ENVIRONMENTAL
SCIENCE

TWELFTH EDITION

ENVIRONMENTAL SCIENCE

A Study of Interrelationships

ELDON D. ENGER

Delta College

BRADLEY F. SMITH

Huxley College of the Environment

Western Washington University

Higher Education

Boston   Burr Ridge, IL   Dubuque, IA   New York   San Francisco   St Louis
Bangkok   Bogotá   Caracas   Kuala Lumpur   Lisbon   London   Madrid   Mexico City
Milan   Montreal   New Delhi   Santiago   Seoul   Singapore   Sydney   Taipei   Toronto

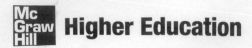 **Higher Education**

ENVIRONMENTAL SCIENCE: A STUDY OF INTERRELATIONSHIPS, TWELFTH EDITION

Published by McGraw-Hill, a business unit of The McGraw-Hill Companies, Inc., 1221 Avenue of the Americas, New York, NY 10020. Copyright © 2010 by The McGraw-Hill Companies, Inc. All rights reserved. Previous editions © 2008, 2006, and 2004. No part of this publication may be reproduced or distributed in any form or by any means, or stored in a database or retrieval system, without the prior written consent of The McGraw-Hill Companies, Inc., including, but not limited to, in any network or other electronic storage or transmission, or broadcast for distance learning.

Some ancillaries, including electronic and print components, may not be available to customers outside the United States.

 This book is printed on recycled, acid-free paper containing 10% postconsumer waste.

1 2 3 4 5 6 7 8 9 0 QPD/QPD 0 9

ISBN 978–0–07–338320–0
MHID 0–07–338320–1

Publisher: *Janice Roerig-Blong*
Executive Editor: *Margaret J. Kemp*
Director of Development: *Kristine Tibbetts*
Marketing Manager: *Heather Chase Wagner*
Project Manager: *Joyce Watters*
Lead Production Supervisor: *Sandy Ludovissy*
Senior Media Project Manager: *Sandra M. Schnee*
Associate Design Coordinator: *Brenda A. Rolwes*
Cover Designer: *Studio Montage, St. Louis, Missouri*
(USE) Cover Image: © *GOODSHOOT/ Alamy*
Senior Photo Research Coordinator: *Lori Hancock*
Compositor: *S4Carlisle Publishing Services*
Typeface: *10/12 Times Roman*
Printer: *Quebecor World Dubuque, IA*

All credits appearing on page or at the end of the book are considered to be an extension of the copyright page.

Library of Congress Cataloging-in-Publication Data

Enger, Eldon D.
  Environmental science : a study of interrelationships / Eldon D. Enger, Bradley S. Smith. — 12th ed.
    p. cm.
  Includes index.
  ISBN 978–0–07–338320–0 — ISBN 0–07–338320–1 (hard copy : alk. paper)
  1. Environmental sciences—Textbooks.   I. Smith, Bradley S.   II. Title.

GE105.E54   2010
363.7—dc22
                                                            2008051006

www.mhhe.com

DEDICATION

*To Judy, my wife and friend,
for sharing life's adventures.*

ELDON ENGER

*To Earl Bohn—teacher, friend,
gentleman, and the embodiment
of the essence of Beavertail Point.*

BRAD SMITH

ABOUT THE AUTHORS

ELDON D. ENGER is an emeritus professor of biology at Delta College, a community college near Saginaw, Michigan. He received his B.A. and M.S. degrees from the University of Michigan. Professor Enger has over 30 years of teaching experience, during which he has taught biology, zoology, environmental science, and several other courses. He has been very active in curriculum and course development. A major contribution to the curriculum was the development of an environmental technician curriculum and the courses that support it. He was also involved in the development of learning community courses in stream ecology, winter ecology, and plant identification. Each of these courses involved students in weekend-long experiences in the outdoors that paired environmental education with physical activity—stream ecology and canoeing, winter ecology and cross-country skiing, and plant identification with backpacking.

Professor Enger is an advocate for variety in teaching methodology. He feels that if students are provided with varied experiences, they are more likely to learn. In addition to the standard textbook assignments, lectures, and laboratory activities, his classes included writing assignments, student presentation of lecture material, debates by students on controversial issues, field experiences, individual student projects, and discussions of local examples and relevant current events. Textbooks are very valuable for presenting content, especially if they contain accurate, informative drawings and visual examples.

Lectures are best used to help students see themes and make connections, and laboratory activities provide important hands-on activities.

Professor Enger received the Bergstein Award for Teaching Excellence and the Scholarly Achievement Award from Delta College and was selected as a Fulbright Exchange Teacher twice—to Australia and Scotland. He has participated as a volunteer in several Earthwatch Research Programs. These include: studying the behavior of a bird known as the long-tailed manakin in Costa Rica, participating in a study to reintroduce endangered marsupials from islands to mainland Australia, and efforts to protect the leatherback turtle in Costa Rica. He also served as a participant in a People to People program, which allowed for an exchange of ideas between U.S. and South African environmental professionals. While traveling he has spent considerable time visiting coral reefs, ocean coasts, mangrove swamps, alpine tundra, prairies, tropical rainforests, cloud forests, deserts, temperate rainforests, coniferous forests, deciduous forests, and many other special ecosystems. This extensive experience provides the background to look at environmental issues from a broad perspective.

Professor Enger is married, has two grown sons, and enjoys a variety of outdoor pursuits such as cross-country skiing, hiking, hunting, fishing, camping, and gardening. Other interests include reading a wide variety of periodicals, beekeeping, singing in a church choir, and preserving garden produce.

BRADLEY F. SMITH is the Dean of Huxley College of the Environment at Western Washington University in Bellingham, Washington. Prior to assuming the position as Dean in 1994, he served as the first Director of the Office of Environmental Education for the U.S. Environmental Protection Agency in Washington, D.C., from 1991 to 1994. Dean Smith also served as the Acting President of the National Environmental Education and Training Foundation in Washington, D.C., and as a Special Assistant to the EPA Administrator.

Before moving to Washington, D.C., Dean Smith was a professor of political science and environmental studies for 15 years, and the executive director of an environmental education center and nature refuge for five years.

Dean Smith has considerable international experience. He was a Fulbright Exchange Teacher to England and worked as a research associate for Environment Canada in New Brunswick. He is a frequent speaker on environmental issues worldwide and serves on the International Scholars Program for the U.S. Information Agency. He also served as a U.S. representative on the Tri-Lateral Commission on environmental education with Canada and Mexico. In 1995, he was awarded a NATO Fellowship to study the environmental problems associated with the closure of former Soviet military bases in Eastern Europe. Dean Smith is an Adjunct Professor at Far Eastern State University in Vladivostok, Russia, and is a member of the Russian Academy of Transport. He also serves as a commissioner for the International Union for the Conservation of Nature (IUCN) and is the President of the World Conservation Learning Network for the IUCN.

Nationally, Dean Smith serves as a member/advisor for many environmental organizations' boards of directors and advisory councils, including the National Environmental Education and Training Foundation. He served as the chair of the Washington State Sustainability Council and is the past President of the Council of Environmental Deans and Directors. He previously served on President Clinton's Council for Sustainable Development (Education Task Force).

Dean Smith holds B.A. and M.A. degrees in Political Science/International Relations and Public Administration and a Ph.D. from the School of Natural Resources and the Environment at the University of Michigan.

Dean Smith and his wife, Daria, live along the shores of Puget Sound in Bellingham, Washington, and spend part of the summer at their summer home on the shores of Lake Huron in the Upper Peninsula of Michigan. He has two grown children and is an avid outdoor enthusiast.

BRIEF CONTENTS

CHAPTER 1   Environmental Interrelationships 1

CHAPTER 2   Environmental Ethics 14

CHAPTER 3   Environmental Risk: Economics, Assessment, and Management 36

CHAPTER 4   Interrelated Scientific Principles: Matter, Energy, and Environment 61

CHAPTER 5   Interactions: Environments and Organisms 79

CHAPTER 6   Kinds of Ecosystems and Communities 108

CHAPTER 7   Populations: Characteristics and Issues 139

CHAPTER 8   Energy and Civilization: Patterns of Consumption 170

CHAPTER 9   Energy Sources 185

CHAPTER 10   Nuclear Energy 213

CHAPTER 11   Biodiversity Issues 234

CHAPTER 12   Land-Use Planning 264

CHAPTER 13   Soil and Its Uses 288

CHAPTER 14   Agricultural Methods and Pest Management 311

CHAPTER 15   Water Management 334

CHAPTER 16   Air Quality Issues 365

CHAPTER 17   Solid Waste Management and Disposal 392

CHAPTER 18   Environmental Regulations: Hazardous Substances and Wastes 409

CHAPTER 19   Environmental Policy and Decision Making 427

APPENDIX 1 451
APPENDIX 2 452
GLOSSARY 454
CREDITS 464
INDEX 467

CONTENTS

Preface xviii
Critical Thinking xxiii
Guided Tour xxiv

CHAPTER 1

ENVIRONMENTAL INTERRELATIONSHIPS 1

The Nature of Environmental Science 2
   Interrelatedness Is a Core Concept 2
   An Ecosystem Approach 3

WATER CONNECTIONS: Social and Biological Interactions in the Management of Keoladeo National Park, India 4

   Political and Economic Issues 4
   The Global Nature of Environmental Concerns 5
Regional Environmental Concerns 6
   The Wilderness North 6

GOING GREEN: Individual Decisions Matter 7

   The Agricultural Middle 7
   The Dry West 8
   The Forested West 9
   The Great Lakes and Industrial Northeast 9
   The Diverse South 10

CAMPUS SUSTAINABILITY INITIATIVE: The Association for the Advancement of Sustainability in Higher Education 12

ISSUES & ANALYSIS: Government Regulation and Personal Property 12

CHAPTER 2

ENVIRONMENTAL ETHICS 14

The Call for a New Ethic 15
Environmental Ethics 15
   Ethics and Laws 16
   Conflicting Ethical Positions 16
   The Greening of Religion 16
   Three Philosophical Approaches to Environmental Ethics 16

CAMPUS SUSTAINABILITY INITIATIVE: Investment Responsibility at Duke University 18

   Other Philosophical Approaches 18
Environmental Attitudes 18
   Development 18
   Preservation 19
   Conservation 19

CASE STUDY 2.1: Early Philosophers of Nature 20

   Sustainable Development 21
Environmental Justice 21
Societal Environmental Ethics 23
Corporate Environmental Ethics 23
   The Legal Status of Corporations 23

CASE STUDY 2.2: Environmental Disasters and Poverty 24

   Waste and Pollution 25
   Profitability and Power 25
   Is There a Corporate Environmental Ethic? 25
   Green Business Concepts 26
Individual Environmental Ethics 27
The Ethics of Consumption 27
   Food 27

GOING GREEN: Do We Consume Too Much? 28

   Energy 28
   Water 29
   Wild Nature 29
Personal Choices 29
Global Environmental Ethics 29

WATER CONNECTIONS: A Global Water Ethic 30

ISSUES & ANALYSIS: Environmental Dissent: Is Ecoterrorism Justified? 33

CHAPTER 3

ENVIRONMENTAL RISK: ECONOMICS, ASSESSMENT, AND MANAGEMENT 36

Characterizing Risk 37
Risk and Economics 37
   Risk Assessment 37
   Risk Management 38

CASE STUDY 3.1: What's in a Number? 39

   Risk Tolerance 40
   True and Perceived Risks 40
Environmental Economics 41
   Resources 41
   Supply and Demand 42

WATER CONNECTIONS: Valuing the Removal of the ELWHA and Glines Dams 43

   Assigning Value to Natural Resources 43
   Environmental Costs 44
   Cost-Benefit Analysis 45

Concerns About the Use of Cost-Benefit Analysis 46
Comparing Economic and Ecological Systems 47
Common Property Resource Problems—The Tragedy
of the Commons 48
Green Economics 49
Using Economic Tools to Address Environmental Issues 49

GOING GREEN: Green-Collar Jobs 50

Subsidies 50
Liability Protection and Grants for Small Business 51
Market-Based Instruments 51
Life Cycle Analysis and Extended Product Responsibility 52

CASE STUDY 3.2: Pollution Prevention Pays! 53

Green Marketing Principles 54
Economics and Sustainable Development 54

CAMPUS SUSTAINABILITY INITIATIVE:
Campus Business Partnership to Reduce Greenhouse Gas
Emissions 55

Economics, Environment, and Developing Nations 57

ISSUES & ANALYSIS: The Economics
and Risks of Mercury Contamination 58

CHAPTER 4

INTERRELATED SCIENTIFIC PRINCIPLES:
MATTER, ENERGY, AND ENVIRONMENT 61

The Nature of Science 62
Basic Assumptions in Science 62
Cause-and-Effect Relationships 62
Elements of the Scientific Method 62
Limitations of Science 66
Pseudoscience 66
The Structure of Matter 66
Atomic Structure 66
The Molecular Nature of Matter 67
A Word About Water 67
Acids, Bases, and pH 67

GOING GREEN: Evaluating Green Claims 68

Inorganic and Organic Matter 69
Chemical Reactions 69

WATER CONNECTIONS: Applying the Scientific
Method—Acid Rain 70

Chemical Reactions in Living Things 71
Chemistry and the Environment 72
Energy Principles 72
Kinds of Energy 72
States of Matter 72
First and Second Laws of Thermodynamics 73
Environmental Implications of Energy Flow 74
Entropy Increases 74
Energy Quality 74

CAMPUS SUSTAINABILITY INITIATIVE: Cooling
Off the University of Arizona 75

Biological Systems and Thermodynamics 75
Pollution and Thermodynamics 75

ISSUES & ANALYSIS: Diesel Engine
Trade-offs 76

CHAPTER 5

INTERACTIONS: ENVIRONMENTS
AND ORGANISMS 79

Ecological Concepts 80
Environment 80

CAMPUS SUSTAINABILITY INITIATIVE: Creek
Restoration at the University of Arkansas–Little Rock 81

Limiting Factors 81
Habitat and Niche 83
The Role of Natural Selection and Evolution 84
Genes, Populations, and Species 84
Natural Selection 85
Evolutionary Patterns 86
Kinds of Organism Interactions 89
Predation 89
Competition 90
Symbiotic Relationships 92
Some Relationships Are Difficult to Categorize 94
Community and Ecosystem Interactions 94
Major Roles of Organisms in Ecosystems 95
Keystone Species 95
Energy Flow Through Ecosystems 96
Food Chains and Food Webs 97
Nutrient Cycles in Ecosystems—Biogeochemical Cycles 98

WATER CONNECTIONS: Changes in the Food
Chain of the Great Lakes 100

Human Impact on Nutrient Cycles 103

GOING GREEN: Phosphorus-Free Lawn Fertilizer 105

ISSUES & ANALYSIS: Phosphate
Mining in Nauru 106

CHAPTER 6

KINDS OF ECOSYSTEMS
AND COMMUNITIES 108

Succession 109
Primary Succession 109
Secondary Succession 111
Modern Concepts of Succession and Climax 112
Biomes Are Determined by Climate 114
Precipitation and Temperature 114
The Effect of Elevation on Climate and Vegetation 115
Major Biomes of the World 115
Desert 116

GOING GREEN: Conservation Easements 117

Temperate Grassland 118

CAMPUS SUSTAINABILITY INITIATIVE:
The Blue Oak Ranch Reserve of the University
of California–Berkeley 119

Savanna 119
Mediterranean Shrublands (Chaparral) 120
Tropical Dry Forest 121
Tropical Rainforest 122

Contents

CASE STUDY 6.1: Grassland Succession 124

Temperate Deciduous Forest 124
Temperate Rainforest 126
Taiga, Northern Coniferous Forest, or Boreal Forest 126
Tundra 127

Major Aquatic Ecosystems 129
Marine Ecosystems 129
Freshwater Ecosystems 133

WATER CONNECTIONS: Varzea Forests—Where the Amazon River and Land Meet 134

ISSUES & ANALYSIS: Ecosystem Loss in North America 137

CHAPTER 7

POPULATIONS: CHARACTERISTICS AND ISSUES 139

Population Characteristics 140
Natality—Birthrate 140
Mortality—Death Rate 140
Population Growth Rate 141
Sex Ratio 141
Age Distribution 142
Population Density and Spatial Distribution 143
Summary of Factors That Influence Population Growth Rates 143

A Population Growth Curve 143
Factors That Limit Population Size 144
Extrinsic and Intrinsic Limiting Factors 144
Density-Dependent and Density-Independent Limiting Factors 144

Categories of Limiting Factors 145
Availability of Raw Materials 145
Availability of Energy 145
Accumulation of Waste Products 145
Interactions Among Organisms 146

Carrying Capacity 146

GOING GREEN: Increasing Populations of Red-Cockaded Woodpeckers 147

Reproductive Strategies and Population Fluctuations 148
K-Strategists and r-Strategists 148
Population Cycles 149

Human Population Growth 149
Human Population Characteristics and Implications 151
Economic Development 151
Measuring the Environmental Impact of a Population 151

CASE STUDY 7.1: Thomas Malthus and His Essay on Population 152

The Ecological Footprint Concept 153

Factors That Influence Human Population Growth 153
Biological Factors 153
Social Factors 155
Economic Factors 156
Political Factors 156

Population Growth Rates and Standard of Living 157
Hunger, Food Production, and Environmental Degradation 158
Environmental Impacts of Food Production 158

CASE STUDY 7.2: The Grameen Bank and Microcredit 159

The Human Energy Pyramid 160

Economics and Politics of Hunger 160
Solving the Problem 161

The Demographic Transition Concept 161
The Demographic Transition Model 161

CAMPUS SUSTAINABILITY INITIATIVE: Auburn University's War on Hunger Initiative 162

Applying the Model 162

The U.S. Population Picture 162

WATER CONNECTIONS: Drinking Water: A Basic Right? 164

What Does the Future Hold? 164
Available Raw Materials 164
Available Energy 164

CASE STUDY 7.3: North America—Population Comparisons 165

Waste Disposal 165
Interaction with Other Organisms 165
Social Factors Influence Human Population 166
Ultimate Size Limitation 166

ISSUES & ANALYSIS: The Lesser Snow Goose—A Problem Population 167

CHAPTER 8

ENERGY AND CIVILIZATION: PATTERNS OF CONSUMPTION 170

History of Energy Consumption 171
Biological Energy Sources 171
Increased Use of Wood 171
Fossil Fuels and the Industrial Revolution 172
The Role of the Automobile 172
Growth in the Use of Natural Gas 173

GOING GREEN: Reducing Automobile Use in Cities 174

How Energy Is Used 174
Residential and Commercial Energy Use 174

CASE STUDY 8.1: Biomass Fuels and the Developing World 175

Industrial Energy Use 175

WATER CONNECTIONS: Heating Water—Saving Energy 176

Transportation Energy Use 177

Electrical Energy 177
The Economics and Politics of Energy Use 179
Fuel Economy and Government Policy 179
Electricity Pricing 179
The Importance of OPEC 180

Energy Consumption Trends 180
Growth in Energy Use 180
Available Energy Sources 181
Political and Economic Factors 181

ISSUES & ANALYSIS: Government Action and Energy Policy 182

CAMPUS SUSTAINABILITY INITIATIVE: Delta College and Energy Efficiency 183

CHAPTER 9

ENERGY SOURCES 185

Energy Sources 186
Resources and Reserves 186
Fossil-Fuel Formation 187
    Coal 187
    Oil and Natural Gas 188
Issues Related to the Use of Fossil Fuels 188
    Coal Use 189
    Oil Use 192
    Natural Gas Use 194

    CASE STUDY 9.1: The Arctic National Wildlife
    Refuge 195

Renewable Sources of Energy 196
    Biomass Conversion 196
    Hydroelectric Power 199
    Solar Energy 201

    WATER CONNECTIONS: Solar Stills and
    Drinking Water 204

    Wind Energy 204
    Geothermal Energy 205

    CAMPUS SUSTAINABILITY INITIATIVE:
    Western Washington University Purchases Green Power
    from North Dakota Wind Farms 206

    Tidal Power 206
Energy Conservation 207

    GOING GREEN: Hybrid Electric Vehicles 209

Are Fuel Cells in the Future? 209

    ISSUES & ANALYSIS: Does Ethanol
    Fuel Make Sense? 210

CHAPTER 10

NUCLEAR ENERGY 213

The Nature of Nuclear Energy 214
    Measuring Radiation 215
    Biological Effects of Ionizing Radiation 216
    Radiation Protection 217
    Nuclear Chain Reaction 217
The History of Nuclear Energy Development 217
Nuclear Fission Reactors 218
    Boiling-Water Reactors 219
    Pressurized-Water Reactors 219
    Heavy-Water Reactors 219
    Gas-Cooled Reactors 220
Investigating Nuclear Alternatives 220
    Breeder Reactors 220
    Nuclear Fusion 220
The Nuclear Fuel Cycle 221
    Mining and Milling 221
    Enrichment and Fuel Fabrication 221
    Use in a Reactor 221
    Reprocessing and Waste Disposal 221

    WATER CONNECTIONS: Water and Nuclear
    Power Plants 222

    Transportation Issues 222
Nuclear Concerns 222
    Reactor Safety 222
    Terrorism 223
    Worker and Public Exposure to Radiation 224
    Contamination from Nuclear Research and Weapons Production 224
    Disposal of Nuclear Weapons 224
    Radioactive Waste Disposal 224
    Thermal Pollution 227
    Decommissioning 228

    GOING GREEN: Returning a Nuclear Plant
    Site to Public Use 229

The Future of Nuclear Power 229
    Social Forces 229

    CAMPUS SUSTAINABILITY INITIATIVE:
    Oregon State University and Passive Nuclear
    Power Plants 230

    Technical Trends 230

    CASE STUDY 10.1: The Hanford Facility: A
    Storehouse of Nuclear Remains 231

    ISSUES & ANALYSIS: Yucca Mountain and Nuclear
    Waste Storage 232

CHAPTER 11

BIODIVERSITY ISSUES 234

Biodiversity Loss and Extinction 235
    Kinds of Organisms Prone to Extinction 235
    Extinction as a Result of Human Activity 236
Describing Biodiversity 236
    Genetic Diversity 236
    Species Diversity 237
    Ecosystem Diversity 239
The Value of Biodiversity 239
    Biological and Ecosystem Services Values 239
    Direct Economic Values 241
    Ethical Values 241
Threats to Biodiversity 242
    Habitat Loss 242

    WATER CONNECTIONS: Freshwater Biodiversity 244

    Overexploitation 248
    Introduction of Exotic Species 250
    Control of Predator and Pest Organisms 251
    Climate Change 253
What Is Being Done to Preserve Biodiversity? 253

    CAMPUS SUSTAINABILITY INITIATIVE:
    University of Kansas Biodiversity Partnership 254

    Legal Protection 254

    CASE STUDY 11.1: Millennium Ecosystem
    Assessment Report and the Millennium
    Declaration 256

    GOING GREEN: Consumer Choices Related
    to Biodiversity 257

    Sustainable Management of Wildlife Populations 258
    Sustainable Management of Fish Populations 260

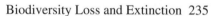

CASE STUDY 11.2: The California Condor 261

ISSUES & ANALYSIS: The Problem of Image 262

CHAPTER 12
LAND-USE PLANNING 264

The Need for Planning 265
Historical Forces That Shaped Land Use 265
    The Rural-to-Urban Shift 265
    Urbanization in the Developing World 265
Migration from the Central City to the Suburbs 265

WATER CONNECTIONS: Waterways and Development 267

Factors That Contribute to Sprawl 269
    Lifestyle Factors 269
    Economic Factors 270
    Planning and Policy Factors 270
Problems Associated with Unplanned Urban Growth 270
    Transportation Problems 271
    Air Pollution 271
    Low Energy Efficiency 271
    Loss of Sense of Community 271
    Death of the Central City 271
    Higher Infrastructure Costs 271
    Loss of Open Space 272
    Loss of Farmland 272
    Water Pollution Problems 272
    Floodplain Problems 272
    Wetlands Misuse 273
    Other Land-Use Considerations 273
    Land-Use Planning and Aesthetic Pollution 274
Land-Use Planning Principles 274
Mechanisms for Implementing Land-Use Plans 276
    Establishing State or Regional Planning Agencies 276
    Purchasing Land or Use Rights 277

CAMPUS SUSTAINABILITY INITIATIVE: Green Building on Campus 278

    Regulating Use 278
Special Urban Planning Issues 279
    Urban Transportation Planning 279
    Urban Recreation Planning 280
    Redevelopment of Inner-City Areas 281
    Smart Growth 281

GOING GREEN: Green Building 282

Federal Government Land-Use Issues 283

ISSUES & ANALYSIS: Smart Communities' Success Stories 285

CHAPTER 13
SOIL AND ITS USES 288

Geologic Processes 289
Soil and Land 291
Soil Formation 292
Soil Properties 292

CAMPUS SUSTAINABILITY INITIATIVE: Composting on Campus 295

Soil Profile 294
Soil Erosion 298

WATER CONNECTIONS: Water and Erosion 301

Soil Conservation Practices 301
    Soil Quality Management Components 302
    Contour Farming 303

GOING GREEN: Green Landscaping 304

    Strip Farming 304
    Terracing 304
    Waterways 304
    Windbreaks 305
Conventional Versus Conservation Tillage 305
Protecting Soil on Nonfarm Land 307

CASE STUDY 13.1: Land Capability Classes 308

ISSUES & ANALYSIS: Soil Fertility and Hunger in Africa 309

CHAPTER 14
AGRICULTURAL METHODS AND PEST MANAGEMENT 311

The Development of Agriculture 312
    Shifting Agriculture 312
    Labor-Intensive Agriculture 312
    Mechanized Agriculture 312
Fossil Fuel Versus Muscle Power 314
The Impact of Fertilizer 314
Agricultural Chemical Use 315
    Insecticides 315

CASE STUDY 14.1: DDT—A Historical Perspective 316

    Herbicides 317
    Fungicides and Rodenticides 318
    Other Agricultural Chemicals 318

WATER CONNECTIONS: The Dead Zone of the Gulf of Mexico 319

CAMPUS SUSTAINABILITY INITIATIVE: Integrated Pest Management at Seattle University 320

Problems with Pesticide Use 320
    Persistence 320
    Bioaccumulation and Biomagnification 320

CASE STUDY 14.2: Economic Development and Food Production in China 322

    Pesticide Resistance 322
    Effects on Nontarget Organisms 323
    Human Health Concerns 323
Why Are Pesticides So Widely Used? 324

GOING GREEN: Organic Farming: Helping to Promote Sustainable Agriculture 325

Contents   xiii

Alternatives to Conventional Agriculture 325
   Sustainable Agriculture 325
   Techniques for Protecting Soil and Water Resources 326
   Integrated Pest Management 327
   Genetically Modified Crops 330

   ISSUES & ANALYSIS: What Does "Certified Organic" Food Mean? 332

CHAPTER 15
WATER MANAGEMENT 334

The Water Issue 335
The Hydrologic Cycle 337
Human Influences on the Hydrologic Cycle 339
Kinds of Water Use 339
   Domestic Use of Water 339

   WATER CONNECTIONS: The Bottled Water Boom 340

   Agricultural Use of Water 342
   Industrial Use of Water 343
   In-Stream Use of Water 344

   CASE STUDY 15.1: Growing Demands for a Limited Supply of Water in the West 345

Kinds and Sources of Water Pollution 346

   GOING GREEN: Water Reuse 349

   Municipal Water Pollution 349
   Agricultural Water Pollution 350
   Industrial Water Pollution 350
   Thermal Pollution 351
   Marine Oil Pollution 351
   Groundwater Pollution 352
Water-Use Planning Issues 353

   CASE STUDY 15.2: Restoring the Everglades 354

   Water Diversion 355
   Wastewater Treatment 356
   Salinization 358
   Groundwater Mining 358
   Preserving Scenic Water Areas and Wildlife Habitats 360

   CAMPUS SUSTAINABILITY INITIATIVE: Conserving Water on Campus 361

   ISSUES & ANALYSIS: Is There Lead in Our Drinking Water? 362

CHAPTER 16
AIR QUALITY ISSUES 365

The Atmosphere 366
Pollution of the Atmosphere 366
Categories of Air Pollutants 367
   Carbon Monoxide 367
   Particulate Matter 368

   CASE STUDY 16.1: Air Pollution in Mexico City 369

   Sulfur Dioxide 370
   Nitrogen Dioxide 370
   Lead 370

   Volatile Organic Compounds 370
   Ground-Level Ozone and Photochemical Smog 371
   Hazardous Air Pollutants 373
Control of Air Pollution 373
   The Clean Air Act 373
   Actions That Have Reduced Air Pollution 374
Acid Deposition 375
   Causes of Acid Precipitation 375
   Effects on Structures 375

   CAMPUS SUSTAINABILITY INITIATIVE: New York University's Co-Generation Plant 376

   Effects on Terrestrial Ecosystems 376
   Effects on Aquatic Ecosystems 377
Ozone Depletion 378
   Why Stratospheric Ozone Is Important 378
   Ozone Destruction 378
   Actions to Protect the Ozone Layer 378
Global Warming and Climate Change 378
   Causes of Global Warming and Climate Change 379
   Potential Consequences of Global Warming and Climate Change 381
Addressing Climate Change 383
   Energy Efficiency 383

   GOING GREEN: Germany's Energy Policy 384

   The Role of Biomass 384
   Technological Approaches 384
   Political and Economic Forces 384
Indoor Air Pollution 385
   Sources of Indoor Air Pollutants 385
   Significance of Weatherizing Buildings 385
   Secondhand Smoke 385

   WATER CONNECTIONS: Decline in Arctic Sea Ice 386

   Radon 387
Noise Pollution 387

   ISSUES & ANALYSIS: Pollution, Policy, and Personal Choice 389

CHAPTER 17
SOLID WASTE MANAGEMENT AND DISPOSAL 392

Kinds of Solid Waste 393
Municipal Solid Waste 394

   GOING GREEN: Garbage Goes Green 395

Methods of Waste Disposal 395
   Landfills 395

   WATER CONNECTIONS: Landfills' Impact on Water 397

   Incineration 398
   Producing Mulch and Compost 399

   CASE STUDY 17.1: Resins Used in Consumer Packaging 400

   Source Reduction 401
   Recycling 402

   CASE STUDY 17.2: Beverage Container Deposit-Refund Programs 403

CAMPUS SUSTAINABILITY INITIATIVE:
Recycling Partnership at Northern Arizona University 404

ISSUES & ANALYSIS: Paper or Plastic
or Plastax? 406

CHAPTER 18

ENVIRONMENTAL REGULATIONS: HAZARDOUS SUBSTANCES AND WASTES 409

Hazardous and Toxic Materials in Our Environment 410
Hazardous and Toxic Substances—Some Definitions 410
Defining Hazardous Waste 411
Determining Regulations 412
 Identification of Hazardous and Toxic Materials 412
 Setting Exposure Limits 413

CASE STUDY 18.1: Determining Toxicity 414

 Acute and Chronic Toxicity 414
 Synergism 414

WATER CONNECTIONS: Dioxins in the
Tittabawassee River Floodplain 415

 Persistent and Nonpersistent Pollutants 415
Environmental Problems Caused by
 Hazardous Wastes 416
Health Risks Associated with Hazardous Wastes 416
Hazardous-Waste Dumps—A Legacy of Abuse 416
 Toxic Chemical Releases 418

GOING GREEN: Guide to Electronics Recycling 419

Hazardous-Waste Management Choices 419
 Reducing the Amount of Waste at the Source 420

CAMPUS SUSTAINABILITY INITIATIVE:
Chemical Exchange at the University of
British Columbia 421

 Recycling Wastes 422
 Treating Wastes 422
 Disposal Methods 422
International Trade in Hazardous Wastes 422
Hazardous-Waste Management Program Evolution 423

ISSUES & ANALYSIS: Household
Hazardous Waste 424

CHAPTER 19

ENVIRONMENTAL POLICY AND DECISION MAKING 427

New Challenges for a New Century 428
 Forces and Trends 428
 Kinds of Policy Responses 428
 Government and Governance 429
 Learning from the Past 430
 Thinking About the Future 430

WATER CONNECTIONS: Shared Water
Resources 431

 Defining the Future 432

Development of Environmental Policy in the United States 43.
 Legislative Action 432
 The Role of Nongovernmental Organizations 434
 The Challenge for U.S. Environmental Policy 434
Environmental Policy and Regulation 435

GOING GREEN: Investing in a Green Future 436

 The Significance of Administrative Law 436
 National Environmental Policy Act—Landmark Legislation 436
 Other Important Environmental Legislation 436
 Role of the Environmental Protection Agency 437
The Greening of Geopolitics 438
 International Aspects of Environmental Problems 438

CASE STUDY 19.1: The Environmental Effects of
Hurricane Katrina 439

 National Security Issues 440
Terrorism and the Environment 441
International Environmental Policy 442
 The Role of the United Nations 442
 Earth Summit on Environment and Development 444
 Environmental Policy and the European Union 445
 New International Instruments 445

CAMPUS SUSTAINABILITY INITIATIVE:
College and University Presidents' Climate
Commitment 446

It All Comes Back to *You* 447

ISSUES & ANALYSIS: Gasoline, Taxes, and the
Environment 448

APPENDIX 1 451
APPENDIX 2 452
GLOSSARY 454
CREDITS 464
INDEX 467

LIST OF CASE STUDIES

2.1 Early Philosophers of Nature 20

2.2 Environmental Disasters and Poverty 24

3.1 What's in a Number? 39

3.2 Pollution Prevention Pays! 53

6.1 Grassland Succession 124

7.1 Thomas Malthus and His Essay on Population 152

7.2 The Grameen Bank and Microcredit 159

7.3 North America—Population Comparisons 165

8.1 Biomass Fuels and the Developing World 175

9.1 The Arctic National Wildlife Refuge 195

10.1 The Hanford Facility: A Storehouse of Nuclear
Remains 231

11.1 Millennium Ecosystem Assessment Report
and the Millennium Declaration 256

11.2 The California Condor 261

Capability Classes 308

—A Historical Perspective 316

onomic Development and Food Production
in China 322

15.1 Growing Demands for a Limited Supply of Water in
the West 345

15.2 Restoring the Everglades 354

16.1 Air Pollution in Mexico City 369

17.1 Resins used in Consumer Packaging 400

17.2 Beverage Container Deposit-Refund Programs 403

18.1 Determining Toxicity 414

19.1 The Environmental Effects of Hurricane Katrina 439

LIST OF CAMPUS SUSTAINABILITY INITIATIVES

CHAPTER 1
The Association for the Advancement of Sustainability in
Higher Education 12

CHAPTER 2
Investment Responsibility at Duke University 18

CHAPTER 3
Campus Business Partnership to Reduce Greenhouse
Gas Emissions 55

CHAPTER 4
Cooling Off the University of Arizona 75

CHAPTER 5
Creek Restoration at the University of Arkansas–Little
Rock 81

CHAPTER 6
The Blue Oak Ranch Reserve of the University of
California–Berkeley 119

CHAPTER 7
Auburn University's War on Hunger Initiative 162

CHAPTER 8
Delta College and Energy Efficiency 183

CHAPTER 9
Western Washington University Purchases Green Power
from North Dakota Wind Farms 206

CHAPTER 10
Oregon State University and Passive Nuclear
Power Plants 230

CHAPTER 11
University of Kansas Biodiversity Partnership 254

CHAPTER 12
Green Building on Campus 278

CHAPTER 13
Composting on Campus 295

CHAPTER 14
Integrated Pest Management at Seattle University 320

CHAPTER 15
Conserving Water on Campus 361

CHAPTER 16
New York University's Co-Generation Plant 376

CHAPTER 17
Recycling Partnership at Northern Arizona University 404

CHAPTER 18
Chemical Exchange at the University of
British Columbia 421

CHAPTER 19
College and University Presidents'
Climate Commitment 446

LIST OF GOING GREEN FEATURES

CHAPTER 1
Individual Decisions Matter 7

CHAPTER 2
Do We Consume Too Much? 28

CHAPTER 3
Green-Collar Jobs 50

CHAPTER 4
Evaluating Green Claims 68

CHAPTER 5
Phosphorus-Free Lawn Fertilizer 105

CHAPTER 6
Conservation Easements 117

CHAPTER 7
Increasing Populations of Red-Cockaded
Woodpeckers 147

CHAPTER 8
Reducing Automobile Use in Cities 174

CHAPTER 9
Hybrid Electric Vehicles 209

CHAPTER 10
Returning a Nuclear Plant Site to Public Use 229

CHAPTER 11
Consumer Choices Related to Biodiversity 257

CHAPTER 12
Green Building 282

CHAPTER 13
Green Landscaping 304

CHAPTER 14
Organic Farming: Helping to Promote Sustainable
Agriculture 325

CHAPTER 15
Water Reuse 349

CHAPTER 16
Germany's Energy Policy 384

CHAPTER 17
Garbage Goes Green 395

CHAPTER 18
Guide to Electronics Recycling 419

CHAPTER 19
Investing in a Green Future 436

LIST OF WATER CONNECTIONS

CHAPTER 1
Social and Biological Interactions in the Management of
Keoladeo National Park, India 4

CHAPTER 2
A Global Water Ethic 30

CHAPTER 3
Valuing the Removal of the ELWHA and Glines Dams 43

CHAPTER 4
Applying the Scientific Method—Acid Rain 70

CHAPTER 5
Changes in the Food Chain of the Great Lakes 100

CHAPTER 6
Varzea Forests—Where the Amazon River and Land
Meet 134

CHAPTER 7
Drinking Water: A Basic Right? 164

CHAPTER 8
Heating Water—Saving Energy 176

CHAPTER 9
Solar Stills and Drinking Water 204

CHAPTER 10
Water and Nuclear Power Plants 222

CHAPTER 11
Freshwater Biodiversity 244

CHAPTER 12
Waterways and Development 267

CHAPTER 13
Water and Erosion 301

CHAPTER 14
The Dead Zone of the Gulf of Mexico 319

CHAPTER 15
The Bottled Water Boom 340

CHAPTER 16
Decline in Arctic Sea Ice 386

CHAPTER 17
Landfills' Impact on Water 397

CHAPTER 18
Dioxins in the Tittabawassee River Floodplain 415

CHAPTER 19
Shared Water Resources 431

THE ROLE OF ENVIRONMENTAL SCIENCE IN SOCIETY

We live in a time of great change and challenge. A quick read of the headlines of any newspaper provides images of disease, hunger, poverty, natural disasters, and pollution. Challenges, however, are also opportunities. Opportunities exist because of the changes the global society must make. Simply put, we cannot continue with business as usual. Such a path is not sustainable. What does that mean? In short, we must do things differently. For example, different farming practices will allow crops to be raised with fewer chemicals and less water. Buildings can be constructed with new, more sustainable methods. Transportation can be provided while using less energy. In other words, we must think differently. Environmental science is a discipline that fosters new ways of thinking. Environmental science is an applied science designed to help address and solve the challenges the world faces. It is also by its very nature a global science. This text, for example, has been translated and published in Spanish, Chinese, and Korean. Therefore, students in Santiago, Shanghai, Seoul, or Seattle are learning the "how's and why's" involved in thinking and acting sustainably. At the end of the day we all share the same air, water, and one not-so-big planet. It's important for all of us to make it last.

WHY "A STUDY OF INTERRELATIONSHIPS"?

Environmental science is an interdisciplinary field. Because environmental problems occur as a result of the interaction between humans and the natural world, we must include both scientific and social aspects when we seek solutions to environmental problems. Therefore, the central theme of this book is interrelatedness. It is important to have a historical perspective, to appreciate economic and political realities, to recognize the role of different social experiences and ethical backgrounds, and to integrate these with the science that describes the natural world and how we affect it. *Environmental Science: A Study of Interrelationships* incorporates all of these sources of information when discussing any environmental issue.

WHAT MAKES THIS TEXT UNIQUE?

We present a balanced view of issues, diligently avoiding personal biases and fashionable philosophies.

It is not the purpose of this textbook to tell readers what to think. Rather, our goal is to provide access to information and the conceptual framework needed to understand complex issues so that readers can comprehend the nature of environmental problems and formulate their own views. Two features of the text encourage readers to think about issues and formulate their own thoughts.

- The **Issues & Analysis** box near the end of each chapter presents real-world examples of environmental problems and prompts students to think about the issues involved and respond to a series of questions.
- The **What's Your Take?** feature found at the end of each chapter asks students to take a stand on a particular issue and develop arguments to support their position.

We recognize that environmental problems are global in nature.

Three features of the text support this concern:

- Throughout the text, the authors have made a point to use **examples** from around the world as well as those from North America.
- **Case Studies** provide examples of specific situations that allow students to see how the concepts discussed in the chapter can be applied to everyday situations.
- The presence of easily accessible **Foldout World Maps** at the back of the text allows students to quickly locate a country or region geographically.

NEW TO THIS EDITION

The twelfth edition of *Environmental Science: A Study of Interrelationships* is the result of extensive analysis of the text and the evaluation of input from environmental science instructors who conscientiously reviewed chapters during the revision. We have used the constructive comments provided by these professionals in our continuing efforts to enhance the strengths of the text. The following is a list of global changes we have made, along with a description of significantly revised chapters. To see a more detailed list of chapter-by-chapter changes, please contact your McGraw-Hill sales representative.

Focus on the Positive Environmental science often seems to focus on the negative, since one of the outcomes of any analysis of an environmental situation is to highlight problems and point out where change is needed. We often overlook the many positive actions of individuals and organizations. Therefore, in this edition three new features call attention to the positive:

- **Going Green** boxes describe actions that are having a positive environmental impact. Some of these actions are taken by governments, some are by corporations, and some are individual efforts.
- **Campus Sustainability Initiatives** highlight some of the many actions of students and the colleges and universities they attend that are making a positive environmental impact.
- **Thinking Green** is an end-of-chapter feature that asks students to consider making changes that will have a positive environmental impact.

Focus on Water Interrelatedness is a core concept in environmental science. Although this concept can be illustrated in many ways, in this edition we have chosen to use water as a theme. A **Water Connections** box appears in every chapter. Sometimes the topic of water is also addressed as a heading in the text.

Revised Art Program About 100 new photos have been added or substituted throughout the text to depict real-life situations. Over 50 illustrations, graphs, and charts are new or revised to present detailed information in a form that is easier to comprehend than if that same material were presented in text form.

Several Significantly Revised Chapters Every chapter has a new Going Green, Water Connections, and Campus Sustainability Initiative box. In addition, many chapters have other significant changes, including:

Chapter 1, Environmental Interrelationships There is a new section entitled *Interrelatedness Is a Core Concept.* It uses John Muir's statement *"Tug on anything at all and you'll find it connected to everything else in the universe," as a theme.* It then highlights the changes brought about by the reintroduction of wolves into Yellowstone National Park to show how one simple change has far ranging impacts. A new illustration accompanies this addition.

Chapter 5, Interactions: Environments and Organisms The section on limiting factors and range of tolerance was rewritten and supported with a new illustration. A new food web illustration was substituted. A new case study discusses the changes in food chains in the Great Lakes.

Chapter 6, Kinds of Ecosystems and Communities A new section on the *temperate rainforest* was added and supported with photographs and a graph. Many new photographs were added or substituted to help better describe the nature of specific biomes.

Chapter 9, Energy Sources The chapter was updated with the most recent energy data on energy supply and consumption. The section on renewable energy was reorganized and greatly revised. A new "Issues and Analysis" feature discusses the pros and cons of corn ethanol production. There are many new and substituted photos.

Chapter 11, Biodiversity Issues The sections *Biological and Ecosystem Services Values* and *Threats to Biodiversity* were rewritten. A new section on the importance of climate change to biodiversity was added. New figures illustrate the concepts of genetic diversity, species diversity, and ecosystem diversity. Many new figures have been added and tables and graphs have been updated. The table of *Estimated Values of Ecosystem Services* was significantly modified.

Chapter 15, Water Management This chapter features a new figure on the global distribution of the world's water and a new map showing areas of the world experiencing water stress. New content has been added on the role of the oceans as the primary regulator of global climate and an important sink for greenhouse gases. There is also new content on pricing of water in countries and expanded coverage on the restoration of the Everglades. Also featured is expanded coverage on groundwater usage.

Chapter 19, Environmental Policy and Decision Making This chapter has gone through a major reorganization, including new material on the challenge for U.S. environmental policy. New material on the complexity of ecological problem solving and China's rising energy consumption has been added.

ACKNOWLEDGMENTS

The creation of a textbook requires a dedicated team of professionals who provide guidance, criticism, and encouragement. It is also important to have open communication and dialogue to deal with the many issues that arise during the development and production of a text. Therefore, we would like to thank Sponsoring Editor Marge Kemp; Developmental Editor Robin Reed of S4Carlisle Publishing Services; Project Manager Joyce Watters; Production Supervisor Sandy Ludovissy; Photo Research Coordinator Lori Hancock; Designer Brenda Rolwes; and Media Project Manager Sandy Schnee for their suggestions and kindnesses. Finally, we'd like to thank our many colleagues who have reviewed all, or part, of *Environmental Science: A Study of Interrelationships.* Their valuable input has continued to shape this text and help it meet the needs of instructors around the world.

Eldon D. Enger
Bradley F. Smith

TEACHING AND LEARNING SUPPLEMENTS

McGraw-Hill offers various tools and technology products to support *Environmental Science*. Students can order supplemental study materials by contacting their local bookstore or by calling 800-262-4729. Instructors can obtain teaching aids by calling the Customer Service Department at 800-338-3987, visiting the McGraw-Hill website at www.mhhe.com, or by contacting their local McGraw-Hill sales representative.

Teaching Supplements for Instructors

Website (www.mhhe.com/enger12e)

The text-specific website offers an extensive array of teaching tools. In addition to all of the student assets available, this site includes:

- Instructor's manual
- Class activities
- Answers to review and critical thinking questions
- Interactive base maps
- Additional Case Studies and Global Perspective features that appeared in previous editions of the book

Presentation Center (found at www.mhhe.com/enger12e)
Build instructional materials wherever whenever and however you want!
McGraw-Hill's Presentation Center is an on-line digital library containing assets such as photos, artwork, animations, PowerPoints, and other media types that can be used to create customized lectures, visually enhanced tests and quizzes, compelling course websites, or attractive printed support materials.

Access to your book, access to all books!
The Presentation Center library includes thousands of assets from many McGraw-Hill titles. This ever-growing resource gives instructors the power to utilize assets specific to an adopted textbook as well as content from all other books in the library.

Nothing could be easier!
Accessed from the instructor side of your textbook's website, Presentation Center's dynamic search engine allows you to explore by discipline, course, textbook chapter, asset type, or keyword. Simply browse, select, and download the files you need to build engaging course materials. All assets are copyright McGraw-Hill Higher Education but can be used by instructors for classroom purposes.

Instructors will find the following digital assets for *Environmental Science*:

- **Color Art** Full-color digital files of illustrations in the text can be readily incorporated into lecture presentations, exams, or custom-made classroom materials. These include all of the 3-D realistic art found in this edition, representing some of the most important concepts in environmental science.
- **Photos** Digital files of photographs from the text can be reproduced for multiple classroom uses.
- **Additional Photos** 317 full-color bonus photographs are available in a separate file. These photos are searchable by content and will add interest and contextual support to your lectures.
- **Tables** Every table that appears in the text is provided in electronic format.
- **Videos** This special collection of 84 underwater video clips displays interesting habitats and behaviors for many animals in the ocean.

- **Animations** 94 full-color animations that illustrate many different cepts covered in the study of environmental science are available for in creating classroom lectures, testing materials, or online course communication. The visual impact of motion will enhance classroom presentations and increase comprehension.
- **Global Base Maps** 88 base maps for all world regions and major sub-regions are offered in four versions: black-and-white and full-color, both with labels and without labels. These choices allow instructors the flexibility to plan class activities, quizzing opportunities, study tools, and PowerPoint enhancements.
- **PowerPoint Lecture Outlines** Ready-made presentations that combine art and photos and lecture notes are provided for each of the 19 chapters of the text. These outlines can be used as they are or tailored to reflect your preferred lecture topics and sequences.
- **PowerPoint Image Slides** For instructors who prefer to create their lectures from scratch, all illustrations, photos, and tables are preinserted by chapter into blank PowerPoint slides for convenience.

Computerized Test Bank

The computerized test bank, also found at www.mhhe.com/enger12e, is a valuable resource for the instructor. The test bank utilizes EZ-Test software for quickly creating customized exams. This flexible and user-friendly program allows instructors to search for questions by topic, format, or difficulty level and edit existing questions or add new ones. Multiple versions of the test can be created, and any test can be exported for use with course management systems such as WebCT, Blackboard, or PageOut. Word files of the test bank are included for those instructors who prefer to work outside of the test-generator software.

Earth and Environmental Science DVD by Discovery Channel Education (ISBN: 978-0-07-352541-9; MHID: 0-07-352541-3)

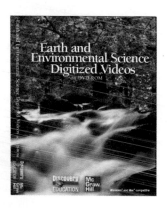

Begin your class with a quick peek at science in action. The exciting NEW DVD by Discovery Channel Education offers 50 short (three- to five-minute) videos on topics ranging from conservation to volcanoes. Search by topic and download into your PowerPoint lecture.

McGraw-Hill's Biology Digitized Videos (ISBN: 978-0-07-312155-0; MHID: 0-07-312155-X)

Licensed from some of the highest quality life-science video producers in the world, these brief video clips on DVD range in length from 15 seconds to two minutes and cover all areas of general biology, from cells to ecosystems. Engaging and informative, McGraw-Hill's digitized biology videos will help capture students' interest while illustrating key biological concepts, applications, and processes.

Course Delivery Systems

With help from WebCT, Blackboard, and other course management systems, professors can take complete control of their course content. Course cartridges containing website content, on-line testing, and powerful student tracking features are readily available for use within these platforms.

Electronic Textbook

CourseSmart is a new way for faculty to find and review eTextbooks. It's also a great option for students who are interested in accessing their course materials digitally and saving money. CourseSmart offers thousands of the most commonly adopted textbooks across hundreds of courses from a wide variety of higher education publishers. It is the only place for faculty to review and compare the full text of a textbook online, providing immediate access without the environmental impact of requesting a print exam copy. At CourseSmart, students can save up to 50 percent off the cost of a print book, reduce their impact on the environment, and gain access to powerful web tools for learning including full text search, notes and highlighting, and email tools for sharing notes between classmates. **www.CourseSmart.com**

Learning Supplements for Students

Website (www.mhhe.com/enger12e)

This text-specific website offers a wide variety of student resources providing many opportunities to master the core concepts in environmental science. Visit the website to learn more about the exciting features provided for students through the *Environmental Science: A Study of Interrelationships* website:

- Practice quizzes
- Labeling exercises
- Weblinks
- Guide to electronic research
- Period table
- Career information
- Guest essays

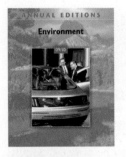

Annual Editions: Environment 09/10 by Sharp (MHID: 0-07-351549-3)

This Twenty-Eighth Edition provides convenient, inexpensive access to current articles selected from some of the most respected magazines, newspapers, and journals published today. Organizational features include: an annotated listing of selected World Wide Web sites; an annotated table of contents; a topic guide; a general introduction; brief overviews for each section; and an instructor's resource guide with testing materials. *Using Annual Editions in the Classroom* is also offered as a practical guide for instructors.

Taking Sides: Clashing Views on Controversial Environmental Issues, Expanded Thirteenth Edition by Easton (MHID: 0-07-351445-4)

This Expanded Thirteenth Edition of *Taking Sides: Environmental Issues* presents two additional current controversial issues in a debate-style format designed to stimulate student interest and develop critical thinking skills. Each issue is thoughtfully framed with an issue summary, an issue introduction, and a postscript. *Taking Sides* readers also feature annotated listings of selected World Wide Web sites. An instructor's resource guide with testing material is available for each volume. *Using Taking Sides in the Classroom* is also an excellent instructor resource.

Field & Laboratory Exercises in Environmental Science, Seventh Edition, by Enger and Smith (ISBN: 978-0-07-290913-5; MHID: 0-07-290913-7)

The major objectives of this manual are to provide students with hands-on experiences that are relevant, easy to understand, applicable to the student's life, and presented in an interesting, informative format. Ranging from field and lab experiments to conducting social and personal assessments of the environmental impact of human activities, the manual presents something for everyone, regardless of the budget or facilities of each class. These labs are grouped by categories that can be used in conjunction with any introductory environmental textbook.

Global Studies: The World at a Glance, Second Edition, by Tessema (ISBN: 978-0-07-340408-0; MHID: 0-07-340408-X)

This book features a compilation of up-to-date data and accurate information on some of the important facts about the world we live in. While it is close to impossible to stay current on every nation's capital, type of government, currency, major languages, population, religions, political structure, climate, economics, and more, this book is intended to help students to understand these essential facts in order to make useful applications.

Sources: Notable Selections in Environmental Studies, Second Edition, by Goldfarb (ISBN: 978-0-07-303186-6; MHID: 0-07-303186-0)

This volume brings together primary source selections of enduring intellectual value—classic articles, book excerpts, and research studies—that have shaped environmental studies and our contemporary understanding of it. The book includes carefully edited selections from the works of the most distinguished environmental observers, past and present. Selections are organized topically around the following major areas of study: energy, environmental degradation, population issues and the environment, human health and the environment, and environment and society.

Student Atlas of Environmental Issues by Allen (ISBN: 978-0-69-736520-0; MHID: 0-69-736520-4)

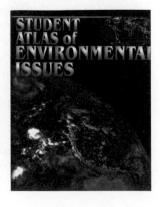

This atlas is an invaluable pedagogical tool for exploring the human impact on the air, waters, biosphere, and land in every major world region. This informative resource provides a unique combination of maps and data that help students understand the dimensions of the world's environmental problems and the geographical basis of these problems.

We live in an age of information. Computers, e-mail, the Internet, CD-ROMs, instant news, and fax machines bring us information more quickly than ever before. A simple search of the Internet will provide huge amounts of information. Some of the information has been subjected to scrutiny and is quite valid, some is well-informed opinion, some is naive misinformation, and some is even designed to mislead. How do we critically evaluate the information we get?

Critical thinking involves a set of skills that helps us to evaluate information, arguments, and opinions in a systematic and thoughtful way. Critical thinking also can help us better understand our own opinions as well as the points of view of others. It can help us evaluate the quality of evidence, recognize bias, characterize the assumptions behind arguments, identify the implications of decisions, and avoid jumping to conclusions.

CHARACTERISTICS OF CRITICAL THINKING

Critical thinking involves skills that allow us to sort information in a meaningful way and discard invalid or useless information while recognizing that which is valuable. Some key components of critical thinking are:

RECOGNIZE THE IMPORTANCE OF CONTEXT

All information is based on certain assumptions. It is important to recognize what those assumptions are. Critical thinking involves looking closely at an argument or opinion by identifying the historical, social, political, economic, and scientific context in which the argument is being made. It is also important to understand the kinds of bias contained in the argument and the level of knowledge the presenter has.

CONSIDER ALTERNATIVE VIEWS

A critical thinker must be able to understand and evaluate different points of view. Often these points of view may be quite varied. It is important to keep an open mind and to look at all the information objectively and try to see the value in alternative points of view. Often people miss obvious solutions to problems because they focus on a certain avenue of thinking and unconsciously dismiss valid alternative solutions.

EXPECT AND ACCEPT MISTAKES

Good critical thinking is exploratory and speculative, tempered by honesty and a recognition that we may be wrong. It takes courage to develop an argument, engage in debate with others, and admit that your thinking contains errors or illogical components. By the same token, be willing to point out what you perceive to be shortcomings in the arguments of others. It is always best to do this with good grace and good humor.

HAVE CLEAR GOALS

When analyzing an argument or information, keep your goals clearly in mind. It is often easy to get sidetracked. A clear goal will allow you to quickly sort information into that which is pertinent and that which may be interesting but not germane to the particular issue you are exploring.

EVALUATE THE VALIDITY OF EVIDENCE

Information comes in many forms and has differing degrees of validity. When evaluating information, it is important to understand that not all the information from a source may be of equal quality. Often content about a topic is a mix of solid information interspersed with less certain speculations or assumptions. Apply a strong critical attitude to each separate piece of information. Often what appears to be a minor, insignificant error or misunderstanding can cause an entire argument to unravel.

CRITICAL THINKING REQUIRES PRACTICE

As with most skills, you become better if you practice. At the end of each chapter in the text, there is a series of questions that allow you to practice critical thinking skills. Some of these questions are straightforward and simply ask you to recall information from the chapter. Others ask you to apply the information from the chapter to other similar contexts. Still others ask you to develop arguments that require you to superimpose the knowledge you have gained from the chapter on quite different social, economic, or political contexts from your own.

Practice, practice, practice.

GUIDED TOUR

UP-TO-DATE COVERAGE OF CURRENT ISSUES

Environmental issues and the facts related to issues constantly change. Therefore, the authors strive to provide the most current information available.

WRITTEN FOR STUDENTS

HEADINGS AND SUBHEADINGS

Numerous headings and subheadings help students follow the organization of the subject matter.

CASE STUDIES

Case studies provide in-depth coverage of current topics.

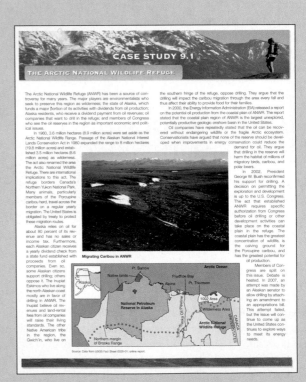

INTERRELATEDNESS IS A CENTRAL THEME

WATER CONNECTIONS

The new Water Connections feature shows how water is involved in most aspects of environmental issues.

TEXT INCLUDES MORE THAN SCIENCE.

Social, political, and economic aspects of environmental issues are included throughout the text.

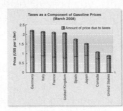

POSITIVE TRENDS ARE HIGHLIGHTED

GOING GREEN

The new Going Green feature shows specific examples of actions that are environmentally friendly.

CAMPUS SUSTAINABILITY INITIATIVES

The new Campus Sustainability Initiative points out actions of students and the institutions they attend that are making a difference.

THINKING GREEN

The new Thinking Green feature points out individual actions that can have an impact and encourage students to be involved.

THINKING GREEN

1. Look for locally grown produce in the supermarket—less energy is used to transport locally grown products.
2. Join a local environmental organization.
3. Volunteer for your local Earth Day event in April.
4. Visit a natural area, nature center, or park typical of your region and learn to identify five plants.

5. Go to the website of the League of Conservation Voters, click on the Scorecard tab, and find out the "environmental score" of your senators and representative.

EXCELLENT ILLUSTRATIONS

Photos, drawings, and tables are used to help students visualize complex ideas and organize their thinking.

Going Green — CONSERVATION EASEMENTS

There are over 1600 private organizations in the United States that are involved in conservation of land. Some are small, single-purpose organizations that protect a small parcel of land with special conservation value. On the other hand, The Nature Conservancy is an international organization that has protected millions of acres.

People often develop an attachment to their land and wish to see it preserved even after they have died. People may have a long family history of using the land for farming or ranching and want to see that use continue. Others may recognize that their land has special conservation value because of its geology, scenic value, or biodiversity and wish to see it protected for the public good. Others may simply have a purely emotional reason for wanting to preserve their land. One of the tools used by land conservation organizations is a legal tool known as a *conservation easement*.

A conservation easement is a legally binding agreement placed on a piece of privately held land that limits the future use or development even when the land is passed to heirs or sold. For example, a conservation easement may prohibit the subdividing of a piece of land or restrict buildings to a specific portion of the property, or an easement may specify that the public must have access to view significant biological or geological features. Alternatively, the easement may restrict access to protect endangered species or archeological sites.

Regardless of their motivation, when people enter into a conservation easement they give up something. In some cases, people donate a conservation easement and receive no financial benefit. In other cases, they may sell a conservation easement to an organization that agrees to provide stewardship of the property into the future. In nearly all cases the placement of a conservation easement on property diminishes its economic value, since its future use is restricted. Yet, thousands of people have entered into such arrangements. As of 2005, in the United States, over 6 million acres of land (an area about the size of Vermont) had been protected by conservation easements.

CAMPUS SUSTAINABILITY INITIATIVE

THE ASSOCIATION FOR THE ADVANCEMENT OF SUSTAINABILITY IN HIGHER EDUCATION

The Association for the Advancement of Sustainability in Higher Education (AASHE) was founded in 2006 as a membership organization of colleges and universities in the United States and Canada. There are currently about 500 member colleges and universities. AASHE's mission is to promote sustainability in all aspects of higher education. Its definition of sustainability includes human and ecological health, social justice, secure livelihoods, and a better world for all generations. A core concept of AASHE is that higher education must be a leader in preparing students and employees to understand the importance of sustainability and to work toward achieving it. Furthermore, campuses should showcase sustainability in their operations and curriculum.

To accomplish its goals, AASHE sponsors conferences and workshops to educate members. It also provides networking opportunities and an e-bulletin to facilitate the exchange of information about sustainable practices on campuses.

AASHE is currently developing a rating system that will allow educational institutions to assess their progress toward achieving sustainability. The Sustainability Tracking, Assessment, and Rating System (STARS) focuses on three major categories of activity: education and research, operations, and administration and finance.

In each chapter of this edition of *Environmental Science: A Study of Interrelationships*, we will highlight the efforts of one of the member colleges of AASHE to achieve sustainability. Is your college a member? Go to the AASHE website and check its membership list.

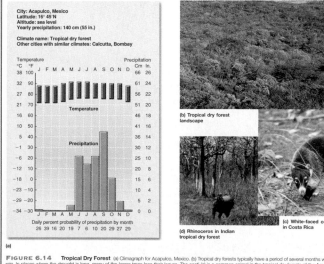

City: Acapulco, Mexico
Latitude: 16° 45´N
Altitude: sea level
Yearly precipitation: 140 cm (55 in.)

Climate name: Tropical dry forest
Other cities with similar climates: Calcutta, Bombay

(b) Tropical dry forest landscape

(c) White-faced coati in Costa Rica

(d) Rhinoceros in Indian tropical dry forest

FIGURE 6.14   Tropical Dry Forest (a) Climagraph for Acapulco, Mexico. (b) Tropical dry forests typically have a period of several months with no rain. In places where the drought is long, many of the larger trees lose their leaves. The coati (c) is a common animal in the tropical dry forests of the Americas. The endangered one-horned rhinoceros (d) is an inhabitant of the tropical dry forest of Asia.

Environmental resistance

Decreasing O_2 supply — Low food supply — Disease — Predators — Limited space

Carrying capacity

Population size

Time

FIGURE 7.7   Carrying Capacity A number of factors in the environment, such as oxygen supply, food supply, diseases, predators, and space, determine the number of organisms that can survive in a given area—the carrying capacity of that area. The environmental factors that limit populations are known collectively as environmental resistance.

CRITICAL THINKING AND APPLICATIONS – VITAL FOR EVERY STUDENT!

CRITICAL THINKING ESSAY

An essay on critical thinking is present in the front matter of the text.

ISSUES & ANALYSIS READINGS

Issues & Analysis readings present real-world, current issues and provide questions that prompt students to think about the complex issues involved.

WHAT'S YOUR TAKE?

This feature presents an issue and asks students to choose one side of the issue and develop arguments that support their position. This activity helps students develop and enhance their critical thinking skills.

CRITICAL THINKING QUESTIONS

Critical Thinking Questions appear at the end of each chapter. The questions require students to evaluate information, recognize bias, characterize the assumptions behind arguments, and organize information.

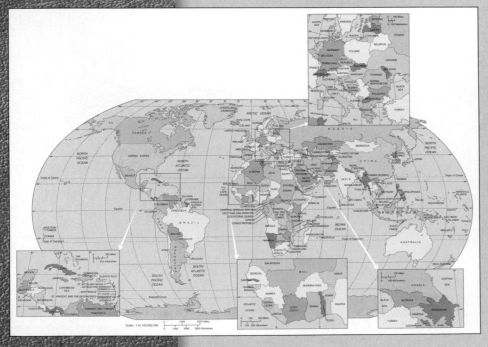

FOLDOUT MAPS

Included at the end of this book as foldouts are two maps: a political map showing the boundaries of the countries throughout the world and a global vegetation map.

CHAPTER 1

ENVIRONMENTAL INTERRELATIONSHIPS

Environmental science is the study of interrelationships between humans and the natural world. These fishermen are interacting with the aquatic environment of fish and if they are successful they will influence the population of fish in this part of the ocean.

CHAPTER OUTLINE

The Nature of Environmental Science
    Interrelatedness Is a Core Concept
    An Ecosystem Approach
    Political and Economic Issues
    The Global Nature of Environmental Concerns
Regional Environmental Concerns
    The Wilderness North
    The Agricultural Middle
    The Dry West
    The Forested West
    The Great Lakes and Industrial Northeast
    The Diverse South

ISSUES & ANALYSIS
Government Regulation and Personal Property 12

CAMPUS SUSTAINABILITY INITIATIVE
The Association for the Advancement of Sustainability in Higher Education 12

GOING GREEN
Individual Decisions Matter 7

WATER CONNECTIONS
Social and Biological Interactions in the Management of Keoladeo National Park, India 4

OBJECTIVES

After reading this chapter, you should be able to:

- Understand why environmental problems are complex and interrelated.

- Realize that environmental problems involve social, ethical, political, and economic issues, not just scientific issues.

- Understand that acceptable solutions to environmental problems often are not easy to achieve.

- Understand that all organisms have an impact on their surroundings.

- Understand what is meant by an ecosystem approach to environmental problem solving.

- Recognize that different geographic regions have somewhat different environmental problems, but the process for resolving them is often the same and involves compromise.

A Global Perspective on "Sustainable Development of the Mekong River Delta" can be found on the book's website
at www.mhhe.com/enger12e along with other interesting readings.

THE NATURE OF ENVIRONMENTAL SCIENCE

Environmental science is an interdisciplinary field that includes both scientific and social aspects of human impact on the world. The word *environment* is usually understood to mean the surrounding conditions that affect organisms. In a broader definition, **environment** is everything that affects an organism during its lifetime. In turn, all organisms including people affect many components in their environment. **Science** is an approach to studying the natural world that involves formulating hypotheses and then testing them to see if the hypotheses are supported or refuted. However, because humans are organized into complex societies, environmental science also must deal with politics, social organization, economics, ethics, and philosophy. Thus, environmental science is a mixture of the traditional science, individual and societal values, economic factors, and political awareness that are important to solving environmental problems. (See figure 1.1.)

Although environmental science as a field of study is evolving, it is rooted in the early history of civilization. Many ancient cultures expressed a reverence for the plants, animals, and geographic features that provided them with food, water, and transportation. These features are still appreciated by many modern people. Although the following quote from Henry David Thoreau (1817–62) is over a century old, it is consistent with current environmental philosophy:

> I wish to speak a word for Nature, for absolute freedom and wildness, as contrasted with a freedom and culture merely civil . . . to regard man as an inhabitant, or a part and parcel of Nature, rather than a member of society.

The current interest in the state of the environment began with philosophers like Thoreau and scientists like Rachel Carson and received emphasis from the organization of the first Earth Day on April 22, 1970. Subsequent Earth Days reaffirmed this commitment. As a result of this continuing interest in the state of the world and how people both affect it and are affected by it, environmental science is now a standard course or program at many colleges. It is also included in the curriculum of high schools. Most of the concepts covered by environmental science courses had previously been taught in ecology, conservation, biology, or geography courses. Environmental science incorporates the scientific aspects of these courses with input from the social sciences, such as economics, sociology, and political science, creating a new interdisciplinary field.

INTERRELATEDNESS IS A CORE CONCEPT

A central factor that makes the study of environmental science so interesting/frustrating/challenging is the high degree of interrelatedness among seemingly unrelated factors. Many naturalists and philosophers have thought about this over the years. The naturalist John Muir captured this interrelatedness theme in the following quote:

Tug on anything at all and you'll find it connected to everything else in the universe.

Charles Darwin exemplified this same kind of thinking when he proposed that the production of seeds in red clover plants in fields in England was directly related to the number of cats in the area. His logic was as follows: Villagers keep cats, cats hunt and kill meadow mice, bumblebees build their nests in the ground, meadow

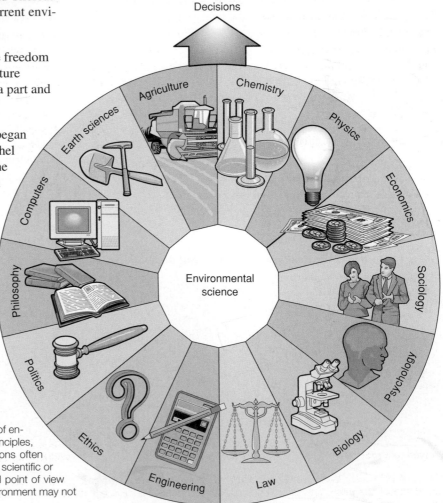

FIGURE 1.1 **Environmental Science** The field of environmental science involves an understanding of scientific principles, economic influences, and political action. Environmental decisions often involve compromise. A decision that may be supportable from a scientific or economic point of view may not be supportable from a political point of view without modification. Often political decisions relating to the environment may not be supported by economic analysis.

mice eat the honeycomb and larvae of bumblebees, and bumblebees have long tongues that allow them to pollinate clover, which other bees have difficulty doing. Therefore, since cats eat mice, more bumblebees survive to pollinate clover plants. While this may seem to be a fanciful story, let's look at a concrete example that illustrates how seemingly distinct things may actually be interconnected.

The reintroduction of wolves to Yellowstone National Park has resulted in many changes to the Yellowstone ecosystem. The initial introduction of 31 wolves in 1995 and 1996 has resulted in a current population of about 320 wolves. Several changes to the Yellowstone ecosystem can be directly attributed to the alterations brought about by wolves.

Wolves kill and eat elk. This has resulted in a significant reduction in the size of the elk herd from about 19,000 prior to wolf reintroduction to about 11,000 now. The presence of wolves also has modified the behavior of elk. Because they must be more vigilant and move about more because of the predatory behavior of wolves, elk spend less time feeding on willow, cottonwood, and aspen. Both the change in behavior and the reduced size of the elk herd have allowed the regeneration of stands of willow and aspen. This has in turn resulted in increased numbers of beavers that use these trees for food. The dams built by beavers tend to slow the flow of water and increase the recharge of groundwater. Furthermore, the stands of willow along the banks of streams cool the water and improve fish habitat. The stands of willow also provide needed habitat for some songbirds.

Wolves directly compete with coyotes and kill them if they have the opportunity. Thus, since the reintroduction of wolves the coyote population has fallen to half its previous level. There is evidence that the populations of the prey of coyotes—voles, mice, and other rodents—have increased. The increased availability of this food source has resulted in an increase in the number of foxes, hawks, and owls.

Thus, it is fair to say that the reintroduction of the wolf has changed how water flows through the landscape, and has led to increased populations of many organisms—willow, aspen, beaver, songbirds, foxes, certain rodent, hawks, and owls; and to the decline in the population of other organisms—coyote and elk (figure 1.2). Truly this is a story that illustrates the point made by Muir—*Tug on*

anything at all and you'll find it connected to everything els[e] universe.

It is important to understand that the reintroduction of wolves to Yellowstone was not purely a scientific undertaking. Many biologists and environmentalists argued that it was important to restore the wolf to its former habitat for biological reasons. Others looked at the issue in terms of ethics and felt that humans had an ethical obligation to restore wolves to their former habitat. While park managers could easily see the problems created by a lack of wolves and a huge elk population, they could not simply make the decision to bring back the wolf. A long history of controlling animals that could prey on livestock had to be overcome. Ranchers strongly opposed the reintroduction of wolves and saw this as an economic issue. If wolves left the park and killed their livestock, they would lose money. The farm lobby in Congress is very strong and fought long and hard to prevent the reintroduction. After a lengthy period of hearings and many compromises— including a fund to pay ranchers for cattle killed by wolves—the U.S. Fish and Wildlife Service was authorized to proceed with the reintroductions. Thus, the interconnectedness theme associated with the reintroduction of wolves to Yellowstone also applies to social, economic, and political realms of human activity.

AN ECOSYSTEM APPROACH

Environmental science involves an understanding that the natural world is organized into interrelated units called ecosystems. An **ecosystem** is a region in which the organisms and the physical environment form an interacting unit. Within an ecosystem there is a complex network of interrelationships. For example, weather affects plants, plants use minerals in the soil and are food for animals, animals spread plant seeds, plants secure the soil, and plants evaporate water, which affects weather.

Some ecosystems have easily recognized boundaries. Examples are lakes, islands, floodplains, watersheds separated by mountains, and many others. Large ecosystems always include smaller ones. A large watershed, for example, may include a number of lakes, rivers, streams, and a variety of terrestrial ecosystems. A forest ecosystem

| Wolves reintroduce | Elk decline | Willows increase | Beavers increase |

FIGURE 1.2 **Wolf Reintroduction Reveals Interrelatedness** The reintroduction of wolves to Yellowstone National Park initiated changes that rippled through the Yellowstone ecosystem. As wolves increased, elk and coyote populations decreased. Decreases in these populations resulted in increases in the populations of willow and aspen trees, beaver, foxes, and songbirds.

Keoladeo National Park is a small (2873 hectares; 7096 acres) artificial wetland system located near Bharatpur on the Ganges plain in India. The wetland was created in 1750 by local royalty to attract migratory birds for hunting. Today, over 350 bird species, including the endangered Siberian crane, inhabit the park seasonally. In 1982, Keoladeo was declared a national park and was designated as a World Heritage Site by the UNESCO in 1985.

Its designation as a park resulted in several changes to the way in which the land was used. Local villagers were prohibited from using the land to graze their cattle and water buffalo, and they could no longer harvest plants to use as food for their animals and for other purposes. It turned out that the traditional uses of the land were important for maintaining habitat suitable for the Siberian crane.

Water buffalo, certain aquatic plants, and the Siberian crane coexisted in a three-way relationship. The buffalo grazed on the aquatic vegetation, which kept it short and maintained open water areas. These conditions made it possible for the cranes to dig up rhizomes and tubers of the aquatic plants for food. In 1983, however, the India Wildlife Protection Act prohibited the grazing of buffalo in the park. As a result, the weeds grew to their full height and in solid masses that created a physical barrier that prevented the cranes from accessing their main food source, which led to a dramatic decrease in the numbers of cranes in the park.

The Wildlife Protection Act was formulated and implemented without consultation with local scientists or local communities. A decade-long study, costing nearly US$ 1 million indicated that grazing buffalo were key to controlling the growth of grasses and water weeds and that control of these plants was needed to support the Siberian crane and other bird populations. Local communities and scientists already knew this.

In recent years, drought and competition with farmers for irrigation water have resulted in many of the ponds drying up. Many of the migratory birds that visited the site have not been doing so because of the lack of

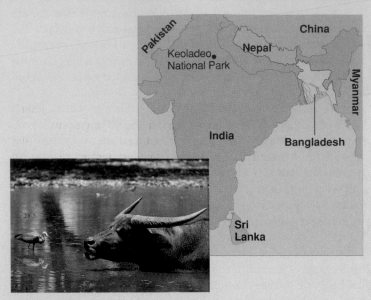

Water buffalo

water. The lack of appropriate water management has caused UNESCO to threaten to remove the World Heritage Site designation. The Indian government responded by constructing a canal to supply water that will maintain adequate water levels during periods of drought. Although the canal has brought additional water, it is not enough to solve the problem of drought. Therefore, ways to provide additional water are being considered.

---

may cover hundreds of square kilometers and include swampy areas, openings, and streams as subsystems within it. Often the boundaries between ecosystems are indistinct, as in the transition from grassland to desert. Grassland gradually becomes desert, depending on the historical pattern of rainfall in an area. Thus, defining an ecosystem boundary is often a matter of practical convenience.

However, an ecosystem approach is important to dealing with environmental problems. The task of an environmental scientist is to recognize and understand the natural interactions that take place and to integrate these with the uses humans must make of the natural world.

POLITICAL AND ECONOMIC ISSUES

Most social and political decisions are made with respect to political jurisdictions, but environmental problems do not necessarily coincide with these artificial political boundaries. For example, air pollution may involve several local units of government, several states or provinces, and even different nations. Air pollution generated in China affects air quality in western coastal states in the United States and in British Columbia, Canada. On a more local level, the air pollution generated in Juarez, Mexico, causes problems in the neighboring city of El Paso, Texas. But the issue is more than air quality and human health. Lower wage rates and less strict environmental laws have influenced some U.S. industries to move to Mexico for economic advantages. Mexico and many other developing nations are struggling to improve their environmental image and need the money generated by foreign investment to improve the conditions and the environment in which their people live.

The issue of declining salmon stocks in the Pacific Northwest of the United States and British Columbia, Canada, illustrates the political and economic friction associated with a resource that crosses political boundaries. From the U.S. perspective alone there are five federal cabinet-level departments, two federal agencies, and five federal laws, and numerous tribal treaties that affect decisions about the use of this resource. Furthermore, commercial fishers from several states and provinces are economically affected by any decisions made concerning the harvesting of these fish. (See figure 1.3.) They are politically active and try to influence the laws and rulings of state, provincial, and national governments.

The freshwater resources of the Great Lakes are a shared resource of eight U.S. states and the Canadian provinces of Ontario and Quebec. Problems associated with the use of this resource are regulated by the International Joint Commission. The International Joint Commission was established in 1909, when the Boundary Waters Treaty was signed between the United States and Canada. The treaty was established in part to provide that the "boundary waters and waters flowing across the boundary shall not be polluted on either side to the injury of health or property of the other." The commission has been instrumental in identifying areas of concern and encouraging the cleanup of polluted sites that affect the quality of the Great Lakes and other boundary waters. In general, the two national governments and the state and provincial governments have listened to the commission's advice and have responded by initiating cleanup activities and regulating the export of water from the region.

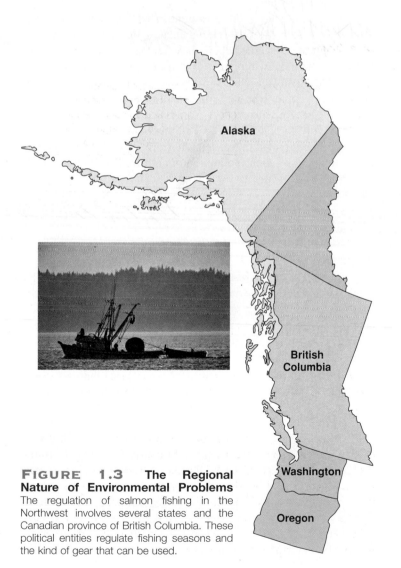

FIGURE 1.3 The Regional Nature of Environmental Problems The regulation of salmon fishing in the Northwest involves several states and the Canadian province of British Columbia. These political entities regulate fishing seasons and the kind of gear that can be used.

THE GLOBAL NATURE OF ENVIRONMENTAL CONCERNS

As the human population has increased, the natural ecosystems of the Earth have been stressed. Recognition of this has led to international activities to address concerns about the Earth's natural systems and how humans are affecting them.

The Earth Summit

The first worldwide meeting of heads of state that was directed to a concern for the environment took place at the Earth Summit, formally known as the United Nations Conference on Environment and Development (UNCED) in Rio de Janeiro in 1992. One of the key outcomes of the conference was a series of policy statements on sustainable development that were identified as Agenda 21.

More than 178 governments at the 1992 conference adopted three documents related to sustainable development: Agenda 21, the Rio Declaration on Environment and Development, and the Statement of Principles for the Sustainable Management of Forests. The United Nations Commission on Sustainable Development was created in 1993 to monitor and report on implementation of the agreements. Follow-up conferences were held in 1997 and 2002 to assess progress.

Climate Change

In 1997, representatives from 125 nations met in Kyoto, Japan, for the Third Conference of the United Nations Framework Convention on Climate Change. This conference, commonly referred to as the Kyoto Conference on Climate Change, resulted in commitments from the participating nations to reduce their overall emissions of six greenhouse gases (linked to global warming) by at least 5 percent below 1990 levels and to do so between the years 2008 and 2012. The Kyoto Protocol, as the agreement was called, was viewed by many as one of the most important steps to date in environmental protection and international diplomacy. It is clear that many countries will not meet their goals, but are making progress toward their goals.

The Millennium Ecosystem Assessment

In 2005, the Millennium Ecosystem Assessment was completed. It was initiated by the United Nations and included input from over 1360 experts from around the world. It looked at the services provided by ecosystems and evaluated the status of each service. Four broad areas of ecosystem services were identified: supporting services, provisioning services, regulating services, and cultural services. Supporting services include such ecosystem functions as: photosynthesis, soil formation, nutrient cycling, and water cycling. Provisioning services include resources provided by ecosystems such as: food, fiber, genetic resources, natural medicines, and freshwater. Regulating services include ecosystem activities that affect air quality, water flow, erosion control, water purification, climate control, disease regulation, pest regulation, pollination, and natural hazards. Cultural services include spiritual, religious, and aesthetic values, and the use of the natural world for recreation.

In general the report is quite negative. As the human population has grown we are putting pressure on the natural ecosystems of the world, and most are being negatively affected. Food production is one bright spot. Production of crops, livestock, and fish from aquaculture have increased. However, this is at the expense of the loss of soil from erosion, the conversion of natural ecosystems to managed agricultural systems, and overconsumption of water resources.

REGIONAL ENVIRONMENTAL CONCERNS

To illustrate the interrelated nature of environmental issues, we will look at several regions of North America and highlight some of the key features and issues of each. (See figure 1.4.) For example, protecting endangered species is a concern in many parts of the world. In the Pacific Northwest, an endangered species known as the northern spotted owl depends on undisturbed mature forests for its survival. Development and logging may conflict with the survival of the owl. In most metropolitan areas, the problem of endangered species is purely historical, since the construction of cities has destroyed the previously existing ecosystem. Here we present a number of regional vignettes to illustrate the complexity

and interrelatedness of environmental issues. Each region has specific environmental issues that capture the attention of the people who live there.

THE WILDERNESS NORTH

Much of Alaska and Northern Canada can be characterized as **wilderness**—areas with minimal human influence. Much of this land is owned by governments, not by individuals, so government policies have a large effect on what happens in these regions. These areas have important economic values in their trees, animals, scenery, and other natural resources. Exploitation of the region's natural resources involves significant trade-offs. Usually, a portion of the natural world is altered permanently, but the area altered is so small that many people consider it insignificant. Because of the severe climate, northern wilderness areas tend to be very sensitive to insults and take a long time to repair damage done by unwise exploitation. Mining, oil exploration, development of hydroelectric projects, and harvesting of timber all require roads and other human artifacts. These activities also alter the culture of native people by introducing new technologies and generating economic benefits.

In the past, many short-term political and economic decisions failed to look at long-term environmental implications. Today, however, people are concerned about these remaining wilderness areas. Politicians are more willing to look at the scientific and recreational values of wilderness as well as the economic value of exploitation.

Native people, who consider much of this region to be their land, have become increasingly sophisticated in negotiating with state, provincial, and federal governments to protect rights they feel they were granted in treaties. They are sensitive to changes in land use or government policy that would force changes in their traditional way of life.

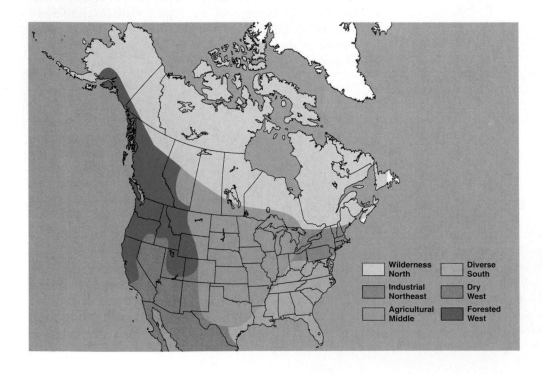

FIGURE 1.4 **Regions of North America** Because of natural features of the land and the uses people make of the land, different regions of North America face different kinds of environmental issues. In each region, people face a large number of specific issues, but certain kinds of issues are more important in some regions than others.

Legend:
- Wilderness North
- Industrial Northeast
- Agricultural Middle
- Diverse South
- Dry West
- Forested West

There is a growing awareness that sustainability needs to be a core value if future generations are to inherit an Earth worth having. Those who support green initiatives are motivated in many different ways. Some are motivated by ethical or moral beliefs that they should "live lightly on the land." Some are motivated by the economic realities of rising energy costs or the costs associated with correcting environmental mistakes. Some simply want to be seen as having green values.

Regardless of their motivation, people around the world are making green decisions. Organizers of conferences and concerts are buying carbon credits to offset the impact of their events. Companies have discovered that consumers seek green products. Governments have passed laws that encourage their citizens to live more sustainably. Ultimately, however, green initiatives depend on individuals making everyday decisions. How many pairs of shoes do I really need? Do I really need the latest electronic gadget? Should I buy products that are produced locally? In the final analysis, most daily decisions have an environmental impact and you have a role to play.

To call attention to these bits of good news, several features in this book will highlight green initiatives. Each chapter will have a "Going Green" feature that highlights a particular green initiative. In addition, specific college campuses will be showcased for their particular green initiatives. Finally, you will be asked to consider changes that you can make that collectively can help lead to a sustainable society.

Walrus harvesting

A clear-cut forest

Grizzly bear fishing for salmon

FIGURE 1.5   The Wilderness North Protection of wilderness is a major issue in this region. The major points of conflict involve the government role in managing these lands and wildlife, the protection of the rights and beliefs of native people, and the desire of many to exploit the mineral and other resources of the region.

Concerned citizens, business interests, and environmental activists have become increasingly sophisticated in influencing decisions made by government. The process of compromise is often difficult and does not always ensure wise decisions, but most governments now realize they must listen to the concerns of their citizens and balance economic benefits with social and cultural benefits. (See figure 1.5.)

THE AGRICULTURAL MIDDLE

The middle of the North American continent is dominated by intensive agriculture. This means that the original, natural ecosystems have been replaced by managed agricultural enterprise. It is important to understand that this area was at one time wilderness. Today, you would need to search very hard to find regions of true wilderness in Iowa, Indiana, or southern Manitoba. Some special areas have been set aside to preserve fragments of the original natural plant and animal associations, but most of the land has been converted to agriculture wherever practical.

The economic value generated by this use of a rich soil resource is tremendous, and most of the land is privately owned. Governments cannot easily control what happens on these privately held lands. But governments indirectly encourage certain activities through departments of agriculture that encourage agricultural research, grant special subsidies to farmers in the form of guaranteed prices for their products and other special payments, and develop markets for products.

One of the major, nonpoint pollution sources (pollution that does not have an easily identified point of origin) is agriculture. Air pollution in the form of dust is an inevitable result of tilling the land. Soil erosion occurs when soil is exposed to wind and moving water and leads to siltation of rivers, impoundments, and lakes. Fertilizers and other agricultural chemicals blow or are washed from the areas where they are applied. Nutrients washed from the land enter rivers and lakes where they encourage the growth of algae,

Dust

Food production

Pesticide use

Erosion

FIGURE 1.6 **The Agricultural Middle** The rich soil resource of this region has been converted to managed agricultural activity. The use of pesticides and fertilizer and exposure of the land to erosion cause concern about pollution of surface and groundwater. Most farmers still maintain that these practices are essential in modern agriculture and that they can be used safely and with minimal pollution.

lowering water quality. The use of pesticides causes concern about human exposure, effects on wildlife that are accidentally exposed, and residues in foods produced.

Since many communities in this region rely on groundwater for drinking water, the use of fertilizers and pesticides, and their potential for entering the groundwater as a result of unwise or irresponsible use, is a consumer issue. In addition, many farmers use groundwater for irrigation, which lowers the water table and leaves less groundwater for other purposes.

In an effort to stay in business and preserve their way of life, farmers must use modern technology. Careful use of these tools can reduce their impact; irresponsible use causes increased erosion, water pollution, and risk to humans. (See figure 1.6.)

THE DRY WEST

Where rainfall is inadequate to support agriculture, ranching and raising livestock are possible. This is true in much of the drier portions of western North America. Because much of the land is of low economic value, most is still the property of the federal government, which encourages its use by providing water for livestock and irrigation at minimal cost, offering low-cost grazing rights, and encouraging mining and other development.

Many people believe that government agencies have seriously mismanaged these lands. They assert that the governmental agencies are controlled by special interest groups and powerful politicians sensitive to the demands of ranchers, that they subsidize ranchers by charging too little for grazing rights, and that they

allow destructive overgrazing because of the economic desires of ranchers. Ranchers argue that they require access to government-owned land, that they cannot afford significantly increased grazing fees, and that changing government policies would destroy a way of life that is important to the regional economy.

Water is an extremely valuable resource in this region. It is needed for municipal use and for agriculture. Many areas, particularly the river valleys, have fertile soils that can be used for intensive agriculture. Cash crops such as cotton, fruits, and vegetables can be grown if water is available for irrigation. Because water tends to evaporate from the soil rapidly, long-term use of irrigated lands often results in the buildup of salts in the soil, thus reducing fertility. Irrigation water flowing from fields is polluted by agricultural chemicals that make it unsuitable for other uses such as drinking. As cities in the region grow, an increasing conflict arises between urban dwellers who need water for drinking and other purposes, and ranchers and farmers who need the water for livestock and agriculture. Increased demand for water will result in shortages, and decisions will have to be made about who will ultimately get the water and at what price. If the urban areas get the water they want, some farmers and ranchers will go out of business. If the agricultural interests get the water, urban growth and development will have to be limited and expensive changes will have to be made to conserve domestic water use.

Because population density is low in most of this region, much of the land has a wilderness character. Increasingly, a conflict has developed between the economic management of the

Grazing

Irrigation

Urban areas

Grazing on federal land

FIGURE 1.7 **The Dry West** Water is a key issue in this region. Both city dwellers and rural ranchers and farmers need water, and conflict results when there is not enough water to satisfy the desires of all. In addition, much of the land in this region is owned by the federal government. This raises concerns about how the government manages the land and how government policy affects the people of the region.

land for livestock production and the desire on the part of many to preserve the "wilderness." Designating an area as wilderness means that certain uses are no longer permitted. This offends individuals and groups who have traditionally used the area for grazing, hunting, and other pursuits. A long history of use and abuse of this land by overgrazing, modification to encourage plants valuable for livestock, and the introduction of grasses for livestock has significantly altered the region so that it cannot truly be called wilderness. The low population density does, however, provide a remoteness and natural character that many seek to preserve. (See figure 1.7.)

THE FORESTED WEST

The coastal areas and mountain ranges of the western United States and Canada receive sufficient rainfall to support extensive coniferous forests. Since most of these areas are not suitable for farmland, they have been maintained as forests with some grazing activity in the more open forests. Governments and large commercial timber companies own large sections of these lands. Government forest managers (U.S. Forest Service, Bureau of Land Management, Environment Canada, and various state and provincial departments) historically have sold timbercutting rights at a loss and are thought by many to be too interested in the production of forest products at the expense of other, less tangible values. In 1993, the U.S. Forest Service was directed to stop below-cost timber sales.

This policy change has become a major issue in the old-growth forests of the Pacific Northwest, where timber interests maintain that they must have access to government-owned forests in order to

remain in business. Many of these areas have significant wilderness, scenic, and recreational value. Environmental interests point out that it makes no sense to complain about the destruction of tropical rainforests in South America while North America makes plans to cut large areas of previously uncut, temperate rainforest. Are the intangible values of preserving an ancient forest ecosystem as important as the economic values provided by timber and jobs?

Environmental organizations are concerned about the consequences logging would have on organisms that require mature, old-growth forests for their survival. Grizzly bear habitat in Alaska and British Columbia could be altered significantly by logging; the northern spotted owl has become a symbol of the conflict between logging and preservation in Oregon and Washington; and preservation of coastal redwood forests has become an issue in northern California. (See figure 1.8.)

THE GREAT LAKES AND INDUSTRIAL NORTHEAST

While much of the West and Central regions of North America are characterized by low population densities and small towns, major portions of the Great Lakes and Northeast are dominated by large metropolitan complexes that generate social and resource needs that are difficult to satisfy. Many of these older cities were formed around industrial centers that have declined, leaving behind poverty, environmental problems in abandoned industrial sites, and difficulties with solid waste disposal, air quality, and land-use priorities. Interspersed among the major metropolitan areas are small towns, farmland, and forests.

Forest resources

Grizzly bear

Cut logs being hauled

FIGURE 1.8   The Forested West The cutting of forested areas for timber production destroys the previous ecosystem. Some see the trees as a valuable resource that provides jobs and building materials. Others see the forest ecosystem as a natural resource that should be preserved. In addition, government ownership of much of this land has generated considerable political debate about what the appropriate use of the land should be.

One of the major resources of the region is water transport. The Great Lakes and eastern seacoast are extremely important to commerce; ships can travel throughout the area by way of the St. Lawrence Seaway and the Great Lakes through a series of locks and canals that bypass natural barriers. Because of the importance of shipping in this region, harbors have been constructed and waterways have been deepened by dredging. The waterways are maintained at considerable government expense.

One of the greatest problems associated with the industrial uses of the Great Lakes and East Coast is the historic problem of the contamination of the water with toxic materials. In some cases, unthinking or unethical individuals dumped toxins directly into the water. In other cases, small, accidental spills or leaks over long periods of time have contaminated the sediments in harbors and bays. The cleanup of contaminated industrial sites and sediments in the adjacent water is a major economic and social cost.

A major concern about these pollutants is that they bioaccumulate (see chapter 14) in the food chain. The concentrations of some chemicals in the fat tissue of top predators, such as lake trout and fish-eating birds, can be a million times higher than the concentration in the water. Because of this, government agencies have issued consumption advisories for some fish and shellfish in contaminated areas. Since many kinds of fish can swim great distances, advisories for the Great Lakes warn against eating certain fish taken anywhere within the lakes, not just from the site of contamination. Similarly, Chesapeake Bay has been subjected to years of thoughtless pollution, resulting in reduced fish and shellfish populations and advisories against consuming some organisms taken from the bay.

Water always generates considerable recreational value. Consequently, conflicts arise between those who want to use the water for industrial and shipping purposes and those who wish to use it for recreation. Due to the fact that so much of the North American population is concentrated in this region, the economic value of recreational use is extremely high. Consumer pressure is great to clean up contaminated sites and prevent future pollution. Contaminated areas do not enhance tourism or quality of life.

Most of these older, large cities had no plan to shape their growth. As a result, open space for people is limited and urban dwellers have few opportunities to interact with the natural world. Children who grow up in these cities often do not know that milk comes from a cow—they have never seen, smelled, or touched a cow. Consequently, urban people have difficulty understanding the feeling rural people have for the land. These urban dwellers may never have an opportunity to experience wilderness. Their major environmental priorities are cleaning up contaminated sites, providing more parks and recreation facilities, reducing air and water pollution, and improving transportation. (See figure 1.9.)

THE DIVERSE SOUTH

In many ways, the South is a microcosm of all the regions previously discussed. The petrochemical industry dominates the economies of Texas and Louisiana, and forestry and agriculture are significant elements of the economy in other parts of the region. Major metropolitan areas thrive, and much of the area is linked to the coast either directly or by the Mississippi River and its tributaries. The environmental issues faced in the South are as diverse as those in the other regions.

Inner-city decay in Chicago

New York harbor

Chicago central city

FIGURE 1.9 **The Great Lakes and the Industrial Northeast** Industry, waterways, and population centers are defining elements of this region. The historically extensive use of the Great Lakes and coastal areas of the Northeast for industry, because of the ease of providing water transportation, has resulted in many older cities with poor land-use practices. Rebuilding cities, providing recreational opportunities for urban dwellers, and repairing previous environmental damage are important issues. The water resources of the region provide transportation, recreation, and industrial opportunities.

Some areas of the South (particularly Florida) have had extremely rapid population growth, which has led to groundwater problems, transportation problems, and concerns about regulating the rate of growth. Growth means money to developers and investors, but it requires municipal services, which are the responsibility of local governments. Too many people and too much development also threaten remaining natural ecosystems.

Poverty has been a problem in many areas of the South. This creates a climate that encourages state and local governments to accept industrial development at the expense of other values. Often, jobs are more important than the environmental consequences of the jobs; low-paying jobs are better than no jobs.

The use of the coastline is of major concern in many parts of the South. The coast is a desirable place to live, which may encourage unwise development on barrier islands and in areas that are subject to flooding during hurricanes and other severe weather events. In addition, industrial activity along the coast has resulted in the loss of wetlands. (See figure 1.10.)

Miami metropolitan area

Everglades

Petrochemical plant

FIGURE 1.10 **The Diverse South** Poverty has been a historically important problem in the region. Often the creation of jobs was considered more important than the environmental consequences of those jobs. The use of coastal areas for industry has resulted in pollution of coastal waters. The heavy use of the Mississippi River for transportation and industry has caused pollution problems. In addition, the desirable climate in the South has resulted in intense pressure to develop new housing for those who want to move to the region. Unwise development of housing on fragile coastal sites has resulted in damage to buildings by storms and the actions of the oceans. This causes intense debate on land use.

Environmental Interrelationships 　　11

THE ASSOCIATION FOR THE ADVANCEMENT OF SUSTAINABILITY IN HIGHER EDUCATION

The Association for the Advancement of Sustainability in Higher Education (AASHE) was founded in 2006 as a membership organization of colleges and universities in the United States and Canada. There are currently about 500 member colleges and universities. AASHE's mission is to promote sustainability in all aspects of higher education. Its definition of sustainability includes human and ecological health, social justice, secure livelihoods, and a better world for all generations. A core concept of AASHE is that higher education must be a leader in preparing students and employees to understand the importance of sustainability and to work toward achieving it. Furthermore, campuses should showcase sustainability in their operations and curriculum.

To accomplish its goals, AASHE sponsors conferences and workshops to educate members. It also provides networking opportunities and an e-bulletin to facilitate the exchange of information about sustainable practices on campuses.

AASHE is currently developing a rating system that will allow educational institutions to assess their progress toward achieving sustainability. The Sustainability Tracking, Assessment, and Rating System (STARS) focuses on three major categories of activity: education and research, operations, and administration and finance.

In each chapter of this edition of *Environmental Science: A Study of Interrelationships,* we will highlight the efforts of one of the member colleges of AASHE to achieve sustainability. Is your college a member? Go to the AASHE website and check its membership list.

ISSUES & ANALYSIS

Government Regulation and Personal Property

There are many ways in which government intrudes into your personal lives. Many kinds of environmental regulations require people to modify their behavior. However, one of the most controversial situations occurs when government infringes on personal property rights. The Endangered Species Act requires that people do no harm to threatened and endangered species. They may not be hunted or harvested and often special areas are established to assure their protection.

Many people have found after they have purchased a piece of land that it has endangered species as inhabitants. They are then faced with a situation in which they cannot use the land as they intended, and the land loses much of its value to them. Some argue that they should be allowed to use the land for their original purpose because they did not know it was habitat for an endangered species. Others argue that they have been deprived of a valuable good by the federal government and that the government should compensate them for their loss.

On the other hand, the people charged with enforcing the regulations say that they are simply following the laws of the land and that the landowner must obey.

The threatened California gnatcatcher is a small grey bird that inhabits coastal sagebrush habitats. This photo shows the results of a compromise between developers and efforts to preserve the California gnatcatcher. Some land that was originally planned for development has been set aside as habitat for the gnatcatcher.

- Do you think landowners should be compensated for their loss by the federal government?
- Should purchasers of land do better research about the land they want to buy?

SUMMARY

Environmental science involves science, economics, ethics, and politics in arriving at solutions to environmental problems. Artificial political boundaries create difficulties in managing environmental problems because most environmental units, or ecosystems, do not coincide with political boundaries. Therefore, a regional approach to solving environmental problems, one that incorporates natural geographic units, is ideal. Furthermore, as the population of the world has grown and the exchange of people and goods between countries has increased, many environmental problems have become global in nature. Each region of the world has certain environmental issues that are of primary concern because of the mix of population, resource use patterns, and culture.

THINKING GREEN

1. Look for locally grown produce in the supermarket—less energy is used to transport locally grown products.
2. Join a local environmental organization.
3. Volunteer for your local Earth Day event in April.
4. Visit a natural area, nature center, or park typical of your region and learn to identify five plants.
5. Go to the website of the League of Conservation Voters, click on the Scorecard tab, and find out the "environmental score" of your senators and representative.

WHAT'S YOUR TAKE?

The governors of the Great Lakes states have signed an agreement that prohibits the export of water from the Great Lakes. They argue that the water is a valuable resource that is needed by the citizens of their states and that export would deprive the states' citizens of the resource.

Regions of the country that are water poor argue that the water in the Great Lakes is a resource that should be shared by all citizens of the country. Develop an argument that supports or refutes the governors' stated policy.

REVIEW QUESTIONS

1. Describe why finding solutions to environmental problems is so difficult. Do you think it has always been as complicated?
2. Describe what is meant by an ecosystem approach to environmental problem solving. Is this the right approach?
3. List two key environmental issues for each of the following regions: the wilderness North, the agricultural middle, the forested West, the dry West, the Great Lakes and industrial Northeast, and the South. How are the issues changing?
4. Define environment and ecosystem and provide examples of these terms from your region.
5. Describe how environmental conflicts are resolved.
6. Select a local environmental issue and write a short essay presenting all sides of the question. Is there a solution to this problem?

CRITICAL THINKING QUESTIONS

1. Imagine you are a U.S. congressional representative from a western state and a new wilderness area is being proposed for your district. Who might contact you to influence your decision? What course of action would you take? Why?
2. How do you weigh in on the issue of jobs or the environment? What limits do you set on economic growth? Environmental protection?
3. Imagine you are an environmentalist in your area who is interested in local environmental issues. What kinds of issues might these be?
4. Imagine that you lived in the urban East and that you were an advocate of wilderness preservation. What disagreements might you have with residents of the wilderness North or the arid West? How would you justify your interest in wilderness preservation to these residents?
5. You are the superintendent of Yellowstone National Park and want to move to an ecosystem approach to managing the park. How might an ecosystem approach change the current park? How would you present your ideas to surrounding landowners?
6. Look at the issue of global warming from several different disciplinary perspectives—economics, climatology, sociology, political science, agronomy. What might be some questions that each discipline could contribute to our understanding of global warming?

CHAPTER 2

ENVIRONMENTAL ETHICS

Human use alters the natural world. Often we must ask what kind of use is ethical? Is clearing a tropical forest for agriculture ethically supportable?

CHAPTER OUTLINE

The Call for a New Ethic
Environmental Ethics
    Ethics and Laws
    Conflicting Ethical Positions
    The Greening of Religion
    Three Philosophical Approaches to Environmental Ethics
    Other Philosophical Approaches
Environmental Attitudes
    Development
    Preservation
    Conservation
    Sustainable Development
Environmental Justice
Societal Environmental Ethics
Corporate Environmental Ethics
    The Legal Status of Corporations
    Waste and Pollution
    Profitability and Power
    Is There a Corporate Environmental Ethic?
    Green Business Concepts
Individual Environmental Ethics
The Ethics of Consumption
    Food
    Energy
    Water
    Wild Nature
Personal Choices
Global Environmental Ethics

ISSUES & ANALYSIS
Environmental Dissent: Is Ecoterrorism Justified? 33

CASE STUDIES
Early Philosophers of Nature 20
Environmental Disasters and Poverty 24

CAMPUS SUSTAINABILITY INITIATIVE
Investment Responsibility at Duke University 18

GOING GREEN
Do We Consume Too Much? 28

WATER CONNECTIONS
A Global Water Ethic 30

OBJECTIVES

After reading this chapter, you should be able to:

- Understand the role of ethics in society.
- Recognize the importance of a personal ethical commitment.
- List three conflicting attitudes toward nature.
- Explain the connection between material wealth and resource exploitation.
- Describe the factors associated with environmental justice.
- Explain how corporate behavior connects to the state of the environment.
- Describe how environmental leaders in industry are promoting more sustainable practices.
- Describe the influence that corporations wield because of their size.
- Explain the relationship among economic growth and environmental degradation.
- Explain some of the relationships between affluence, poverty, and environmental degradation.
- Explain the importance of individual ethical commitments toward environment.
- Explain why global action on the environment is necessary.

Global Perspectives on "The Gray Whales of Neah Bay," "The Environmental Cost of Rapid Industrialization," and "Connecting Peace, Social Justice, and Environmental Quality" can be found on the book's website at www.mhhe.com/enger12e along with other interesting readings.

THE CALL FOR A NEW ETHIC

The most beautiful object I have ever seen in a photograph in all my life is the planet Earth seen from the distance of the moon, hanging in space, obviously alive. Although it seems at first glance to be made up of innumerable separate species of living things, on closer examination every one of its things, working parts, including us, is interdependently connected to all the other working parts. It is, to put it one way, the only truly closed ecosystem any of us know about.

—Lewis Thomas

When we try to pick out anything by itself, we find it hitched to everything else in the Universe.

—John Muir

One of the marvels of recent technology is that we can see the Earth from the perspective of space, a blue sphere unique among all the planets in our solar system. (See figure 2.1.) Looking at ourselves from space, it becomes obvious that a lot of what we do on our home planet connects us to something or somebody else. This means, as Harvard University ecologist William Clark points out, that only as a global species, "pooling our knowledge, coordinating our actions, and sharing what the planet has to offer—do we have any prospect for managing the planet's transformation along pathways of sustainable development."

Some people see little value in an undeveloped river and feel it is unreasonable to leave it flowing in a natural state. It could be argued that rivers throughout the world ought to be controlled to provide power, irrigation, and navigation for the benefit of humans. It could also be argued that to not use these resources would be wasteful.

In the U.S. Pacific Northwest, there is a conflict over the value of old-growth forests. Economic interests want to use the forests for timber production and feel that to not do so would cause economic hardship. They argue that trees are simply a resource to be used in any way deemed necessary for human economic benefit. An opposing view is that all the living things that make up the forest have a kind of value beyond their economic utility. Removing the trees would destroy something ethically significant that took hundreds of years to develop and may be almost impossible to replace.

Interactions between people and their environment are as old as human civilization. The problem of managing those interactions, however, has been transformed today by unprecedented increases in the rate, scale, and complexity of the interactions. At one time, pollution was viewed as a local, temporary event. Today pollution may involve several countries and may affect multiple generations. The debates over chemical and radioactive waste disposal are examples of the increasingly international nature of pollution. Many European countries are concerned about the transportation of radioactive and toxic wastes across their borders. What were once straightforward confrontations between ecological preservation and economic growth have today become complex balancing acts containing multiple economic, political, and ethical dimensions.

FIGURE 2.1 **The Earth as Seen from Space** Political, geographic and national differences among humans do not seem so important from this perspective. In reality, we all share the same planet.

The character of the environmental changes that today's technology makes possible and the increased public awareness of the importance of the natural environment mean that we have entered a new age of environmental challenges. Across the world, thousands of people believe that part of what is needed to meet these challenges is the development of a new and more robust environmental ethic.

ENVIRONMENTAL ETHICS

Ethics is one branch of philosophy. Ethics seeks to define what is right and what is wrong. For example, most cultures are ethically committed to the idea that it is wrong to needlessly take life. Many cultures ground this belief on the existence of a right to life. It is considered unethical to deprive humans of this right to their life. Ethics can help us to understand *what* actions are wrong and *why* they are wrong. Many of the issues discussed throughout this book (energy, population, environmental risk, biodiversity, land-use planning, air quality, etc.) have ethical dimensions.

Across the world, not all cultures share the same ethical commitments. **Cultural relativism** in ethics acknowledges that these differences exist. On some occasions, it is appropriate to show sensitivity to legitimate differences in ethical commitments. However, despite the presence of some differences, there are also many cases in which ethical commitments can and should be globally agreed upon. The rights to life, liberty, and security of person, for example, are judged to be important across the globe. The 1948 Universal Declaration on Human Rights issued by the United Nations expressed a commitment to these basic human rights. Given the importance of the planetary ecosystem to all of Earth's inhabitants, an area that shows potential for similar global agreement is the question of the proper treatment of the natural environment.

ETHICS AND LAWS

Ideally, the **laws** of a particular nation or community should match the ethical commitments of those living there. Sometimes laws are changed to match ethical commitments only after a long period of struggle and debate. In the United States, the abolition of slavery, women's suffrage, civil rights laws, and regulations that protect the welfare of animals are all examples of changes in legislation that came about only after long periods of public debate and struggle. As a result of these struggles, the country's laws now fall more in line with the ethical commitments of its people.

Not every action that is ethically right can have a law supporting it. There are no laws saying that you have to help your elderly neighbor unload her groceries from the car. But even without a law this still might be the ethically right thing to do. In the case of environmental issues, care needs to be taken over when it is appropriate to legislate something and when action should be left up to the individual's sense of right and wrong. For example, most people today agree that knowingly putting harmful pollutants into the water and into the air is unethical. But while it may be appropriate to legislate against deliberately dumping toxic chemicals into a river, it may not be appropriate to legislate against driving a car more than a certain amount per week. Similarly, it may not be appropriate to legislate how many material goods people can purchase, or how much food they can waste, or how many kilowatts of electricity they can use, or how large a family they can have. On these issues, individual environmental action tends to be determined more by custom, habit, and certain social and economic pressures. In addition to these factors, a **strong personal ethical commitment** can help guide behavior in the absence of supporting laws.

CONFLICTING ETHICAL POSITIONS

Even when people have strong personal ethical commitments, they might find that some of their commitments conflict. For example, a mayor might have an ethical commitment to preserving the land around a city but at the same time have an ethical commitment to bringing in the jobs associated with the construction of a new factory on the outskirts of town. There are often difficult balances to be struck between multiple ethical values.

As you can see, ethics can be very complicated. Ethical issues dealing with the environment are especially complex because sometimes it appears that what is good for people conflicts with

what is good for the environment. Saving the forest might mean the loss of some logging jobs. While recognizing that there are some real conflicts involved, it is also important to see that it is not necessarily the case that when the environment wins people lose. In a surprising number of cases it turns out that what is good for the environment is also good for people. For example, even when forest protection reduces logging jobs, a healthier forest might lead to new jobs in recreation, fisheries, and tourism. Searching for genuine "win-win" situations has become a priority in environmental decision making.

THE GREENING OF RELIGION

For many years, environmental issues were considered to be the concern of scientists, lawyers, and policy makers. Now the ethical dimensions of the environmental crisis are becoming more evident. What is our moral responsibility toward future generations? How can we ensure equitable development that does not destroy the environment? Can religious and cultural perspectives be considered in creating viable solutions to environmental challenges?

Until recently, religious communities have been so absorbed in internal sectarian affairs that they were unaware of the magnitude of the environmental concerns facing the world. Certainly, the natural world figures prominently in the world's major religions: God's creation of material reality in Judaism, Christianity, and Islam; the manifestation of the divine in the karmic processes underlying the recycling of matter in Hinduism and Jainism; the interdependence of life in Buddhism; and the Tao (the Way) that courses through nature in Confucianism and Taoism.

Today, many religious leaders recognize that religions, as enduring shapers of culture and values, can make major contributions to the rethinking of our current environmental impasse. Religions have developed ethics for homicide, suicide, and genocide; now they are challenged to respond to biocide and ecocide. Moreover, the environment presents itself as one of the most compelling concerns for robust inter-religious dialogue. The common ground is the Earth itself, along with a shared sense among the world's religions of the interdependence of all life.

THREE PHILOSOPHICAL APPROACHES TO ENVIRONMENTAL ETHICS

Given the complexity of the issues, environmental philosophers have developed a number of theoretical approaches to help us see more clearly our ethical responsibilities concerning the environment. In these environmentally conscious times, most people agree that we need to be environmentally responsible. Toxic waste contaminates groundwater, oil spills destroy shorelines, and fossil fuels produce carbon dioxide, thus adding to global warming. The goal of environmental ethics, then, is not simply to convince us that we should be concerned about the environment—many already are. Instead, environmental ethics focuses on the moral foundation of environmental responsibility and how far this responsibility extends. There are three primary theories of moral responsibility regarding the environment. Although each can support environmental responsibility, their approaches are different. (See figure 2.2.)

Anthropocentrism

The first of these theories is **anthropocentrism** or human-centered ethics. Anthropocentrism is the view that all environmental responsibility is derived from human interests alone. The assumption here is that only human beings are morally significant and have direct moral standing. Since the environment is crucial to human well-being and human survival, we have an indirect duty toward the environment, that is, a duty derived from human interests. We must ensure that the Earth remains environmentally hospitable for supporting human life and even that it remains a pleasant place for humans to live. Nevertheless, according to this view, the value of the environment lies in its *instrumental worth* for humans. Nature is fundamentally an instrument for human manipulation. Some anthropocentrists have argued that our environmental duties are derived both from the immediate benefit that people receive from the environment and from the benefit that future generations of people will receive. But critics have maintained that since future generations of people do not yet exist, then, strictly speaking, they cannot have rights any more than a dead person can have rights. Nevertheless, both parties to this dispute acknowledge that environmental concern derives solely from human interests.

Biocentrism

A second theory of moral responsibility to the environment is **biocentrism** or life-centered environmental ethics. According to the broadest version of the biocentric theory, all forms of life have an inherent right to exist. A number of biocentrists recognize a hierarchy of values among species. Some, for example, believe that we have a greater responsibility to protect animal species than plant species and a greater responsibility to protect mammals than invertebrates. Another group of biocentrists, known as "biocentric egalitarians," take the view that all living organisms have an exactly equal right to exist. Since the act of survival inevitably involves some killing (for food and shelter) it is hard to know where biocentric egalitarians can draw the lines and still be ethically consistent.

Ecocentrism

The third approach to environmental responsibility, called **ecocentrism,** maintains that the environment deserves direct moral consideration and not consideration that is merely derived from human or animal interests. In ecocentrism it is suggested that the environment itself, not just the living organisms that inhabit it, has moral worth. Some ecocentrists talk in terms of the systemic value that a particular ecosystem possesses as the matrix that makes biological life possible. Others, go beyond particular ecosystems and suggest that the biological system on Earth as a whole has an integrity to it that gives it moral standing. Another version goes even further and ascribes personhood to the planet, suggesting that Mother Earth or "Gaia" should have the same right to life as any mother.

One of the earliest and most well known spokespersons for ecocentrism was the ecologist Aldo Leopold. Leopold is most famous for his 1949 book *A Sand County Almanac,* which was written in response to the relentless destruction of the landscape that Leopold had witnessed during his life. *A Sand County Almanac* redefined the

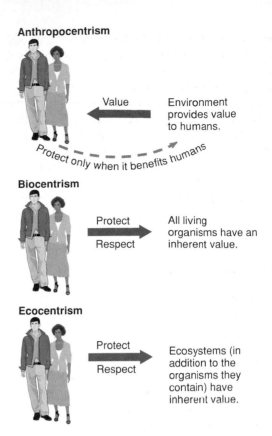

FIGURE 2.2 **Philosophical Approaches** Of the three major approaches only anthropocentrism refers all value back to human needs and interests.

relationship between humankind and the Earth. Leopold devoted an entire chapter of his book to "The Land Ethic."

> All ethics so far evolved rest upon a single premise: that the individual is a member of a community of interdependent parts. . . . The land ethic simply enlarges the boundaries of the community to include soils, waters, plants, and animals, or collectively the land . . . a land ethic changes the role of *Homo sapiens* from conqueror of the land-community to plain member and citizen of it. It implies respect for his fellow-members, and also respect for the community as such.
>
> It is inconceivable to me that an ethical relation to land can exist without love, respect, and admiration for land, and a high regard for its value. By value, I of course mean something far broader than mere economic value; I mean value in the philosophical sense.

What Leopold put forth in "The Land Ethic" was viewed by many as a radical shift in how humans perceive themselves in relation to the environment. Originally we saw ourselves as conquerors of the land. Now, according to Leopold, we need to see ourselves as members of a community that also includes the land and the water.

Leopold went on to claim that "a thing is right when it tends to preserve the integrity, stability, and beauty of the biotic community. It is wrong when it tends otherwise. . . . We abuse land because we regard it as a commodity belonging to us. When we see land as a community to which we belong, we may begin to use it with love and respect."

INVESTMENT RESPONSIBILITY AT DUKE UNIVERSITY

Duke University passed a resolution in 2008 that prohibits it from making future direct investment in companies engaged in business with the government of Sudan. The resolution, in protest against that government's human rights violations in the war-torn region of Darfur, covers the $8.2 billion in endowments and assets of Duke University. This is the first time that Duke invoked its investment responsibility policy, which was established by the Board of Trustees in 2004.

The policy has a two-step protocol for handling issues related to socially responsible investment. When an issue is presented by a member of the university community, it is prescreened by the President's Special Committee on Investment Responsibility. If that group finds the issue merits further investigation, it turns the matter over to the President's Advisory Committee on Investment Responsibility for a decision.

The Sudan divestment will remain in effect until the U.S. government lifts sanctions against Sudan. The sanctions were established in 1997 by the Clinton Administration and added to by an executive order of President Bush in 2006.

The resolution does not list any specific businesses in which Duke will have no direct investment, but instead uses a list of "heavily engaged companies" compiled by the international Sudan Divestment Task Force. Duke joins a growing list of universities that are divesting or prohibiting future investments in Sudan. Among the universities that have taken similar actions are Yale, Princeton, Harvard, and Stanford.

Does your university or college have a social or environmental investment policy?

Do you believe that such policies can influence change?

What issues do you believe should be covered by such policies?

OTHER PHILOSOPHICAL APPROACHES

As traditional political and national boundaries fade or shift in importance, new variations of environmental philosophy are fast emerging. Many of these variations are founded on an awareness that humanity is part of nature and that nature's component parts are interdependent. Beyond the three ethical positions discussed previously, other areas of thought recently developed by philosophers to address the environmental crisis include:

Ecofeminism—the view that there are important theoretical, historical, and empirical connections between how society treats women and how it treats the environment.

Social ecology—the view that social hierarchies are directly connected to behaviors that lead to environmental destruction. Social ecologists are strong supporters of the environmental justice movement (see pp. 21–23).

Deep ecology—the generally ecocentric view that a new spiritual sense of oneness with the Earth is the essential starting point for a more healthy relationship with the environment. Many deep ecologists are environmental activists (see p. 33).

Environmental pragmatism—an approach that focuses on policy rather than ethics. Environmental pragmatists think that a human-centered ethic with a long-range perspective will come to many of the same conclusions about environmental policy as an ecocentric ethic. Consequently they find the emphasis on ethical theories unhelpful.

Environmental aesthetics—the study of how to appreciate beauty in the natural world. Some environmental aesthetics advocates think that the most effective philosophical ground for protecting the natural environment is to think in terms of protecting natural beauty.

Animal rights/welfare—this position asserts that humans have a strong moral obligation to nonhuman animals. Strictly speaking, this is not an environmental position because the commitment is to individual animals and not to ecosystems or ecological health. Animal rights advocates are particularly concerned about the treatment of farm animals and animals used in medical research.

ENVIRONMENTAL ATTITUDES

Even for those who have studied environmental ethics carefully, it is never easy to act in accordance with one particular ethic in everything. Ethical commitments pull in different directions at different times. Because of these difficulties, it is sometimes easier to talk in terms of general *attitudes* or *approaches* to the environment rather than in terms of particular *ethics*. The three most common approaches are (a) the *development* approach, (b) the *preservation* approach, and (c) the *conservation* approach. (See figure 2.3.)

DEVELOPMENT

The **development** approach tends to be the most anthropocentric of the three. It assumes that the human race is and should be master of nature and that the Earth and its resources exist solely for our benefit and pleasure. This approach is reinforced by the capitalist work

Development

Conservation

Preservation

FIGURE 2.3   Environmental Attitudes
Development, preservation, and conservation are different attitudes toward nature. These attitudes reflect a person's ethical commitments.

ethic, which historically dictated that humans should create value for themselves by putting their labor into both land and materials in order to convert them into marketable products. In its unrestrained form, the development approach suggests that improvements in the human condition require converting ever more of nature over to human use. The approach thinks highly of human creativity and ingenuity and holds that continual economic growth is itself a moral ideal for society. In the development approach, the environment has value only insofar as human beings economically utilize it. This mindset has very often accompanied the process of industrialization and modernization in a country.

PRESERVATION

The **preservationist** approach tends to be the most ecocentric of the three common attitudes toward the environment. Rather than seek to convert all of nature over to human uses, preservationists want to see large portions of nature preserved intact. Nature, they argue, has intrinsic value or inherent worth apart from human uses. Preservationists have various ways of articulating their position.

During the nineteenth century, preservationists often gave openly religious reasons for protecting the natural world. John Muir condemned the "temple destroyers, devotees of ravaging commercialism" who, "instead of lifting their eyes to the God of the mountains, lift them to the Almighty dollar." This was not a call for better cost-benefit analysis: Muir described nature not as a commodity but as a companion. Nature is sacred, Muir held, whether or not resources are scarce.

Philosophers such as Emerson and Thoreau thought of nature as full of divinity. Walt Whitman celebrated a leaf of grass as no less than the "journeywork of the stars." "After you have exhausted what there is in business, politics, conviviality, love, and so on and found that none of these finally satisfy, or permanently wear—what remains? Nature remains," Whitman wrote. These philosophers thought of nature as a refuge from economic activity, not as a resource for it.

While many preservationists adopt an ecocentric ethic, some also include anthropocentric principles in their arguments. These preservationists wish to keep large parts of nature intact for aesthetic or recreational reasons. They believe that nature is beautiful and restorative and should be preserved to ensure that wild places exist for future humans to hike, camp, fish, or just enjoy some solitude.

CONSERVATION

The third environmental approach is the **conservationist** approach. Conservationism tends to strike a balance between unrestrained development and preservationism. Conservationism is anthropocentric in the sense that it is interested in promoting human well-being. But conservationists tend to consider a wider range of long-term human goods in their decisions about environmental management. Conservationist Gifford Pinchot, for example, arguing with preservationist John Muir at the start of the twentieth century about how to manage American forests, thought that the forests should primarily be managed to serve human needs and interests. But Pinchot could see that timber harvest rates should be kept low enough for the forest to have time to regenerate itself. He also realized that water quality for nearby humans was best preserved by keeping large patches of forest intact.

Many hunters are conservationists. Hunters throughout North America and Northern Europe have protected large amounts of habitat for waterfowl, deer, and elk. Hunters have both biocentric and anthropocentric elements to their thinking. Ethically sensitive hunters often see the value of nonhuman animal species and put ethical

EARLY PHILOSOPHERS OF NATURE

The philosophy behind the environmental movement had its roots in the nineteenth century. Among many notable conservationist philosophers, several stand out: Ralph Waldo Emerson, Henry David Thoreau, John Muir, Aldo Leopold, and Rachel Carson.

In an essay entitled "Nature" published in 1836, Emerson wrote that "behind nature, throughout nature, spirit is present." Emerson was an early critic of rampant economic development, and he sought to correct what he considered to be the social and spiritual errors of his time. In his *Journals,* published in 1840, Emerson stated that "a question which well deserves examination now is the Dangers of Commerce. This invasion of Nature by Trade with its Money, its Credit, its Steam, its Railroads, threatens to upset the balance of Man and Nature."

Henry David Thoreau was a writer and a naturalist who held beliefs similar to Emerson's. Thoreau's bias fell on the side of "truth in nature and wilderness over the deceits of urban civilization." The countryside around Concord, Massachusetts, fascinated and exhilarated him as much as the commercialism of the city depressed him. It was near Concord that Thoreau wrote his classic, *Walden,* which describes the two years he lived in a small cabin on the edge of Walden pond in order to have direct contact with nature's "essential facts of Life." In his later writings and journals, Thoreau summarized his feelings toward nature with prophetic vision:

> But most men, it seems to me, do not care for Nature and would sell their share in all her beauty, as long as they may live, for a stated sum—many for a glass of rum. Thank God, man cannot as yet fly, and lay waste the sky as well as the earth! We are safe on that side for the present. It is for the very reason that some do not care for these things that we need to continue to protect all from the vandalism of a few. (1861)

John Muir combined the intellectual ponderings of a philosopher with the hard-core, pragmatic characteristics of a leader. Muir believed that "wilderness mirrors divinity, nourishes humanity, and vivifies the spirit." Muir tried to convince people to leave the cities for a while to enjoy the wilderness. However, he felt that the wilderness was threatened. In the 1876 article entitled, "God's First Temples: How Shall We Preserve Our Forests?" published in the *Sacramento Record Union,* Muir argued that only government control could save California's finest sequoia groves from the "ravages of fools." In the early 1890s, Muir organized the Sierra Club to "explore, enjoy, and render accessible the mountain regions of the Pacific Coast" and to enlist the support of the government in preserving these areas. His actions in the West convinced the federal government to restrict development in the Yosemite Valley, which preserved its beauty for generations to come.

Aldo Leopold was a thinker as well as an activist in the early conservation movement. As a philosopher, Leopold summed up his feelings in *A Sand County Almanac:*

> Wilderness is the raw material out of which man has hammered the artifact called civilization. No living man will see again the long grass prairie, where a sea of prairie flowers lapped at the stirrups of the pioneer. No living man will see again the virgin pineries of the Lake States, or the flatwoods of the coastal plain, or the giant hardwoods.

While serving in the U.S. Forest Service in New Mexico in the 1920s, Leopold worked for the protection of parts of the forest as early wilderness areas. An avid hunter himself, Leopold learned a great deal about attitudes toward the control of predators as part of his Forest Service job.

Ralph Waldo Emerson

Henry David Thoreau

John Muir

Aldo Leopold

Rachel Carson

When he left the Forest Service he pioneered the field of game management. As a professor of game management at the University of Wisconsin, he argued that regulated hunting should be used to maintain a proper balance of wildlife on that landscape. His *Sand County Almanac,* published shortly after his death in 1949, laid down the principles of his land ethic.

While most people talk about what's wrong with the way things are, few actually go ahead and change it. Rachel Carson ranks among those few. A distinguished naturalist and best-selling nature writer, Rachel Carson published in the *New Yorker* in 1960 a series of articles that generated widespread discussion about pesticides. In 1962, she published *Silent Spring,* which dramatized the potential dangers of pesticides to food, wildlife, and humans and eventually led to changes in pesticide use in the United States.

Although some technical details of her book have been shown to be in error by later research, her basic thesis that pesticides can contaminate and cause widespread damage to the ecosystem has been established. Unfortunately, Carson's early death from cancer came before her book was recognized as one of the most important events in the history of environmental awareness and action in the twentieth century.

constraints on the way they hunt them. Even though a hunter tends to think that the human interest in harvesting the meat ultimately overrides the animal's interest in staying alive, he or she often believes that the animal has a place on the landscape and that the world is a better place if it contains healthy populations of wild animals.

SUSTAINABLE DEVELOPMENT

Sustainable development, a term first used in 1987 in a UN-sponsored document called the Brundtland Report, is often defined as "meeting the needs of current generations without compromising the ability of future generations to meet theirs." Like conservationism, sustainable development is a middle ground that seeks to promote appropriate development in order to alleviate poverty while still preserving the ecological health of the landscape. Sustainable development does not focus solely on environmental issues. The United Nations 2005 World Summit Document refers to the "interdependent and mutually reinforcing pillars" of sustainable development as economic development, social development, and environmental protection. (See figure 2.4.) Indigenous peoples have argued that there are four pillars of sustainable development—the fourth being cultural.

Green development is generally differentiated from sustainable development in that green development prioritizes what its proponents consider to be environmental sustainability over economic and cultural considerations. Proponents of sustainable development argue that it provides a context in which to improve overall sustainability where green development is unattainable. For example, a cutting-edge wastewater treatment plant with extremely high maintenance costs may not be sustainable in regions of the world with fewer financial resources. An environmentally ideal plant that is shut down due to bankruptcy is obviously less sustainable than one that is maintainable by the community, even if it is somewhat less effective from an environmental standpoint.

Still other researchers view environmental and social challenges as opportunities for development action. This is particularly true in the concept of sustainable enterprise that frames these global needs as opportunities for private enterprise to provide innovative and entrepreneurial solutions. This view is now being taught at many business schools.

The United Nations Conference on Environment and Development in Rio de Janeiro, Brazil, in 1992 produced a document entitled Agenda 21, which set out a roadmap for sustainable development. A follow-up conference in South Africa in 2002 drew up a Plan of Implementation for sustainable development. Many observers of the 2002 conference questioned why there had been such a lack of international progress in alleviating poverty and protecting the environment. Part of the problem is that people differ in their opinions on how to strike the right balance between the *development* and *preservation* aspects of sustainable development.

ENVIRONMENTAL JUSTICE

In 1982, African-American residents of Warren County, North Carolina, protested the use of a landfill in their community as a dumpsite for PCB-contaminated waste. (See figure 2.5.) Some

(a)

(b)

FIGURE 2.5 **Environmental Justice** (a) Residents of Warren County, North Carolina, protest the dumping of hazardous PCBs in their local landfill. (b) The direct action in Warren County marked the birth of the environmental justice movement in the United States.

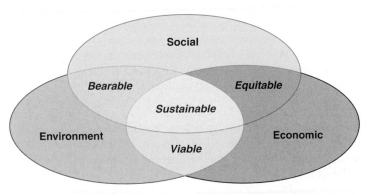

FIGURE 2.4 **Scheme of Sustainable Development** This occurs at the confluence of three constituent parts.

observers asserted that the proposal constituted a form of environmental racism that took advantage of the poorest and least politically influential people in the state. Residents of Warren County took direct action by lying down on the road in front of the trucks headed to the landfill. Hundreds were arrested including Walter Fauntroy, a U.S. Congressional delegate for Washington, D.C., who had traveled to North Carolina in support of the protests.

Following the protests, a number of studies quickly confirmed that toxic waste facilities were indeed disproportionately located in minority neighborhoods. Not only that, the studies also revealed that enforcement of laws was slower and fines levied against violators were lower in areas where residents were made up of poor minorities. By the early 1990s, **environmental justice** (EJ) was recognized throughout the environmental movement as being a critical component of environmental protection.

In 1998, the Environmental Protection Agency (EPA) in the United States characterized environmental justice as a simple matter of fair treatment. The EPA outlined the federal government's commitment to the principle that "no group of people, including racial, ethnic, or socioeconomic groups, should bear a disproportionate share of the negative environmental consequences resulting from industrial, municipal, and commercial operations or the execution of federal, state, local, and tribal programs and policies." The EPA's website on environmental justice (http://www.epa.gov/compliance/environmentaljustice) states that environmental justice also involves "the meaningful involvement of all people regardless of race, color, national origin, or income with respect to the development, implementation, and enforcement of environmental laws, regulations, and policies." Environmental justice is therefore closely related to civil rights. According to the EPA's definition, deliberate discrimination need not be involved. Any action that affects one social group disproportionately is in violation of EPA rules. The difficulty arises in defining what to measure and what should be the standard of comparison.

As a first step in evaluating whether a group is unduly disadvantaged, a policy maker must consider who is affected. Most ethnic data relate to census tracts, zip codes, city boundaries, and counties. If the facility is to be located in a wealthy county but near the county border close to a poor community, how does the policy maker draw the line? Should prevailing winds be considered? Many industrial sites are located where land is cheap; people of low income may choose to live in those areas to minimize living expenses. How are these decisions weighed?

Another difficulty arises in determining whether and how particular groups will be disadvantaged. Landfills, chemical plants, and other industrial works bring benefits to some, although they may harm others. They create jobs, change land values, and generate revenues that are spent in the community. How do officials compare the benefits with the losses? How should potential health risks from a facility be compared with the overall health benefits that jobs and higher incomes bring?

Despite the complications, there are clearly many ways in which governments can act to ensure fairness in how environmental costs and benefits affect their citizens. Governments have established laws and directives to eliminate discrimination in housing, education, and employment, but until the rise of the environmental justice movement they had done very little to address discrimination in environmental practices. One factor that makes environmental justice issues especially troubling is the inverse relationship between who generates the problem and who bears the burden. Studies show that the affluent members of society generate most of the waste, while the impoverished members tend to bear most of the burden of this waste.

Environmental justice encompasses a wide range of issues. In addition to the question of where to place hazardous and polluting facilities, environmental justice questions arise in relation to transportation, safe housing, lead poisoning, water quality, access to recreation, exposure to noise pollution, the viability of subsistence fisheries, access to environmental information, hazardous waste cleanup, and exposure to natural disasters such as Hurricane Katrina (see p. 24). In the United States, environmental justice issues also arise in relation to pollution on Native American Indian reservations. Further questions of environmental justice result from international trade policies that tend to congregate polluting factories in particular countries and their border regions.

The environmental movement in America began as the concern of middle-class and affluent white people. Environmental justice has broadened the demographic of the movement and has raised the profile of many important environmental issues that were simply being ignored. Minorities, indigenous people, and people of color have forced a dialogue about race, class, discrimination, and equity in relation to the environment. They have established beyond doubt that patterns of environmental destruction have important social dimensions.

Environmental Justice Highlights

1982
National attention focused on a series of protests in the low-income, minority community of Warren County, North Carolina, over a landfill filled with PCB-contaminated soil from 14 other counties throughout the state. Over 500 people were arrested. The protest prompted the U.S. General Accounting Office (GAO) to initiate a study of hazardous waste landfills in eight southern states. The GAO study concluded that three out of every four landfills were located in minority communities.

1987
The United Church of Christ Commission on Racial Justice published a report showing that race was the most significant factor in the siting of toxic waste facilities throughout the nation. More than 60 percent of African-American and Hispanic people lived in a neighborhood near a hazardous waste site. A similar study by the *National Law Journal* found that polluters in minority communities paid 54 percent lower fines and the EPA took 20 percent longer to place toxic sites on the national priority action list.

1992
EPA created the Office of Environmental Justice to examine the issue of environmental justice in all agency policies and programs. EPA reported that low-income and minority communities were more likely to be exposed to lead, contaminated fish, air pollution, hazardous waste, and agricultural pesticides.

1993

Environmental justice became one of EPA's top priorities, and an independent advisory group, the National Environmental Justice Advisory Council, was formed from industry experts, activists, and officials.

1994

President Clinton signed Executive Order 12898, a directive requiring all federal agencies to begin taking environmental justice into account. The order specified that "Each Federal agency shall make achieving environmental justice part of its mission by identifying and addressing, as appropriate, disproportionately high and adverse human health or environmental effects of its programs, policies, and activities on minority populations and low-income populations." In a separate memo, the order called for enforcement through three federal statutes: the Civil Rights Act of 1964, the National Environmental Policy Act of 1969, and the Clean Air Act of 1972.

2000

EPA issued a memo outlining additional federal statutes in which environmental justice issues might be involved, including: the Resource Conservation and Recovery Act, the Clean Water Act, the Safe Drinking Water Act, and the Marine Protection, Research and Sanctuaries Act.

2001

A special task force was created within EPA to provide additional resources to investigate the backlog of environmental justice complaints.

2003

The U.S. Commission on Civil Rights, an independent group charged with monitoring federal civil rights enforcement, issued to Congress its report titled "Not in My Backyard," which found that several federal agencies (EPA, DOT, HUD, and DOI) have failed to fully implement the 1994 Environmental Justice Executive Order.

SOCIETAL ENVIRONMENTAL ETHICS

The environmental ethic expressed by a society is a product of the decisions and choices made by a diverse range of social actors that includes individuals, businesses, and national leaders. Western, developed societies have long acted as if the Earth has unlimited reserves of natural resources, an unlimited ability to assimilate wastes, and a limitless ability to accommodate unchecked growth.

The economies of developed nations have been based on a rationale that favors continual growth. Unfortunately, this growth has not always been carefully planned or even desired. This "growth mania" has resulted in the unsustainable use of nonrenewable resources for comfortable homes, well-equipped hospitals, convenient transportation, fast-food outlets, VCRs, home computers, and battery-operated toys, among other things. In economic terms, such

"growth" measures out as "productivity." But the question arises, "What is enough?" Poor societies have too little, but rich societies never say, "Halt! We have enough." The Indian philosopher and statesman Mahatma Gandhi said, "The Earth provides enough to satisfy every person's need, but not every person's greed."

Until the last quarter of the twentieth century, **economic growth** and **resource exploitation** were by far the dominant orientations toward the natural environment in industrialized societies. Developing countries were encouraged to follow similar anthropocentric paths. Since the rise of the modern environmental movement in the last 40 years, things have started to change. Some of the most dramatic changes have occurred in corporate business practices.

CORPORATE ENVIRONMENTAL ETHICS

The enormous effects of business on the state of the environment highlight the important need for corporate environmental ethics.

THE LEGAL STATUS OF CORPORATIONS

Corporations are legal entities designed to operate at a profit. They possess certain rights and privileges, such as the right to own property and the limited liability of their shareholders. Though in some legal senses corporations operate as "artificial persons," a corporation's primary purpose is neither to benefit the public nor to protect the environment but to generate a financial return for its shareholders. This does not, however, mean that a corporation has no ethical obligations to the public or to the environment. Shareholders can demand that their directors run the corporation ethically. (See figure 2.6.)

In business, incorporation allows for the organization and concentration of wealth and power far surpassing that of individuals or partnerships. Some of the most important decisions affecting our environment are made not by governments or the public but by executives who wield massive corporate power.

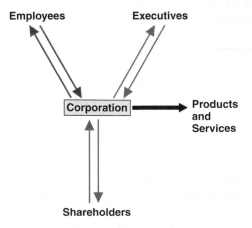

FIGURE 2.6 Corporate Obligations Corporate obligations do not involve the environment unless the shareholders, employees, or executives demand it.

ENVIRONMENTAL DISASTERS AND POVERTY

On August 29, 2005, Hurricane Katrina came ashore on the U.S. Gulf Coast between Mobile, Alabama, and New Orleans, Louisiana. Katrina was an enormous hurricane, just one of the 26 named storms that hit the Americas in the worst Atlantic hurricane season in history. A few hours after the hurricane made landfall, the combination of the storm surge and the torrential rain falling inland overwhelmed the levees that were supposed to protect New Orleans. Up to 80 percent of the city flooded. Close to 1000 people died in Louisiana alone, with most of those deaths occurring in New Orleans. Mandatory evacuation orders were issued for New Orleans' 500,000 residents in the days that followed the storm.

The devastation caused by Hurricane Katrina starkly illustrated the way in which environmental destruction can cause particular hardship for the poor. Twenty-eight percent of New Orleans residents lived below the poverty line. Information about the hurricane and about evacuation options was harder for the city's poorer residents to access. Compounding the problem, a large number of these poorer households did not own cars. As environmental justice advocate Robert Bullard pointed out, without a car, a driver's license, or a credit card even a timely evacuation order can be extremely hard to obey. As Hurricane Katrina battered the city, poor and minority residents were generally left behind, forced to flee to crowded and unsanitary temporary shelters like the New Orleans Superdome.

Outside of the New Orleans city center, some of the areas hardest hit by the hurricane were home to many chemical and petrochemical plants. The location of those plants on the Mississippi River close to African-American communities is a legacy of the area's social and economic history that raises its own environmental justice concerns. In the wake of Hurricane Katrina, damage to the industrial infrastructure threatened homes with floodwaters that were laced with a toxic stew of chemicals. Those people left behind were often forced to wade through the contaminated waters to safety. Their saturated homes required demolition.

The images of hardship, suffering, and death in Louisiana that were broadcast worldwide after the disaster illustrated how, even in wealthy countries, the burden of environmental disasters falls particularly hard on the poor, the sick, and the elderly. Bullard claims that Hurricane Katrina might have finally brought environmental justice out of the closet.

Hurricane Katrina heading toward New Orleans on August 28, 2005.

Flooding in New Orleans immediately after Hurricane Katrina.

WASTE AND POLLUTION

The daily tasks of industry, such as procuring raw materials, manufacturing and marketing products, and disposing of wastes, cause large amounts of pollution. This is not because any industry or company has adopted pollution as a corporate policy. It is simply a fact that industries consume energy and resources to make their products and that they must sell those products profitably to exist.

When raw materials are processed, some waste is usually inevitable. It is often hard to completely control all the by-products of a manufacturing process. Some of the waste material may simply be useless. For example, the food-service industry uses energy to prepare meals. Much of the energy is lost as waste heat. Smoke and odors are released into the atmosphere and spoiled food items must be discarded. Heat, smoke, and food wastes appear to be part of the cost of doing business in the food industry. The cost of controlling this waste can be very important in determining a company's profit margin.

The cheaper it is to produce an item, the greater the possible profit. Ethics are clearly involved when a corporation cuts corners in production quality or waste disposal to maximize profit without regard for public or environmental well-being. Often it is cheaper in the short run to dump wastes into a river than to install a waste-water treatment facility, and it is cheaper to release wastes into the air than it is to trap them in filters. Actions such as these **externalize the costs** of doing business so that the public or the environment, rather than the corporation, pays those costs. Many people consider such pollution unethical and immoral, but on a corporate balance sheet it can look like just another of the factors that determine **profitability.** (See figure 2.7.) Because stockholders expect a return on an investment, corporations are often drawn toward making decisions based on short-term profitability rather than long-term benefit to the environment or society.

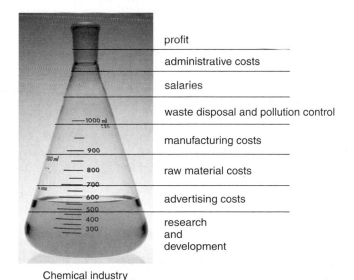

profit

administrative costs

salaries

waste disposal and pollution control

manufacturing costs

raw material costs

advertising costs

research
and
development

Chemical industry

FIGURE 2.7 **Corporate Decision Making** Corporations must make a profit. When they look at pollution control and waste disposal they view its cost like any other cost: and reductions in cost increase profits.

PROFITABILITY AND POWER

The amount of profit a corporation realizes determines how much it can expand. To expand continually, a corporation must increase the demand for its products through advertising. The more demand there is for its products, the more financial resources and power it has. The more power it has, the greater its influence over decision makers who can create conditions favorable to its expansion plans. The process becomes a seemingly never-ending spiral.

New technologies, expanding markets, and the increasingly international nature of commerce have dramatically increased both corporate power and the global exploitation of natural resources. Political and economic institutions such as the World Trade Organization (WTO) generally support the development of private and corporate enterprise in the service of international trade. In order to keep up with the internationalizing of commerce, institutions on a national level often defend and promote private property rights rather than social and environmental interests. As part of the same cycle, corporations and individual business owners find themselves with financial incentives to use loopholes, political pressure, and the time-consuming nature of legal action to circumvent or delay compliance with social or environmental regulations.

IS THERE A CORPORATE ENVIRONMENTAL ETHIC?

Corporations have certainly made more frequent references to environmental issues over the past several years. Is such concern only rhetoric and social marketing, or is it the beginning of a new corporate environmental ethic?

Corporations face real choices between using environmentally friendly or harmful production processes. As the idea of an environmental ethic has become more firmly established within society, corporations are being increasingly pressured to adopt more environmentally and socially responsible practices. Real improvements have been made. For example, Ray Anderson of Interface Incorporated has led the way in greening the carpet industry by reducing the amount of waste produced by his company by 75 percent. At the same time, this measure has saved his company more than $200 million since 1995. The International Organization for Standardization (www.iso.org) has developed a program it calls ISO 14000 to encourage industries to adopt the most environmentally sensitive production practices.

Reaction by the business community to the 1989 *Exxon Valdez* oil spill in Alaska is a good example of the mixed responses of corporations to the ethical challenges. (See figure 2.8.) The Oil Pollution Act (OPA) was passed in 1990 to reduce the environmental impact of future oil spills and has resulted in a 94 percent reduction in spills. One of the new regulations in OPA was that all large tankers must have double hulls or be phased out of service by 2010. To get around the law, however, many oil carriers shifted their oil transport operations to lightly regulated oil barges. This reduction in oil spill safety led to several barge oil spills, including one in January 1996 in Rhode Island's Moonstone Bay and another in March 1997 in Texas's Galveston Bay.

(a)

(b)

FIGURE 2.8 **Oil Spill Response** The 1989 oil spill in Prince William Sound, Alaska, led to the development of the CERES Principles. (a) The otter is a victim of the spill. (b) The *Exxon Valdez* aground on Bligh Reef.

In stark contrast, another group of environmentalists, investors, and companies motivated by the *Valdez* spill and other environmental incidents formed the Coalition for Environmentally Responsible Economics (CERES) group in 1989 and created a set of 10 environmental standards by which their business practices could be measured called the CERES Principles (they were first called the Valdez Principles). CERES companies pledge to voluntarily go beyond the requirements set by the law to strive for environmental excellence through business practices that: (1) protect the biosphere, (2) sustainably use natural resources, (3) reduce and dispose of waste safely, (4) conserve energy, (5) minimize environmental risks through safe technologies, (6) reduce the use, manufacture, and sale of products and services that cause environmental damage, (7) restore environmental damage, (8) inform the public of any health, safety, or environmental conditions, (9) consider environmental policy in management decisions, and (10) report the results of an annual environmental audit to the public. Today, over 70 companies have publicly endorsed the CERES Principles, including 13 Fortune 500 firms. In addition, CERES coordinates an investor network and worldwide, CERES firms have won environmental awards from many organizations in recognition of their approach.

In 1997, the Global Reporting Initiative (GRI) was established. Convened by CERES in partnership with the United Nations Environment Programme, the GRI encourages the active participation of corporations, nongovernmental organizations, accountancy organizations, business associations, and other stakeholders from around the world. The mission of the GRI is to develop globally applicable guidelines for reporting on economic, environmental, and social performance, initially for corporations and eventually for any business, governmental, or nongovernmental organization.

The GRI's *Sustainability Reporting Guidelines* were released in draft form in 1999. The GRI guidelines represent the first global framework for comprehensive sustainability reporting, encompassing the "triple bottom line" of economic, environmental, and social issues.

Improved disclosure of sustainability information is an essential ingredient in the mix of approaches needed to meet the governance challenges in the globalizing economy. Today, at least 2000 companies around the world voluntarily report information on their economic, environmental, and social policies, practices, and performance. Unfortunately, this information can sometimes be inconsistent and incomplete. Measurement and reporting practices vary widely according to industry, location, and regulatory requirements. The GRI's *Sustainability Reporting Guidelines* are designed to address some of these challenges. Currently, 392 organizations in 33 countries follow GRI's guidelines.

GREEN BUSINESS CONCEPTS

There will be little political enthusiasm for preserving the environment if preservation results in national economic collapse. Neither does it make sense to maintain industrial productivity at the cost of breathable air, drinkable water, wildlife species, parks, and wildernesses. Environmental advocates should consider the corporation's need to make a profit when they demand that businesses take more account of the environmental consequences of their actions. Corporations should recognize that profits must come neither at the cost of the health of current and future generations nor at the cost of species extinctions. **Natural capitalism** is the idea that businesses can both expand their profits and take good care of the environment. Natural capitalism works. The 3M Company is estimated to have saved up to US $500 million over the last 20 years through its Pollution Prevention Pays (3P) program. Innovations in **ecological** and **environmental economics** promote accounting techniques that make visible all of the social and environmental costs of doing business so that these costs can no longer be externalized by corporate decision makers.

In the mid-1990s, a concept called **industrial ecology** emerged that links industrial production and environmental quality. One of the most important elements of industrial ecology is that it models industrial production on biological production. Industrial ecology forces industry to account for where waste is going. Dictionaries define *waste* as useless or worthless material. In nature, however, nothing is discarded. All materials ultimately get re-used. Industrial ecology makes it clear that discarding or wasting materials taken from the Earth at great cost is a short-sighted view. Materials and products that are no longer in use could be termed *residues* rather than *wastes*. Residues are materials that our economy has not yet learned to use efficiently. In this view, a pollutant is a resource out of place. In industrial ecology, good environmental practices are good economics. It forces us to view pollution and waste in a new way.

The environmental movement has effectively influenced public opinion and started to move the business community toward greater environmental responsibility. More complex and stringent environmental and public safety demands will increasingly influence corporate decisions throughout this century. Business itself will expand its horizons and find new ways to make profits while promoting environmental benefits and minimizing environmental costs. However, as population increases and as consumption levels rise around the globe, the environmental burden on the planet will inevitably continue to rise. Changes in business practices alone will not be enough.

INDIVIDUAL ENVIRONMENTAL ETHICS

Ethical changes in society and business must start with individuals. We have to recognize that our individual actions have a bearing on environmental quality and that each of us bears some personal responsibility for the quality of the environment in which we live. In other words, environmental ethics must express themselves not only in new national laws and in better business practices but also in significant changes in the ways in which we all live.

Various public opinion polls conducted over the past decade have indicated that Americans think environmental problems can often be given a quick technological fix. The Roper polling organization found that the public believes that "cars, not drivers, pollute, so business should invent pollution-free autos. Coal utilities, not electricity consumers, pollute, so less environmentally dangerous generation methods should be found." It appears that many individuals want the environment cleaned up, but they do not want to make major lifestyle changes to make that happen.

While new technologies will certainly play a major role in the future in lessening the environmental impact of our lifestyles, individual behavioral choices today can also make a significant difference to the health of ecological systems. Environmental ethics must therefore take hold not only at the level of government and business but also at the level of personal choices about consumption.

THE ETHICS OF CONSUMPTION

In 1994, when delegates from around the world gathered in Cairo for the International Conference on Population and Development, representatives from developing countries protested that a baby born in the United States will consume during its lifetime at least 20 times as much of the world's resources as an African or Indian baby. The problem for the world's environment, they argued, is overconsumption in the Northern Hemisphere, not just overpopulation in the Southern Hemisphere. Do we in the Northern Hemisphere consume too much?

North Americans, only 5 percent of the world's population, consume one-fourth of the world's oil. They use more water and own more cars than anybody else. They waste more food than most people in sub-Saharan Africa eat. It has been estimated that if the rest of the world consumed at the rate at which people in the United States consume, we would need five more Planet Earths to supply the resources.

Ever since he wrote a book called *The Population Bomb* in 1968, ecologist Paul Ehrlich has argued that the American lifestyle is driving the global ecosystem to the brink of collapse. But others, including the economist Julian Simon, have argued that Ehrlich couldn't be more wrong. It is not resources that limit economic growth and lifestyles, Simon has insisted, but human ingenuity.

In 1980, the two wagered money on their competing world-views. They picked something easily measurable—the value of metals—to put their theories to the test. Ehrlich predicted that world economic growth would make copper, chrome, nickel, tin, and tungsten scarcer and thus drive the prices up. Simon argued that human ingenuity is always capable of finding technological fixes for scarcity and that the price would go down. By 1990 all five metals had decreased in value. Simon had won the bet. Ehrlich claimed the decrease in price was partly the result of a global recession. He also claimed that many of the costs of producing these metals were still being externalized, with the result that their market price was lower than it should be. But Simon argued that the metals decreased in price because superior materials such as plastics, fiber optics, and ceramics had been developed to replace them. The Ehrlich-Simon argument is actually an old one, and, despite the outcome of their bet, it remains unsettled. What do you think? With the question of consumption in mind, let's look at how consumption could affect several areas in the future—food, energy, water, and wild nature.

FOOD

Two centuries ago, Thomas Malthus declared that worldwide famine was inevitable as human population growth outpaced food production. In 1972, a group of scholars known as the Club of Rome predicted much the same thing for the waning years of the twentieth century. It did not happen because—so far, at least—human ingenuity has outpaced population growth.

Fertilizers, pesticides, and high-yield crops have more than doubled world food production in the past 40 years. The reason 850 million people go hungry today and 6 million children under the age

The desire to consume is nothing new. It has been around for millennia. People need to consume to survive. However, consumption has evolved as people have found new ways to help make their lives simpler and/or to use their resources more efficiently. We consume a variety of resources and products today as we move beyond meeting basic needs to include luxury items and technological innovations to improve efficiency. Such consumption beyond minimal and basic needs is not necessarily a bad thing in and of itself—throughout history we have always sought to find ways to make our lives a bit easier to live. However, increasing, there are important issues around consumerism that need to be understood. For example:

- How are the products and resources we consume actually produced?
- What are the impacts of that process of production on the environment, on society, and on individuals?

- What are the impacts of certain forms of consumption on the environment, on society, and on individuals?
- What is a necessity and what is a luxury?
- Businesses and advertising are major engines in promoting the consumption of products so that they may survive. How much of what we consume is influenced by their needs versus our needs?
- How do material values influence our relationships with other people?
- What impact does that have on our personal values?

We can likely think of numerous other questions as well. We can, additionally, see that consumerism and consumption are at the core of many, if not most, societies. The impacts of consumerism—positive and negative—are very significant in all aspects of our lives, as well as on our planet.

of 5 die each year from hunger-related causes is not that there is not enough food in the world but that social, economic, and political conditions make it impossible for those who need the food to get it. This tragedy is made more troubling by the fact that in 2000 the world reached the historic landmark of there being the same number of overweight people as those that were malnourished.

Norman Borlaug, who won the Nobel Peace Prize in 1970 for his role in developing high-yield crops, predicts that genetic engineering and other new technologies will keep food production ahead of population increases over the next half century. New technologies, however, are not free from controversy. The Mexican government recently confirmed that genetically modified corn has escaped into native corn populations, and the European Union ended a five-year ban on U.S. imports by requiring labeling of all foods containing more than 0.9 percent materials from genetically modified organisms.

Adding to the uncertainty about future food production are factors that include decreasing soil fertility caused by repeated chemical applications, desertification and erosion caused by poor farming techniques, and the loss of available cropland as a result of urbanization. The increasing evidence of rapid global climate change also makes extrapolations from past harvests increasingly unreliable indicators of the future. With global population set to peak at around 9 billion people in the middle of the twenty-first century, it remains unclear whether there will be enough food to go around. Even if it turns out that enough food can be produced for the world in the twenty-first century, whether everybody will get a fair share is much less certain.

ENERGY

If everybody on Earth consumed as much oil as the average American, the world's known reserves would be gone in about 40 years. Even at current rates of consumption, known reserves

will not last through the current century. Technological optimists, however, tell us not to worry. New technologies, they say, will avert a global energy crisis.

Already oil companies have developed cheaper and more efficient ways to find oil and extract it from the ground, possibly extending the supply into the twenty-second century. In many regions of the world, natural gas is replacing oil as the primary source of domestic and industrial power. New coal gasification technologies also hold promise for cleaner and extended fossil fuel power. However, it is impossible to ignore the fact that there is a finite amount of fossil fuel on the planet. These fuels cannot be the world's primary power source forever. Even before fossil fuels run out, concerns about global warming may compel the world to stop burning them. Furthermore, since fossil fuels are found in abundance only in particular parts of the world, geopolitical events can suddenly cause fuel prices to spike in ways that can be disastrous for national economies. Natural disasters such as Hurricane Katrina can also destroy infrastructure and add to the uncertainty surrounding fossil fuel supply.

The more foresighted energy companies are already looking ahead by investing in the technologies that will replace fossil fuels. In some countries, the winds of political change have brought nuclear power back onto the table. In others, solar, wind, wave, and biomass technologies are already meeting increasing proportions of national energy needs. A great deal of optimism is placed on the development of fuel cell technologies. A fuel cell is essentially a refillable, hydrogen-powered battery that produces zero pollution. Since hydrogen is the most abundant element in the universe, there is no shortage of supply. The problem is that most of this hydrogen exists in unusable forms, already combined with other elements in more stable molecules. Separating the usable hydrogen is not technically difficult, but the process takes energy itself, raising questions about whether it might take more energy to make the hydrogen available than the fuel cell will end up producing. With accelerating global demand, it remains unclear whether there will be enough clean energy supply to meet the world's needs in the years ahead.

WATER

The world of the future may not need oil, but without water, humanity could not last more than a few days. Right now, humans use about half the planet's accessible supply of renewable freshwater—the supply regenerated each year and available for human use. A simple doubling of agricultural production with no efficiency improvements would push that fraction to about 85 percent. Unlike fossil fuels, which could eventually be replaced by other energy sources, there is no substitute for water.

Technologies such as desalination, which removes salt from seawater, can be used in rare circumstances. But removing salt from water takes a lot of energy and is expensive. In the Persian Gulf, one place that uses desalination, wealth makes it possible. It has been said that in the Persian Gulf, they turn "oil into water."

Some regions have already reached their water limits, with massive dams and aqueducts diverting almost every drop of water for human use. In the U.S. Southwest, the diversion is so complete that by the time the Colorado River reaches its mouth in the Sea of Cortez, it has no water in it. Much of Los Angeles's water comes from more than 300 kilometers (186 miles) away.

More than any other resource, water may limit consumerism during the next century. "In the next century," World Bank Vice President Ismail Serageldin predicted a few years ago, "wars will be fought over water."

WILD NATURE

Every day in the United States, somewhere from 1000 to 2000 hectares of farmland and natural areas are permanently lost to development. As more and more people around the world achieve modern standards of living, the land area converted to houses, shopping malls, roads, and industrial parks will continue to increase. The planet will labor under rising levels of resource extraction and pollution. Tropical rainforests will be cut and wild lands will become entombed under pavement. Mighty rivers like the Yangtze and Nile, already dammed and diverted, will become even more canal-like. As the new century progresses more and more of us will live urbanized lives. The few pockets of wild nature that remain will be biologically isolated from each other by development. We will increasingly live in a world of our own making.

PERSONAL CHOICES

Threats to supplies of food, energy, water, and wild nature certainly require action on national and international levels. However, in each of these cases ethically responsible action can also begin with the individual. Individuals committed to a strong environmental ethic can make many lifestyle changes to significantly reduce their personal impact on the planet. Food choice is one place to start. Eating food that is produced locally, is low on the food chain, and is grown with a minimum of chemical fertilizers and pesticides not only reduces the environmental impact of food production; it might also lead to better health. Heart disease and certain cancers are increasingly being linked to diets high in animal fats. Buying durable consumer products and reusing or repairing products that still have useful life in them reduces the amount of raw materials that have to be extracted from the ground to meet your needs. Conserving energy at home and on the road can help lessen the amount of fossil fuels you use to support your lifestyle. Living in town rather than out in the suburbs, lobbying for the protection of wild areas, and voting for officials who take environmental issues seriously are all ways that your own environmental ethic can directly contribute to a reduced environmental impact. Consumer behavior is a vote for things you believe in. Lifestyle choices are an expression of your ethical commitments.

The concept of an **ecological footprint** has been developed to help individuals measure their environmental impact on the Earth. One's ecological footprint is defined as "the area of Earth's productive land and water required to supply the resources that an individual demands, as well as to absorb the wastes that the individual produces." Websites exist that allow you to estimate your ecological footprint and to compare it to the footprint of others by answering a few questions about your lifestyle. Running through one of these exercises is a good way to gain a sense of personal responsibility for your own environmental impact. To learn more about ecological footprints visit:

http://www.earthday.org/Footprint/info.asp
http://www.rprogress.org/newprojects/ecolFoot

Finally, think about the words of anthropologist Margaret Mead, who vividly drew attention to the importance of individual action when she stated, "Never doubt that a small group of thoughtful, committed citizens can change the world. Indeed, it's the only thing that ever has."

GLOBAL ENVIRONMENTAL ETHICS

As human stresses on the environment increase, the stability of the planet's ecological systems becomes more and more uncertain. Small environmental changes can create large-scale and unpredictable disruptions. Increased atmospheric carbon dioxide and methane, whether caused by humans or not, are leading to changes in surface temperatures that will result in major ecological effects. Feedback loops add to the urgency. For example, just a small reduction in seasonal snow and ice coverage in Arctic regions due to global warming can greatly increase the amount of solar energy the Earth absorbs. This additional energy itself raises atmospheric temperature, leading to a further reduction in snow coverage. Once established, this feedback loop continually reinforces itself. Some models predict that ocean currents, nutrient flows, and hydrologic cycles could make radical shifts from historic patterns in a matter of

Scientists have long stated the vital importance of a truth we all know on some level to be common sense: Without water, there is no life. With water, there is at least the possibility of life—both here on Earth and, perhaps, as some speculate, elsewhere in the cosmos.

Given its fundamental role for all human survival and the antiquity of our cultural reflection on its importance, one might have expected humans to have developed a broad consensus of thought or a measure of cumulative wisdom about water usage in the ecosystem. But what we might call a *water ethic*—a set of common understandings, shared values, and widely accepted norms governing how humans ought to behave with reference to water—does not appear to be widely thought of in contemporary human affairs. Water itself is far from uniformly appreciated. Some cultures extol its value as priceless, while others behave as if it were worthless.

We live on what has been called the "water planet," yet over 99 percent of Earth's water is either saline or frozen. Humankind depends upon the remaining 1 percent for its survival. Competition for that 1 percent has already become intense in many parts of the world, and even those who live in water-abundant regions are becoming conscious of water as a precious asset. Beyond valuing water sources, however, we are only just beginning to become aware of the downstream impact that our water habits are having upon whole communities of life-forms that inhabit lake, river, estuary, and marine environments, some of which may prove vital for our own survival.

Regarding the world's major rivers, the world atlas no longer tells the truth. Today, dozens of the greatest rivers are dry long before they reach the sea. They include the Nile in Egypt, the Yellow River in China, the Indus in Pakistan, the Rio Grande and the Colorado in the United States, the Murray in Australia, and the Jordan, which is emptied before it can even reach the country that bears its name.

Recent droughts in parts of the United States have focused interest on allocation of water resources for basic needs. During the California drought, when many cities were passing emergency water rationing and increasing basic costs to consumers, other cities were permitting potable water to irrigate golf courses and allowing new swimming pools to be built. Who makes the decisions on how water is allocated? How can a state as diverse as California make ethical decisions about water that treat everyone equally and fairly?

In addition to its role as the substance and medium for all life-forms, water needs to be respected as a geologic force. In the face of recent extreme weather events, cultures around the world are realizing that as humans we did not create, nor can we control, the hydrologic cycle. This can be difficult for technological cultures accustomed to the illusion of dominance over nature. The multiple roles that water plays in the evolving climate system are being investigated with an eye to the future. Any shift in climate is likely to involve new patterns of global water distribution, and these new circumstances will in turn require humankind to devise new habits, policies, and ethical norms to govern water usage. Perhaps now is the time that we begin to discuss a new global water ethic.

---

months. Such disruptions would cause catastrophic environmental change by shifting agricultural regions, threatening species with extinction, decimating crop harvests, and pushing tropical diseases into areas where they are currently unknown. Glaciers will continue to melt and ocean waters will rise, flooding heavily populated low-lying places like Bangladesh, the Netherlands, and even parts of Florida and the U.S. Gulf Coast. Millions of people would be displaced by famine, flood, and drought.

As environmental justice advocates point out, these changes will hit the poor and those least able to respond to them first. However, the changes predicted are of such magnitude that even the very wealthy countries will suffer environmental consequences that they cannot hope to avoid. These scenarios are not distant, future worries. They are here now. The year 2005 was one of the top two hottest years on record. The record hurricane season of 2005 in the United States may be just a small indication of what lies in store.

Many of these problems require global solutions. In 1990, Noel Brown, the director of the United Nations North American Environmental Programme, stated:

> Suddenly and rather uniquely the world appears to be saying the same thing. We are approaching what I have termed a consensual moment in history, where suddenly from most quarters we get a sense that the world community

is now agreeing that the environment has become a matter of global priority and action.

This new sense of urgency and common cause about the environment is leading to unprecedented cooperation in some areas. Despite their political differences, Arab, Israeli, Russian, and American environmental professionals have been working together for several years. Ecological degradation in any nation almost inevitably impinges on the quality of life in others. For years, acid rain has been a major irritant in relations between the United States and Canada. Drought in Africa and deforestation in Haiti have resulted in waves of refugees. From the Nile to the Rio Grande, conflicts flare over water rights. The growing megacities of the Third World are time bombs of civil unrest.

Much of the current environmental crisis is rooted in and exacerbated by the widening gap between rich and poor nations. Industrialized countries contain only 20 percent of the world's population, yet they control 80 percent of the world's goods and create most of its pollution. (See figure 2.9.) The developing countries are hardest hit by overpopulation, malnutrition, and disease. As these nations struggle to catch up with the developed world and improve the quality of life for their people, a vicious circle begins: Their efforts at rapid industrialization poison their cities, while their attempts to boost agricultural production often result in the

Less developed   **More developed**

Approximate relative ecological footprint of one person
in the developing world.

Approximate relative ecological footprint of one person
in the developed world.

FIGURE 2.9 **Lifestyle and Environmental Impact** Significant differences in lifestyles and their environmental impact exist between the rich and poor nations of the world. The ecological footprint of those in the industrialized world can be 20 to 40 times higher than the footprint of those in the developing world. What would be the environmental impact on the Earth if the citizens of China and India and other less-developed countries enjoyed the standard of living in North America? Can we deny them that opportunity?

destruction of their forests and the depletion of their soils, which lead to greater poverty.

Perhaps one of the most important questions for the future is, "Will the nations of the world be able to set aside their political differences to work toward a global environmental course of action?" The United Nations Conference on Human Environment held in Stockholm, Sweden, in 1972 was a step in the right direction. Out of that international conference was born the UN Environment Programme, a separate department of the United Nations that deals with environmental issues. A second world environmental conference was held in 1992 in Brazil and a third in 2002 in South Africa. Each followed up the Stockholm conference with many new international initiatives. A major world conference on climate change was held in Kyoto, Japan, in 1997. (See Global Perspective: The Kyoto Protocol in chapter 16.) Through organizations and conferences such as these, nations can work together to solve common environmental problems. Several major international treaties are listed in Table 2.1.

Even those who doubt the accuracy of the gloomiest predictions about our environmental future can hardly deny that we live on a changing planet. Uncertainty about the ecological baseline that we will be dealing with 50—or even five—years from now means that it would be wise to think hard about the environmental consequences of our actions. As John Muir and Lewis Thomas expressed in their remarks quoted at the beginning of this chapter, humans should expect that many of their actions on planet Earth will in time affect something or somebody else. Anytime our actions may harm another, we face serious ethical questions. Environmental ethics suggests that we may have an obligation beyond simply minimizing the harm that we cause to our families, our neighbors, our fellow human citizens, and future generations of people that will live on Planet Earth. It suggests that we may also have an obligation to minimize the harm we cause to the ecological systems and the biodiversity of the Earth itself. The ecological systems, many now believe, deserve moral consideration for what they are in themselves, quite apart from their undeniable importance to human beings. Recognizing that our treatment of the natural environment is an ethical issue is a good start on the challenges that lie ahead.

Nobel Peace Prize winner Wangari Maathai captured these sentiments about ethics in the speech that she gave shortly after receiving her prize:

> Today we are faced with a challenge that calls for a shift in our thinking, so that humanity stops threatening its life-support system. We are called to assist the Earth to heal her wounds and in the process heal our own—indeed, to embrace the whole creation in all its diversity, beauty and wonder. This will happen if we see the need to revive our sense of belonging to a larger family of life, with which we have shared our evolutionary process. In the course of history, there comes a time when humanity is called to shift to a new level of consciousness, to reach a higher moral ground. A time when we have to shed our fear and give hope to each other. That time is now.

TABLE 2.1 Major International Environmental Treaties and Their Impact

Treaty	Year	What This Treaty Accomplished
Convention on International Trade in Endangered Species of Wild Fauna and Flora (CITES)	1973	Regulated trade of endangered species. Today, 164 countries have signed and trade in 30,000 plant and animals species is monitored. Enforcement varies by country; however, not one CITES-listed species has gone extinct since the treaty was fully initiated in 1975. See chapter 12 or the CITES website: http://www.cites.org
International Tropical Timber Agreement	1983	Created the International Tropical Timber Organization to set trade and conservation policies for tropical forest products. ITTO has 59 member organizations, which represent 90% of global timber trade and 80% of tropical forest resources. See the ITTO website: http://www.itto.org.jp/live/index.jsp
Montreal Protocol on Substances That Deplete the Ozone Layer	1987	Controlled the production and consumption of chemicals that cause ozone depletion in the atmosphere. (Ozone shields the Earth from harmful ultraviolet radiation.) This treaty successfully phased out the production of these chemicals and promoted the production of alternatives. See chapter 16 or the Montreal Protocol website: http://www.afeas.org/montreal_protocol.html
UN Framework Convention on Climate Change, Kyoto Protocol	1992, 1997	These two agreements have created an international commitment to reduce greenhouse gases. In all, 189 countries have signed the framework convention. The Kyoto Protocol for the first time set specific targets for greenhouse gas reduction. At a December 2005 meeting in Montreal, implementation and compliance mechanisms for the Kyoto Protocol were established. By the end of 2005, 157 countries had ratified the 1997 Protocol. The United States, the world's largest emitter of greenhouse gases, had not. Neither had Australia. http://unfccc.int/resource/convkp.html
Stockholm Convention on Persistent Organic Pollutants	2001	An agreement to work toward restricting and ultimately eliminating the production, use, and storage of persistent organic pollutants (POPS) such as DDT, furans, dieldrin, dioxins, and polychlorinated biphenyls. These chemicals do not break down quickly and they move easily through water and air. The chemicals accumulate to dangerous levels in the fatty tissue of animals that are higher up the food chain (including humans). Traces of these chemicals are now found all over the globe with especially high concentrations in polar regions. See the POPS website: http://www.pops.int

Environmental Dissent:
Is Ecoterrorism Justified?

At the first Earth Day in 1970, 20 million people from grassroots organizations throughout the United States came together in many different ways to voice their concern about the destruction of the environment. Since that first protest and celebration, environmental activists have used many creative strategies to bring about change, including letter writing, sit-ins, media campaigns, voting, lobbying, legislation, and economic tools such as conservation easements. Civil disobedience is also one of the strategies that has been used to call attention to environmental problems. For example, 25-year-old Julia Butterfly Hill spent two years living in a 600-year-old redwood tree to protest the logging of ancient forests on private land in California by Pacific Lumber. One of the results of the media attention created by her tree-sitting and the hard work of many in government and industry was the creation of a 10,000-acre public preserve that cost $480 million.

Environmental protest is taking on a new meaning, however, as radical ecoterrorist groups increasingly use violence to halt activity that they believe destroys or exploits the natural environment. In the name of environmental protection, ecoterrorist groups have burned, bombed, and sabotaged government offices, suburban subdivisions, ski resorts, logging operations, mink farms, restaurants, fur companies, universities, agricultural sites, and animal research laboratories. In 1995, following the Oklahoma City bombing, the Federal Bureau of Investigation established teams aimed at combating domestic terrorism. In 2002, the FBI testified before Congress that two ecoterrorist groups alone, the Earth Liberation Front (ELF) and the Animal Liberation Front (ALF), had committed over 600 criminal acts in the United States costing more than $43 million in a six-year period. In January 2006, the U.S. Attorney General, Alberto Gonzales, indicted 11 environmental activists for their role in "domestic terrorism activities."

Earth Liberation Front is believed to have been founded in Brighton, England, in 1992 by Earth First! members who felt that criminal acts such as bombing, arson, and tree-spiking should be used to protest environmental issues. Tree-spiking is the practice of driving a spike or metal object into tree trunks to destroy logging equipment when the trees are harvested. ELF's mission is to ensure that all forms of life on the planet can flourish. It holds that anyone participating in earning wealth by "threatening the ability of all life on the planet to exist" should be stopped. In 1993 ELF joined forces with the Animal Liberation Front, a radical animal rights group whose mission it is to discourage any acts that "harm any animal, human, or nonhuman." Jointly, ELF and ALF are credited with raiding and burning the Bureau of Land Management wild horse corrals in Oregon and burning the U.S. Department of Agriculture Animal Damage Control Building in Washington. ELF alone is credited with arsons in Colorado, Oregon, Washington, Michigan, Indiana, and New York. Both groups employ extensive and sophisticated surveillance systems, operate nonhierarchically in strict secrecy, and leave little, if any, evidence at the scene. Federal, state, and local agencies have collaborated to arrest and prosecute several ELF members for such crimes as arson, tree-spiking, extortion, destruction of property, and attacks on businesses that work with animals.

- How should we as a society protect the rights of all life on the planet? Do we have a moral obligation to do so?
- Is this method of environmental protest effective? Does it result in protection?
- What are the ethical issues involved in using violent methods compared to other approaches to environmental conservation?

SUMMARY

People of different cultures view their place in the world from different perspectives. Among the things that shape their views are religious understandings, economic pressures, geographic location, and fundamental knowledge of nature. Because of this diversity of backgrounds, different cultures put different values on the natural world and the individual organisms that compose it. Environmental ethics investigates the justifications for these different positions.

Three common attitudes toward nature are the development approach, which assumes that nature is for people to use for their own purposes; the preservationist approach, which assumes that nature has value in itself and should be preserved intact; and the conservationist approach, which recognizes that we must use nature to meet human needs but encourages us to do so in a sustainable manner. The conservationist approach is generally known today as "sustainable development."

Ethical obligations to the environment are usually closely connected to ethical obligations toward people, particularly poor people and minority groups. Environmental justice is about ensuring that no group is made to bear a disproportionate burden of environmental harm. Environmental justice is also about ensuring that governments develop and enforce environmental regulations fairly across different segments of society. The environmental justice movement has forced environmentalists to recognize that you cannot think about protecting nature without also thinking about people.

Recognition that there is an ethical obligation to protect the environment can be made by corporations, by individuals, by nations, and by international bodies. Corporate environmental ethics are complicated by the existence of a corporate obligation to its shareholders to make a profit. Corporations often wield tremendous economic power that can be used to influence public opinion and political will. Many corporations are now being driven to include environmental ethics in their business practices by their shareholders. Natural capitalism and industrial ecology are ideas that promote ways of doing profitable business while also protecting the environment.

Corporations are composed of individuals. An increasing sensitivity of individual citizens to environmental concerns can change the political and economic climate for the whole of society. Individuals must demonstrate strong commitments to environmental ethics in their personal choices and behaviors. The concept of an ecological footprint has been developed to help individuals gauge their personal environmental impact.

Global commitments to the protection of the environment are enormously important. Accelerating international trade and communication technologies mean that the world is getting smaller while the potential impact of humanity on the planet is getting larger and more uncertain. Tens of millions of people are added to the world's population each year while economic development increases the environmental impact of those already here. Opportunities for global cooperation and agreement are of critical importance in facing these real and increasing challenges. Environmental ethics has a role to play in shaping human attitudes toward the environment from the smallest personal choice to the largest international treaty.

THINKING GREEN

1. Calculate your ecological footprint.
2. Participate in sustainability activities in your community or university.
3. Give up the use of your car for part of each week. Walk, bicycle, or take public transportation.
4. Read *Silent Spring, A Sand County Almanac*, and *Walden*.
5. Work with a local business in helping it apply green business concepts.

WHAT'S YOUR TAKE?

The grizzly bear (*Ursus arctos horribilis*) has been receiving federal protection under the U.S. Endangered Species Act for over 30 years. The federal government has now proposed removing that protection on the basis of increasing numbers of bears in the Greater Yellowstone Ecosystem and elsewhere. There is considerable disagreement among conservation biologists about how many bears are needed for the species to be "recovered."

When a species is delisted, management is handed over to individual states. If it proposes an acceptable plan, a state may introduce a management plan that includes the hunting of the previously listed species. What kind of ethic underlies the Endangered Species Act? What kind of ethic underlies a management plan that includes hunting? Develop an ethical argument for or against the delisting of the grizzly bear.

REVIEW QUESTIONS

1. Why does the environmental crisis demand a new ethic?
2. What is the relationship between ethics and law?
3. Describe three types of environmental ethics developed by philosophers.
4. Describe three common attitudes toward the environment found in modern society.
5. Why is environmental justice part of the environmental movement?
6. What are the conflicts between corporate behavior and environmental ethics?
7. How can individuals direct business toward better environmental practices?
8. How can individuals implement environmental ethics in their own lives?
9. Where does global environmental ethics fit in the broad scheme of environmental protection?

1. Give three different ethical justifications for protecting a forest using an anthropocentric, a biocentric, and an ecocentric viewpoint.

2. Which approach to the environment—development, preservation, or conservation—do you think you adopt in your own life? Do you think it appropriate for everybody in the world to share the same attitude you hold?

3. What ethical obligations do you personally feel toward wolves and whales? What ethical obligations do you feel toward future generations of people? What ethical obligations do you feel toward future generations of wolves and whales?

4. Until recently, it was generally believed that growth and development were unquestionably good. Now, at the beginning of the twenty-first century, some are beginning to question that belief. Are those questions appropriate ones to address to a citizen of the developing world? Would you ever make the argument that development has gone far enough?

5. Imagine you are a business executive who wants to pursue an environmental policy for your company that limits pollution and uses fewer raw materials but would cost more. What might be the discussion at your next board of directors meeting? How would you make your case to your directors and your shareholders?

6. In 1997 Ojibwa Indians in northern Wisconsin sat on the railroad tracks to block a shipment of sulfuric acid from crossing their reservation on its way to a controversial copper mine in Michigan. Try to put yourself in their position. What values, beliefs, and perspectives might have contributed to their actions? Was it the right thing to do? How would you try and mediate a heated conversation between an Ojibwa protestor and a Michigan copper miner?

7. Wangari Maathai led a protest movement against the stripping of African forests. Are environmental issues important enough for citizens to become activists and perhaps break the law? How could an environmental activist respond to the pro-development position that it is more important to feed people and lift them out of poverty than it is to save a few trees? Are there any ethical principles that you think an environmental activist and a pro-development advocate share?

8. Reread the Case Study about early philosophers of nature. Is the environmental crisis in certain respects a problem in ethics? Does philosophy have a role to play in helping to solve the problem? Should scientists, business leaders, and politicians study environmental ethics? What role is environmental ethics going to play in your own life from this point on?

CHAPTER 3

ENVIRONMENTAL RISK:
ECONOMICS, ASSESSMENT,
AND MANAGEMENT

Many environmental questions are framed by the interrelated concepts of risk and cost. Although mountain biking carries substantial risk, proper training, appropriate behavior, and use of the correct equipment reduce the risk. Similarly, the risk of environmental damage can be lessened by making wise choices.

CHAPTER OUTLINE

Characterizing Risk
Risk and Economics
  Risk Assessment
  Risk Management
  Risk Tolerance
  True and Perceived Risks
Environmental Economics
  Resources
  Supply and Demand
  Assigning Value to Natural Resources
  Environmental Costs
  Cost-Benefit Analysis
  Concerns About the Use of Cost-Benefit Analysis
  Comparing Economic and Ecological Systems
  Common Property Resource Problems—The Tragedy of the Commons
  Green Economics
Using Economic Tools to Address Environmental Issues
  Subsidies
  Liability Protection and Grants for Small Business
  Market-Based Instruments
  Life Cycle Analysis and Extended Product Responsibility
  Green Marketing Principles
Economics and Sustainable Development
Economics, Environment, and Developing Nations

ISSUES & ANALYSIS
The Economics and Risks of Mercury Contamination 58

CASE STUDIES
What's in a Number? 39
Pollution Prevention Pays! 53

CAMPUS SUSTAINABILITY INITIATIVE
Campus Business Partnership to Reduce Greenhouse Gas Emissions 55

GOING GREEN
Green-Collar Jobs 50

WATER CONNECTIONS
Valuing the Removal of the ELWHA and Glines Dams 43

OBJECTIVES

After reading this chapter, you should be able to:

- Describe why the analysis of risk has become an important tool in environmental decision making.
- Understand the difference between risk assessment and risk management.
- Describe the issues involved in risk management.
- Understand the difference between true and perceived risks.
- Define what an economic good or service is.
- Understand the relationship between the available supply of a commodity or service and its price.
- Understand how and why cost-benefit analysis is used.
- Understand the concept of sustainable development.
- Understand environmental external costs and the economics of pollution prevention.
- Understand market approaches to solving environmental problems.
- Describe RBCA and Eco-RBCA.
- Understand what is meant by risk tolerance.
- Understand the concept of perceived versus actual risk.

The following additional Case Studies can be found on the book's website at www.mhhe.com/enger12e along with other interesting readings: "ASTM International E2205-02 Standard Guide for Risk-Based Corrective Action (RBCA) for Protection of Ecological Resources," "Great Lakes Fisheries Resources: Who Cares? A Role-Play Exercise," and "Energy Savings Through Light Replacement, Cost-Benefit Analysis."

CHARACTERIZING RISK

Risk is the probability that a condition or action will lead to an injury, damage, or loss. When we consider any activity or situation that poses a risk, we generally think about three factors: the probability of a bad outcome, the consequences of a bad outcome, and the cost of dealing with a bad outcome. **Probability** is a mathematical statement about how likely it is that something will happen. Probability is often stated in terms like, "The probability of developing a particular illness is 1 in 10,000," or "The likelihood of winning the lottery is 1 in 5 million." It is important to make a distinction between *probability* and *possibility*. When we say something is *possible,* we are just saying that it could occur. It is a very inexact term. *Probability* specifically defines in mathematical terms how likely it is that a *possible* event will occur.

The *consequences* of a bad outcome resulting from the acceptance of a risk may be minor or catastrophic. For example, ammonia is a common household product. Exposure to ammonia will result in 100 percent of people reacting with watery eyes and other symptoms. The probability of an exposure and the probability of an adverse effect are high; however, the consequences are not severe, and there are no lasting effects after the person recovers. Therefore, we are willing to use ammonia in our homes and accept the high probability of an exposure. By contrast, if a large dam were to fail, it would cause extensive property damage and the deaths of thousands of people downstream. Because the consequences of a failure are great, we insist on very high engineering standards so that the probability of a failure is extremely low.

One of the consequences of accepting a risk is the *economic cost* of dealing with bad outcomes. If people become ill or are injured, health care costs are likely to be associated with the acceptance of the risk. If a dam fails and a flood occurs downstream, there will be loss of life and property, which ultimately is converted to an economic cost. The assessment and management of risk involve an understanding of the probability and the consequences of decisions. (See figure 3.1.)

RISK AND ECONOMICS

Most decisions in life involve an analysis of two factors: risk and cost. We commonly ask such questions as "How likely is it that someone will be hurt?" and "What is the cost of this course of action?" Furthermore, these two factors are often interrelated. When we make economic decisions, we may be risking our hard earned money. Risky decisions that lead to physical harm are often reduced to economic terms when medical care costs or legal fees are incurred. Environmental decision making is no different. Most environmental decisions involve finding a balance between the perceived cost of enduring the risk and the economic cost of eliminating the conditions that pose the risk. If a new air-pollution regulation is proposed to reduce the public's exposure to a chemical that is thought to cause disease in a small percentage of the exposed public, industry will be sure to point out that it will cost a considerable amount of money to put these controls in place and will reduce profitability. Citizens also may point out that their tax money will have to support a larger governmental bureaucracy to ensure that the regulations are followed. On the other side, advocates will point out the new regulations will lead to reduced risk of illness and reduced health care costs for people who live in areas affected by the pollutant.

RISK ASSESSMENT

Environmental **risk assessment** is the use of facts and assumptions to estimate the probability of harm to human health or the environment that may result from particular management decisions. An environmental risk assessment process provides environmental decision makers with an orderly, clearly stated, and consistent way to deal with scientific issues when evaluating whether a risk exists, the magnitude of the risk, and the consequences of the negative outcome of accepting the risk. Voluntary organizations such as the International Standardization Organization (ISO) and ASTM have tried to develop acceptable industry-wide standards for guidance in determining environmental and human health risks based on actual or perceived risks. Many industrial and regulatory agencies have joined in this somewhat difficult exercise.

Calculating the risk to humans of a particular activity, chemical, technology, or policy is difficult, and several tools are used to help clarify the risk. If a situation is well known, scientists use probabilities based on past experience to estimate risks. For example, the risk of developing black lung disease from coal dust in mines is well established, and people can be informed of the risks involved and of actions that can reduce the risk.

There are also environmental risks that do not directly affect human health. If human activities cause the extinction of species, there is a negative environmental

FIGURE 3.1   **Decision-Making Process**   The assessment, cost, and consequences of risks are all important to the decision-making process.

Figure labels: % Probability of risk; Consequences of risk; Economics of risk; Decision-making process and prioritization; Decisions

impact, although direct human impact may not be obvious. Similarly, unwise policy decisions may lead to the unsustainable harvest of forest products, fish, wildlife, or other resources that will deplete the resource for future generations.

To estimate the risks associated with new technologies or policies for which there is no established history, models must be used. One common method for modeling the risk of chemical exposure to human health is to expose animals to known quantities of a chemical to gain some insight into how dangerous a material or situation may be. However, a rat or rabbit may not react in the same way as a human. Therefore, animal studies are only indicators of human risk.

In other situations, the impact of a new policy initiative may be modeled using computer simulations. For example, in an attempt to understand the risks associated with global climate change, complex computer models of climate have been used to assess the effects resulting from current energy policy, which contributes to climate change. In the final analysis, most risk assessments are statistical statements that are estimates of the probability of negative effects, as in the examples listed in table 3.1. Such estimates typically are modified to ensure that a lack of complete knowledge does not result in an underestimation of the risk. People may be more or less sensitive to the effects of certain chemicals than the laboratory animals studied. Also, people vary in their sensitivity to compounds. Thus, what may present no risk to one person may be a high risk to others. Persons with breathing difficulties are more likely to be adversely affected by high levels of air pollutants than are healthy individuals. In addition, the estimate of human risk is based on extrapolation from animal tests in which high, chronic doses are used. Human exposure is likely to be lower or infrequent.

Because of all these uncertainties, government regulators have decided to err on the side of safety to protect the public health. For example, the decisions to continue registration of pesticides, to list substances as hazardous air pollutants under the Clean Air Act, and to regulate water contaminants under the Safe Drinking Water Act set conditions of use and allowable exposure limits that provide a large margin of safety. Thus, if animal studies show an effect from the presence of a chemical at a certain dose, the allowable dose for humans is set at a lower level. It can be difficult to precisely measure the toxicity of pollutants once they contact or react with air, water, or soils. This leads to debates over the actual risks and to criticism from those who say this approach carries protection to the extreme, usually at the expense of industry. Conversely, others suggest that this method of setting regulations often underestimates the risks to humans of chronic, low-level exposures.

Risk assessment is also being used to help set regulatory priorities and support regulatory action. Those chemicals, technologies, or situations that have the highest potential to cause damage to health or the environment receive attention first, while those perceived as having minor impacts receive less immediate attention. Medical waste is perceived as high-risk, and laws have been enacted to minimize the risk, while the risk associated with the use of fertilizer on lawns is considered minimal and is not regulated.

Many of the most important threats to human health and the environment are highly uncertain. In addition to quantifying risk, a risk assessment process can state the uncertainty associated with alternative approaches to dealing with environmental issues. This can help institutions to determine research priorities and plan in a way that is consistent with scientific and public concern for environmental protection.

RISK MANAGEMENT

Risk management is a decision-making process that involves weighing policy alternatives and selecting the most appropriate regulatory action by integrating the results of risk assessment with engineering data and with social, economic, and political concerns. It is included as part of all good environmental management systems within business and industry. The purpose of risk management is to reduce the probability or magnitude of a negative outcome. This process involves understanding the probability and consequences of the risk and the factors that contribute to increasing or decreasing the risk. For example, automobile accidents are a leading cause of accidental death. Recognizing that the probability is high that a person will be involved in an automobile accident leads to management of the risk so that the consequences of the accident are minimized. Some management activities are designed to reduce the number of accidents. Traffic lights, warning signs, speed limits, and laws against drunk driving are all designed to reduce the number of accidents. Other activities are designed to reduce the trauma to people who are involved in accidents. Air bags, seat

TABLE 3.1	Estimates of Selected Environmental Causes of Death	
Risk Factor	**Approximate Lifetime Risk of Death (per 1000)**
Smoking 1–2 packs of cigarettes per day	38–175
Having 200 chest X rays per year	7–30
Driving a motor vehicle	17
Eating one 8-ounce meal per week of Great Lakes salmon with 1984 contaminant levels	11–12
Eating one 8-ounce meal per week of Great Lakes salmon with 1987 contaminant levels	3–6
Breathing air in U.S. urban areas at 1980 contaminant levels	0.1–6
Recreational boating	3.5
Drinking one 12-ounce beer per day	1–2
Recreational hunting	1.5
Complications from insect bites and stings	0.014

Source: Data from Indiana State Department of Health.

Risk values are often stated as numbers. When the risk concern is cancer, the risk number represents the probability of additional cancer cases occurring. For example, such an estimate for pollutant X might be expressed as 1×10^{-6}, or simply 10^{-6}. This number can also be written as 0.000001, or one in a million—meaning one additional case of cancer projected in a population of 1 million people exposed to a certain level of pollutant X over their lifetimes. Similarly, 5×10^{-7}, or 0.0000005, or five in *10 million,* indicates a potential risk of five additional cancer cases in a population of 10 million people exposed to a certain level of the pollutant. These numbers signify additional cases above what normally occurs in the general population. The normal rate is referred to as the background cancer incidence. American Cancer Society statistics indicate that the background cancer incidence in the general population is one in three over a lifetime. (One-third of the population will develop some form of cancer during their lifetime.)

If the effect associated with pollutant X is a health effect other than cancer, such as neurotoxicity (nerve damage) or birth defects, then numbers are typically given as the levels of exposure below which no harm is estimated to occur. This often takes the form of a reference dose (RfD). A reference dose is typically expressed in terms of milligrams (of pollutant) per kilogram of body weight per day; for example, 0.004 mg/kg/day. A reference dose typically has a large uncertainty associated with it. It may be too high or too low by several orders of magnitude (i.e., multiples of 10).

An important point to remember is that numbers by themselves don't tell the whole story. For instance, a cancer risk value of 10^{-6} for the "average exposed person" is not the same thing as a cancer risk of 10^{-6} for a "most exposed individual" (perhaps someone exposed from living or working in a highly contaminated area), even though the numbers are identical. It's important to know the difference. Omitting the qualifier "average" or "most exposed" incompletely describes the risk and could result in an inappropriate assessment of the risk. A numerical estimate is only as good as the quality of the data it is based on. You need to ask yourself the following kinds of questions: What data exist to support the risk assessment? Do the data include human epidemiological as well as animal studies? Do the laboratory studies include data on more than one species? If multiple species were tested, did they all respond similarly to the test substance? Are there pieces of information that you would like to have but do not? What assumptions underlie the risk assessment? What is the overall confidence level in the risk assessment? All of these qualitative considerations are essential when deciding how confident you are that the "numbers" used to characterize a risk are meaningful.

Source: Data from *EPA Journal.*

belts, and car designs that absorb the energy of an impact are examples. A risk management plan includes:

1. Evaluating the scientific information regarding various kinds of risks
2. Deciding how much risk is acceptable
3. Deciding which risks should be given the highest priority
4. Deciding where the greatest benefit would be realized by spending limited funds
5. Deciding how the plan will be enforced and monitored

The process of developing a risk management plan begins with an evaluation of the scientific evidence that quantifies the magnitude of a risk. The scientific basis can be thought of as a kind of problem definition. Science determines that some threat or hazard exists but does not specify which risks are most important. With environmental concerns such as hazardous waste, climate change, ozone depletion, and acid rain, the scientific basis for regulatory decisions is often controversial. Hazardous substances are tested on animals. Are animal tests appropriate for determining impacts on humans? There is no easy answer to this question. Dealing with climate change, ozone depletion, and acid rain require projecting into the future and estimating the magnitude of future effects. Will the sea level rise? How many additional skin cancers will be caused by depletion of the ozone layer? How many lakes will become acidified? Estimates from equally reputable sources vary widely. Which ones do we believe? For example, it is a fact that dioxin is a highly toxic material known to cause cancer in laboratory animals. It is also very difficult to prove that human exposure to dioxin has led to the development of cancer, although high exposures have resulted in acne in exposed workers.

From a risk management standpoint, whether one is dealing with a site-specific situation or a national standard, the deciding question ultimately is: What degree of risk is acceptable? In general, we are not talking about a "zero risk" standard but rather the concept of **negligible risk:** At what point is there really no significant health or environmental risk? At what point is there an adequate safety margin to protect public health and the environment?

Once the scientific evidence has been evaluated, it is possible to integrate economic and political factors to determine how much risk is acceptable and to prioritize the assignment of economic and personnel resources needed to solve the problems. This is why problem definition is so important. Defining the problem helps to determine the rest of the policy process (making rules, passing laws, or issuing statements) and the appropriate enforcement actions.

Even after a policy has been developed and regulations have been put in place, however, there is often still controversy. For example, some observers believe specific chemicals such as herbicides pose many threats that need to be addressed. Others believe these chemicals pose little threat; instead, they see scare tactics and government regulations as unnecessary attacks on businesses. The commercial logging of forests poses risks of soil erosion and the loss of resident animal species. The timber industry sees these risks as minimal, while many environmentalists consider the risks unacceptable. These and similar disagreements are often serious public-relations problems for both government and business because most of the public have a poor understanding of the risks they accept daily.

RISK TOLERANCE

Business and industry must have management policies or risk tolerance programs, either formal or understood. Each entity has some level of risk it is willing to accept. In the environmental arena, this can take many forms. Depending upon the situation, policies and/or tolerance for environmental health and safety risks can vary greatly. Generally the more familiar or well understood the issues are, the greater the level of risk that is acceptable. It is the unknown risks that cause problems for most. Questions need to be asked that address the variables encountered: How will the information be processed? What results will be used for making business decisions? Who will perform the tasks and interpret the results? Following an initial review, further inquiry may be necessary. Decisions about the use of property, or the handling of chemical waste, or hazardous substances, can have a material impact on a business.

TRUE AND PERCEIVED RISKS

Perceptions play a large role in all things environmental. For example, in the recent past, there was a great deal of fear about asbestos in our public schools. There was a public outcry to rid the buildings of this "dangerous, carcinogenic material" immediately to protect our vulnerable children. The perception of many was that our children were at risk. It seemed as if the asbestos would leap off the walls and sicken our children. Simply being near asbestos or in a building that had asbestos-containing material would result in lung cancer and death. This simply was not the case. Asbestos can be dangerous and can cause lung and other cancers as well as asbestosis. However, to become ill with these or other diseases, you must first be exposed. That is, you must receive a regular, chronic dose (over many years). To receive a dose, you must inhale (breathe in) or ingest (swallow) fibers of asbestos. Asbestos is a naturally occurring mineral that comes in several forms, and a person could work in a room made from solid asbestos and never become ill.

Asbestos does not become a problem unless it is disturbed or removed during renovation or demolition. The worst thing we could have done was to start ripping it out of our schools. The best practice was to leave it in place and encapsulate it with a coating. Simple painting will do in many cases. Only during removal should abatement measures be required. Spraying with water is sometimes all that is needed. The perceived risk was much greater than the actual risk. Becoming educated about the risks enables us to make better decisions about environmental and human health issues. This saves time and money, thus maximizing the resources available to reduce actual risks to ourselves and the environment.

Risk estimates by "experts" and by the "public" on many environmental problems differ significantly. Almost every daily activity—driving, walking, or working—involves some element of risk. (See table 3.2.) People often overestimate the frequency and seriousness of dramatic, sensational, well-publicized causes of death and underestimate the risks from more familiar causes that claim lives one by one.

This discrepancy and the reasons for it are extremely important because the public generally does not trust experts to make important risk decisions alone. The public generally perceives involuntary risks, such as nuclear power plants or nuclear weapons, as greater than voluntary risks, such as drinking alcohol or smoking. In addition, the public perceives newer technologies, such as genetic engineering or toxic-waste incinerators, as greater risks than more familiar technologies, such as automobiles and dams. Many people are afraid of flying for fear of crashing; however, motor vehicle accidents account for a far greater number of deaths—over 40,000 in the United States each year, compared to less than a thousand from plane crashes.

One of the most profound dilemmas facing decision makers and public health scientists is how to address the discrepancy between the scientific and public perceptions of environmental risks. Numerous studies have shown that in the last 20 years, environmental hazards truly affecting health status in the country are not those receiving the highest attention, whether measured by public

TABLE 3.2 Causes of Accidental Death in the United States, 2007

Cause	Number
Motor vehicle accidents (cars, trucks, buses)	43,620
Falls	16,321
Accidental poisoning	15,100
Unspecified accidents	7,484
Suffocation	6,254
Fires	4,321
Drowning	4,103
Aircraft accidents	1,007
Lightning	48

Source: National Center for Health Statistics.

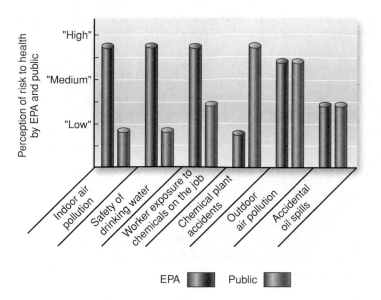

FIGURE 3.2 **Perception of Risk** Professional regulators and the public do not always agree on what risks are.

opinion polls, news coverage, congressional actions, or government expenditures. (See figure 3.2.)

Indoor air pollution, in its various forms, receives relatively little attention compared with outdoor sources and yet probably accounts for as much, if not more, poor health. Hazardous waste dumps, on the other hand, which are difficult to associate with any measurable ill health, attract much attention and resources. The same chemicals in the form of common consumer products, such as household cleaners, pesticides, and fuel (gasoline), account for much more exposure and ill health and yet raise comparatively little concern from the public.

Several explanations exist for this difference in perception, the major ones relating to the fact that the public uses a number of criteria other than health risk to establish its concerns. However, this mismatch between real and perceived risks has significant consequences. In a world of finite financial resources, when money is used to reduce risks that have little measurable health impact, there is less to spend on interventions that address more significant risks.

Some researchers argue that the public is frequently misled by the politics of public health and environmental safety. This is understandable since many prominent people become involved in such issues and use their public image to encourage people to look at issues from a particular point of view.

Whatever the issue, it is hard to ignore the will of the people, particularly when sentiments are firmly held and not easily changed. A fundamental issue surfaces concerning the proper role of government and other organizations in a democracy when it comes to matters of risk. Should the government focus available resources and technology where they can have the greatest tangible impact on human and ecological well-being, or should it focus them on problems about which the public is most upset? What is the proper balance? For example, would adequate prenatal health care for all pregnant women have a greater effect on the health of children than removing asbestos from all school buildings?

Obviously, there are no clear answers to these questions. Experts and the public, however, are both beginning to realize that they each have something to offer to the debate. Many risk experts who have been accustomed to looking at numbers and probabilities are now conceding that a rationale exists for looking at risk in broader terms. At the same time, the public is being supplied with more data to enable them to make more informed judgments.

Throughout this discussion of risk assessment and management, we have made numerous references to costs and economics. It is not economically possible to eliminate all risk. Sometimes risk identification is all that is possible or required. A risk elimination process can be desirable but not always beneficial. As risk is eliminated, the cost of the product or service increases. Many environmental issues are difficult to evaluate from a purely economic point of view, but economics is one of the tools used to analyze any environmental problem.

ENVIRONMENTAL ECONOMICS

Economics is the study of how people choose to use resources to produce goods and services and how these goods and services are distributed to the public. In other words, economics is an allocation process that determines the purposes to which resources are put. In many respects, environmental problems are primarily economic problems. While this may be an overstatement, it is not possible to view environmental issues outside the normal economic process that is central to our way of life. Pollution prevention often takes on a purely economic aspect when we look at "waste in, waste out" or mass-balance equations to determine the costs associated with waste. Businesses today cannot simply ignore the economics of environmental considerations. They are sometimes required to maintain environmental management systems even to do business with certain companies or even countries. To appreciate the interplay between environmental issues and economics, it is important to have an understanding of some basic economic concepts.

RESOURCES

Economists look at **resources** as the available supply of something that can be used. Classically, there are three kinds of resources: labor, capital, and land. Labor is commonly referred to as a human resource. Capital is anything that enables the efficient production of goods and services (technology and knowledge are examples). Land can be thought of as the natural resources of the planet. **Natural resources** are structures and processes that humans can use for their own purposes but cannot create. The agricultural productivity of the soil, rivers, minerals, forests, wildlife, and weather (wind, sunlight, rainfall) are all examples of natural resources. The landscape is also a natural resource, as we see in countries with a combination of mountainous terrain and high rainfall that can be used to generate hydroelectric power or in those that have beautiful scenery or biotic resources that foster tourism.

Natural resources are usually categorized as either renewable or nonrenewable. **Renewable resources** can be formed or regenerated by natural processes. Soil, vegetation, animals, air, and water are renewable primarily because they naturally undergo processes that repair, regenerate, or cleanse them when their quality or quantity is reduced. Just because a resource is renewable, however, does not mean that it is inexhaustible. Overuse of renewable resources can result in their irreversible degradation. **Nonrenewable resources** are not replaced by natural processes, or the rate of replacement is so slow as to be ineffective. For example, iron ore, fossil fuels, and mountainous landscapes are nonrenewable on human timescales. Therefore, when nonrenewable resources are used up, they are gone, and a substitute must be found or we must do without.

SUPPLY AND DEMAND

An economic good or service can be defined as anything that is scarce. Scarcity exists whenever the demand for anything exceeds its supply. We live in a world of general scarcity, where resources are limited relative to the desires of humans to consume them. The mechanism by which resources are allocated involves the establishment of a price for a good or service. The price describes how we value goods and services and is set by the relationship among the supply of a good or service and society's demand for it.

The **supply** is the amount of a good or service people are willing to *sell* at a given price. **Demand** is the amount of a good or service that consumers are willing and able to *buy* at a given price. The **price** of a good or service is its monetary value. One of the important mechanisms that determines the price is the relationship between the supply and demand, which is often illustrated with a **supply/demand curve.** For any good or service, there is a constantly shifting relationship among supply, demand, and price. The price of a product or service reflects the strength of the demand for and the availability of the commodity. When demand exceeds supply, the price rises. The increase in price results in a chain of economic events. Price increases cause people to seek alternatives or to decide not to use a product or service, which results in a lower quantity demanded.

For example, prices for recycled paper materials, such as old corrugated cardboard, fluctuate significantly based on their supply and demand. (See figure 3.3.) The supply of old corrugated cardboard does not vary much, because there are well-established recycling programs in place that capture over 70 percent of discarded corrugated cardboard. However, the demand fluctuates significantly depending on several factors. A primary factor that determines demand is the general state of the economy. When the economy is strong, people buy things and those things are typically shipped in corrugated cardboard containers. This results in an increase in demand and an increase in the price cardboard manufacturers are willing to pay for old corrugated cardboard. Conversely, when people are not buying things, less packaging is needed, demand falls, and the price falls as well. A second factor that determines demand is the strength of the export market. When other countries are buying old corrugated cardboard, less is available for

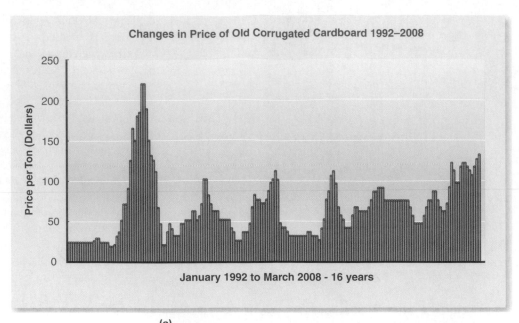

(a)

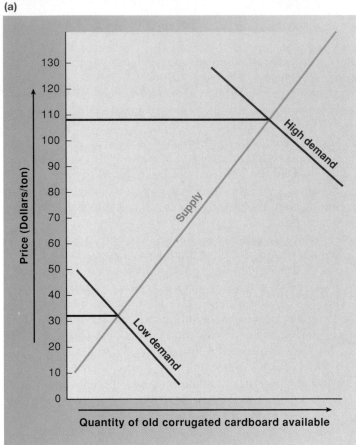

(b)

FIGURE 3.3   Supply and Demand for Old Corrugated Cardboard Graph *a* shows that the price for old corrugated cardboard varies considerably. The supply of old corrugated cardboard is relatively fixed because about 70 percent is captured for recycling. However, the demand varies. Graph *b* shows a typical supply/demand diagram. Demand for old corrugated cardboard is high when the U.S. economy is doing well or exports of old corrugated cardboard are high. Because demand is high the price is high. When the U.S. economy is not doing well or exports of old corrugated cardboard are low, the demand falls and so does the price.

Source: (a) Data from USEPA.

More than half of the major North American rivers have been dammed, diverted, or otherwise controlled. Although the structures provide hydropower, control floods, supply irrigation, and increase navigation, they have changed the hydrological regime—damaging aquatic life, recreational opportunities, and livelihoods of some indigenous peoples. The ecological and economic costs of dams are being increasingly evaluated in comparison to their anticipated benefits, and some have been removed. At least 465 dams have been decommissioned in the United States, with about 100 more planned for removal. There has also been a trend toward river restoration in the United States since 1900, with most projects directed to enhancing water quality, managing riparian zones, removing in-stream habitat, allowing fish passage, and stabilizing stream banks.

An environmental impact analysis using CVM (contingent valuation method) was conducted in the 1990s to explore the removal of the Elwha and Glines dams in the state of Washington. These two 30- and 60-meter-high dams, respectively, are old and block the migration of salmon to 110 km of pristine water located in the Olympic National Park.

The dams also impact the Lower Elwha Klallam Tribe, which relies on the salmon and the river for its physical, spiritual, and cultural well-being. Dam removal could bring substantial fishing benefits, more than tripling the salmon populations. The cost of removing the dams, especially the sediment build-up, is estimated at about $125 million. Recreational and commercial fishing benefits resulting from dam removal would not be sufficient to cover these costs.

A CVM survey was conducted and yielded a 68 percent response in Washington State, and a 55 percent response for the rest of the United States. Willingness to pay for dam removal ranged from $73 per household for Washington to $68 for the rest of the United States. If every household in Washington State were to pay $73, the cost of the dam removal and river restoration could be covered.

After years of negotiations it has been decided that the dams will be removed, and the Elwha Restoration Project will go forward. This is the biggest dam-removal project in history, and an event of national significance. It is expected that the two dams will be removed in stages over the course of three years, between 2010 and 2013.

the domestic market and the price increases. Finally, when the price of old corrugated cardboard approaches $125 per ton, cardboard producers can buy pulpwood at about the same price and begin to switch from using old corrugated cardboard to pulpwood.

Similarly, food production depends heavily on petroleum for the energy to plant, harvest, and transport food crops. In addition, petrochemicals are used to make fertilizer and chemical pest-control agents. If the demand for energy exceeds the supply, the price of petroleum increases. As petroleum prices rise, farmers reduce their petroleum use. Perhaps they farm less land or use less fertilizer or pesticide. Because farmers are using less energy, they will produce less food, so the supply of food decreases. Thus, an increase in petroleum prices results in an increase in food prices. As the prices of certain foods rise, consumers seek less costly foods.

When the supply of a commodity exceeds the demand, producers must lower their prices to get rid of the product, and eventually, some of the producers go out of business. Ironically, this happens to farmers when they have a series of good years. Production is high, prices fall, and some farmers go out of business.

ASSIGNING VALUE TO NATURAL RESOURCES

We assign value to natural resources based on our perception of their relative scarcity. We are willing to pay for goods or services we value highly and are unwilling to pay for things we think there is plenty of. For example, we will readily pay for a warm, safe place to live but would be offended if someone suggested that we pay for the air we breathe.

If a natural resource has always been rare, it is expensive. Pearls and precious metals are expensive because they have always been rare. If the supply of a resource is very large and the demand for it is low, the resource may be thought of as free. Sunlight, oceans, and air are often not even thought of as natural resources because their supply is so large. However, modern technologies have allowed us to exploit natural resources to a much greater extent than our ancestors were able to achieve and resources that were once considered limitless are now rare. For example, in the past, land and its covering of soil was considered a limitless natural resource, but as the population grew and the demand for food, lodging, and transportation increased, we began to realize that land is a finite, nonrenewable resource. The economic value of land is highest in metropolitan areas, where open land is unavailable. Unplanned, unwise, or inappropriate use can result in severe damage to the land and its soil. (See figure 3.4.)

Even renewable resources can be overexploited. If the overexploitation is severe and prolonged, the resource itself may be destroyed. For example, overharvesting of fish, wildlife, or forests can change the natural ecosystem so much that it cannot recover, and a resource that should have been renewable becomes a depleted nonrenewable resource.

Valuing natural resources and evaluating policies where institutions such as markets do not exist, and where there is a lack of individual property rights, pose challenges. Under such uncertainties, and where divergent sets of values exist, the economic value of common resources can be measured by the maximum amount of other goods and services that individuals are willing to give up to obtain a given good or service. Therefore, it is possible to weigh the benefits from an activity such as the construction of a dam against its negative impacts on fishing, livelihoods of nearby communities, and changes to aesthetic values. This method of valuation is called the contingent valuation method (CVM). (See Water Connections box above.)

FIGURE 3.4   Mismanagement of a Renewable Resource
Although soil is a renewable resource, extensive use can permanently damage it. Many of the world's deserts were formed or extended by unwise use of farmland. This photograph shows a once-productive farm now abandoned to the wind and sand because the soil was mistreated and allowed to erode.

Valuation presents a set of challenges beyond conflicting value systems or lack of existing market institutions. It uses national and local measures to estimate the economic values of tangible and intangible services provided by the environment. Valuation work has been undertaken on areas such as the value of nontimber forest products, forestry, and the health impacts of air pollution and water-borne diseases. However, studies on less tangible but yet important services, such as water purification and the prevention of natural disasters, in addition to recreational, aesthetic, and cultural services, have been hard to get. To get objective monetary estimates of these services remains a challenge. Market data are limited to a small number of services provided by ecosystems. Furthermore, methodologies such as cost-benefit analysis and CVM may raise problems of bias.

ENVIRONMENTAL COSTS

Air pollution, water pollution, plant and animal extinctions, depletion of a resource, and loss of scenic quality are all examples of the **environmental costs** of resource exploitation. Often environmental costs are difficult to assess, since they are not easily converted to monetary values. This is especially true with the loss of "aesthetics" such as a beautiful scene, relaxing surroundings, or recreational opportunities. These losses are often calculated in "man-hours" spent or lost due to environmental degradation. In addition, since they may not be recognized immediately, environmental costs are often **deferred costs,** which must be paid at a later date. For example, when dams were built on the Colorado River to provide electric power and irrigation water, planners did not anticipate that the changes in the flow of the river would reduce habitat for endangered bird species, lead to the loss of fish species because the water is colder, and result in increased salinity in lower regions of the river. Soil erosion is another example

of a deferred cost. The damage done by practices that increase soil erosion may not be felt immediately, but eventually as the amount of damage accumulates, the cost becomes obvious to future generations.

Many of the important environmental problems facing the world today arise because modern production techniques and consumption patterns transfer waste disposal, pollution, and health costs to society. Such expenses, whether they are measured in monetary terms or in diminished environmental quality, are borne by someone other than the individuals who use a resource. They are referred to as **external costs.** For example, when a logging operation removes so many trees from a hillside that runoff from the hillside destroys streams and causes mudslides, the logging operation has transferred a cost to the public. Another example is the thousands of hazardous waste sites produced by industries that have since ceased to exist. The cleanup of these abandoned hazardous waste sites became the responsibility of government and the taxpayers. The entities that created the sites avoided paying for their cleanup. Similarly, when a new shopping complex is built, many additional external costs are paid for by the public and the municipality. Additional roads, police and fire protection, sewer and water services, runoff from parking lots, and pressure to convert the remaining adjacent land to shopping are all external costs typically borne by the taxpayer.

The extraction of mineral resources is a good example of the variety of environmental costs that accompany resource use. All mining operations involve the separation of the valuable mineral from the surrounding rock. The surrounding rock must then be disposed of in some way. These pieces of rock are usually piled on the surface of the Earth, where they are known as mine tailings and present an eyesore. It is difficult to get vegetation to grow on these deposits. Some mine tailings contain materials (such as asbestos, arsenic, lead, and radioactive materials) that can be harmful to humans and other living things.

Many types of mining operations require vast quantities of water for the extraction process. The quality of this water is degraded, so it is unsuitable for drinking, irrigation, or recreation. Since mining disturbs the natural vegetation in an area, water may carry soil particles into streams and cause erosion and siltation. Some mining operations, such as strip mining, rearrange the top layers of the soil, which lessens or eliminates its productivity for a long time. (See figure 3.5.) Strip mining has disturbed approximately 75,000 square kilometers (30,000 square miles) of U.S. land, an area equivalent to the state of Maine.

Probably most environmental costs have both deferred and external aspects. A good example of a problem that is both a deferred and an external cost is the damage caused by the use of high-sulfur coal as an inexpensive way to produce electricity. The sulfur compounds released into the atmosphere resulted in acid rain that caused a decline in the growth of forests and damage to buildings and other structures. The damage accumulated over time so the cost of acid rain was a deferred cost. The cost of the damage was paid for by the public as fewer scenic vistas, by forest products industries with fewer trees to harvest, and by property owners as repair costs for buildings and other structures, so it was an external cost not paid for by the electric utilities directly.

FIGURE 3.5   **A Strip-Mining Operation** It is easy to see the important impact a mine of this type has on the local environment. Unfortunately, many mining operations are located in areas that are also known for their scenic beauty.

Throughout history, humans have sought to improve living conditions and eliminate the misery caused by hunger and disease. In general, we rely on science and technology to improve our quality of life. While technological progress can improve the quality of life, it can also generate new sources of pollution. The development of the steam engine allowed machines to replace animal power and human labor but increased the amount of smoke and other pollutants in the air as well as the need for fuel. The modern chemical industry has produced many extremely valuable synthetic materials (plastics, pesticides, medicines), but it has also produced toxic pollutants.

It is not always easy to agree on what constitutes pollution. To some, the smell of a little wood smoke in the air is pleasant; others do not like the odor. A business may consider advertising signs valuable and necessary; others consider them to be visual pollution. Finally, it is important to recognize that it is impossible to eliminate all the negative effects produced by humans and our economic processes. The difficult question is to determine the levels of pollution that are acceptable. (See figure 3.6.)

As people recognize the significance of environmental costs, these costs are being converted to economic costs as stricter controls on pollution and environmental degradation are enforced. It takes money to clean up polluted water and air or to reclaim land that has been degraded, and the people who cause the damage should not be allowed to defer the cost or escape paying for the necessary cleanup or remediation.

Pollution-control costs include pollution costs and pollution-prevention costs. **Pollution costs** include such things as the private or public expenditures to correct pollution damage once pollution has occurred, the increased health costs because of pollution, and the loss of the use of public resources because of pollution. **Pollution-prevention costs (P2)** are those incurred either in the private sector or by government to prevent, either entirely or partially, the pollution that would otherwise result from some production or consumption activity. The cost incurred by local government to treat its sewage before releasing it into a river is a P2 cost; so is the cost incurred by an electric utility to prevent air pollution by installing new equipment.

Environmental costs also may include lost opportunities or values because the resource could not be used for another purpose. For example, if houses are built in a forested region, its possible use as a natural area for hiking or hunting is lost. Similarly, when land is converted to roads and parking lots, the opportunity is lost to use the land for farming or other purposes.

A primary environmental cost is pollution. **Pollution** is any addition of matter or energy that degrades the environment for humans and other organisms. When we think about pollution, however, we usually mean something that people produce in large enough quantities that it interferes with our health or well-being. Two primary factors that affect the amount of damage done by pollution are the size of the population and the development of technology that "invents" new forms of pollution.

When the human population was small and people lived in a simple manner, the wastes produced were biological and so dilute that they usually did not constitute a pollution problem. People used what was naturally available and did not manufacture many products. Humans, like any other animal, fit into their natural ecosystems. Their waste products were **biodegradable** materials that were broken down into simpler chemicals, such as water and carbon dioxide, by the action of decomposer organisms.

Human-initiated pollution became a problem when human populations became so concentrated that their waste materials could not be broken down as fast as they were produced. As the population increased, people began to congregate and establish villages, towns, and cities. The release of large amounts of smoke, biological waste, and trash faster than they could be absorbed and dispersed resulted in pollution, which led to unhealthy living conditions.

COST-BENEFIT ANALYSIS

Because resources are limited and there are competing uses for most resources, it is essential that a process be used to help decide the most appropriate use of a scarce resource. **Cost-benefit analysis** is a formal quantitative method of assessing the costs and benefits of competing uses of a resource or solutions to a problem and deciding which is the most effective. It has long been the case in many developed countries that major projects, especially those undertaken by the government, require some form of cost-benefit

Water pollution. This sign indicates it is unsafe to swim in this area because of high levels of bacteria.

Smog. Smog that develops when air pollution is trapped is a serious health hazard.

Health hazard

Smoke. Smoke contains small particles that can cause lung problems.

Solvents. Solvents evaporate and cause localized pollution.

Odors. Feedlots create an odor problem that many people find offensive.

Thermal pollution. Cooling towers release heat into the atmosphere and can cause localized fog.

Annoyance

Visual pollution. Unsightly surroundings are annoying but not hazardous to your health.

Junkyard. This is unsightly but constitutes only a minor safety hazard.

FIGURE 3.6 **Examples of Pollution** There are many kinds of pollution. Some are major health concerns, whereas others merely annoy.

People use cost-benefit analysis to determine whether a policy generates more social costs than social benefits and, if benefits outweigh costs, how much activity would obtain optimal results. Steps in cost-benefit analysis include:

1. Identification of the project to be evaluated
2. Determination of all impacts, favorable and unfavorable, present and future, on all of society
3. Determination of the value of those impacts, either directly through market values or indirectly through price estimates
4. Calculation of the net benefit, which is the total value of positive impacts less the total value of negative impacts

For example, the cost of reducing the amount of lead in drinking water in the United States to acceptable limits is estimated to be about $125 million a year. The benefits to the nation's health from such a program are estimated at nearly $1 billion per year. Thus, under a cost-benefit analysis, the program is economically sound. Table 3.3 gives examples of the kinds of costs and benefits involved in improving air quality. Although it is not a complete list, the table indicates the kinds of considerations that go into a cost-benefit analysis. Some of these are easy to measure in monetary terms; others are not.

CONCERNS ABOUT THE USE OF COST-BENEFIT ANALYSIS

Critics of cost-benefit analysis often raise the question of whether everything can be analyzed from an economic point of view. Some people argue that if the only measure of value is economic, many simple noneconomic values such as beauty or cleanliness can be justified only if they are given economic value. (See figure 3.7.)

There are clearly benefits to requiring such analysis. Although environmental issues must be considered at some point during project evaluation, efforts to do so are hampered by the difficulty of assigning specific value to environmental resources. In cases of Third World development projects, these already difficult environmental issues are made more difficult by cultural and socioeconomic differences. A less-developed country, for example, may be less inclined to insist on expensive emissions-treatment technology on a project that will provide jobs and economic development because it is unable to afford the treatment technology and it values the jobs very highly.

analysis with respect to environmental impacts and regulations. In the United States, for example, such requirements were established by the National Environmental Policy Act of 1969, which mandates environmental impact statements for major government-supported projects. Increasingly, similar analyses are required for projects supported by national and international lending institutions such as the World Bank.

TABLE 3.3 Costs and Benefits of Improving Air Quality

Costs	Benefits
Installation and maintenance of new technology:	Reduced deaths and disease
Scrubbers on smokestacks	Fewer respiratory problems
Automobile emissions control	Reduced plant and animal damage
Redesign of industries and machines	Lower cleaning costs for industry and public
Additional energy costs to industry and public	More clear, sunny days; better visibility
Retraining of employees to use new technology	Less eye irritation
Costs associated with monitoring and enforcement	Fewer odor problems

Cost-Benefit Analysis

Economic costs and benefits

- Proposed action
- Alternate actions
- What are the total monetary costs of the project?
- What are monetary benefits?
- Compare economic costs and benefits
- Evaluate and compare costs and benefits
- Who will cover costs? Who will reap benefits?
- Final decision—consider economic and noneconomic factors

Environmental costs and benefits

- What environmental elements and systems will be affected?
- What will be the consequences to human health and welfare?
- Identify and quantify
- Identify and quantify
- Establish monetary values, if possible
- Establish monetary values, if possible
- What elements and systems cannot be given a monetary value?
- What consequences cannot be given a monetary value?

One particularly compelling critique of cost-benefit analysis is that for analysis to be applied to a specific policy, the analyst must decide which preferences count—that is, which preferences are the most important for cost-benefit analysis. In theory, cost-benefit analysis should count all benefits and costs associated with the policy under review, regardless of who benefits or bears the costs. In practice, however, this is not always done. For example, if a cost is spread thinly over a large population, it may not be recognized as a cost at all. The cost of air pollution in many parts of the world could fall into such a category. Debates over how to count benefits and costs for future generations, inanimate objects such as rivers, and nonhumans, such as endangered species, are also common.

COMPARING ECONOMIC AND ECOLOGICAL SYSTEMS

For most natural scientists, current crises such as biodiversity loss, climate change, and many other environmental problems are symptoms of an imbalance between the socioeconomic system and the natural world. While it is true that humans have always changed the natural world, it is also clear that this imprint is much greater now than anything experienced in the past. One reason for the profound effect of human activity on the natural world is the fact that there are so many of us.

One of the problems associated with matching economic processes with environmental resources is the great differences in the way economic systems and ecological systems function. The loss of biodiversity is an example that illustrates the conflicting frameworks of economics and ecology. Market decisions fail to account for the context of a species or the interconnections between resource quality and ecosystem functions. For example, from an economic point of view, the value of land used for beef production is measured according to its contribution to its economic output (beef). Yet long before economic output and the use value of land decrease, the diversity of grass varieties, microorganisms in the soil, or groundwater quality may be affected by intensive beef production. As long as yields are maintained, these environmental changes go unnoticed by economic measurements and are unimportant to land-use decisions. This does not need to be the case. It should be pointed out that in Zimbabwe and other African nations, some ranchers now earn more money managing native species of

FIGURE 3.7 **Assigning Economic Values to Resource Use** The way we use resources is based on the perceived value of the resource. Not all people see the same value for a resource, and values are not always easy to measure.

wildlife for ecotourism in a biodiverse landscape than they would from raising cattle in a landscape with reduced biodiversity.

Another obvious difference between economics and ecology is the great difference in the time frame of markets and ecosystems. Many ecosystem processes take place over tens of thousands and even millions of years. The time frame for market decisions is short. It may be as short as minutes for stock trades or as long as a few years for the development and construction of a factory. Where U.S. economic policy is concerned, two- to four-year election cycles are the frame of reference. For investors and dividend earners, performance time frames of three months to one year are the rule.

Space or place is another issue. For ecosystems, place is critical. Take groundwater as an example. Soil quality, hydro geological conditions, regional precipitation rates, plants that live in the region, and losses from evaporation, transpiration, and groundwater flow all contribute to the size and location of groundwater reservoirs. These capacities are not simply transferable from one location to another. For economic activities, place is increasingly irrelevant. Topography, location, and function within a bioregion or local ecological features do not enter into economic calculations except as simple functions of transportation costs or comparative advantage. Production is transferable, and the preferred location is anywhere production costs are the lowest.

Another difference between economics and ecology is that they are measured in different units. The unifying measure of market economics is money. Progress is measured in monetary units that everyone uses and understands to some degree. Ecological systems are measured in physical units such as calories of energy, carbon dioxide absorption, centimeters of rainfall, or parts per million of nitrate contamination. Focusing only on the economic value of resources while ignoring environmental health may mask serious changes in environmental quality or function. (See Case Study on RBCA and protection of ecological resources.)

COMMON PROPERTY RESOURCE PROBLEMS—THE TRAGEDY OF THE COMMONS

Economists have stated that when everybody shares ownership of a resource, there is a strong tendency to overexploit and misuse that resource. Thus, common public ownership could be better described as effectively having no owner. The problems inherent in common ownership of resources were outlined by biologist Garrett Hardin in a classic essay entitled "The Tragedy of the Commons" (1968). The original "commons" were areas of pastureland in England that were provided free by the king to anyone who wished to graze cattle.

There are no problems on the commons as long as the number of animals is small in relation to the size of the pasture. From the point of view of each herder, however, the optimal strategy is to enlarge his or her herd as much as possible: If my animals do not eat the grass, someone else's will. Thus, the size of each herd grows, and the density of stock increases until the commons becomes overgrazed. The result is that everyone eventually loses as the animals die of starvation. The tragedy is that even though the eventual result should be perfectly clear, no one acts to avert disaster.

The ecosphere is one big commons stocked with air, water, and irreplaceable mineral resources—a "people's pasture," to be used in common, but it is a pasture with very real limits. Each nation attempts to extract as much from the commons as possible without regard to other countries. Furthermore, the United States and other industrial nations consume far more than their fair share of the total world resource harvest each year, much of it imported from less-developed nations.

One clear modern example of this problem involves the overharvest of marine organisms. Since no one owns the oceans, many countries feel they have the right to exploit the fisheries resources present. As in the case of grazing cattle, individuals seek to get as many fish as possible before someone else does. Currently, the UN estimates that nearly all of the marine fisheries of the world are being fished at or above capacity.

Another example of this is the idea of shared fisheries in the Great Lakes region. Commercial fisheries, recreational fishers, Native American tribal fishing, and regulatory agencies in both the United States and Canada have tried for many years to deal with overexploitation of the Great Lakes fishery. Tribal customs and treaties that date back hundreds of years complicate the issue because traditional fishing methods have changed from simple spearing, trapping, and hook-and-line taking to modern gill netting that takes fish indiscriminately.

Invasive and exotic species, pollution, and overfishing are just a few of the issues that must be taken into consideration when dealing with declining fisheries. Fishing regulations and zones were not designed from an ecosystem approach, but rather by political means and sometimes haphazardly with complete disregard for spawning grounds or other important biological realities. These issues have been addressed somewhat in recent years through the Sea Grant program, the International Joint Commission, and other groups.

Finally, common ownership of land resources, such as parks and streets, is the source of other environmental problems. People who litter in public parks do not generally dump trash on their own property. The lack of enforceable property rights to commonly owned resources explains much of what economist John Kenneth Galbraith has termed "public squalor amid private affluence." Common ownership of the ocean makes it inexpensive for ships and oil drilling platforms to use the ocean as a dump for their wastes. (See figure 3.8.)

The tragedy of the commons also operates on an individual level. Most people are aware of air pollution, but they continue to drive their automobiles. Many families claim to need a second or third car. It is not that these people are antisocial; most would be willing to drive smaller or fewer cars if everyone else did, and they could get along with only one small car if public transport were adequate. But people frequently get "locked into" harmful situations, waiting for others to take the first step, and many unwittingly contribute to tragedies of the commons. After all, what harm can be done by the birth of one more child, the careless disposal of one more beer can, or the installation of one more air conditioner?

GREEN ECONOMICS

The world has witnessed three economic transformation in the past century: First came the industrial revolution, then the technology revolution, then our modern era of globalization. The world now stands at the threshold of another great change: the age of green economics.

The evidence is all around us, often in unexpected places. Brazil, for example has become one of the biggest players in green economics, drawing 44 percent of its energy needs from renewable fuels. The world average is 13 percent. Much is made of the fact that China is poised to surpass the United States as the world's largest emitter of greenhouse gases. Less well known, however, are its more recent efforts to confront grave environmental problems. China is on track to invest $10 billion in renewable energy—second only to Germany. It has become a world leader in solar and wind power. China has pledged to reduce energy consumption (per unit of gross domestic product) by 20 percent by 2014—not far removed from Europe's commitment to a 20 percent reduction in greenhouse gas emissions by 2020.

Some estimates show that growth in global energy demand could be cut in half over the next 15 years simply by deploying existing technologies yielding a return on investment of 10 percent or more. The Intergovernmental Panel on Climate Change (IPCC) reports such demand cuts in very practical ways, from tougher standards for air conditioners and refrigerators to improved efficiency in industry, building, and transport. It estimates that overcoming serious climate change may cost as little 0.1 percent of global GDP a year over the next three decades.

Growth need not suffer and, in fact, may accelerate. Research by the University of California at Berkeley indicates that the United States could create 300,000 jobs if 20 percent of electricity needs were met by renewable energy. A German firm predicts that more people will be employed in Germany's environmental technology industry than in the auto industry by the end of the next decade. The U.N. Environment Program estimates that global investment in zero greenhouse energy will reach $1.9 trillion by 2020.

USING ECONOMIC TOOLS TO ADDRESS ENVIRONMENTAL ISSUES

The traditional way of dealing with environmental issues is to develop regulations that prohibit certain kinds of behavior. This is often called a "command and control" approach. It has been very

FIGURE 3.8   **The Ocean Is a Common Property Resource**
Since the oceans of the world are a shared resource that nobody owns, there is a tendency to use the resource unwisely. Growing populations in coastal areas lead to more marine pollution and destruction of coastal habitats. Many countries use the ocean as a dump for unwanted trash. Some 6.5 million metric tons of litter find their way into the sea each year.

Everyone knows what blue-collar and white-collar jobs are, but now we have a job of another hue—a green-collar job. It's hard to exactly say what a green-collar job is; however, it has been reported that the green-collar sector is growing in the United States and could include more than 14 million workers by 2017. According to the American Solar Energy Society, there are 8.5 million U.S. jobs that involve Earth-friendly enterprises and renewable energy sources. That figure could grow by 5 million in the next decade. Another study by the Cleantech Network, a venture capital firm for green business, showed that up to half a million new jobs in ecologically responsible trades will develop by 2012. These new jobs will be at every income level and include titles such as:

- Green product designer—designs products that use less energy and raw materials to produce and consume less energy and resources to use.

- Energy rating auditor—performs a comprehensive analysis of a building's energy efficiency.

- Environmental manager—coordinates management of organization's environmental performance to protect and conserve natural resources.

- Biological systems engineer—designs, manages, and develops systems and equipment that produce, package, process, and distribute the world's food and fiber supplies.

- Permaculture specialist—analyzes land use and community building to create a harmonious blend of buildings, plants, animals, soils, and water.

In addition, professionals will find opportunities by adding green to their skill sets, from accountants who can manage corporate carbon emission offsets to zookeepers who must maintain environmentally sensitive and ecologically friendly animal habitats.

Part of the growth in green-collar jobs will come from government initiatives. In 2007, the U.S. House of Representatives passed the Green Jobs Act, which provides $125 million annually to train people for green vocational fields that offer living wages and upward mobility for low-income communities.

In the private sector, Bank of America launched a $20 billion initiative to support environmentally sustainable business activity to address global climate change, and Citigroup plans to commit $50 billion to environmental projects over the next decade. Many companies such as Dow Chemical and DuPont have added Chief Sustainability Officer (CSO) to their list of top-level executives.

Is there a green-collar job in your future? Listed next are some links to job boards that you may want to look at as you think about your career plans or a career change.

- Ecojobs
- Greenjobs
- Greenbiz

---

effective at reducing air and water pollution, protecting endangered species, and requiring that environmental concerns be addressed by environmental impact statements. However, there are also tools that use economic incentives to encourage environmental stewardship.

SUBSIDIES

A **subsidy** is a gift from government to individuals or private enterprise to encourage actions considered important to the public interest. Subsidies may include consumer rebates for purchases of environmentally friendly goods, loans for businesses planning to implement environmental products, and other monetary incentives designed to reduce the costs of improving environmental performance.

Governments frequently subsidize agriculture, transportation, space technology, and communication. These gifts, whether loans, favorable tax situations, or direct grants, are all paid for by taxes on the public, so in effect they are an external cost.

When subsidy programs have a clear purpose and are used for short periods to move to new ways of doing business, they can be very useful. Government payments to farmers that encourage them to permanently take highly erodable land from production reduces erosion and the buildup of sediment in local streams. The

better water quality benefits fish, and the return of land to more natural vegetation benefits wildlife. Government programs that purchase the fishing boats of fishers who are displaced when fishing quotas are reduced are a form of subsidy to the fishing industry. The cost of government management of federal forests is a subsidy to the forest products industry, which can protect forest resources while ensuring a livelihood for loggers at the same time.

Subsidies are often used inappropriately, however, and when they are, they can lead to economic distortions. One of the effects of a subsidy is to keep the price of a good or service below its true market price. The actual cost of a subsidized good or service is higher than the subsidized market price because subsidy costs must be added to the market price to arrive at the product's true cost. Agricultural subsidies greatly distort the price of food. One common agricultural subsidy is a program that guarantees a price to a farmer for the products produced. If the market price is below the guaranteed price, the government buys the products at the guaranteed price or pays the farmer the difference between the market price and the guaranteed price. On average, U.S. farmers receive about 20 percent of their income from government payments. Other developed countries have programs that support their farmers in similar fashion. In addition, a huge bureaucracy is needed to manage the complex program. One of the unintended outcomes of such subsidies is that farmers are encouraged to produce more on less land. This results in the use of more fertilizer and pesticides that can damage the environment, and typically, there is overproduction of agricultural products.

Once subsidies become a part of the economic fabric of a country, they are very difficult to eliminate. In 1996, the U.S. Congress passed the Freedom to Farm Act, which eliminated many agricultural subsidies and was hailed as the end of agricultural subsidies. It did not work, however, and in 2002, a new federal farm bill abandoned the 1996 goal of reducing farm payments and authorized an 80 percent increase in expenditures.

Conversely, China has successfully reduced its subsidy for coal. In China, subsidy rates for coal declined from an estimated 61 percent in 1985 to 9 percent in 2000. Private mines now account for about half of all production, and some 80 percent of the coal is now sold at international prices. These reforms have had numerous benefits. Energy intensity in China has fallen by about 50 percent since 1980, and the government's total subsidy for fossil fuels fell from about $25 billion in 1990–91 to $9 billion in 2000.

The building of roads and bridges is a major part of the U.S. federal budget. In 2002, over $31 billion was allocated to the building and improvement of roads. This constitutes a subsidy for automobile transportation. Higher taxes on automobile use to cover the cost of building and repairing highways would encourage the use of more energy-efficient public transport.

LIABILITY PROTECTION AND GRANTS FOR SMALL BUSINESS

On January 11, 2002, President George W. Bush signed an important piece of environmental legislation into law: the Small Business Liability Relief and Brownfield Revitalization Act (SBLRBRA). This law provided incentives for small businesses and other entities to develop so-called **brownfields** (those areas perceived to have environmental liabilities), most of which are in urban areas. Prior to SBLRBRA, brownfield areas were considered too risky to purchase and develop since purchasers could potentially acquire the environmental liabilities associated with the property. Liability protection was provided in Title I and Title II, Subtitle B. In Subtitle A, funding was provided for small businesses and other entities to revitalize these areas. Working together, many states and local industries have used tax incentives and other methods to further encourage development in these previously undesirable or unusable areas.

Businesses must conduct "all appropriate inquiry" into the property prior to acquisition. They must conduct no activities that contribute to the environmental impact, and they must take "due care" while conducting business. They must also utilize "use limitation" criteria—that is, if they say they will build and use a plating operation, they had better do so and not open a day care center where children might be exposed to contaminated soil or groundwater. This program has resulted in many successful projects that have brought business back to where it once was, minimizing impact on the green belts outside urban areas.

MARKET-BASED INSTRUMENTS

With the growing interest in environmental protection during the past three decades, policy makers are examining new methods to reduce harm to the environment. One area of growing interest is market-based instruments. Market-based instruments provide an alternative to the common command-and-control legislation because they use economic forces and the ingenuity of entrepreneurs to achieve a high degree of environmental protection at a low cost. One of the benefits of market-based instruments is that they can be used to determine fair prices for environmental resources. Because of subsidies and external costs, many environmental resources are under-priced. Instead of inflexible, top down government directives, market-based policies take advantage of price signals and give entrepreneurs the freedom to choose the solution most economically efficient for them. For example, a price can be established for pollution causing activities. Companies are then allowed to decide for themselves how best to achieve the required level of environmental protection. To date, most of these market-based policies have been implemented in developed nations and in some rapidly growing developing nations. In virtually all cases, they have been introduced as supplements to, not substitutes for, traditional government regulations.

Several kinds of market-based instruments are currently in use:

Information programs provide consumers with information about the environmental consequences of purchasing decisions. Information about the environmental consequences of choices make clear to consumers that it is in their personal interest to change their decisions or behaviors. Examples include information tags on electric appliances that inform the public about the energy efficiency of the product, the mileage ratings of various automobiles, and labeling on pesticide products that describes safe use and disposal. Another type of program, such as the Toxic Release Inventory in the United States, discloses information on environmental releases by polluters. This provides corporations with incentives to improve their environmental performance to enhance their public image.

Tradable emissions permits give companies the right to emit specified quantities of pollutants. Companies that emit less than the specified amounts can sell their permits to other firms or "bank" them for future use. Thus, businesses responsible for pollution have an incentive to internalize the external cost they were previously imposing on society: If they clean up their pollution sources, they can realize a profit by selling their permit to pollute. Once a business recognizes the possibility of selling its permit, it sees that reducing pollution can have an economic benefit. The establishment of tradable sulfur dioxide pollution permits for coal-fired power plants has resulted in huge reductions in the amount of sulfur dioxide released.

Emissions fees and taxes provide incentives for environmental improvement by making environmentally damaging activity or products more expensive. Businesses and individuals reduce their level of pollution wherever it is cheaper to reduce the pollution than to pay the charges. Emissions fees can be useful when pollution is coming from many small sources, such as vehicular emissions or agricultural runoff, where direct regulation or trading schemes are impractical. Taxes and fees contribute to government revenue and thus offset some of the indirect costs incurred by government to protect environmental resources.

China, a pollution tax system is intended to raise revenue ~~~vestment in industrial pollution control, help pay for regula-~~~ activities, and encourage enterprises to comply with emission and effluent standards. The system imposes noncompliance fees on discharges that exceed standards and assesses fines and other charges on violations of regulations.

In the Netherlands, an effluent tax on industrial wastewater has been viewed as successful. Especially among larger companies, the tax worked as an incentive to reduce pollution. In a survey of 150 larger companies, about two-thirds said the tax was the main factor in their decision to reduce discharges. As the volume of pollution from industrial sources dropped, rates were increased to cover the fixed costs of sewage water treatment plants. Rising rates are providing a further incentive for more companies to start purifying their sewage water.

Deposit-refund programs place a surcharge on the price of a product that is refunded when the used product is returned for reuse or recycling. Deposit refund schemes have been widely used to encourage recycling. In Japan, deposits are made for the return of bottles. In 2002, the German government imposed a deposit of 0.25 euros on drink cans and disposable glass and plastic (PET) bottles. Eleven states in the United States have similar laws, but so far Congress has been unwilling to pass a national bill.

Performance bonds are fees that are collected to ensure that proper care is taken to protect environmental resources. Some nations—including Indonesia, Malaysia, and Costa Rica—use performance bonds to ensure that reforestation takes place after timber harvesting. The United States also has used this approach to ensure that strip-mined lands are reclaimed. Before a mining permit can be granted, a company must post a performance bond sufficient to cover the cost of reclaiming the site in the event the company does not complete reclamation. The bond is not fully released until all performance standards have been met and full reclamation of the site, including permanent revegetation, is successful—a five-year period in the East and Midwest and 10 years in the arid West. The bond can be partially released as various phases of reclamation are successfully completed.

Although each of these economic incentives can be effective by itself, they also can be used effectively in combination. For example, deposit-refund and tradable emissions programs work better if supported by information programs. Communities that adopted pay-by-the-bag systems of trash disposal had fewer problems if households were given adequate information well in advance. Environmental tax systems can incorporate emissions trading so that taxes can be levied on net emissions after trades.

LIFE CYCLE ANALYSIS AND EXTENDED PRODUCT RESPONSIBILITY

Life cycle analysis is the process of assessing the environmental effects associated with the production, use, reuse, and disposal of a product over its entire useful life. Life cycle analysis can help us understand the full cost of new products and their associated technologies.

The various stages in the product chain include raw material acquisition, manufacturing processes, transportation, use by the consumer, and ultimately disposal of the used product. When this approach is used, it is possible to identify changes in product design and process technology that would reduce the ultimate environmental impact of the production, use, and disposal of the product. All factors along the product chain share responsibility for the life cycle environmental impacts of the product, from the upstream impacts inherent in the selection of materials and impacts from the manufacturing process itself to downstream impacts from the use and disposal of the product. (See figure 3.9.)

Because the relationships among industrial processes are complex, life cycle analysis requires an understanding of material flows, resource reuse, and product substitution. Shifting to an approach that considers all resources, products, and waste as an interdependent system will take time, but governments can encourage this kind of thinking by establishing regulations that prevent industries from externalizing their pollution costs and providing economic incentives for those that use life cycle analysis in their product development and planning.

A logical extension of life cycle analysis is extended product responsibility. **Extended product responsibility** is the concept that the producer of a product is responsible for all the negative effects involved in its production, including the ultimate disposal of the product when its useful life is over. The logic behind extended product responsibility is that if manufacturers pay for the post consumer impacts of products, they will design them differently to reduce waste.

Many people identify the German packaging ordinance as one of the first instances of extended product responsibility. Under the

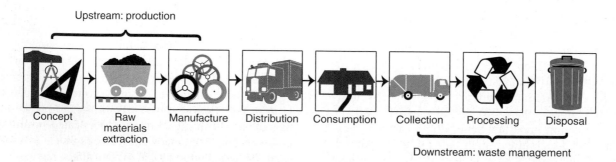

FIGURE 3.9 **The Life Cycle of a Typical Product** When life cycle analysis is undertaken, it is important to identify all the steps in the process— from obtaining raw materials, through the manufacturing process, to the final disposal of the item.
Source: *Environment*, vol. 39, no. 7, September 1997.

POLLUTION PREVENTION PAYS!

The philosophy of pollution prevention is that pollution should be prevented or reduced at the source whenever feasible. It is increasingly being shown that preventing pollution can cut business costs and thus increase profits. Pollution prevention, then, does make cents!

For example, several years ago, 3M's European chemical plant in Belgium switched from a polluting solvent to a safer but more expensive water-based substance to make the adhesive for its Scotch™ Brand Magic™ Tape. The switch was not made to satisfy any environmental law in Belgium or the European Union. 3M managers were complying with company policy to adopt the strictest pollution-control regulations that any of its subsidiaries is subject to—even in countries that have no pollution laws at all. Part of the policy is founded on corporate public relations, a response to growing customer demand for "green" products and environmentally responsible companies. But as many North American multinationals with similar global environmental policies are discovering, cleaning up waste, whether voluntarily or as required by law, can cut costs dramatically. Since 1975, 3M's "Pollution Prevention Pays" program—or 3P—has cut the company's air, water, and waste pollution by nearly 900,000 tons and saved the company almost $900 million. Less waste has meant less spending to comply with pollution control laws. But, in many cases, 3M actually has made money selling wastes it formerly hauled away. And because of recycling prompted by the 3P program, it has saved money by not having to buy as many raw materials.

AT&T followed a similar path. In 1990, it set voluntary goals for the company's 40 manufacturing and 2500 nonmanufacturing sites worldwide. According to its latest estimates, AT&T has (1) reduced toxic air emissions, many caused by solvents used in the manufacture of computer circuit boards, by 73 percent; (2) reduced emissions of chlorofluorocarbons—gases blamed for destroying the ozone layer in the Earth's atmosphere—by 76 percent; and (3) reduced manufacturing waste 39 percent.

Xerox Corporation had focused on recycling materials in its global environment efforts. It provides buyers of its copiers with free United Parcel Service pickup of used copier cartridges, which contain metal-alloy parts that otherwise would wind up in landfills. The cartridges and other parts are now cleaned and used to make new ones. Other recycled Xerox copier parts include power supplies, motors, paper transport systems, printed wiring boards, and metal rollers. In all, 1 million parts per year are remanufactured. The initial design and equipment investment was $10 million. Annual savings total $200 million.

EXAMPLES OF COMMON POLLUTION PREVENTION TECHNIQUES

- Improved process control to use energy and materials more efficiently
- Improved catalysis or reactor design to reduce by-products, increase yield, and save energy in chemical processes
- Alternative processes (e.g., low- or no-chlorine pulping for paper)
- In-process material recovery (e.g., vapor recovery, water reuse, and heavy metals recovery)
- Alternatives to chlorofluorocarbons and other organic solvents
- High-efficiency paint and coating application
- Substitutes for heavy metals and other toxic substances
- Cleaner or alternative fuels and renewable energy
- Energy-efficient motors, lighting, heat exchangers, etc.
- Water conservation
- Improved "housekeeping" and maintenance in industry

German packaging ordinance, consumers, retailers, and packaging manufacturers all share this responsibility, with the financial burden of waste management falling on the retailers and packaging manufacturers. This is thought to be one of the first instances of the concept of "take back," or taking the product back for disposal to the place that made it. This made companies look very hard at how they would manufacture something. The thought was: "This thing is coming back to me someday; how will I recycle/reuse or dispose of it?"

Although no national legislation in the United States mandates extended product responsibility, there are several instances in which manufacturers of specific products have implemented it. When several states passed legislation that required manufacturers of nickel-cadmium batteries to take back the worn-out batteries, manufacturers instituted a national take back program. Kodak has established a program of taking back and recycling single-use cameras. The chemical industry has instituted a program known as Responsible Care®. The concept originated in Canada and has since spread to 46 countries. The primary goals of Responsible Care® are to improve chemical processes and ensure the safe production, transport, use, and disposal of the products of the industry.

Specific benefits of extended product responsibility include:

Cost savings result when manufacturers take back used products because manufacturers recover valuable materials, reuse them, and save money.

Consideration of extended product responsibility has led to companies redesigning products to facilitate disassembly and recycling.

There are more efficient environmental protections, since it is easier to design environmental safety into the product than to try to clean up the problems created by products after they have been dispersed to consumers.

This concept is designated by the U.S. Environmental Protection Agency as Design for Environment (DfE). In other words, prior to going into full production, a product is designed with a consideration of the environmental impacts or aspects that will result from manufacturing it. The company sees the cost-benefit analysis prior to making the product. It can then make rational decisions as to the environmental "liabilities" associated with the manufacture of the product. Production changes, reformulations, source reductions, or complete redesign may be necessary to make the product more environmentally sound, and therefore more cost effective. Some industries have found that the product they designed was a great thing until they examined the environmental costs associated with its manufacture—after they had already gone to full production. This resulted in their product costs being too high to compete in the marketplace (they had to pay for the cost of hazardous waste by-products).

Despite the benefits of extended product responsibility, obstacles remain. These include:

The cost of instituting extended product responsibility programs

The lack of information and tools to assess all impacts of the production, use, and disposal of a product

Difficulty in building relationships among individuals and institutions involved in different stages in the life cycle of a product

Hazardous waste regulations that require hazardous waste permits for collection and disposal of certain products

Antitrust laws that make it difficult for companies to cooperate

Nevertheless, extended product responsibility can be an important tool of industry and policy makers that can lead to less costly and more flexible ways of dealing with the environmental costs of manufacturing and consuming goods.

GREEN MARKETING PRINCIPLES

Evidence indicates that successful green products have three marketing principles in common: consumer value positioning, calibration of consumer knowledge, and the credibility of product claims.

Consumer Value Positioning

- Design environmental products to perform as well as (or better than) alternatives.
- Promote and deliver the consumer desired value of environmental products and target relevant consumer market segments (such as market health benefits among health-conscious consumers).

Calibration of Consumer Knowledge

- Educate consumers with marketing messages that connect environmental product attributes with desired consumer value (for example, "energy efficiency saves money" or "pesticide free produce is healthier").
- Frame environmental product attributes as "solutions" for consumer needs such as "rechargeable batteries offer longer performance."

- Create engaging and educational Internet sites about environmental products' desired consumer value (for example, Tide Coldwater's interactive site allows visitors to calculate their likely annual money savings based on their laundry habits, utility source (gas or electricity), and zip code location.

Credibility of Product Claims

- Use environmental product and consumer benefit claims that are specific, meaningful, unpretentious, and qualified.
- Obtain product endorsements or eco-certification from reliable third parties, and educate consumers about the meaning behind those endorsements and certifications.
- Encourage consumer evangelism via consumers' social and Internet communication networks with compelling, interesting, and/or entertaining information about environmental products (for example, Tide's "Coldwater Challenge" website includes a map of the United States so visitors can track and watch their personal influence spread when their friends request a free sample).

ECONOMICS AND SUSTAINABLE DEVELOPMENT

Sustainable development has become an important policy priority for the world. The most commonly used definition of the term *sustainable development* is one that originated with the 1987 report, *Our Common Future,* by the World Commission on Environment and Development (known as the Bruntland Commission). It states that "**sustainable development** is development that meets the needs of the present without compromising the ability of future generations to meet their own needs." This definition reflects the dual societal objectives of economic development and environmental stewardship.

However, similar terms such as *sustainable growth* and *sustainable use* have been used interchangeably with *sustainable development,* as if their meanings were the same. They are not. *Sustainable growth* is a contradiction in terms: Nothing physical can grow indefinitely. *Sustainable use* is applicable only to renewable resources: it means using them at rates within their capacity for renewal.

The concept of *sustainability* has gained usage because of increasing concern over the exploitation of natural resources for economic development at the expense of environmental quality. Although disagreement exists as to the precise meaning of the term beyond respect for the quality of life of future generations, most definitions refer to the viability of natural resources and ecosystems over time and to the maintenance of human living standards and economic growth.

In the United States, a biology textbook (*New Essentials of Biology* published in 1911 by the American Book Company) described the "destruction of the forests by waste cutting" and the impact of forests on our economy and the need for replanting of trees after lumbering. It also described how "forests are of much

CAMPUS SUSTAINABILITY INITIATIVE

CAMPUS BUSINESS PARTNERSHIP TO REDUCE GREENHOUSE GAS EMISSIONS

In 2008, Chevron Energy Solutions, a unit of Chevron Corporation, and the Contra Costa Community College District (CCCCD) in Martinez, California announced the completion of the first phase of the largest solar power installation ever constructed for an institution of higher learning in North America. The project is part of a multi-facility energy efficiency and solar program that is expected to save CCCCD more than $70 million over 25 years.

The state of the art energy infrastructure upgrades—designed, engineered, and constructed by Chevron Energy Solutions—make CCCCD's three college campuses (enrolling more than 58,000 students) more energy efficient, reliable, and environmentally friendly.

The program includes three types of improvements:

- A 3.2-megawatt solar power generation system comprising photo-voltaic panels mounted on 34 parking canopies in six parking lots of the three college campuses

- High-efficiency lighting and energy management systems installed at the three campuses, as well as high-efficiency heating, ventilation, and air conditioning equipment

- High-voltage electrical system replacements on the campuses

The solar installation is expected to generate about 4 million kilowatt hours of power each year, supplying up to half of CCCCD's peak electricity needs. This renewable power will offset the production of about 5.6 million pounds of carbon dioxide emissions annually—equivalent to removing 629 cars from the road or planting 400 hectares of trees.

The $35.2 million project cost is being offset by about $8.5 million in rebates and other incentives administered by Pacific Gas and Electric Company under the State of California's Solar Initiative, Self Generation Incentive Program, and the Community College Partnership Program. The net amount of $26.7 million, supported by California bond funds, will be recovered over time by the annual cost savings achieved as a result of the new systems.

The California Lieutenant Governor commented on the partnership by stating, "Thinking green can no longer be a choice when looking toward the future. Smart businesses and colleges are looking over the horizon, building partnerships, and understanding that the risks and opportunity associated with this critical issue must be part of their overall plan to grow and to be successful in the future."

---

importance because they 1. regulate our water supplies 2. prevent erosion 3. change climate 4. are of great commercial importance. Man is responsible for the destruction of one of this nation's most valuable assets. This is primarily due to wrong and wasteful lumbering." The chapter went on to describe the loss of trees as a monetary issue and continually mentioned "waste" from "forest to finished product." Even at that relatively early time it was recognized that humans may be overharvesting and not maintaining a renewable resource. The book suggests that "forests may be artificially planted . . . two seedlings planted for every tree cut is the rule followed in Europe." Recognition of a problem is one thing; implementation of these ideas is another.

A sustainable agricultural system, for example, can be defined as one that can indefinitely meet the demands for food and fiber at socially acceptable economic costs and environmental impacts. Gaylord Nelson, the founder of the first Earth Day, listed five characteristics that define sustainability:

1. *Renewability:* A community must use renewable resources, such as water, topsoil, and energy sources no faster than they can replace themselves. The rate of consumption of renewable resources cannot exceed the rate of regeneration.

2. *Substitution:* Whenever possible, a community should use renewable resources instead of nonrenewable resources. This can be difficult because of barriers to substitution. To be sustainable, a community has to make the transition before the nonrenewable resources become prohibitively scarce.

3. *Interdependence:* A sustainable community recognizes that it is a part of a larger system and that it cannot be sustainable unless the larger system is also sustainable. A sustainable community does not import resources in a way that impoverishes other communities, nor does it export its wastes in a way that pollutes other communities.

4. *Adaptability:* A sustainable community can absorb shocks and adapt to take advantage of new opportunities. This requires a diversified economy, educated citizens, and a spirit of solidarity. A sustainable community invests in and uses research and development.

5. *Institutional commitment:* A sustainable community adopts laws and political processes that mandate sustainability. Its economic system supports sustainable production and consumption. Its educational systems teach people to value and practice sustainable behavior.

Some people assume that a slowdown of economic growth is needed to prevent further deterioration of the environment. Whether or not a slowdown is necessary provokes sharp differences of opinion. One school of thought argues that economic growth is essential to finance the investments necessary to prevent pollution and to improve the environment by a better allocation of

(a)

(b)

FIGURE 3.10 Indian Deforestation Causes Floods in Bangladesh (a) Because the Ganges River drains much of India and the country of Bangladesh is at the mouth of the river, deforestation and poor land use in India can result in (b) devastating floods in Bangladesh.

resources. Another school of thought, which is also progrowth, stresses the great potential of science and technology to solve problems and promotes technological advances as the way to solve environmental problems. Neither of these schools of thought sees any need for fundamental changes in the nature and foundation of economic policy. Environmental issues are viewed mainly as a matter of setting priorities in the allocation of resources.

A newer school of economic thought believes that economic and environmental well-being are mutually reinforcing goals that must be pursued simultaneously if either one is to be reached. Economic growth will create its own ruin if it continues to undermine the healthy functioning of Earth's natural systems or to exhaust natural resources. It is also true that healthy economies are most likely to provide the necessary financial investments to support environmental protection. For this reason, one of the principal objectives of environmental policy must be to ensure a decent standard of living for all. The solution, at least in the broad scope, would be for a society to manage its economic growth in such a way as to do no irreparable damage to its environment.

If sustainable development is to become feasible, it will be necessary to transform our approach to economic policy. Historically, rapid exploitation of resources has provided only short-term economic growth, and the environmental consequences in some cases have been incurable. For example, 40 years ago, forests covered 30 percent of Ethiopia. Today, forests cover only 1 percent, and deserts are expanding. Trees once covered one-half of India; today, only 14 percent of the land is in forests. As the Indian trees and topsoil disappear, the citizens of Bangladesh drown in India's runoff. (See figure 3.10.) One of the steps necessary to move economies toward sustainable development is to change the definition of gross national product (GNP) to include environmental improvement or decline.

Sustainable development is a worthy goal, but many changes are needed for the concept to be viable. One of these involves the transfer of modern, environmentally sound technology to developing nations. Tan Sri Razali, former chairman of the UN Commission on Sustainable Development, has stated that this transfer is the "key global action to sustainable development." Another major obstacle to sustainable development in many countries is a social structure that gives most of the nation's wealth to a tiny minority of its people. It has been said that a person who is worrying about his or her next meal is not going to listen to lectures on protecting the environment. What seem like some of the worst environmental outrages to residents in the Northern Hemisphere—cutting rainforests to make charcoal for sale as cooking fuel, for example—are often committed by people who have no other source of income.

The disparities that mark individual countries are also reflected in the planet as a whole. Most of the wealth is concentrated in the Northern Hemisphere. From the Southern Hemisphere's point of view, it is the rich world's growing consumption patterns—big cars, refrigerators, and climate-controlled shopping malls—that are the problem. The problem for the long term is that people in developing countries now want those consumer items that make life in the industrial world so comfortable—and these items use large amounts of energy and raw materials for their production and use. If the standard of living in China and India were to rise to that of Germany or the United States, the environmental impact on the planet would be significant.

Sustainable development requires choices based on values. To make intelligent choices, the public must have information about the way economic decisions affect the environment. A. W. Clausen, in his final address as president of the World Bank, noted the

> increasing awareness that environmental precautions are essential for continued economic development over the long run. Conservation, in its broadest sense, is not a luxury for people rich enough to vacation in scenic parks. It is not just a motherhood issue. Rather, the goal of economic growth itself dictates a serious and abiding concern for resource management.

High-income developed nations with high educational levels, such as the United States, Japan, and much of Europe, are in a position to promote sustainable development. They have the resources to invest in research and the technologies to implement research findings. Some believe that the world should not impose environmental protection standards on poorer nations without also helping them to move into the economic mainstream.

ECONOMICS, ENVIRONMENT, AND DEVELOPING NATIONS

As previously mentioned, the Earth's "natural capital," on which humankind depends for food, security, medicines, and machines, includes both nonrenewable resources such as minerals, oil, and mountains and renewable resources such as soil, sunlight, and biological diversity. Many countries in the developing world have resources that they wish to develop in order to improve the economic conditions of their inhabitants. To pay for development projects, many economically poor nations are forced to borrow money from banks in the developed world.

The debt they have incurred is a perverse incentive to overexploit their resources. Because they have borrowed money and creditors expect repayment of the debt, many countries must divert a major part of their gross domestic product to debt payment. In 2001, the debt in the developing nations had risen to over US $1900 billion, a figure equal to about half their collective gross national product. The burden of external debt is so great that many developing nations feel forced to overexploit their natural resources rather than manage them sustainably. The debt burdens have led to investments in programs absolutely necessary for immediate survival and in projects with safe, short-term returns. Environmental impacts are often neglected because severely indebted countries feel they cannot afford to pay attention to environmental costs until other problems are resolved. These "other problems" include the reality that simple day-to-day survival issues take precedence over environmental protection. This held true for the United States as it first struggled to survive in its formative years. All countries go through different phases during their development, and they often must work toward the time when they can literally afford to worry about the environment. Nevertheless, often environmental impacts cause international problems.

One new method of helping manage a nation's debt crisis is referred to as debt-for-nature exchange. **Debt for-nature exchanges** are an innovative mechanism for addressing the debt issue while encouraging investment in conservation and sustainable development. Three players are involved in debt-for-nature exchanges: the debtor nation, the creditor, and a third party interested in conservation initiatives. The exchange works as follows:

1. The conservation organization buys the debt from the creditor at a discount.
2. Although the creditor receives only partial payment of the initial loan, some return is better than a total loss.
3. The debtor country has the debt removed and is relieved of the huge burden of paying interest on the debt.
4. In exchange, the conservation organization requires the debtor country to spend money on appropriate conservation and sustainable development projects.

Debt-for-nature exchanges originated in 1987, when a nonprofit organization, Conservation International, bought $650,000 of Bolivia's foreign debt in exchange for Bolivia's promise to establish a national park. By 2005, at least 16 debtor countries—in the Caribbean, Africa, Eastern Europe, and Latin America—had made similar deals with official and nongovernmental organizations. By 2005, nearly US $160 million of debt around the world had been purchased at a cost of some US $35 million, but the debtor countries spent the equivalent of US $78 million. This money was used to establish biosphere reserves and national parks, develop watershed protection programs, build inventories of endangered species, and develop environmental education.

The primary goal of debt-for-nature exchanges has not been debt reduction but the funding of natural-resource management investment. The contribution made by exchanges could increase, as in the case of the Dominican Republic, where 10 percent of the country's outstanding foreign commercial debt is to be redeemed by exchanges. Although eliminating the debt crisis alone is no guarantee of investment in environmentally sound projects, instruments such as debt-for-nature exchanges can, on a small scale, reduce the mismanagement of natural resources and encourage sustainable development.

Attitudes of banks in the industrialized nations also seem to be changing. For example, the World Bank, which lends money for Third World development projects, has long been criticized by environmental groups for backing large, ecologically unsound programs, such as a cattle-raising project in Botswana that led to overgrazing. During the past few years, however, the World Bank has been factoring environmental concerns into its programs. One product of this new approach is an environmental action plan for Madagascar. The 20-year plan, which has been drawn up jointly with the World Wildlife Fund, is aimed at heightening public awareness of environmental issues, setting up and managing protected areas, and encouraging sustainable development.

Another problem associated with resource exploitation in developing countries is the short-term exploitation of a country's resources by foreign corporations. Because poor countries do not have the financial means to develop their natural resources themselves, they often contract with corporations to do the development. Many of these corporations have no long-term commitment to the project and withdraw as soon as the project becomes unprofitable. For example, many forest resources have been logged with no interest in sustaining the resource. Companies are involved for the short-term economic gain and have little interest in the long-term economic needs of the people of the country.

The Economics and Risks of Mercury Contamination

Mercury is a chemical element that is used in many industrial processes and products. Since it is a liquid metal, it is used in many kinds of electrical applications. Common uses of mercury include fluorescent light-bulbs, mercury vapor lights, electrical tilt switches, and certain kinds of small batteries. It also is alloyed with silver and other metals to produce fillings in teeth. Elemental mercury by itself is poisonous, but so few people have direct access to elemental mercury that it constitutes a minor risk. Certain bacteria, however, are able to convert elemental mercury to methylmercury, which can be easily taken up by living things and stored in the body.

Some of the mercury available for conversion to methylmercury is from natural rock. Thus, some parts of the world naturally have high mercury levels compared to others. A common human-generated source of mercury is the combustion of coal in electric power plants and similar facilities. The mercury is released into the air and distributed over the landscape downwind from the power plant.

Contamination with methylmercury is a particular problem in aquatic food chains where fish at higher trophic levels accumulate high quantities of methylmercury in their tissues. Fish are an important source of omega-3 fatty acids that are associated with reductions in heart disease. If fish are declared unfit for human consumption because of methylmercury levels, there is a major economic impact on the people who catch and process fish. There can also be an increased health risk if people switch to eating other foods that do not have omega-3 fatty acids. One study of the risks and benefits of Alaskan natives eating fish determined that the risk to health from ingesting small amounts of methyl mercury in fish was less than the risk associated with changing to a diet that included other sources of protein.

In 2000, under the Clinton administration, the EPA announced that it would require reductions in the amount of mercury released from coal fired power plants. This would require expensive additions to power plants to capture the mercury so that it was not released into the atmosphere. The power industry protested that changes would be too expensive, and the program was never initiated. In December 2003, under the George W. Bush administration, the EPA announced a mercury emission trading proposal to deal with the problem. Under this proposal, coal fired power plants will be issued permits to release specific amounts of mercury. If they do not release much mercury, they can sell their permit to other power plants that are not able to reduce the amount of mercury they release. Proponents point out that a similar emissions trading program for sulfur dioxide resulted in major reductions in sulfur dioxide release. Critics of the mercury emissions program say that the program is a concession to the power industry.

- Should power companies be able to buy rights to pollute?
- Should all pollution be prevented regardless of the cost, or should risk assessment and economic analysis be a part of the decision-making process?

SUMMARY

Risk is the probability that a condition or action will lead to an injury, damage, or loss. Risk assessment is the use of facts and assumptions to estimate the probability of harm to human health or the environment that may result from exposures to pollutants or toxic agents. While it is difficult to calculate risks, risk assessment is used in risk management, which analyzes the risk factors in decision making. The politics of risk management focus on the adequacy of scientific evidence, which is often open to divergent interpretations. In assessing risk, people often overestimate new and unfamiliar risks while underestimating familiar ones.

To a large degree, environmental problems can be viewed as economic problems that revolve around decisions about how to use resources. Many environmental costs are deferred (paid at a later date) or external (paid by someone other than the entity that causes the problem). Pollution is a good example of both a deferred and an external cost. Cost-benefit analysis is concerned with whether a policy generates more social benefits than social costs. Criticism of cost-benefit analysis is based on the question of whether everything has an economic value. It has been argued that if economic thinking dominates society, then noneconomic values, such as beauty, can survive only if a monetary value is assigned to them. There is a strong tendency to overexploit and misuse resources that are shared by all. This concept was developed by Garrett Hardin in his essay, "The Tragedy of the Commons."

Economic policies and concepts, such as supply and demand and subsidies, play important roles in environmental decision making. The balance between the amount of a good or service available for purchase and the demand for that commodity determines the price. Subsidies are gifts from government to encourage desired behaviors. Recently, several kinds of market-based approaches have been developed to deal with the economic costs of environmental problems. These approaches include information programs, tradable emissions, emissions fees, deposit refund programs, and performance bond programs. The goal of all these mechanisms is to introduce a profit motive for institutions and individuals to use resources wisely.

A newer school of economic thought is referred to as sustainable development. Sustainable development has been defined as actions that address the needs of the present without compromising the ability of future generations to meet their own needs. Sustainable development requires choices based on values. Economic concepts are also being applied to the debt-laden developing countries. One such approach is the debt-for-nature exchange. This program, which involves transferring loan payments for land that is later turned into parks and wildlife preserves, is gaining popularity.

1. If you smoke, quit! If you do not smoke, help a friend who does smoke to quit.
2. Research the "green marketing" claims for several products that you use.
3. Work on sustainability projects in your college.
4. Develop a cost-benefit analysis for a local issue.
5. Buy products that come in a container that is reusable or that requires a deposit.

WHAT'S YOUR TAKE?

Mercury contamination of our lakes, streams, oceans, soil, and groundwater is an ongoing issue. Because of surface water contamination, many state and federal agencies have issued health advisories to limit the amounts of certain fish and shellfish that people eat. If you like to eat fish or shellfish, would you limit your consumption based on warnings from state or federal health agencies? Develop an argument for or against consuming fish that potentially contain mercury.

REVIEW QUESTIONS

1. How is risk assessment used in environmental decision making?
2. What is incorporated in a cost-benefit analysis?
3. What are some of the concerns about the use of cost-benefit analysis in environmental decision making?
4. What concerns are associated with sustainable development?
5. What are some examples of external environmental costs?
6. Define what is meant by pollution-prevention costs.
7. Define the problem of common property resource ownership. Provide some examples.
8. Describe the concept of debt-for-nature exchanges.
9. Give examples of subsidies, market based instruments, and life cycle analysis.
10. What kinds of risks are willingly accepted by people?
11. Give examples of renewable and nonrenewable resources.
12. Why are environmental costs often deferred costs?
13. What is meant by "take back"?
14. Why would a small business be interested in SBLRBRA?
15. Why do some companies have different levels of risk tolerance?
16. What does Eco-RBCA have to do with ecological assessment and cleanup?
17. Why is the perception of risk so important to how we look at environmental health and safety issues?
18. How could DfE have an effect on our environment?
19. What impact could governments have on sustainability of natural resources through political means?

CRITICAL THINKING QUESTIONS

1. If you were a regulatory official, what kind of information would you require to make a decision about whether a certain chemical was "safe" or not? What level of risk would you deem acceptable for society? For yourself and your family?
2. Why do you suppose some carcinogenic agents, such as those in cigarettes, are so difficult to regulate?
3. Imagine you were assessing the risk of a new chemical plant being built along the Mississippi River in Louisiana. Identify some of the risks that you would want to assess. What kinds of data would you need to assess whether or not the risk was acceptable? Do you think that some risks are harder to quantify than others? Why?
4. Granting polluting industries or countries the right to buy and sell emissions permits is a controversial idea. Some argue that the market is the best way to limit pollution. Others argue that trade in permits allows polluting industries to continue to pollute and concentrates that pollution. What do you think?
5. Imagine you are an independent economist who is conducting a cost-benefit analysis of a hydroelectric project. What might be the costs of this project? The benefits? How would you quantify the costs of the project? The benefits? What kinds of costs and benefits might be hard to quantify or might be too tangential to the project to figure into the official estimates?
6. Do you think environmentalists should or should not stretch traditional cost-benefit analysis to include how development affects the environment? What are the benefits to this? The risks?
7. Looking at your own life, what kinds of risks do you take? What kinds are you unwilling to take? What criteria do you use to make a decision about acceptable and unacceptable risk?
8. Is current worldwide growth and development sustainable? If there were less growth, what would be the effect on developing countries? How could we achieve a just distribution of resources and still limit growth?

9. Should our policies reflect an interest in preserving resources for future generations? If so, what level of resources should be preserved? What would you be willing to do without in order to save for the future?

10. If you owned a small business in the United States and were looking to expand your business within your home state, would you consider purchasing a contaminated piece of property in a downtown or urban area? Why or why not? Would you be concerned about the environmental liabilities associated with it? Would you be worried about the health and safety of your workers if you located there? Why? What might be your concerns, or how could you find out? Would you conduct an environmental assessment prior to applying for financing the purchase and construction? Knowing what you know now, what type of risk tolerance do you have concerning your finances, your workers' health and safety, and the environment?

11. If you were developing a new product ("super" sprocket), and found that you had to use chrome plating in the manufacturing process, how might you utilize DfE concepts in the development stage of design? What types of questions might you ask the production engineer? What type of P2 techniques might you employ? Would you consider "take back" at your facility? Why or why not?

INTERRELATED SCIENTIFIC PRINCIPLES: MATTER, ENERGY, AND ENVIRONMENT

Many of the processes that occur in the natural world involve interactions between matter and energy. Fire occurs when the chemical energy in wood is converted to heat and light. Control of fire was a major technological achievement in early human history.

CHAPTER OUTLINE

The Nature of Science
   Basic Assumptions in Science
   Cause-and-Effect Relationships
   Elements of the Scientific Method
Limitations of Science
Pseudoscience
The Structure of Matter
   Atomic Structure
   The Molecular Nature of Matter
   A Word About Water
   Acids, Bases, and pH
   Inorganic and Organic Matter
   Chemical Reactions
   Chemical Reactions in Living Things
   Chemistry and the Environment
Energy Principles
   Kinds of Energy
   States of Matter
   First and Second Laws of Thermodynamics
Environmental Implications of Energy Flow
   Entropy Increases
   Energy Quality
   Biological Systems and Thermodynamics
   Pollution and Thermodynamics

ISSUES & ANALYSIS
Diesel Engine Trade-offs 76

CAMPUS SUSTAINABILITY INITIATIVE
Cooling Off the University of Arizona 75

GOING GREEN
Evaluating Green Claims 68

WATER CONNECTIONS
Applying the Scientific Method—Acid Rain 70

OBJECTIVES

After reading this chapter, you should be able to:

- Understand that science is usually reliable because information is gathered in a manner that requires impartial evaluation and continuous revision.

- Understand that matter is made up of atoms that have a specific subatomic structure of protons, neutrons, and electrons.

- Recognize that each element is made of atoms that have a specific number of protons and electrons and that isotopes of the same element may differ in the number of neutrons present.

- Recognize that atoms may be combined and held together by chemical bonds to produce molecules.

- Understand that rearranging chemical bonds results in chemical reactions and that these reactions are associated with energy changes.

- Recognize that matter may be solid, liquid, or gas, depending on the amount of kinetic energy contained by the molecules.

- Realize that energy can be neither created nor destroyed, but when energy is converted from one form to another, some energy is converted into a less useful form.

- Understand that energy can be of different qualities.

THE NATURE OF SCIENCE

Since environmental science involves the analysis of data, it is useful to understand how scientists gather and evaluate information. It is also important to understand some chemical and physical principles as a background for evaluating environmental issues. An understanding of these scientific principles will also help you to appreciate the ecological concepts in the chapters that follow.

The word *science* creates a variety of images in the mind. Some people feel that it is a powerful word and are threatened by it. Others are baffled by scientific topics and have developed an unrealistic belief that scientists are brilliant individuals who can solve any problem. For example, there are those who believe that the conservation of fossil fuels is unnecessary because scientists will soon "find" a replacement energy source. Similarly, many are convinced that if government really wanted to, it would allocate sufficient funds to allow scientists to find a cure for AIDS. Such images do not accurately portray what science is really like.

Science is a process used to solve problems or develop an understanding of nature that involves testing possible answers. Science is distinguished from other fields of study by how knowledge is acquired rather than by what is studied. The process has become known as the *scientific method.* The **scientific method** is a way of gaining information (facts) about the world by forming possible solutions to questions, followed by rigorous testing to determine if the proposed solutions are valid.

BASIC ASSUMPTIONS IN SCIENCE

When using the scientific method, scientists make several fundamental assumptions. They presume that:

1. There are specific causes for events observed in the natural world;
2. The causes can be identified;
3. There are general rules or patterns that can be used to describe what happens in nature;
4. An event that occurs repeatedly probably has the same cause each time;
5. What one person perceives can be perceived by others; and
6. The same fundamental rules of nature apply regardless of where and when they occur.

For example, we have all observed lightning associated with thunderstorms. According to the assumptions just stated, we should expect that there is an explanation that would account for all cases of lightning regardless of where or when they occur and that all people could make the same observations. We know from scientific observations and experiments that lightning is caused by a difference in electrical charge, that the behavior of lightning follows general rules that are the same as those seen with static electricity, and that all lightning that has been measured has the same cause wherever and whenever it occurred.

CAUSE-AND-EFFECT RELATIONSHIPS

Scientists distinguish between situations that are merely correlated (happen together) and those that are correlated and show **cause-and-effect relationships.** Many events are correlated, but not all correlations show cause-and-effect. When an event occurs as a direct result of a previous event, a cause-and-effect relationship exists. For example, lightning and thunder are correlated—thunder follows lightning—but they also have a cause-and-effect relationship—lightning causes thunder.

The relationships between autumn and trees dropping their leaves is more difficult to sort out. Because autumn brings colder temperatures, many people assume that the cold temperature causes leaves to turn color and fall. Cold temperatures are correlated with falling leaves. However there is no cause-and-effect relationship. The cause of the change in trees is actually the shortening of days that occurs in the autumn. Experiments have shown that artificially shortening the length of days in a greenhouse will cause trees to drop their leaves with no change in temperature. Knowing that a cause-and-effect relationship exists enables us to predict what will happen should that same set of circumstances occur in the future.

ELEMENTS OF THE SCIENTIFIC METHOD

The scientific method requires a systematic search for information and a continual checking and rechecking to see if previous ideas are still supported by new information. If the new evidence is not supportive, scientists discard or change their original ideas. Scientific ideas undergo constant reevaluation, criticism, and modification. The scientific method involves several important identifiable components, including:

* careful observation,
* asking questions about observed events,
* the construction and testing of hypotheses,
* an openness to new information and ideas, and
* a willingness to submit one's ideas to the scrutiny of others.

Underlying all of these activities is constant attention to accuracy and freedom from bias.

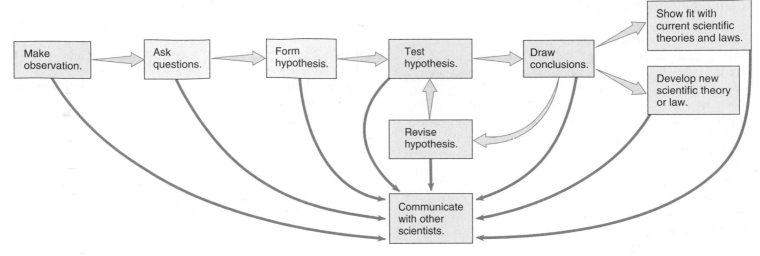

FIGURE 4.1 **Elements of the Scientific Method** The scientific method consists of several kinds of activities. Observation of a natural phenomenon is usually the first step. Observation often leads people to ask questions about the observation they have made or to try to determine why the event occurred. This questioning is typically followed by the construction of a hypothesis that attempts to explain why the phenomenon occurred. The hypothesis is then tested to see if it is supported. Often this involves experimentation. If the hypothesis is not substantiated, it is modified and tested in its new form. It is important at all times that others in the scientific community be informed by publishing observations of unusual events, their probable cause, and the results of experiments that test hypotheses. Occasionally, this method of inquiry leads to the development of theories that tie together many bits of information into broad statements that state why things happen in nature and serve to guide future thinking about a specific area of science. Scientific laws are similar broad statements that describe how things happen in nature.

The scientific method is not, however, an inflexible series of steps that must be followed in a specific order. Figure 4.1 shows how these steps may be linked.

Observation

Scientific inquiry often begins with an observation that an event has occurred. An **observation** occurs when we use our senses (smell, sight, hearing, taste, touch) or an extension of our senses (microscope, tape recorder, X-ray machine, thermometer) to record an event. Observation is more than a casual awareness. You may hear a sound or see an image without really observing it. Do you know what music was being played in the shopping mall? You certainly heard it, but if you are unable to tell someone else what it was, you didn't "observe" it. If you had prepared yourself to observe the music being played, you would be able to identify it. When scientists talk about their observations, they are referring to careful, thoughtful recognition of an event—not just casual notice. Scientists train themselves to improve their observational skills, since careful observation is important in all parts of the scientific method. (See figure 4.2.)

Because many of the instruments used in scientific investigations are complicated, we might get the feeling that science is incredibly complex, when in reality these sophisticated tools are being used simply to answer questions that are relatively easy to understand. For example, a microscope has several knobs to turn and a specially designed light source. It requires considerable skill to use properly, but it is essentially a fancy magnifying glass that allows small objects to be seen more clearly. The microscope has enabled scientists to answer some relatively fundamental questions such as: Are there living things in pond water? and Are living things made up of smaller subunits? Similarly, chemical tests allow us to determine the

FIGURE 4.2 **Observation** Careful observation is an important part of the scientific method. This technician is making observations on the characteristics of the soil and recording the results.

amounts of specific materials dissolved in water, and a pH meter allows us to determine how acidic or basic a solution is. Both are simple activities, but if we are not familiar with the procedures, we might consider the processes hard to understand.

Questioning and Exploring

Observations often lead one to ask questions about the observations. Why did this event happen? Will it happen again in the same circumstances? Is it related to something else? Some questions may be simple speculation, but others may inspire you to further investigation. The formation of the questions is not as simple as it might seem because the way the questions are asked will determine how you go about answering them. A question that is too broad or too complex may be impossible to answer; therefore, a great deal of effort is put into asking the question in the right way. In some situations, this can be the most time-consuming part of the scientific method; asking the right question is critical to how you look for answers. For example, you observe that robins eat the berries of many plants but avoid others. You could ask the following questions:

1. Do the robins dislike the flavor of some berries?
2. Will robins eat more of one kind of berry if given a choice between two kinds of berries?

The second question is obviously easier to answer.

Once a decision has been made about what question to ask, scientists *explore other sources of knowledge* to gain more information. Perhaps the question already has been answered by someone else or several possible answers already have been rejected. Knowing what others have already done saves one time. This process usually involves reading appropriate science publications, exploring information on the Internet, or contacting fellow scientists interested in the same field of study. Even if the particular question has not already been answered, scientific literature and other scientists can provide insights that may lead to a solution. After exploring the appropriate literature, a decision is made about whether to continue to explore the question. If the scientist is still intrigued by the question, a formal hypothesis is constructed, and the process of inquiry continues at a different level.

Constructing Hypotheses

A **hypothesis** is a statement that provides a possible answer to a question or an explanation for an observation that can be tested. A good hypothesis must be logical, account for all the relevant information currently available, allow one to predict future events relating to the question being asked, and be testable. Furthermore, if one has the choice of several competing hypotheses, one should use the simplest hypothesis with the fewest assumptions. Just as deciding which questions to ask is often difficult, the formation of a hypothesis requires much critical thought and mental exploration. If the hypothesis is not logical or does not account for all the observed facts in the situation, it must be rejected. If a hypothesis is not testable it is mere speculation.

Testing Hypotheses

Keep in mind that a hypothesis is based on observations and information gained from other knowledgeable sources. It predicts how an event will occur under specific circumstances. Scientists test the predictive ability of a hypothesis to see if the hypothesis is supported or is disproved. If you disprove the hypothesis, it is rejected, and a new hypothesis must be constructed. However, if you cannot disprove a hypothesis, it increases your confidence in the hypothesis, but it does not prove it to be true in all cases and for all time. Science always allows for the questioning of ideas and the substitution of new ones that more completely describe what is known at a particular point in time. It could be that an alternative hypothesis you haven't thought of explains the situation or that you have not made the appropriate observations to indicate that your hypothesis is wrong.

The test of a hypothesis can take several forms. It may simply involve the collection of pertinent information that already exists from a variety of sources. For example, if you visited a cemetery and observed from reading the tombstones that an unusually large number of people of different ages died in the same year, you could hypothesize that there was an epidemic of disease or a natural disaster that caused the deaths. Consulting historical newspaper accounts would be a good way to test this hypothesis.

In other cases, a hypothesis may be tested by simply making additional observations. For example, if you hypothesized that a certain species of bird used cavities in trees as places to build nests, you could observe many birds of the species and record the kinds of nests they built and where they built them.

Another common method for testing a hypothesis involves devising an experiment. An **experiment** is a re-creation of an event or occurrence in a way that enables a scientist to support or disprove a hypothesis. This can be difficult because a particular event may involve a great many separate happenings called **variables.** The best experimental design is a **controlled experiment** in which two groups differ in only one way. For example, tumors of the skin and liver occur in the fish that live in certain rivers (*observation*). This raises the question: What causes the tumors? Many people feel that the tumors are caused by toxic chemicals that have been released into the rivers by industrial plants (*hypothesis*). However, it is possible that the tumors are caused by a virus, by exposure to natural substances in the water, or are the result of genes present in the fish. The following experiment could be conducted to test the hypothesis that industrial contaminants cause the tumors: Fish could be collected from the river and placed in one of two groups. One group (the control group) would be raised in a container through which the normal river water passes. The second group (experimental group) would be raised in an identical container through which water from the industrial facility passes. There would need to be large numbers of fish in both groups. This kind of experiment is called a controlled experiment. If the fish in the experimental group develop a significantly larger number of tumors than the control group, something in the water from the plant is the probable cause of the tumors. This is particularly true if the chemicals present in the water are already known to cause tumors. After the data have been evaluated, the results of the experiment would be published.

The results of a well-designed experiment should be able to support or disprove a hypothesis. However, this does not always occur. Sometimes the results of an experiment are inconclusive. This means that a new experiment must be conducted or that more information must be collected. Often, it is necessary to have large amounts of information before a decision can be made about the validity of a hypothesis. The public often fails to appreciate why it is necessary to perform experiments on so many subjects or to repeat experiments again and again.

The concept of **reproducibility** is important to the scientific method. Because it is often not easy for scientists to eliminate unconscious bias, independent investigators must be able to reproduce the experiment to see if they get the same results. To do this, they must have a complete and accurate written document to work from. That means the scientists must publish the methods and results of their experiment. This process of publishing one's work for others to examine and criticize is one of the most important steps in the process of scientific discovery. The results of experiments are only considered reliable if they are supported by many experiments and by different investigators.

The Development of Theories and Laws

When broad consensus exists about an area of science, it is known as a theory or law. A **theory** is a widely accepted, plausible generalization about fundamental concepts in science that explains *why* things happen. An example of a scientific theory is the **kinetic molecular theory,** which states that all matter is made up of tiny, moving particles. As you can see, this is a very broad statement, and it is the result of years of observation, questioning, experimentation, and data analysis. Because we are so confident that the theory explains the nature of matter, we use this concept to explain why materials disperse in water or air, why materials change from solids to liquids, and why different chemicals can interact during chemical reactions.

Theories and hypotheses are different. A hypothesis provides a possible explanation for a specific question; a theory is a broad concept that shapes how scientists look at the world and how they frame their hypotheses. Because they are broad, unifying statements, there are few theories. However, just because a theory exists does not mean that testing stops. As scientists continue to gain new information, they may find exceptions to a theory or even in rare cases disprove a theory.

It is important to recognize that the word *theory* is often used in a much less restrictive sense. Often it is used in ordinary conversation to describe a vague idea or a hunch. This is not a theory in the scientific sense. So when you see or hear the word *theory,* you must look at the context to see if the speaker or writer is referring to a theory in the scientific sense.

A **scientific law** is a uniform or constant fact of nature that describes *what* happens in nature. An example of a scientific law is the **law of conservation of mass,** which states that matter is not gained or lost during a chemical reaction. While laws describe what happens and theories describe why things happen, in a way, laws and theories are similar. They have both been examined repeatedly and are regarded as excellent predictors of how nature behaves.

Communication

At several points in the discussion of the scientific method the significance of communication has arisen. Communication is a central characteristic of the scientific method. Science is conducted openly, under the critical eyes of others who are interested in the same questions. An important part of the communication process involves the publication of articles in scientific journals about one's research, thoughts, and opinions. This communication can occur at any point during the process of scientific discovery.

People may ask questions about unusual observations. They may publish preliminary results of incomplete experiments. They may publish reports that summarize large bodies of material. And they often publish strongly held opinions that may not always be supportable with current data. This provides other scientists with an opportunity to criticize, make suggestions, or agree. (See figure 4.3.) Scientists also talk to one another at conferences and by phone, e-mail, and the Internet. The result is that science is subjected to examination by many minds as it is discovered, discussed, and refined.

FIGURE 4.3 **Communication** One important way in which scientists communicate is through publication in scientific journals.

LIMITATIONS OF SCIENCE

Science is a powerful tool for developing an understanding of the natural world, but it cannot analyze international politics, decide if family-planning programs should be instituted, or evaluate the significance of a beautiful landscape. These tasks are beyond the scope of scientific investigation. This does not mean that scientists cannot comment on such issues. They often do. But they should not be regarded as more knowledgeable on these issues just because they are scientists. Scientists may know more about the scientific aspects of these issues, but they struggle with the same moral and ethical questions that face all people, and their judgments on these matters can be just as biased as anyone else's. Consequently, major differences of opinion often exist among lawmakers, regulatory agencies, special interest groups, and members of scientific organizations about the significance or value of specific scientific information.

It is important to differentiate between the scientific data collected and the opinions scientists have about what the data mean. Scientists form and state opinions that may not always be supported by fact, just as other people do. Equally reputable scientists commonly state opinions that are in direct contradiction. This is especially true in environmental science, where predictions about the future must be based on inadequate or fragmentary data. The issue of climate change (covered in chapter 16) is an example of this.

It is important to recognize that scientific knowledge can be used by different people to support opinions that may not be valid. For example, the following statements are all factual.

1. Many of the kinds of chemicals used in modern agriculture are toxic to humans and other animals.
2. Small amounts of agricultural chemicals have been detected in some agricultural products.
3. Low levels of some toxic materials have been strongly linked to a variety of human illnesses.

This does not mean that *all* foods grown with the use of *any* chemicals are less nutritious or are dangerous to health or that "organically grown" foods are necessarily more nutritious or healthful because they have been grown without agricultural chemicals. The idea that something that is artificial is necessarily bad and something natural is necessarily good is an oversimplification. After all, many plants such as tobacco, poison ivy, and rhubarb leaves naturally contain toxic materials. Furthermore, the use of chemical fertilizers has contributed to the health and well-being of the human population because increased crop yields due to fertilizer use account for about one-third of the food grown in the world, thus reducing malnutrition. However, it is appropriate to question if the use of agricultural chemicals is always necessary or if trace amounts of specific agricultural chemicals in food are dangerous.

PSEUDOSCIENCE

Pseudoscience (*pseudo* = false) is a deceptive practice that uses the appearance or language of science to convince, confuse, or mislead people into thinking that something has scientific validity when it does not. When pseudoscientific claims are closely examined, it is found that they are not supported by unbiased tests.

Often facts are selected to support a particular point of view. For example, certain kinds of ionizing radiation are known to cause cancer. Electrical devices and power lines emit electromagnetic radiation. These facts have caused many people to see a link between exposure to electromagnetic radiation and health, when there is none. They maintain that living near electric power lines causes cancer or other health effects. However, careful scientific studies, which compared people who lived near power lines with matched groups of people who did not live near power lines, show that those who live near power lines have no more health problems than those that do not. Often it is interesting to look at the motivation of those making pseudoscientific claims. Are they selling something? Are they seeking money for presumed injuries?

THE STRUCTURE OF MATTER

Now that we have an appreciation for the methods of science, it is time to explore some basic information and theories about the structure and function of various kinds of matter. **Matter** is anything that takes up space and has mass. Air, water, trees, cement, and gold are all examples of matter. As stated earlier, the kinetic molecular theory is a central theory that describes the structure and activity of matter. This theory states that all matter is made up of tiny objects that are in constant motion. Although different kinds of matter have different properties, they all are similar in one fundamental way. They are all made up of one or more kinds of smaller subunits called atoms.

ATOMIC STRUCTURE

Atoms are fundamental subunits of matter. There are 92 kinds of atoms found in nature. Each kind forms a specific type of matter known as an **element.** Gold (Au), oxygen (O), and mercury (Hg) are examples of elements. All atoms have a central region known as a **nucleus,** which is composed of two kinds of relatively heavy particles: positively charged particles called **protons** and uncharged particles called **neutrons.** Surrounding the nucleus of the atom is a cloud of relatively lightweight, fast-moving, negatively charged particles called **electrons.**

The atoms of each element differ from one another in the number of protons, neutrons, and electrons present. For example, a typical mercury atom contains 80 protons and 80 electrons; gold has 79 of each, and oxygen only eight of each. (See figure 4.4.) (Appendix 1, page 451, contains a periodic table of the elements.) All atoms of an element always have the same number of protons and electrons, but the number of neutrons may vary from one atom to the next.

Atoms of the same element that differ from one another in the number of neutrons they contain are called **isotopes.** For example, there are three isotopes of the element hydrogen. All hydrogen atoms have one proton and one electron, but one isotope of hydrogen has no neutrons, one has one neutron, and one has two neutrons. These isotopes behave the same chemically but have different masses since they contain different numbers of neutrons. (See figure 4.5.)

THE MOLECULAR NATURE OF MATTER

The kinetic molecular theory states that all matter is made of tiny particles that are in constant motion. However, there are several kinds of these tiny particles. In some instances, atoms act as individual particles. In other instances, atoms bond to one another chemically to form stable units called **molecules.** In still other cases, atoms or molecules may gain or lose electrons and thus become electrically charged particles called **ions.** Atoms or molecules that lose electrons are positively charged because they have more protons (+) than electrons (−). Those that gain electrons have more electrons (−) than protons (+) and are negatively charged.

Oppositely charged ions are attracted to one another and may form stable units similar to molecules; however, they typically split into their individual ions when dissolved. For example, table salt (NaCl) is composed of sodium ions (Na^+) and chloride ions (Cl^-). It is a white, crystalline material when dry but separates into individual ions when placed in water.

When two or more atoms or ions are bonded chemically, a new kind of matter called a **compound** is formed. While only 92 kinds of atoms are commonly found, there are millions of ways atoms can be combined to form compounds. Water (H_2O), table sugar ($C_6H_{12}O_6$), table salt (NaCl), and methane gas (CH_4) are examples of compounds.

Many other kinds of matter are **mixtures,** variable combinations of atoms, ions, or molecules. Honey is a mixture of several sugars and water; concrete is a mixture of cement, sand, gravel, and reinforcing rods; and air is a mixture of several gases, of which the most common are nitrogen and oxygen. Table 4.1 summarizes the various kinds of matter and the subunits of which they are composed.

A WORD ABOUT WATER

The temperature of the Earth is such that water can exist in all three phases—solid, liquid, and gas. If we look at Earth from space, it is a blue planet with patches of white. The blue color is caused by the three-quarters of the surface of the Earth covered with water in the oceans. The wisps of white are caused by water droplets in clouds. Water in all its forms determines the weather and climate of regions of the Earth. Earth's surface is shaped by the flow of water and ice. Life on Earth is based on water. Most kinds of organisms live in water and the most common molecule found in living things is water.

Water molecules are polar molecules with positive and negative ends. Since unlike charges attract each other, water molecules tend to stick together. This contributes to several of water's unusual properties. Because water molecules are polar, it takes a great deal of energy to cause water molecules to separate from one another—to go from the liquid form to the water vapor. Thus, the evaporation of water has the ability to cool the surroundings. The polar nature of water molecules also contributes to its ability to dissolve most substances. It is often described as the universal solvent. Almost everything dissolves, to some extent, in water. Many important solutions are formed as a result of water.

ACIDS, BASES, AND pH

Acid and bases are two classes of compounds that are of special interest. Their characteristics are determined by the nature of their chemical bonds. When acids are dissolved in water, hydrogen ions

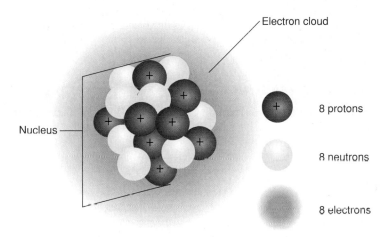

Electron cloud

Nucleus

+ 8 protons

8 neutrons

8 electrons

FIGURE 4.4 **Diagrammatic Oxygen Atom** Most oxygen atoms are composed of a nucleus containing eight positively charged protons and eight neutrons without charges. Eight negatively charged electrons move in a cloud around the nucleus.

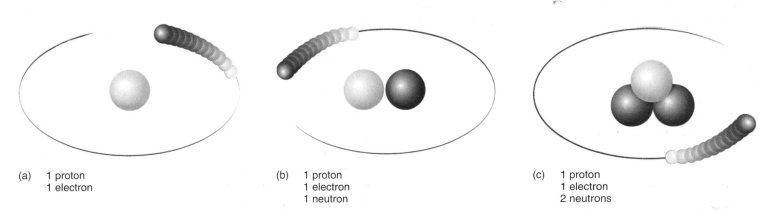

(a) 1 proton
     1 electron

(b) 1 proton
     1 electron
     1 neutron

(c) 1 proton
     1 electron
     2 neutrons

FIGURE 4.5 **Isotopes of Hydrogen** (a) The most common form of hydrogen is the isotope that has one proton and no neutrons in its nucleus. (b) The isotope deuterium has one proton and one neutron. (c) The isotope tritium has two neutrons and one proton. Each of these isotopes of hydrogen also has one electron. Most scientists use the term *hydrogen* in a generic sense—that is, the term is not specific but might refer to any or all of these isotopes.

Businesses have discovered that there is money to be made by marketing their products as ecologically friendly, Earth friendly, or green. But how do you know when a product is truly green? Most of us do not have the expertise to evaluate highly technical statements about products, so it is useful to have some general guidelines that help separate the charlatans from those that have a sincere interest in marketing a green product. Be a skeptical consumer.

1. Look for statements that are specific. Statements like "contains all natural ingredients," or "biodegradable," or "nontoxic" are not useful unless there is other evidence provided to support the claim. A "natural ingredients" statement should be backed up with a list of the actual ingredients and their sources. A claim of being biodegradable is not a very useful term, since most things will biodegrade if given enough time and the proper microorganisms. However, if there are statements about how quickly materials break down or that they break down in septic systems or sewage treatment plants, you can have more confidence. A claim of being nontoxic is meaningless—nearly everything is toxic in high enough concentrations, even table salt, ethanol, and vinegar.

2. Look for evidence that companies have a general environmental commitment. Do they market in recycled containers? Do they sell concentrated products that can be diluted at home, thus saving energy? Do they sell with a minimum of packaging?

3. Look for statements about the use of recycled materials in the product. Saying a product or container is recyclable is not very useful if there is no easy way to recycle it. Does the business have a program for recycling its products? If a claim is made that the product is made from recycled materials, evaluate the claim. What percent is post-consumer waste? Many products that claim to be made from recycled materials are made from materials salvaged during the manufacturing process.

4. Look for evidence that the claims are highlighting one positive element while ignoring other important issues. A statement that a product contains no chlorine is not useful if the other ingredients are as bad or worse.

5. Look for statements that take credit for things they are legally required to do. "Meets all FDA requirements" and "contains no CFCs" are meaningless statements because it would be illegal to sell the product if it didn't meet FDA requirements and it is illegal to sell products containing CFCs.

TABLE 4.1 Relationships Between the Kinds of Subunits Found in Matter

Category of Matter	Subunits	Characteristics
Subatomic particles	protons	Positively charged
		Located in nucleus of the atom
	neutrons	Have no charge
		Located in nucleus of the atom
	electrons	Negatively charged
		Located outside the nucleus of the atom
Elements	atoms	Atoms of an element are composed of specific arrangements of protons, neutrons, and electrons. Atoms of different elements differ in the number of protons, neutrons, and electrons present.
Compounds	molecules or ions	Compounds are composed of two or more atoms or ions chemically bonded together. Different compounds contain specific atoms or ions in specific proportions.
Mixtures	atoms, molecules, or ions	The molecular particles in mixtures are not chemically bonded to each other. The number of each kind of molecular particle present is variable.

(H^+) are set free. A *hydrogen ion* is positive because it has lost its electron and now has only the positive charge of its proton. Therefore, a hydrogen ion is a proton. An **acid** is any compound that releases hydrogen ions (protons) in a solution. Some familiar examples of common acids are sulfuric acid (H_2SO_4) in automobile batteries and acetic acid (HCH_2COOH) in vinegar.

A **base** is the opposite of an acid in that it accepts hydrogen ions in solution. Many common bases release **hydroxide ions** (OH^-). This ion is composed of an oxygen atom and a hydrogen atom bonded together but with an additional electron. The hydroxide ion is negatively charged. It is a base because it is able to accept hydrogen ions in solution to form water ($H^+ + OH^- \rightarrow H_2O$). A very strong base often used in oven cleaners is sodium hydroxide (NaOH). Sometimes people refer to a base as an alkali and solutions that are basic often are called alkaline solutions.

The concentration of an acid or base solution is given by a number called its **pH.** The pH scale is a measure of hydrogen ion concentration. However, the pH scale is different from what you might expect. First, it is an inverse scale, which means that the lower the pH, the greater the number of hydrogen ions present. Second, the scale is logarithmic, which means that a difference between two consecutive pHs is really a difference by a factor of 10. For example, a pH of 7 indicates that the solution is neutral and has an equal number of H^+ ions and OH^- ions, but a pH of 6 means that the solution has 10 times more hydrogen ions than it would at a pH of 7.

As the number of hydrogen ions in the solution increases, the pH gets smaller. A number higher than 7 indicates that the solution has more OH^- than H^+. As the number of hydroxide ions increases, the pH gets larger. The higher the pH, the more concentrated the hydroxide ions. (See figure 4.6.)

INORGANIC AND ORGANIC MATTER

Inorganic and organic matter are usually distinguished from one another by one fact: Organic matter consists of molecules that contain carbon atoms that are usually bonded to form chains or rings. Consequently, organic molecules can be very large. Many different kinds of organic compounds exist. Inorganic compounds generally consist of small molecules and combinations of ions, and relatively few kinds exist. All living things contain molecules of organic compounds. They must either be able to manufacture organic compounds from inorganic compounds or to modify organic compounds they obtain from eating organic material. Typically, chemical bonds in organic molecules contain a large amount of chemical energy that can be released when the bonds are broken and new inorganic compounds are produced. Salt, water, metals, sand, and oxygen are examples of inorganic matter. Sugars, proteins, and fats are examples of organic compounds that are produced and used by living things. Natural gas, oil, and coal are all examples of organic substances that were originally produced by living things but have been modified by geologic processes.

CHEMICAL REACTIONS

When atoms or ions combine to form compounds, they are held together by chemical bonds, as in the water molecule in figure 4.7. **Chemical bonds** are attractive forces between atoms resulting from the interaction of their electrons. Each chemical bond contains a certain amount of energy. When chemical bonds are broken or formed, a chemical reaction occurs. During chemical reactions, the amount of energy within the chemical bonds changes. If the chemical bonds in the new compounds have less chemical energy than the previous compounds, some of the energy is released, often as heat and light. These kinds of reactions are called **exothermic reactions.** In other cases, the newly formed chemical bonds contain more energy than was present in the compounds from which they were formed. Such reactions are called **endothermic reactions.** For such a reaction to occur, the additional energy must come from an external source.

A common example of an exothermic reaction is the burning of natural gas. The primary ingredient in natural gas is the compound methane. When methane and oxygen are mixed together and a small amount of energy is used to start the reaction, the chemical bonds in the methane and oxygen (reactants) are rearranged to form two

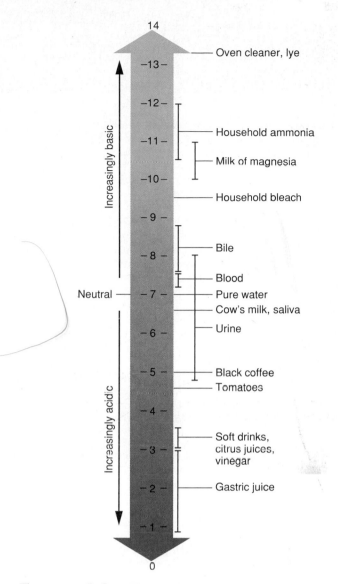

FIGURE 4.6 **The pH Scale** The concentration of hydrogen ions (H^+) is greatest when the pH is lowest. At a pH of 7.0, the concentrations of H^+ and OH^- are equal. We usually say as the pH gets smaller, the solution becomes more acidic. As the pH gets larger, the solution becomes more basic or alkaline.

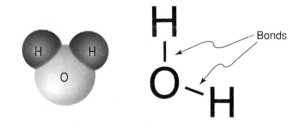

FIGURE 4.7 **Water Molecule** A water molecule consists of an atom of oxygen bonded to two atoms of hydrogen. The molecule is not symmetrical. The hydrogen atoms are not on opposite sides of the oxygen atom.

different compounds, carbon dioxide and water (products). In this kind of reaction, the products have less energy than the reactants. The leftover energy is released as light and heat. (See figure 4.8.) In every reaction, the amount of energy in the reactants and in the products can be compared and the differences accounted for by energy loss to, or gain from, the surroundings.

In the 1970s people in North America began to *observe* that in the northeastern United States and adjacent Canada, certain bodies of water and certain forested areas were changing. Fish were disappearing and trees were dying. They began to ask *questions* about why this was occurring. At the same time people began to make *observations* about the pH of rain and snow. They found that in many places in the Northeast the pH of rain was very low—it was acid. Normal rain has a pH of about 5.7, and measurements of rain indicated that the pH was often much more acid than normal. People began to ask if there was a *cause-and-effect relationship* between the acid precipitation and the decline of forests and certain lakes.

When the pH of the water in many northeastern lakes was measured it was determined that they were acid and that there was probably a *cause-and-effect relationship* between the acid condition of the water in the lakes and the decline of fish. (Although forest decline was a more difficult problem to understand, research eventually established a *cause-and-effect relationship* between forest decline and acid rain.) The next *question* that was asked was, what was the cause of the acidification of the lakes? Since many of the lakes were downwind of industrial centers, people asked the *question* Is the low pH of the rain the result of industrial activities? In particular, what activities might cause the acid rain?

Experimental evidence linking industrial emissions to acid rain came from several kinds of studies. Much of this research took place during the 1980s.

1. Measurements of the emissions from industrial smokestacks revealed that there were a variety of compounds in the emissions that were acidic in nature. In particular, sulfur compounds from the burning of high-sulfur coal provided compounds that formed acids.

2. The rain in areas that were not downwind of industrial settings had a more normal pH.

3. Reducing emissions resulted in less acid rain.

Since this is a complex problem and involves many scientists working on particular aspects of the problem, there was a great deal of *communication* among scientists as these studies were occurring.

Ultimately, most scientists *concluded* that the cause of the acidification of lakes and the decline of forests was acid rain and the cause of the acid rain was industrial emissions. During the 1990s the U.S. Congress passed legislation that led to the reduction of acid emissions from industrial sources. Efforts to reduce emissions of acid-forming substances continue today. However, as sulfur emissions have been reduced the emphasis today is on nitrogen compounds that result from the burning of fuels by industry, automobiles, and homes.

Testing for Acid Rain

Methane + Oxygen $\longrightarrow$ Carbon dioxide + Water + Heat + Light

CH_4 + $2O_2$ $\longrightarrow$ CO_2 + $2H_2O$ + Heat + Light

$$H-\underset{\underset{H}{|}}{\overset{\overset{H}{|}}{C}}-H + 2\ [O{=}O] \longrightarrow O{=}C{=}O + 2\left[\begin{array}{c} O \\ /\backslash \\ H\ H \end{array}\right] + \text{Heat} + \text{Light}$$

FIGURE 4.8   A Chemical Reaction When methane is burned, chemical bonds are rearranged, and the excess chemical bond energy is released as light and heat. The same atoms are present in the reactants as in the products, but they are bonded in different ways, resulting in molecules of new substances.

Even energy-yielding reactions usually need an input of energy to get the reaction started. This initial input of energy is called **activation energy.** In certain cases, the amount of activation energy required to start the reaction can be reduced by the use of a catalyst. A **catalyst** is a substance that alters the rate of a reaction, but the catalyst itself is not consumed or altered in the process. Catalysts are used in catalytic converters, which are attached to automobile exhaust systems. The purpose of the catalytic converter is to bring about more complete burning of the fuel, thus resulting in less air pollution. Most of the materials that are not completely burned by the engine require high temperatures to react further; with the presence of catalysts, these reactions can occur at lower temperatures.

Endothermic reactions require an input of energy in order to occur. For example, nitrogen gas and oxygen gas can be combined to form nitrous oxide ($O_2 + N_2 + heat \rightarrow 2NO$). Another important endothermic reaction is the process of photosynthesis, which is discussed in the following section.

CHEMICAL REACTIONS IN LIVING THINGS

Living things are constructed of cells that are themselves made up of both inorganic and organic matter in very specific arrangements. The chemical reactions that occur in living things are regulated by protein molecules called **enzymes** that reduce the activation energy needed to start the reactions. Enzymes are important since the high temperatures required to start these reactions without enzymes would destroy living organisms. Many enzymes are arranged in such a way that they cooperate in controlling a chain of reactions, as in photosynthesis and respiration.

Photosynthesis is the process plants use to convert inorganic material into organic matter, with the assistance of light energy. Light energy enables the smaller inorganic molecules (water and carbon dioxide) to be converted into organic sugar molecules. In the process, molecular oxygen is released.

$$6CO_2 + 6H_2O \xrightarrow[\text{in chloroplasts}]{\text{light energy}} C_6H_{12}O_6 + 6O_2$$

$$\text{Carbon dioxide} + \text{Water} \longrightarrow \text{Sugar} + \text{Oxygen}$$

Molecules of the green pigment chlorophyll are found in cellular structures called chloroplasts. Chlorophyll is responsible for trapping the sunlight energy needed in the process of photosynthesis. Therefore, photosynthesis takes place in the green portions of the plant, usually the leaves. (See figure 4.9.) The organic molecules

FIGURE 4.9 **Photosynthesis** This reaction is an example of one that requires an input of energy (sunlight) to combine low-energy molecules (CO_2 and H_2O) to form sugar ($C_6H_{12}O_6$) with a greater amount of chemical bond energy. Molecular oxygen (O_2) is also produced.

produced as a result of photosynthesis can be used as an energy source by the plants and by organisms that eat the plants.

Respiration involves the use of atmospheric oxygen to break down large, organic molecules (sugars, fats, and proteins) into smaller, inorganic molecules (carbon dioxide and water). This process releases energy the organisms can use.

$$C_6H_{12}O_6 + 6O_2 \longrightarrow 6CO_2 + 6H_2O + Energy$$

$$Sugar + Oxygen \longrightarrow Carbon\ dioxide + Water + Energy$$

(See figure 4.10.) All organisms, including plants, must carry on some form of respiration, since all organisms need a source of energy to maintain life.

CHEMISTRY AND THE ENVIRONMENT

Chemistry is extremely important in discussions of environmental problems. The chemicals of fertilizer and pesticides are extremely important in food production, but they also present environmental problems. Photochemical smog is a problem created by chemical reactions that take place between pollutants and other components of the atmosphere. Persistent organic chemicals are those that are not broken down by microorganisms and remain in the environment for long periods. Organic molecules in water reduce the amount of oxygen available to animals that need it to stay alive. These and other examples of environmental chemistry will be discussed in later chapters on agriculture, air pollution, and water pollution.

ENERGY PRINCIPLES

The "Chemical Reactions in Living Things" section started out with a description of matter, yet it used the concept of energy to describe chemical bonds, chemical reactions, and molecular motion. That is because energy and matter are inseparable. It is difficult to describe one without the other. **Energy** is the ability to do work. Work is done when an object is moved over a distance. This occurs even at the molecular level.

KINDS OF ENERGY

There are several kinds of energy. Heat, light, electricity, and chemical energy are common forms. The energy contained by moving objects is called **kinetic energy.** The moving molecules in air have kinetic energy, as does water running downhill or a dog chasing a ball. In contrast, **potential energy** is the energy matter has because of its position. The water behind a dam has potential energy by virtue of its elevated position. (See figure 4.11.) An electron moved to a position farther from the nucleus has increased potential energy due to the increased distance between the electron and the nucleus.

STATES OF MATTER

Depending on the amount of energy present, matter can occur in three common states: solid, liquid, or gas. The physical nature of matter changes when a change occurs in the amount of kinetic energy its molecular particles contain, but the chemical nature of matter and the kinds of chemical reactions it will undergo remain the same. For example, water vapor, liquid water, and ice all have the same chemical

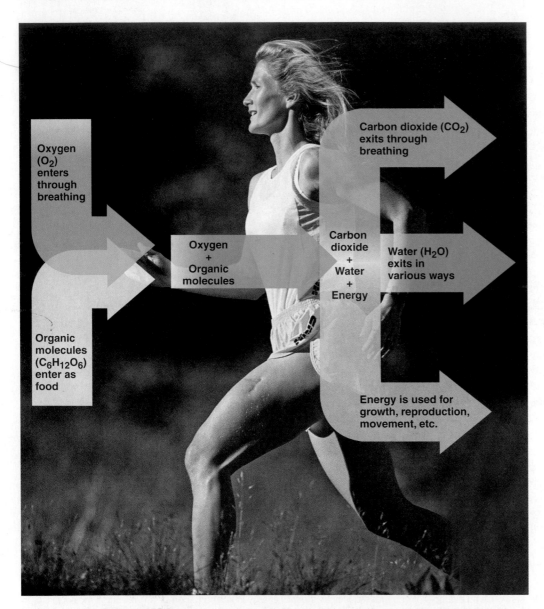

FIGURE 4.10 **Respiration** Respiration involves the release of energy from organic molecules when they react with oxygen. In addition to providing energy in a usable form, respiration produces carbon dioxide and water.

FIGURE 4.11 Kinetic and Potential Energy Kinetic and potential energy are interconvertible. The potential energy possessed by the water behind a dam is converted to kinetic energy as the water flows to a lower level.

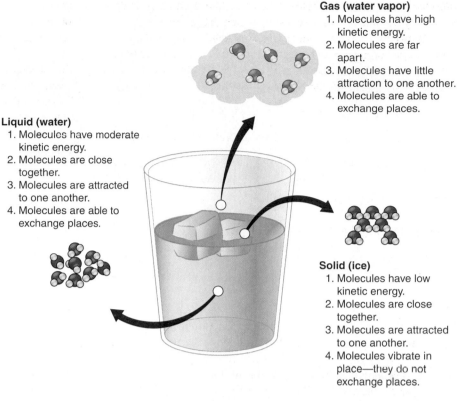

Gas (water vapor)
1. Molecules have high kinetic energy.
2. Molecules are far apart.
3. Molecules have little attraction to one another.
4. Molecules are able to exchange places.

Liquid (water)
1. Molecules have moderate kinetic energy.
2. Molecules are close together.
3. Molecules are attracted to one another.
4. Molecules are able to exchange places.

Solid (ice)
1. Molecules have low kinetic energy.
2. Molecules are close together.
3. Molecules are attracted to one another.
4. Molecules vibrate in place—they do not exchange places.

FIGURE 4.12 States of Matter Matter exists in one of three states, depending on the amount of kinetic energy the molecules have. The higher the amount of energy, the greater the distance between molecules and the greater their degree of freedom of movement.

composition but differ in the arrangement and activity of their molecules. The amount of kinetic energy molecules have determines how rapidly they move. (See figure 4.12.) In solids, the molecular particles have comparatively little energy, and they vibrate in place very close to one another. In liquids, the particles have more energy, are farther apart from one another, and will roll, tumble, and flow over each other. In gases, the molecular particles move very rapidly and are very far apart. All that is necessary to change the physical nature of a substance is an energy change. Heat energy must be added or removed.

When two forms of matter have different temperatures, heat energy will flow from the one with the higher temperature to the one with the lower temperature. The temperature of the cooler matter increases while that of the warmer matter decreases. You experience this whenever you touch a cold or hot object. This is referred to as a **sensible heat** transfer. When heat energy is used to change the state of matter from solid to liquid at its melting point or liquid to gas at its boiling point, heat is transferred, but the temperature of the matter does not change. This is called a **latent heat** transfer. You have experienced this effect when water evaporates from your skin. Your body supplies the heat necessary to convert liquid water to water vapor. While the temperature of the water does not change, the physical state of the water does. The heat that was transferred to the water caused it to evaporate, and your body cooled. When substances change from gas to liquid at the boiling point or liquid to solid at the freezing point, there is a corresponding release of heat energy without a change in temperature.

FIRST AND SECOND LAWS OF THERMODYNAMICS

Energy can exist in several different forms, and it is possible to convert one form of energy into another. However, the total amount of energy remains constant.

The **first law of thermodynamics** states that energy can neither be created nor destroyed; it can only be changed from one form into another. From a human perspective, some forms of energy are more useful than others. We tend to make extensive use of electrical energy for a variety of purposes, but there is very little electrical energy present in nature. Therefore, we convert other forms of energy into electrical energy.

The **second law of thermodynamics** states that whenever energy is converted from one form to another, some of the useful energy is lost. The energy that cannot be used to do useful work is called **entropy.** Therefore, another way to state the second law of thermodynamics is to say that when energy is converted from one form to another, entropy increases. An alternative way to look at the idea of entropy is to say that entropy is a measure of disorder and that the amount of disorder (entropy) typically increases when energy conversions take place. Obviously, it is possible to generate greater order in a system (living things are a good example of highly ordered things), but when things become more highly ordered, the disorder of the surroundings must increase. For example, all living things release heat to their surroundings.

It is important to understand that when energy is converted from one form to another, there is no loss to *total* energy, but there is a loss of *useful* energy. For example, gasoline, which contains chemical energy, can be burned in an automobile engine to provide the kinetic energy to turn the wheels. The heat from the burning gasoline is used to move the pistons, which turns the crankshaft and eventually causes the wheels to turn. At each step in the process, some heat energy is lost from the system. During

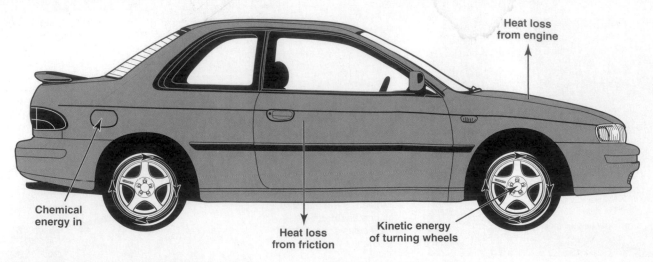

FIGURE 4.13 **Second Law of Thermodynamics** Whenever energy is converted from one form to another, some of the useful energy is lost, usually in the form of heat. The burning of gasoline produces heat, which is lost to the atmosphere. As the kinetic energy of moving engine parts is transferred to the wheels, friction generates some additional heat. All of these steps produce low-quality heat in accordance with the second law of thermodynamics.

combustion, heat is lost from the engine. Friction results in further loss of energy as heat. Therefore, the amount of useful energy (kinetic energy of the wheels turning) is much less than the total amount of chemical energy present in the gasoline that was burned. (See figure 4.13).

Within the universe, energy is being converted from one form to another continuously. Stars are converting nuclear energy into heat and light. Animals are converting the chemical potential energy found in food into kinetic energy that allows them to move. Plants are converting sunlight energy into the chemical bond energy of sugar molecules. In each of these cases, some energy is produced that is not able to do useful work. This is generally in the form of heat lost to the surroundings.

ENVIRONMENTAL IMPLICATIONS OF ENERGY FLOW

If we wish to understand many kinds of environmental problems, we must first understand the nature of energy and energy flow.

ENTROPY INCREASES

The heat produced when energy conversions occur is dissipated throughout the universe. This is a common experience. All machines and living things that manipulate energy release heat. It is also true that organized matter tends to become more disordered unless an external source of energy is available to maintain the ordered arrangement. Houses fall into ruin, automobiles rust, and appliances wear out unless work is done to maintain them. In reality, all of these phenomena involve the loss of heat. The organisms that decompose the wood in our houses release heat. The chemical reaction that causes rust releases heat. Friction, caused by the movement of parts of a machine against each other, generates heat and causes the parts to wear.

Ultimately, orderly arrangements of matter, such as clothing, automobiles, or living organisms, become disordered. There is an increase in entropy. Eventually, nonliving objects wear out and living things die and decompose. This process of becoming more disordered coincides with the constant flow of energy toward a dilute form of heat. This dissipated, low-quality heat has little value to us, since we are unable to use it.

ENERGY QUALITY

It is important to understand that some forms of energy are more useful to us than others. Some forms, such as electrical energy, are of high quality because they can be easily used to perform a variety of useful actions. Other forms, such as the heat in the water of the ocean, are of low quality because we are not able to use them for useful purposes. Although the total *quantity* of heat energy in the ocean is much greater than the total amount of electrical energy in the world, little useful work can be done with the heat energy in the ocean because it is of *low quality*. Therefore, it is not as valuable as other forms of energy that can be used to do work for us.

The reason the heat of the ocean is of little value is related to the small temperature difference between two sources of heat. When two objects differ in temperature, heat will flow from the warmer to the cooler object. The greater the temperature difference, the more useful the work that can be done. For example, fossil-fuel power plants burn fuel to heat water and convert it to steam. High-temperature steam enters the turbine, while cold cooling water condenses the steam as it leaves the turbine. This steep temperature gradient also provides a steep pressure gradient as heat energy flows from the steam to the cold water, which causes a turbine to turn, which generates electricity. Because the average temperature of the ocean is not high, and it is difficult to find another object that has a greatly lower temperature than the ocean, it is difficult to use the huge heat content of the ocean to do useful work for us.

CAMPUS SUSTAINABILITY INITIATIVE

COOLING OFF THE UNIVERSITY OF ARIZONA

In hot climates, air conditioning makes the hot part of the year bearable. However, air conditioning has its costs—both monetary and environmental. The electricity needed to provide air conditioning costs money and the production of that electricity results in carbon emissions.

The University of Arizona's water chilling and ice storing project is reducing the amount of electricity needed to cool the campus, which is both cost effective and environmentally conscious. The system involves the use of water chillers, cooling towers, pumps, and underground pipes that connect to buildings.

Because daytime and nighttime temperatures in Arizona can be drastically different, the university turns the water chillers on during the cooler evening hours to produce ice. The project can produce more than 900 tons of ice per hour. Currently, this process begins at 7:30 P.M. In the morning around 10 A.M., the university turns the chillers off and the ice begins to melt. The cold water is then delivered all over campus to cool buildings. This process moves the bulk of the university's electrical demand for air conditioning from the hot daytime hours to the cooler nighttime hours, which means that less energy is needed to provide air conditioning. The system has been called one of the most energy efficient and environmentally friendly chilled water systems in the world. Annual savings are expected to be about half a million dollars.

It is important to understand that energy that is of low quality from our point of view may still have significance to the world in which we live. For example, the distribution of heat energy in the ocean tends to moderate the temperature of coastal climates, contributes to weather patterns, and causes ocean currents that are extremely important in many ways. It is also important to recognize that we can sometimes figure new ways to convert low-quality energy to high-quality energy. For example, it is possible to use the waste heat from power plants to heat cities if the power plants are located in the cities. Scientists have recently made major improvements in wind turbines and photovoltaic cells that allow us to economically convert low-quality light or wind to high-quality electricity.

BIOLOGICAL SYSTEMS AND THERMODYNAMICS

From the point of view of ecological systems, organisms such as plants do photosynthesis and are able to convert low-quality light energy to high-quality chemical energy in the organic molecules they produce. Eventually, they will use this stored energy for their needs, or it will be used by some other organism that has eaten the plant. In accordance with the second law of thermodynamics, all organisms, including humans, are in the process of converting high-quality energy into low-quality energy. Waste heat is produced when the chemical-bond energy in food is converted into the energy needed to move, grow, or respond. The process of releasing chemical-bond energy from food by organisms is known as cellular respiration. From an energy point of view, it is comparable to the process of **combustion,** which is the burning of fuel to obtain heat, light, or some other form of useful energy. The efficiency of cellular respiration is relatively high. About 40 percent of the energy contained in food is released in a useful form. The rest is dissipated as low-quality heat.

POLLUTION AND THERMODYNAMICS

An unfortunate consequence of energy conversion is pollution. The heat lost from most energy conversions is a pollutant. The wear of the brakes used to stop cars results in pollution. The emissions from power plants pollute. All of these are examples of the effect of the second law of thermodynamics. If each person on Earth used less energy, there would be less waste heat and other forms of pollution that result from energy conversion. The amount of energy in the universe is limited. Only a small portion of that energy is of high quality. The use of high-quality energy decreases the amount of useful energy available, as more low-quality heat is generated. All life and all activities are subject to these important physical principles described by the first and second laws of thermodynamics.

Diesel Engine Trade-offs

In the United States nearly all automobiles have gasoline engines. Only about 1 percent have diesel engines. In Europe more than 50 percent of new automobiles sold have diesel engines. Automobile diesel engines are much more efficient than gasoline engines. They have an efficiency of 35 to 42 percent compared to gasoline engines with an efficiency of 25 to 30 percent. This means that they go much farther on a liter of fuel than do equivalent gasoline engines. Furthermore diesel engines last longer than gasoline engines.

Since gasoline engines are less efficient than diesel engines, gasoline engines produce 20 to 40 percent more carbon dioxide than diesel engines for the same distance driven. Increased carbon dioxide is known to cause a warming of the Earth's atmosphere. One of the reasons for the popularity of diesel engines in Europe is concern about the effect of carbon dioxide on global climate.

Current diesel engines produce more particulate matter and nitrogen oxides than do gasoline engines. However, current air quality guidelines for particulate matter and nitrogen oxides in Europe are more stringent than those in the United States. The World Health Organization has estimated that thousands of people die each year because of particulate air pollution. Changes are being made in fuels, engine design, and pollution control devices that reduce the amount of particulate matter and nitrogen oxides produced.

- Should U.S. automakers switch to diesel engines as the Europeans have?
- Is global climate change an important reason to use diesel engines?
- Should problems with particulate emissions prevent the development of diesel engines for passenger vehicles?

SUMMARY

Science is a method of gathering and organizing information. It involves observation, asking questions, exploring alternative sources of information, hypothesis formation, the testing of hypotheses, and publication of the results for others to evaluate. A hypothesis is a logical prediction about how things work that must account for all the known information and be testable. The process of science attempts to be careful, unbiased, and reliable in the way information is collected and evaluated. This often involves conducting experiments to test the validity of a hypothesis. If a hypothesis is continually supported by the addition of new facts, it may be incorporated into a theory. A theory is a broadly written, widely accepted generalization that ties together large bodies of information and explains why things happen. Similarly, a law is a broad statement that describes what happens in nature.

The fundamental unit of matter is the atom, which is made up of protons and neutrons in the nucleus surrounded by a cloud of moving electrons. The number of protons for any one type of atom is constant, but the number of neutrons in different atoms of the same type of atom may vary. The number of electrons is equal to the number of protons. Protons have a positive charge, neutrons lack a charge, and electrons have a negative charge.

Molecules are units made of a combination of two or more atoms bonded to one another. Chemical bonds are physical attractions between atoms resulting from the interaction of their electrons. When chemical bonds are broken or formed, a chemical reaction occurs, and the amount of energy within the chemical bonds is changed. Chemical reactions require activation energy to get the reaction started.

Matter that is composed of only one kind of atom is known as an element. Matter that is composed of small units containing different kinds of atoms bonded in specific ratios is known as a compound. An atom or molecule that has gained or lost electrons so that it has an electric charge is known as an ion.

Matter can occur in three states: solid, liquid, and gas. These three differ in the amount of energy the molecular units contain and the distance between the units. Kinetic energy is the energy contained by moving objects. Potential energy is the energy an object has because of its position.

The first law of thermodynamics states that the amount of energy in the universe is constant, that energy can neither be created nor destroyed. The second law of thermodynamics states that when energy is converted from one form to another, some of the useful energy is lost (entropy increases). Some forms of energy are more useful than others. The quality of the energy determines how much useful work can be accomplished by expending the energy. Low-temperature heat sources are of poor quality, since they cannot be used to do useful work.

THINKING GREEN

1. Replace household chemicals with those that are more environmentally friendly.
2. Find out when the next household hazardous waste cleanup is in your community and get rid of unwanted hazardous materials.
3. Read the labels on your household cleaning products.
4. Use baking soda as a substitute for toothpaste.
5. Reduce your consumption of electricity by 10 percent.

WHAT'S YOUR TAKE?

The following chart gives the typical light output per watt of electricity used for various kinds of lighting devices:

Kind of Light	Light Output per Watt of Electricity Used
100-watt incandescent lightbulb	About 17 lumens/watt
White LED	About 17 lumens/watt
Compact fluorescent bulb	About 50 lumens/watt
Standard 40-watt fluorescent tube	About 60 lumens/watt
T8 fluorescent with electronic Ballast (32 watt)	About 90 lumens/watt

Compact fluorescent bulbs can be screwed into standard light fixtures. They cost about 10 times more than an incandescent lightbulb but last about 10 times longer.

Based on this information, should you replace incandescent lightbulbs with compact fluorescents? Develop an argument supporting your position.

REVIEW QUESTIONS

1. How do scientific disciplines differ from nonscientific disciplines?
2. What is a hypothesis? Why is it an important part of the way scientists think?
3. Why are events that happen only once difficult to analyze from a scientific point of view?
4. What is the scientific method, and what processes does it involve?
5. How are the second law of thermodynamics and pollution related?
6. Diagram an atom of oxygen and label its parts.
7. What happens to atoms during a chemical reaction?
8. State the first and second laws of thermodynamics.
9. How do solids, liquids, and gases differ from one another at the molecular level?
10. List five kinds of energy.
11. Are all kinds of energy equal in their capacity to bring about changes? Why or why not?

CRITICAL THINKING QUESTIONS

1. You observe that a high percentage of frogs, which are especially sensitive to environmental poisons, in small ponds in your agricultural region have birth defects. Suspecting agricultural chemicals present in runoff to be the culprit, state the hypothesis in your own words. Next, devise an experiment that might help you support or reject your hypothesis.
2. Given the experiment you proposed in Critical Thinking Question 1, imagine some results that would support that hypothesis. Now imagine you are a different scientist, one who is very skeptical of the initial hypothesis. How convincing do you find these data? What other possible explanations (hypotheses) might there be to explain the results? Devise a different experiment to test this new hypothesis.
3. Increasingly, environmental issues such as global climate change are moving to the forefront of world concern. What role should science play in public policy decisions? How should we decide between competing scientific explanations about an environmental concern such as global climate change? What might be some of the criteria for deciding what is "good science" and what is "bad science"?
4. How important are the first and second laws of thermodynamics to explaining environmental issues? Using the concepts in these laws of thermodynamics, try to explain a particular environmental issue. How does an understanding of thermodynamics change your conceptual framework regarding this issue?

5. The text points out that incandescent lightbulbs are only 5–10 percent efficient at using energy to accomplish their task, while new, initially more expensive, compact fluorescent lighting uses significantly less electricity to provide the same quantity of light. Examine the contextual framework of those who advocate for new lighting methods and the contextual framework of those who continue to design and build using the old methods. What are the major differences in perspective?

What could you suggest be done to help bring these different perspectives closer together?

6. Some scientists argue that living organisms constantly battle against the principles of the second law of thermodynamics using the principles of the first law of thermodynamics. What might they mean by this? Do you think this is accurate? What might be some of the implications of this for living organisms?

CHAPTER 5

INTERACTIONS: ENVIRONMENTS AND ORGANISMS

There are many kinds of organism interactions. Fungi like this mushroom are decomposers that break down dead organic matter. The mushrooms may, in turn, become food for other organisms.

CHAPTER OUTLINE

Ecological Concepts
  Environment
  Limiting Factors
  Habitat and Niche
The Role of Natural Selection and Evolution
  Genes, Populations, and Species
  Natural Selection
  Evolutionary Patterns
Kinds of Organism Interactions
  Predation
  Competition
  Symbiotic Relationships
  Some Relationships Are Difficult to Categorize
Community and Ecosystem Interactions
  Major Roles of Organisms in Ecosystems
  Keystone Species
  Energy Flow Through Ecosystems
  Food Chains and Food Webs
  Nutrient Cycles in Ecosystems—Biogeochemical Cycles
  Human Impact on Nutrient Cycles

ISSUES & ANALYSIS
Phosphate Mining in Nauru 106

CAMPUS SUSTAINABILITY INITIATIVE
Creek Restoration at the University of Arkansas–Little Rock 81

GOING GREEN
Phosphorus-Free Lawn Fertilizer 105

WATER CONNECTIONS
Changes in the Food Chain of the Great Lakes 100

OBJECTIVES

After reading this chapter, you should be able to:

- Identify and list abiotic and biotic factors in an ecosystem.
- Define *niche.*
- Describe the process of natural selection as it operates to refine the fit among organism, habitat, and niche.
- Describe predator-prey, parasite-host, competitive, mutualistic, and commensalistic relationships.
- Differentiate between a community and an ecosystem.
- Define the roles of producer, herbivore, carnivore, omnivore, scavenger, parasite, and decomposer.
- Describe energy flow through an ecosystem.
- Relate the concepts of food webs and food chains to trophic levels.
- Explain the cycling of nutrients such as nitrogen, carbon, and phosphorus through an ecosystem.

ECOLOGICAL CONCEPTS

The science of **ecology** is the study of the ways organisms interact with each other and with their nonliving surroundings. Ecology deals with the ways in which organisms are adapted to their surroundings, how they make use of these surroundings, and how an area is altered by the presence and activities of organisms. These interactions involve energy and matter. Living things require a constant flow of energy and matter to assure their survival. If the flow of energy and matter ceases, the organisms die.

All organisms are dependent on other organisms in some way. One organism may eat another and use it for energy and raw materials. One organism may temporarily use another without harming it. One organism may provide a service for another, such as when animals distribute plant seeds or bacteria break down dead organic matter for reuse. The study of ecology can be divided into many specialties and be looked at from several levels of organization. (See figure 5.1.) Before we can explore the field of ecology in greater depth, we must become familiar with some of the standard vocabulary of this field.

ENVIRONMENT

Everything that affects an organism during its lifetime is collectively known as its **environment.** Environment is a very broad concept. For example, during its lifetime, an animal such as a raccoon is likely to interact with millions of other organisms (bacteria, food organisms, parasites, mates, predators), drink copious amounts of water, breathe huge quantities of air, and respond to daily changes in temperature and humidity. This list only begins to describe the various components that make up the raccoon's environment. Because of this complexity, it is useful to subdivide the concept of environment into abiotic (nonliving) and biotic (living) factors.

Abiotic Factors

Abiotic factors are nonliving things that influence an organism. They can be organized into several broad categories: energy, nonliving matter, living space, and processes that involve the interactions of nonliving matter and energy.

FIGURE 5.1 **Levels of Organization in Ecology** Ecology is the science that deals with the study of interactions between organisms and their environment. This study can take place at several different levels, from the broad ecosystem level through community interactions to population studies and the study of the niche of individual organisms. Ecology also involves study of the physical environment and the atoms and molecules that make up both the living and nonliving parts of an ecosystem.

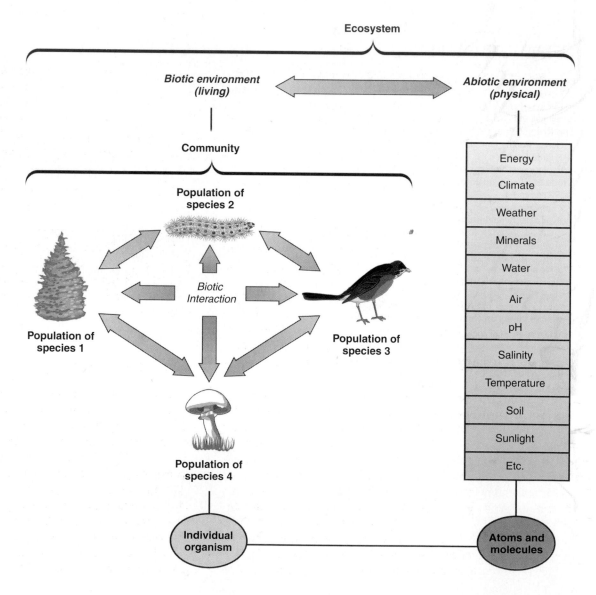

CAMPUS SUSTAINABILITY INITIATIVE

CREEK RESTORATION AT THE UNIVERSITY OF ARKANSAS–LITTLE ROCK

A partnership that involves the University of Arkansas–Little Rock, the City of Little Rock, and the Audubon Society of Arkansas is turning a 5-acre urban concrete and asphalt part of the campus into a creek shoreline of native trees and grasses. The site is located in a floodplain of Coleman Creek.

The first part of the project includes the removal of concrete pilings, asphalt, and five unusable buildings. When the demolition is completed, landscaping will begin. The landscaping will include planting of native grasses and trees. Foot paths will be constructed that will create a "circle of life" in which historical markers will be placed to identify the creek as a stop-over along the old Southwest Trail. The old Southwest Trail was part of the "Trail of Tears" over which members of the Chickasaw and Choctaw tribes were forced to migrate to reservations farther west.

This project is the first step in a 47-acre greenway reaching the full length of the campus that will include plantings, the construction of bicycle and walking trails, and other structures like benches and bridges. The restoration project also will provide an outdoor laboratory for biologists, Earth scientists, and hydrologists for teaching and research activities. The reclamation of the area is also part of the strategic plan for the University District aimed at improving life and business in the neighborhood surrounding the campus and will connect to other open space in the area.

Energy is required by all organisms to maintain themselves. The ultimate source of energy for almost all organisms is the sun; in the case of plants, the sun directly supplies the energy necessary for them to maintain themselves. Animals obtain their energy by eating plants or other animals that eat plants. Ultimately, the amount of living material that can exist in an area is determined by the amount of energy plants, algae, and bacteria can trap.

Matter in the form of atoms such as carbon, nitrogen, and phosphorus and molecules such as water provides the structural framework of organisms. Organisms constantly obtain these materials from their environment. The atoms become part of an organism's body structure for a short time, and eventually all of them are returned to the environment through respiration, excretion, or death and decay.

The *space* organisms inhabit has a particular structure and location that is also an important abiotic aspect of their environment. Some organisms live in the ocean; others live on land at sea level; still others live on mountaintops or fly through the air. Some spaces are homogeneous and flat; others are a jumble of rocks of different sizes. Some are close to the equator; others are near the poles.

Ecological processes involve interactions between matter and energy. The climate (average weather patterns over a number of years) of an area involves energy in the form of solar radiation interacting with the matter that makes up the Earth. The kind of climate present is determined by many different factors, including the amount of solar radiation, proximity to the equator, prevailing wind patterns, and closeness to water. The intensity and duration of sunlight in an area cause daily and seasonal changes in temperature. Differences in temperature generate wind. Solar radiation is also responsible for generating ocean currents and the evaporation of water into the atmosphere that subsequently falls as precipitation. Depending on the climate, precipitation may be of several forms: rain, snow, hail, or fog. Furthermore, there may be seasonal precipitation patterns.

Soil-building processes are influenced by prevailing weather patterns, local topography, and the geologic history of the region. These factors interact to produce a variety of soils that range from those that are coarse, sandy, dry, and infertile to fertile soils composed of fine particles that hold moisture.

Biotic Factors

The **biotic factors** of an organism's environment include all forms of life with which it interacts. Some broad categories are: plants that carry on photosynthesis; animals that eat other organisms; bacteria and fungi that cause decay; bacteria, viruses, and other parasitic organisms that cause disease; and other individuals of the same species.

LIMITING FACTORS

Although organisms interact with their surroundings in many ways, certain factors may be critical to a particular species' success. A shortage or absence of a factor can restrict the success of the species; thus, it is known as a **limiting factor.** Limiting factors may be either abiotic or biotic and can be quite different from one species to another. Many plants are limited by scarcity of water, light, or specific soil nutrients such as nitrogen or phosphorus. Monarch butterflies are limited by the number of available milkweed plants, since their developing caterpillars use this plant as their only food source. (See figure 5.2.)

Climatic factors such as temperature range, humidity, periods of drought, or length of winter are often limiting factors.

Adult monarch on milkweed

Caterpillar feeding on milkweed

FIGURE 5.2 **Milkweed Is a Limiting Factor for Monarch Butterflies** Monarch butterflies lay their eggs on various kinds of milkweed plants. The larvae eat the leaves of the milkweed plant. Thus, monarchs are limited by the number of milkweeds in the area.

For example, many species of snakes and lizards are limited to the warmer parts of the world because they have difficulty maintaining their body temperature in cold climates and cannot survive long periods of cold. If we look at the number of species of snakes and lizards, we see that the number of species declines as one moves from warmer to colder climates. While this is a general trend, there is much variation among species. The **range of tolerance** of a species is the degree to which it is able to withstand environmental variation. Some species have a broad range of tolerance, whereas others have a narrow range of tolerance. The common garter snake (*Thamnophis sirtalis*) consists of several subspecies and is found throughout the United States and Southern Canada and has scattered populations in Mexico. Thus, it is able to tolerate a wide variety of climates. Conversely, the green anole (*Anolis carolinensis*) has a much narrower range of tolerance and is found only in the southeastern part of the United States where temperatures are mild. (See figure 5.3.)

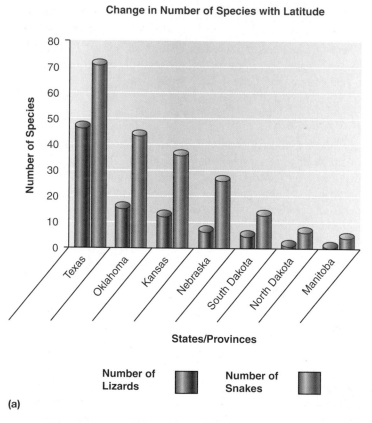

(b) Garter snake

(c) Green anole

(a)

Number of Lizards

Number of Snakes

FIGURE 5.3 **Temperature Is a Limiting Factor** Cold temperature is a limiting factor for many kinds of reptiles. Snakes and lizards are less common in cold regions than in warm regions. (a) The graph shows the number of species of snakes and lizards in regions of central North America. Note that the number of species declines as one proceeds from south to north. (b) Some species, like the common garter snake (*Thamnophis sirtalis*), have a broad range of tolerance and are not limited by cold temperature. It is found throughout the United States and several Canadian provinces. (c) However, the green anole (*Anolis carolinensis*) has a very narrow range of tolerance and is found only in the warm, humid southeastern states.

HABITAT AND NICHE

As we have just seen, it is impossible to understand an organism apart from its environment. The environment influences the organism, and organisms affect the environment. To focus attention on specific elements of this interaction, ecologists have developed two concepts that need to be clearly understood: habitat and niche.

Habitat—Place

The **habitat** of an organism is the space that the organism inhabits, the place where it lives (its address). We tend to characterize an organism's habitat by highlighting some prominent physical or biological feature of its environment such as soil type, availability of water, climatic conditions, or predominant plant species that exist in the area. For example, mosses are small plants that must be covered by a thin film of water in order to reproduce. In addition, many kinds dry out and die if they are exposed to sunlight, wind, and drought. Therefore, the typical habitat of moss is likely to be cool, moist, and shady. (See figure 5.4.) Likewise, a rapidly flowing, cool, well-oxygenated stream with many bottom-dwelling insects is good trout habitat, while open prairie with lots of grass is preferred by bison, prairie dogs, and many kinds of hawks and falcons. The particular biological requirements of an organism determine the kind of habitat in which it is likely to be found.

Niche—Role

The **niche** of an organism is the functional role it has in its surroundings (its profession). A description of an organism's niche

FIGURE 5.4 **Moss Habitat** The habitat of mosses is typically cool, moist, and shady, since many mosses die if they are subjected to drying. In addition, mosses must have a thin layer of water present in order to reproduce sexually.

includes all the ways it affects the organisms with which it interacts as well as how it modifies its physical surroundings. In addition, the description of a niche includes all of the things that happen to the organism. For example, beavers frequently flood areas by building dams of mud and sticks across streams. (See figure 5.5.) The flooding has several effects. It provides beavers with a larger area of deep water, which they need for protection; it

Beavers eat woody plants

Beaver gnawing

Beaver dam

Beaver ponds provide habitat

Beaver ponds provide habitat

FIGURE 5.5 **The Ecological Niche of a Beaver** The ecological niche of an organism is a complex set of interactions between an organism and its surroundings, which includes all of the ways an organism influences its surroundings as well as all of the ways the organism is affected by its environment. A beaver's ecological niche includes building dams and flooding forested areas, killing trees, providing habitat for waterbirds and other animals, serving as food for predators, and many other effects.

FIGURE 5.6   **The Niche of a Dandelion** A dandelion is a familiar plant that commonly invades disturbed sites because it produces many seeds that are blown easily to new areas. It serves as food for various herbivores, supplies nectar to bees, and can regrow quickly from its root if its leaves are removed.

provides a pond habitat for many other species of animals such as ducks and fish; and it kills trees that cannot live with their roots under water. The animals that are attracted to the pond and the beavers often become food for predators. After the beavers have eaten all the suitable food, such as aspen, they abandon the pond, migrate to other areas along the stream, and begin the whole process over again.

In this recitation of beaver characteristics, we have listed several effects that the animal has on its local environment. It changes the physical environment by flooding, it kills trees, it enhances the environment for other animals, and it is a food source for predators. This is only a superficial glimpse of the many aspects of the beaver's interaction with its environment. A complete catalog of all aspects of its niche would make up a separate book.

Another familiar organism is the dandelion. (See figure 5.6.) Its niche includes the fact that it is an opportunistic plant that rapidly becomes established in sunny, disturbed sites. It can produce thousands of parachutelike seeds that are easily carried by the wind over long distances. (You have probably helped this process by blowing on the fluffy, white collections of seeds of a mature dandelion fruit.) Furthermore, it often produces several sets of flowers per year. Since there are so many seeds and they are so easily distributed, the plant can easily establish itself in any sunny, disturbed site, including lawns. Since it is a plant, one major aspect of its niche is the ability to carry on photosynthesis and grow. It uses water and nutrients from the soil to produce new plant parts. Since dandelions need direct sunlight to grow successfully, mowing lawns helps provide just the right conditions for dandelions because the vegetation is never allowed to get so tall that dandelions are shaded. Many kinds of animals, including some humans, use the plant for food. The young leaves may be eaten in a salad, and the blossoms can be used to make dandelion wine. Bees visit the flowers regularly to obtain nectar and pollen.

THE ROLE OF NATURAL SELECTION AND EVOLUTION

Since organisms generally are well adapted to their surroundings and fill a particular niche, it is important that we develop an understanding of the processes that lead to this high degree of adaptation.

Furthermore, since the mechanisms that result in adaptation occur within a species, we need to understand the nature of a species.

GENES, POPULATIONS, AND SPECIES

We can look at an organism from several points of view. We can consider an individual, groups of individuals of the same kind, or groups that are distinct from other groups. This leads us to discuss three interrelated concepts.

Genes are distinct pieces of DNA that determine the characteristics an individual displays. There are genes for structures such as leaf shape or feather color, behaviors such as cricket chirps or migratory activity, and physiological processes such as photosynthesis or muscular contractions. Each individual has a particular set of genes.

A **population** is considered to be all the organisms of the same kind found within a specific geographic region. (See figure 5.7.) The individuals of a population will have very similar sets of genes, although there will be some individual variation. Because some genetic difference exists among individuals

FIGURE 5.7   **A Penguin Population** This population of King penguins is from South Georgia Island near Antarctica.

in a population, a population contains more kinds of genes than any individual within the population. Reproduction also takes place among individuals in a population so that genes are passed from one generation to the next. The concept of a species is an extension of this thinking about genes, groups, and reproduction.

A **species** is a population of all the organisms potentially capable of reproducing naturally among themselves and having offspring that also reproduce. Therefore, the concept of a species is a population concept. *An individual organism is not a species but a member of a species.* It is also a genetic concept, since individuals that are of different species are not capable of exchanging genes through reproduction. (See figure 5.8.)

Some species are easy to recognize. We easily recognize humans as a distinct species. Most people recognize a dandelion when they see it and do not confuse it with other kinds of plants that have yellow flowers. Other species are not as easy to recognize. Most of us cannot tell one species of mosquito from another or identify different species of grasses. Because of this, we tend to lump organisms into large categories and do not recognize the many subtle niche differences that exist among the similar-appearing species. However, species of mosquitoes are quite distinct from one another genetically and occupy different niches. Only certain species of mosquitoes carry and transmit the human

disease malaria. Other species transmit the dog heartworm parasite. Each mosquito species is active during certain portions of the day or night. And each species requires specific conditions to reproduce.

NATURAL SELECTION

As we have seen, each species of organism is specifically adapted to a particular habitat in which it has a very specific role (niche). But how is it that each species of plant, animal, fungus, or bacterium fits into its environment in such a precise way? Since most of the structural, physiological, and behavioral characteristics organisms display are determined by the genes they possess, these characteristics are passed from one generation to the next when individuals reproduce. The process that leads to this close fit between the characteristics organisms display and the demands of their environment is known as natural selection.

Charles Darwin is generally credited with developing the concept of natural selection. Although he did not understand the gene concept, he understood that characteristics were passed from parent to offspring. He also observed the highly adaptive nature of the relationship between organisms and their environment and developed the concept of natural selection to explain how this adaptation came about. (See figure 5.9.)

Natural selection is the process that determines which individuals within a species will reproduce and pass their genes to the next generation. The changes that we see in the genes and the characteristics displayed by successive generations of a population of organisms over time is known as **evolution.** Thus, natural

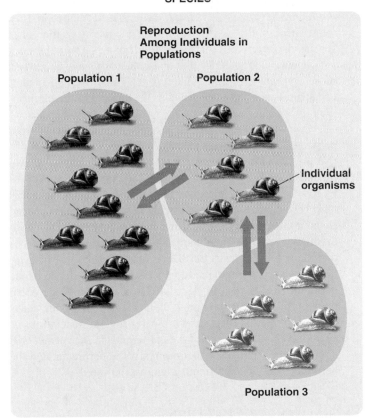

SPECIES

Reproduction Among Individuals in Populations

Population 1   Population 2

Individual organisms

Population 3

FIGURE 5.8 **The Species Concept** A species is all the organisms of a particular kind capable of interbreeding and producing fertile offspring. Often species consist of many distinct populations that show genetic differences from one another.

FIGURE 5.9 **Charles Darwin** Charles Darwin observed that each organism fit into its surroundings. He developed the theory of natural selection to explain how adaptation came about.

FIGURE 5.10 **Genetic Variation** Since genes are responsible for many facial characteristics, the differences in facial characteristics shown in this photograph are a good visual illustration of genetic variation. This population of students also differs in genes for height, susceptibility to disease, and many other characteristics.

selection is the mechanism that causes evolution to occur. Several conditions and steps are involved in the process of natural selection.

1. *Individuals within a species show genetically determined variation; some of the variations are useful and others are not.* (See figure 5.10.) For example, individual animals that are part of the same species show variation in color, size, or susceptibility to disease. Some combinations of genes are more valuable to the success of the individual than others.

2. *Organisms within a species typically produce many more offspring than are needed to replace the parents when they die. Most of the offspring die.* One blueberry bush may produce hundreds of berries with several seeds in each berry, or a pair of rabbits may have three to four litters of offspring each summer, with several young in each litter. Few of the seeds or baby rabbits become reproducing adults. (See figure 5.11.)

3. *The excess number of individuals results in a shortage of specific resources.* Individuals within a species must compete with each other for food, space, mates, or other requirements that are in limited supply. If you plant 100 bean seeds in a pot, many of them will begin to grow, but eventually, some will become taller and get the majority of the sunlight while the remaining plants are

FIGURE 5.11 **Excessive Reproduction** One blueberry bush produces thousands of seeds (offspring). Only a few will actually germinate and only a few of the seedlings will become mature plants.

shaded and die. Great horned owls typically produce two young at a time, but if food is in short supply, the larger of the two young will get the majority of the food.

4. *Because of variation among individuals, some have a greater chance of obtaining needed resources and, therefore, have a greater likelihood of surviving and reproducing than others.* Individuals that have genes that allow them to obtain needed resources and avoid threats to their survival will be more likely to survive and reproduce. Even if less well-adapted individuals survive, they may mature more slowly and not be able to reproduce as many times as the more well-adapted members of the species.

The degree to which organisms are adapted to their environment influences their reproductive success and is referred to as fitness. It is important to recognize that fitness does not necessarily mean the condition of being strong or vigorous. In this context, it means how well the organism fits in with all the aspects of its surroundings so that it successfully passes its genes to the next generation. For example, in a lodgepole pine forest, many lodgepole pine seedlings become established following a fire. Some grow more rapidly and obtain more sunlight and nutrients. Those that grow more rapidly are more likely to survive. These are also likely to reproduce for longer periods and pass more genes to future generations than those that die or grow more slowly.

5. *As time passes and each generation is subjected to the same process of natural selection, the percentage of individuals showing favorable variations will increase and those having unfavorable variations will decrease.* Those that reproduce more successfully pass on to the next generation the genes for the characteristics that made them successful in their environment, and the genes that made them successful become more common in future generations. Thus, each species of organism is continually refined to be adapted to the environment in which it exists.

One modern example of genetic change resulting from natural selection involves the development of pest populations that are resistant to the pesticides previously used to control them. Figure 5.12 shows a graph of the number of species of weeds that have populations that are resistant to commonly used herbicides. When an herbicide is first used against weed pests, it kills most of them. However, in many cases, some individual weed plants within the species happen to have genes that allow them to resist the effects of the herbicide. These individuals are better adapted to survive in the presence of the herbicide and have a higher likelihood of surviving. When they reproduce, they pass on to their offspring the same genes that contributed to their survival. After several generations of such selection, a majority of the individuals in the species will contain genes that allow them to resist the herbicide, and the herbicide is no longer effective against the weed.

EVOLUTIONARY PATTERNS

When we look at the effects of natural selection over time, we can see considerable change in the characteristics of a species and kinds of species present. Some changes take thousands or millions of years

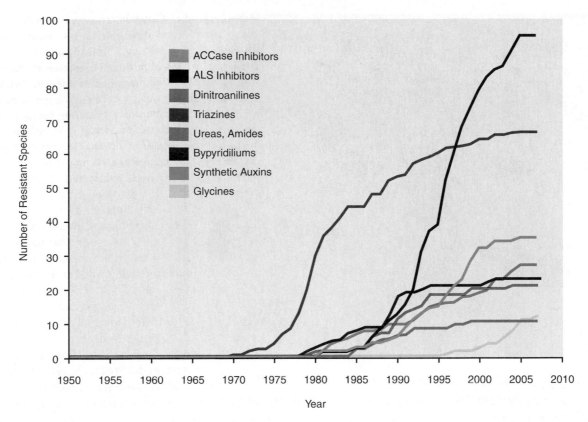

FIGURE 5.12 **Change Resulting from Natural Selection** Populations of weed plants that have been subjected repeatedly to herbicides often develop resistant populations. Those individual weed plants that were able to resist the effects of the herbicide lived to reproduce and pass on their genes for resistance to their offspring; thus, resistant populations of weeds develop. This graph shows the number of weeds with resistance to commonly used herbicides.

Source: Data from Ian Heap, "The International Survey of Herbicide Resistant Weeds," [cited 2 April 2008], Internet.

to occur. Others, such as resistance to pesticides, can occur in a few years. (See figure 5.12.) Natural selection involves the processes that bring about change in species, and the end result of the natural selection process observable in organisms is called evolution.

Scientists have continuously shown that this theory of natural selection can explain the development of most aspects of the structure, function, and behavior of organisms. It is the central idea that helps explain how species adapt to their surroundings. When we discuss environmental problems, it is helpful to understand that species change and that as the environment is changed, either naturally or by human action, some species will adapt to the new conditions while others will not.

There are many examples that demonstrate the validity of the process of natural selection and the evolutionary changes that result from natural selection. In recent times, we have become aware that many species of insects, weeds, and bacteria have become resistant to the insecticides, herbicides, and antibiotics that formerly were effective against them.

When we look at the evolutionary history of organisms in the fossil record over long time periods, it becomes obvious that new species come into being while other species disappear.

Speciation

Speciation is the production of new species from previously existing species. It is thought to occur as a result of a species

dividing into two isolated subpopulations. If the two subpopulations contain some genetic differences and their environments are somewhat different, natural selection will work on the two groups differently and they will begin to diverge from each other. Eventually, the differences may become so great that the two subpopulations are no longer able to interbreed. At this point, they are two different species. This process has resulted in millions of different species. Figure 5.13 shows a collection of a few of the thousands of species of beetles.

Among plants, another common mechanism known as polyploidy results in new species. **Polyploidy** is a condition in which the number of sets of chromosomes in the cells of the plant is increased. Many organisms are diploid; that is, they have two sets of chromosomes. They got a set from each parent, one set in the egg and one set in the sperm. Polyploid organisms may have several sets of chromosomes. The details of how polyploidy comes about are not important for this discussion. It is sufficient to recognize that many species of plants appear to have extra sets of chromosomes, and they are not able to reproduce with closely related species that have a different number of sets of chromosomes.

Extinction

Extinction is the loss of an entire species and is a common feature of the evolution of organisms. The environment in which

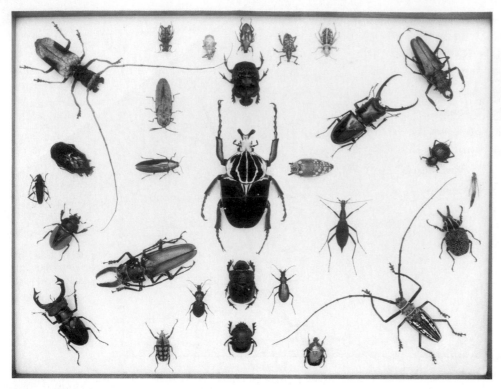

FIGURE 5.13 **Speciation** This photo shows a few of the many different species of beetles.

Natural selection is constantly at work shaping organisms to fit a changing environment. It is clear that humans have had a significant impact on the extinction of many kinds of species. Wherever humans have modified the environment for their purposes (farming, forestry, cities, hunting, and introducing exotic organisms), species are typically displaced from the area. If large areas are modified, entire species may be displaced. Ultimately, humans are also subject to evolution and the possibility of extinction as well.

Although extinction is a common event, there have been several instances of past mass extinctions in which major portions of the living world went extinct in a relatively short time. The background rate of extinction is estimated to be about 10 species per year. However, the current rate is much higher than that, leading many people to suggest that we may be in the middle of a mass extinction caused by human influences on the Earth. See chapter 11 on biodiversity for a more complete discussion.

organisms exist does not remain constant over long time periods. Those species that lack the genetic resources to cope with a changing environment go extinct. Of the estimated 500 million species of organisms that are believed to have ever existed on Earth since life began, perhaps only 5 million to 10 million are currently in existence. This represents an extinction rate of 98 to 99 percent. Obviously, these numbers are estimates, but the fact remains that extinction has been the fate of most species of organisms. In fact, studies of recent fossils and other geologic features show that only thousands of years ago, huge glaciers covered much of Europe and the northern parts of North America. Humans coexisted with mammoths, saber-toothed tigers, and giant cave bears. As the climate became warmer and the glaciers receded, and humans continued to prey on these animals, new pressures affected the organisms in the area. Some, including the mammoths, saber-toothed tigers, and giant cave bears, did not adapt and became extinct. (See figure 5.14.) Others, such as humans, horses, and many kinds of plants, adapted to the new conditions and so survive to the present.

It is also possible to have the local extinction of specific populations of a species. Most species consist of many different populations that may differ from one another in significant ways. Often some of these populations have small local populations that can easily be driven to extinction. While these local extinctions are not the same as the extinction of an entire species, local extinctions often result in the loss of specific gene combinations. Many of the organisms listed on the endangered species list are really local populations of a more widely distributed species.

Coevolution

Coevolution is the concept that two or more species of organisms can reciprocally influence the evolutionary direction of the other. In other words, organisms affect the evolution of other organisms. Since all organisms are influenced by other organisms, this is a common pattern.

FIGURE 5.14 **Extinction** This figure shows a scene that would have been typical at the end of the last ice age. Nearly all of these large mammals are extinct.

For example, grazing animals and the grasses they consume have coevolved. Grasses that are eaten by grazing animals grow from the base of the plant near the ground rather than from the tips of the branches as many plants do. Furthermore, grasses have hard materials in their cell walls that make it difficult for animals to crush the cell walls and digest them. Grazing animals have different kinds of adaptations that overcome these deterrents. Many grazers have teeth that are very long or grow continuously to compensate for the wear associated with grinding hard cell walls. Others, such as cattle, have complicated digestive tracts that allow microorganisms to do most of the work of digestion.

Similarly, the red color and production of nectar by many kinds of flowers is attractive to hummingbirds, which pollinate the flowers at the same time as they consume nectar from the flower. In addition to the red color that is common for many flowers that are pollinated by hummingbirds, many such flowers are long and tubular. The long bill of the hummingbird is a matching adaptation to the shape of the flower. (See figure 5.15.)

The "Kinds of Organism Interactions" section will explore in more detail the ways that organisms interact and the results of long periods of coevolution.

KINDS OF ORGANISM INTERACTIONS

Ecologists look at organisms and how they interact with their surroundings. Perhaps the most important interactions occur between organisms. Ecologists have identified several general types of organism-to-organism interactions that are common in all ecosystems. When we closely examine how organisms interact, we see that each organism has specific characteristics that make it well suited to its role. An understanding of the concept of natural selection allows us to see how interactions between different species of organisms can result in species that are finely tuned to a specific role.

As you read this section, notice how each species has special characteristics that equip it for its specific role (niche). Because these interactions involve two kinds of organisms interacting, we should expect to see examples of coevolution. If the interaction between two species is the result of a long period of interaction, we should expect to see that each species has characteristics that specifically adapt it to be successful in its role.

PREDATION

One common kind of interaction called **predation** occurs when one organism, known as a **predator,** kills and eats another, known as the **prey.** (See figure 5.16.) The predator benefits from killing and eating the prey and the prey is harmed. Some examples of predator-prey relationships are lions and zebras, robins and earthworms, wolves and moose, and toads and beetles. Even a few plants show predatory behavior. The Venus flytrap has specially modified leaves that can quickly fold together and trap insects that are then digested.

FIGURE 5.15 **Coevolution** The red tubular flowers and the nectar they produce are attractive to hummingbirds. The long bill of the hummingbird allows it to reach the nectar at the base of the flower. The two kinds of organisms have coevolved.

FIGURE 5.16 **Predator-Prey Relationship** Cheetahs are predators of impalas. Cheetahs can run extremely fast to catch the equally fast impalas. Impalas have excellent eyesight, which helps them see predators. The chameleon hunts by remaining immobile and ambushing prey. The long, sticky tongue is its primary means of capturing unsuspecting insects as they fly past.

FIGURE 5.17 **Intraspecific Competition** Whenever a needed resource is in limited supply, organisms compete for it. Intraspecific competition for sunlight among these pine trees has resulted in the tall, straight trunks. Those trees that did not grow fast enough died.

To succeed, predators employ several strategies. Some strong and speedy predators (cheetahs, lions, sharks) chase and overpower their prey; other species lie in wait and quickly strike prey that happen to come near them (many lizards and hawks); and some (spiders) use snares to help them catch prey. At the same time, prey species have many characteristics that help them avoid predation. Many have keen senses that allow them to detect predators, others are camouflaged so they are not conspicuous, and many can avoid detection by remaining motionless when predators are in the area.

An adaptation common to many prey species is a high reproductive rate compared to predators. For example, field mice may have 10 to 20 offspring per year, while hawks typically have two to three. Because of this high reproductive rate, prey species can endure a high mortality rate and still maintain a viable population. Certainly, the *individual* organism that is killed and eaten is harmed, but the prey *species* is not, since the prey individuals that die are likely to be the old, the slow, the sick, and the less well-adapted members of the population. The healthier, quicker, and

better-adapted individuals are more likely to survive. When these survivors reproduce, their offspring are more likely to have characteristics that help them survive; they are better adapted to their environment.

At the same time, a similar process is taking place in the predator population. Since poorly adapted individuals are less likely to capture prey, they are less likely to survive and reproduce. The predator and prey species are both participants in the natural selection process. This dynamic relationship between predator and prey species is a complex one that continues to intrigue ecologists.

COMPETITION

A second type of interaction between species is **competition,** in which two organisms strive to obtain the same limited resource. Ecologists distinguish two different kinds of competition. **Intraspecific competition** is competition between members of the same species. **Interspecific competition** is competition between members of different species. In either case organisms are harmed to some extent. However, this does not mean that there is no winner or loser. If two organisms are harmed to different extents, the one that is harmed less is the winner in the competition, and the one that is harmed more is the loser.

Examples of Intraspecific Competition

Lodgepole pine trees release their seeds following a fire. Thus, a large number of lodgepole pine seedlings begin growing close to one another and compete for water, minerals, and sunlight. None of the trees grows as rapidly as it could because its access to resources is restricted by the presence of the other trees. Eventually, because of differences in genetics or specific location, some of the pines will grow faster and will get a greater share of the resources. The taller trees will get more sunlight and the shorter trees will receive less. In time, some of the smaller trees die. They lost the competition. (See figure 5.17.)

Similarly, when two robins are competing for the same worm, only one gets it. Both organisms were harmed because they had to expend energy in fighting for the worm, but one got some food and was harmed less than the one that fought and got nothing. Other examples of intraspecific competition include corn plants in a field competing for water and nutrients, male elk competing with one another for the right to mate with the females, and wood ducks competing for the holes in dead trees to use for nesting sites.

Examples of Interspecific Competition

Many species of predators (hawks, owls, foxes, coyotes) may use the same prey species (mice, rabbits) as a food source. If the supply of food is inadequate, intense competition for food will occur and certain predator species may be more successful than others. (See figure 5.18.)

In grasslands, the same kind of competition for limited resources occurs. Rapidly growing, taller grasses get more of the water, minerals, and sunlight, while shorter species are less successful.

FIGURE 5.18 **Interspecific Competition** The lion and vultures are involved in interspecific competition for the carcass of the zebra. At this point, the lion is winning in the competition.

Often the shorter species are found to be more abundant when the taller species are removed by grazers, fire, or other activities.

Competition and Natural Selection

Competition among members of the same species is a major force in shaping the evolution of a species. When resources are limited, less well-adapted individuals are more likely to die or be denied mating privileges. Because the most successful organisms are likely to have larger numbers of offspring, each succeeding generation will contain more of the genetic characteristics that are favorable for survival of the species in that particular environment. Since individuals of the same species have similar needs, competition among them is usually very intense. A slight advantage on the part of one individual may mean the difference between survival and death.

As with intraspecific competition, one of the effects of interspecific competition is that the species that has the larger number of successful individuals emerges from the interaction better adapted to its environment than its less successful rivals.

The **competitive exclusion principle** is the concept that no two species can occupy the same ecological niche in the same place at the same time. The more similar two species are, the more intense will be the competition between them. If one of the two competing species is better adapted to live in the area than the other, the less-fit species must evolve into a slightly different niche, migrate to a different geographic area, or become extinct.

When the niche requirements of two similar species are examined closely, we usually find significant differences between the niches of the two species. The difference in niche requirements reduces the intensity of the competition between the two species. For example, many small forest birds eat insects. However, they may obtain them in different ways; a flycatcher sits on a branch and makes short flights to snatch insects from the air, a woodpecker excavates openings to obtain insects in rotting wood, and many warblers flit about in the foliage capturing insects.

Even among these categories there are specialists. Different species of warblers look in different parts of trees for their insect food. Because of this niche specialization, they do not directly compete with one another. (See figure 5.19).

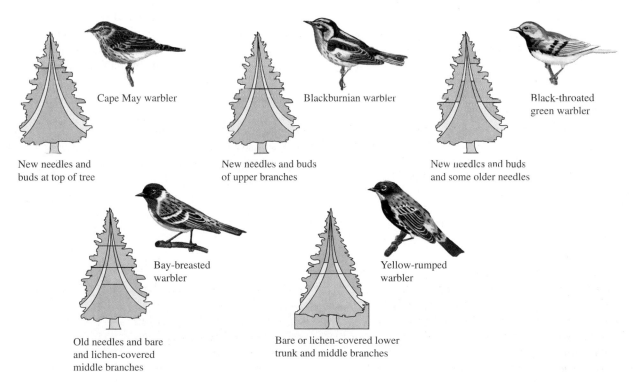

Cape May warbler

Blackburnian warbler

Black-throated green warbler

New needles and buds at top of tree

New needles and buds of upper branches

New needles and buds and some older needles

Bay-breasted warbler

Yellow-rumped warbler

Old needles and bare and lichen-covered middle branches

Bare or lichen-covered lower trunk and middle branches

FIGURE 5.19 **Niche Specialization** Although all of these warblers have similar feeding habits, the intensity of competition is reduced because they search for insects on different parts of the tree.

SYMBIOTIC RELATIONSHIPS

Symbiosis is a close, long-lasting, physical relationship between two different species. In other words, the two species are usually in physical contact and at least one of them derives some sort of benefit from this contact. There are three different categories of symbiotic relationships: parasitism, commensalism, and mutualism.

Parasitism

Parasitism is a relationship in which one organism, known as the **parasite,** lives in or on another organism, known as the **host,** from which it derives nourishment. Generally, the parasite is much smaller than the host. Although the host is harmed by the interaction, it is generally not killed immediately by the parasite, and some host individuals may live a long time and be relatively little affected by their parasites. Some parasites are much more destructive than others, however.

Newly established parasite-host relationships are likely to be more destructive than those that have a long evolutionary history. With a long-standing interaction between the parasite and the host, the two species generally evolve in such a way that they can accommodate one another. It is not in the parasite's best interest to kill its host. If it does, it must find another. Likewise, the host evolves defenses against the parasite, often reducing the harm done by the parasite to a level the host can tolerate.

Many parasites have complex life histories that involve two or more host species for different stages in the parasite's life cycle. Many worm parasites have their adult, reproductive stage in a carnivore (the definitive host), but they have an immature stage that reproduces asexually in another animal (the intermediate host) that the carnivore uses as food. Thus, a common dog tapeworm is found in its immature form in certain internal organs of rabbits.

Other parasite life cycles involve animals that carry the parasite from one host to another. These carriers are known as **vectors.** For example, many blood-feeding insects and mites can transmit parasites from one animal to another when they obtain blood meals from successive hosts. Malaria, Lyme disease, and sleeping sickness are transmitted by these kinds of vectors.

Parasites that live on the surface of their hosts are known as **ectoparasites.** Fleas, lice, mites, and some molds and mildews are examples of ectoparasites. Many other parasites, such as tapeworms, malaria parasites, many kinds of bacteria, and some fungi, are called **endoparasites** because they live inside the bodies of their hosts. A tapeworm lives in the intestines of its host, where it is able to resist being digested and makes use of the nutrients in the intestine. If a host has only one or two tapeworms, it can live for some time with little discomfort, supporting itself and its parasites. If the number of parasites is large, the host may die.

Even plants can be parasites. Mistletoe is a flowering plant that is parasitic on trees. It establishes itself on the surface of a tree when a bird transfers the seed to the tree. It then grows down into the water-conducting tissues of the tree and uses the water and minerals it obtains from these tissues to support its own growth.

Parasitism is a very common life strategy. If we were to categorize all the organisms in the world, we would find many more parasitic species than nonparasitic species. Each organism, including you, has many others that use it as a host. (See figure 5.20.)

Commensalism

Commensalism is a relationship between organisms in which one organism benefits while the other is not affected. It is possible to visualize a parasitic relationship evolving into a commensal one. Since parasites generally evolve to do as little harm to their host as possible and the host is combating the negative effects of the parasite, they might eventually evolve to the point where the host is not harmed at all.

Many examples of commensal relationships exist. Many orchids use trees as a surface upon which to grow. The tree is not harmed or helped, but the orchid needs a surface upon which to establish itself and also benefits by being close to the top of the tree,

Varroa mite (ectoparasite)

Tapeworm (endoparasite)

Mistletoe

FIGURE 5.20 **Parasitism** Varroa mites are ectoparasites on honeybees. They suck the body fluids of bees and severely weaken them. Tapeworms are endoparasites that live inside the intestines of their hosts, where they absorb food from their hosts' intestines. Mistletoe is a plant that is parasitic on other plants.

(a)

FIGURE 5.21 **Commensalism** Remoras hitch a ride on sharks and feed on the scraps of food lost by the sharks. This is a benefit to the remoras. The sharks do not appear to be affected by the presence of the remoras.

where it can get more sunlight and rain. Some mosses, ferns, and many vines also make use of the surfaces of trees in this way.

In the ocean, many sharks have a smaller fish known as a remora attached to them. Remoras have a sucker on the top of their heads that they can use to attach to the shark. In this way, they can hitch a ride as the shark swims along. When the shark feeds, the remora frees itself and obtains small bits of food that the shark misses. Then, the remora reattaches. The shark does not appear to be positively or negatively affected by remoras. (See figure 5.21.)

Many commensal relationships are rather opportunistic and may not involve long-term physical contact. For example, many birds rely on trees of many different species for places to build their nests but do not use the same tree year after year. Similarly, in the spring, bumblebees typically build nests in underground mouse nests that are no longer in use.

Mutualism

Mutualism is another kind of symbiotic relationship and is actually beneficial to both species involved. In many mutualistic relationships, the relationship is obligatory; the species cannot live without each other. In others, the species can exist separately but are more successful when they are involved in a mutualistic relationship. Some species of *Acacia,* a thorny tree, provide food in the form of sugar solutions in little structures on their stems. Certain species of ants feed on the solutions and live in the tree, which they will protect from other animals by attacking any animal that begins to feed on the tree. Both organisms benefit; the ants receive food and a place to live, and the tree is protected from animals that would use it as food.

One soil nutrient that is usually a limiting factor for plant growth is nitrogen. Many kinds of plants, such as legumes (beans, clover, and acacia trees) and alder trees, have bacteria that live in their roots in little nodules. The roots form these nodules when they are infected with certain kinds of bacteria. The bacteria do not cause

(b)

FIGURE 5.22 **Mutualism** (a) The growths on the roots of this plant contain beneficial bacteria that make nitrogen available to the plant. The relationship is also beneficial to the bacteria, since the bacteria obtain necessary raw materials from the plant. It is a mutually beneficial relationship. (b) The oxpecker (bird) and the impala have a mutualistic relationship. The oxpecker obtains food by removing parasites from the surface of the impala and the impala benefits from a reduced parasite load.

disease but provide the plants with nitrogen-containing molecules that the plants can use for growth. The nitrogen-fixing bacteria benefit from the living site and nutrients that the plants provide, and the plants benefit from the nitrogen they receive. (See figure 5.22.)

Similarly, many kinds of fungi form an association with the roots of plants. The root-fungus associations are called **mycorrhizae.** The fungus obtains organic molecules from the roots of the plant, and the branched nature of the fungus assists the plant in obtaining nutrients such as phosphates and nitrates. In many cases, it is clear that the relationship is obligatory.

SOME RELATIONSHIPS ARE DIFFICULT TO CATEGORIZE

Sometimes it is not easy to categorize the relationships that organisms have with each other. For example, it is not always easy to say whether a relationship is a predator-prey relationship or a host-parasite relationship. How would you classify a mosquito or a tick? Both of these animals require blood meals to live and reproduce. They don't kill and eat their prey. Neither do they live in or on a host for a long period of time. This question points out the difficulty encountered when we try to place all kinds of organism interactions into a few categories. However, we can eliminate this problem if we call them temporary parasites or blood predators.

Another relationship that doesn't fit well is the relationship that certain birds such as cowbirds and European cuckoos have with other birds. Cowbirds and European cuckoos do not build nests but lay their eggs in the nests of other species of birds, who are left to care for a foster nestling at the expense of their own nestlings, who generally die. This situation is usually called nest parasitism or brood parasitism. (See figure 5.23.)

What about grazing animals? Are they predators or parasites on the plants that they eat? Sometimes they kill the plant they eat, while at other times they simply remove part of the plant and the rest continues to grow. In either case, the plant has been harmed by the interaction and the grazer has benefited.

There are also mutualistic relationships that do not require permanent contact between the participants in the relationship. Bees and the flowering plants they pollinate both benefit from their interactions. The bees obtain pollen and nectar for food and the plants are pollinated. But the active part of the relationship involves only a part of the life of any plant, and the bees are not restricted to any one species of plant for the food. They must actually switch to different flowers at different times of the year.

COMMUNITY AND ECOSYSTEM INTERACTIONS

Thus far, we have discussed specific ways in which individual organisms interact with one another and with their physical surroundings. However, often it is useful to look at ecological relationships from a broader perspective. Two concepts that focus on relationships that involve many different kinds of interactions are community and ecosystem.

A **community** is an assemblage of all the interacting populations of different species of organisms in an area. Some species play minor roles, while others play major roles, but all are part of the community. For example, the grasses of the prairie have a major role, since they carry on photosynthesis and provide food and shelter for the animals that live in the area. Grasshoppers, prairie dogs, and bison are important consumers of grass. Meadowlarks consume many kinds of insects, and though they are a conspicuous and colorful part of the prairie scene, they have a relatively minor role and little to do with maintaining a prairie community. Bacteria and fungi in the soil break down the bodies of dead plants and animals and provide nutrients to plants. Communities consist of interacting populations of different species, but these species interact with their physical world as well.

An **ecosystem** is a defined space in which interactions take place between a community, with all its complex interrelationships, and the physical environment. The physical world has a major impact on what kinds of plants and animals can live in an area. We do not expect to see a banana tree in the Arctic or a walrus in the Mississippi River. Banana trees are adapted to warm, moist, tropical areas, and walruses require cold ocean waters. Some ecosystems, such as grasslands and certain kinds of forests, are shaped by periodic fires. The kind of soil and the amount of moisture also influence the kinds of organisms found in an area.

While it is easy to see that the physical environment places limitations on the kinds of organisms that can live in an area, it is also important to recognize that organisms impact their physical surroundings. Trees break the force of the wind, grazing animals form paths, and earthworms create holes that aerate the soil. While the concepts of community and ecosystem are closely related, an ecosystem is a broader concept because it involves physical as well as biological processes.

Every system has parts that are related to one another in specific ways. A bicycle has wheels, a frame, handlebars, brakes, pedals, and a seat. These parts must be organized in a certain way or the system known as a bicycle will not function. Similarly, ecosystems have parts that must be organized in specific ways or the systems will not operate. To more fully develop the concept of ecosystem, we will look at ecosystems from three points of view: the major roles played by organisms, the way energy is utilized within ecosystems, and the way atoms are cycled from one organism to another.

FIGURE 5.23 **Nest (Brood) Parasitism** This red-eyed vireo is feeding the nestling of a brown-headed cowbird. A female cowbird laid its egg in the vireo's nest. The vireo is harmed because it is not raising its own young, and the cowbird benefits because it did not need to expend energy to build and defend a nest or collect food for its own young.

MAJOR ROLES OF ORGANISMS IN ECOSYSTEMS

Ecologists have traditionally divided organisms' roles in ecosystems into three broad categories: producers, consumers, and decomposers.

Producers

Producers are organisms that are able to use sources of energy to make complex, organic molecules from the simple inorganic substances in their environment. In nearly all ecosystems, energy is supplied by the sun, and organisms such as plants, algae, and tiny aquatic organisms called phytoplankton use light energy to carry on photosynthesis. Since producers are the only organisms in an ecosystem that can trap energy and make new organic material from inorganic material, all other organisms rely on producers as a source of food, either directly or indirectly.

Consumers

Consumers are organisms that require organic matter as a source of food. They consume organic matter to provide themselves with energy and the organic molecules necessary to build their own bodies. An important part of their role is the process of respiration in which they break down organic matter to inorganic matter.

However, consumers can be further subdivided into categories based on the kinds of things they eat and the way they obtain food.

Primary consumers, also known as **herbivores,** are animals that eat producers (plants or phytoplankton) as a source of food. Herbivores, such as leaf-eating insects and seed-eating birds, are usually quite numerous in ecosystems, where they serve as food for the next organisms in the chain.

Secondary consumers or **carnivores** are animals that eat other animals. Secondary consumers can be further subdivided into categories based on what kind of prey they capture and eat. Some carnivores, such as ladybird beetles, primarily eat herbivores, such as aphids; others, such as eagles, primarily eat fish that are themselves carnivores. While these are interesting conceptual distinctions, most carnivores will eat any animal they can capture and kill.

In addition, many animals, called **omnivores,** include both plants and animals in their diet. Even animals that are considered to be carnivores (foxes, bears) regularly include large amounts of plant material in their diets. Conversely, animals often thought of as herbivores (mice, squirrels, seed-eating birds) regularly consume animals as a source of food. Parasites are also consumers that have a special way of obtaining their food.

Decomposers

Decomposers are organisms that use nonliving organic matter as a source of energy and raw materials to build their bodies. Whenever an organism sheds a part of itself, excretes waste products, or dies, it provides a source of food for decomposers. Since decomposers carry on respiration, they are extremely important in recycling matter by converting organic matter to inorganic material. Many small animals, fungi, and bacteria fill this niche. (See table 5.1.)

KEYSTONE SPECIES

Ecosystems typically consist of many different species interacting with each other and their physical surroundings. However, some species have more central roles than others. In recognition of this idea, ecologists have developed the concept of keystone species.

A **keystone species** is one that has a critical role to play in the maintenance of specific ecosystems. In prairie ecosystems, grazing animals are extremely important in maintaining the mix of species typical of a grassland. Without the many influences of the grazers, the nature of the prairie changes. A study of the American tallgrass prairie indicated that when bison are present, they increase the biodiversity of the site. Bison typically eat grasses and, therefore, allow smaller plant species that would normally be shaded by tall grasses to be successful. In ungrazed plots, the tall grasses become the dominant vegetation and biodiversity decreases. Bison also dig depressions in the soil, called wallows, to provide themselves with dust or mud with which they can coat themselves. These wallows retain many species of plants that typically live in disturbed areas. Bison urine has also been shown to be an important source of nitrogen for the plants.

The activities of bison even affect the extent and impact of fire, another important feature of grassland ecosystems. Since bison prefer to feed on recently burned sites and revisit these sites several times throughout the year, they tend to create a patchwork

TABLE 5.1 Roles in an Ecosystem

Category	Major Role or Action	Examples
Producer	Converts simple inorganic molecules into organic molecules by the process of photosynthesis	Trees, flowers, grasses, ferns, mosses, algae
Consumer	Uses organic matter as a source of food	Animals, fungi, bacteria
Herbivore	Eats plants directly	Grasshopper, elk, human vegetarian
Carnivore	Kills and eats animals	Wolf, pike, dragonfly
Omnivore	Eats both plants and animals	Rats, raccoons, most humans
Scavenger	Eats meat but often gets it from animals that died by accident or illness, or were killed by other animals	Coyote, vulture, blowflies
Parasite	Lives in or on another living organism and gets food from it	Tapeworm, many bacteria, some insects
Decomposer	Returns organic material to inorganic material; completes recycling of atoms	Fungi, bacteria, some insects and worms

FIGURE 5.24 **A Keystone Species** Bison are a keystone species in prairie ecosystems. They affect the kinds of plants that live in an area, provide nutrients in the form of urine and manure, and produce wallows that provide special habitats for certain plants and animals.

of grazed and ungrazed areas. The grazed areas are less likely to be able to sustain a fire, and fires likely will be more prevalent in ungrazed patches. (See figure 5.24.)

The concept of keystone species has also been studied in marine ecosystems. The relationship among sea urchins, sea otters, and kelp forests suggests that sea otters are a keystone species. Sea otters eat sea urchins, which eat kelp. A reduction in the number of otters results in an increase in the number of sea urchins. Increased numbers of sea urchins lead to heavy grazing of the kelp by sea urchins. When the amount of kelp is severely reduced, fish and many other animals that live within the kelp beds lose their habitat and biodiversity is significantly reduced.

The concept of keystone species is useful to ecologists and resource managers because it helps them to realize that all species cannot be treated equally. Some species have pivotal roles, and their elimination or severe reduction can significantly alter ecosystems. In some cases, the loss of a keystone species can result in the permanent modification of an ecosystem into something considerably different from the original mix of species.

ENERGY FLOW THROUGH ECOSYSTEMS

An ecosystem is a stable, self-regulating unit. This does not mean that an ecosystem is unchanging. The organisms within it are growing, reproducing, dying, and decaying. In addition, an ecosystem must

have a continuous input of energy to retain its stability. The only significant source of energy for most ecosystems is sunlight. Producers are the only organisms that are capable of trapping solar energy through the process of photosynthesis and making it available to the ecosystem. The energy is stored in the form of chemical bonds in large organic molecules such as carbohydrates (sugars, starches), fats, and proteins. The energy stored in the molecules of producers is transferred to other organisms when the producers are eaten.

Trophic Levels

Each step in the flow of energy through an ecosystem is known as a **trophic level.** Producers (plants, algae, phytoplankton) constitute the first trophic level, and herbivores constitute the second trophic level. Carnivores that eat herbivores are the third trophic level, and carnivores that eat other carnivores are the fourth trophic level. Omnivores, parasites, and scavengers occupy different trophic levels, depending on what they happen to be eating at the time. If we eat a piece of steak, we are at the third trophic level; if we eat celery, we are at the second trophic level. (See figure 5.25.)

Energy Relationships

The second law of thermodynamics states that whenever energy is converted from one form to another, some of the energy is converted to a nonuseful form (typically, low-quality heat). Thus, there is always less useful energy following an energy conversion. When energy passes from one trophic level to the next, there is less useful energy left with each successive trophic level. This loss

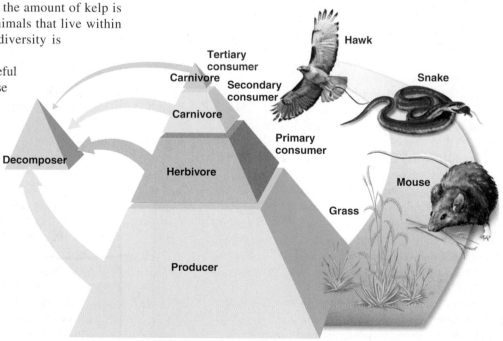

FIGURE 5.25 **Categories of Organisms Within an Ecosystem** Organisms within ecosystems can be placed in several broad categories based on how they obtain the energy they need for their survival.

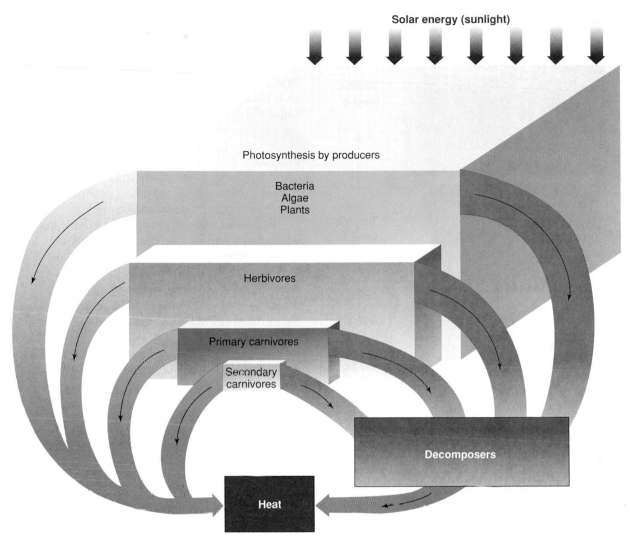

Solar energy (sunlight)

Photosynthesis by producers

Bacteria
Algae
Plants

Herbivores

Primary carnivores

Secondary carnivores

Decomposers

Heat

FIGURE 5.26 **Energy Flow Through Ecosystems** As energy flows through an ecosystem, it passes through several levels known as trophic levels. Each trophic level contains a certain amount of energy. Each time energy flows to another trophic level, approximately 90 percent of the useful energy is lost, usually as heat to the surroundings. Therefore, in most ecosystems, higher trophic levels contain less energy and fewer organisms.

of low-quality heat is dissipated to the surroundings and warms the air, water, or soil. In addition to this loss of heat, organisms must expend energy to maintain their own life processes. It takes energy to chew food, defend nests, walk to waterholes, or produce and raise offspring. Therefore, the amount of energy contained in higher trophic levels is considerably less than that at lower levels. Approximately 90 percent of the useful energy is lost with each transfer to the next highest trophic level. So in any ecosystem, the amount of energy contained in the herbivore trophic level is only about 10 percent of the energy contained in the producer trophic level. The amount of energy at the third trophic level is approximately 1 percent of that found in the first trophic level. (See figure 5.26.)

Because it is difficult to actually measure the amount of energy contained in each trophic level, ecologists often use other measures to approximate the relationship between the amounts of energy at each level. One of these is the biomass. The **biomass** is the weight of living material in a trophic level. It is often possible in a simple ecosystem to collect and weigh all the producers, herbivores, and

carnivores. The weights often show the same 90 percent loss from one trophic level to the next as happens with the amount of energy.

FOOD CHAINS AND FOOD WEBS

A **food chain** is a series of organisms occupying different trophic levels through which energy passes as a result of one organism consuming another. For example, willow trees grow well in very moist soil, perhaps near a pond. The trees' leaves capture sunlight and convert carbon dioxide and water into sugars and other organic molecules. The leaves serve as a food source for insects, such as caterpillars, grasshoppers, and leaf beetles, that have chewing mouthparts and a digestive system adapted to plant food. Some of these insects are eaten by spiders, which fall from the trees into the pond below, where they are consumed by a frog. As the frog swims from one lily pad to another, a large bass consumes the frog. A human may use an artificial frog as a lure to entice the bass from its hiding place. A fish dinner is the final step in this chain of events that began with the leaves of a willow tree. (See figure 5.27.) This food chain has six trophic

BIOMES ARE DETERMINED BY CLIMATE

Biomes are terrestrial climax communities with wide geographic distribution. (See figure 6.7.) Although the concept of biomes is useful for discussing general patterns and processes, it is important to recognize that when different communities within a particular biome are examined, they will show differences in the exact species present. However, in broad terms, the general structure of the ecosystem and the kinds of niches and habitats present are similar.

PRECIPITATION AND TEMPERATURE

Patterns of precipitation and temperature are two primary nonbiological factors that have major impacts on the kind of climax community that develops in any part of the world. Several aspects of precipitation are important: the total amount of precipitation per year, the form in which it arrives (rain, snow, sleet), and its seasonal distribution. Precipitation may be evenly spaced throughout the year, or it may be concentrated at particular times so that there are wet and dry seasons.

The temperature patterns are also important and can vary considerably in different parts of the world. Tropical areas have warm, relatively unchanging temperatures throughout the year. Areas near the poles have long winters with extremely cold temperatures and relatively short, cool summers. Other areas are more evenly divided between cold and warm periods of the year. (See figure 6.8.)

Although temperature and precipitation are of primary importance, several other factors may influence the kind of climax community present. Periodic fires are important in maintaining some grassland and shrub climax communities because the fires prevent the establishment of larger, woody species. Some parts of the world have frequent, strong winds that prevent the establishment of trees and cause rapid drying of the soil. The type of soil present is also very important. Sandy soils tend to dry out quickly and may not allow the establishment of more water-demanding species such as trees, while extremely wet soils

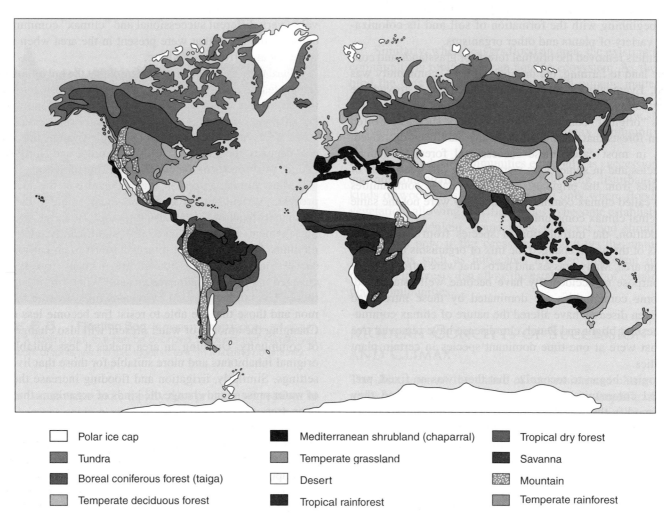

☐ Polar ice cap	■ Mediterranean shrubland (chaparral)	■ Tropical dry forest
■ Tundra	■ Temperate grassland	■ Savanna
■ Boreal coniferous forest (taiga)	☐ Desert	▦ Mountain
■ Temperate deciduous forest	■ Tropical rainforest	■ Temperate rainforest

FIGURE 6.7 **Biomes of the World** Although most biomes are named for a major type of vegetation, each also includes a specialized group of animals adapted to the plants and the biome's climatic conditions.

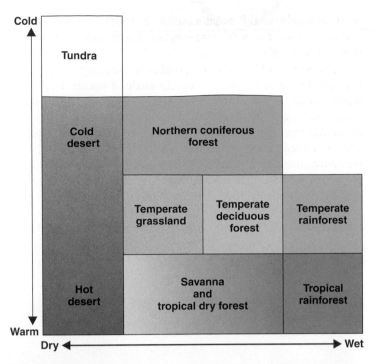

FIGURE 6.8 **Influence of Precipitation and Temperature on Vegetation** Temperature and moisture are two major factors that influence the kind of vegetation that can occur in an area. Areas with low moisture and low temperatures produce tundra; areas with high moisture and freezing temperatures during part of the year produce deciduous or coniferous forests; dry areas produce deserts; moderate amounts of rainfall or seasonal rainfall support temperate grasslands or savannas; and areas with high rainfall and high temperatures support tropical rainforests.

may allow only certain species of trees to grow. Obviously, the kinds of organisms currently living in the area are also important, since their offspring will be the ones available to colonize a new area.

THE EFFECT OF ELEVATION ON CLIMATE AND VEGETATION

The distribution of terrestrial ecosystems is primarily related to precipitation and temperature. The temperature is warmest near the equator and becomes cooler toward the poles. Similarly, as the height above sea level increases, the average temperature decreases. This means that even at the equator, it is possible to have cold temperatures on the peaks of tall mountains. As one proceeds from sea level to the tops of mountains, it is possible to pass through a series of biomes that is similar to what would be encountered as one traveled from the equator to the North Pole. (See figure 6.9.)

MAJOR BIOMES OF THE WORLD

In the next sections, we will look at the major biomes of the world and highlight the abiotic and biotic features typical of each biome.

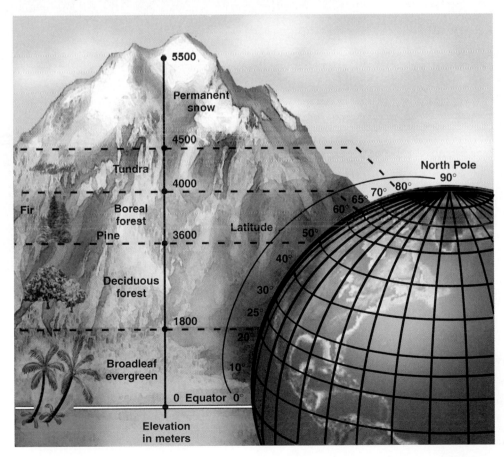

FIGURE 6.9 **Relationship Among Height above Sea Level, Latitude, and Vegetation** As one travels up a mountain, the climate changes. The higher the elevation, the cooler the climate. Even in the tropics, tall mountains can have snow on the top. Thus, it is possible to experience the same change in vegetation by traveling up a mountain as one would experience traveling from the equator to the North Pole.

DESERT

Deserts are found throughout the world.

Climate

A lack of water is the primary factor that determines that an area will be a desert. **Deserts** are areas that generally average less than 25 centimeters (10 inches) of precipitation per year. (See figure 6.10.) When and how precipitation arrives is quite variable in different deserts. Some deserts receive most of the moisture as snow or rain in the winter months, while in others rain comes in the form of thundershowers at infrequent intervals. If rain comes as heavy thundershowers, much of the water does not sink into the ground but runs off into gullies. Also, since the rate of evaporation is high,

plant growth and flowering usually coincide with the periods when moisture is available. Deserts are also likely to be windy.

We often think of deserts as hot, dry wastelands devoid of life. However, many deserts are quite cool during a major part of the year. Certainly, the Sahara Desert and the deserts of the southwestern United States and Mexico are hot during much of the year, but the desert areas of the northwestern United States and the Gobi Desert in Central Asia can be extremely cold during winter months and have relatively cool summers. Furthermore, the temperature can vary greatly during a 24-hour period. Since deserts receive little rainfall, it is logical that most will have infrequent cloud cover. With no clouds to block out the sun, during the day the soil surface and the air above it tend to heat up rapidly. After the sun has set, the absence of an insulating

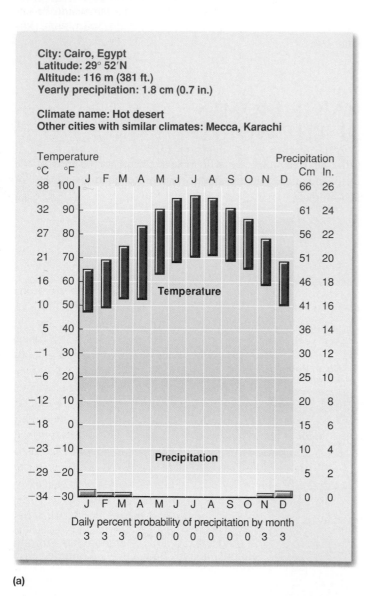

(a)

(b) Desert landscape

(c) Coyote

(d) Collared lizard

FIGURE 6.10 **Desert** (a) Climagraph for Cairo, Egypt. (b) The desert receives less than 25 centimeters (10 inches) of precipitation per year, yet it teems with life. Cactus, sagebrush, lichens, snakes, small mammals, birds, and insects inhabit the desert. (c) Coyotes are common in North American deserts. (d) Collared lizards are common reptiles in many deserts of the United States. Because daytime temperatures are often high, most animals are active only at night, when the air temperature drops significantly. Cool deserts also exist in many parts of the world, where rainfall is low but temperatures are not high.

There are over 1600 private organizations in the United States that are involved in conservation of land. Some are small, single-purpose organizations that protect a small parcel of land with special conservation value. On the other hand, The Nature Conservancy is an international organization that has protected millions of acres.

People often develop an attachment to their land and wish to see it preserved even after they have died. People may have a long family history of using the land for farming or ranching and want to see that use continue. Others may recognize that their land has special conservation value because of its geology, scenic value, or biodiversity and wish to see it protected for the public good. Others may simply have a purely emotional reason for wanting to preserve their land. One of the tools used by land conservation organizations is a legal tool known as a *conservation easement.*

A conservation easement is a legally binding agreement placed on a piece of privately held land that limits the future use

or development even when the land is passed to heirs or sold. For example, a conservation easement may prohibit the subdividing of a piece of land or restrict buildings to a specific portion of the property, or an easement may specify that the public must have access to view significant biological or geological features. Alternatively, the easement may restrict access to protect endangered species or archeological sites.

Regardless of their motivation, when people enter into a conservation easement they give up something. In some cases, people donate a conservation easement and receive no financial benefit. In other cases, they may sell a conservation easement to an organization that agrees to provide stewardship of the property into the future. In nearly all cases the placement of a conservation easement on property diminishes its economic value, since its future use is restricted. Yet, thousands of people have entered into such arrangements. As of 2005, in the United States, over 6 million acres of land (an area about the size of Vermont) had been protected by conservation easements.

layer of clouds allows heat energy to be reradiated from the Earth, and the area cools off rapidly. Cool to cold nights are typical even in "hot" deserts, especially during the winter months.

Organisms

Another misconception about deserts is that few species of organisms live in the desert. There are many species, but they typically have low numbers of individuals. For example, a conspicuous feature of deserts is the dispersed nature of the plants. There is a significant amount of space between them. Similarly, animals do not have large, dense populations. However, those species that are present are specially adapted to survive in dry, often hot environments. For example, water evaporates from the surfaces of leaves. As an adaptation to this condition, many desert plants have very small leaves that allow them to conserve water. Some even lose their leaves entirely during the driest part of the year. Some, such as cactus, have the ability to store water in their spongy bodies or their roots for use during drier periods. Other plants have parts or seeds that lie dormant until the rains come. Then they germinate, grow rapidly, reproduce, and die, or become dormant until the next rains. Even the perennial plants are tied to the infrequent rains. During these times, the plants are most likely to produce flowers

and reproduce. Many desert plants are spiny. The spines discourage large animals from eating the leaves and young twigs.

The desert has many kinds of animals. However, they are often overlooked because their populations are low, numerous species are of small size, and many are inactive during the hot part of the day. They also aren't seen in large, conspicuous groups. Many insects, lizards, snakes, small mammals, grazing mammals, carnivorous mammals, and birds are common in desert areas. All of the animals that live in deserts are able to survive with a minimal amount of water. Some receive nearly all of their water from the moisture in the food they eat. They generally have an outer skin or cuticle that resists water loss, so they lose little water by evaporation. They also have physiological adaptations, such as extremely efficient kidneys, that allow them to retain water. They often limit their activities to the cooler part of the day (the evening), and small mammals may spend considerable amounts of time in underground burrows during the day, which allows them to avoid extreme temperatures and to conserve water.

Human Impact

Throughout history, deserts have been regions where humans have had little impact. The harshness of the climate does not allow for agriculture. Therefore, hunter-gatherer societies were the most common ones associated with deserts. Some deserts support nomadic herding in which herders move their livestock to find patches of vegetation for grazing. Modern technology allows for the transport of water to the desert. This has resulted in the development of cities in some desert areas and some limited agriculture as a result of irrigation.

TEMPERATE GRASSLAND

Temperate grasslands, also known as **prairies** or **steppes,** are widely distributed over temperate parts of the world.

Climate

As with deserts, the major factor that contributes to the establishment of a temperate grassland is the amount of available moisture. Grasslands generally receive between 25 and 75 centimeters (10 to 30 inches) of precipitation per year. These areas are windy with hot summers and cold-to-mild winters. In many grasslands, fire is an important force in preventing the invasion of trees and releasing nutrients from dead plants to the soil.

Organisms

Grasses make up 60 to 90 percent of the vegetation. Many other kinds of flowering plants are interspersed with the grasses. (See figure 6.11.) Typically, the grasses and other plants are very

City: Tehran, Iran
Latitude: 35° 41′N
Altitude: 1220 m (4002 ft.)
Yearly precipitation: 26 cm (10.1 in.)

Climate name: Midaltitude dryland
Other cities with similar climates: Salt Lake City, Ankara

Daily percent probability of precipitation by month
13 14 16 10 6 3 3 0 0 3 10 13

(a)

(b) Prairie landscape

(c) Pronghorn

(d) Grasshopper

(e) Meadowlark

FIGURE 6.11 **Temperate Grassland** (a) Climagraph for Tehran, Iran. (b) Grasses are better able to withstand low water levels than are trees. Therefore, in areas that have moderate rainfall, grasses are the dominant plants. (c & d) Pronghorns and grasshoppers are common herbivores and (e) meadowlarks are common consumers of insects in North American grasslands.

THE BLUE OAK RANCH RESERVE OF THE UNIVERSITY OF CALIFORNIA—BERKELEY

The Blue Oak Ranch Reserve is located in the Mount Hamilton Range near San Jose, California. The 3260-acre (1300 hectare) reserve is protected by a conservation easement held by The Nature Conservancy and is the newest of 36 California reserves overseen by the University of California's Natural Reserve System. The reserve has special research and education value because it has remained undeveloped for its entire history.

The reserve helps protect many species of plants and wildlife and provides many research opportunities. There are more than 430 species of plants, 130 species of birds, 41 species of mammals, 14 species of reptiles, 7 species of amphibians—including the rare Foothill yellow-legged frogs and the endangered California tiger salamander—7 species of fish, and hundreds of species of invertebrates that make use of the reserve. The reserve contains mature blue oaks, valley oaks, and two species of live oak. The reserve has many mature oaks, but few oak seedlings. One avenue of research would be to try to find out why these oak woodlands are not regenerating.

Because the reserve is part of a much larger area of protected land, it is an important link for migratory wildlife. Thus, its location provides opportunities for major research projects that address large-scale conservation and land management issues of regional, state, and national concern.

The reserve's location near a fast-growing urban area also makes it ideal for studying issues like nitrogen deposition and loss of biodiversity that are particularly acute at the interface between wild land and urban development.

---

close together, and their roots form a network that binds the soil together. Trees, which generally require greater amounts of water, are rare in these areas except along watercourses.

The primary consumers are animals that eat the grasses, such as large herds of migratory, grazing mammals such as bison, wildebeests, wild horses, and various kinds of sheep, cattle, and goats. While the grazers are important as consumers of the grasses, they also supply fertilizer from their dung and discourage invasion by woody species of plants because they eat the young shoots.

In addition to grazing mammals, many kinds of insects, including grasshoppers and other herbivorous insects, dung beetles (which feed on the dung of grazing animals), and several kinds of flies are common. Some of these flies bite to obtain blood. Others lay their eggs in the dung of large mammals. Some feed on dead animals and lay their eggs in carcasses. Small herbivorous mammals, such as mice and ground squirrels, are also common. Birds are often associated with grazing mammals. They eat the insects stirred up by the mammals or feed on the insects that bite them. Other birds feed on seeds and other plant parts. Reptiles (snakes and lizards) and other carnivores such as coyotes, foxes, and hawks feed on small mammals and insects.

Human Impact

Most of the moist grasslands of the world have been converted to agriculture, since the rich, deep soil that developed as a result of the activities of centuries of soil building is useful for growing cultivated grasses such as corn (maize) and wheat. The drier grasslands have been converted to the raising of domesticated grazers such as cattle, sheep, and goats. Therefore, little undisturbed grassland is left, and those fragments that remain need to be preserved as refuges for the grassland species that once occupied huge portions of the globe.

SAVANNA

Savannas are found in tropical parts of Africa, South America, and Australia and are characterized by extensive grasslands spotted with occasional trees or patches of trees. (See figure 6.12.) Although savannas receive 50 to 150 centimeters (20 to 60 inches) of rain per year, the rain is not distributed evenly throughout the year. Typically, a period of heavy rainfall is followed by a prolonged drought. This results in a very seasonally structured ecosystem.

Organisms

The plants and animals time their reproductive activities to coincide with the rainy period, when limiting factors are less severe. The predominant plants are grasses, but many drought-resistant, flat-topped, thorny trees are common. As with grasslands, fire is a common feature of the savanna, and the trees present are resistant to fire damage. Many of these trees are particularly important because they are legumes that are involved in nitrogen fixation. They also provide shade and nesting sites for animals. As with grasslands, the predominant mammals are the grazers. Wallabies in Australia, wildebeests, zebras, elephants, and various species of antelope in Africa, and capybaras (rodents) in South America are examples. In Africa, the large herds of grazing animals provide food for many different kinds of large carnivores (lions, hyenas, leopards). Many kinds of rodents, birds, insects, and reptiles are associated with this biome. Among the insects, mound-building termites are particularly common.

City: Nairobi, Kenya
Latitude: 1° 18′S
Altitude: 1,661 m (5,450 ft.)
Yearly precipitation: 96 cm (38 in.)

Climate name: Savanna
Other cities with similar climates: Dakar, Kampala

Daily percent probability of precipitation by month
17 20 37 53 57 30 20 23 20 27 50 37

(a)

(b) Savanna landscape

(c) Lion

(d) Secretary birds

(e) Thomson's gazelles

FIGURE 6.12 **Savanna** (a) Climagraph for Nairobi, Kenya. (b) Savannas develop in tropical areas that have seasonal rainfall. They typically have grasses as the dominant vegetation with drought- and fire-resistant trees scattered through the area. Grazing animals such as elephants and gazelles (b, e) are common herbivores and lions and secretary birds (c, d) are common carnivores in African savannas.

Human Impact

Savannas have been heavily impacted by agriculture. Farming is possible in the more moist regions, and the drier regions are used for the raising of livestock. Because of the long periods of drought, the raising of crops is often difficult without irrigation. Some areas support nomadic herding. In Africa there are extensive areas set aside as parks and natural areas and ecotourism is an important source of income. However, there is a constant struggle between the people who want to use the land for agriculture or grazing and those who want to preserve it in a more natural state.

MEDITERRANEAN SHRUBLANDS (CHAPARRAL)

The **Mediterranean shrublands** are located near oceans and are dominated by shrubby plants.

Climate

Mediterranean shrublands have a climate with wet, cool winters and hot, dry summers. Rainfall is 40 to 100 centimeters (15 to 40 inches) per year. As the name implies, this biome is typical of the Mediterranean coast and is also found in coastal southern

California, the southern tip of Africa, a portion of the west coast of Chile, and southern Australia.

Organisms

The vegetation is dominated by woody shrubs that are adapted to withstand the hot, dry summer. (See figure 6.13.) Often the plants are dormant during the summer. Fire is a common feature of this biome, and the shrubs are adapted to withstand occasional fires. The kinds of animals vary widely in the different regions of the world with this biome. Many kinds of insects, reptiles, birds, and mammals are found in these areas. In the chaparral of California, rattlesnakes, spiders, coyotes, lizards, and rodents are typical inhabitants.

Human Impact

Very little undisturbed Mediterranean shrubland still exists. The combination of moderate climate and closeness to the ocean has resulted in all Mediterranean shrublands being heavily altered by human activity. Agriculture is common, often with the aid of irrigation, and many major cities are located in this biome.

TROPICAL DRY FOREST

The **tropical dry forest** is another biome that is heavily influenced by seasonal rainfall. Tropical dry forests are found in parts of Central and South America, Australia, Africa, and Asia (particularly India and Myanmar).

Climate

Many of the tropical dry forests have a monsoon climate in which several months of heavy rainfall are followed by extensive dry periods ranging from a few to as many as eight months. (See figure 6.14.) The rainfall may be as low as 50 centimeters (20 inches) or as high as 200 centimeters (80 inches).

City: Rome, Italy
Latitude: 41° 48′N
Altitude: 115 m (377 ft.)
Yearly precipitation: 85 cm (33.3 in.)

Climate name: Mediterranean
Other cities with similar climates:
    Athens, Los Angeles, Valparaiso

Daily percent probability of precipitation by month
26 39 16 20 19 7 6 10 20 29 27 29

(a)

(b) Chaparral landscape

(c) California quail

(d) Black-tailed jack rabbit

FIGURE 6.13 **Mediterranean Shrubland** (a) Climagraph for Rome, Italy. (b) Mediterranean shrublands are characterized by a period of winter rains and a dry, hot summer. The dominant plants are drought-resistant, woody shrubs. (c,d) Common animals in the Mediterranean shrubland (chaparral) of California are the California quail and black-tailed jack rabbit.

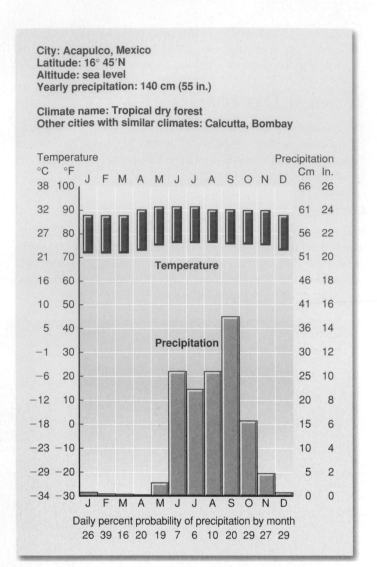

City: Acapulco, Mexico
Latitude: 16° 45'N
Altitude: sea level
Yearly precipitation: 140 cm (55 in.)

Climate name: Tropical dry forest
Other cities with similar climates: Calcutta, Bombay

Temperature
°C   °F
Precipitation
Cm   In.

Temperature

Precipitation

Daily percent probability of precipitation by month
26 39 16 20 19 7 6 10 20 29 27 29

(a)

(b) Tropical dry forest landscape

(c) White-faced coati in Costa Rica

(d) Rhinoceros in Indian tropical dry forest

FIGURE 6.14   Tropical Dry Forest (a) Climagraph for Acapulco, Mexico. (b) Tropical dry forests typically have a period of several months with no rain. In places where the drought is long, many of the larger trees lose their leaves. The coati (c) is a common animal in the tropical dry forests of the Americas. The endangered one-horned rhinoceros (d) is an inhabitant of the tropical dry forest of Asia.

Organisms

Since the rainfall is highly seasonal, many of the plants have special adaptations for enduring drought. In many of the regions that have extensive dry periods, many of the trees drop their leaves during the dry period. Many of the species of animals found here are also found in more moist tropical forests of the region. However, there are fewer kinds in dry forests than in rainforests.

Human Impacts

Many of these forests occur in areas of very high human population. Therefore, the harvesting of wood for fuel and building materials has heavily affected these forests. In addition, many of these forests have been converted to farming or the grazing of animals.

TROPICAL RAINFOREST

Tropical rainforests are located near the equator in Central and South America, Africa, Southeast Asia, and some islands in the Caribbean Sea and Pacific Ocean. (See figure 6.15.)

Climate

The temperature is normally warm and relatively constant. There is no frost, and it rains nearly every day. Most areas receive in excess of 200 centimeters (80 inches) of rain per year. Some receive 500 centimeters (200 inches) or more. Because of the warm temperatures and abundant rainfall, most plants grow very rapidly; however, soils are usually poor in nutrients because water tends to carry away any nutrients not immediately taken up by plants. Many of the trees have extensive root networks, associated with fungi (mycorrhizae), near

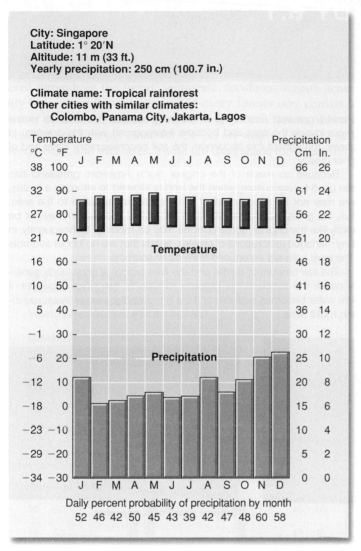

City: Singapore
Latitude: 1° 20′N
Altitude: 11 m (33 ft.)
Yearly precipitation: 250 cm (100.7 in.)

Climate name: Tropical rainforest
Other cities with similar climates:
  Colombo, Panama City, Jakarta, Lagos

| Temperature | | Precipitation | |
| °C | °F | Cm | In. |

Daily percent probability of precipitation by month
52 46 42 50 45 43 39 42 47 48 60 58

(a)

(b) Tropical rainforest landscape

(c) Blue morpho butterfly

(d) Squirrel monkey

FIGURE G.15 Tropical Rainforest (a) Climagraph for Singapore. (b-d) Tropical rainforests develop in areas with high rainfall and warm temperatures. They have an extremely diverse mixture of plants and animals such as birds, butterflies, and monkeys.

the surface of the soil that allow them to capture nutrients from decaying vegetation before the nutrients can be carried away.

Organisms

Tropical rainforests have a greater diversity of species than any other biome. More species are found in the tropical rainforests of the world than in the rest of the world combined. A small area of a few square kilometers is likely to have hundreds of species of trees. Furthermore, it is typical to have distances of a kilometer or more between two individuals of the same species. Balsa, teakwood, and many other ornamental woods are from tropical trees.

Each of those trees is home to a set of animals and plants that use it as food, shelter, or support. The canopy, which forms a solid wall of leaves between the sun and the forest floor, consists of two or three levels. A few trees, called emergent trees, protrude above the canopy. Below the canopy is a layer of understory tree species.

Since most of the sunlight is captured by the trees, only shade-tolerant plants live beneath the trees' canopy. In addition, the understory has many vines that attach themselves to the tall trees and grow toward the sun. When the vines reach the canopy, they can compete effectively with their supporting tree for available sunlight. In addition to supporting various vines, each tree serves as a surface for the growth of ferns, mosses, and orchids.

Recently, biologists discovered a whole new community of organisms that live in the canopy of these forests. Rainfall is a source of new nutrients, since atmospheric particles and gases dissolve as the rain falls. The canopy contains many kinds of epiphytic plants (plants that live on the surfaces of other plants) that trap many of these nutrients in the canopy before they can reach the soil.

Pelagic Marine Ecosystems

In the open ocean, many kinds of organisms float or swim actively. Crustaceans, fish, and whales swim actively as they pursue food. Organisms that are not attached to the bottom are called **pelagic organisms,** and the ecosystem they are a part of is called a **pelagic ecosystem.**

The term **plankton** is used to describe aquatic organisms that are so small and weakly swimming that they are simply carried by currents. As with all ecosystems, the organisms at the bottom of the energy pyramid carry on photosynthesis.

Phytoplankton are planktonic organisms that carry on photosynthesis. In the open ocean, a majority of these organisms are small, microscopic, floating algae and bacteria. The upper layer of the ocean, where the sun's rays penetrate, is known as the **euphotic zone.** It is in this euphotic zone where phytoplankton are most common. The thickness of the euphotic zone varies with the degree of clarity of the water. In clear water it can be up to 150 meters (500 feet) in depth.

Zooplankton are small, weakly swimming animals of many kinds that feed on the phytoplankton. Zooplankton are often located at a greater depth in the ocean than the phytoplankton but migrate upward at night and feed on the large population of phytoplankton. The zooplankton are in turn eaten by larger animals such as fish and larger shrimp, which are eaten by larger fish such as salmon, tuna, sharks, and mackerel. (See figure 6.20.)

A major factor that influences the nature of a marine community is the kind and amount of material dissolved in the water. Of particular importance is the amount of dissolved, inorganic nutrients available to the organisms that carry on photosynthesis. Phosphorus, nitrogen, and carbon are all required for the construction of new living material. In water, these are often in short supply. Therefore, the most productive aquatic ecosystems are

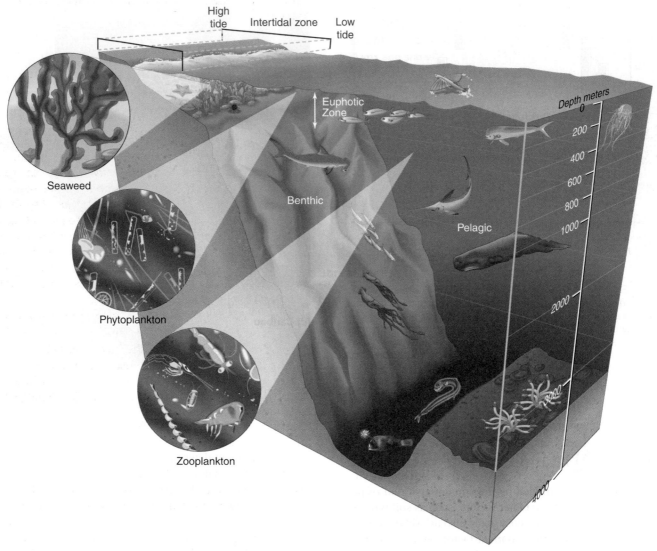

FIGURE 6.20 **Marine Ecosystems** All of the photosynthetic activity of the ocean occurs in the shallow water called the euphotic zone, either in attached algae near the shore or in minute phytoplankton in the upper levels of the open ocean. Consumers are either free-swimming pelagic organisms or benthic organisms that live on the bottom. Small animals that feed on phytoplankton are known as zooplankton.

those in which these essential nutrients are most common. These areas include places in oceans where currents bring up nutrients that have settled to the bottom and places where rivers deposit their load of suspended and dissolved materials.

Benthic Marine Ecosystems

Organisms that live on the ocean bottom, whether attached or not, are known as **benthic** organisms, and the ecosystem of which they are a part is called a **benthic ecosystem.** Some fish, clams, oysters, various crustaceans, sponges, sea anemones, and many other kinds of organisms live on the bottom. In shallow water, sunlight can penetrate to the bottom, and a variety of attached photosynthetic organisms commonly called seaweeds are common. Since they are attached and some, such as kelp, can grow to very large size, many other bottom-dwelling organisms, such as sea urchins, worms, and fish, are associated with them.

The substrate is very important in determining the kind of benthic community that develops. Sand tends to shift and move, making it difficult for large plants or algae to become established, although some clams, burrowing worms, and small crustaceans find sand to be a suitable habitat. Clams filter water and obtain plankton and detritus or burrow through the sand, feeding on other inhabitants. Mud may provide suitable habitats for some kinds of rooted plants, such as mangrove trees or sea grasses. Although mud usually contains little oxygen, it still may be inhabited by a variety of burrowing organisms that feed by filtering the water above them or that feed on other animals in the mud. Rocky surfaces in the ocean provide a good substrate for many kinds of large algae. Associated with this profuse growth of algae is a large variety of animals. (See figure 6.21.)

Temperature also has an impact on the kind of benthic community established. Some communities, such as coral reefs or mangrove swamps, are found only in areas where the water is warm.

Coral reef ecosystems are produced by coral animals that build cup-shaped external skeletons around themselves. Corals protrude from their skeletons to capture food and expose themselves to the sun. Exposure to sunlight is important because corals contain single-celled algae within their bodies. These algae carry on photosynthesis and provide both themselves and the coral animals with the nutrients necessary for growth. This mutualistic relationship between algae and coral is the basis for a very productive community of organisms.

The skeletons of the corals provide a surface upon which many other kinds of animals live. Some of these animals feed on corals directly, while others feed on small plankton and bits of algae that establish themselves among the coral organisms. Many kinds of fish, crustaceans, sponges, clams, and snails are members of coral reef ecosystems.

Because they require warm water, coral ecosystems are found only near the equator. Coral ecosystems also require shallow, clear water, since the algae must have ample sunlight to carry on photosynthesis. Coral reefs are considered to be among the most productive ecosystems on Earth. (See figure 6.22.)

Mangrove swamp ecosystems are tropical forest ecosystems that occupy shallow water near the shore and the adjacent land. The dominant organisms are special kinds of trees that can tolerate the high salt content of the ocean because the trees can excrete salt from their leaves. In areas where the water is shallow and wave action is not too great, the trees can become established. They have seeds that actually begin to germinate on the tree. When the germinated seed falls from the tree it floats in the water. When the seeds become trapped in mud, they take root.

The trees also have extensively developed roots that extend above the water, where they can obtain oxygen and prop up the plant. The trees trap sediment and provide places for oysters, crabs, jellyfish, sponges, and fish to live. The trapping of sediment and the continual extension of mangroves into shallow areas result in the development of a terrestrial ecosystem in what was once shallow ocean. Mangroves are found in south Florida, the Caribbean, Southeast Asia, Africa, and other parts of the world where tropical mudflats occur. (See figure 6.23.)

An **abyssal ecosystem** is a benthic ecosystem that occurs at great depths in the ocean. In such deep regions of the ocean there is no light to support photosynthesis. Therefore, the animals must rely on a continuous rain of organic matter from the euphotic zone. Essentially, all of the organisms in this environment are scavengers that feed on whatever drifts their way. Many of the animals are small and generate light that they use for finding or attracting food.

Estuaries

An **estuary** is a special category of aquatic ecosystem that consists of shallow, partially enclosed areas where freshwater enters the ocean. The saltiness of the water in the estuary changes with tides and the flow of water from rivers. The organisms that live here are specially adapted to this set of physical conditions, and the number of species is less than in the ocean or in freshwater.

Estuaries are particularly productive ecosystems because of the large amounts of nutrients introduced into the basin from the rivers that run into them. This is further enhanced by the fact that the shallow water allows light to penetrate to most of the water in the basin. Phytoplankton and attached algae and plants are able to use the sunlight and the nutrients for rapid growth. This photosynthetic activity supports many kinds of organisms in the estuary. Estuaries are especially important as nursery sites for fish and crustaceans such as flounder and shrimp.

The adults enter these productive, sheltered areas to reproduce and then return to the ocean. The young spend their early life in the estuary and eventually leave as they get larger and are more able to survive in the ocean. Estuaries also trap sediment. This activity tends to prevent many kinds of pollutants from reaching the ocean and also results in the gradual filling in of the estuary, which may eventually become a salt marsh and then part of a terrestrial ecosystem.

Human Impact on Marine Ecosystems

Since the oceans cover about 70 percent of the Earth's surface, it is hard to imagine that humans can have a major impact on them. However, we use the oceans in a wide variety of ways. The oceans provide a major source of protein in the form of fish, shrimp, and other animals. However, overfishing has destroyed many of the traditional fishing industries of the world. Fish farming results in the addition of nutrients and has caused diseases to because we tend to introduce nutrients from agriculture and organic wastes from a variety of industrial, agricultural, and municipal sources. These topics are discussed in greater depth in chapter 15.

Streams and Rivers

Streams and rivers are a second category of freshwater ecosystem. Since the water is moving, planktonic organisms are less important than are attached organisms. Most algae grow attached to rocks and other objects on the bottom. This collection of attached algae, animals, and fungi is called the **periphyton.** Since the water is shallow, light can penetrate easily to the bottom (except for large or extremely muddy rivers). Even so, it is difficult for photosynthetic organisms to accumulate the nutrients necessary for growth, and most streams are not very productive. As a matter of fact, the major input of nutrients is from organic matter that falls into the stream from terrestrial sources. These are primarily the leaves from trees and other vegetation, as well as the bodies of living and dead insects.

Within the stream is a community of organisms that are specifically adapted to use the debris as a source of food. Bacteria and fungi colonize the organic matter, and many kinds of insects shred and eat this organic matter along with the fungi and bacteria living on it. The feces (intestinal wastes) of these insects and the tiny particles produced during the eating process become food for other insects that build nets to capture the tiny bits of organic matter that drift their way. These insects are in turn eaten by carnivorous insects and fish.

Organisms in larger rivers and muddy streams, which have less light penetration, rely in large part on the food that drifts their way from the many streams that empty into the river. These larger rivers tend to be warmer and to have slower-moving water. Consequently, the amount of oxygen is usually less, and the species of plants and animals change. Any additional organic matter added to the river system adds to the BOD, further reducing the oxygen in the water. Plants may become established along the river bank and contribute to the ecosystem by carrying on photosynthesis and providing hiding places for animals.

Just as estuaries are a bridge between freshwater and marine ecosystems, swamps and marshes are a transition between aquatic and terrestrial ecosystems. **Swamps** are wetlands that contain trees that are able to live in places that are either permanently flooded or flooded for a major part of the year. **Marshes** are wetlands that are dominated by grasses and reeds. Many swamps and marshes are successional states that eventually become totally terrestrial communities.

Human Impact on Freshwater Ecosystems

Freshwater resources in lakes and rivers account for about 0.02 percent of the world's water. Most freshwater ecosystems have been heavily affected by human activity. Any activity that takes place on land ultimately affects freshwater because of runoff from the land. Agricultural runoff, sewage, sediment, and trash all find their way into streams and lakes. Chapter 15 covers these issues in greater detail.

SUMMARY

Ecosystems change as one kind of organism replaces another in a process called succession. Ultimately, a relatively stable stage is reached, called the climax community. Succession may begin with bare rock or water, in which case it is called primary succession, or may occur when the original ecosystem is destroyed, in which case it is called secondary succession. The stages that lead to the climax are called successional stages.

Major regional terrestrial climax communities are called biomes. The primary determiners of the kinds of biomes that develop are the amount and yearly distribution of rainfall and the yearly temperature cycle. Major biomes are desert, grassland, savanna, Mediterranean shrublands, tropical dry forest, tropical rainforest, temperate deciduous forest, taiga, and tundra. Each has a particular set of organisms that is adapted to the climatic conditions typical for the area. As one proceeds up a mountainside, it is possible to witness the same kind of change in biomes that occurs if one were to travel from the equator to the North Pole.

Aquatic ecosystems can be divided into marine (saltwater) and freshwater ecosystems. In the ocean, some organisms live in open water and are called pelagic organisms. Light penetrates only the upper layer of water; therefore, this region is called the euphotic zone. Tiny photosynthetic organisms that float near the surface are called phytoplankton. They are eaten by small animals known as zooplankton, which in turn are eaten by fish and other larger organisms.

The kind of material that makes up the shore determines the mixture of organisms that live there. Rocky shores provide surfaces to which organisms can attach; sandy shores do not. Muddy shores are often poor in oxygen, but marshes and swamps may develop in these areas. Coral reefs are tropical marine ecosystems dominated by coral animals. Mangrove swamps are tropical marine shoreline ecosystems dominated by trees. Estuaries occur where freshwater streams and rivers enter the ocean. They are usually shallow, very productive areas. Many marine organisms use estuaries for reproduction.

Insects are common in freshwater and absent in marine systems. Lakes show a structure similar to that of the ocean, but the species are different. Deep, cold-water lakes with poor productivity are called oligotrophic, while shallow, warm-water, highly productive lakes are called eutrophic. Streams differ from lakes in that most of the organic matter present in them falls into them from the surrounding land. Thus, organisms in streams are highly sensitive to the land uses that occur near the streams.

Ecosystem Loss in North America

North America contains a variety of species and ecosystems, including temperate rainforests, grasslands, wetlands, deserts, and more. Species in the United States include grizzly bears, spotted owls, ghost-faced bats, horned puffins, and redwood trees, as only a few examples.

As in the tropics, North America's storehouse of biodiversity is being threatened. As of May 2002, the U.S. Fish and Wildlife Service and the National Marine Fisheries Service combined had listed 1231 species (496 animal species and 735 plant species) as endangered or threatened in the United States. Hundreds of other species are being considered as possible additions to the list. According to The Nature Conservancy, one-third of all U.S. plant and animal species are in need of protection. Many freshwater fishes and wetland species such as mussels, crayfish and amphibians are particularly vulnerable. Nearly 500 species in the United States may be nearly extinct.

Canada's endangered species list included 353 species as of May 2002. Among them are the wolverine, killer whale, eastern barn owl, western rattlesnake, tailed frog, white-throated swift, peregrine falcon, and whooping crane. Many of Canada's ecosystems are also in danger. According to the Canadian Nature Federation, 240 hectares (593 acres) of

Most prairie has been converted to agriculture.

wildlife habitat are converted or fragmented every hour in Canada, and habitat destruction threatens more than 80 percent of Canada's endangered species with extinction.

Mexico's rich biodiversity is also being lost. Home to nearly 10 percent of the world's terrestrial species, Mexico has a high number of endemic species, the richest diversity of reptiles and cacti, and the second richest diversity of mammals in the world. But almost half of Mexico's 25 million hectares (62 million acres) of tropical dry and humid forests have been cleared for agriculture and grazing, leaving only 10 percent in stable condition. More than 50 percent of Mexico is dry coastal sage scrub or desert, and overgrazing and human-caused fires have degraded much of this land.

Consider the following facts compiled by the World Wildlife Fund:

North American Ecosystem	Percent of Ecosystem Lost
Original North American tallgrass prairie	More than 99 percent transformed
Original primary forest in the 48 contiguous United States	More than 95 percent lost
Midwest oak savanna	More than 98 percent altered
Old-growth forest in the Pacific Northwest	About 90 percent cleared
Wild or scenic rivers in the United States	Between 90 percent and 98 percent degraded
Coastal sage scrub in the United States	Between 70 percent and 90 percent disturbed
Original wetlands in the United States	More than 50 percent drained and filled

- Can you give examples of ecosystem lost in your area?
- What were the circumstances that led to the loss in your area?
- Was there an alternative to the loss of the ecosystem?
- Were any endangered or threatened species affected?

Source: President's Committee of Advisors on Science and Technology, *Biodiversity: Connecting with the Tapestry of Life.* Washington, D.C., 2002.

THINKING GREEN

1. Learn to identify five plants native to your area.
2. Observe the behavior of an insect, reptile, amphibian, bird, or mammal in its natural habitat.
3. Participate in a local program to eliminate invasive species.
4. Participate in Earth Day (April 22) and Arbor Day (in the spring but the date varies by state) activities in your community.
5. Visit the National Wildlife Federation website and learn about its Backyard Wildlife Habitat Program.
6. Visit a disturbed site—vacant lot, road side, abandoned farmland. What evidence do you see that succession is taking place?

WHAT'S YOUR TAKE?

If you review the Issues & Analysis: Ecosystem Loss in North America, you will find that in the United States, 99 percent of the tallgrass prairie has been transformed, 95 percent of the original forest is gone, and 50 percent of wetlands have been drained and filled. Much of this occurred because people used the resources or converted them to farmland. Many people and conservation organizations in economically advanced countries feel that poor countries should preserve their natural ecosystems (particularly tropical rainforests) rather than use them for economic development. Many people in these developing countries feel they have a right to use their own resources the way people in other countries have.

Choose to support either preservation or development of these resources, and prepare arguments to support your position.

REVIEW QUESTIONS

1. Describe the process of succession. How does primary succession differ from secondary succession?

2. How does a climax community differ from a successional community?

3. List three characteristics typical of each of the following biomes: tropical rainforest, desert, tundra, taiga, savanna, Mediterranean shrublands, tropical dry forest, temperate grassland, temperate rainforest, and temperate deciduous forest.

4. What two primary factors determine the kind of terrestrial biome that will develop in an area?

5. How does height above sea level affect the kind of biome present?

6. What areas of the ocean are the most productive?

7. How does the nature of the substrate affect the kinds of organisms found at the shore?

8. What is the role of each of the following organisms in a marine ecosystem: phytoplankton, zooplankton, algae, coral animals, and fish?

9. List three differences between freshwater and marine ecosystems.

10. What is an estuary? Why are estuaries important?

CRITICAL THINKING QUESTIONS

1. Does the concept of a "climax community" make sense? Why or why not?

2. What do you think about restoring ecosystems that have been degraded by human activity? Should it be done or not? Why? Who should pay for this reconstruction?

3. Identify the biome in which you live. What environmental factors are instrumental in maintaining this biome? What is the current health of your biome? What are the current threats to its health? How might your biome have looked 100, 1000, 10,000 years ago?

4. Imagine you are a conservation biologist who is being asked by local residents what the likely environmental outcomes of development would be in the tropical rainforest in which they live. What would you tell them? Why do you give them this evaluation? What evidence can you cite for your claims?

5. The text says that 90 percent of the old-growth temperate rainforest in the Pacific Northwest has been logged. What to do with the remaining 10 percent is still a question. Some say it should be logged, and others say it should be preserved. What values, beliefs, and perspectives are held by each side? What is your ethic regarding logging old-growth in this area? What values, beliefs, and perspectives do you hold regarding this issue?

6. Much of the old-growth forest in the United States has been logged, economic gains have been realized, and second-growth forests have become established. This is not the case in the tropical rainforests, although they are being lost at alarming rates. Should developed countries, which have already "cashed in" on their resources, have anything to say about what is happening in developing countries? Why do you think the way you do?

CHAPTER 7

POPULATIONS: CHARACTERISTICS AND ISSUES

Populations are collections of organisms of the same species. This group of Magellanic penguins constitutes a population with certain characteristics that may differ somewhat from the characteristics of penguin groups that inhabit other parts of South America.

CHAPTER OUTLINE

Population Characteristics
   Natality—Birthrate
   Mortality—Death Rate
   Population Growth Rate
   Sex Ratio
   Age Distribution
   Population Density and Spatial Distribution
   Summary of Factors that Influence Population Growth Rates
A Population Growth Curve
Factors that Limit Population Size
   Extrinsic and Intrinsic Limiting Factors
   Density-Dependent and Density-Independent Limiting Factors
Categories of Limiting Factors
   Availability of Raw Materials
   Availability of Energy
   Accumulation of Waste Products
   Interactions Among Organisms
Carrying Capacity
Reproductive Strategies and Population Fluctuations
   K-Strategists and r-Strategists
   Population Cycles
Human Population Growth
Human Population Characteristics and Implications
   Economic Development
   Measuring the Environmental Impact of a Population
   The Ecological Footprint Concept
Factors that Influence Human Population Growth
   Biological Factors
   Social Factors
   Economic Factors
   Political Factors
Population Growth Rates and Standard of Living
Hunger, Food Production, and Environmental Degradation
   Environmental Impacts of Food Production
   The Human Energy Pyramid
   Economics and Politics of Hunger
   Solving the Problem
The Demographic Transition Concept
   The Demographic Transition Model
   Applying the Model
The U.S. Population Picture
What Does the Future Hold?
   Available Raw Materials
   Available Energy
   Waste Disposal
   Interaction with Other Organisms
   Social Factors Influence Human Population
   Ultimate Size Limitation

ISSUES & ANALYSIS
The Lesser Snow Goose—A Problem Population 167

CASE STUDIES
Thomas Malthus and His Essay on Population 152
The Grameen Bank and Microcredit 159
North America—Population Comparisons 165

CAMPUS SUSTAINABILITY INITIATIVE
Auburn University's War on Hunger Initiative 162

GOING GREEN
Increasing Populations of Red-Cockaded Woodpeckers 147

WATER CONNECTIONS
Drinking Water: A Basic Right? 164

OBJECTIVES

After reading this chapter, you should be able to:

- Understand that birthrate and death rate are both important in determining the population growth rate.
- Define the following characteristics of a population: natality, mortality, sex ratio, age distribution, biotic potential, and spatial distribution.
- Explain the significance of biotic potential to the rate of population growth.
- Describe the lag, exponential growth, deceleration, and stable equilibrium phases of a population growth curve. Explain why each of these stages occurs.
- Describe how limiting factors determine the carrying capacity for a population.
- List the four categories of limiting factors.
- Recognize that humans are subject to the same forces of environmental resistance as are other organisms.
- Understand the implications of overreproduction.
- Explain how human population growth is influenced by social, theological, philosophical, and political thinking.
- Explain why the age distribution and the status and role of women affect population growth projections.
- Recognize that countries in the more-developed world are experiencing an increase in the average age of their populations.
- Recognize that most countries of the world have a rapidly growing population.
- Describe the implications of the demographic transition concept.
- Recognize that rapid population growth and poverty are linked.

POPULATION CHARACTERISTICS

A **population** can be defined as a group of individuals of the same species inhabiting an area. Just as individuals within a population are recognizable, different populations of the same species have specific characteristics that distinguish them from one another. Some important ways in which populations differ include natality (birthrate), mortality (death rate), sex ratio, age distribution, growth rates, density, and spatial distribution.

NATALITY—BIRTHRATE

Natality refers to the number of individuals added to the population through reproduction over a particular time period. There are two ways in which new individual organisms are produced—asexual reproduction and sexual reproduction.

Asexual Reproduction

Bacteria and other tiny organisms reproduce primarily asexually when they divide to form new individuals that are identical to the original parent organism. Even plants and many kinds of animals, such as sponges, jellyfish, and many kinds of worms, reproduce asexually by dividing into two parts or by budding off small portions of themselves that become independent individuals. Even some insects and lizards have a special kind of asexual reproduction in which the females lay unfertilized eggs that are genetically identical to the female.

Sexual Reproduction

However, most species have some stage in their life cycle in which they reproduce sexually. In plant populations, sexual reproduction results in the production of numerous seeds, but the seeds must land in appropriate soil conditions before they will germinate to produce a new individual. Animal species also typically produce large numbers of offspring as a result of sexual reproduction.

In human populations, natality is usually described in terms of the **birthrate,** the number of individuals born per 1000 individuals per year. For example, if a population of 2000 individuals produced 20 offspring during one year, the birthrate would be 10 per thousand per year. The natality for most species is typically quite high. Most species produce many more offspring than are needed to replace the parents.

MORTALITY—DEATH RATE

It is important to recognize that the growth of a population is not determined by the birthrate (natality) alone. **Mortality,** the number of deaths in a population over a particular time period, is also important. For most species, mortality rates are very high, particularly among the younger individuals. For example, of all the seeds that plants produce, very few will result in a mature plant that itself will produce offspring. Many seeds are eaten by animals, some do not germinate because they never find proper soil conditions, and those that germinate must compete with other organisms for nutrients and sunlight.

In human population studies, mortality is usually discussed in terms of the **death rate,** the number of people who die per 1000 individuals per year. Compared to the high mortality of the young of most species, the infant death rate of long-lived animals such as humans is relatively low. In order for the size of a population to grow, the number of individuals added by reproduction must be greater than the number leaving it by dying. (See figure 7.1.)

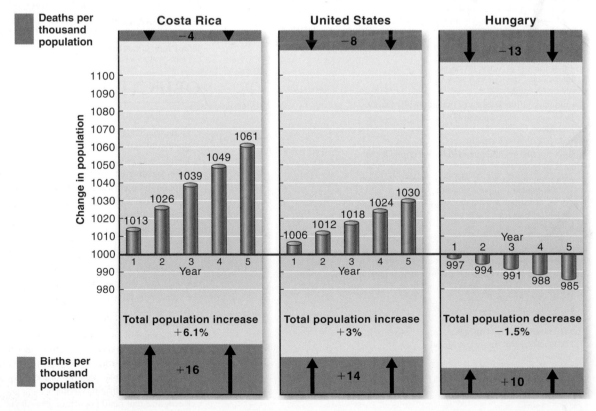

FIGURE 7.1 Effect of Birthrate and Death Rate on Population Size For a population to grow, the birthrate must exceed the death rate for a period of time. These three human populations illustrate how the combined effects of births and deaths would change population size if birthrates and death rates were maintained for a five-year period.
Source: Data from World Population Data Sheet 2008, Population Reference Bureau, Inc., Washington, D.C.

(a)

(b)

(c)

FIGURE 7.2 **Types of Survivorship Curves** (a) The Dall sheep is a large mammal that produces relatively few young. Most of the young survive, and survival is high until individuals reach old age, when they are more susceptible to predation and disease. (b) The curve shown for the white-crowned sparrow is typical of that for many kinds of birds. After a period of high mortality among the young, the mortality rate is about equal for all ages of adult birds. (c) Many small animals and plants, such as the Mediterranean shrub *Cleome droserifolia,* produce enormous numbers of offspring. Mortality is very high in the younger individuals, and few individuals reach old age.

Another way to view mortality is to view how likely it is that an offspring will survive to a specific age. One way of visualizing this is with a survivorship curve. A **survivorship curve** shows the proportion of individuals likely to survive to each age. While each species is different, three general types of survivorship curves can be recognized: species that have high mortality among their young, species in which mortality is evenly spread over all age groups, and species in which survival is high until old age, when mortality is high. Figure 7.2 gives examples of species that fit these three general categories.

POPULATION GROWTH RATE

The **population growth rate** is the birthrate minus the death rate. In human population studies, the population growth rate is usually expressed as a percentage of the total population. For example, in the United States, the birthrate is 14 births per thousand individuals in the population. The death rate is 8 per thousand. The difference between the two is 6 per thousand, which is equal to an annual population increase of 0.6 percent (6/1000).

SEX RATIO

The population growth rate is greatly influenced by the sex ratio of the population. The **sex ratio** refers to the relative numbers of males and females. (Many kinds of organisms, such as earthworms and most plants, have both kinds of sex organs in the same body; sex ratio has no meaning for these species.) The number of females

is very important, since they ultimately determine the number of offspring produced in the population. In polygamous species, one male may mate with many females. Therefore, the number of males is less important to the population growth rate than the number of females. In monogamous species, a male and female pair up, mate, and raise their young together. Unpaired females are not likely to be fertilized and raise young. Even if an unpaired female is fertilized, she will be less successful in raising young.

It is typical in most species that the sex ratio is about 1:1 (one female to one male). However, there are populations in which this is not true. In populations of many species of game animals, the males are shot (have a higher mortality) and the females are not. This results in an uneven sex ratio in which the females outnumber the males. In many social insect populations (bees, ants, and wasps), the number of females greatly exceeds the number of males at all times, though most of the females are sterile. In humans, about 106 males are born for every 100 females. However, in the United States, by the time people reach their mid-twenties, a higher death rate for males has equalized the sex ratio. The higher male death rate continues into old age, when women outnumber men.

AGE DISTRIBUTION

The **age distribution** is the number of individuals of each age in the population. Age distribution greatly influences the population growth rate. As you can see in figure 7.3, some are prereproductive juveniles, some are reproducing adults, and some are postreproductive adults. If the population has a large number of prereproductive juveniles, it would be expected to grow in the future as the young become sexually mature. If the majority of a population is made up of reproducing adults, the population should be growing. If the population is made up of old individuals whose reproductive success is low, the population is likely to fall.

Many species, particularly those that have short life spans, have age distributions that change significantly during the course of a year. Species typically produce their young during specific parts of the year. Annual plants (those that live for only one year) produce seeds that germinate in the spring or following a rainy period of the year. Therefore, during one part of the year, most of the individuals are newly germinated seeds and are prereproductive.

As time passes, nearly all of those seedlings that survive become reproducing adults and produce seeds. Later in the year, they all die. A similar pattern is seen in many insects that go through their entire life cycle in a year. They emerge from eggs as larvae, transform into adults, mate and lay eggs, and die. Animals that live for several years typically produce their young at a time when food is abundant. In northern climates, this is generally in the spring of the year. In regions where rainfall is sporadic (deserts) or highly seasonal (savannas and some forests), the production of offspring usually occurs following rain. Thus, there is a surge in the number of prereproductive individuals at specific times of the year.

In species that live a long time, it is possible for a population to have an age distribution in which the proportion of individuals in these three categories is relatively constant. Since mortality is generally higher among young individuals, such populations typically have more prereproductive individuals than reproductive individuals and more reproductive individuals than postreproductive individuals.

Human populations exhibit several types of age distribution. (See figure 7.3.) Kenya's population has a large prereproductive and reproductive component. This means that it will continue to increase rapidly for some time. The United States has a very large reproductive component with a declining number of prereproductive individuals. Eventually, if there were no immigration, the U.S. population would begin to decline if current trends in birthrates and death rates continued. Italy has an age distribution with high postreproductive

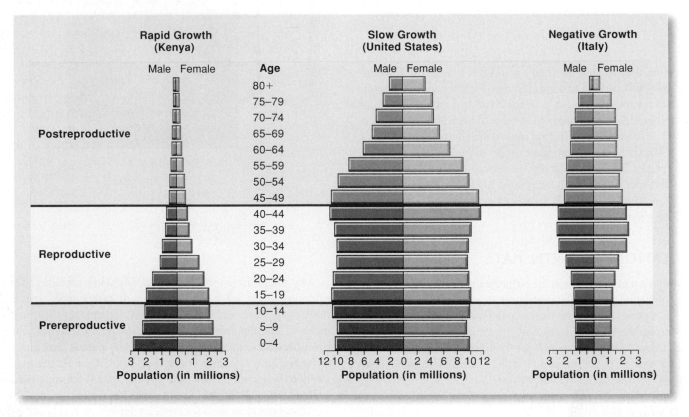

FIGURE 7.3 **Age Distribution in Human Populations** The relative numbers of individuals in each of the three categories (prereproductive, reproductive, and postreproductive) are good clues to the future growth of a population. Kenya has a large number of young individuals who will become reproducing adults. Therefore, this population is likely to grow rapidly. The United States has a large proportion of reproductive individuals and a moderate number of prereproductive individuals. Therefore, the population is likely to grow slowly. Italy has a declining number of reproductive individuals and a very small number of prereproductive individuals. Therefore, its population has begun to decline.
Source: Data from United States Census Bureau International Data Base.

and low prereproductive portions of the population. With low numbers of prereproductive individuals entering their reproductive years, the population of Italy has begun to decline.

POPULATION DENSITY AND SPATIAL DISTRIBUTION

Because of such factors as soil type, quality of habitat, and availability of water, organisms normally are distributed unevenly. Some populations have many individuals clustered into a small space, while other populations of the same species may be widely dispersed. **Population density** is the number of organisms per unit area. For example, fruitfly populations are very dense around a source of rotting fruit, while they are rare in other places. Similarly, humans are often clustered into dense concentrations we call cities, with lower densities in rural areas.

When the population density is too great, all individuals within the population are injured because they compete severely with each other for necessary resources. Plants may compete for water, soil nutrients, or sunlight. Animals may compete for food, shelter, or nesting sites. In animal populations, overcrowding might cause some individuals to explore and migrate into new areas. This movement from densely populated locations to new areas is called **dispersal.** It relieves the overcrowded conditions in the home area and, at the same time, increases the population in the places to which they migrate. Often, it is juvenile individuals that relieve overcrowding by leaving. The pressure to migrate from a population **(emigration)** may be a result of seasonal reproduction leading to a rapid increase in population size or environmental changes that intensify competition among members of the same species. For example, as water holes dry up, competition for water increases, and many desert birds emigrate to areas where water is still available.

The organisms that leave one population often become members of a different population. This migration into an area **(immigration)** may introduce characteristics that were not in the population originally. When Europeans immigrated to North America, they brought genetic and cultural characteristics that had a tremendous impact on the existing Native American population. Among other things, Europeans brought diseases that were foreign to the Native Americans. These diseases increased the death rate and lowered the birthrate of Native Americans, resulting in a sharp decrease in the size of their populations.

SUMMARY OF FACTORS THAT INFLUENCE POPULATION GROWTH RATES

Populations have an inherent tendency to increase in size. However, as we have just seen, many factors influence the rate at which a population can grow. At the simplest level, the rate of increase is determined by subtracting the number of individuals leaving the population from the number entering. Individuals leave the population either by death or emigration. Individuals enter the population by birth or immigration. Birthrates and death rates are influenced by several factors, including the number of females in the population and their age. In addition, the density of a population may encourage individuals to leave because of intense competition for a limited supply of resources.

A POPULATION GROWTH CURVE

Each species has a **biotic potential** or inherent reproductive capacity, which is its biological ability to produce offspring. Reproducing individuals of some species, such as watermelon plants or moths, may produce hundreds or thousands of offspring (seeds or caterpillars) per year, while others, such as geese, may produce 10 to 12 young per year. (See figure 7.4.) Some large animals, such as bears or elephants, may produce one young every two to three years. Although there are large differences among species, generally, adults produce many more offspring during their lifetimes than are needed to replace themselves when they die. However, among organisms that produce large numbers of offspring, most of the young die, so only a few survive to become reproductive adults themselves.

Because most species have a high biotic potential, there is a natural tendency for populations to increase. If we consider a hypothetical situation in which mortality is not a factor, we could have the following situation. If two mice produced four offspring and they all lived, eventually they would produce offspring of their own, while their parents continued to reproduce as well.

Watermelon offspring (seeds)

Moth offspring (caterpillars)

Geese

FIGURE 7.4 **Biotic Potential** The ability of a species to reproduce greatly exceeds the number necessary to replace those who die. Here are some examples of the prodigious reproductive abilities of some species.

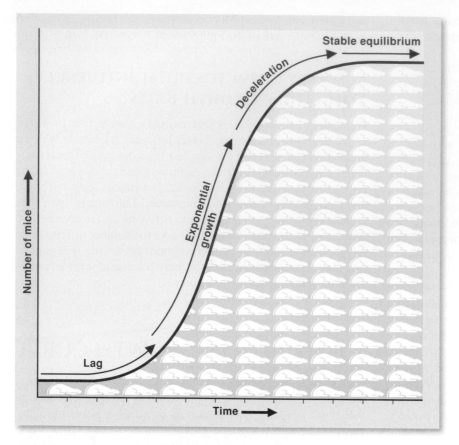

FIGURE 7.5 **A Typical Population Growth Curve** In this mouse population, there is little growth during the lag phase. During the exponential growth phase, the population rises rapidly as increasing numbers of individuals reach reproductive age. Eventually, the population growth rate begins to slow during the deceleration phase and the population reaches a stable equilibrium phase, during which the birthrate equals the death rate.

Under these conditions, the population will grow exponentially. Exponential growth results in a population increasing by the same percentage each year. For example, if the population were to double each year, we would have 2, 4, 8, 16, 32, etc. individuals in the population. While populations cannot grow exponentially forever, they often have an exponential period of growth.

Population growth often follows a particular pattern, consisting of a lag phase, an exponential growth phase, a deceleration phase, and a stable equilibrium phase. Figure 7.5 shows a typical population growth curve. During the first portion of the curve, known as the **lag phase,** the population grows very slowly because there are few births, since the process of reproduction and growth of offspring takes time. Organisms must mature into adults before they can reproduce. While the offspring begin to mate and have young, the parents may be producing a second set of offspring. Since more organisms now are reproducing, the population begins to increase at an accelerating rate. This stage is known as the **exponential growth phase (log phase).** The population will continue to grow as long as the birthrate exceeds the death rate. Eventually, however, the population growth rate will begin to slow as the death rate and the birthrate come to equal one another. This is the **deceleration phase.** When the birthrate and death rate become equal, the population will stop growing and reach a relatively

stable population size. This stage is known as the **stable equilibrium phase.**

It is important to recognize that although the size of the population may not be changing, the individuals are changing. As new individuals enter by birth or immigration, others leave by death or emigration. For most organisms, the first indication that a population is entering a stable equilibrium phase is an increase in the death rate. A decline in the birthrate may also contribute to the stabilizing of population size. Usually, this occurs after an increase in the death rate.

FACTORS THAT LIMIT POPULATION SIZE

Populations cannot continue to increase indefinitely. Eventually, some factor or set of factors acts to limit the size of a population. The factors that prevent unlimited population growth are known as **limiting factors.** All of the different limiting factors that act on a population are collectively known as **environmental resistance.**

EXTRINSIC AND INTRINSIC LIMITING FACTORS

Some factors that control populations come from outside the population and are known as **extrinsic limiting factors.** Predators, loss of a food source, lack of sunlight, or accidents of nature are all extrinsic factors. However, the populations of many kinds of organisms appear to be regulated by factors from within the populations themselves. Such limiting factors are called **intrinsic limiting factors.** For example, a study of rats under crowded living conditions showed that as conditions became more crowded, abnormal social behavior became common. There was a decrease in litter size, fewer litters per year were produced, mothers were more likely to ignore their young, and adults killed many young. Thus changes in the behavior of the members of the rat population itself resulted in lower birthrates and higher death rates that limit population size. Among populations of white-tailed deer, it is well known that reproductive success is reduced when the deer experience a series of severe winters. When times are bad, the female deer are more likely to have single offspring than twins.

DENSITY-DEPENDENT AND DENSITY-INDEPENDENT LIMITING FACTORS

Density-dependent limiting factors are those that become more effective as the density of the population increases. For example, the larger a population becomes, the more likely it is that predators will have a chance to catch some of the individuals. A prolonged period of increasing population allows the size of the predator population to increase as well. Disease epidemics are also more common in large, dense populations because dense populations allow for the easy spread of parasites from one individual to another. The rat example discussed previously is another good example of a density-dependent factor operating because the amount of abnormal behavior increased as the density of the population increased. In general, whenever there is competition among members of a population, its intensity increases as the population density increases. Large organisms that tend to live a long time and have relatively few young are most likely to be controlled by density-dependent factors.

Density-independent limiting factors are population-controlling influences that are not related to the density of the population. They are usually accidental or occasional extrinsic factors in nature that happen regardless of the density of a population. A sudden rainstorm may drown many small plant seedlings and soil organisms. Many plants and animals are killed by frosts that come late in spring or early in the fall. A small pond may dry up, resulting in the death of many organisms. The organisms most likely to be controlled by density-independent factors are small, short-lived organisms that can reproduce very rapidly.

CATEGORIES OF LIMITING FACTORS

For most populations, limiting factors recognized as components of environmental resistance can be placed into four broad categories: (1) the availability of raw materials, (2) the availability of energy, (3) the accumulation of waste products, and (4) interactions among organisms.

AVAILABILITY OF RAW MATERIALS

Raw materials come in many forms. For example, plants need nitrogen and magnesium from the soil as raw materials for the manufacture of chlorophyll. If these minerals are not present in sufficient quantities, the plant population cannot increase. The application of fertilizers is a way of preventing certain raw materials from being a limiting factor. In effect, the carrying capacity has been increased because this limiting factor has been removed. A carrying capacity still exists, but it is set at a new level, and some new primary limiting factor will emerge. Perhaps it will be the amount of water, the number of insects that feed on the plants, or competition for sunlight.

Animals also require certain minerals as raw material, which they obtain in their diets. They may also require objects with which to build nests, to provide places for escape, or to serve as observation sites.

AVAILABILITY OF ENERGY

Energy sources are important to all organisms. Plants require energy in the form of sunlight for photosynthesis, so the amount of light can be a limiting factor for many plants. When small plants are in the shade of trees, they often do not grow well and have small populations because they do not receive enough sunlight. Animals require energy in the form of the food they eat, and if food is scarce, many die.

ACCUMULATION OF WASTE PRODUCTS

The accumulation of waste products is not normally a limiting factor for plants, since they produce few wastes, but it can be for other kinds of organisms. Bacteria, other tiny organisms, and many kinds of aquatic organisms that live in small ecosystems such as puddles, pools, or aquariums may be limited by wastes. When a small number of a species of bacterium are placed on a petri plate with nutrient agar (a jellylike material containing food substances), the population growth follows a curve shown in figure 7.6. As expected, it begins with a lag phase, continues through an exponential growth phase, and eventually levels off in a stable equilibrium phase. However, in this small, enclosed space, there is no way to get rid of the toxic waste products, which accumulate, eventually killing the bacteria. This decline in population size is known as the **death phase.**

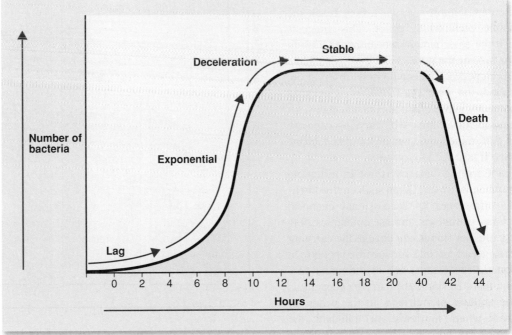

FIGURE 7.6 **A Bacterial Growth Curve** The initial change in population size follows a typical population growth curve until waste products become lethal. The buildup of waste products lowers the carrying capacity. When a population begins to decline, it enters the death phase.

INTERACTIONS AMONG ORGANISMS

Interactions among organisms are also important in determining population size. For example, white-tailed deer and cottontail rabbits eat the twigs of many species of small trees and shrubs. Repeated browsing by herbivores retards growth and can cause death of trees and shrubs. Thus, damage by herbivores can limit the size of some tree and shrub populations. Many single-celled aquatic organisms produce waste products that build up to toxic levels and result in the death of fish. Parasites and predators weaken or cause the premature death of individuals, thus limiting the size of the population.

Some studies indicate that populations can be controlled by interaction among individuals within the population. A study of laboratory rats shows that crowding causes a breakdown in normal social behavior, which leads to fewer births and increased deaths. The changes observed include abnormal mating behavior, decreased litter size, fewer litters per year, lack of maternal care, and increased aggression in some rats or withdrawal in others. Thus, limiting factors can reduce birthrates as well as increase death rates. Many other kinds of animals have shown similar reductions in breeding success when population densities were high.

CARRYING CAPACITY

The populations of many organisms are at their maximum size when they reach the stable equilibrium phase. This suggests that the environment sets an upper limit to the size of the population. Ecologists have developed a concept for this observation, called the *carrying capacity*. **Carrying capacity** is the maximum sustainable population for an area. The carrying capacity is determined by a set of limiting factors. (See figure 7.7.)

Carrying capacity is not an inflexible number, however. Often such environmental differences as successional changes, climate variations, disease epidemics, forest fires, or floods can change the carrying capacity of an area for specific species. In aquatic ecosystems one of the major factors that determine the carrying capacity is the amount of nutrients in the water. In areas where nutrients are abundant, the numbers of various kinds of organisms are high. Often nutrient levels fluctuate with changes in current or runoff from the land, and plant and animal populations fluctuate

as well. In addition, a change that negatively affects the carrying capacity for one species may increase the carrying capacity for another. For example, the cutting down of a mature forest followed by the growth of young trees increases the carrying capacity for deer and rabbits, which use the new growth for food, but decreases the carrying capacity for squirrels, which need mature, fruit-producing trees as a source of food and old, hollow trees for shelter.

Wildlife management practices often encourage modifications to the environment that will increase the carrying capacity for the designated game species. The goal of wildlife managers is to have the highest sustainable population available for harvest by hunters. Typical habitat modifications include creating water holes, cutting forests to provide young growth, planting food plots, and building artificial nesting sites.

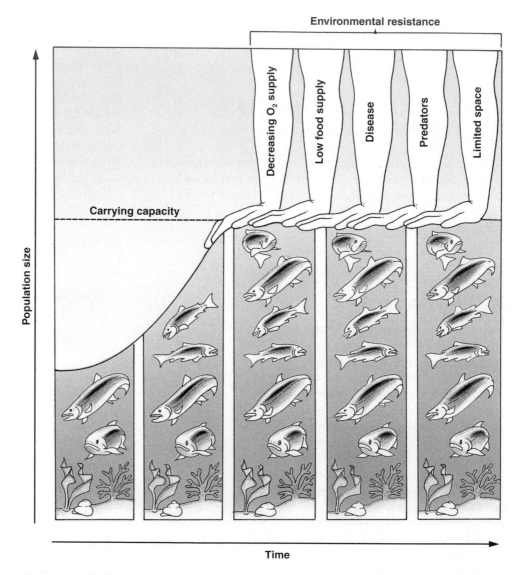

FIGURE 7.7 **Carrying Capacity** A number of factors in the environment, such as oxygen supply, food supply, diseases, predators, and space, determine the number of organisms that can survive in a given area—the carrying capacity of that area. The environmental factors that limit populations are known collectively as environmental resistance.

The red-cockaded woodpecker (*Picoides borealis*) is listed as an endangered species. This medium-sized bird (about the size of a cardinal) is a cooperative colony nester—the dominant male and female raise young with the support of nonbreeding members of the colony. They are only found in the southeastern United States—southern Virginia to eastern Texas—where native southern yellow pine forests occur. Several pine species, including slash pine, shortleaf pine, loblolly pine, and longleaf pine, are typical of this region. The original forests were fire-adapted in that mature trees were able to withstand moderate ground fires. This resulted in a rather open forest type. The woodpeckers typically construct their nesting cavities in older, diseased longleaf pine trees.

The trees these birds use for nesting are also commercially important. Thus, the amount of suitable breeding habitat has been severely reduced as older trees are harvested and natural stands of pines have been replaced with plantations, where large tracts are planted to a single species and the trees are harvested before they reach old age.

Since much of the suitable habitat is privately owned, protecting populations of red-cockaded woodpeckers requires the cooperation of private landowners, conservation organizations, state and federal governments, and commercial forest products companies.

In 1998, International Paper entered into an agreement with the U.S Fish and Wildlife Service, which is responsible for monitoring the status of endangered species, to increase the amount of suitable nesting habitat on its lands. International Paper agreed to set aside particular parcels of forest to maintain colonies of red-cockaded woodpeckers. One of those parcels was the Southlands Experimental Forest near Bainbridge, Georgia. When the agreement was signed in 1998, there were three male red-cockaded woodpeckers at the site. By 2008, there were over 50 individuals. The increase is attributable to protection and improvement of the birds' habitat and transfer of birds to the area from other locations. In 2006, the company decided to sell nearly all of its land holdings in the United States. Many environmentally sensitive lands were sold to conservation organizations such as The Nature Conservancy and the Conservation Fund, as well as state governments. The Southlands Experimental Forest was sold to the state of Georgia with some funding assistance from the Conservation Fund. This land transfer protects the population gains made by this population of red-cockaded woodpeckers.

Red-cockaded woodpecker habitat

Red-cockaded woodpecker

REPRODUCTIVE STRATEGIES AND POPULATION FLUCTUATIONS

So far, we have talked about population growth as if all organisms reach a stable population when they reach the carrying capacity. That is an appropriate way to begin to understand population changes, but the real world is much more complicated.

K-STRATEGISTS AND r-STRATEGISTS

Species can be divided into two broad categories based on their reproductive strategies. **K-strategists** are organisms that typically reach a stable population as the population reaches the carrying capacity. K-strategists usually occupy relatively stable environments and tend to be large organisms that have relatively long lives, produce few offspring, and provide care for their offspring. Their reproductive strategy is to invest a great deal of energy in producing a few offspring that have a good chance of living to reproduce. Deer, lions, and swans are examples of this kind of organism. Humans generally produce single offspring, and even in countries with high infant mortality, 80 percent of the children survive beyond one year of age, and the majority of these will reach adulthood. Generally, populations of K-strategists are controlled by density-dependent limiting factors that become more severe as the size of the population increases. For example, as the size of the hawk population increases, the competition among hawks for available food becomes more severe. The increased competition for food is a density-dependent limiting factor that leads to less food for the young in the nest. Therefore, many of the young die, and population growth rate slows as the carrying capacity for the area is reached.

The **r-strategists** are typically small organisms that have a short life, produce many offspring, exploit unstable environments, and do not reach a carrying capacity. Examples are bacteria, protozoa, many insects, and some small mammals. The reproductive strategy of r-strategists is to expend large amounts of energy producing many offspring but to provide limited care (often none) for them. Consequently, there is high mortality among the young. For example, one female oyster may produce a million eggs, but of those that become fertilized and grow into larvae, only a few find suitable places to attach themselves and grow into mature oysters. Typically, populations of r-strategists are limited by density-independent limiting factors. These factors can include changing weather conditions that kill large numbers of organisms, habitat loss such as occurs when a pond dries up or fire destroys a forest, or an event such as a deep snow or flood that buries sources of food and leads to the death of entire populations. The population

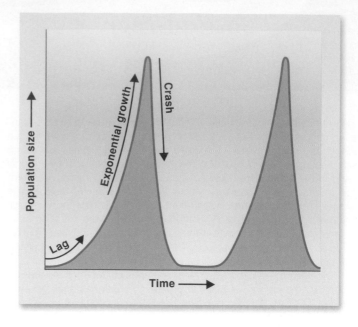

FIGURE 7.8   A Population Growth Curve for Short-Lived Organisms Organisms that are small and only live a short time often have the kind of population growth curve shown here. There is a lag phase followed by an exponential growth phase. However, instead of entering into a stable equilibrium phase, the population reaches a maximum and crashes.

size of r-strategists is likely to fluctuate wildly. They reproduce rapidly, and the size of the population increases until some density-independent factor causes the population to crash; then they begin the cycle all over again. (See figure 7.8.)

The concepts of K- or r-strategists describe idealized situations. (See table 7.1.) (The letters K and r in *K-strategists* and *r-strategists* come from a mathematical equation in which K represents the carrying capacity of the environment and r represents the biotic potential of the species.) In the real world many organisms don't fit clearly into either category. For example, many kinds of mammals provide care for their offspring but have short life spans. On the other hand, many reptiles such as turtles may live for many years, but they produce large numbers of eggs and do not care for them.

TABLE 7.1	**A Comparison of Life History Characteristics of Typical K- and r-Strategists**	
Characteristic	K-Strategist	r-Strategist
Environmental stability	Stable	Unstable
Size of organism	Large	Small
Length of life	Long, most live to reproduce	Short, most die before reproducing
Number of offspring	Small number produced, parental care provided	Large number produced, no parental care
Primary limiting factors	Density-dependent limiting factors	Density-independent limiting factors
Population growth pattern	Exponential growth followed by a stable equilibrium stage at the carrying capacity	Exponential growth followed by a population crash
Examples	Alligators, humans, redwood trees	Protozoa, mosquitoes, annual plants

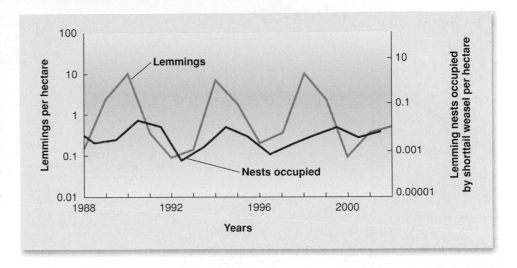

FIGURE 7.9 **Population Cycles** In many northern regions of the world, population cycles are common. In the case of collared lemmings and shorttail weasels on Greenland, interactions between the two populations result in population cycles of about four years. The graph shows the population of lemmings per hectare and the number of lemming nests occupied by weasels per hectare. A hectare is 10,000 square meters—about 2.47 acres. The researchers used the number of lemming nests occupied by weasels as an indirect measure of the size of the weasel population because of difficulties in measuring the size of the weasel population by other means.

Source: Data from O. Gilg, I. Hamski, and B. Sittler, "Cyclic Dynamics in a Simple Predator-Prey Community," *Transpol'air* Online Magazine, www.transpolair.com.

Since humans are K-strategists, it may be difficult for us to appreciate that the r-strategy can be viable from an evolutionary point of view. Resources that are present only for a short time can be exploited most effectively if many individuals of one species monopolize the resource, while denying other species access to it. Rapid reproduction can place a species in a position to compete against other species that are not able to increase numbers as rapidly. Obviously, most of the individuals will die, but not before they have left some offspring or resistant stages that will be capable of exploiting the resource should it become available again.

Even K-strategists, however, have some fluctuations in their population size for a variety of reasons. One reason is that even in relatively stable ecosystems, there will be variations from year to year. Floods, droughts, fires, extreme cold, and similar events may affect the carrying capacity of an area, thus causing fluctuations in population size. Epidemic disease or increased predation may also lead to populations that vary in size from year to year. Many endangered species have reduced populations because their normal environment has been altered either naturally or as a result of human activity. (See chapter 11 on biodiversity issues.)

POPULATION CYCLES

In northern regions of the world, many kinds of animals show distinct population cycles—periods of relatively large populations followed by periods of small populations. This is generally thought to be the case because the ecosystems are relatively simple, with few kinds of organisms affecting one another. Many of these cycles are quite regular. Biologists have been studying these cycles since the 1920s, and they have developed several theories about why northern populations cycle.

One idea is that heavy feeding by large populations of herbivores causes the plants to produce increased amounts of chemicals, which taste bad or are toxic. A second thought is that when an herbivore population is large, many different predators shift to eating them and the herbivore population crashes. Another idea is that interactions between a prey organism and a specialized predator naturally lead to population cycles. The length of the population cycle depends on the reproductive biology of the prey and their predators.

A study of the population biology of the collared lemming *(Dicrostonyx groenlandicus)* on Greenland illustrates the population interactions between lemmings and four different predators. Lemmings have a very high biotic potential. They produce two to three litters of young per year. Their population is held in check by four different predators. Three of these predators—the snowy owl, arctic fox, and longtailed skua (a bird that resembles a gull)—are generalist predators whose consumption of lemmings is directly related to the size of the lemming population. They constitute a density-dependent limiting factor for the lemming population. When lemming numbers are low, these predators seek other prey. For example, the snowy owl often migrates to other regions when lemming numbers are low in a particular region. The fourth predator is the shorttail weasel *(Mustela ermina)*, a specialist predator on lemmings. The weasels are much more dependent on lemmings for food than the other predators. Since the weasels mate once a year, their populations increase at a slower rate than those of the lemmings. As weasel populations increase, however, they eventually become large enough that they drive down the lemming populations. The resulting decrease in lemmings leads to a decline in the number of weasels, which allows greater survival of lemmings, which ultimately leads to another cycle of increased weasel numbers. (See figure 7.9.)

HUMAN POPULATION GROWTH

The human population growth curve has a long lag phase followed by a sharply rising exponential growth phase that is still rapidly increasing. (See figure 7.10.) A major reason for the continuing increase in the size of the human population is that the human species has lowered its death rate. When various countries reduce environmental resistance by increasing food production or controlling disease, they share this technology throughout the world. Developed countries send health care personnel to all parts of the globe to improve the quality of life for people in less-developed countries. Physicians offer advice on nutrition, and engineers develop wastewater treatment systems. Improved sanitary facilities in India and

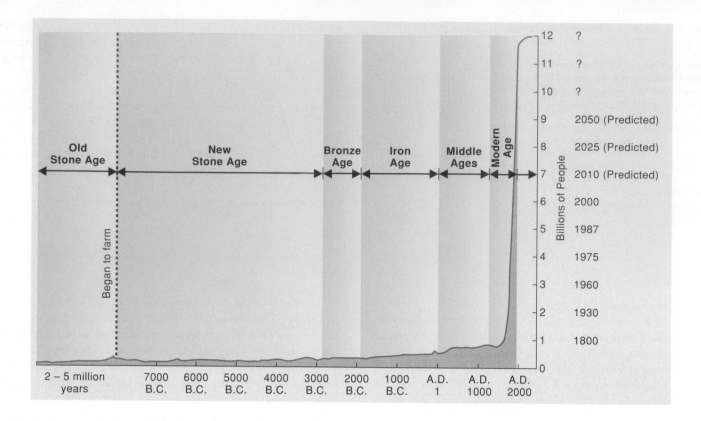

FIGURE 7.10 **The Historical Human Population Curve** From A.D. 1800 to A.D. 1930, the number of humans doubled (from 1 billion to 2 billion) and then doubled again by 1975 (4 billion) and is projected to double again (8 billion) by the year 2025. How long can this pattern continue before the Earth's ultimate carrying capacity is reached?

Source: Data from Jean Van Det Tak, et al., "Our Population Predicament: A New Look," *Population Bulletin,* vol. 34, no. 5 (December 1979), Population Reference Bureau, Washington, D.C.: and more recent data from the Population Reference Bureau.

Indonesia, for example, decreased deaths caused by cholera. These advancements tend to reduce death rates while birthrates remain high. Thus, the size of the human population increases rapidly.

Let us examine the human population situation from a different perspective. The world population is currently increasing at an annual rate of 1.2 percent. That may not seem like much, but even at 1.2 percent, the population is growing rapidly. It can be difficult to comprehend the impact of a 1 or 2 percent annual increase. Remember that a growth rate in any population compounds itself, since many of the additional individuals eventually reproduce, thus adding more individuals. One way to look at this growth is to determine how much time is needed to double the population. This is a valuable method because most of us can appreciate what life would be like if the number of people in our locality were doubled, particularly if the doubling were to occur within our lifetime.

Figure 7.11 shows the relationship between the rate of annual increase for the human population and the number of years it would take to double the population if that rate were to continue.

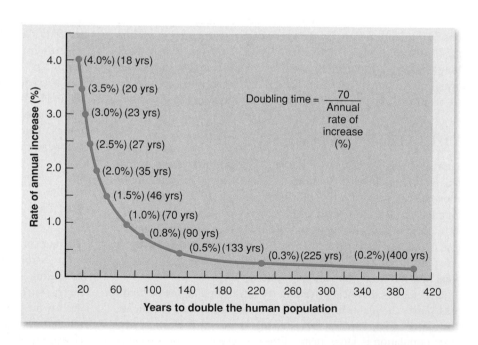

FIGURE 7.11 **Doubling Time for the Human Population** This graph shows the relationship between the rate of annual increase in percent and doubling time. A population growth rate of 1 percent per year would result in the doubling of the population in about 70 years. A population growth rate of 3 percent per year would result in a population doubling in about 23 years.

The doubling time for the human population is easily calculated by dividing the number 70 by the annual rate of increase. Thus, at a 1 percent rate of annual increase, the population will double in 70 years (70/1). At a 2 percent rate of annual increase, the human population will double in 35 years (70/2). The current worldwide annual increase of about 1.2 percent will double the world human population in about 58 years.

HUMAN POPULATION CHARACTERISTICS AND IMPLICATIONS

The human population dilemma is very complex. To appreciate it, we must understand current population characteristics and how they are related to social, political, and economic conditions.

ECONOMIC DEVELOPMENT

The world can be divided into two segments based on the state of economic development of the countries. The **more-developed countries** of the world typically have a per capita income that exceeds US $10,000; they include all of Europe, Canada, the United States, Australia, New Zealand, and Japan, with a combined population of about 1.2 billion people. The remaining countries of the world are referred to as **less-developed countries** and typically have a per capita income of less than US $5000. The population

of these countries totals almost 5.5 billion people, nearly 3 billion of whom live on less than US $2 per day. While these definitions constitute an oversimplification and some countries are exceptions, basically this means that the majority of Asian, Latin American, and African citizens are much less well off economically than those who live in the more-developed countries. Collectively, the more-developed countries of the world have relatively stable populations and are expected to grow by about 5 percent between 2008 and 2050. The less-developed regions of the world, however, have high population growth rates and are expected to grow by about 47 percent between 2008 and 2050. (See figure 7.12.) If these trends continue, the total population of the less-developed world will increase from the current 5.5 billion to over 8 billion by 2050, when this region will contain over 86 percent of the world's people.

MEASURING THE ENVIRONMENTAL IMPACT OF A POPULATION

Human population growth is tied to economic development and is a contributing factor to nearly all environmental problems. Current population growth has led to

1. famine in areas where food production cannot keep pace with increasing numbers of people;

2. political unrest in areas with great disparities in the availability of resources (jobs, goods, food);

3. environmental degradation (erosion, desertification, strip mining, oil spills, groundwater mining) caused by poor agricultural

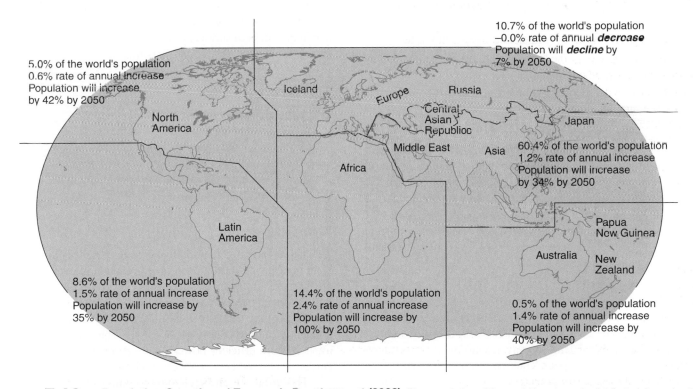

FIGURE 7.12 **Population Growth and Economic Development (2008)** The population of the world is not evenly distributed. It can be divided into the economically more-developed and less-developed nations. The more-developed nations are indicated in green and the less-developed in tan. Currently, about 83 percent of the world's population is in the less-developed nations of Latin America, Africa, and Asia. These areas also have the highest rates of population increase. Because of the high birthrates, they are likely to remain less developed and will constitute about 86 percent of the world's population by the year 2050.

In 1798, Thomas Robert Malthus, an Englishman, published an essay on human population. In it, he presented an idea that was contrary to popular opinion. His basic thesis was that human population increased in a geometric or exponential manner (2, 4, 8, 16, 32, 64, etc.), while the ability to produce food increased only in an arithmetic manner (1, 2, 3, 4, 5, 6, etc.). The ultimate outcome of these different rates would be that population would outgrow the ability of the land to produce food. He concluded that wars, famines, plagues, and natural disasters would be the means of controlling the size of the human population. His predictions were hotly debated by the intellectual community of his day. His assumptions and conclusions were attacked as erroneous and against the best interest of society. At the time he wrote the essay, the popular opinion was that human knowledge and "moral constraint" would be able to create a world that would supply all human needs in abundance. One of Malthus's basic postulates was that "commerce between the sexes" (sexual intercourse) would continue unchanged,

Thomas Robert Malthus

while other philosophers of the day believed that sexual behavior would take less procreative forms and human population would be limited. Only within the past 50 years, however, have really effective conception-control mechanisms become widely accepted and used, and they are used primarily in developed countries.

Malthus did not foresee the use of contraception, major changes in agricultural production techniques, or the exporting of excess people to colonies in the Americas, Australia, and other parts of the world. These factors, as well as high death rates, prevented the most devastating of his predictions from coming true. However, in many parts of the world today, people are experiencing the forms of population control (famine, epidemic disease, wars, and natural disasters) predicted by Malthus in 1798. Many people feel that his original predictions were valid—only his timescale was not correct—and that we are seeing his predictions come true today.

practices and the destructive effects of exploitation of natural resources;

4. water pollution caused by human and industrial waste;

5. air pollution caused by the human need to use energy for personal and industrial applications; and

6. extinctions caused by people converting natural ecosystems to managed agricultural ecosystems.

Several factors interact to determine the impact of a society on the resources of its country. These include the land and other natural resources available, the size of the population, the amount of resources consumed per person, and the environmental damage caused by using resources. The following equation is often used as a shorthand way of stating these relationships: $I = P \times A \times T$ (*Impact* on the environment = *Population* size $\times$ *Affluence* (amount of resources consumed per person) $\times$ *Technology* (effects of methods used to provide items consumed).

Population

As the population of a country increases, it puts a greater demand on its resources. Some countries have abundant natural resources, such as good agricultural land, energy resources, or mineral resources. Others are resource poor. Thus, some countries can sustain high populations while others cannot.

Population density, the number of people per unit of land area, relates the size of the population to the resources available. A million people spread out over the huge area of the Amazon Basin have much less impact on resources than that same million people in a small island country because the impact is distributed over a greater land surface. Countries with abundant resources can sustain higher population densities than resource-poor countries.

Affluence

People in highly developed countries consume huge amounts of resources. Citizens of these countries eat more food, particularly animal protein, which requires larger agricultural inputs than does a vegetarian diet. They have more material possessions and consume vast amounts of energy compared to people in the less-developed world. (See figure 7.13.)

Technology

The technology used to provide the things people consume and use is an important contributor to environmental impact. Some methods are efficient and have minor impacts; others are very damaging. For example, the use of firewood to heat homes and provide fuel for cooking can lead to deforestation. Similarly, the use of inefficient, polluting, coal-fired power plants has a high environmental impact. More efficient power plants or the use of wind or solar energy to provide energy lowers environmental impact.

The affluence and technology components of this equation are very difficult to tease apart. Often the per-capita energy consumption is used as a measure of the combined effects of affluence and technology.

Possessions of an Indian family

Possessions of an American family

FIGURE 7.13 **Differences in Affluence** These two families have greatly different numbers of possessions. Those with more possessions have a greater ecological footprint.

THE ECOLOGICAL FOOTPRINT CONCEPT

The environmental impact of the developed world is often underestimated because the population in these countries is relatively stable and local environmental conditions are good. However, developed countries purchase goods and services from other parts of the world, often degrading environmental conditions in less-developed countries. Thus, the environmental impact of highly developed regions such as North America, Japan, Australia, New Zealand, and Europe is often felt in distant places, while the impact on resources in the developed region may be minimal.

This has led to the development of the concept of the ecological footprint of a society. The **ecological footprint** is a measure of the land area required to provide the resources and absorb the wastes of a population. Most of the more-developed countries of the world have a much larger ecological footprint than is represented by their land area. For example, Japan has a highly developed economy but few resources. Thus, it must import most of the materials it needs. One study calculated that the ecological impact of Japan is nearly five times larger than its locally available resources. The same study estimated that the ecological footprint of the United States is 1.5 times locally available resources.

It is clear that as the world human population continues to increase, it will become more difficult to limit the environmental degradation that accompanies it. Since much of the population growth will occur in the less-developed areas of the world that have weak economies, the money to invest in pollution control, health programs, and sustainable agricultural practices will not be present.

While controlling world population growth would not eliminate all environmental problems, it could reduce the rate at which environmental degradation is occurring. It is also generally believed that the quality of life for many people in the world would improve if their populations grew less rapidly. Why, then, does the human population continue to grow at such a rapid rate?

FACTORS THAT INFLUENCE HUMAN POPULATION GROWTH

Human populations are subject to the same biological factors discussed earlier in this chapter. There is an ultimate carrying capacity for the human population. Eventually, limiting factors will cause human populations to stabilize. However, unlike other kinds of organisms, humans are also influenced by social, political, economic, and ethical factors. We have accumulated knowledge that allows us to predict the future. We can make conscious decisions based on the likely course of events and adjust our lives accordingly. Part of our knowledge is the certainty that as populations continue to increase, death rates and birthrates will become equal. This can happen by allowing the death rate to rise or by choosing to limit the birthrate. Controlling human population would seem to be a simple process. Once people understand that lowering the birthrate is more humane than allowing the death rate to rise, they should make the "correct" decision and control their birthrates; however, it is not quite that simple.

BIOLOGICAL FACTORS

The scientific study of human populations, their characteristics, how these characteristics affect growth, and the consequences of that growth is known as **demography.** Demographers can predict the future growth of a population by looking at several biological indicators.

Birthrate and Death Rate

Currently, in most countries of the world, the birthrate exceeds the death rate. Therefore, the size of the population must increase. (See table 7.2.) Some countries that have high birthrates and high

Populations: Characteristics and Issues 153

TABLE 7.2 Population Characteristics of the 20 Most Populous Countries, 2008

Country	Current Population (Millions)	Births per 1000 Individuals	Deaths per 1000 Individuals	Infant Mortality Rate (Deaths per 1000 Live Births)	Total Fertility Rate (Children per Woman per Lifetime)	% Married Women Using Birth Control	Rate of Natural Increase (Annual %)	Projected Population Changes 2008–50 (%)
World	6705	21	8	49	2.6	62	1.2	39
Russia	141.9	12	15	9	1.4	67	(−0.3)	(−22)
Germany	82.2	8	10	3.9	1.3	75	(−0.2)	(−13)
Japan	127.7	9	9	2.8	1.3	52	0.0	(−25)
China	1324.7	12	7	23	1.6	90	0.5	8
Thailand	66.1	13	8	16	1.6	72	0.5	4
United States	304.5	14	8	6.6	2.1	73	0.6	44
Vietnam	86.2	17	5	16	2.1	78	1.2	31
Turkey	74.8	19	6	23	2.2	71	1.2	19
Brazil	195.1	20	6	24	2.3	76	1.3	33
Iran	72.2	20	5	32	2.1	79	1.4	39
Indonesia	239.9	21	6	34	2.6	61	1.5	43
India	1149.3	24	8	57	2.8	31	1.6	53
Mexico	107.7	20	5	19	2.3	71	1.6	22
Bangladesh	147.8	24	7	52	2.7	56	1.7	46
Egypt	74.9	27	6	33	3.1	59	2.0	57
Philippines	90.5	26	5	25	3.3	51	2.1	66
Pakistan	172.8	31	8	75	4.1	30	2.2	71
Nigeria	148.1	43	18	100	5.9	12	2.5	91
Ethiopia	79.1	40	15	77	5.3	15	2.5	87
Democratic Republic of Congo	66.5	44	13	92	6.5	21	3.1	185

Source: Population Reference Bureau, 2008 Population Data Sheet.

death rates—with birthrates greatly exceeding the death rates—will grow rapidly (Nigeria and Ethiopia). Such countries usually have an extremely high mortality rate among children because of disease and malnutrition; but because the birthrate still greatly exceeds the death rate, the populations will grow rapidly.

Some countries have high birthrates and low death rates and will grow rapidly (Mexico and Indonesia). Infant mortality rates are moderately high in these countries. Other countries have low birthrates and death rates that closely match the birthrates; they will grow slowly (Japan and the United States). These and other more-developed countries typically have very low infant mortality rates. The disruption caused by the political upheaval in the former Soviet Union and Eastern Europe has resulted in several countries (e.g., Russia and Germany) having death rates that are equal to or exceed birthrates, causing their populations to decline. Because of these countries and the generally low rates of growth in the rest of Europe, the European region as a whole has a declining population.

Total Fertility Rate

The most important determinant of the rate at which human populations grow is related to how many women in the population are having children and the number of children each woman will have. The **total fertility rate** of a population is the number of children born per woman in her lifetime. A total fertility rate of 2.1 is known as **replacement fertility,** since parents produce 2 children who will replace the parents when they die. Eventually, if the total fertility rate is maintained at 2.1, population growth will stabilize. A rate of 2.1 is used rather than 2.0 because some children do not live very long after birth and therefore will not contribute to the population for very long. When a population is not growing, and the number of births equals the number of deaths, it is said to exhibit **zero population growth.**

For several reasons, however, a total fertility rate of 2.1 will not necessarily immediately result in a stable population with zero growth. First, the death rate may fall as living conditions improve and people live longer. If the death rate falls faster than the birthrate, there will still be an increase in the population even though it is reproducing at the replacement rate.

Age Distribution

The **age distribution,** the number of people of each age in the population, also has a great deal to do with the rate of population growth. If a population has many young people who are raising families or who will be raising families in the near future, the population will continue to increase even if the families limit themselves to two children. Depending on the number of young people in a population, it may take 20 years to a century for the population of a country to stabilize so that there is no net growth.

SOCIAL FACTORS

It is clear that populations in economically developed countries of the world have low fertility rates and low rates of population growth and that the less-developed countries have high fertility rates and high population growth rates. It also appears obvious that reducing fertility rates would be to everyone's advantage; however, not everyone in the world feels that way. Several factors influence the number of children a couple would like to have. Some are religious, some are traditional, some are social, and some are economic.

Culture and Traditions

The major social factors that determine family size are the status and desires of women in the culture. In many male-dominated cultures, the traditional role of women is to marry and raise children. Often this role is coupled with strong religious input as well. Typically, little value is placed on educating women, and early marriage is encouraged. In these cultures, women are totally dependent on their husbands and children in old age. Because early marriage is encouraged, fertility rates are high, since women are exposed to the probability of pregnancy for more of their fertile years. Lack of education reduces options for women in these cultures. They do not have the option to not marry or to delay marriage and thus reduce the number of children they will bear.

By contrast, in much of the developed world, women are educated, delay marriage, and have fewer children. It has been said that the single most important activity needed to reduce the world population growth rate is to educate women. Whenever the educational level of women increases, fertility rates fall. Figure 7.14 compares total fertility rates and educational levels of women in the 20 most populous countries of the world. The educational level of women is strongly correlated with the total fertility rate and economic well-being of a population.

Data on the age at which women marry and have children are strongly correlated with total fertility rate. Early marriage results in women being exposed to the probability of pregnancy for a longer period of their lives and results in a higher total fertility rate. Table 7.3 compares the age of marriage and total fertility rate.

Even childrearing practices have an influence on population growth rates. In countries where breast feeding is practiced, several benefits accrue. Breast milk is an excellent source of nutrients for the infant as well as a source of antibodies against some diseases. Furthermore, since many women do not return to a normal reproductive cycle until after they have stopped nursing, during the months a woman is breast feeding her child, she is less likely to become pregnant. Since in many cultures, breast feeding may continue for two years, it serves to increase the time between successive births. Increased time between births results in a lower mortality among women of childbearing age.

Attitudes Toward Birth Control

As women become better educated and obtain higher-paying jobs, they become financially independent and can afford to marry later and consequently have fewer children. Better-educated women are also more likely to have access to and use birth control. In economically advanced countries, a high proportion of women typically use contraception. In the less-developed countries, contraceptive use is much lower—about 28 percent in Africa, about 71 percent in Latin America, and about 67 percent in Asia (about 56 percent if China is excluded).

It is important to recognize that access to birth control alone will not solve the population problem. What is most important is the desire of women to limit the size of their families. In developed countries, the use of birth control is extremely important in regulating the birthrate. This is true regardless of religion and previous historical birthrates. For example, Italy and Spain are both traditionally Catholic countries and have low total fertility rates (Spain 1.4 and Italy 1.3). The average for the developed countries of the world is 1.6. Obviously, women in these countries make use of birth control to help them regulate the size of their families. By contrast, Mexico, which is also a traditionally Catholic country, has a total fertility rate of 2.3, which is somewhat less than 2.8, which is typical for the less-developed world regardless of religious tradition.

Women in the less-developed world typically have more children than they think is ideal, and the number of children they have is higher than the replacement fertility rate of 2.1 children. Access to birth control will allow them to limit the number of children they have to their desired number and to space their children at more convenient intervals, but they still desire more children than the 2.1 needed for replacement. Why do they desire large families? There are several reasons. In areas where infant mortality is high, it is traditional to have large families, since several of a woman's children may die before they reach adulthood. This is particularly important in the less-developed world, where there is

TABLE 7.3 Age of Marriage, Births to Young Mothers, and Total Fertility Rate (2007)

	Africa	Latin America	Asia Excluding China	Asia	Developed World
% women 15–19 years old who have married	23	17	24	15	3
% of women 15–19 years old who give birth in any year	11	8	8	5	2
Total Fertility Rate	5.0	2.5	2.8	2.4	1.6

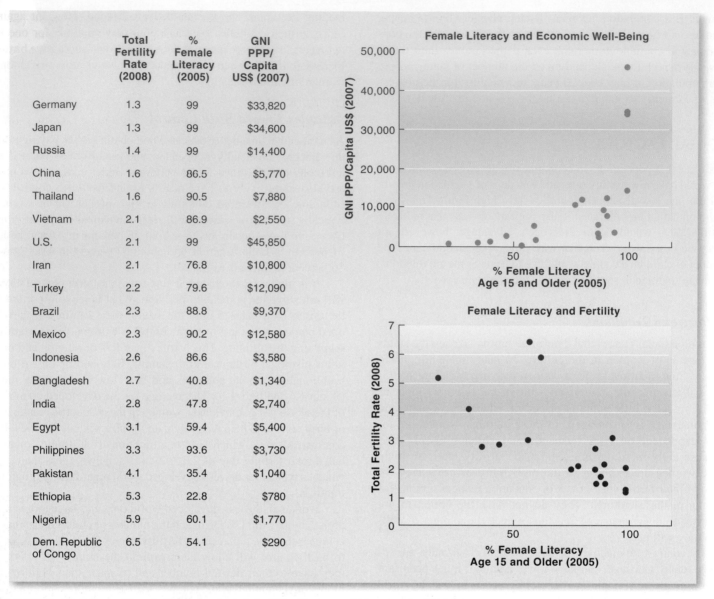

	Total Fertility Rate (2008)	% Female Literacy (2005)	GNI PPP/ Capita US$ (2007)
Germany	1.3	99	$33,820
Japan	1.3	99	$34,600
Russia	1.4	99	$14,400
China	1.6	86.5	$5,770
Thailand	1.6	90.5	$7,880
Vietnam	2.1	86.9	$2,550
U.S.	2.1	99	$45,850
Iran	2.1	76.8	$10,800
Turkey	2.2	79.6	$12,090
Brazil	2.3	88.8	$9,370
Mexico	2.3	90.2	$12,580
Indonesia	2.6	86.6	$3,580
Bangladesh	2.7	40.8	$1,340
India	2.8	47.8	$2,740
Egypt	3.1	59.4	$5,400
Philippines	3.3	93.6	$3,730
Pakistan	4.1	35.4	$1,040
Ethiopia	5.3	22.8	$780
Nigeria	5.9	60.1	$1,770
Dem. Republic of Congo	6.5	54.1	$290

FIGURE 7.14 **Female Literacy, Total Fertility Rate, and Economic Well-Being** These graphs show the relationship among female literacy, total fertility, and economic well-being of the 20 most populous countries of the world. In general the higher the female literacy rate the lower the total fertility rate and the higher the per capita income.

Source: Data from Population Reference Bureau, 2008 Data Sheet, and *Human Development Report 2007/2008*, United Nations Development program.

no government program of social security. Parents are more secure in old age if they have several children to contribute to their needs when they become elderly and can no longer work.

ECONOMIC FACTORS

In less-developed countries, the economic benefits of children are extremely important. Even young children can be given jobs that contribute to the family economy. They can protect livestock from predators, gather firewood, cook, or carry water. In the developed world, large numbers of children are an economic drain. They are prevented by law from working, they must be sent to school at great expense, and they consume large amounts of the family income. Many parents in the developed world make an economic decision about having children in the same way they buy a house

or car: "We are not having children right away. We are going to wait until we are better off financially."

POLITICAL FACTORS

Two other factors that influence the population growth rate of a country are government policies on population growth and immigration.

Government Population Policy

Many countries in Europe have official policies that state their population growth rates are too low. As their populations age and there are few births, they are concerned about a lack of working-age people in the future and have instituted programs that are meant to encourage people to have children. For example, Hungary, Sweden, and several other European countries provide paid maternity leave for mothers during the early months of a child's life and the guarantee of a job when the mother returns to work. Many countries provide childcare facilities and other services that make it possible for both parents to work. This removes some of the economic barriers that tend to reduce the birthrate. The tax system in many countries, such as the United States, provides an indirect payment for children by allowing a deduction for each child. Canada pays a bonus to couples on the birth of a child.

By contrast, most countries in the developing world publicly state that their population growth rates are too high. To reduce the birthrate, they have programs that provide information on maternal and child health and on birth control. The provision of free or low-cost access to contraceptives is usually a part of their population-control effort as well.

China and India are the two most populous countries in the world, each with over a billion people. China has taken steps to control its population and now has a total fertility rate of 1.6 children per woman while India has a total fertility rate of 2.8. This difference between these two countries is the result of different policy decisions over the last 50 years. The history of China's population policy is an interesting study of how government policy affects reproductive activity among its citizens. When the People's Republic of China was established in 1949, the official policy of the government was to encourage births, because more Chinese would be able to produce more goods and services, and production was the key to economic prosperity. The population grew from 540 million to 614 million between 1949 and 1955, while economic progress was slow. Consequently, the government changed its policy and began to promote population control.

Since 1955 China has had a series of family planning programs all aimed at reducing the number of births. Although many believe these programs include human rights violations, the birthrate has fallen steadily, and the current total fertility rate is 1.6, well below the replacement rate. Ninety percent of couples use contraception; the most commonly used forms are male and female sterilization and the intrauterine device. Abortion is also an important aspect of this program, with a ratio of over 600 abortions per 1000 live births.

By contrast, during the same 50 years, India has had little success in controlling its population. In 2000, a new plan was unveiled that had the goal of bringing the total fertility rate from 3.1 children per woman to 2 (replacement rate) by 2010. (In 2008, the total fertility rate was 2.8, so it is highly unlikely that the goal will be met.) In the past, the emphasis of government programs was on meeting goals of sterilization and contraceptive use, but this has not been successful. Today, about 55 percent of couples use contraceptives.

This new plan emphasizes improvements in the quality of life of the people. The major thrusts are to reduce infant and maternal death, immunize children against preventable disease, and encourage girls to attend school. It is hoped that improved health will remove the perceived need for large numbers of births. Currently, less than 50 percent of the women in India can read and write. The

emphasis on improving the educational status of women to the experiences of other developing countries. In m countries, it has been shown that an increase in the education of women is linked to lower fertility rates.

Immigration

The immigration policies of a country also have a significant impact on the rate at which the population grows. Birthrates are currently so low in several European countries, Japan, and China that these countries will likely have a shortage of working-age citizens in the near future. One way to solve this problem is to encourage immigration from other parts of the world.

The developed countries are under tremendous pressure to accept immigrants. The standard of living in these countries is a tremendous magnet for refugees or people who seek a better life than is possible where they currently live. In the United States, approximately one-third of the population increase experienced each year is the result of immigration. Canada encourages immigrants and has set a goal of accepting 300,000 new immigrants each year. This is about 1 percent of its current population.

POPULATION GROWTH RATES AND STANDARD OF LIVING

There appears to be an inverse relationship between the rate at which the population of a country is growing and its standard of living. The **standard of living** is an abstract concept that attempts to quantify the quality of life of people. Standard of living is a difficult concept to quantify, since various cultures have different attitudes and feelings about what is desirable. However, several factors can be included in an analysis of standard of living: economic well-being, health conditions, and the ability to change one's status in the society. Figure 7.15 lists several factors that are important in determining standard of living and compares three countries with very different standards of living (the United States, Argentina, and Kenya). One important economic measure of standard of living is the average purchasing power per person. One index of purchasing power is the **gross national income (GNI)**. The GNI is an index that measures the total goods and services generated within a country as well as income earned by citizens of the country who are living in other countries. Since the prices of goods and services vary from one country to another, a true comparison of purchasing power requires some adjustments. Therefore, a technique used to compare economic well-being across countries is a measure called the GNI PPP (gross national income purchasing power parity). Finally, the GNI PPP can be divided by the number of people in the country to get a per capita (per person) GNI PPP. As you can see from figure 7.15, a wide economic gap exists between economically advanced countries and those that are less developed. Yet the people of less-developed countries aspire to the same standard of living enjoyed by people in the developed world.

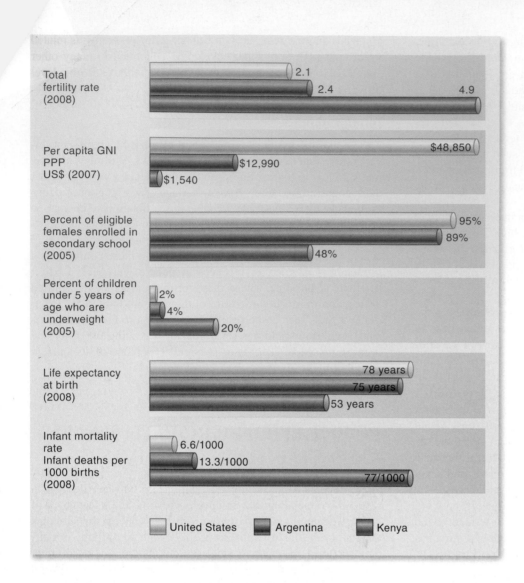

FIGURE 7.15 **Standard of Living and Population Growth in Three Countries** Standard of living is a measure of how well one lives. It is not possible to get a precise definition, but when we compare the United States, Argentina, and Kenya, it is obvious that there are great differences in how the people in these countries live. Kenya has a high population growth rate, a low life expectancy, a high infant mortality rate, and many people without adequate food. Furthermore, their incomes are low and they are poorly educated. The United States has a low population growth rate, a high life expectancy, a low infant mortality rate, and many people who eat too much. People in the United States have high educational levels and high incomes. Argentina is intermediate in all of these characteristics.

Source: Data from The Population Reference Bureau, 2008 Population Data Sheet; and Human Development Report 2007/ 2008, United Nations Development Program.

Health criteria reflect many aspects of standard of living. Access to such things as health care, safe drinking water, and adequate food are reflected in life expectancy, infant mortality, and growth rates of children. The United States and Argentina have similar life expectancies (over 75 years) and adequate nutrition. Kenya has a low life expectancy (53 years), many undernourished children (20 percent are underweight), and a high infant mortality rate (77 per 1000). The United States has a low infant mortality rate (6.6 per 1000). Argentina has an intermediate infant mortality rate (13.3 per 1000).

Finally, the educational status of people determines the kinds of jobs that are available and the likelihood of being able to improve one's status. In general, men are more likely to receive an education than women, but the educational status of women has a direct bearing on the number of children they will have and, therefore, on the economic well-being of the family. In Argentina and the United States over 89 percent of girls of high school age attend high school but only 48 percent do so in Kenya. Obviously, tremendous differences exist in the standard of living among these three countries. What the average U.S. citizen would consider poverty level would be considered a luxurious life for the average person in Kenya.

HUNGER, FOOD PRODUCTION, AND ENVIRONMENTAL DEGRADATION

As the human population increases, the demand for food rises. People must either grow food themselves or purchase it. Most people in the developed world purchase what they need and have more than enough food to eat. Most people in the less-developed world must grow their own food and have very little money to purchase additional food. Typically, these farmers have very little surplus. If crops fail, people starve. Even in countries with the highest population (China and India), the majority of the people live on the land and farm. (Fifty-five percent of Chinese and 72 percent of Indians live in rural areas.)

ENVIRONMENTAL IMPACTS OF FOOD PRODUCTION

The human population can increase only if the populations of other kinds of plants and animals decrease. Each ecosystem has a maximum biomass that can exist within it. There can be

Poverty and environmental degradation go together. Today, more than half of the people of the world live on less than US $2 a day. Impoverished people do not worry about the environmental impacts of their actions when they are trying to provide food and other basic necessities for themselves and their families. Therefore, the easing of poverty can have a positive effect on environmental conditions. Typically, the poor are unable to get loans from traditional sources and must go to moneylenders who charge extremely high rates of interest. This locks them into a cycle of poverty.

The Grameen Bank is a creation of Muhammad Yunus, a professor of economics at Chittagong University in Bangladesh, that provides an alternative source of credit for poor people. It is an outgrowth of an experiment during the late 1970s that showed that it was possible to make loans to the poor with a high rate of repayment and that the loans would improve the standard of living of the borrowers. In 1983, the Bangladeshi government established the Grameen Bank (*grameen* means village or rural in the Bengali language) as an independent bank. The bank provides tiny loans without any collateral (most are for less than US $200) to the rural poor to begin a variety of businesses that can generate income that will raise the standard of living of the people. About 90 percent of the borrowers are women. The large proportion of loans to women is based on several factors. Unmarried women (widows, divorced, abandoned) are the poorest segments of society and have the greatest need. In addition, lenders feel that women are more likely to use the money to improve the conditions of their families and to repay the loan.

One of the primary reasons the bank has been successful is that the borrowers are helped to manage the repayment of their loans. Many of the conditions of the loans may be considered intrusive, but the system works. Some of the loan conditions are:

1. There is close supervision by bank personnel.

2. Borrowers must become members of a small group of their peers that advises the borrower and supervises the repayment of loans. Because most decisions about loans are made in meetings with other borrowers present, peer pressure is strong to use the loan wisely.

3. Loans are paid back on a weekly basis, which means that loans must be for activities that provide a quick return on investment and rely on skills already possessed by the borrower.

4. Compulsory saving is required of those who receive loans. This reduces the likelihood that poor decisions by the borrower will result in squandering of the new income generated by the business started with the aid of the loan.

Today, the Grameen Bank is owned by the rural poor whom it serves. Borrowers own 90 percent of the bank's shares, while the Bangladeshi government owns the remaining 10 percent. Through its loans, about 5 percent of its borrowers come out of poverty every year. In 2006, Muhammad Yunus was awarded the Nobel Peace Prize for his efforts to help the world's poor.

The success of the Grameen Bank has led to replication of the concept in other countries. In 1997, a Microcredit Summit was held in Washington, D.C., that attracted 2900 delegates from 137 countries. They adopted a goal of reaching 100 million of the world's poorest families with microcredit and other financial services, preferably through the women in those families, by 2005. By the end of 2004, about 2500 microcredit institutions existed in nearly 100 countries. They had served over 92 million clients, over 83 percent of whom were women. The year 2005 was declared the International Year of Microcredit by the United Nations. The goal of reaching 100 million people with microcredit by the end of 2005 was not met. But by the end of 2007, 130 million people had been reached with microcredit worldwide.

shifts within ecosystems to allow an increase in the population of one species, but this always adversely affects certain other populations because they are competing for the same basic resources.

When humans need food, they convert natural ecosystems to artificially maintained agricultural ecosystems. The natural mix of plants and animals is destroyed and replaced with species useful to humans. If these agricultural ecosystems are mismanaged, the region's total productivity may fall below that of the original ecosystem. The desertification in Africa and destruction of tropical rainforests are well-known examples. In countries where food is in short supply and the population is growing, pressure is intense to convert remaining natural ecosystems to agriculture. Typically, these areas are the least desirable for agriculture and will not be productive. However, to a starving population, the short-term gain is all that matters. The long-term health of the environment is sacrificed for the immediate needs of the population.

THE HUMAN ENERGY PYRAMID

A consequence of the basic need for food is that people in less-developed countries generally feed at lower trophic levels than do those in the developed world. (See figure 7.16.) Converting the less-concentrated carbohydrates of plants into more nutritionally valuable animal protein and fat is an expensive process. During the process of feeding plants to animals and harvesting animal products, approximately 90 percent of the energy in the original plants is lost. (See chapter 5.) Although many modern agricultural practices in the developed world obtain better efficiencies than this, most of the people in the developing world are not able to use such sophisticated systems. Thus, their conversion rates approach the 90 percent loss characteristic of natural ecosystems. Therefore, in terms of economics and energy, people in less-developed countries must consume the plants themselves rather than feed the plants to animals and then consume the animals. In most cases, if the plants were fed to animals, many people would starve to death. On the other hand, a lack of protein in diets that consist primarily of plants can lead to malnutrition. It is possible to get adequate protein from a proper mixture of plant foods. In regions where food is in short supply, however, the appropriate mixture of foods is often not available. Thus, many people in the less-developed world suffer from a lack of adequate protein, which stunts their physical and mental development.

In contrast, in most of the developed world, meat and other animal protein sources are important parts of the diet. Many people suffer from overnutrition (they eat too much); they are "malnourished" in a different sense. About 66 percent of North Americans are overweight and 30 percent are obese. The ecological impact of one person eating at the carnivore level is about 10 times that of a person eating at the herbivore level. If people in the developed world were to reduce their animal protein intake, they would significantly reduce their demands on world resources. Almost all of the corn and soybeans grown in the United States are used as animal feed or to produce biofuels. Instead, if these grains were used to feed people, less grain would have to be grown and the impact on farmland would be reduced.

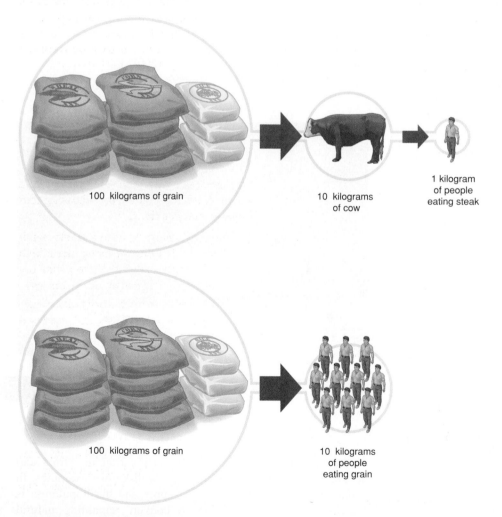

100 kilograms of grain → 10 kilograms of cow → 1 kilogram of people eating steak

100 kilograms of grain → 10 kilograms of people eating grain

FIGURE 7.16 **Population and Trophic Levels** The larger a population, the more energy it takes to sustain the population. Every time one organism is eaten by another organism, approximately 90 percent of the energy is lost. Therefore, when countries are densely populated, they usually feed at the herbivore trophic level because they cannot afford the 90 percent energy loss that occurs when plants are fed to animals. The same amount of grain can support 10 times more people at the herbivore level than at the carnivore level.

ECONOMICS AND POLITICS OF HUNGER

In countries where food is in short supply, agricultural land is already being exploited to its limit, and there is still a need for more food. This makes the United States, Canada, Australia, Argentina, New Zealand, and the European Union net food exporters. Many countries, such as India and China, are able to grow enough food for their people but do not have any left for export. Others, including many nations of the former Soviet Union, are not able to grow enough to meet their own needs and, therefore, must import food.

Improved plant varieties, irrigation, and improved agricultural methods have dramatically increased food production in some parts of the world. In recent years, India, China, and much of southern Asia have moved from being food importers to being self-sufficient and, in some cases, food exporters.

A country that is a net food importer is not necessarily destitute. Japan and some European countries are net food importers but have enough economic assets to purchase what they need. Hunger occurs when countries do not produce enough food to feed their people and cannot obtain food through purchase or humanitarian aid.

The current situation with respect to world food production and hunger is very complicated. It involves the resources needed to produce food, such as arable land, labor, and machines; appropriate crop selection; and economic incentives. It also involves the maldistribution of food within countries. This is often an economic problem, since the poorest in most countries have difficulty finding the basic necessities of life, while the rich have an excess of food and other resources. In addition, political activities often determine food availability. War, payment of foreign debt, corruption, and poor management often contribute to hunger and malnutrition.

The areas of greatest need are in sub-Saharan Africa. Africa is the only major region of the world where per capita grain production has decreased over the past few decades. People in these regions are trying to use marginal lands for food production, as forests, scrubland, and grasslands are converted to agriculture. Often, this land is not able to support continued agricultural production. This leads to erosion and desertification.

SOLVING THE PROBLEM

What should be done about countries that are unable to raise enough food for their people and are unable to buy the food they need? This is not an easy question. A simple humanitarian solution to the problem is for the developed countries to supply food. Many religious and humanitarian organizations do an excellent service by taking food to those who need it and save many lives. However, the aim should always be to provide temporary help and insist that the people of the country develop mechanisms for solving their own problem. Often, emergency food programs result in large numbers of people migrating from their rural (agricultural) areas to cities, where they are unable to support themselves. They become dependent on the food aid and stop working to raise their own food, not because they do not want to work but because they need to leave their fields to go to the food distribution centers. Many humanitarian organizations now recognize the futility of trying to feed people with gifts from the developed world. They try to provide food aid in local villages rather than in large cities and support projects that provide incentives for improving the local agricultural economy. The emphasis must be on self-sufficiency.

THE DEMOGRAPHIC TRANSITION CONCEPT

Clearly there is a relationship between the standard of living and the population growth rate. Countries with the highest standard of living have the lowest population growth rate, and those with the lowest standard of living have the highest population growth rate. This has led many people to suggest that countries naturally go through a series of stages called **demographic transition.**

THE DEMOGRAPHIC TRANSITION MODEL

The demographic transition model is based on the historical, social, and economic development of Europe and North America. In a demographic transition, the following four stages occur (see figure 7.17):

1. Initially, countries have a stable population with a high birthrate and a high death rate. Death rates often vary because of famine and epidemic disease.

2. Improved economic and social conditions (control of disease and increased food availability) bring about a period of rapid population growth as death rates fall. Birthrates remain high.

3. As countries develop an industrial economy, birthrates begin to drop because people desire smaller families and use contraceptives.

4. Eventually, birthrates and death rates again become balanced. However, the population now has *low* birthrates and *low* death rates.

This is a very comfortable model because it suggests that if a country can develop a modern industrial economy, then social, political, and economic processes will naturally cause its population to stabilize.

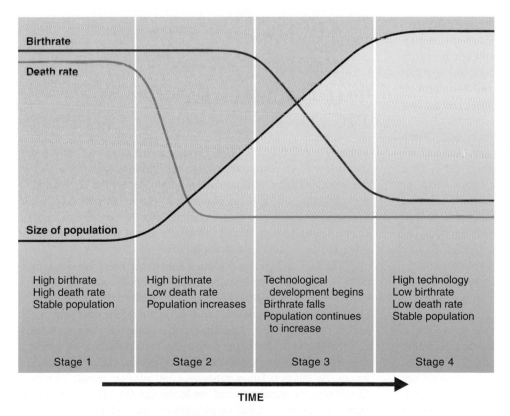

FIGURE 7.17 **Demographic Transition** The demographic transition model suggests that as a country develops technologically, it automatically experiences a drop in the birthrate. This certainly has been the experience of the developed countries of the world. However, the developed countries make up less than 20 percent of the world's population. It is doubtful whether the less-developed countries can achieve the kind of technological advances experienced in the developed world.

APPLYING THE MODEL

However, the model leads to some serious questions. Can the historical pattern exhibited by Europe and North America be repeated in the less-developed countries of today? Europe, North America, Japan, and Australia passed through this transition period when world population was lower and when energy and natural resources were still abundant. It is doubtful whether these supplies are adequate to allow for the industrialization of the major portion of the world currently classified as less developed.

Furthermore, when the countries of Europe and North America passed through the demographic transition, they had access to large expanses of unexploited lands, either within their boundaries or in their colonies. This provided a safety valve for expanding populations during the early stages of the transition. Without this safety valve, it would have been impossible to deal adequately with the population while simultaneously encouraging economic development. Today, less-developed countries may be unable to accumulate the necessary capital to develop economically, since they do not have uninhabited places to which their people can migrate and an ever-increasing population is a severe economic drain.

A second concern is the time element. With the world population increasing as rapidly as it is, industrialization probably cannot occur fast enough to have a significant impact on population growth. As long as people in less-developed countries are poor, there is a strong incentive to have large numbers of children. Children are a form of social security because they take care of their elderly parents. Only people in developed countries can save money for their old age. They can choose to have children, who are expensive to raise, or to invest money in some other way.

Today, most people feel that this model provides important insight into why some populations stabilize, but that most countries will require assistance in the form of economic development funds, education, and birth control information and technology if they are to be able to make the transition.

THE U.S. POPULATION PICTURE

In many ways, the U.S. population, which has a total fertility rate that is low (2.1), is similar to those of other developed countries of the world with low birthrates. One might expect the population to be stabilizing under these conditions. However, two factors are operating to cause significant change over the next 50 years. One factor has to do with the age structure of the population, and the other has to do with immigration policy.

The U.S. population includes a **postwar baby boom** component, which has significantly affected population trends. These baby boomers were born during an approximately 15-year period (1947–61) following World War II, when birthrates were much higher than today, and constitute a bulge in the age distribution profile. (See figure 7.18.) As members of this group have raised families, they have had a significant influence on how the U.S. population has grown. As this population bulge ages and younger people limit their family size, the population will gradually age. By 2030, about 20 percent of the population will be 65 years of age or older.

Both legal and illegal immigration significantly influence future population growth trends. Even with the current total fertility rate of 2.0 children per woman, the population is still growing by about 1.1 percent per year. About 0.6 percent is the result of

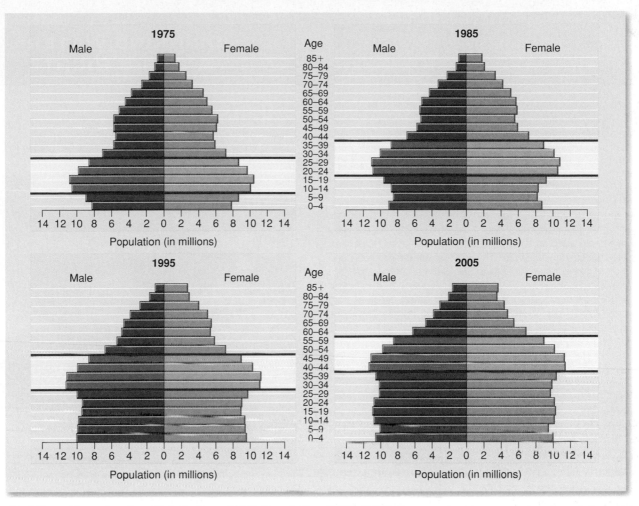

FIGURE 7.18 **Changing Age Distribution of U.S. Population (1975–2005)** These graphs show the number of people in the United States at each age level. Notice that in the year 2005 a bulge begins to form at age 40 to 44 and ends at about age 55 to 59. These people represent the "baby boom" that followed World War II. As you compare the age distribution for 1975, 1985, and 1995 with 2005, you can see that this group of people moves through the population. As this portion of the population has aged, it has had a large impact on the nature of the U.S. population. In the 1970s and 1980s, baby boomers were in school. In the 1990s, they were in their middle working years. In 2005, some of them were beginning to retire, and many more will do so throughout the decade.

Source: Data from the U.S. Department of Commerce, U.S. Census Bureau, International Data Base.

natural increases due to the difference between birthrates and death rates. The remainder is the result of immigration into the United States. The U.S. Census Bureau projects that immigration will increase significantly and will account for 50 percent of population growth by the year 2050.

Current immigration policy in the United States is difficult to characterize. Strong measures are being taken to reduce illegal immigration across the southern border. This is in part due to pressures placed on Congress by states that receive large numbers of illegal immigrants. Illegal immigrants add to the education and health care costs that states must fund. At the same time, some segments of the U.S. economy (agriculture, tourism) maintain that they are unable to find workers to do certain kinds of work. Consequently, special guest workers are allowed to enter the country for limited periods to serve the needs of these segments of the economy. There is also a consistent policy of allowing immigration that reunites families of U.S. residents. Obviously, the families that fall into this category are likely to include U.S. citizens who were recent immigrants themselves. Most immigration policy is

the result of political decisions rather than decisions that relate to population policy or a concern about the rate at which the U.S. population is growing.

Projections based on the 2000 census indicate that the population will continue to grow and does not seem to be moving toward zero growth despite the fact that the total fertility rate is at the replacement rate of 2.1. Two reasons account for this. Immigration adds about a million people per year, and new immigrants typically are young and have larger numbers of children than nonimmigrants. This is likely to result in an increase from about 305 million people in 2008 to about 438 million by 2050.

Differences in family size exist between different segments of the U.S. population. The Hispanic and Asian-American portions of the population tend to have larger families. Furthermore, many of the people in these groups are recent immigrants, so these portions of the population will grow rapidly, and the caucasian portion will decline. Thus, the U.S. population will be much more ethnically diverse in the future.

Death rates in many countries of the developing world are directly related to sanitation and access to safe drinking water. Target 10 of the United Nations Millennium Development Goals recognizes this connection and states that the goal is to

Halve by 2015, the proportion of people without sustainable access to safe drinking water and basic sanitation.

As of the year 2004—the latest year for which there are data—about one-sixth of the world's population (over 1.1 billion people) did not have access to safe drinking water. Those who live in rural areas are least likely to have access to safe drinking water. The regions with the greatest need to improve safe drinking water access are Sub-Saharan Africa, East Asia (China, North and South Korea, and Taiwan), and South Asia (India, Pakistan, and Bangladesh). Even those who did have access to safe drinking water typically did not have access in their homes. Someone from the household—usually women or children—must walk to a safe water source and carry water back to their homes. If people must walk more than 30 minutes round trip to get water, they are not likely to get enough water to satisfy the basic needs of washing, cooking, and drinking—about 20 liters/person/day (about 5 gallons/person/day).

Diarrheal diseases caused by bacteria, viruses, or protozoa in drinking water or contaminated food cause about 2.2 million deaths a year and nearly 90 percent of these cases are attributable to a lack of safe drinking water. Most of these deaths occur in children under 5 years of age. Children who are subject to repeated bouts of severe diarrhea are also likely to suffer from malnutrition because they are unable to absorb nutrients during their illness.

Children in India obtaining water from an unsafe source.

WHAT DOES THE FUTURE HOLD?

Humans are subject to the same limiting factors as other species. We cannot increase beyond our ability to acquire raw materials and energy and safely dispose of our wastes. We also must remember that interactions with other species and with other humans will help determine our carrying capacity. Furthermore, when we think about the human population and carrying capacity, we need to distinguish between the biological carrying capacity, which describes how many people the Earth can support, and a cultural carrying capacity, which describes how many people the Earth can support with a reasonable standard of living. Regardless of which of these two concepts we are thinking about, the same four basic factors are involved. Let us look at these four factors in more detail.

AVAILABLE RAW MATERIALS

Many of us think of raw materials simply as the amount of food available. However, we have become increasingly dependent on technology, and our lifestyles are directly tied to our use of other kinds of resources, such as irrigation water, genetic research, and antibiotics. Food production is becoming a limiting factor for some segments of the world's human population. Malnutrition is a serious problem in many parts of the world because sufficient food is not available. Currently, about 1 billion people (one-sixth of the world's population) suffer from a lack of adequate food.

AVAILABLE ENERGY

The second factor, available energy, involves problems similar to those of raw materials. Essentially, all species on Earth are ultimately dependent on sunlight for their energy. Currently, the world's human population depends on fossil fuels to raise food, modify the environment, and move from place to place. When energy prices increase, much of the world's population is placed in jeopardy because incomes are not sufficient to pay the increased costs for energy and other essentials. New, less disruptive methods of harnessing this energy must be developed to support an increasing population. Increases in the efficiency with which energy is used could reduce demands on fossil fuels. In addition, the development of more efficient solar and wind energy conversion systems could reduce the need for fossil fuels.

The three countries that make up North America (Canada, the United States, and Mexico) interact politically, socially, and economically. The characteristics of their populations determine how they interact. Canada and the United States are both wealthy countries with similar age structures and low total fertility rates. The United States has a total fertility rate of 2.1, and Canada has a total fertility rate of 1.6. Both countries have a relatively small number of young people (20 percent or less of the population is under 15 years of age) and a relatively large number of older people (about 13 to 14 percent are 65 or older). Without immigration, these countries would have stable or falling populations. Therefore, they must rely on immigration to supply additional people to fill the workplace. This is particularly true for jobs with low pay. The United States currently receives about 1 million immigrants per year. Canada currently receives about 230,000 immigrants per year but is planning to increase that to about 300,000 in the future.

Mexico, on the other hand, has a young, rapidly growing population. About 31 percent of the population is under 15 years of age, and the total fertility rate is 2.6. At that rate, the population will double in about 32 years. Furthermore, the average Mexican has a purchasing power about 36 percent of that in Canada and about 27 percent of that in the United States. These conditions create a strong incentive for individuals to migrate from Mexico to other parts of the world. The United States is the usual country of entry. The number is hard to evaluate, since many enter the United States illegally or as seasonal workers and eventually plan to return to Mexico. Current estimates are that about 250,000 people enter the United States from Mexico each year.

In response to the large number of illegal immigrants from Mexico, the United States has erected fences and increased surveillance. In 2008, about 725,000 people were apprehended attempting to cross the border from Mexico to the United States. Thus, instead of crossing at normal points of entry, illegal immigrants are likely to cross the border in remote desert areas, which has resulted in numerous deaths due to exposure.

The economic interplay between Mexico and the United States has several components. Working in the United States allows Mexican immigrants to improve their economic status. Furthermore, their presence in the United States has a significant effect on the economy of Mexico, since many immigrants send much of their income to Mexico to support family members. The North American Free Trade Agreement (NAFTA) allows for relatively free exchange of goods and services among Canada, the United States, and Mexico. Consequently, many Canadian, U.S., and European businesses have built assembly plants in Mexico to make use of the abundant inexpensive labor, particularly along the U.S.-Mexican border. Labor leaders in Canada and the United States are concerned about the effect access to low-cost labor will have on their membership and complain that many high-paying jobs have been moved to Mexico, where labor costs are lower.

Country	Population Size 2008 (Millions)	Birthrate per 1000	Death Rate per 1000	Rate of Natural Increase	Total Fertility Rate (Children per woman per lifetime)	Life Expectancy (Years)	Infant Mortality Rate (Deaths per 1000 births)	Per Capita GNI PPP 2007 US$	% Under 15 Years of Age	% Over 65 Years of Age	Annual Immigration/ Emigration (Estimate)
Canada	33.3	11	7	0.3	1.6	80	5.4	35,310	17	14	+230,000
United States	304.5	14	8	0.6	2.1	78	6.6	45,850	20	13	+1,000,000
Mexico	107.7	20	5	1.9	2.3	75	19	12,580	32	6	−650,000

Source: Population Reference Bureau, 2008 Population Data Sheet.

WASTE DISPOSAL

Waste disposal is the third factor determining the carrying capacity for humans. Most pollution is, in reality, the waste product of human activity. Lack of adequate sewage treatment and safe drinking water causes large numbers of deaths each year. Some people are convinced that disregard for the quality of our environment will be a major limiting factor. In any case, it makes good sense to control pollution and to work toward cleaning our environment.

INTERACTION WITH OTHER ORGANISMS

The fourth factor that determines the carrying capacity of a species is interaction with other organisms. We need to become aware that we are not the only species of importance. When we convert land to meet our needs, we displace other species from their habitats. Many of these displaced organisms are not able to compete with us successfully and must migrate or become extinct. Unfortunately, as humans expand their domain, the areas available to these displaced organisms become rarer. Parks and natural areas have become tiny refuges for the plants and animals that once occupied vast expanses of land. If these refuges fall to the developer's bulldozer or are converted to agricultural use, many organisms will become extinct. What today seems like an unimportant organism, one that we could easily do without, may someday be seen as an important link to our very survival. It is also important to recognize that many organisms provide services that we enjoy without thinking about them. Forest trees release water and moderate temperature changes, bees and other insects pollinate crops, insect predators eat pests, and decomposers recycle dead organisms. All of these activities illustrate how we rely on other organisms. Eliminating the services of these valuable organisms would be detrimental to our way of life.

SOCIAL FACTORS INFLUENCE HUMAN POPULATION

Human survival depends on interaction and cooperation with other humans. Current technology and medical knowledge are available to control human population growth and to improve the health of the people of the world. Why, then, does the population continue to increase, and why do large numbers of people continue to live in poverty, suffer from preventable diseases, and endure malnutrition? Humans are social animals who have freedom of choice and frequently do not do what is considered "best" from an unemotional, uninvolved, biological point of view. People make decisions based on history, social situations, ethical and religious considerations, and personal desires. The biggest obstacles to controlling human population are not biological but are the province of philosophers, theologians, politicians, and sociologists. People in all fields need to understand that the cause of the population problem has both biological and social components if they are to successfully develop strategies for addressing it.

ULTIMATE SIZE LIMITATION

The human population is subject to the same biological constraints as are other species of organisms. We can say with certainty that our population will ultimately reach its carrying capacity and stabilize. There is disagreement about how many people can exist when the carrying capacity is reached. Some people suggest we are already approaching the carrying capacity, while others maintain that we could more than double the population before the carrying capacity is reached. Furthermore, uncertainty exists about what the primary limiting factors will be and about the quality of life the inhabitants of a more populous world would have. If the human population continues to grow at its current rate of 1.2 percent per year, the population will reach 9.4 billion by 2050.

As with all K-strategist species, when the population increases, density-dependent limiting factors will become more significant. Some people suggest that a lack of food, a lack of water, or increased waste will ultimately control the size of the human population. Still others suggest that, in the future, social controls will limit population growth. These social controls could be either voluntary or involuntary. In the economically developed portions of the world, families have voluntarily lowered their birthrates to fewer than two children per woman. Most of the poorer countries of the world have higher birthrates. What kinds of measures are needed to encourage them to limit their populations? Will voluntary compliance with stated national goals be enough, or will enforced sterilization and economic penalties become the norm? Others are concerned that countries will launch wars to gain control of limited resources or to simply eliminate people who compete for the use of those resources.

It is also important to consider the age structure of the world population. In most of the world, there are many reproductive and prereproductive individuals. Since most of these individuals are currently reproducing or will reproduce in the near future, even if they reduce their rate of reproduction, there will be a sharp increase in the number of people in the world in the next few years.

No one knows what the ultimate human population size will be or what the most potent limiting factors will be, but most agree that we are approaching the maximum sustainable human population. If the human population continues to increase, eventually the amount of agricultural land available will not be able to satisfy the demand for food.

The Lesser Snow Goose—A Problem Population

Lesser snow geese use breeding grounds in the Arctic and subarctic regions of Canada and wintering grounds on salt marshes of the Gulf of Mexico, California, and Mexico. (See map.) One population of the lesser snow goose has breeding grounds around Hudson Bay and wintering grounds on the Gulf of Mexico. Because these birds migrate south from Hudson Bay through the central United States west of the Mississippi River to the west coast of the Gulf of Mexico, they are called the midcontinent population. Populations of the midcontinent lesser snow goose have grown from about 800,000 birds in 1969 to an estimated 4.5 million in 2005. (Some experts believe the number may be closer to 6 million.) These numbers are so large that the nesting birds are destroying their breeding habitat.

What has caused this drastic population increase? A primary reason appears to be a change in feeding behavior. Originally, during the winter snow geese fed primarily on the roots and tubers of aquatic vegetation in the salt marshes of the Gulf of Mexico. Since this habitat and the food source it provided is limited, there was a natural control on the size of the population. However, many of these wetland areas have been destroyed. Simultaneously, agriculture (particularly rice farming) in the Gulf Coast region provided an alternative food source. As the geese began to use agricultural areas for food, they had an essentially unlimited source of food during the winter months, which led to very high winter survival. Furthermore, concerns about erosion have resulted in farmers using no-till and reduced tillage practices in the grain fields along the geese's flyway. The fields provided food for the fall migration that improved the survival of young snow geese. During the spring migration, geese used the same resources to store food as body fat that they took with them to their northern breeding grounds. Other factors that may have played a role are the decline in the number of people who hunt geese and a warming of the climate that leads to greater breeding success on the cold northern breeding grounds. The large goose population is causing habitat destruction on their Canadian breeding grounds.

Habitat destruction occurring in Canada is a result of several factors. Snow geese feed by ripping plants from the ground, and they nest in large colonies. Therefore, they can have a powerful local impact on the vegetation. This coupled with a huge increase in the breeding population has resulted in the destruction of large areas of the coastal vegetation around Hudson Bay. (See photo.)

What can be done to control the snow goose population and protect their breeding habitat? There are two possibilities: let nature take its course or use population management tools. If nothing is done, the geese will eventually destroy so much of their breeding habitat that they will be unable to breed successfully, and the population will crash. It could take decades for the habitat to recover, however. The alternative is to use population management tools to try to resolve the problem. Several management activities have been instituted. These include allowing hunters to kill more geese, increasing the length of the hunting season, allowing hunts during the spring as well as the fall migration seasons, and allowing more harvest by Canadian native communities. In recent years, the harvest in the United States and Canada has increased, varying between 1 million and 1.5 million geese killed. Preliminary evidence suggests that the increased harvest is starting to have an effect. It will take several more years, however, to determine if the population can be controlled.

- Do you think the killing of geese is justified to protect the coastal habitat of Hudson Bay?
- Do you think it is better to have the population crash as a result of natural forces or to reduce the size of the population by increasing the harvest?

Map Legend

- Wintering grounds
- Nesting grounds
- Principal staging areas
- Pacific flyway migration
- Greater snow goose migration
- Midcontinental spring migration
- Midcontinental fall migration
- Western Canadian Arctic population

The map shows the migration routes between the traditional nesting and wintering areas of different snow goose populations.

This photo shows the kind of damage done by geese. The exclosures were established by Dawn R. Bazely and Robert L. Jefferies of the Hudson Bay Project.

Populations: Characteristics and Issues

SUMMARY

A population is a group of organisms of the same species that inhabits an area. The birthrate (natality) is the number of individuals entering the population by reproduction during a certain period. The death rate (mortality) measures the number of individuals that die in a population during a certain period. Population growth is determined by the combined effects of the birthrate and death rate.

The sex ratio of a population is a way of stating the relative number of males and females. Age distribution and the sex ratio have a profound impact on population growth. Most organisms have a biotic potential much greater than that needed to replace dying organisms.

Interactions among individuals in a population, such as competition, predation, and parasitism, are also important in determining population size. Organisms may migrate into (immigrate) or migrate out of (emigrate) an area as a result of competitive pressure.

A typical population growth curve shows a lag phase followed by an exponential growth phase, a deceleration phase, and a stable equilibrium phase at the carrying capacity. The carrying capacity is determined by many limiting factors that are collectively known as environmental resistance. The four major categories of environmental resistance are available raw materials, available energy, disposal of wastes, and interactions among organisms. Some populations experience a death phase following the stable equilibrium phase.

K-strategists typically are large, long-lived organisms that reach a stable population at the carrying capacity. Their population size is usually controlled by density-dependent limiting factors. Organisms that are r-strategists are generally small, short-lived organisms that reproduce very quickly. Their populations do not generally reach a carrying capacity but crash because of some density-independent limiting factor.

Currently, the world's population is growing very rapidly. The causes for human population growth are not just biological but also social, political, philosophical, and theological. Many of the problems of the world are caused or made worse by an increasing human population. Most of the growth is occurring in the less-developed areas of the world (Africa, Asia, and Latin America), where people have a low standard of living. The more-developed regions of the world, with their high standard of living, have relatively slow population growth and, in some instances, declining populations.

Demography is the study of human populations and the things that affect them. Demographers study the sex ratio and age distribution within a population to predict future growth. Population growth rates are determined by biological factors such as birthrate, which is determined by the number of women in the population and the age of the women, and death rate. Sociological and economic conditions are also important, since they affect the number of children desired by women, which helps set the population growth rate. In more-developed countries, women usually have access to jobs. Couples marry later, and they make decisions about the number of children they will have based on the economic cost of raising children. In the less-developed world, women marry earlier, and children have economic value as additional workers, as future caregivers for the parents, and as status for either or both parents.

The demographic transition model suggests that as a country becomes industrialized, its population begins to stabilize. However, there is little hope that the Earth can support the entire world's population in the style of the industrialized nations. It is doubtful whether there are enough energy resources and other natural resources to develop the less-developed countries or whether there is enough time to change trends of population growth. Highly developed nations should anticipate increased pressure in the future to share their wealth with less-developed countries.

THINKING GREEN

1. Participate in a local effort to eliminate alien species.
2. Make a donation to an organization devoted to improving living conditions in the developing world.
3. Express your views on population policy to your congressional representatives.
4. Eat at the herbivore trophic level for one week.
5. Trace your family back three generations. Start with your grandparents and list all of their descendants.
6. Determine the number of siblings for each of the following: each of your grandparents, each parent, you.

Like much of the developed world, the United States has an aging population and would cease to grow without immigration. Immigrants (both legal and illegal) have larger families than non-immigrants. Immigrants often take low-paying jobs that the rest of the population does not want. Illegal immigrants from Mexico constitute a major problem, and the United States spends over a billion dollars each year to try to control illegal immigration.

Consequently, many people support a guest-worker program in which immigrants could enter the country for specific time periods but must eventually return to their home countries. Choose to support or oppose the concept of a guest-worker program, and develop arguments to support your point of view.

REVIEW QUESTIONS

1. How is biotic potential related to the rate at which a population will grow?
2. List three characteristics populations might have.
3. Why do some populations grow? What factors help to determine the rate of this growth?
4. Under what conditions might a death phase occur?
5. List four factors that could determine the carrying capacity of an animal species.
6. How do the concepts of birthrate and population growth differ?
7. How does the population growth curve of humans compare with that of bacteria on a petri dish?
8. How do K-strategists and r-strategists differ?
9. As the human population continues to increase, what might happen to other species?
10. All successful organisms overproduce. What advantage does this provide for the species? What disadvantages may occur?
11. What is demography?
12. What is demographic transition? What is it based on?
13. What does the age distribution of a population mean?
14. List 10 differences between your standard of living and that of someone in a less-developed country.
15. Why do people who live in overpopulated countries use plants as their main source of food?
16. Which three areas of the world have the highest population growth rate? Which three areas of the world have the lowest standard of living?
17. What role does the status of women play in determining population growth rates?
18. Describe three reasons why women in the less-developed world might desire more than two children.

CRITICAL THINKING QUESTIONS

1. Why do you suppose some organisms display high natality and others display lower natality? For example, why do cottontail rabbits show high natality and wolves relatively low natality? Why wouldn't all organisms display high natality?
2. Consider the differences between K-strategists and r-strategists. What costs are incurred by adopting either strategy? What evolutionary benefits does each strategy enjoy?
3. Do you think it is appropriate for developed countries to persuade less-developed countries to limit their population growth? What would be appropriate and inappropriate interventions, according to your ethics? Now imagine you are a citizen of a less-developed country. What might be your reply to those who live in more-developed countries? Why?
4. Population growth causes many environmental problems. Identify some of these problems. What role do you think technology will play in solving these problems? Are you optimistic or pessimistic about these problems being solved through technology? Why?
5. Do you think that demographic transition will be a viable option for world development? What evidence leads you to your conclusions? What role should the developed countries play in the current demographic transition of developing countries? Why?
6. Imagine a debate between an American and a Sudanese person about human population and the scarcity of resources. What perspectives do you think the American might bring to the debate? What perspectives do you think the Sudanese would bring? What might be their points of common ground? On what might they differ?
7. Many people in developing countries hope to achieve the standard of living of those in the developed world. What might be the effect of this pressure on the environment in developing countries? On the political relationship between developing countries and already developed countries? What ethical perspective do you think should guide this changing relationship?
8. The demographic changes occurring in Mexico have an influence on the United States. What problems does Mexico face regarding its demographics? Should the United States be involved in Mexican population policy?

CHAPTER 8

ENERGY AND CIVILIZATION: PATTERNS OF CONSUMPTION

Modern societies are highly dependent on the use of energy. This scene of Rosslyn, Virginia, shows the use of energy for transportation and lighting.

CHAPTER OUTLINE

History of Energy Consumption
    Biological Energy Sources
    Increased Use of Wood
    Fossil Fuels and the Industrial Revolution
    The Role of the Automobile
    Growth in the Use of Natural Gas
How Energy Is Used
    Residential and Commercial Energy Use
    Industrial Energy Use
    Transportation Energy Use
Electrical Energy
The Economics and Politics of Energy Use
    Fuel Economy and Government Policy
    Electricity Pricing
    The Importance of OPEC
Energy Consumption Trends
    Growth in Energy Use
    Available Energy Sources
    Political and Economic Factors

ISSUES & ANALYSIS
Government Action and Energy Policy 182

CASE STUDY
Biomass Fuels and the Developing World 175

CAMPUS SUSTAINABILITY INITIATIVE
Delta College and Energy Efficiency 183

GOING GREEN
Reducing Automobile Use in Cities 174

WATER CONNECTIONS
Heating Water–Saving Energy 176

OBJECTIVES

After reading this chapter, you should be able to:

- Explain why all organisms require a constant input of energy.
- Describe how per capita energy consumption increased as civilization developed from hunting and gathering to primitive agriculture to advanced cultures.
- Describe how advanced modern civilizations developed as new fuels were used to run machines.
- Correlate the Industrial Revolution with social and economic changes.
- Explain how cheap oil and natural gas led to a consumption-oriented society.
- Explain how the automobile changed people's lifestyles.
- Explain why energy consumption is growing more rapidly in developing countries than in the industrialized world.
- Describe the role of OPEC in determining oil prices.

HISTORY OF ENERGY CONSUMPTION

Every form of life and all societies require a constant input of energy. If the flow of energy through organisms or societies ceases, they stop functioning and begin to disintegrate. Some organisms and societies are more energy efficient than others. In general, complex industrial civilizations use more energy than simple hunter-gatherer or primitive agricultural cultures. If modern societies are to survive, they must continue to expend energy. However, they may need to change their pattern of energy consumption as current sources become limited.

BIOLOGICAL ENERGY SOURCES

Energy is essential to maintain life. In every ecosystem, the sun provides that energy. (See chapter 5.) The first transfer of energy occurs during photosynthesis, when plants convert light energy into the chemical energy in the organic molecules they produce. Herbivorous animals utilize the food energy in the plants. The herbivores, in turn, are a source of energy for carnivores.

Because nearly all of their energy requirements were supplied by food, primitive humans were no different from other animals in their ecosystems. In such hunter-gatherer cultures, nearly all human energy needs were met by using plants and animals as food, tools, and fuel. (See figure 8.1.)

Early in human history, people began to use additional sources of energy to make their lives more comfortable. They domesticated plants and animals to provide a more dependable supply of food. They no longer needed to depend solely on gathering wild plants and hunting wild animals for sustenance. Domesticated animals also furnished a source of energy for transportation, farming, and other tasks. (See figure 8.2.) Wood or other plant material provided a source of fuel for heating and cooking. Eventually, this biomass energy was used in simple technologies, such as shaping tools and extracting metals.

INCREASED USE OF WOOD

Early civilizations, such as the Aztecs, Chinese, Indians, Greeks, Egyptians, and Romans were culturally advanced, but their societies used human muscle, animal muscle, and burned biomass as sources of energy. Except for limited use of some wind-powered and water-powered devices such as ships and canoes, the controlled use of fire was the first use of energy in a form other than food. Wood was the primary fuel. (Wood was also used for building materials and other cultural uses.) The energy provided by wood enabled people to cook their food, heat their dwellings, and develop a primitive form of metallurgy. Such advances separated humans from other animals. When dense populations of humans made heavy use of wood for fuel and building materials, they eventually used up the readily available sources and had to import wood or seek alternative forms of fuel.

Because of a long history of high population density, India and some other parts of the world experienced a wood shortage

FIGURE 8.1 **Hunter-Gatherer Society** In a hunter-gatherer society, people obtain nearly all of their energy from the collection of wild plants and the hunting of animals.

FIGURE 8.2 **Animal Power** The domestication of animals was an important stage in the development of human civilization. Domesticated animals provided people with a source of power other than their own muscles.

hundreds of years before Europe and North America did. In many of these areas, animal dung replaced wood as a fuel source. It is still used today in some parts of the world.

Western Europe and North America were able to use wood as a fuel for a longer period of time. The forests of Europe supplied sufficient fuel until the thirteenth century. In North America, vast expanses of virgin forests supplied adequate fuel until the late nineteenth century. Fortunately, when local supplies of wood declined in Europe and North America, coal, formed from fossilized plant remains, was available as an alternative energy source. By 1890, coal had replaced wood as the primary energy source in North America.

FOSSIL FUELS AND THE INDUSTRIAL REVOLUTION

Fossil fuels are the remains of plants, animals, and microorganisms that lived millions of years ago. (The energy in these fuels is stored sunlight, just as the biomass of wood represents stored sunlight.) During the Carboniferous period, 286 million to 362 million years ago, when the Earth's climate was warmer and wetter than it is now, conditions were conducive to the formation of large deposits of coal.

Oil and natural gas formed primarily from one-celled marine organisms whose bodies accumulated in large quantities on the seafloor and became compressed over millions of years. Heat and pressure from overlying sediment layers eventually converted the organic matter into oil and gas. Ever since machines replaced muscle power, the major energy sources for the world have been fossil remains from the distant past.

Historically, the first fossil fuel to be used extensively was coal. In the early eighteenth century, regions of the world that had readily available coal deposits were able to switch to this new fuel and participate in a major cultural change known as the **Industrial Revolution.** (See figure 8.3.) The Industrial Revolution began in England and spread to much of Europe and North America. It involved the invention of machines that replaced human and animal labor in manufacturing and transporting goods. Central to this change was the invention of the steam engine, which could convert heat energy into the energy of motion. The steam engine made possible the large-scale mining of coal. Before steam engines, coal mines flooded and thus were not economical. The source of energy for steam engines was either wood or coal; wood was quickly replaced by coal in most cases. Nations without a source of coal or those possessing coal reserves that were not easily exploited did not participate in the Industrial Revolution.

Prior to the Industrial Revolution, Europe and North America were predominately rural. Goods were manufactured on a small scale in the home. However, as machines and the coal to power them became increasingly available, the factory system of manufacturing products replaced the small home-based operation. Because expanding factories required a constantly increasing labor supply, people left the farms and congregated in areas surrounding the factories. Villages became towns, and towns became cities. Widespread use of coal in cities resulted in increased air pollution. In spite of these changes, the Industrial Revolution was viewed as progress. Energy consumption increased, economies grew, and people prospered. Within a span of 200 years, the daily per capita energy consumption of industrialized nations increased eightfold.

THE ROLE OF THE AUTOMOBILE

Although the Chinese used some oil and natural gas as early as 1000 B.C., the oil well that Edwin L. Drake, an early oil prospector, drilled in Pennsylvania in 1859 was the beginning of the modern petroleum era. For the first 60 years of production, the principal use of oil was to make kerosene, a fuel burned in lamps to provide lighting. The gasoline produced was discarded as a waste product. During this time, oil was abundant relative to its demand, and thus, it had a low price.

The invention of the automobile dramatically increased the demand for oil products. In 1900, the United States had only 8000 automobiles. By 1950, it had over 40 million cars, and by 2000, over 200 million. More oil was needed to make automobile fuel and lubricants. During this period, the percentage of energy provided by oil increased from about 2 percent in 1900 to about 40 percent by 1950 and has remained at about 40 percent to the present. (See figure 8.4.)

The growth of the automobile industry, first in the United States and then in other industrialized countries, led to roadway construction, which required energy. Thus, the energy costs of driving a car were greater than just the fuel consumed in travel. As roads improved, higher speeds were possible. Bigger and faster cars required more fuel and even better roads. So roads were continually being improved, and better cars were being produced. A cycle of *more chasing more* had begun. In North America and much of Europe, the convenience of the automobile encouraged two-car families, which created a demand for more energy.

More cars meant more jobs in the automobile industry, the steel industry, the glass industry, and hundreds of other industries. Constructing thousands of kilometers of roads created additional jobs. Thus, the automobile industry played a major role in the economic development of the industrialized world. All this wealth gave people more money for cars and other necessities of life. The car, originally a luxury, was now considered a necessity.

FIGURE 8.3 **The Industrial Revolution** The invention of the steam engine led to the development of many machines during the Industrial Revolution.

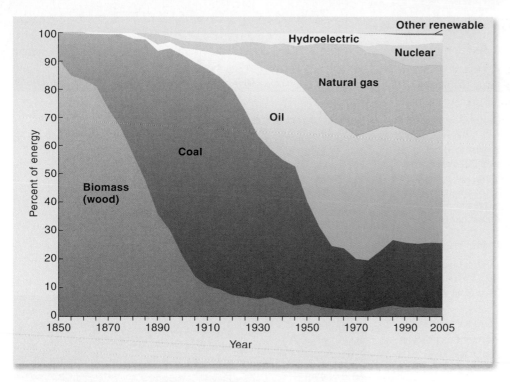

FIGURE 8.4 **Changes in Energy Sources** This graph shows how the percentage of energy obtained from various sources in the United States has changed over time. Until 1900, wood and coal were the predominant sources of energy. These sources declined in importance as oil and natural gas increased. Since 1950, oil and natural gas have provided over 60 percent of the energy, while coal has fallen to about 20 percent and wood to about 3 percent. Nuclear power has grown to about 8 percent, and hydroelectric power provides less than 3 percent. Other renewable sources (geothermal, solar, and wind) collectively account for less than 1 percent.

Source: Data from *Annual Energy Review, 2007,* Energy Information Administration.

As people moved to the suburbs, they also changed their buying habits. Labor-saving, energy-consuming devices became essential in the home. The vacuum cleaner, dishwasher, garbage disposal, and automatic garage door opener are only a few of the ways human power has been replaced with electrical power. About 25 percent of the electrical energy produced in North America is used in homes to operate lights and appliances. Other aspects of our lifestyles illustrate our energy dependence. Regardless of where we live, we expect Central American bananas, Florida oranges, California lettuce, Texas beef, Hawaiian pineapples, Ontario fruit, and Nova Scotia lobsters to be readily available at all seasons. What we often fail to consider is the amount of energy required to process, refrigerate, and transport these items. The car, the modern home, the farm, and the variety of items on our grocery shelves are only a few indications of how our lifestyles are based on a continuing supply of cheap, abundant energy.

GROWTH IN THE USE OF NATURAL GAS

The car also altered people's lifestyles. Vacationers could travel greater distances. New resorts and chains of motels, restaurants, and other businesses developed to serve the motoring public, creating thousands of new jobs. Because people could live farther from work, they began to move to the suburbs. (See figure 8.5.)

Initially, natural gas was a waste product of oil production that was burned at oil wells because it was difficult to store and transport. Before 1940, natural gas accounted for less than 10 percent of energy consumed in the United States. In 1943, in response to World War II, a federally financed pipeline was constructed to transport oil within the United States.

(a)

(b)

FIGURE 8.5 **Energy-Demanding Lifestyle** (a) Building private homes on large individual lots some distance from shopping areas and places of employment is directly related to the heavy use of the automobile as a mode of transportation. (b) Heating and cooling a large, enclosed shopping mall, along with the gasoline consumed in driving to the shopping center, increase the demand for energy.

Energy and Civilization: Patterns of Consumption

Although people like the convenience of the personal automobile, cars are the least energy-efficient means of transportation in cities. Ironically, the desire to use cars has resulted in traffic congestion in most large cities that has further reduced efficiency and eliminated the convenience of cars. To convince people to use other, more efficient public transport and to relieve congestion, many cities have passed regulations that use financial or other means to discourage automobile traffic in highly congested areas. Several different mechanisms have been successful.

LONDON

Automobiles entering central London have their license plates photographed. Drivers of cars in the central city must pay a fee of £5 (about US $20) a day. Those who do not pay by 10 P.M. face a doubled fee.

HONG KONG

Electronic sensors on cars record highway travel and time of day. Drivers are issued a monthly bill (commuter hours are the most expensive).

SINGAPORE

Automobiles and motorcycles are equipped with an in-vehicle unit that automatically deducts money from a cash card when the vehicle passes through a sensor and enters the restricted zone of downtown Singapore between 7:30 A.M. and 7 P.M.

GOTHENBURG, SWEDEN

To encourage pedestrian traffic, the central business district has been divided like a pie into zones, with cars prohibited from moving directly from one zone to another. Autos move from zone to zone by way of a peripheral ring road.

ROME AND FLORENCE

All traffic except buses, taxis, delivery vehicles, and cars belonging to area residents have been banned between 7:30 A.M. and 7:30 P.M.

TOKYO

Before closing the sale, the buyer of a standard-size vehicle must show evidence that a permanent parking space is available for the car. To comply with the law, some drivers have constructed home garages with lifts to permit parking one car above the other.

After the war, the federal government sold these pipelines to private corporations. The corporations converted the pipelines to transport natural gas. Thus, a direct link was established between the natural gas fields in the Southwest and the markets in the Midwest and East. By 1970, about 30 percent of energy needs were being met by natural gas. Currently, about 23 percent of the energy consumed in the United States is from natural gas, primarily for home heating and industrial purposes.

HOW ENERGY IS USED

The amount of energy consumed by countries of the world varies widely. (See figure 8.6.) The highly industrialized countries consume much more energy than less-developed countries. However, even among countries with the same level of development, great differences exist in the amount of energy they use as well as in how they use it. To maintain their style of living, individuals in the United States use about twice as much energy as people in France, Germany, or Japan, 5.5 times more energy than the people of China, and about 20 times more energy than the people of India.

Differences also exist in the purposes for which people use energy. Industrialized nations use energy about equally for three purposes: (1) residential and commercial uses, (2) industrial uses, and (3) transportation. Less-developed nations with little industry use most of their energy for residential purposes (cooking and heating). Countries that are making the transition from less-developed to industrial economies use large amounts of energy to develop their industrial base.

RESIDENTIAL AND COMMERCIAL ENERGY USE

The amount of energy required for residential and commercial use varies greatly throughout the world. For example, about 22 percent of the energy used in North America is for residential purposes and 18 percent for commercial purposes, while in India, 44 percent

BIOMASS FUELS AND THE DEVELOPING WORLD

Although most of the world uses fossil fuels as energy sources, much of the developing world relies on *biomass* as its source of energy. The biomass can be wood, grass, agricultural waste, or dung. According to the United Nations, 2 billion people (30 percent of the world's population) use biomass as fuel for cooking and heating dwellings. In developing countries, nearly 40 percent of energy used comes from biomass. In some regions, however, the percentage is much higher. For example, in sub-Saharan Africa, fuelwood provides about 80 percent of energy consumed. Worldwide, about 60 percent of wood removed from the world's forests is used for fuel.

This dependence on biomass has several major impacts:

- Often women and children must walk long distances and spend long hours collecting firewood and transporting it to their homes.

- Because the fuel is burned in open fires or inefficient stoves, smoke contaminates homes and affects the health of the people. The World Health Organization estimates that in the developing world, 40 percent of acute respiratory infections are associated with poor indoor air quality related to burning biomass. A majority of those who become ill are women and children because the children are in homes with their mothers who spend time cooking food for their families.

- Often the fuel is harvested unsustainably. Thus, the need for an inexpensive source of energy is a cause of deforestation. Furthermore, deforested areas are prone to soil erosion.

- When dung or agricultural waste is used for fuel, it cannot be used as an additive to improve the fertility or organic content of the soil. Thus, use of these materials for fuel negatively affects agricultural productivity.

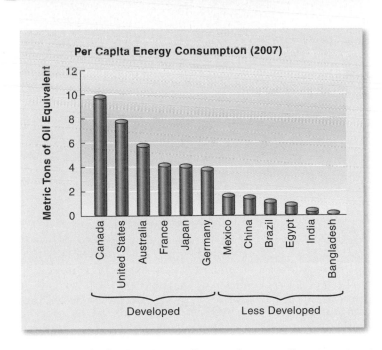

FIGURE 8.6 Per Capita Energy Consumption Countries of the developed world use much more energy per person than do countries of the less-developed world. However, there are great differences in energy use among developed nations. France, Germany, and Japan use about half the energy per person as does the United States.

Source: Data from *BP Statistical Review of World Energy*, June 2008; and *Population Reference Bureau, 2007 Population Data Sheet*.

of the energy is for residential uses and 3.4 percent for commercial purposes. The ways that residential energy is used also vary widely. In the United States, 47 percent is used for space heating. In Canada, which has a cold climate, 60 percent of the residential energy is used for heating. However, in many parts of Africa and Asia much of the energy used in the home is for cooking, and much of that energy comes from firewood.

In much of the less-developed world, cooking is done over open fires. Using fuel-efficient stoves instead of open fires could reduce these energy requirements by 50 percent. Improving efficiency would protect wood resources, reduce the time or money needed to obtain firewood, and improve the health of people because they would breathe less wood smoke. (See figure 8.7.)

Computer systems and the Internet are a relatively new segment of the economy that consumes energy. Although early estimates suggested that this segment of the economy would consume over 10 percent of the U.S. electrical energy supply, more recent estimates put the energy consumed at about 2 percent of the electrical energy supply.

INDUSTRIAL ENERGY USE

The amount of energy countries use for industrial processes varies considerably. Nonindustrial countries use little energy for industry. Countries that are developing new industries dedicate a high percentage of their energy use to them. They divert energy to

Water is a substance with a high heat capacity. That means that it takes a great deal of heat energy to increase its temperature. Conversely, a great deal of heat needs to be transferred from water for it to cool.

This property of water has several significant connections to the way we produce and use energy. Access to hot water for washing clothes and dishes and for bathing is considered a necessity in modern society, but this convenience carries a cost. About 17 percent of the energy used in households is used to heat water.

There are several simple things you can do to reduce the cost of heating water:

1. Use less hot water by installing low-flow fixtures on showers and faucets and take shorter showers.

2. Reduce the thermostat setting on the water heater to 120°F (49°C). This saves energy because energy is not used to raise the temperature above that which is needed.

3. Insulate the hot water tank so that the heated water does not lose heat through the surface of the tank.

4. Insulate the pipes that carry the hot water to reduce the rate at which the water loses heat.

(a)

(b)

FIGURE 8.7 **Wood Use for Cooking** Many people in the developing world use wood as a primary fuel for cooking. Reliance on wood is a major cause of deforestation. (a) In many parts of the world such as Africa, cooking is often done over open fires. This is a very inefficient use of fuel. (b) The use of simple mud stoves as in this home in Nepal greatly increases efficiency.

the developing industries at the expense of other sectors of their economy. Highly industrialized countries use a significant amount of their energy in industry, but their energy use is high in other sectors as well. In the United States, industry claims about 32 percent of the energy used.

The amount of energy required in a country's industrial sector depends on the types of industrial processes used. Many countries use inefficient processes and could reduce their energy consumption by converting to more energy-efficient ones. However, they need capital investment to upgrade their industries and reduce energy

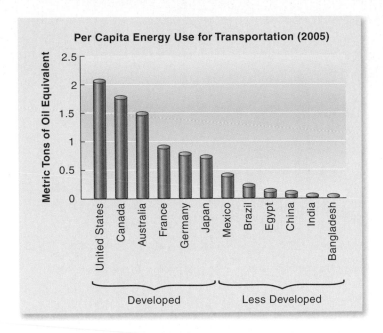

FIGURE 8.8 **Per Capita Energy Use for Transportation (2005)** Per capita energy use for transportation is highest among developed countries. However, there are great differences among countries. Most European countries and Japan use less than half the energy per person compared to the United States and Canada.
Source: International Energy Agency on-line statistics.

FIGURE 8.9 **Public Transportation** People will use public transportation in situations where automobile travel is inconvenient, time consuming, or expensive. Public transportation is more energy efficient than travel by private automobile.

consumption. Some countries cannot afford the upgrade. For example, many countries of the former Soviet Union still use outdated, inefficient open-hearth furnaces to produce steel. These furnaces require nearly double the worldwide energy average to produce a metric ton of steel compared to more efficient processes. By contrast, China, the world's largest steel producer, has closed all its open-hearth steelmaking operations in favor of more efficient processes.

TRANSPORTATION ENERGY USE

As with residential, commercial, and industrial uses, the amount of energy used for transportation varies widely throughout the world. In some of the less-developed nations, transportation uses are very small. Per capita energy use for transportation is larger in developing countries and highest in highly developed countries. (See figure 8.8.) There are also great differences in the percentage of energy devoted to transportation. Bangladesh, India, and China use about 10 percent or less of their energy for transportation, while most developing and industrialized countries use 25 to 40 percent for the same purpose.

Once a country's state of development has been taken into account, the specific combination of bus, rail, waterways, and private automobiles is the main factor in determining a country's energy use for transportation. In Europe, Latin America, and many other parts of the world, rail and bus transport are widely used because they are more efficient than private automobile travel, governments support these transportation methods, or a large part of the populace is unable to afford an automobile. In countries with high population densities, rail and bus transport is particularly efficient. (See figure 8.9.)

In general, automobiles require about twice the energy per passenger kilometer than does bus or rail transport. In addition, most of these countries have high taxes on fuel, which raise the cost to the consumer and encourage the use of public transport.

In North America, the situation is different. Government policy has kept the cost of energy artificially low and supported the automobile industry while removing support for bus and rail transport. Consequently, the automobile plays a dominant role, and public transport is used primarily in metropolitan areas. (See figure 8.10.) Rail and bus transport are about twice as energy efficient as private automobiles. Private automobiles in North America, with about 5 percent of the world's population, consume about 40 percent of the gasoline produced in the world. Air travel is relatively expensive in terms of energy, although it is slightly more efficient than an automobile carrying a single passenger. Passengers, however, are paying for the convenience of rapid travel over long distances.

ELECTRICAL ENERGY

Electrical energy is such a large proportion of energy consumed in most countries that it deserves special comment. Electricity is both a way that energy is consumed and a way that it is supplied. Almost all electrical energy is produced as a result of burning fossil fuels. Thus, we can look at electrical energy as a use to which fossil fuel energy is put. In the same way we use natural gas to heat homes, we can use natural gas to produce electricity. Because the transportation of electrical energy is so simple and the uses to

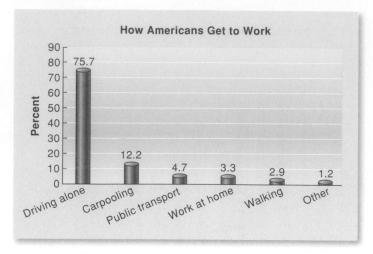

FIGURE 8.10 **How Americans Get to Work** The vast majority of Americans commute to work in an automobile with one person in it.

which it can be put are so varied, electricity is a major form in which energy is supplied to people of the world.

Electrical energy can be produced in many ways. The primary methods of generating electricity are burning fossil fuels, nuclear power plants, hydroelectric plants, and other renewable methods (geothermal, wind, tidal, solar). The combination of methods used to generate electricity in any country depends on the natural resources of the country and government policy. Figure 8.11 shows several countries that were selected to show how individual countries use different combinations of technologies to produce electricity. Norway gets nearly all of its electricity from hydroelectric plants. Iceland gets its electricity from a combination of hydroelectricity and geothermal energy. France relies heavily on nuclear power. Most countries have a substantial amount of their energy produced from fossil fuels, primarily coal.

As with other forms of energy use, electrical consumption in different regions of the world varies widely. The industrialized countries of the world, with about 20 percent of the world's population, consume 60 percent of the world's electricity. Less-developed nations of the world, which have about 80 percent of the world's population, use 40 percent of the world's electricity. The per capita use of electricity in North America is 10 times greater than average per capita use in the less-developed countries. In Bangladesh, the annual per capita use of electricity is about 147 kilowatt hours, which is enough to light a 100-watt lightbulb for less than two months. The per capita consumption of electricity in North America is about 100 times greater than in Bangladesh. Electrical consumption among developed countries also varies widely. For example, the per capita use in Japan and the 15 countries of the European Union is about 60 percent of that in North America.

The production and distribution of electricity is a major step in the economic development of a country. In developed nations, about a quarter of the electricity is used by industry. The remainder is used primarily for residential and commercial

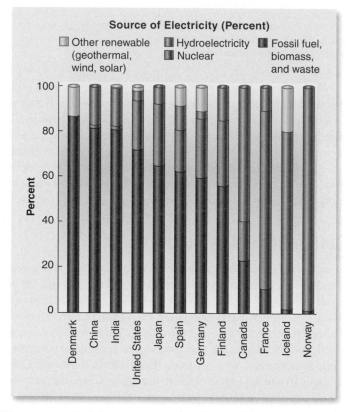

FIGURE 8.11 **Sources of Electricity** Electricity is generated from fossil fuels, nuclear power, hydropower, and other renewable forms of energy (wind, solar, geothermal, tidal). However, countries differ in the combination of technologies used to produce electricity. The differences are based on available natural resources and government policy. The countries shown here were selected to show these differences. Norway and Iceland have abundant hydroelectricity resources. In addition, Iceland has much geothermal energy. France has made the political decision to generate most of its electricity from nuclear power plants. Denmark, Spain, and Germany have made commitments to produce electricity from renewable technologies—particularly wind.

Sources: *BP Statistical Review of World Energy 2007* and other sources.

purposes. In nations that are developing their industrial base, over half of the electricity is used by industry. For example, industries consume about 50 percent of the electricity used in South Korea.

THE ECONOMICS AND POLITICS OF ENERGY USE

A direct link exists between economic growth and the availability of inexpensive energy. The replacement of human and animal energy with fossil fuels began with the Industrial Revolution and was greatly accelerated by the supply of cheap, easy-to-handle, and highly efficient fuels. Because the use of inexpensive fossil fuels allows each worker to produce more goods and services, productivity increased. The result was unprecedented economic growth in Europe, North America, and the rest of the industrialized world.

Because of this link between energy and productivity, most industrial societies want to ensure a continuous supply of affordable energy. The higher the price of energy, the more expensive goods and services become. To keep costs down, many countries have subsidized their energy industries and maintained energy prices at artificially low levels. International trade in fossil fuels has a major influence on the world economy and politics. The emphasis on low-priced fuels has encouraged high rates of consumption.

FUEL ECONOMY AND GOVERNMENT POLICY

Governments fashion policies that influence how people use energy. Automobile fuel efficiency is one area in which government policy has had significant impact. For example, the price of a liter of gasoline is determined by two major factors; (1) the cost of purchasing and processing crude oil into gasoline and (2) various taxes. Most of the differences in gasoline prices among countries are a result of taxes and reflect differences in government policy toward motor vehicle transportation. The cost of taxes to the U.S. consumer is about 12 percent of the retail gasoline price, and in Canada, about 30 percent of the price of gasoline is taxes, while in Japan and many European countries, taxes account for 40 percent to 60 percent of the cost of gasoline. (See figure 8.12.) When we compare the kinds of automobiles driven, we find a direct relationship between the cost of fuel and fuel efficiency. In the United States and Canada, the average fleet fuel economy is about 8.8 liters per 100 kilometers (26.7 miles per gallon) and about 8.6 liters per 100 kilometers (27 miles per gallon), respectively. This compares with a European average of about 6 liters per 100 kilometers (40 miles per gallon). The average European car driver pays more than twice as much for fuel as U.S. and Canadian drivers and uses about 26 percent less fuel to drive the same distance as a U.S. driver. Since taxes make up the majority of the price of gasoline in Europe, government tax policy has provided an incentive for people to purchase fuel-efficient automobiles.

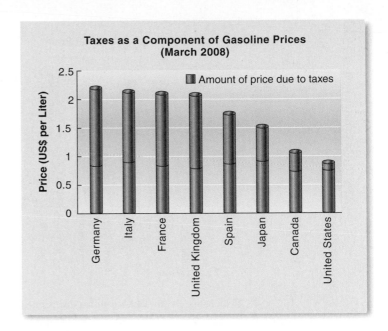

FIGURE 8.12 **Gasoline Taxes and Fuel Efficiency** The price paid for fuel is greatly influenced by the amount of tax paid. High fuel prices cause consumers to choose automobiles with greater fuel efficiency. Source: International Energy Agency.

Another objective of governments is to have a mechanism for generating the money needed to build and repair roads. Many European countries raise more money from fuel taxes than they spend on building and repairing roads. The United States, on the other hand, raises approximately 60 percent of the monies needed for roads from fuel taxes. The relatively low cost of fuel in the United States encourages more travel, which increases road repair costs.

The European Union has also taken steps to comply with the Kyoto Treaty that mandates reductions in the amount of carbon dioxide released into the atmosphere. To achieve carbon dioxide reduction targets, European automobile manufacturers had voluntarily agreed to reduce their carbon dioxide emissions to 140 grams of carbon dioxide per kilometer by 2008. They did not meet the goal and have expressed doubts that they will meet the 130 gram level set by the European Union for 2012. Many European countries have instituted taxes based on the amount of carbon dioxide emitted by a car. People who own cars that emit higher amounts pay much higher taxes. By contrast, the United States has not signed the Kyoto Treaty and had falling fuel efficiency throughout the 1990s as more people bought and drove SUVs until about 2005. There have been tiny gains in fuel efficiency in the United States since 2005.

ELECTRICITY PRICING

Since electricity is such an important source of energy for many uses, it is interesting to look at how different countries structure electricity prices. Because of the nature of the electrical industry, most countries regulate the industry and influence the price utilities

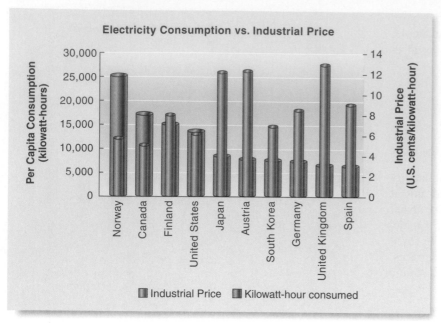

(a)

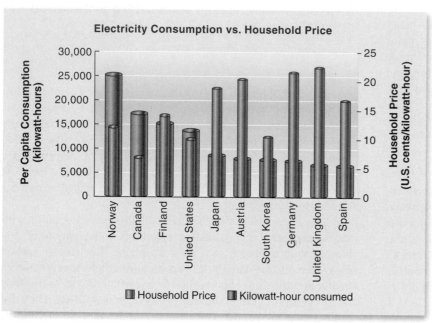

(b)

FIGURE 8.13 **Electricity Price and Consumption** The price of electricity is directly related to the amount consumed. In most countries, electric utilities are monopolies and governments regulate the price charged to the consumer. Most countries have separate prices for industrial and household use with industrial use prices being lower. Graph (a) shows the relationship between the price charged to industry and national per capita use. Graph (b) shows the relationship between the price charged to households and national per capita use.
Source: Data from International Energy Agency, Key Energy Statistics 2007.

are able to charge. Furthermore, the cost to industrial users is typically about half that charged to residential customers. Figure 8.13 shows industrial and residential costs per kilowatt-hour (including charges per kilowatt-hour used, fees, surcharges, and taxes) for electricity in several countries. Obviously, higher prices discourage use. The United States, Canada, Norway, and Finland, which have some of the lowest prices for electricity, have the highest rates of consumption. Canada, Norway, and Finland have high proportions of their electricity generated by hydroelectric plants, which contributes to the low cost of electricity in those countries. Countries such as Japan, which must import fossil fuels to generate electricity, have a higher cost to generate the electricity, which is reflected in the price.

THE IMPORTANCE OF OPEC

The Organization of Petroleum Exporting Countries (OPEC) began in September 1960, when the governments of five of the world's leading oil-exporting countries agreed to form a cartel. Three of the original members—Saudi Arabia, Iraq, and Kuwait—were Arab countries, while Venezuela and Iran were not. Today, 13 countries belong to OPEC. These include seven Arab states—Saudi Arabia, Kuwait, Libya, Algeria, Iraq, Qatar, and United Arab Emirates—and six non-Arab members—Iran, Indonesia, Nigeria, Ecuador, Angola, and Venezuela. OPEC nations control over 75 percent of the world's estimated oil reserves of 1200 billion barrels of oil. Middle Eastern OPEC countries control over 60 percent of this total, which makes OPEC and the Middle East important world influences. Today, OPEC countries control more than 40 percent of the world's oil production and are a major force in determining price. (See table 8.1.)

Increased solidarity among OPEC countries, continuing political instability in the Middle East, and increased demand by countries such as China and India coupled with changes in the value of the dollar and activities by oil speculators, caused oil prices to peak at over US $147 per barrel in mid-2008 before falling at the end of 2008 as the world economy entered a recession.

ENERGY CONSUMPTION TRENDS

From a historical point of view, it is possible to plot changes in energy consumption. Economics, politics, public attitudes, and many other factors must be incorporated into an analysis of energy use trends.

GROWTH IN ENERGY USE

In 2007, world energy consumption was around 11,099 million metric tons of oil equivalent, an increase of 25 percent over 10 years. Of this total, conventional fossil fuels—oil, natural gas, and coal—accounted for nearly 90 percent.

TABLE 8.1 Major World Oil-Producing Countries

	Country	Production (thousand barrels/day) 2007	Percent of Total
OPEC	**Saudi Arabia**	**10,233.9**	**12.1**
	Russian Federation	9875.8	11.7
	United States	8481.1	10.0
OPEC	**Iran**	**4043.4**	**4.8**
	China	3901.0	4.6
	Mexico	3501.4	4.1
	Canada	3358.5	4.0
OPEC	**United Arab Emirates**	**2947.7**	**3.5**
OPEC	**Venezuela**	**2666.6**	**3.2**
OPEC	**Kuwait**	**2613.2**	**3.1**
	Norway	2565.3	3.0
OPEC	**Nigeria**	**2352.4**	**2.8**
	Brazil	2279.0	2.7
OPEC	**Algeria**	**2173.2**	**2.6**
OPEC	**Iraq**	**2093.8**	**2.5**
OPEC	**Libya**	**1844.6**	**2.2**
OPEC	**Angola**	**1768.7**	**2.1**
	United Kingdom	1690.0	1.9
	Kazakhstan	1445.0	1.7
OPEC	**Qatar**	**1136.0**	**1.3**
OPEC	**Indonesia**	**1043.7**	**1.2**
OPEC	**Ecuador**	**300.7**	**0.4**
	World	**84,600.6**	

Source: U.S. Energy Information Administration.

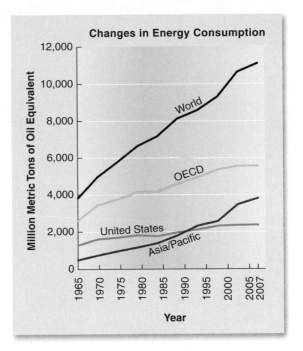

FIGURE 8.14 **Changes in World Energy Consumption by Region** World energy consumption has increased steadily. Currently the energy consumption of the economically developed nations (OECD) accounts for about half of world consumption. In recent years the fastest growth in energy use has been in the Asia Pacific region.
Source: Data from *BP Statistical Review of World Energy,* 2008.

AVAILABLE ENERGY SOURCES

Oil remains the world's major source of energy, accounting for about 36 percent of primary energy demand. Coal accounts for 28 percent and natural gas for 24 percent; the remainder is supplied mainly by nuclear energy and hydropower. The current percentages are likely to remain the same into the future.

POLITICAL AND ECONOMIC FACTORS

Political and economic factors have a great deal of influence on energy consumption. The two primary factors that determine energy use are political stability in parts of the world that supply oil and the price of that oil. Since OPEC and countries of the Middle East control over 40 percent of the world's oil production and 75 percent of the oil reserves, political stability in this region is very important. (See figure 8.15.)

The energy consumption behavior of most people is motivated by economics rather than by a desire to use energy resources wisely. When the price of energy increases, the price of everything else increases as well and eventually consumption falls. Conversely, when energy prices fall, consumption increases. During the 1980s, energy costs declined and people in North America and Europe became less concerned about their energy consumption. They used more energy to heat and cool their homes and buildings, bought and used more home appliances, and bought bigger cars. Governments can manipulate energy prices by increasing or decreasing taxes on energy, granting subsidies to energy producers, and using other means.

Over half of world energy is consumed by the 25 countries that are members of the Organization for Economic Cooperation and Development (OECD). These countries (Australia, New Zealand, Japan, Canada, Mexico, the United States, and the countries of Europe) are the developed nations of the world. In the last decade, energy consumption in OECD countries has risen moderately (less than 1 percent per year) while economic growth has continued. There has also been a shift toward service-based economies, with energy-intensive industries moving to non-OECD countries. In contrast, in countries that are becoming more economically advanced (particularly parts of Asia), energy consumption is increasing at a faster rate (about 5 percent per year). Currently China's energy consumption is growing at 7.7 percent per year and India's is growing at nearly 7 percent per year. Since these two countries contain over 1/3 of the world's population, they are having a major impact on energy use in the world. Growing demand for oil from countries like China and India has led to higher oil prices. We should expect to see this pattern continue and countries with emerging economies increasingly demanding more energy. (See figure 8.14.)

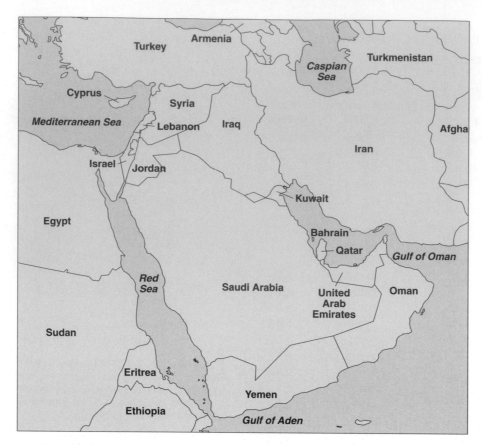

FIGURE 8.15   **Persian Gulf OPEC Countries**   The six OPEC countries of the Persian Gulf (Saudi Arabia, Iraq, Iran, Kuwait, Qatar, and United Arab Emirates) control a major portion of the oil reserves and oil production of the world. Political instability in this region leads to increased oil prices.

Over the past several years, world oil prices have been extremely volatile. As recently as 1999, consumers benefited from oil prices that fell to under US $13 per barrel—a result of oversupply caused by lower demand for oil in both southeast Asia, which was suffering from an economic recession, and North America and Western Europe, which had warmer than expected winters. Since then there has been an increase in the price of oil due to increased demand from China, India, and other rapidly industrializing countries, political instability in the Middle East—in part the result of the wars in Iraq and Afghanistan—and increased solidarity among OPEC countries. In 2003, the price of oil was about $32 per barrel. In 2008, it exceeded $147 per barrel—a 360 percent increase in five years before falling at the end of 2008.

ISSUES & ANALYSIS

Government Action and Energy Policy

Governments use economic tools to encourage desired behaviors. Taxes on goods or services artificially increase the price and encourage a search for alternatives. Subsidies are a gift from government to encourage individuals or corporations to pursue certain kinds of activities. When regulations specify particular standards or actions, fines or fees are charged to those who do not meet the standards.

With respect to energy policy, there are several ways in which these tools are used:

1. The issue of global warming has generated the idea of a carbon tax. Carbon dioxide is released when fossil fuels are used, so a carbon tax would be a tax on energy. This tax would artificially raise the price of energy and cause people to change their behavior to use less energy or shift to forms of energy that release less carbon dioxide.

2. Taxes on gasoline and diesel fuel can be used to encourage consumers to change their driving habits or the kinds of vehicles they drive.

3. Oil companies receive large subsidies to encourage them to explore for oil. Small subsidies are also provided for the exploration of certain alternative energy sources such as solar energy and wind energy.

4. Fuel economy standards for motor vehicles require manufacturers to meet standards by a certain date or they must pay a fine.

All of these activities have their supporters and detractors. How do you feel?

- Do you support subsidies to companies to explore for oil?
- Would you support increased taxes on gasoline?
- Should government impose a carbon tax on all fossil fuels?
- Should automobile manufacturers be forced to make more fuel-efficient vehicles?

CAMPUS SUSTAINABILITY INITIATIVE

DELTA COLLEGE AND ENERGY EFFICIENCY

Delta College is a community college that serves, central Michigan. The campus is situated on a square mile of mixed farmland and woodlots. In the 1960s, the board of trustees agreed to preserve the woodlots as natural areas that could be used by the public and as a resource for biology, art, physical education, and other classes.

Delta College is one of 90 member colleges involved in the Sustainability Tracking, Assessment, and Rating System (STARS) of the Association for the Advancement of Sustainability in Higher Education (AASHE). The STARS project is designed to develop a rating system that will allow colleges to gauge their progress toward meeting sustainability goals. A "triple bottom" approach that weighs financial, environmental, and social benefits in departmental decision-making is a core philosophy.

The original campus was built in the early 1960s when energy efficiency was not a priority. Thus, energy conservation has been a long-term goal of the college for financial as well as environmental reasons. Whenever renovations occur, new energy-saving features are added. Some of the projects that support this goal include:

- An energy management system is programmed to automatically turn off heating/ventilation/air-conditioning systems and lighting during unoccupied hours.

- Heat recovery systems installed in renovated buildings capture the heat from air being exhausted from buildings and return it to the buildings.

- Energy-efficient lighting has been installed throughout the campus.

- Occupancy sensors in all classrooms and offices automatically turn off lights when the room is unoccupied.

- In corridors and rooms with windows, photocells record light levels and automatically shut off lights when incoming natural light meets minimum lighting levels.

- A system that allows the volume of air being supplied to a space to match the usage needs was built into renovated buildings.

- Single-glazed windows were replaced with energy-efficient, low-E insulated windows.

- Photovoltaic panels were installed on the roof of one of the buildings. This installation will serve as a demonstration and training site for several curricula.

- A chilled water system produces ice during the evening hours that is used to cool the building during the following day. This mechanism shifts electricity demand from peak daytime hours to off peak nighttime hours.

SUMMARY

A constant supply of energy is required by all living things. Energy has a major influence on society. A direct correlation exists between the amount of energy used and the complexity of civilizations.

Wood furnished most of the energy and construction materials for early civilizations. Heavy use of wood in densely populated areas eventually resulted in shortages, so fossil fuels replaced wood as a prime source of energy. Fossil fuels were formed from the remains of plants, animals, and microorganisms that lived millions of years ago. Fossil-fuel consumption in conjunction with the invention of labor-saving machines resulted in the Industrial Revolution, which led to the development of technology-oriented societies today in the developed world.

The invention of the automobile caused major changes in the lifestyles of people that led to greater consumption of energy.

Because of the high dependence of modern societies on oil as a source of energy, OPEC countries, which control a majority of the world's oil, can set the price of oil through collective action.

Throughout the world, residential and commercial uses, industries, transportation, and electrical utilities require energy. Because of financial, political, and other factors, nations vary in the amount of energy they use as well as in how they use it. In general, rich countries use large amounts of energy and poor countries use much less. Analysts expect the worldwide demand for energy to increase steadily and the growth in energy usage by those countries that are becoming industrialized to be greater than that of the countries that are already industrialized.

THINKING GREEN

1. List 10 activities you engage in each day that require fossil fuel or electrical energy. Choose one that you will abstain from for a week.

2. Keep a record of your transportation activities for one week. Record miles driven, subway fares paid, taxi fares, etc. How many of the trips were really necessary?

3. Write your governmental representative expressing your opinion on energy policy.

4. If a trip is less than 1 kilometer (0.6 mile), you can walk the round trip (2 km) in about half an hour and not use any fossil fuel energy.

WHAT'S YOUR TAKE?

The price of gasoline is a hot topic. Although Americans pay less than half as much for gasoline as people who live in other economically developed countries, many feel that U.S. prices are too high. Other people feel that increasing the price of gasoline is necessary to stimulate change in the way Americans use energy. Choose to support either the idea of keeping gasoline prices low or the idea of artificially increasing prices. Develop arguments that support your position.

REVIEW QUESTIONS

1. Why was the sun able to provide all energy requirements for human needs before the Industrial Revolution?

2. In addition to food, what energy requirements does a civilization have?

3. Why were some countries unable to use the technologies developed during the Industrial Revolution?

4. What factors caused a shift from wood to coal as a source of energy?

5. How were energy needs in World War II responsible for the subsequent increased consumption of natural gas?

6. What part does government regulation play in changing the consumption of natural gas and oil?

7. Why was much of the natural gas that was first produced wasted?

8. What was the initial use of oil? What single factor was responsible for a rapid increase in oil consumption?

9. List the three purposes for which a civilization uses energy.

10. Why is OPEC important in the world's economy?

11. Give examples of how political and economic events affect energy prices and usage.

CRITICAL THINKING QUESTIONS

1. Imagine you are a historian writing about the Industrial Revolution. Imagine that you also have your new knowledge of environmental science and its perspective. What kind of a story would you tell about the development of industry in Europe and the United States? Would it be a story of triumph or tragedy, or some other story? Why?

2. What might be some of the effects of raising gasoline taxes in the United States to the rate that most Europeans pay for gasoline? Why? What do you think about this possibility?

3. Some argue that the price of gasoline in the United States is artificially low because it does not take into account all of the costs of producing and using gasoline. If you were to figure out the "true" cost of gasoline, what kinds of factors would you want to take into account?

4. How has the ubiquitous nature of automobiles changed the United States? Do you feel these changes are, on balance, positive or negative? What should the future look like regarding automobile use in the United States? How can this be accomplished?

5. The Organization of Petroleum Exporting Countries (OPEC) controls over 75 percent of the known oil reserves. What political and economic effects do you think this has? Does this have any effect on energy use?

6. How do you think projected energy consumption will affect world politics and economics, given current concerns about global warming?

ENERGY SOURCES

The three primary sources of energy are coal, oil, and natural gas. Natural gas pipelines connect production sites with industrial, commercial, and residential customers.

CHAPTER OUTLINE

Energy Sources
Resources and Reserves
Fossil-Fuel Formation
    Coal
    Oil and Natural Gas
Issues Related to the Use of Fossil Fuels
    Coal Use
    Oil Use
    Natural Gas Use
Renewable Sources of Energy
    Biomass Conversion
    Hydroelectric Power
    Solar Energy
    Wind Energy
    Geothermal Energy
    Tidal Power
Energy Conservation
Are Fuel Cells in the Future?

ISSUES & ANALYSIS
Does Ethanol Fuel Make Sense? 210

CASE STUDY
The Arctic National Wildlife Refuge 195

CAMPUS SUSTAINABILITY INITIATIVE
Western Washington University Purchases Green Power from
    North Dakota Wind Farms 206

GOING GREEN
Hybrid Electric Vehicles 209

WATER CONNECTIONS
Solar Stills and Drinking Water 204

OBJECTIVES

After reading this chapter, you should be able to:

- Differentiate between resources and reserves.
- Identify peat, lignite, bituminous coal, and anthracite coal as steps in the process of coal formation.
- Recognize that natural gas and oil are formed from ancient marine deposits.
- Explain how various methods of coal mining can have negative environmental impacts.
- Explain why surface mining of coal is used in some areas and underground mining in other areas.
- Explain why it is more expensive to find and produce oil today than it was in the past.
- Recognize that secondary recovery methods have been developed to increase the proportion of oil and natural gas obtained from deposits.
- Recognize that transport of natural gas is still a problem in some areas of the world.
- Explain why the amount of energy supplied by hydroelectric power is limited.
- Describe how wind, geothermal, and tidal energy are used to produce electricity.
- Recognize that wind, geothermal, and tidal energy can be developed only in areas with the proper geologic or geographical features.
- Describe the use of solar energy in passive heating systems, active heating systems, and the generation of electricity.
- Recognize that fuelwood is a major source of energy in many parts of the less-developed world and that fuelwood shortages are common.
- Describe the potential and limitations of biomass conversion and waste incineration as sources of energy.
- Recognize that energy conservation can significantly reduce our need for additional energy sources.

ENERGY SOURCES

Chapter 8 outlined the historical development of energy consumption and how advances in civilizations were closely linked to the availability and exploitation of energy. New manufacturing processes relied on dependable sources of energy. Technology accelerated in the twentieth century. Between 1900 and 2007, world energy consumption increased by a factor of about 16 and economic activity increased by more than 70 times, but population increased only slightly more than four times. (See table 9.1.)

The energy sources most commonly used by industrialized nations are the fossil fuels: oil, coal, and natural gas, which supply about 80 percent of the world's energy. Fossil fuels were formed hundreds of millions of years ago. They are the accumulation of energy-rich organic molecules produced by organisms as a result of photosynthesis over millions of years. We can think of fossil fuels as concentrated, stored solar energy. The rate of formation of fossil fuels is so slow that no significant amount of fossil fuels will be formed over the course of human history. Since we are using these resources much faster than they can be produced and the amount of these materials is finite, they are known as **nonrenewable energy sources.** Eventually, human demands will exhaust the supplies of coal, oil, and natural gas.

In addition to nonrenewable fossil fuels, there are several renewable energy sources. **Renewable energy sources** replenish themselves or are continuously present as a feature of the solar system. Some forms of renewable energy can also be referred to as *perpetual energy.* For example, in plants, photosynthesis converts light energy into chemical energy.

$$\text{Carbon dioxide} + \text{Water} + \text{Sunlight energy} \rightarrow \text{Biomass (Chemical energy)} + \text{Oxygen}$$

This energy is stored in the organic molecules of the plant as wood, starch, oils, or other compounds. Any form of biomass—plant, animal, alga, or fungus—can be traced back to the energy of the sun. Since biomass is constantly being produced, it is a form of renewable energy. Solar, geothermal, and tidal energy are renewable energy sources because they are continuously available. Anyone who has ever lain in the sun, seen a geyser or hot springs, or been swimming in the surf has experienced these forms of energy. Progress has been made in the use of renewable sources of energy, as in heating and cooling homes and businesses with solar energy. However, technical, economic, and cultural challenges must be

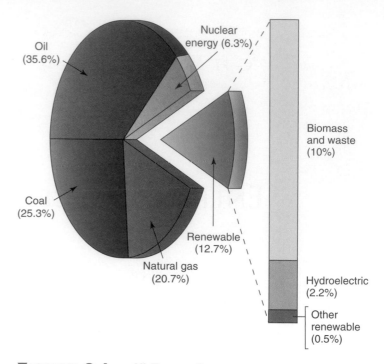

FIGURE 9.1 **All Energy Sources** The use of nonrenewable energy sources, including fossil fuels and nuclear energy, far outweighs the use of renewable sources in the industrialized world today.
Data from International Energy Agency, 2007.

addressed before renewable energy will meet a significant percentage of humans' energy demands. Figure 9.1 shows the proportions of renewable and nonrenewable energy sources used in the world today.

RESOURCES AND RESERVES

When discussing deposits of nonrenewable resources, such as fossil fuels, we must differentiate between deposits that can be extracted and those that cannot. From a technical point of view, a **resource** is a naturally occurring substance of use to humans that can *potentially* be extracted using current technology. **Reserves** are known deposits from which materials *can* be extracted profitably with existing technology *under prevailing economic conditions.* It is important to recognize that the concept of *reserves* is an economic idea and is only loosely tied to the total quantity of a material present in the world. Therefore, reserves are smaller than resources. (See figure 9.2.)

Both terms are used when discussing the amount of fossil-fuel deposits a country has at its disposal. This can cause considerable confusion if the difference between these concepts is not understood. The total amount of a *resource* such as coal or oil changes only by the amount used each year. The amount of a *reserve* changes as technology advances, new deposits are discovered, and economic conditions vary. Furthermore, countries often restate the amount of their reserves for political reasons. Thus, there can be large increases in the amount of *reserves,* while the total amount of the *resource* falls.

When we read about the availability of fossil fuels, we must remember that if the cost of removing and processing a fuel is greater than the fuel's market value, no one is going to produce it. Also, if the amount of energy used to produce,

TABLE 9.1 World Population, Economic Output, and Fossil-Fuel Consumption

	Population (Billions)	Gross World Product (Trillion 2000 US$)	Fossil-Fuel Consumption (Billion Metric Tons Coal Equivalent)
1900	1.6	0.6	1
1950	2.5	2.9	3
2007	6.6	43.3	16

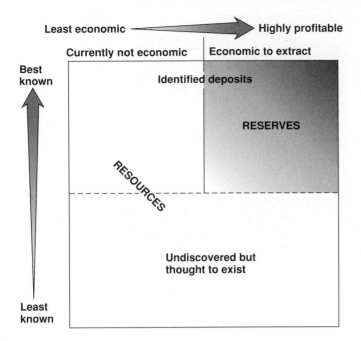

FIGURE 9.2 **Resources and Reserves** Each term describes the amount of a natural resource present. Reserves are those known deposits that can be profitably obtained using current technology under current economic conditions. Reserves are shown in the box in the upper right-hand corner in this diagram. The darker the color, the better known the deposit and the more profitable it is to extract. Resources are much larger quantities that include undiscovered deposits and deposits that currently cannot be profitably used, although it might be feasible to do so in the future If technology or market conditions change.
Source: Adapted from the U.S. Bureau of Mines.

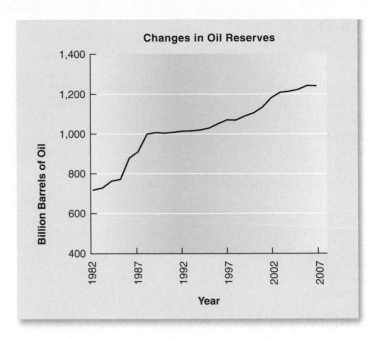

FIGURE 9.3 **Changes in Proved Oil Reserves** The figure shows the changes in proved oil reserves over a 25-year period. The major changes in the 1980s were primarily due to more accurate reporting rather than new discoveries.
Source: Data from *BP Statistical Review of World Energy,* June 2008.

TABLE 9.2 Coal Formation—Changes in Carbon Content

	Carbon Content	Physical Characteristics
Peat	5%	Recognizable plant material
Lignite	25–46%	Brown and crumbly
Subbituminous	46–60%	Black and crumbly
Bituminous	60–86%	Black and soft
Anthracite	86–98%	Black and hard

refine, and transport a fuel is greater than its potential energy, the fuel will not be produced. A net useful energy yield is necessary to exploit the resource. However, in the future, new technology or changing prices may permit the profitable removal of some fossil fuels that currently are not profitable. If so, those resources will be reclassified as reserves.

To further illustrate the concept of reserves and how technology and economics influence their magnitude, let us look at the history of oil. When the first oil well in North America was drilled in Pennsylvania in 1859, it greatly expanded the estimate of the amount of oil in the Earth. There was a sudden increase in the known oil reserves. In the years that followed, new deposits were discovered. Better drilling techniques led to the discovery of deeper oil deposits, and offshore drilling established the location of oil under the ocean floor. At the time of their discovery, these deep deposits and the offshore deposits added to the estimated size of the world's oil resources. But they did not necessarily add to the reserves because it was not always profitable to extract the oil. With advances in drilling and pumping methods and increases in oil prices, it eventually became profitable to obtain oil from many of these deposits. As it became economical to extract them, they were reclassified as reserves. (See figure 9.3.)

FOSSIL-FUEL FORMATION

Fossil fuels are the remains of once-living organisms that were preserved and altered as a result of geologic forces. Significant differences exist in the formation of coal from that of oil and natural gas.

COAL

Coal was formed from plant material that has been subjected to heat and pressure. Tropical freshwater swamps covered many regions of the Earth 300 million years ago. Conditions in these swamps favored extremely rapid plant growth, resulting in large accumulations of plant material. Because this plant material collected under water, decay was inhibited, and a spongy mass of organic material formed. Similar deposits are being formed today and are known as peat.

Due to geologic changes in the Earth, some of these organic deposits were submerged by seas. The plant material that had collected in the swamps was then covered by sediment. The weight of the plant material plus the weight of the sediment on top of it compressed it into coal.

Depending on the amount of time the organic matter has been subjected to geologic processes, several qualities of material are produced. (See table 9.2.) Most parts of the world have coal deposits. (See figure 9.4.)

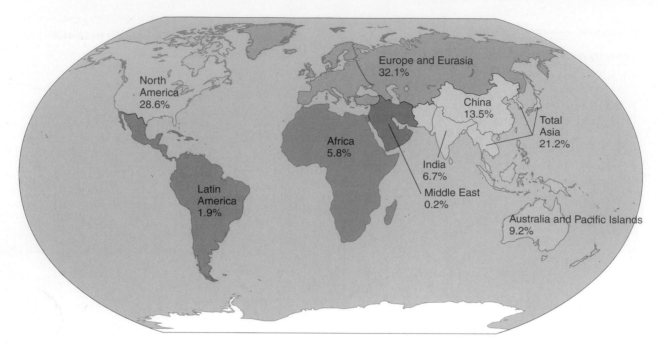

FIGURE 9.4 **Recoverable Coal Reserves of the World 2007** The percentage indicates the coal reserves in different parts of the world. This coal can be recovered under present local economic conditions using available technology.

Source: Data from *BP Statistical Review of World Energy,* 2008.

OIL AND NATURAL GAS

Oil and natural gas, like coal, are products from the past. They probably originated from microscopic marine organisms. When these organisms died and accumulated on the ocean bottom and were buried by sediments, their breakdown released oil droplets. Gradually, the muddy sediment formed rock called shale, which contained dispersed oil droplets. Although shale is common and contains a great deal of oil, extraction from shale is difficult because the oil is not concentrated. However, in instances where a layer of porous sandstone formed on top of the oil-containing shale and an impermeable layer of rock formed on top of the sandstone, concentrations of oil often form. Usually, the trapped oil does not exist as a liquid mass but rather as a concentration of oil within sandstone pores, where it accumulates because water and gas pressure force it out of the shale. (See figure 9.5.) These accumulations of oil are more likely to occur if the rock layers were folded by geological forces.

Natural gas, like coal and oil, forms from fossil remains. If the heat generated within the Earth reached high enough temperatures, natural gas could have formed along with or instead of oil. This would have happened as the organic material changed to lighter, more volatile (easily evaporated) hydrocarbons than those found in oil. The most common hydrocarbon in natural gas is the gas methane (CH_4). Water, liquid hydrocarbons, and other gases may be present in natural gas as it is pumped from a well.

The conditions that led to the formation of oil and gas deposits were not evenly distributed throughout the world. Figure 9.6 illustrates the geographic distribution of oil reserves. The Middle East has over 60 percent of the world's oil reserves. Figure 9.7 shows the geographic distribution of natural gas reserves. The Middle East and Eurasia (primarily Russia) have about 76 percent of the world's natural gas reserves.

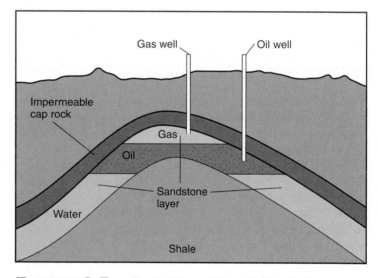

FIGURE 9.5 **Crude Oil and Natural Gas Pool** Water and gas pressure force oil and gas out of the shale and into sandstone capped by impermeable rock.

Source: Adapted with permission from Arthur N. Strahler, *Planet Earth.* Copyright © 1972 by Arthur N. Strahler.

ISSUES RELATED TO THE USE OF FOSSIL FUELS

As previously mentioned, of the world's commercial energy, about 80 percent is furnished by the three nonrenewable fossil-fuel resources: coal, oil, and natural gas. Coal supplies about 25 percent, oil supplies about 36 percent, and natural gas supplies about 21 percent. Each fuel has advantages and disadvantages and requires special techniques for its production and use.

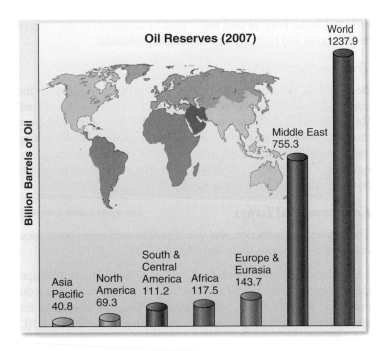

FIGURE 9.6 **World Oil Reserves 2007** The world's supply of oil is not distributed equally. The Middle East controls over 60 percent of the world's oil reserves.

Source: Data from *BP Statistical Review of World Energy,* 2008.

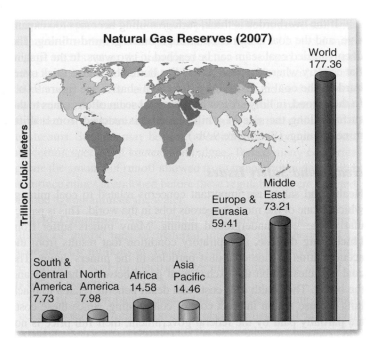

FIGURE 9.7 **World Natural Gas Reserves 2007** Natural gas reserves, like oil and coal, are concentrated in certain regions of the world. The Middle East and Eurasia have about 75 percent of the world's natural gas reserves.

Source: Data from *BP Statistical Review of World Energy,* 2008.

COAL USE

Coal is the world's most abundant fossil fuel, but it supplies only about 25 percent of the energy used in the world. It varies in quality and is generally classified in four categories: lignite, subbituminous, bituminous, and anthracite. Lignite (brown) coal has a high moisture content and is crumbly in nature, which makes it the least desirable form. It has a low energy content that makes transportation over long distances uneconomic. Therefore, most lignite is burned in power plants built near the coal mine. Over 60 percent of the lignite used is from Europe. Subbituminous coal has a lower moisture content and a higher carbon content (46–60 percent) than lignite and is typically used as fuel for electric power plants. Bituminous (soft) coal has a low moisture content and a high carbon content (60–86 percent). It is primarily used in electrical power generation but is also used in other industrial uses such as cement production and steel making. Bituminous coal is the most widely used because it is the easiest to mine and the most abundant. It supplies about 20 percent of the world's energy requirements. Anthracite (hard) coal is 86–98 percent carbon. It is relatively rare and is used primarily in heating of buildings and for specialty uses.

Extraction Methods

Because coal was formed as a result of plant material being buried under layers of sediment, it must be mined. There are two methods of extracting coal: surface mining and underground mining. **Surface mining** (strip mining) involves removing the material located on top of a vein of coal, called **overburden,** to get at the coal beneath. (See figure 9.8.) Coal is usually surface mined when the overburden is less than 100 meters (328 feet) thick. This type of mining operation

FIGURE 9.8 **Surface Mining** Aerial view F&M Coal strip mining site, Preston County, West Virginia.

is efficient because it removes most of the coal in a vein and can be profitably used for a seam of coal as thin as half a meter. For these reasons, surface mining results in the best utilization of coal reserves. Advances in the methods of surface mining and the development of better equipment have increased surface mining activity in the United States from 30 percent of the coal production in 1970 to more than 60 percent today. This trend toward increased surface mining has also occurred in Canada, Australia, and the former Soviet Union.

FIGURE 9.12 **Acid Mine Drainage** The red color of the river is a common characteristic of acid mine drainage.

OIL USE

Oil has several characteristics that make it superior to coal as a source of energy. Its extraction causes less environmental damage than does coal mining. It is a more concentrated source of energy than coal, it burns with less pollution, and it can be

moved easily through pipes. These characteristics make it an ideal fuel for automobiles. However, it is often difficult to find.

Extraction Methods

Today, geologists use a series of tests to locate underground formations that may contain oil. When a likely area is identified, a test well is drilled to determine if oil is actually present. Since the many easy-to-reach oil fields have already been tapped, drilling now focuses on smaller amounts of oil in less accessible sites, which means that the cost of oil from most recent discoveries is higher than that from the large, easy-to-locate sources of the past. As oil deposits located below land have become more difficult to find, geologists have widened the search to include the ocean floor. Building an offshore drilling platform can cost millions of dollars. To reduce the cost, as many as 70 wells may be sunk from a single platform. (See figure 9.13.)

Once a source of oil has been located, the greatest technological problems involve techniques used to extract the oil and transport it to the surface If the water or gas pressure associated with an oil deposit is great enough, the oil is forced to the surface when

FIGURE 9.13 **Offshore Drilling** Once the drilling platform is secured to the ocean floor, a number of wells can be sunk to obtain the gas or oil.
Source: (line art) American Petroleum Institute.

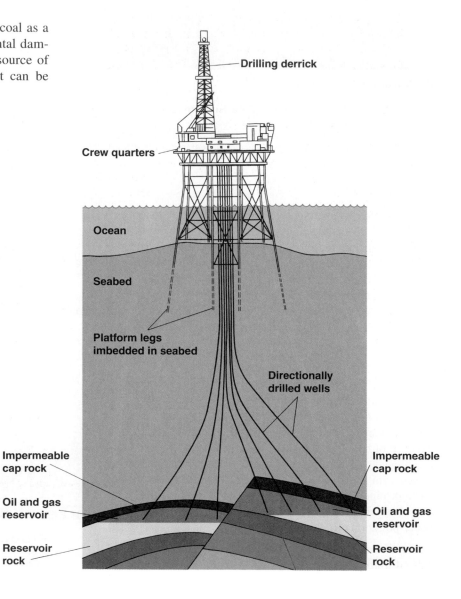

Drilling derrick

Crew quarters

Ocean

Seabed

Platform legs imbedded in seabed

Directionally drilled wells

Impermeable cap rock

Oil and gas reservoir

Reservoir rock

Impermeable cap rock

Oil and gas reservoir

Reservoir rock

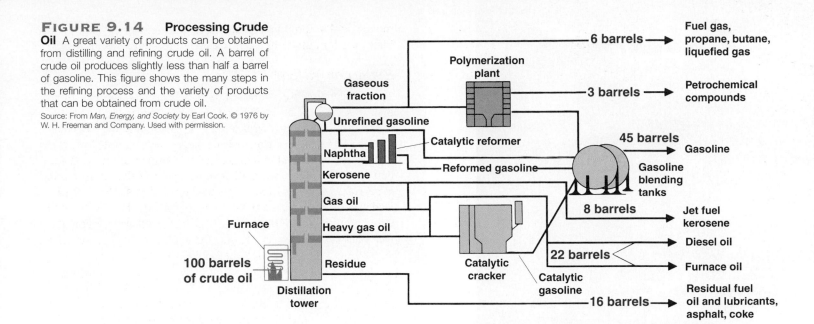

FIGURE 9.14 **Processing Crude Oil** A great variety of products can be obtained from distilling and refining crude oil. A barrel of crude oil produces slightly less than half a barrel of gasoline. This figure shows the many steps in the refining process and the variety of products that can be obtained from crude oil.

Source: From *Man, Energy, and Society* by Earl Cook. © 1976 by W. H. Freeman and Company. Used with permission.

a well is drilled. When the natural pressure is not great enough, the oil must be pumped to the surface. These techniques are often referred to as *primary recovery methods* and can extract 5 to 30 percent of the oil depending on geologic characteristics of the source and the viscosity of the oil. In most oil fields, secondary recovery is used to recover more of the oil. *Secondary recovery methods* include pumping water or gas into the well to drive the oil out of the pores in the rock. These techniques typically result in up to 40 percent of the oil being extracted. As oil prices increase, more expensive and aggressive recovery methods will need to be used. *Tertiary recovery methods* include pumping steam into the well to lower the viscosity of the oil and allow it to flow more readily. Other techniques include more aggressive pumping of gases or chemicals into wells. All of these methods are expensive and are only used if the price of oil is high and the likelihood of getting significant additional production is great.

Processing Crude Oil

Oil, as it comes from the ground, is not in a form suitable for use. It must be refined. The various components of crude oil can be separated and collected by heating the oil in a distillation tower. (See figure 9.14.) After distillation, the products may be further refined by "cracking." In this process, heat, pressure, and catalysts are used to produce a higher percentage of volatile chemicals, such as gasoline, from less volatile liquids, such as diesel fuel and furnace oils. It is possible, within limits, to obtain many products from one barrel of oil. In addition, petrochemicals from oil serve as raw materials for a variety of synthetic compounds. (See figure 9.15.)

Environmental Issues

Liquids are much easier to transport than are solids or gases. Oil pipelines are the primary methods by which oil is transported on continents. When the oil must cross the ocean, giant supertankers

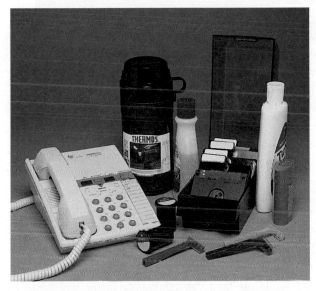

FIGURE 9.15 **Oil-Based Synthetic Materials** These common household items are produced from chemicals derived from oil. Although petrochemicals represent only about 3 percent of each barrel of oil, they are extremely profitable for the oil companies.

carry huge amounts of oil. (See figure 9.16.) The primary problems associated with transportation are leaks and spills. All of the extraction, transportation, and refining activities create opportunities for accidental or routine releases that may cause air or water pollution.

Oil spills in the oceans have been widely reported by the news media. Because of new regulations, changes in tanker hull design, and greater attention to safety, the number of tanker spills has declined over the last few decades, while the amount of oil being transported has increased. (See table 9.3.) Although major shipping accidents are spectacular and release large amounts of oil, it is estimated that nearly 60 percent of the oil pollution in the oceans is the result of routine shipping operations unrelated to oil tankers.

(a) Trans-Alaska pipeline

(b) Oil tanker

FIGURE 9.16 **Transportation of Oil** (a) Oil pipelines and (b) oil tankers are the primary methods used to transport oil. When accidents occur, oil leaks contaminate the soil or water.

TABLE 9.3 Average Annual Oil Spills (over 7 metric tons) from Tankers

Years	Metric Tons of Oil Spilled per Year	Number of Spills per Year
1970s	314,200	25.2
1980s	117,600	9.3
1990s	113,800	7.8
2000–2007	24,000	3.6

Oil spills on land can contaminate soil and underground water. The evaporation of oil products and the incomplete burning of oil fuels contribute to air pollution. These problems are discussed in chapter 16.

NATURAL GAS USE

Natural gas, the third major source of fossil-fuel energy, supplies about 21 percent of the world's energy.

Extraction Methods

The drilling operations to obtain natural gas are similar to those used for oil. In fact, a well may yield both oil and natural gas. As with oil, secondary recovery methods that pump air or water into a well are used to obtain the maximum amount of natural gas from a deposit. After processing, the gas is piped to the consumer for use.

Transport Methods

Transport of natural gas still presents a problem in some parts of the world. In the Middle East, Mexico, Venezuela, and Nigeria, wells are too far from consumers to make pipelines practical, so much of the natural gas is burned as a waste product at the wells. However, new methods of transporting natural gas and converting it into other products are being explored. At −162°C (−126°F), natural gas becomes a liquid and has only 1/600 of the volume of its gaseous form. Tankers have been designed to transport **liquefied natural gas** from the area of production to an area of demand. In 2007, over 211 billion cubic meters (about 7500 billion cubic feet) of natural gas were shipped between countries as liquefied natural gas. This is over 7.5 percent of the natural gas consumed in the world. Of that amount, Japan alone imported 80 billion cubic meters (2800 billion cubic feet). As the demand for natural gas increases, the amount of it wasted will decrease and new methods of transportation will be employed. Higher prices will make it profitable to transport natural gas greater distances between the wells and the consumers.

A major public concern about liquefied natural gas is the safety at loading and unloading facilities. When new ports are suggested, there is concern about explosions that could result from accidents or the actions of terrorists. Because of these concerns, the loading and unloading facilities are often located several kilometers off-shore.

Environmental Issues

Of the three fossil fuels, natural gas is the least disruptive to the environment. A natural gas well does not produce any unsightly waste, although there may be local odor problems. Except for the

The Arctic National Wildlife Refuge (ANWR) has been a source of controversy for many years. The major players are environmentalists who seek to preserve this region as wilderness; the state of Alaska, which funds a major portion of its activities with dividends from oil production; Alaska residents, who receive a dividend payment from oil revenues; oil companies that want to drill in the refuge; and members of Congress who see the oil reserves in the region as important economic and political issues.

In 1960, 3.6 million hectares (8.9 million acres) were set aside as the Arctic National Wildlife Range. Passage of the Alaskan National Interest Lands Conservation Act in 1980 expanded the range to 8 million hectares (19.8 million acres) and established 3.5 million hectares (8.6 million acres) as wilderness. The act also renamed the area the Arctic National Wildlife Refuge. There are international implications to this act. The refuge borders Canada's Northern Yukon National Park. Many animals, particularly members of the Porcupine caribou herd, travel across the border on a regular yearly migration. The United States is obligated by treaty to protect these migration routes.

Alaska relies on oil for about 80 percent of its revenue and has no sales or income tax. Furthermore, each Alaskan citizen receives a yearly dividend check from a state fund established with proceeds from oil companies. Even so, some Alaskan citizens support drilling; others oppose it. The Inupiat Eskimos who live along the north Alaskan coast mostly are in favor of drilling in ANWR. The Inupiat believe oil revenues and land-rental fees from oil companies will raise their living standards. The other Native American tribe in the region, the Gwich'in, who live on the southern fringe of the refuge, oppose drilling. They argue that the drilling will impact the caribou migration through the area every fall and thus affect their ability to provide food for their families.

In 2000, the Energy Information Administration (EIA) released a report on the potential oil production from the coastal plain of ANWR. The report stated that the coastal plain region of ANWR is the largest unexplored, potentially productive geologic onshore basin in the United States.

Oil companies have repeatedly stated that the oil can be recovered without endangering wildlife or the fragile Arctic ecosystem. Conservationists have argued that none of the reserve should be developed when improvements in energy conservation could reduce the demand for oil. They argue that drilling in the reserve will harm the habitat of millions of migratory birds, caribou, and polar bears.

In 2002, President George W. Bush reconfirmed his support for drilling. A decision on permitting the exploration and development is up to the U.S. Congress. The act that established ANWR requires specific authorization from Congress before oil drilling or other development activities can take place on the coastal plain in the refuge. The coastal plain has the greatest concentration of wildlife, is the calving ground for the Porcupine caribou, and has the greatest potential for oil production.

Members of Congress are split on this issue. Debate is heated. In 2007, an attempt was made by an Alaskan senator to allow drilling by attaching an amendment to an appropriations bill. This attempt failed, but the issue will continue to come up as the United States continues to explore ways to meet its energy needs.

Migrating Caribou in ANWR

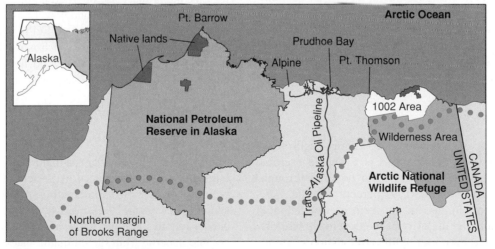

Source: Data from USGS Fact Sheet 0028-01: online report.

danger of an explosion or fire, natural gas poses no harm to the environment during transport. Since it is clean burning, it causes almost no air pollution. The products of its combustion are carbon dioxide and water. Although the burning of natural gas produces carbon dioxide, which contributes to global warming, it produces less carbon dioxide than does coal or oil. Global warming is discussed in chapter 16.

Although natural gas is used primarily for heat energy, it does have other uses, such as the manufacture of petrochemicals and fertilizer. Methane contains hydrogen atoms that are combined with nitrogen from the air to form ammonia, which can be used as fertilizer.

RENEWABLE SOURCES OF ENERGY

The burning of fossil fuels (oil, natural gas, and coal) provides over 80 percent of the energy used in the world. Nuclear energy provides an additional 6.3 percent. The burning of fossil fuels is also responsible for the most of the human-caused carbon dioxide emissions. Energy consumption has been growing at a rate of over 2 percent per year and has nearly doubled in the past 20 years. If growth continues at this rate, we can expect a further doubling of energy consumption in the next 20 years. Since fossil fuels are nonrenewable, they will eventually become scarce and the price will rise. This has led to increased investment in renewable sources of energy.

Currently, alternative energy sources—biomass, hydroelectricity, wind turbines, solar energy, geothermal energy, and tidal energy—supply about 12.7 percent of the world's total energy. Biomass accounts for about 10 percent of the energy used in the world, since firewood and other plant materials are the primary source of energy in much of the developing world.

Hydroelectric power accounts for over 2 percent and the remaining renewable technologies account for about 0.5 percent. (See figure 9.17.) Some optimistic studies suggest that these sources could provide half of the world's energy needs by 2050. It is unlikely that that will occur, but renewable sources certainly will become much more important as fossil fuel supplies become more expensive.

BIOMASS CONVERSION

Biomass fulfilled almost all of humankind's energy needs prior to the Industrial Revolution. All biomass is traceable back to green plants that convert sunlight into plant material through photosynthesis. As recently as 1850, 91 percent of total U.S. energy consumption was biomass in the form of wood. Since the Industrial Revolution, the majority of the developed world's energy requirements have been met by the combustion of fossil fuels such as coal, oil, and natural gas. Biomass, however, is still the predominant form of energy used by people in the less-developed countries, accounting for 10 percent of world energy use.

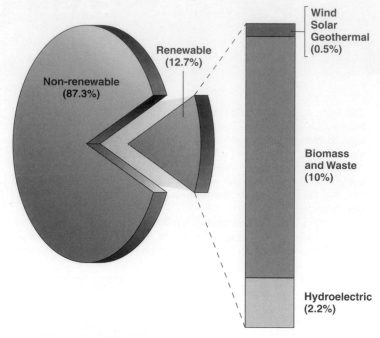

FIGURE 9.17 **Renewable Energy as a Share of Total Energy Consumption (World 2006)** Of the energy consumed in the world, over 87 percent is from nonrenewable fossil fuels and nuclear power. Renewable energy sources provide 12.7 percent, and of that about 80 percent is from the burning of biomass.
Source: International Energy Agency.

Major Types of Biomass

There are several distinct sources of biomass energy: fuelwood, municipal and industrial wastes, agricultural crop residues and animal waste, and energy plantations.

Fuelwood In less-developed countries, wood has been the major source of fuel for centuries. In fact, wood is still the primary source of energy for nearly half of the world's population. In these regions, the primary use of wood is for cooking.

Because of its bulk and low level of energy compared to equal amounts of coal or oil, wood is not practical to transport over a long distance, so most of it is used locally. In the United States, Norway, and Sweden, wood furnishes 10 percent of the energy for home heating. Canada obtains 3 percent of its total energy, not just home heating energy, from wood. Most of this energy is used in forest product industries, such as lumbering and paper mills.

Solid Waste Solid waste is a major source of biomass and other burnable materials produced by society. About 80 percent of this waste is combustible and, therefore, represents a potential energy source. (See figure 9.18.) However, to use waste to produce energy requires that the waste be sorted to separate the burnable organic material from the inorganic material. The sorting is done most economically by those who produce the waste, which means that residents and businesses must separate their trash into garbage, burnable materials, glass, and metals. The trash must be gathered by compartmentalized collection trucks.

FIGURE 9.18 **Waste to Energy** Municipal trash can be burned to produce heat and electricity. This refuse pit is used to feed hoppers of high-temperature furnaces.

The burning of solid waste to produce energy only makes economic sense when the cost of waste disposal is taken into account. In other words, although energy from solid waste is expensive, one can deduct the avoided landfill costs from the cost of producing energy from waste. Where landfill costs are high, waste-to-energy plants make economic sense. In the United States, about 15 percent of solid waste is burned in about 90 plants resulting in about 2500 megawatts of electricity. Europe and Japan have much less available land and have placed restrictions on landfills. Thus, these countries have a much higher rate of burning of solid waste. Countries in Western Europe have over 400 waste-to-energy plants. Japan burns about 80 percent of its waste and Germany burns nearly all of its waste that is not recyclable.

Crop Residues and Animal Wastes The materials that are left following the harvest of a crop can be used as a biomass fuel. In many parts of the world the straw and stalks left on the field are collected and used to provide fuel for heat and cooking. Animal wastes are also used for energy. Animal dung is dried and burned or processed in anaerobic digesters to provide a burnable gas.

Energy Plantations Many crops can be grown for the express purpose of energy production. Crops that have been used for energy include forest plantations, sugarcane, corn, sugar beets, grains, kelp, palm oil, and many others. Two main factors determine whether a crop is suitable for energy use. Good energy crops have a very high yield of dry material per unit of land (dry metric tons per hectare). A high yield reduces land requirements and lowers the cost of producing energy from biomass. Similarly, the amount of energy that can be produced from a biomass crop must be more than the amount of energy required to grow the crop. In some circumstances such as the heavily mechanized corn farms of the U.S. Midwest, the amount of energy in ethanol produced from corn is not much greater than the energy used for tractors, to manufacture fertilizer, and to process the grain into ethanol.

Biomass Conversion Technologies

There are several technologies capable of converting biomass into energy. These include direct combustion and cogeneration, ethanol production, anaerobic digestion, and pyrolysis.

Direct Combustion The most common way that biomass and waste are used for energy production is by burning them. In much of the developing world the primary use of energy derived from biomass is as fires to provide heat for cooking and heating homes.

Large-scale operations are used to power industrial processes or to generate electricity. Large biomass power-generation systems can have efficiencies that are comparable to those of fossil-fuel systems, but this comes at a higher cost due to the design of the burner to handle the higher moisture content of biomass. However, using the biomass in a combined heat- and electricity-production system (or cogeneration system) significantly improves the economics.

Worldwide, about 1 percent of electricity is generated from biomass. This compares to about 16 percent for hydroelectricity. Although the United States produces more total electrical energy from biomass than any other country, its production is only about 1.3 percent of total electrical generation. Finland, which has abundant forest resources, produces nearly 11 percent of its electricity from biomass, although its total production is less than that of the United States.

Biofuels Production Ethanol can be produced from certain biomass materials that contain sugars, starch, or cellulose. The best-known feedstock for ethanol production is sugarcane, but other materials can be used, including wheat, corn, other cereals, and sugar beets.

Ethanol is produced by a process known as fermentation. Typically, sugar or starch is extracted from the biomass crop by crushing and mixing with water and yeast and then keeping the mixture warm in large tanks called fermenters. The yeast breaks down the sugar and converts it to ethanol and carbon dioxide. A distillation process is required to remove the water and other impurities from the dilute alcohol product. The low price of sugar coupled with the high price of oil has prompted Brazil to use its large crop of sugarcane to produce ethanol. Ethanol is sold in a variety of mixtures for automobile fuel, from 100 percent ethanol to 20 percent ethanol and 80 percent gasoline. In total, ethanol provides 40 percent of Brazil's automobile fuel.

In the United States, corn is used for ethanol production and then blended with gasoline to produce E85, a fuel that is 85 percent ethanol and 15 percent gasoline. (See figure 9.19.)

Biodiesel can be produced from the oils in a variety of crops, including soybeans, rapeseed, and palm oil as well as animal fats. These raw materials need to be modified chemically before they can be used as fuel. Currently, about 2 percent of the diesel fuel consumed in the world is biodiesel. Germany leads the world in production of biodiesel fuel with about 36 percent of the total world production.

Anaerobic Digestion Anaerobic digestion involves the decomposition of wet and green biomass or animal waste through bacterial action in the absence of oxygen. This process produces a

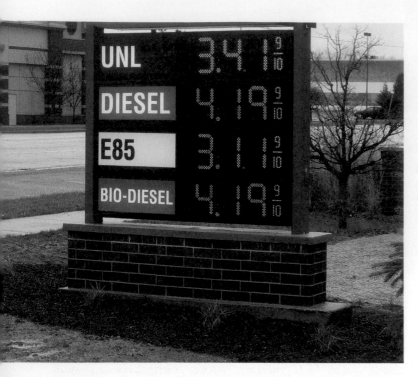

Biofuels Biofuels (E85 and biodiesel) are available for purchase in many parts of the world.

FIGURE 9.21 **Anaerobic Bioreactor** This bioreactor is used to produce methane from municipal sewage sludge.

mixture, consisting primarily of methane and carbon dioxide, known as biogas. The most commonly used technology involves small digesters on farms that generate gas for use in the home or for farm-related activities. China has 500,000 small methane digesters in homes and on farms; India has 100,000; and Korea has 50,000. (See figure 9.20.) Anaerobic digestion can also be used with sewage treatment plants to produce methane. The methane collected can be used to provide heat or run machinery in the plant. (See figure 9.21.) Anaerobic digestion also occurs in landfills. In many landfills the methane gas produced eventually escapes into the atmosphere. However, the gas can be extracted by inserting perforated pipes into the landfill. In this way, the gas will travel through the pipes, under natural pressure, to be used as an energy source, rather than simply escaping into the atmosphere to contribute to greenhouse gas emissions. Some newer landfills have even been designed to encourage anaerobic digestion, which reduces the volume of the waste and provides a valuable energy by-product.

Pyrolysis Pyrolysis is a thermochemical process for converting solid biomass to a more useful fuel. Biomass is heated or partially burned in an oxygen-poor environment to produce a hydrocarbon-rich gas mixture, an oil-like liquid, and a carbon-rich solid residue. Traditionally, in developing countries, the solid residue produced is charcoal, which has a higher energy density than the original fuel.

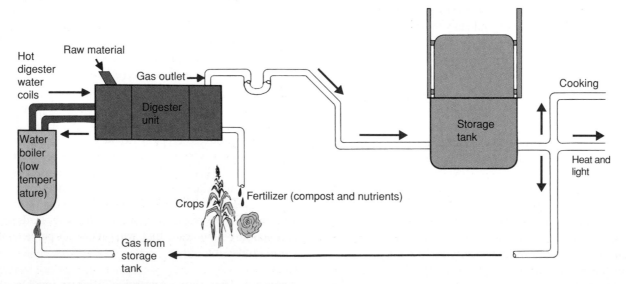

FIGURE 9.20 **Methane Digester** In the digester unit, anaerobic bacteria convert animal waste into methane gas. This gas is then used as a source of fuel. The sludge from this process serves as a fertilizer. In many less-developed countries, this type of digester has the advantages of providing a source of energy and a supply of fertilizer and managing animal wastes, which helps reduce disease.

The traditional charcoal kilns are simply mounds of wood or wood-filled pits in the ground that are covered with earth. However, the process of carbonization is very slow and inefficient in these kilns, and more sophisticated kilns are replacing the traditional ones. The liquid residue or "bio-oil" produced can be easily transported and refined into other products. The process is similar to refining crude oil.

Gasification is a form of pyrolysis, carried out with more air and at high temperatures, to optimize the gas production. The resulting gas, known as producer gas, is a mixture of carbon monoxide, hydrogen, and methane, together with carbon dioxide and nitrogen. The gas is more versatile than the original solid biomass, and it can be used as a source of heat or used in internal combustion engines or gas turbines to produce electricity. During the Second World War, countries such as Australia and Germany even used it to power vehicles.

Environmental Issues

Although the use of biomass and waste as a source energy is often thought of as being environmentally benign, it has many significant environmental impacts.

Habitat and Biodiversity Loss It is estimated that throughout the world there are 1.3 billion people who cannot obtain enough wood or must harvest wood at a rate that exceeds its growth. This has resulted in the destruction of much forest land in Asia and Africa and has hastened the rate of desertification in these regions. (See figure 9.22.)

Another issue associated with biomass energy is the loss of biodiversity. Destroying natural ecosystems to plant sugarcane, grains, palm oil, or other plants can reduce the biodiversity of a region. The plantations lack the complexity of a natural ecosystem and are susceptible to widespread damage by pests or disease.

Air Pollution Burning wood is a source of air pollution. Often the people in developing countries use wood in open fires or poorly designed, inefficient stoves. This results in the release of high amounts of smoke (particulate matter) and other products of incomplete combustion, such as carbon monoxide and hydrocarbons, which contribute to ill health and death.

Respiratory illnesses are particularly common among women and children who spend the most time in the home. Even in the developed world, air pollution from the burning of biomass is a problem. Some cities, such as London, England, have a total ban on burning wood. Vail, Colorado, permits only one wood-burning stove per dwelling. Many areas require woodstoves to have special pollution controls that reduce the amount of particulates and other pollutants released.

The burning of solid waste presents some additional problems. Because solid waste is likely to contain a mixture of materials, including treated paper and plastic, there are additional air pollutants not found in other forms of biomass.

Carbon Dioxide and Global Warming A consensus exists among scientists that biomass fuels and wastes used in a sustainable manner result in no net increase in atmospheric carbon dioxide. Some scientists would even go as far as to declare that sustainable use of biomass would result in a net decrease in atmospheric carbon dioxide. This is based on the assumption that all the carbon dioxide given off by the use of biomass fuels was recently taken in from the atmosphere by photosynthesis. Increased substitution of biomass fuels for fossil fuels would therefore help reduce the potential for global warming, which is caused by increased atmospheric concentrations of carbon dioxide.

Effects on Food Production Although the use of marginal or underutilized land to grow energy crops may make sense, using fertile cropland does not. Since there are millions of people in the world who do not have enough food to eat, the conversion of land from food crops to energy crops presents ethical issues.

The use of crop residues and animal waste as a source of energy also presents some problems. These materials supply an important source of organic matter and soil nutrients for farmers. This is particularly true among subsistence farmers in the developing world. They cannot afford fertilizer and rely on these materials to maintain soil fertility. However, they also need energy. Thus, they must make difficult decisions about how to use this biomass resource.

HYDROELECTRIC POWER

People have long used water to power a variety of machines. Some early uses of water power were to mill grain, saw wood, and run machinery for the textile industry. Flowing water creates energy that can be captured and turned into electricity. This is called hydroelectric power, or *hydropower.*

Technology for Obtaining Hydropower

The most common type of hydroelectric power plant uses a dam on a river to store water in a reservoir. (See figure 9.23.) Water released from the reservoir flows through a turbine, spinning it, which in turn activates a generator to produce electricity. But hydroelectric power does not necessarily require a large dam. In some areas of the world where the streams have steep gradients and a constant flow of water, hydroelectricity may be generated

FIGURE 9.22 **Desertification** The demand for fuelwood in many regions has resulted in the destruction of forests. This is a major cause of desertification.

(a) Glen canyon dam

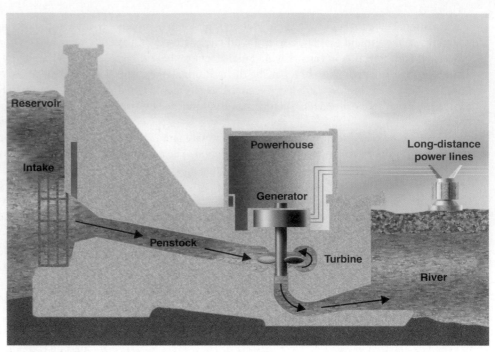

(b) Hydroelectric power plant

FIGURE 9.23 **Hydroelectric Power Plant** (a) The water impounded in this reservoir is used to produce electricity. In addition, this reservoir serves as a means of flood control and provides an area for recreation. (b) This figure shows how a hydroelectric dam produces electricity.
Source: (b) Tennessee Valley Authority.

without a reservoir. Such sites are usually found in mountainous regions and can support only small power-generating stations. Some hydroelectric power plants just use a small canal to channel the river water through a turbine. A small microhydroelectric power system can produce enough electricity for a home, farm, or ranch.

At present, hydroelectricity produces about 2.2 percent of the world's energy supply, or about 16 percent of the world's electricity. In some areas of the world, hydroelectric power is the main source of electricity. More than 35 nations already obtain more than two-thirds of their electricity from falling water. In South and Central America, 65 percent of the electricity used comes from hydroelectric power, compared to 44 percent in the developing world as a whole. Norway gets 99 percent of its electricity and over 65 percent of all its energy from hydroelectricity.

Potential for Additional Hydropower

The potential for developing hydroelectric power is best in mountainous regions and large river valleys. Some areas of the world, such as Canada, the United States, Europe, and Japan, have already developed most of their hydroelectric potential. About 50 percent of the U.S. hydroelectric capacity has been developed. In contrast, Africa has developed only 5 percent of its potential, half of which comes from only three dams: Kariba in East Africa, Aswan on the Nile, and Akosombo in Ghana.

Over the past 10 years, the energy furnished by hydroelectricity worldwide increased by about 17 percent. The World Energy Council estimates that it would be technically possible to triple the electricity produced by hydropower with current technology. The less-developed countries, which have developed about 10 percent of their hydropower, will experience most of this growth.

The projected increase will come mainly from the development of plants on large reservoirs. However, the construction of "minihydro" (less than 10 megawatts) and "microhydro" (less than 1 megawatt) plants is also increasing. Such plants can be built in remote places and can supply electricity to small areas. China has over 80,000 such small stations, and the United States has nearly 1500.

Today's large dams rank among humanity's greatest engineering feats. Table 9.4 lists the locations and sizes of the largest hydroelectric facilities.

Hydroelectric dam projects figure prominently in the economic and investment plans of many developing countries. Egypt electrified virtually all of its villages with power from Aswan. In 2006 China completed construction of a huge hydroelectric dam, known as the Three Gorges Dam, on the Yangtze River. It is the largest hydroelectric dam in the world. Although the dam is complete, it is not expected to reach its full generating capacity of 22,500 megawatts until 2011.

Environmental Issues

It is important to recognize that the construction of a reservoir for a hydroelectric plant causes environmental and social problems. These impacts, however, must be weighed against the environmental

TABLE 9.4 World's Largest Hydroelectric Plants

Name	Country	Rated Capacity Megawatts
Three Gorges Dam	China	22,500
Itaipu	Brazil/Paraguay	14,000
Guri (Simón Bolívar)	Venezuela	10,200
Tucuruí	Brazil	7,960
Grand Coulee	United States	6,809
Sayano Shushenskaya	Russia	6,400
Krasnoyarskaya	Russia	6,000
Robert-Bourassa	Canada	5,616
Churchill Falls	Canada	5,429
Bratskaya	Russia	4,500
Ust Ilimskaya	Russia	4,320
Yaciretá	Argentina/Paraguay	4,050

impacts of alternative sources of electricity. Hydroelectric power plants do not emit any of the standard atmospheric pollutants, such as carbon dioxide or sulfur dioxide given off by fossil fuel–fired power plants. In this respect, hydropower is better than burning coal, oil, or natural gas to produce electricity because it does not contribute to global warming or acid rain.

The most obvious impact of hydroelectric dams is the flooding of vast areas of land, much of it previously forested or used for agriculture. The size of reservoirs created can be extremely large. The Robert-Bourassa project on the Le Grande River in the James Bay region of Quebec has already submerged over 10,000 square kilometers (3861 square miles) of land, and if future plans are carried out, the eventual area of flooding in northern Quebec will be larger than the country of Switzerland. The construction of the Three Gorges Dam in China inundated 153 towns and 4500 villages and caused the displacement of over a million people. In addition, numerous archeological sites were submerged and the nature of the scenic canyons of the Three Gorges was changed.

Large dams and reservoirs can have other impacts on a watershed. Damming a river can alter the amount and quality of water in the river downstream of the dam as well as prevent fish from migrating upstream to spawn. These impacts can be reduced by requiring minimum flows downstream of a dam and by creating fish ladders that allow fish to move upstream past the dam. Silt, normally carried downstream to the lower reaches of a river, is trapped by a dam and deposited on the bed of the reservoir. This silt slowly fills a reservoir, decreasing the amount of water that can be stored and used for electrical generation. The river downstream of the dam is also deprived of silt, which normally fertilizes the river's floodplain during high-water periods. Bacteria present in decaying vegetation can also change mercury, which is sometimes present in rocks underlying a reservoir, into a form that is soluble in water. The mercury accumulates in the bodies of fish and poses a health hazard to those who depend on these fish for food.

SOLAR ENERGY

The sun is often mentioned as the ultimate answer to the world's energy problems. It provides a continuous supply of energy that far exceeds the world's demands. In fact, the amount of energy received from the sun each day is 600 times greater than the amount of energy produced each day by all other energy sources combined. The major problems with solar energy are its intermittent and diffuse nature. It is available only during the day when it is sunny, and it is spread out over the entire Earth, falling on many places like the oceans where it is difficult to collect. All systems that use solar energy must store energy or use supplementary sources of energy when sunlight is not available. Because of differences in the availability of sunlight, some parts of the world are more suited to the use of solar energy than others.

Solar energy is utilized in three ways:

1. In a passive heating system, the sun's energy is converted directly into heat for use at the site where it is collected.

2. In an active heating system, the sun's energy is converted into heat, but the heat must be transferred from the collection area to the place of use.

3. The sun's energy also can be used to generate electricity by heating water to turn turbines or by using photovoltaic cells.

Passive Solar Systems

Anyone who has walked barefoot on a sidewalk or blacktopped surface on a sunny day has experienced the effects of passive solar heating. In a **passive solar system,** light energy is transformed to heat energy when it is absorbed by a surface. Some of the earliest uses of passive solar energy were to dry food and clothes and to evaporate seawater to produce salt. Homes and buildings can be designed to use passive solar energy for heating. (See figure 9.24.)

In the Northern Hemisphere, the south side of a building always receives the most sunlight. Therefore, buildings designed for passive solar heating usually have large south-facing windows. Materials that absorb and store the sun's heat can be built into the sunlit floors and walls. The floors and walls heat up during the day and slowly release heat at night, when the heat is needed most. This passive solar design feature is called *direct gain.*

Other passive solar heating design features include sunspaces and trombe walls. A *sunspace* (which is much like a greenhouse) is built on the south side of a building. As sunlight passes through glass or other glazing, it warms the sunspace. Proper ventilation allows the heat to circulate into the building. On the other hand, a *trombe wall* is a very thick, south-facing wall painted black and made of a material that absorbs a lot of heat. A pane of glass or plastic glazing, installed a few centimeters in front of the wall, helps hold in the heat. The wall heats up slowly during the day; then, as it cools gradually during the night, it gives off its heat inside the building.

Many of the passive solar heating design features also provide daylighting. *Daylighting* is simply the use of natural sunlight to brighten a building's interior, reducing the need for electricity to light the interior of a building. To lighten north-facing rooms and upper levels, a *clerestory*—a row of windows near the peak of the roof—is often used along with an open floor plan inside that allows the light to bounce throughout the building.

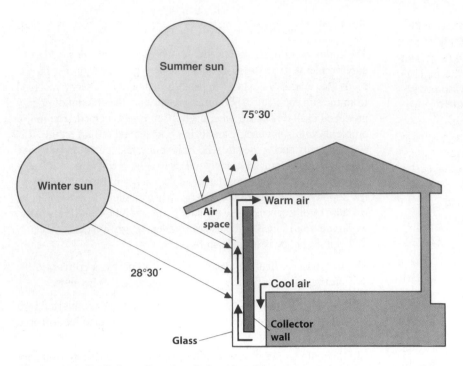

FIGURE 9.24 **Passive Solar Heating** The length of overhang in this home is designed for solar heating at the latitude of St. Louis, Missouri (38°N). In this design, a wall 30 to 40 centimeters (12 to 16 inches) thick is used to collect and store heat. The collector wall is located behind a glass wall and faces south. During a midwinter day, when the sun's angle is 28 degrees, light energy is collected by the wall and stored as heat. At night, the heat stored in the wall is used to warm the house. Natural convection causes the air to circulate past the wall, and the house is heated. During a midsummer day, when the sun's angle is 75 degrees, the overhang shades the collector wall from the sun.

Of course, too much solar heating and daylighting can be a problem during hot summer months. There are design features that can help keep passive solar buildings cool in the summer. For instance, overhangs can be designed to shade windows when the sun is high in the summer. Sunspaces can be closed off from the rest of the building. And a building can be designed to use fresh-air ventilation in the summer.

Active Solar Systems

An **active solar system** requires a solar collector, a pump, and a system of pipes to transfer the heat from the site of production to the area to be heated. (See figure 9.25.)

Active solar collector systems take advantage of the sun to provide energy for domestic water heating, pool heating, ventilation air preheating, and space heating. Water heating for domestic use is generally the most economical application of active solar systems. The demand for hot water is fairly constant throughout the year, so the solar system provides energy savings year-round. Successful use of solar water heating systems requires careful selection of components and proper sizing.

An active solar water heating system can be designed with components sized large enough to provide heating for pools or a combined function of heating both domestic water and space. Space heating requires

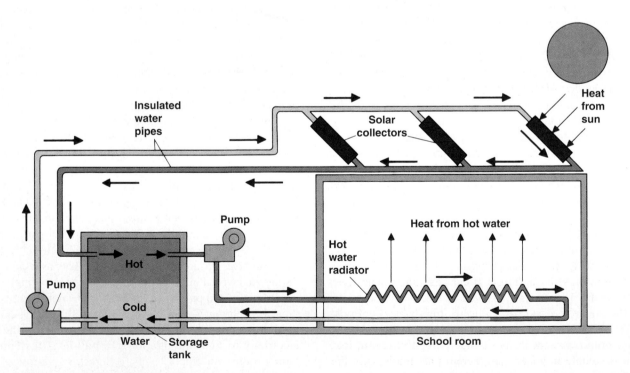

FIGURE 9.25 **Solar Heating Designs** An active solar system requires a solar collector, a pump, a heat storage system, and a system of pipes to convey the heat from one place to another.

a heat-storage system and additional hardware to connect with a heat distribution system. An active solar space heating system makes economic sense if it can offset considerable amounts of heating energy from conventional systems over the life of the building or the system. Rock, water, or specially produced products are used to store heat. The hot liquid in the pipes heats the storage medium, which is used to release heat when the sun is not shining.

Active solar systems are most easily installed in new buildings, but in some cases they can be installed in existing structures. A major consideration in the use of an active solar system is the initial cost of installation.

Solar-Generated Electricity

Solar energy can be used to generate electricity in two different ways. It can be used to create steam that is used to run a turbine similar to that of a conventional power plant, or photovoltaic cells can be used to generate electricity directly from sunlight.

Conventional Electric Generation   To produce electricity using a turbine, the energy from the sun must be collected and concentrated to heat water to steam. There are basically two designs used. One design, called a *solar furnace,* uses mirrors to focus the light at a central point that raises the temperature and allows for the production of steam. An 11-megawatt plant known as the PS10 solar power tower is currently operating in Spain and another 17-megawatt plant known as the Solar Tres Power Tower is currently being built. Several other projects are being planned throughout the world.

Currently, the most successful commercial design is the parabolic trough, which can heat oil in pipes to 390°C (734°F). (See figure 9.26.) This heat can be transferred to water, which is turned into steam that is used to run conventional electricity-generating turbines. The 354-megawatt Solar Energy Generating System (SEGS) in the Mojave Desert in California is the largest solar electric generation facility in the world. The 64-megawatt Nevada Solar One plant opened in 2007 and several plants are being built in Spain.

Photovoltaics   Photovoltaics (PV) are solid-state semiconductor devices that convert light directly into electricity. Photovoltaics are usually made of silicon with traces of other elements. Although making PV cells and modules requires advanced technology, they are very simple to use. PV modules are generally low-voltage DC devices, although arrays of PV modules can be wired for higher voltages, with no moving or wearing parts. Once installed, a PV array generally requires no maintenance other than an occasional cleaning. Most PV systems contain storage batteries, which require some water and maintenance similar to that required by the battery in an automobile.

Thin-film solar cells use layers of semiconductor materials only a few micrometers thick. Thin-film technology has now made it possible for solar cells to double as rooftop shingles, roof tiles, building façades, or the glazing for skylights or atria. (See figure 9.27.) The solar cell version of shingles offers the same protection and durability as ordinary asphalt shingles.

FIGURE 9.26   **Solar Generation of Electricity** This solar-powered electricity-generating plant is capable of generating electricity at a cost that is competitive with other methods of generating electricity.

FIGURE 9.27   **Photovoltaic Shingles** The shingles that will be installed on this roof will produce electricity and protect the occupants from the weather.

Many people who live in remote areas have a problem obtaining safe and adequate drinking water. Many remote arid regions have an abundance of solar energy and little potable water. Because of their remoteness, they must rely on local sources of energy to purify their water. Local sources of water may be contaminated or the groundwater may be too salty. However, these sources of water can be converted to drinking water by using a solar still, a simple device that can be constructed of readily available materials. Energy from the sun is used to evaporate the water, which then condenses on a glass surface and runs to a collecting tank. The impurities are left behind in the still. The still needs to be flushed periodically to remove the impurities.

Three factors drive the photovoltaic industry: cost of the solar installation, efficiency of the system, and government policy. As the cost of the system is reduced and efficiency increases, the price per kilowatt-hour of electricity falls. Currently the price is about 20 US cents per kilowatt-hour, which is much more than people pay for electricity from the power company. A typical commercial solar cell has an efficiency of 15 percent, but new designs suggest that efficiencies up to 40 percent are possible. If these new systems can be produced economically, they will compete effectively with conventional power plants.

In recent years, the amount of PV power installed worldwide has increased dramatically, from 314 megawatts in 1997 to about 5700 megawatts in 2006—an 18-fold increase in 10 years. Three countries dominate in the amount of photovoltaics installed—Germany has 50 percent, Japan has 30 percent, and the United States has 10 percent. Their dominance is the result of government policies that encourage the use of photovoltaics with tax advantages and other incentives.

Environmental Issues

Since solar energy is renewable, it has minimal environmental impact. Thermal systems that use mirrors require large amounts of land to position the mirrors. The SEGS system in California covers 6.4 km² (2.5 mi²). The installation of photovoltaics on buildings does not require additional space and is often incorporated into the design of the building.

WIND ENERGY

As the sun's radiant energy strikes the Earth, that energy is converted into heat, which warms the atmosphere. The Earth is unequally heated because various portions receive different amounts of sunlight. Since warm air is less dense and rises, cooler, denser air flows in to take its place. This flow of air is wind. For centuries, wind has been used to move ships, grind grains, pump water, and do other forms of work. In more recent times, wind has been used to generate electricity. (See figure 9.28.)

Some areas are better suited for producing wind energy than others. Figure 9.29 shows the wind energy potential of regions within the United States. However, location can be a problem.

FIGURE 9.28 **Wind Energy** Fields of wind-powered generators can produce large amounts of electricity.

Although places such as the Dakotas have the strongest winds, they are remote from energy-using population centers, and large losses in the amount of electricity would occur as it is transmitted through electric lines.

Because winds are variable, so is the amount of energy generated by each wind turbine. This means that electrical energy from wind must be coupled with other, more reliable sources of energy.

Future Development

Since the technology to generate electricity from wind is relatively easy to install, sizable increases in capacity occur each year. Europe is the leader in the amount of installed capacity, with a total installed capacity of 56,535 megawatts at the end of 2007. This is an increase of 18 percent in one year. Germany, Denmark, and Spain get more than 10 percent of their electricity from wind generation. Although there has been rapid development of new

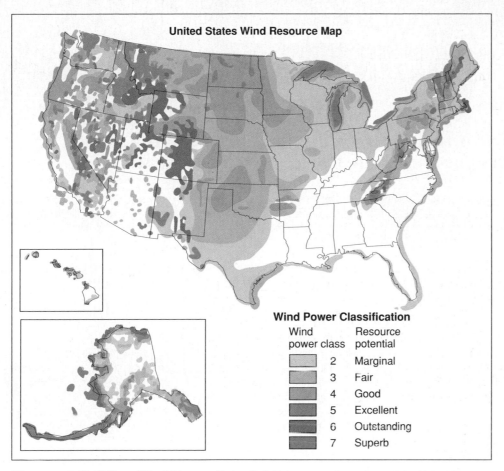

United States Wind Resource Map

Wind Power Classification

Wind power class	Resource potential
2	Marginal
3	Fair
4	Good
5	Excellent
6	Outstanding
7	Superb

FIGURE 9.29 **Wind Energy Potential** This map ranks regions of the United States in terms of their potential to supply electricity from wind energy
Source: U.S. Department of Energy.

annoying. Newer windmills, however, have slower-moving rotors that many birds such as the golden eagle find easier to avoid. Vibrations from the generators can also cause structural problems. In addition, some people consider the sight of a large number of wind generators to be visual pollution.

GEOTHERMAL ENERGY

Geothermal energy is obtained in two different ways. In geologically active areas where hot magma approaches the surface, the heat from the underlying rock can be used to heat water. The heated water can then be used directly either to heat buildings or to generate electricity by way of a steam turbine.

The United States produces about 30 percent of the world's geothermally generated electricity. The Pacific Gas and Electric Company (PG&E) has been producing electricity from geothermal energy since 1960. PG&E's complex of generating units located north of San Francisco is the largest in the world and provides 700 megawatts of power, enough for 700,000 households, or 2.9 million people. However, to put this in perspective, geo-thermal electricity accounts for less than 1 percent of total electricity consumption in the United States. Other countries that produce significant amounts of geothermal electricity are the Philippines, Italy, Mexico, Japan, New Zealand, Indonesia, and Iceland. In Iceland, half of the geothermal energy is used to produce electricity and half is used for heating. In the capital, Reykjavik, all of the buildings are heated with geothermal energy at a cost that is less than 25 percent of what it would be if oil were used

It is also possible to use heat pumps to obtain geothermal energy from areas that are not geologically active. All objects contain heat energy, which can be extracted from and transferred to other locations. Geothermal heat pumps act in a manner similar to a refrigerator, which extracts heat from its interior and exports it to the coils on the back of the unit. A heat pump can extract heat from the Earth and transfer it to a building. Typically the amount of heat energy harvested is three to four times the amount of electrical energy used to run the system. Over 50 percent of heating of buildings due to geothermal energy is the result of increased use of heat pumps.

Technology for Obtaining Geothermal Energy

In areas where a hot mass is near the surface, geothermal energy is tapped by drilling wells to obtain steam. The steam is then used to power electrical generators. (See figure 9.30.)

Geothermal heat pump systems utilize a closed loop of underground pipes. A water-antifreeze solution is circulated through the

wind power capacity, it is important to recognize that the total electrical energy produced by wind today is about 0.5 percent of total worldwide electricity consumption.

In the United States, a push for energy deregulation and concerns about smog, acid rain, and global warming are driving policy makers to require electric utilities to sell electricity from renewable sources. Twenty-nine states have established renewable portfolio standards that require that a certain percentage of electricity be produced from renewable sources by specified dates. The federal government also has provided economic incentives to utilities to construct wind and other renewable electrical-generating facilities. Since wind turbines are easy to site and install, wind energy projects have benefited from these policy decisions. Currently, 35 states have wind power installations.

According to the American Wind Energy Association, wind energy capacity expanded by 46 percent in 2007. However, wind still is responsible for only about 1 percent of all electricity generated in the United States. In 2008, the U.S. Department of Energy published a report that stated that it was technically feasible to generate 20 percent of electricity in the United States from wind by 2030.

Environmental Issues

Wind generators do have some negative effects. The moving blades are a hazard to birds and produce a noise that some find

WESTERN WASHINGTON UNIVERSITY PURCHASES GREEN POWER FROM NORTH DAKOTA WIND FARMS

In 2008, Western Washington University (WWU) in Bellingham, Washington, entered into an agreement with Puget Sound Energy (PSE) to purchase green power from North Dakota wind farms instead of renewable projects in the Pacific Northwest.

Since 2002 PSE has been recognized as offering one of the top 10 green power programs in the United States. In order to provide its customers with renewable purchase options, PSE buys energy from a variety of renewable energy suppliers. About 86 percent of its renewable power is from wind and 14 percent from the burning of biomass. Customers pay a premium of 1.25 cents per kWh to purchase green energy. In 2007, about 200,000 customers participated and purchased about 200 million kWh of electrical energy from renewable sources. This is enough energy to supply about 17,000 homes.

For several years, the U.S. Environmental Protection Agency has ranked WWU among the top 10 green energy purchasers in higher education. Since the fall of 2005, WWU has purchased all of its electrical energy through PSE from a 100 percent renewable source in the Pacific Northwest. This was made possible by Western students voting in 2004 to implement a student fee to offset the cost of purchasing renewable energy—a first in the United States.

Until 2008, WWU purchased renewable energy credits, also known as green tags, from projects located in the Pacific Northwest through PSE. These green tags replace traditional sources of electricity with sustainable sources of energy such as wind, biomass, or solar power. However, Western students continued their research into renewable energy and sought how best to optimize their purchase of green power in terms of combating global warming. That led to a Memorandum of Agreement between WWU and PSE, in which the utility agreed to purchase power from two North Dakota wind projects. WWU selected North Dakota based on new standards that show that North Dakota is one of the most carbon-intensive regions of the United States in terms of energy generation because of the number of coal-fired power plants in the region. Because North Dakota produces most of its electricity from coal, it has a large carbon footprint. Supporting green energy options in North Dakota encourages renewable energy production in North Dakota and results in a larger ultimate carbon offset than if the power were purchased from renewable sources in Washington. Because hydroelectric power plants supply large amounts of electricity in the Pacific Northwest, use of green power from local sources does not offset as much carbon.

---

pipes and heat is extracted from the solution and transferred to the building.

Environmental Issues

The use of geothermal energy from geologically active areas creates some environmental problems. The steam contains hydrogen sulfide gas, which has the odor of rotten eggs and is an unpleasant form of air pollution. (The sulfides from geothermal sources can, however, be removed.) The minerals in the steam corrode pipes and equipment, causing maintenance problems. The minerals are also toxic to fish if waste water is discharged into local bodies of water.

TIDAL POWER

Tides are caused by the gravitational force exerted by the moon and the sun. The magnitude of the gravitational attraction between two objects depends on the masses of the objects and the distance between them. The moon exerts a larger gravitational force on the Earth because, although it is much smaller in mass than the sun, it is a great deal closer than the sun. This force of attraction causes the oceans, which make up 71 percent of the Earth's surface, to bulge along an axis pointing toward the moon. (There is actually a bulge on the side of the Earth farthest from the moon also because

the moon is pulling the Earth away from the water on its surface.) Tides are produced by the rotation of the Earth beneath this bulge in its watery coating, resulting in the rhythmic rise and fall of water levels that can be observed along coasts. Thus, there are two high and two low tides each day. When the sun, moon, and Earth are in a line, the combined effects of sun and moon generate higher tides.

Certain coastal regions experience higher tides than others. This is a result of the amplification of tides caused by local geographical features such as bays and inlets. To produce practical amounts of power, a difference between high and low tides of at least 5 meters (16 feet) is required. About 40 sites around the world have this magnitude of tidal range. The higher the tides, the more electricity can be generated from a given site and the lower the cost of electricity produced. Due to the constraints just described, it has been estimated that only 2 percent, or 60 gigawatts, can potentially be recovered for electricity generation.

Technology for Obtaining Tidal Energy

The technology required to convert tidal energy into electricity is very similar to that used in traditional hydroelectric power plants. The first requirement is a dam or "barrage" across a tidal bay or estuary. Building such dams is expensive. Therefore, the best tidal sites are those where a bay has a narrow opening, thus reducing the

A new style of tidal power generator is being developed in the Dalupiri Ocean Power Plant in the Philippines. This new system is a submerged turbine referred to as a *hydro turbine*. Changing tides in the area generate ocean currents that can turn a hydro turbine. It is estimated that a 1-kilometer (0.62-mile) "tidal bridge" made up of a series of these turbines could generate more electricity than a large nuclear plant. The completed project in the Philippines will span 4 kilometers (2.5 miles) and will include 274 turbines with a generating capacity of 2.3 gigawatts during peak tidal flow.

Environmental Issues

Tidal energy is a renewable source of electricity and does not contribute to global warming. However, changing tidal flows by damming a bay or estuary could result in negative impacts on aquatic and shoreline ecosystems, as well as affecting navigation and recreation.

Studies undertaken to identify the environmental impacts of tidal power have determined that each site is different and the impacts depend greatly on local geography. Local tides changed only slightly due to the La Rance barrage, and the environmental impact has been negligible, but this may not be the case for all other sites. In all cases the barriers and turbines will affect the migration of fish and other marine species.

ENERGY CONSERVATION

Conservation is not a way of generating energy, but it is a way of reducing the need for additional energy, and it saves money for the consumer. Some conservation technologies are sophisticated,

FIGURE 9.30 **Geothermal Power Plant** Steam obtained from geothermal wells is used to produce electricity.

length of dam required. At certain points along the dam, gates and turbines are installed. When the difference in the elevation of the water on the two sides of the barrage is adequate, the gates are opened. This "hydrostatic head" that is created causes water to flow through the turbines, turning an electric generator to produce electricity.

Although the technology required to harness tidal energy is well established, tidal power is expensive, and only one major tidal generating station is in operation. This is a 240-megawatt facility (1 megawatt = 1 million watts) at the mouth of the La Rance river estuary on the northern coast of France (a large coal or nuclear power plant generates about 1000 megawatts of electricity). La Rance generating station has been in operation since 1966 and has been a very reliable source of electricity for France. (See figure 9.31.) Elsewhere, there is a 20-megawatt facility at Annapolis Royal in Nova Scotia, Canada, a 0.4-megawatt tidal power plant near Murmansk, Russia, and a 0.5-megawatt facility on Jangxia Creek in the East China Sea. A 252-megawatt facility is currently being built in South Korea and other sites throughout the world are being evaluated for their potential.

FIGURE 9.31 **Tidal Generating Station** La Rance River Estuary Power Plant in France is the world's largest tidal electrical-generating station.

while others are quite simple. For example, if a small, inexpensive wood-burning stove were developed and used to replace open fires in the less-developed world, energy consumption in these regions could be reduced by 50 percent.

Many observers have pointed out that demanding more energy while failing to conserve is like demanding more water to fill a bathtub while leaving the drain open. To be sure, conservation and efficiency strategies by themselves will not eliminate demands for energy, but they can make the demands much easier to meet, regardless of what options are chosen to provide the primary energy.

Energy efficiency improvements have significantly reduced the need for additional energy sources. Consider these facts:

- Total primary energy use per capita in the United States has changed very little in the past 10 years and the real per capita gross domestic product increased by nearly 20 percent.

- National energy intensity (energy use per unit of GDP) fell 20 percent between 1997 and 2006.

Even though the United States is much more energy-efficient today, the potential is still enormous for additional cost-effective energy savings.

Many conservation techniques are relatively simple and highly cost effective. More efficient and less energy-intensive industry and domestic practices could save large amounts of energy. Improved automobile efficiency, better mass transit, and increased railroad use for passenger and freight traffic are simple and readily available means of conserving transportation energy.

Several technologies that reduce energy consumption are now available. (See figure 9.32.) Highly efficient fluorescent lightbulbs that can be used in regular incandescent fixtures give the same amount of light for 25 percent of the energy, and they produce less heat. Since lighting and air conditioning (which removes the heat from inefficient incandescent lighting) account for 25 percent of U.S. electricity consumption, widespread use of these lights could significantly reduce energy consumption. Low-emissive glass for windows can reduce the amount of heat entering a building while allowing light to enter. The use of this glass in new construction and replacement windows could have a major impact on the energy picture. Many other technologies, such as automatic dimming devices or automatic light-shutoff devices, are being used in new construction.

The shift to more efficient use of energy needs encouragement. Often, poorly designed, energy-inefficient buildings and machines can be produced inexpensively. The short-term cost is low, but the long-term cost is high. The public needs to be educated to look at the long-term economic and energy costs of purchasing poorly designed buildings and appliances.

Electric utilities have recently become part of the energy conservation picture. In some states, they have been allowed to make money on conservation efforts; previously, they could make money only by building more power plants. This encourages them to become involved in energy conservation education, because teaching their customers how to use energy more efficiently allows them to serve more people without building new power plants.

(a)

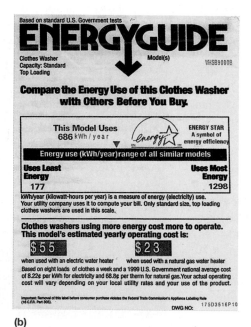

(b)

(c)

FIGURE 9.32 **Energy Conservation** The use of (a) fluorescent lightbulbs, (b) energy-efficient appliances, and (c) low-emissive glass could reduce energy consumption significantly.

In many metropolitan areas, emissions from automobiles are a major contributor to air-quality problems. An increase in the efficiency with which the chemical energy of fuel is converted to the motion of automobiles would greatly reduce air pollution. Because sources of fossil fuels are limited, they will become less available and greater efficiency will extend the limited supplies. There has been an ongoing process of modifying vehicles to improve performance and fuel efficiency. These include: reducing vehicle weight by using lighter materials, streamlining the shape to reduce air resistance, using better tires, and many other modifications. To further increase energy efficiency, more innovative techniques are required.

A great deal of energy is needed to get a vehicle to begin moving from a stop (accelerate), and it takes an equal amount of energy to stop a vehicle once it is moving (decelerate). Internal combustion engines operate most efficiently when they are running at a specific speed (rpm) and work at less than peak efficiency when the vehicle is accelerating or decelerating. Thus, using an internal combustion engine to accelerate is not efficient and contributes significantly to air pollution. When the brakes are applied to stop the vehicle, the kinetic energy possessed by the moving vehicle is converted to heat in the braking system. So the energy that has just been used to accelerate the vehicle is lost as heat when the vehicle is brought to a stop. Furthermore, in metropolitan areas, an automobile sits in traffic with its engine running a significant amount of the time. Thus, the stop-and-go traffic common in city driving provides conditions that significantly reduce the efficient transfer of the chemical energy of fuel to the kinetic energy of turning wheels.

Current hybrid electric vehicles use highly efficient electric motors to provide the energy to start the car moving (accelerate) and when the car is being driven at low speed. An internal combustion engine provides the extra power needed to reach higher speeds on the highway. In conventional vehicles, a great deal of energy is wasted while the car idles in traffic or at stoplights. In hybrid vehicles, only the electric motors powered by batteries are used in stop-and-go traffic, so no fuel is used. The battery is recharged when the car switches back to the gas engine, and in a process called *regenerative braking,* the kinetic energy of the car is captured during braking and stored as electrochemical energy in the battery. Since the internal combustion engine is only used to assist at high speeds, it is smaller, weighs less, and consumes less fuel than engines of conventional vehicles.

Although Toyota and Honda have sold hybrid electric vehicles since 1999, U.S. automakers did not produce a hybrid until 2004. Even then, their focus was on large vehicles that produced only modest fuel economy improvements. In 2008, several standard-sized, hybrid cars produced by U.S manufactures became available on a limited basis.

ARE FUEL CELLS IN THE FUTURE?

Fuel cells use hydrogen as a fuel to produce electrical energy that can be used to do a variety of jobs. Although numerous types of fuel cells are available, the most common is the proton exchange membrane (PEM) fuel cell. Although the specific reactions within the fuel cell involve several steps, they can be summarized as follows: Pressurized hydrogen gas enters the fuel cell and comes in contact with a platinum catalyst that causes the hydrogen molecules (H_2) to split into hydrogen ions (H^+) and electrons. Hydrogen ions are also known as *protons.* A proton exchange membrane allows protons (H^+)—but not the electrons—to flow through it. The electrons instead flow through an electric circuit to do work such as generating light or running a motor. The hydrogen ions flow through the proton exchange membrane and recombine with the electrons that have passed through the circuit and with oxygen to form water.

The advantages of fuel cells over other power sources are numerous. With a low operating temperature and minimal noise generation, fuel cells are safe to install in semiexposed areas. Cells are also self-sustaining, making them ideal for locations far removed from traditional electricity production facilities. Because they can operate separately from the power lines associated with most electricity distribution systems, the risk of weather-related power outages is also diminished. Their most important benefit, however, is that they do not pollute. When they are fueled by pure hydrogen, water, heat, and electricity are the only by-products. Although currently powered by modified fossil fuels, fuel cells are capable of extracting far greater quantities of energy from each unit of fuel than other methods of power generation, thereby reducing the need for fossil fuel extraction.

The automobile industry is working hard to produce fuel cells that are small enough and powerful enough to power cars. They are currently used in some bigger vehicles like buses. (See figure 9.33.)

FIGURE 9.33 **Experimental Fuel Cell Bus** Citaro fuel cell bus is the latest model with fuel cells from DaimlerChrysler AG in Germany. This picture shows one of 30 buses that were delivered to 10 European cities in 2003.

However, there are significant technological challenges to the use of fuel cells. One of the most significant is the need to produce hydrogen. Using fossil fuels to produce hydrogen does not improve the energy picture and actually produces more greenhouse gases. Therefore, it is necessary that nuclear or renewable sources of energy such as solar energy be used to produce the hydrogen. Further technological problems include the need to develop a hydrogen infrastructure so that fuel cell automobiles will be able to obtain fuel on an as-needed basis.

ISSUES & ANALYSIS

Does Ethanol Fuel Make Sense?

To evaluate the feasibility of using ethanol as fuel it is necessary to do some accounting from two points of view: energy gain and economic competitiveness.

Energy Analysis

Many studies have assessed the amount of energy necessary to produce ethanol. Currently in North America, ethanol is made by fermenting the starch in corn. It is essentially the same process as that used to make beer and wine. Therefore, the energy input to make ethanol includes fuel used by farmers to till, plant, harvest, and transport corn. In addition, fertilizer, herbicides, and other agricultural chemicals require energy to produce and apply. Finally, the fermenting of corn to make ethanol and the distillation of the ethanol from a dilute solution both require energy. Various attempts have been made to account for all of these energy inputs and the results range from zero net energy gain to a net gain of 30 percent. If it takes 1 unit of energy to produce about 1.3 units of energy as ethanol, the net gain is 0.3 units.

    The amount of ethanol produced in 2007 was about 7 billion gallons. If the amount of energy need to produce the energy is subtracted, the net gain in energy is equivalent to about 1.6 billion gallons of ethanol. This is less than 1 percent of the amount of gasoline used in 2007.

Economic Analysis

A gallon of ethanol has 67.2 percent of the energy of a gallon of gasoline, so in order to compete economically, a gallon of ethanol must cost about 67.2 percent of a gallon of gasoline. Since a gallon of ethanol has less energy than gasoline, vehicles will travel fewer miles on a gallon of ethanol than on a gallon of gasoline. Most ethanol is used in E85 fuel, which contains 85 percent ethanol and 15 percent gasoline. The following charts show the amount of fuel needed to provide the same amount of energy and the price per gallon needed to provide equivalent amounts of energy per dollar spent.

Gallons Needed to Provide Equivalent Energy		
Gasoline	Ethanol	E85 fuel
1	1.49	1.42

Price/Gallon to Provide Equivalent Energy per Dollar		
Gasoline	Ethanol	E85 fuel
$3.00	$2.00	$2.15

Other Economic Issues

- Currently in the United States those who sell ethanol for fuel receive a $0.51/gallon tax credit, which effectively reduces the price of ethanol fuel by 51 cents per gallon at the taxpayer's expense. If that subsidy were removed, the price of ethanol fuel would be 51 cents higher, and ethanol would not be competitive with gasoline. In nearly all markets, if the 51 cent/gallon subsidy for ethanol were removed, the price of ethanol fuel would be more expensive than gasoline.

- Most of the corn produced in the United States is used for animal feed. Higher demand for corn to make ethanol has caused an increase in the price of corn, which results in higher feed prices for livestock and leads to higher prices for meat products.

- High prices for corn encourages farmers to plant more corn (2007 was a record crop). If more farmland is planted to corn, then less land is planted to wheat, soybeans, and other crops, which results in increased prices for those commodities and higher food prices in general.

- Some studies attempt to assign economic values to soil erosion, increased cost of food, and other factors. Many of those studies show no net gain from the production of ethanol.

What Do You Think?

- Is it important to provide ethanol as a fuel?
- Would you use ethanol?
- Should the production of ethanol be subsidized?

SUMMARY

A resource is a naturally occurring substance of use to humans, a substance that can potentially be extracted using current technology. Reserves are known deposits from which materials can be extracted profitably with existing technology under present economic conditions.

Coal is the world's most abundant fossil fuel. Coal is obtained by either surface mining or underground mining. Problems associated with coal extractions are disruption of the landscape due to surface mining and subsidence due to underground mining. Black lung disease, waste heaps, water and air pollution, and acid mine drainage are additional problems. Oil was originally chosen as an alternative to coal because it was more convenient and less expensive. However, the supply of oil is limited. As oil becomes less readily available, multiple offshore wells, secondary recovery methods, and increased oil exploration will become more common. Natural gas is another major source of fossil-fuel energy. The primary problem associated with natural gas is transport of the gas to consumers.

Fossil fuels are nonrenewable: The amounts of these fuels are finite. When the fossil fuels are exhausted, they will have to be replaced with other forms of energy, probably renewable forms. Hydroelectric power can be increased significantly, but its development requires flooding areas and thus may require the displacement of people. The use of geothermal and tidal energy is limited by geographic locations. Wind power may be used to generate electricity but may require wide, open areas and a large number of wind generators. Solar energy can be collected and used in either passive or active systems and can also be used to generate electricity. Lack of a constant supply of sunlight is solar energy's primary limitation. Fuelwood is a minor source of energy in industrialized countries but is the major source of fuel in many less-developed nations. Biomass can be burned to provide heat for cooking or to produce electricity, or it can be converted to alcohol or used to generate methane. In some communities, solid waste is burned to reduce the volume of the waste and also to supply energy.

Energy conservation can reduce energy demands without noticeably changing standards of living.

THINKING GREEN

1. Contact your local electric utility or visit its website and determine what percentage of the electricity produced comes from fossil fuels, nuclear, hydroelectric, or other renewable energy sources.

2. Does your electric utility offer an opportunity to purchase green energy? If so, what does it cost?

3. Calculate your monthly carbon footprint. The average carbon dioxide produced per person in the United States is 1.7 tons of CO_2 per month (20 tons per year).

   Note: If there are several persons in your household, divide the total energy used in each category by the number of people in the household to determine your individual energy usage.

Energy Use	Tons of CO_2
Kilowatt-hours of electricity $\times$ 0.0006 =	
Therms of natural gas $\times$ 0.00591 =	
Gallons of heating oil $\times$ 0.01015 =	
Gallons of gasoline $\times$ 0.0087 =	

WHAT'S YOUR TAKE?

Alcohol fuel is cheaper than gasoline because it is subsidized by the federal government. In addition, farmers receive subsidies to grow corn from which alcohol is made. Should the government subsidize the production of alcohol for fuel? Choose one side or the other and develop arguments to support your point of view.

REVIEW QUESTIONS

1. Why are fossil fuels important?
2. Distinguish between reserves and resources.
3. What are the advantages of surface mining of coal compared to underground mining? What are the disadvantages of surface mining?
4. Compare the environmental impacts of the use of coal and the use of oil.
5. What are some limiting factors in the development of new hydroelectric generating sites?
6. What factors limit the development of tidal power as a source of electricity?
7. In what parts of the world is geothermal energy available? What creates it?
8. Why can wind be considered a form of solar energy?
9. Compare a passive solar-heating system with an active solar-heating system.
10. What problems are associated with the use of solid waste as a source of energy?
11. List three energy conservation techniques.

CRITICAL THINKING QUESTIONS

1. Given what you know about the economic and environmental costs of different energy sources, would you recommend that your local utility company use hydroelectricity or coal to supplement electric production? What criteria would you use to make your recommendation?
2. Coal-burning electric power plants in the Midwest have contributed to acid rain in the eastern United States. Other energy sources would most likely be costlier than coal, thereby raising electricity rates. Should citizens of another state be able to pressure these utility companies to change the method of generating electricity? What mechanisms might be available to make these changes? How effective are these mechanisms?
3. Imagine you are an official with the Department of Energy and are in the budgeting process for alternative energy research. Where would you put the money? Why?
4. Given your choices from question 3, what do you think the political repercussions of your decision would be? Why?
5. Do you believe that large dam projects like the Three Gorges Dam project in China are, on the whole, beneficial or not? What alternatives would you recommend? Why?
6. Energy conservation is one way to decrease dependence on fossil fuels. What are some things you can do at home, work, or school that would reduce fossil-fuel use and save money?
7. What alternative energy resources that the text has outlined are most useful in your area? How might these be implemented?

CHAPTER **10**

NUCLEAR ENERGY

Nuclear energy can be used to produce electricity. In the process steam must be cooled. Thus, giant cooling towers are common structures associated with nuclear power plants.

CHAPTER OUTLINE

The Nature of Nuclear Energy
  Measuring Radiation
  Biological Effects of Ionizing Radiation
  Radiation Protection
  Nuclear Chain Reaction
The History of Nuclear Energy Development
Nuclear Fission Reactors
  Boiling-Water Reactors
  Pressurized-Water Reactors
  Heavy-Water Reactors
  Gas-Cooled Reactors
Investigating Nuclear Alternatives
  Breeder Reactors
  Nuclear Fusion
The Nuclear Fuel Cycle
  Mining and Milling
  Enrichment and Fuel Fabrication
  Use in a Reactor
  Reprocessing and Waste Disposal
  Transportation Issues
Nuclear Concerns
  Reactor Safety
  Terrorism
  Worker and Public Exposure to Radiation
  Contamination from Nuclear Research and Weapons Production
  Disposal of Nuclear Weapons
  Radioactive Waste Disposal
  Thermal Pollution
  Decommissioning
The Future of Nuclear Power
  Social Forces
  Technical Trends

ISSUES & ANALYSIS
Yucca Mountain and Nuclear Waste Storage 232

CASE STUDY
The Hanford Facility: A Storehouse of Nuclear Remains 231

CAMPUS SUSTAINABILITY INITIATIVE
Oregon State University and Passive Nuclear Power Plants 230

GOING GREEN
Returning a Nuclear Plant Site to Public Use 229

WATER CONNECTIONS
Water and Nuclear Power Plants 222

OBJECTIVES

After reading this chapter, you should be able to:

- Explain how nuclear fission has the potential to provide large amounts of energy.
- Describe how a nuclear reactor produces electricity.
- Describe the basic types of nuclear reactors.
- Explain the steps involved in the nuclear fuel cycle.
- List concerns about the use of nuclear power.
- Explain the problem of decommissioning a nuclear plant.
- Describe how high-level radiation waste is stored.
- Describe the accident at Chernobyl.
- Explain how a breeder reactor differs from other nuclear reactors.

A Global Perspective on "The Nuclear Legacy of the Soviet Union" can be found on the book's website at www.mhhe.com/enger12e along with other interesting readings.

THE NATURE OF NUCLEAR ENERGY

Energy from disintegrating atomic nuclei has tremendous potential. (See figure 10.1.) In terms of benefiting the people of the world, radioactive isotopes are used to assist in diagnosis and to treat certain kinds of cancer such as thyroid and prostate cancer. About 15 percent of the electrical energy generated in the world comes from nuclear power plants. This is slightly less than 6 percent of all the energy consumed per year in the world. Engineers in the former Soviet Union used nuclear explosions to move large amounts of earth and rock to construct dams, canals, and underground storage facilities. The U.S. government briefly considered the possibility of using nuclear explosions to build a new Panama Canal.

On the other hand, nuclear energy has the potential to do great harm. The Japanese cities of Hiroshima and Nagasaki were destroyed by nuclear bombs. The manufacture and storage of nuclear weapons have left a legacy of radioactive wastes. In many cases, these wastes have been mismanaged or carelessly disposed of.

To understand where nuclear energy comes from, it is necessary to review some aspects of atomic structure presented in chapter 4. All atoms are composed of a central region called the nucleus, which contains positively charged protons and neutrons that have no charge. Moving around the nucleus are smaller, negatively charged electrons. Since like charges repel one another, something is required to hold the positively charged protons together in the nucleus. This is called the *nuclear force*. In most atoms, the various forces in the nucleus are balanced and the nucleus is stable. However, some isotopes of atoms are **radioactive;** that is, the nuclei of these atoms are unstable and spontaneously decompose. Neutrons, electrons, protons, and other larger particles are released during nuclear disintegration, along with a great deal of energy. The rate of decomposition is consistent for any given isotope. It is measured and expressed as **radioactive half-life,** which is the time it takes for one-half of the radioactive material to spontaneously decompose. Table 10.1 lists

(a) Bomb blast

(b) Nuclear power plant

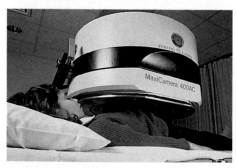

(c) Nuclear medicine

FIGURE 10.1 **Uses of Nuclear Energy** The energy available from the splitting of atoms has been put to many uses: (a) nuclear test detonation. Bikini Atoll, May 1956; (b) a nuclear power plant in Palo Verde, Arizona; (c) a nuclear medicine procedure used to diagnose medical conditions.

TABLE 10.1 Half-Lives and Significance of Some Radioactive Isotopes

Radioactive Isotope	Half-Life	Significance
Uranium-235	700 million years	Fuel in nuclear power plants
Plutonium-239	24,110 years	Nuclear weapons Fuel in some nuclear power plants
Carbon-14	5730 years	Establish age of certain fossils
Americium-241	432.2 years	Used in smoke detectors
Cesium-137	30.17 years	Treat prostate cancer Used to measure thickness of objects in industry
Cobalt-60	5.27 years	Sterilize food by irradiation Cancer therapy Inspect welding seams
Strontium-90	29.1 years	Power source in space vehicles Treat bone tumors
Iridium-192	73.82 days	Inspect welding seams Treat certain cancers
Phosphorus-32	14.3 days	Radioactive tracer in biological studies
Iodine-131	8.06 days	Diagnose and treat thyroid cancer
Radon-222	3.8 days	Naturally occurs in atmosphere of some regions where it causes lung cancers
Radon-220	54.5 seconds	Naturally occurs in atmosphere of some regions where it causes lung cancers

the half-lives of several radioactive isotopes. Figure 10.2 shows a typical radioactive decay curve.

Nuclear disintegration releases energy from the nucleus as **radiation,** of which there are three major types:

Alpha radiation consists of a moving particle composed of two neutrons and two protons. Alpha radiation usually travels through air for less than a meter and can be stopped by a sheet of paper or the outer, nonliving layer of the skin.

Beta radiation consists of moving electrons released from nuclei. Beta particles travel more rapidly than alpha particles and will travel through air for a couple of meters. They are stopped by a layer of clothing, glass, or aluminum.

Gamma radiation is a type of electromagnetic radiation that does not consist of particles. Other forms of electromagnetic radiation are X rays, light, and radio waves. Gamma radiation can pass through your body, several centimeters of lead, or nearly a meter of concrete.

When a radioactive isotope disintegrates and releases particles, it becomes a different kind of atom. For example, uranium-238 ultimately produces lead—but it goes through several steps in the process. (See figure 10.3.)

MEASURING RADIATION

There are several different kinds of radiation measurements for different purposes. Table 10.2 shows some of the common units and their values.

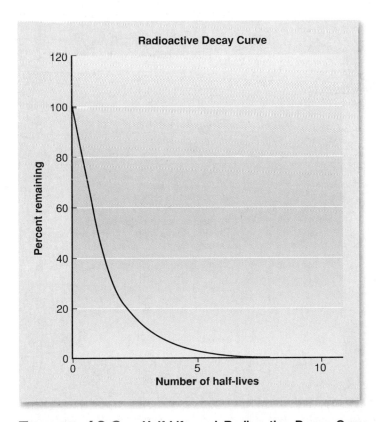

FIGURE 10.2   Half-Life and Radioactive Decay Curve
This curve shows how the amount of a radioactive material decreases over time.

Isotope of Element	Type of Radiation	Half-Life
Uranium-238 (U-238) Protons = 92 Neutrons = 146		4.5 billion years
	Alpha (2 protons and 2 neutrons)	
Thorium-234 (Th-234) Protons = 90 Neutrons = 144		24.5 days
	Beta (electron)	
Protactinium-234 (Pa-234) Protons = 91 Neutrons = 143		1.14 minutes
	Beta (electron)	
Uranium-234 (U-234) Protons = 92 Neutrons = 142		233,000 years
	Alpha (2 protons and 2 neutrons)	
Thorium-230 (Th-230) Protons = 90 Neutrons = 140		83,000 years
	Alpha (2 protons and 2 neutrons)	
Radium-226 (Ra-226) Protons = 88 Neutrons = 138		1590 years
	Alpha (2 protons and 2 neutrons)	
Radon-222 (Rn-222) Protons = 86 Neutrons = 136		3.825 days
	Alpha (2 protons and 2 neutrons)	
Polonium-218 (Po-218) Protons = 84 Neutrons = 134		3.05 minutes
	Alpha (2 protons and 2 neutrons)	
Lead-214 (Pb-214) Protons = 82 Neutrons = 132		26.8 minutes
	Beta (electron)	
Bismuth-214 (Bi-214) Protons = 83 Neutrons = 131		19.7 minutes
	Beta (electron)	
Polonium-214 (Po-214) Protons = 84 Neutrons = 130		0.00015 seconds
	Alpha (2 protons and 2 neutrons)	
Lead-210 (Pb-210) Protons = 82 Neutrons = 128		22 years
	Beta (electron)	
Bismuth-210 (Bi-210) Protons = 83 Neutrons = 127		5 days
	Beta (electron)	
Polonium-210 (Po-210) Protons = 84 Neutrons = 126		140 days
	Alpha (2 protons and 2 neutrons)	
Lead-206 (Pb-206) Protons = 82 Neutrons = 124		Stable

FIGURE 10.3   Radioactive Decay Path of U-238 When an unstable isotope disintegrates, it ultimately leads to a stable isotope of a different element. In the process it is likely to go through several unstable isotopes of other elements before it becomes a stable element.

TABLE 10.2 Radiation Measurement Units

What Is Measured	International Scientific Units	U.S. Commonly Used Units	Application
Number of nuclear disintegrations	becquerel (Bq) 1 Bq = 1 disintegration/ second	curie (Ci) 1 Ci = 37 billion disintegration/second 1 Ci = 37 billion Bq	Quantify the strength of a radiation source
Absorbed dose	gray (Gy) 1 Gy = 1 joule/kilogram of matter	rad 1 rad = 0.01 joule/kilogram of matter 1 rad = 0.01 Gy	Quantify the amount of energy absorbed
Dose equivalent	sievert (Sv) 1 Sv = Gy X quality factor*	rem 1 rem = rad X quality factor* 1 rem = 0.01 Sv	Quantify the potential biological effect of a dose

*For beta and gamma radiation, the quality factor = 1. For alpha radiation, the quality factor = 20.

The amount of radiation absorbed by the human body is a way to quantify the potential damage to tissues from radiation. The **absorbed dose** is the amount of energy absorbed by matter. It is measured in grays or rads. However, the damage caused by alpha particles is 20 times greater than that caused by beta particles or gamma rays. Therefore, a dose equivalent is used. The **dose equivalent** is the absorbed dose times a quality factor. The units for dose equivalents are seiverts or rems. The quality factor for beta and gamma radiation is 1. Therefore, the dose equivalent is the same as the absorbed dose. The quality factor for alpha radiation is 20. Therefore, the dose equivalent for alpha radiation is 20 times the absorbed dose.

Because ionizing radiation affects DNA, it can cause mutations, which are changes in the genetic messages within cells. Mutations can cause two quite different kinds of problems. Mutations that occur in the ovaries or testes can form mutated eggs or sperm, which can lead to abnormal offspring. Care is usually taken to shield these organs from unnecessary radiation. Mutations that occur in other tissues of the body may manifest themselves as abnormal tissue growths known as cancer. Two common cancers that are strongly linked to increased radiation exposure are leukemia and breast cancer. Because mutations are essentially permanent, they may accumulate over time. Therefore, the accumulated effects of radiation over many years may result in the development of cancer later in life.

The effects of large doses (1000 to 1 million rems) are clearly evident because people become ill and die. However, demonstrating known harmful biological effects from smaller doses is much more difficult. (See table 10.3.) Moderate doses (10 to 100 rems) are known to increase the likelihood of cancer and birth defects. The higher the dose, the higher the incidence of abnormality. Lower doses may cause temporary cellular changes, but it is difficult to demonstrate long-term effects.

BIOLOGICAL EFFECTS OF IONIZING RADIATION

When an alpha or beta particle or gamma radiation interacts with atoms, it can dislodge electrons from the atoms and cause the formation of ions. Therefore, these kinds of radiation are called **ionizing radiation.** (X rays are also a form of ionizing radiation, although they are not formed as a result of nuclear disintegration.) When ionization occurs in living tissue it can result in damage to DNA or other important molecules in cells. The degree and kind of damage vary with the kind of radiation, the amount of radiation, the duration of the exposure, and the types of cells irradiated.

TABLE 10.3 Radiation Effects

Source or Benchmark	Dose	Biological Effects
Nuclear bomb blast or accidental exposure in a nuclear facility	100,000 rems/incident	Immediate death
Nuclear accident or accidental exposure to X rays	10,000 rems/incident	Coma, death in 1–2 days
	1000 rems/incident	Death in 2–3 weeks
	800 rems/incident	100% death eventually
	500 rems/incident	50% survival with good medical care
	100 rems/incident	Increased probability of leukemia
	50 rems/incident	Changes in numbers of blood cells observed
	10 rems/incident	Early embryos may show abnormalities
X ray of intestine	1 rem/procedure	Damage or effects difficult to demonstrate
Upper limit for occupationally exposed persons	5 rems/year	
Upper limit for release from nuclear facilities that are not nuclear power plants	0.5 rem/year	
Natural background radiation	0.2–0.3 rem/year	
Upper limit for exposure of general public to radiation above background	0.1 rem/year	
Upper limit for release from nuclear power plants	0.005 rem/year	

Current research is trying to assess the risks associated with repeated exposure to low-level radiation. Thus, the effects of low-level, chronic radiation generate much controversy. Some people feel that all radiation is harmful, that there is no safe level, and that special care must be taken to prevent exposure. Others feel that there is a threshold level of exposure below which there are no biological effects. One problem is that it is likely that there are delayed effects of radiation exposure. Since it is assumed that radiation affects DNA, the effect may not be evident at the time of exposure but would show up later when the affected gene becomes active. This complicates our attempts to link certain health problems such as cancers with previous exposures to radiation. It should be pointed out that no one experiences zero exposure; the average person is exposed to 0.2 to 0.3 rem per year from natural and medical sources.

RADIATION PROTECTION

For the sake of our health, it is important that we know something about the cumulative dose of ionizing radiation we receive. The more radiation a person receives, the more likely it is that there will be biological consequences. Because in most cases people cannot sense ionizing radiation, protection requires specialized sensors. These can be as simple as film badges that record the amount of radiation hitting them. These kinds of devices measure *past exposure* to radiation, which can be useful in determining whether a person should avoid *future exposure*. This is most important for people who work in environments that have a radiation hazard. Because there are limits to the amount of radiation exposure an employee can receive, monitoring allows for changes in job assignments or procedures that limit total exposure.

Time, distance, and shielding are the basic principles of radiation protection. The basic idea behind these three principles is that the total cumulative dose should be minimized. Time is important because the longer one is exposed to a source of radiation, the more radiation is absorbed. Distance is important because the farther a person is from the source, the less likely that person will be hit by radiation. In addition, alpha and beta radiation will only travel a short distance through air. Shielding is important because it can stop radiation and prevent a person from being exposed. Because gamma radiation can travel many meters through air and can penetrate the body, shielding is very important. Water, lead, and concrete are common materials used for shielding from gamma radiation. Another kind of shielding is needed to prevent materials that produce radiation from entering the body. Even though alpha and beta radiation are easily stopped, if radioactive isotopes that emit alpha or beta radiation are ingested or inhaled, the radiation they emit can damage the cells they are in contact with. Thus, shielding from isotopes involves such devices as masks, gloves, and protective clothing. (See figure 10.4.)

NUCLEAR CHAIN REACTION

In addition to releasing alpha, beta, and gamma radiation when they disintegrate, the nuclei of a few kinds of atoms release neutrons. When moving neutrons hit the nuclei of certain other atoms, they

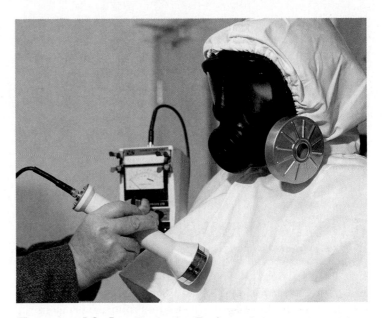

FIGURE 10.4 **Protective Equipment** Persons who work in an area that is subject to radiation exposure must take steps to protect themselves. This worker is wearing protective clothing and a respirator. The surface of his protective suit is being checked with a Geiger counter to determine if he has any radioactive dust particles on his clothing.

can cause those nuclei to split as well. This process is known as **nuclear fission.** If these splitting nuclei also release neutrons, they can strike the nuclei of other atoms, which also disintegrate, resulting in a continuous process called a **nuclear chain reaction.** Only certain kinds of atoms are suitable for the development of a nuclear chain reaction. The two materials commonly used in nuclear reactions are uranium-235 and plutonium-239. In addition, there must be a certain quantity of nuclear fuel (a critical mass) in order for a nuclear chain reaction to occur. It is this process that results in the large amounts of energy released from bombs or nuclear reactors.

THE HISTORY OF NUCLEAR ENERGY DEVELOPMENT

The first controlled nuclear chain reaction occurred at Stagg Field in Chicago in 1942. That event led directly to the development of the atomic bombs dropped on the Japanese cities of Hiroshima and Nagasaki in 1945. The incredible devastation of these two cities demonstrated the potential of nuclear energy for destruction and led to Japan's surrender. For many years, most atomic research involved military applications of nuclear energy as bombs and as power sources for ships. During the years following World War II, the two major military powers of the world—the United States and the former Soviet Union—conducted secret nuclear research projects related to the building and testing of bombs. This continued to be a primary focus of nuclear research until the disintegration of the Soviet Union and the end of the Cold War. A legacy of this military research is a great deal of soil, water, and air contaminated with radioactive material. Many of these contaminated sites have come to light recently and require major cleanup efforts.

After World War II, people began to see the potential for using nuclear energy for peaceful purposes rather than as weapons. The world's first electricity-generating reactor was constructed in the United States in 1951, and the Soviet Union built its first reactor in 1954.

As of 2008, there were 439 nuclear power reactors in operation and there are 36 nuclear power plants under construction in 12 countries. A further 93 are in the planning stages. (See table 10.4.) The United States has 104 nuclear power plants, nearly 25 percent of the world total. Figure 10.5 shows the locations of the nuclear power plants in North America.

NUCLEAR FISSION REACTORS

A **nuclear reactor** is a device that permits a controlled nuclear fission chain reaction. **Uranium-235 (U-235)** is an isotope used to fuel most nuclear fission reactors. When the nucleus of a U-235 atom is struck by a slowly moving neutron from another atom, the nucleus of the atom of U-235 is split into several smaller particles.

TABLE 10.4 Nuclear Reactor Statistics (2008)

Region	Reactors Operating	Reactors Under Construction	Reactors Planned
World	*439*	*36*	*93*
United States	104	0	12
France	59	1	0
Japan	55	2	11
Russia	31	7	10
South Korea	20	3	5
United Kingdom	19	0	0
Canada	18	2	3
Germany	17	0	0
Ukraine	15	0	2
India	15	6	10
China	11	7	24
Sweden	10	0	0
Spain	8	0	0
Rest of World	57	8	16

Source: Data from World Nuclear Association.

Because the nucleus will split, it is said to be **fissionable.** When the nucleus is split, two to three rapidly moving neutrons are released, along with large amounts of energy, that can be harnessed to do work. The neutrons released strike the nuclei of other atoms of U-235 and cause them to undergo fission, which in turn releases more energy and more neutrons, thus resulting in a chain reaction. (See figure 10.6.) Once begun, this chain reaction continues to release energy until the fuel is spent or the neutrons are prevented from striking other nuclei.

In addition to fuel rods containing uranium, reactors contain control rods of cadmium, boron, graphite, or other nonfissionable materials used to control the rate of fission by absorbing neutrons. When control rods are lowered into a reactor, they absorb the neutrons produced by fissioning uranium. There are then fewer neutrons to continue the chain reaction, and the rate of fission decreases. If the control rods are withdrawn, more fission occurs, and more particles, radiation, and heat are produced.

The fuel rods housed in a reactor are surrounded by water or some other type of moderator. A **moderator** absorbs some of the energy of the neutrons, which slows neutrons, enabling them to split the nuclei of other atoms more effectively. Fast-moving neutrons are less effective at splitting atoms than slow-moving neutrons. As U-235 undergoes fission, the energy of the fast-moving neutrons is transferred to water; the neutrons slow down, and the water is heated.

FIGURE 10.5 **Distribution of Nuclear Power Plants in North America**
The United States has the largest number of nuclear power plants of any country in the world—104. Canada has 18 nuclear power plants, and Mexico has two.

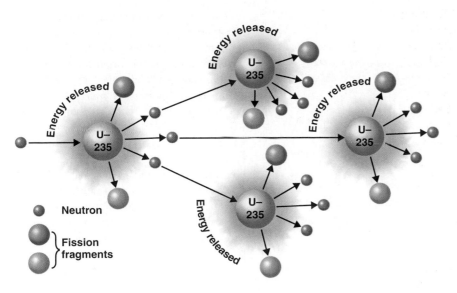

FIGURE 10.6 **Nuclear Fission Chain Reaction** When a neutron strikes a nucleus of U-235, energy is released, and several fission fragments and neutrons are produced. The newly released neutrons may strike other atoms of U-235, causing their nuclei to split with the release of additional neutrons. This series of events is called a *nuclear fission chain reaction*.

In the production of electricity, a nuclear-powered reactor serves the same function as any fossil-fueled boiler. It produces heat, which converts water to steam to operate a turbine that generates electricity. After passing through the turbine, the steam must be cooled, and the water is returned to the reactor to be heated again. Various types of reactors have been constructed to furnish heat for the production of steam. They differ in the moderator used, in how the reactor core is cooled, and in how the heat from the core is used to generate steam. Water is the most commonly used reactor-core coolant and also serves as a neutron moderator. The three most common kind of nuclear reactors are boiling water reactors, pressurized-water reactors, and heavy-water reactors.

BOILING-WATER REACTORS

In a **boiling-water reactor,** water functions as both a moderator and a reactor-core coolant. (See figure 10.7.) Steam is formed within the reactor and transferred directly to the turbine, which turns to generate electricity. A disadvantage of the boiling-water reactor is that the steam passing to the turbine must be treated to remove any radiation. Even then, some radioactive material is left in the steam; therefore, the generating building must be shielded. About 20 percent of the nuclear reactors in the world are boiling-water reactors.

PRESSURIZED-WATER REACTORS

In a **pressurized-water reactor,** water is kept under high pressure so that steam is not allowed to form in the reactor. (See figure 10.8.) A secondary loop transfers the heat from the pressurized water in the reactor to a steam generator. The steam is used to turn the turbine and generate electricity. Such an arrangement reduces the risk of radiation in the steam but adds to the cost of construction by requiring a secondary loop for the steam generator. About 60 percent of the nuclear reactors in the world are pressurized-water reactors. Most of the reactors currently under construction are of this type.

HEAVY-WATER REACTORS

A third type of reactor that uses water as a coolant is a **heavy-water reactor,** developed by Canadians. It uses water that contains the hydrogen isotope deuterium in its molecular structure as the reactor-core coolant and moderator. Since the deuterium atom is twice as heavy as the more common hydrogen isotope, the water that contains deuterium weighs slightly more than ordinary water. Heavy-water reactors are similar to pressurized-water reactors in that they use a steam generator to convert regular water to steam in a secondary loop. The major advantage of a heavy-water reactor is that naturally occurring uranium isotopic mixtures serve as a suitable fuel while other

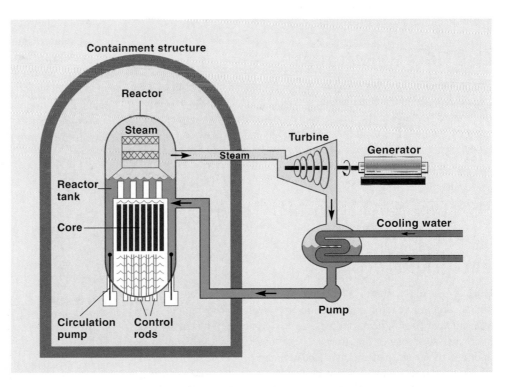

FIGURE 10.7 **Boiling-Water Reactor** A boiling-water reactor is a type of light-water reactor that produces steam to directly power the turbine and produce electricity. Water is used as a moderator and as a reactor-core coolant.

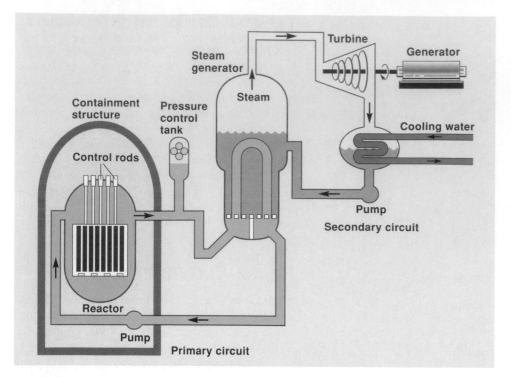

FIGURE 10.8 **Pressurized-Water Reactor** A pressurized-water reactor is a type of light-water reactor that uses a steam generator in a secondary loop to produce steam, which is transferred to the turbine.

function without a moderator. Because it is necessary to move heat away from the reactor core very efficiently, most breeder reactors use liquid metal (often liquid sodium) as a core coolant. Thus, they are often called *liquid metal fast-breeder reactors*. When a fast-moving neutron hits a nonfissionable uranium-238 (U-238) nucleus and is absorbed, an atom of fissionable **plutonium-239 (Pu-239)** is produced. Remember that U-235 is a nuclear fuel and U-238 is not. Thus, a breeder reactor converts a nonfuel (U-238) into a fuel (Pu-239). (See figure 10.9.)

Most breeder reactors are considered experimental and because they produce plutonium-239, which can be used to produce nuclear weapons, they are politically sensitive. Russia has one that is in commercial operation and plans to build several more, but most other countries have discontinued their breeder reactor programs following accidents or political decisions. Japan, the United Kingdom, Germany, the United States, and France have breeder reactors that are currently not in operation. India has a small research reactor operating and is building a larger facility. China and South Korea have expressed interest in developing breeder reactors.

reactors require that the amount of U-235 be enriched to get a suitable fuel. This is possible because heavy water is a better neutron moderator than is regular water. Since it does not require enriched fuel, the cost of producing fuel for a heavy-water reactor is less than that for other reactors. About 10 percent of nuclear reactors are of this type.

GAS-COOLED REACTORS

The **gas-cooled reactor** was developed by atomic scientists in the United Kingdom. Carbon dioxide serves as a coolant for a graphite-moderated core. As in the heavy-water reactor, natural isotopic mixtures of uranium are used as a fuel. However, this is not a popular type of reactor and no new plants of this type are being constructed.

NUCLEAR FUSION

When two lightweight atomic nuclei combine to form a heavier nucleus, a large amount of energy is released. This process is known as **nuclear fusion.** Nuclear fusion is often mentioned as the answer to energy needs. Theoretically, fusion could produce huge amounts of energy. Although fusion occurs in stars, like our sun, it has been technically impossible to create conditions that would allow fusion to provide a reliable source of energy in the foreseeable future. Governments are providing little money for fusion research.

INVESTIGATING NUCLEAR ALTERNATIVES

BREEDER REACTORS

During the early stages of the development of nuclear power plants, breeder reactor construction was seen as the logical step after nuclear fission development. A **nuclear breeder reactor** is a nuclear fission reactor that forms a new supply of nuclear fuel as it operates to produce electricity. Because the process requires fast-moving neutrons, water cannot be used as a moderator because it slows the neutrons too much and most models of breeder reactors

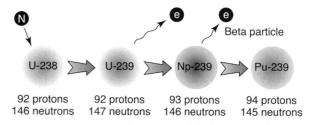

FIGURE 10.9 **Formation of Pu-239 in a Breeder Reactor** When a fast-moving neutron (N) is absorbed by the nucleus of a U-238 atom, a series of reactions results in the formation of Pu-239 from U-238. Two intermediate atoms are U-239 and Np-239, which release beta particles (electrons) from their nuclei. (A neutron in the nucleus can release a beta particle and become a proton.) This series of reactions is important because, while the U-238 does not disintegrate readily and therefore is not a nuclear fuel, the Pu-239 is fissionable and can serve as a nuclear fuel.

THE NUCLEAR FUEL CYCLE

To appreciate the consequences of using nuclear fuels to generate energy, it is important to understand the nuclear fuel cycle, which follows the process from the mining of uranium to the disposal of the waste from power plants.

MINING AND MILLING

The nuclear fuel cycle begins with the mining operation. (See figure 10.10.) Low-grade uranium ore is obtained by underground or surface mining. The ore contains about 0.2 percent uranium by weight. After it is mined, the ore goes through a milling process. It is crushed and treated with a solvent to concentrate the uranium. Milling produces yellow-cake, a material containing 70 to 90 percent uranium oxide. Over half of the world's production of uranium comes from mines in Canada, Australia, and Kazakhstan.

ENRICHMENT AND FUEL FABRICATION

Naturally occurring uranium contains about 99.3 percent nonfissionable U-238 and only 0.7 percent fissionable U-235. This concentration of U-235 is not high enough for most types of reactors, so the amount of U-235 must be increased by enrichment. Since the masses of the isotopes U-235 and U-238 vary only slightly, and the chemical differences are very slight, enrichment is a difficult and expensive process. The enrichment process involves the use of centrifuges to separate the two isotopes of uranium by their slight differences in mass. Enrichment increases the U-235 content from 0.7 percent to 3 percent.

Fuel fabrication involves converting the enriched material into a powder, which is then compacted into pellets about the size of a pencil eraser. These pellets are sealed in metal fuel rods about 4 meters (13 feet) in length, which are loaded into a reactor.

USE IN A REACTOR

As fission occurs in the reactor, the concentration of U-235 atoms decreases. After about three years, a fuel rod does not have enough radioactive material to sustain a chain reaction, and the spent fuel rods must be replaced by new ones. The spent rods are still very radioactive, containing about 1 percent U-235 and 1 percent plutonium.

REPROCESSING AND WASTE DISPOSAL

Because spent fuel rods are radioactive, they must be managed carefully to prevent environmental damage and risks to health. There are two alternatives used to deal with these wastes: reprocessing and long-term storage. Reprocessing involves extracting the remaining U-235 and plutonium from the spent fuel, separating the U-235 from the plutonium, and using both to manufacture new fuel rods. The plutonium is mixed with spent uranium (mostly U-238) to form a mixed oxide fuel that can be used as a fuel in some kinds of nuclear power plants. Besides providing new fuel, reprocessing

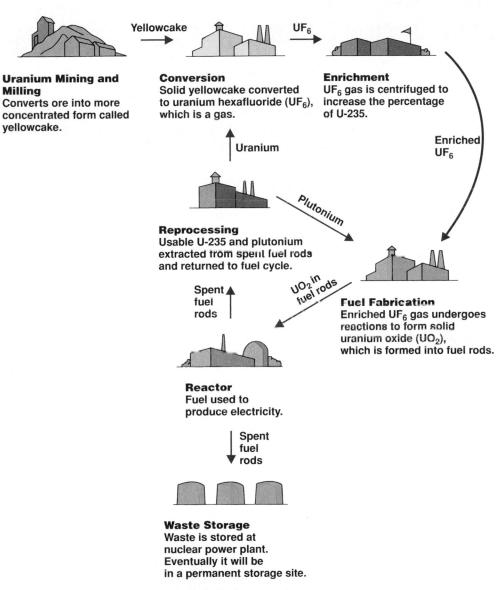

Uranium Mining and Milling
Converts ore into more concentrated form called yellowcake.

Conversion
Solid yellowcake converted to uranium hexafluoride (UF$_6$), which is a gas.

Enrichment
UF$_6$ gas is centrifuged to increase the percentage of U-235.

Reprocessing
Usable U-235 and plutonium extracted from spent fuel rods and returned to fuel cycle.

Fuel Fabrication
Enriched UF$_6$ gas undergoes reactions to form solid uranium oxide (UO$_2$), which is formed into fuel rods.

Reactor
Fuel used to produce electricity.

Waste Storage
Waste is stored at nuclear power plant. Eventually it will be in a permanent storage site.

FIGURE 10.10   Steps in the Nuclear Fuel Cycle The process of obtaining nuclear fuel involves mining, extracting the uranium from the ore, concentrating the U-235, fabricating the fuel rods, installing and using the fuel in a reactor, and disposing of the waste. Some countries reprocess the spent fuel as a way of reducing the amount of waste they must deal with.
Source: U.S. Department of Energy.

Water is involved in nearly all aspects of nuclear power plant operations. Water is the moderator in most nuclear power plants. The water slows down the neutrons so that a nuclear chain reaction can be sustained and controlled. Because water slows rapidly moving particles, it also is an effective shield against radiation. Therefore, spent fuel rods are typically stored in water-filled facilities. Many reactors rely on flooding the reactor core with water as a safety measure in the case of a nuclear accident.

In all of the common forms of nuclear power reactors, water serves as the cooling agent in the core. It carries the heat away from the core to be used to generate steam and turn the turbine. The conversion of liquid water to steam is a necessary step in production of electricity, since steam turns the turbines to produce electricity. Finally, all power plants require water to cool the steam and convert it back into liquid water. Thus, most nuclear power plants are located at sites with ready access to water.

In the summer of 2007, an extended drought in the southeastern part of the United States threatened to cause the shut-down of several nuclear power plants because they were in danger of not having enough water to cool the plant. Although this did not actually happen, it points out the tight link between nuclear generation of electricity and water.

reduces the amount of nuclear waste. At present, India, Japan, Russia, France, and the United Kingdom operate reprocessing plants that reprocess spent fuel rods as an alternative to storing them as a nuclear waste. Just less than half of the nuclear power plants in the world dispose of spent fuels rods by reprocessing.

Those countries that do not reprocess nuclear waste have made the decision to store it. Initially, the waste is stored onsite at the nuclear power plant, but the long term plan is to bury the waste in stable geologic formations. These and other issues related to nuclear waste disposal will be disscussed in the section on nuclear power concerns.

TRANSPORTATION ISSUES

Each step in the nuclear fuel cycle involves the transport of radioactive materials. The uranium mines are some distance from the processing plants. The fuel rods must be transported to the power plants, and the spent rods must be moved to a reprocessing plant or storage area. Each of these links in the fuel cycle presents the possibility of an accident or mishandling that could release radioactive material. Therefore, the methods of transport are very carefully designed and tested before they are used. Many people are convinced that the transport of radioactive materials is hazardous, while others are satisfied that the utmost care is being taken and that the risks are extremely small.

NUCLEAR CONCERNS

Nuclear power provides over 16 percent of the world's electricity. Many new nuclear power plants are currently being built and there are plans to build more in the future. However, a few serious accidents at nuclear facilities, the possibility of terrorist use of nuclear materials, concern about worker and public exposure to radiation, the potential for weapons production, problems with nuclear waste disposal, and other issues continue to raise concerns about the use of nuclear power.

REACTOR SAFETY

Although nearly all nuclear power plants have operated without serious accidents, two accidents have generated concern about nuclear power plant safety: Three Mile Island in 1979 and Chernobyl in 1986.

Three Mile Island

The Three Mile Island nuclear plant is located in the Susquehanna River near Middletown, Pennsylvania. On March 28, 1979, the main pump that supplied cooling water to the reactor broke down. At this point, an emergency coolant should have flooded the reactor and stabilized the temperature. The coolant did start to flow into the reactor core, but a pressure relief valve was struck in the open position so that water also was flowing from the core. In addition, there was no sensor to tell the operator if the reactor was flooded with coolant. Relying on other information, the operator assumed the reactor core was flooded and overrode the automatic emergency cooling system. Without the emergency coolant, the reactor temperature rose rapidly. The control rods eventually stopped fission, but because of the loss of coolant, a partial core meltdown had occurred. Thus, the accident was caused by a combination of equipment failures, lack of appropriate information to the operator, and decisions by the operator.

However, in retrospect, the containment structure worked as designed and prevented the release of radioactive materials from the core. Later that day, radioactive steam was vented into the atmosphere. The dose of radiation that would have been received by a person in the area was estimated to have been about

0.001 rem. This compares to normal background radiation of 0.2–0.3 rem.

The crippled reactor was eventually defueled in 1990 at a cost of about $1 billion. It has been placed in monitored storage until the companion reactor, which is still operating, reaches the end of its useful life. At that time, both reactors will be decommissioned.

Chernobyl

Chernobyl is a small city in Ukraine north of Kiev that became infamous in the spring of 1986, when it became the site of the world's largest nuclear power plant accident. (See figure 10.11.)

At 1 A.M. on April 25, 1986, at Chernobyl Nuclear Power Station-4, a test was begun to measure the amount of electricity that the still-spinning turbine would produce if the steam were shut off. This was important information since the emergency core cooling system required energy for its operation, and the coasting turbine could provide some of that energy until another source became available.

During the experiment, operators violated six important safety rules. They shut off all automatic warning systems, automatic shutdown systems, and the emergency core cooling system for the reactor.

As the test continued, the power output of the reactor rose beyond its normal level and continued to rise. The operators activated the emergency system that was designed to put the control rods back into the reactor and stop the fission. But it was too late. The core had already been deformed, and the control rods would not fit properly; the reaction could not be stopped. In 4.5 seconds, the energy level of the reactor increased 2000 times. The cooling water in the reactor converted to steam and blew the 1000-metric ton (1102-ton) concrete roof from the reactor, and the graphite that was part of the reactor core caught fire.

In less than 10 seconds, Chernobyl became the scene of the world's worst nuclear power plant accident. (See figure 10.12.) It took 10 days to bring the burning reactor under control. The immediate consequences were 37 fatalities; 500 persons hospitalized, including 237 with acute radiation sickness; and 116,000 people evacuated. Of the evacuees, 24,000 received high doses of radiation. Currently, some of these people are experiencing health problems attributable to their radiation exposure. In particular, children or fetuses exposed to fallout are showing increased frequency of thyroid cancer because of exposure to radioactive iodine 131 released from Chernobyl.

One important impact of Chernobyl is that it deepened public concern about the safety of nuclear reactors and many countries halted construction of new plants. However, by 2008 attitudes had changed and there was a renewed interest in nuclear power plants. There were 36 reactors under construction in 13 countries. And a further 93 are in the planning stages, including 12 in the United States.

TERRORISM

After the terrorist attacks on New York and Washington, D.C., on September 11, 2001, fear arose regarding nuclear plants as potential targets for terrorist attacks.

The consensus by nuclear experts is that damage to a nuclear power plant by an aircraft would not significantly damage the containment building or reactor and that normal emergency and containment functions would prevent the release of radioactive materials.

Since most spent nuclear fuel is stored at the nuclear plant, these facilities are a concern as being vulnerable to terrorist attacks. Again, analysis by nuclear experts leads to the conclusion that, although damage to these facilities would result in some radioactive material being released, it would be localized to the plant site. The most difficult contamination to deal with would be through the dispersal of airborne radioactive material into the natural environment.

Probably the greatest terrorism-related threat involving radioactive material is from dirty bombs. Dirty bombs are also known as Radiological Dispersal Devices (RDDs). They are not true nuclear weapons or devices but are simply a combination of conventional explosives and radioactive materials. The explosion of such a device is designed to scatter radioactive material about the environment to cause disruption and create panic rather than to kill large numbers of people. The most easily obtainable radioactive materials that could be used in fashioning such a bomb

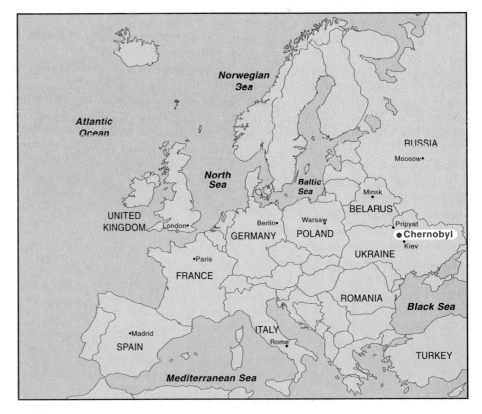

FIGURE 10.11 **Chernobyl** The town of Chernobyl became infamous as the location of the world's worst nuclear power plant accident.

FIGURE 10.12   The Accident at Chernobyl An uncontrolled chain reaction in the reactor of unit 4 resulted in a series of explosions and fires. See the circled area in the photograph.

would be low-level nuclear materials from medical facilities and similar institutions. Highly radioactive materials from power plants or nuclear weapons facilities are more difficult to obtain but would scatter more dangerous nuclear material.

WORKER AND PUBLIC EXPOSURE TO RADIATION

Each step in the nuclear fuel cycle poses a radiation exposure problem, beginning with the mining of uranium. Although the radiation level in the ore is low, miners' prolonged exposure to low-level radiation increases their rates of certain cancers, such as lung cancer. Uranium miners who smoke have an even higher lung cancer rate. After mining, the ore must be crushed in a milling process. This releases radioactive dust into the atmosphere, so workers are chronically exposed to low levels of radioactivity. The crushed rock is left on the surface of the ground as mine tailings. There are approximately 200 million metric tons of low-level radioactive mine tailings in the United States and at least as much in the rest of the world. These tailings constitute a hazard because they are dispersed into the environment.

In the enrichment and fabrication processes, the main dangers are exposure to radiation and accidental release of radioactive material into the environment. Transport also involves exposure risks. Most radioactive material is transported by highways or railroads. If a transporting vehicle were involved in an accident, radioactive material could be released into the environment. Even after fuel rods have been loaded into the reactor, people who work in the area of the reactor risk radiation exposure. Countries that reprocess spent fuel rods must be concerned about the exposure of workers and the possible theft of rods or reprocessed material by terrorist organizations that would use the radioactive material to construct a "dirty" bomb. Finally, the wastes produced present a possible radiation exposure hazard if they are not managed properly.

CONTAMINATION FROM NUCLEAR RESEARCH AND WEAPONS PRODUCTION

Producing nuclear materials for weapons and other military uses involves many of the same steps used to produce nuclear fuel for power reactors. In the United States, the Department of Energy (DOE) is responsible for nuclear research for both weapons and and energy production. Therefore, it is responsibile for the stewardship of the facilities used for both nuclear energy research and weapons production. Some facilities were used for both processing nuclear fuel and providing materials for weapons. In both cases, uranium was mined, concentrated, and transported to sites of use. Furthermore, military uses involved the production and concentration of plutonium. The production and storage facilities invariably became contaminated, as did the surrounding land.

Early in the history of nuclear research, the dangers from radioactive wastes were underappreciated or considered unimportant because of the necessity to produce weapons. Many of the research and production facilities dealt with hazardous chemicals and *minor* radioactive wastes by burying them, pumping them into the ground, storing them in ponds, or releasing them into rivers. As a result, the DOE has become the steward of a large number of sites that are contaminated with both hazardous chemicals and radioactive materials. The magnitude of the problem is huge and includes over 3000 contaminated sites containing hundreds of underground storage tanks, millions of 55-gallon drums of waste, and thousands of sites with contaminated soils.

Several major sites have been cleaned up. Each year there are fewer sites and the remaining sites are smaller.

DISPOSAL OF NUCLEAR WEAPONS

An additional problem has arisen as a result of the reduced importance of nuclear weapons. The political disintegration of the Soviet Union and Eastern Europe made large numbers of nuclear weapons, in both the East and West, unnecessary. But how are nations of the world to dispose of their nuclear weapons? Some nuclear material can be diverted to fuel use in nuclear reactors, and some reactors can be modified to accept enriched uranium or plutonium. The security of these materials, particularly in the former Soviet Union, is a concern.

RADIOACTIVE WASTE DISPOSAL

Although there are differences in the way in which nuclear wastes are defined by different countries, there are four general categories of wastes. Each presents a different level of risk and requires particular methods of handling and disposal.

Transuranic Waste Disposal

Transuranic nuclear waste is highly radioactive waste that contains large numbers of atoms that are larger than uranium with half-lives greater than 20 years. Most of these wastes come from processes involved in the production of nuclear weapons. As the cleanup of former nuclear weapons sites in the United States takes place, transuranic waste is transported to the Waste Isolation Pilot Plant near Carlsbad, New Mexico, for storage. This facility began accepting waste in March 1999. (See figure 10.13.)

Uranium Mining and Milling Waste

Uranium mining and milling waste is produced from the preparation of uranium for both weapons and nuclear power uses. These have low levels of radioactivity but are above background levels. The most serious problems associated with these wastes are direct gamma radiation, windblown dust, the release of radon gas, the erosion and dispersal of materials into surface waters, and the contamination of groundwater. (See figure 10.14.) Waste management activities include simple activities such as building fences, putting up warning signs, and establishing land- use restrictions that keep the public away from the sites. However, covering the wastes with a thick layer of soil and rock prevents erosion, windblown particles, and groundwater contamination. The covering also greatly reduces the gamma radiation and release of radon gas.

High-Level Radioactive Waste Disposal

High-level radioactive waste consists of spent fuel rods and highly radioactive materials from the reprocessing of fuel rods. The disposal of high-level nuclear waste has been a major problem for the nuclear power industry. In several countries—France, Japan, Russia, the United Kingdom, China, India, Switzerland, Belgium, and Bulgaria—accounting for just less than half of nuclear power facilities, the spent fuel rods are reprocessed to reduce the volume of the waste. Other countries,

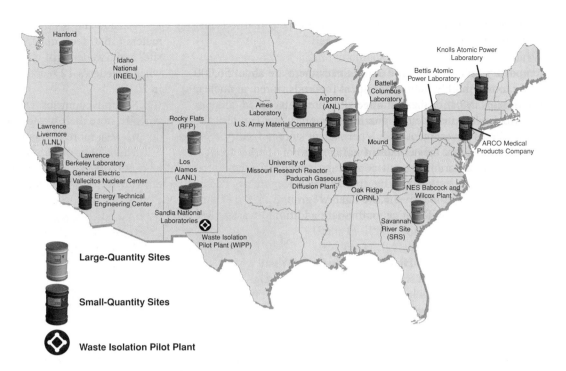

FIGURE 10.13 **Department of Energy High-Level Radioactive Transuranic Waste Sites** The high-level transuranic radioactive waste generated by the U.S. Department of Energy is shipped to the Waste Isolation Pilot Plant near Carlsbad, New Mexico, for storage.
Source: Nuclear Regulatory Commission.

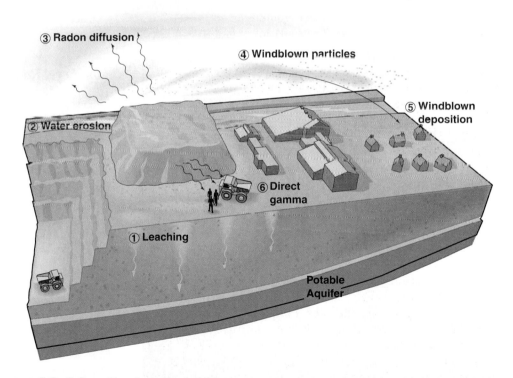

FIGURE 10.14 **Uranium Mine Tailings** Even though the amount of radiation in the tailings is low, the radiation still represents a threat to human health. Radioactivity may be dispersed throughout the environment and may come in contact with humans in several ways: (1) radioactive materials may leach into groundwater; (2) radioactive materials may enter surface water through erosion; (3) radon gas may diffuse from the tailings and enter the air; or (4) persons living near the tailings may receive particles in the air, (5) come in contact with objects that are coated with particulate, or (6) receive direct gamma radiation.
Source: U.S. Environmental Protection Agency.

Nuclear Energy   225

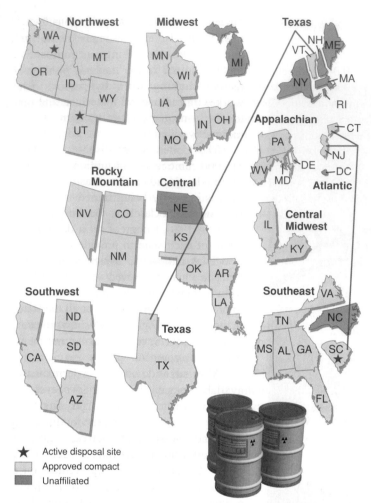

Northwest Midwest Texas

Appalachian

Rocky Mountain Central

Atlantic

Central Midwest

Southwest Texas Southeast

★ Active disposal site
▫ Approved compact
▪ Unaffiliated

FIGURE 10.18 **Low-Level Radioactive Waste Sites** Each state is responsible for the disposal of its low-level radioactive waste. Many states have formed compacts and selected a state to host the disposal site. None of the proposed sites is currently accepting waste. Many other states do not have an acceptable site and are temporarily relying on a site in South Carolina to accept their waste. Eventually, they will need to find other acceptable sites. The map shows the current compacts and active disposal sites.
Source: Nuclear Regulatory Commission.

FIGURE 10.19 **Storage of Low-Level Radioactive Waste** Low-level radioactive waste is stored in specially designated land burial sites.

FIGURE 10.20 **Cooling Towers** Cooling towers draw air over wet surfaces. The evaporation of water cools the surface and removes heat.

DECOMMISSIONING

All industrial facilities have a life expectancy—that is, the number of years they can be profitably operated. The life expectancy for an electrical generating plant, whether fossil-fuel or nuclear out, is about 30 to 40 years, after which time the plant is taken out of service. With a fossil-fuel plant, demolition is relatively simple and quick. A wrecking ball and bulldozers reduce the plant to rubble, which is trucked off to a landfill. The only harm to the environment is usually the dust raised by the demolition.

Demolition of a nuclear plant is not so simple. In fact, nuclear plants are not demolished; they are decommissioned. **Decommissioning** involves removing the fuel, cleaning surfaces, and permanently preventing people from coming into contact with the contaminated buildings or equipment.

The decommissioning of a plant is a two-step process:

Stage 1 involves removing fuel rods and water used in the reactor and properly storing or disposing of them. This removes 99 percent of the radioactivity. The radioactive material remaining consists of contaminants and activated materials. Activated materials are atoms that have been converted to radioactive isotopes as a result of exposure to radiation during the time the plant was in operation. These materials must be dealt with before the plant can be fully decommissioned.

Big Rock Point nuclear power plant was the fifth nuclear power plant built in the United States. It was a small (67-megawatt) boiling water reactor. When the plant was shut down in 1997, it was the longest operating (35 years) nuclear power plant in the country and had operated safely for that entire period. During its last 20 years of operation, Big Rock Point operated without a lost time accident. The plant was licensed in 1962 and, in its early years, was used for research related to kinds of fueling and related issues. In 1965, its primary purpose became that of producing electricity, although for 10 years during the time it was producing electricity, it was also used to produce colbalt-60 for medical purposes.

In 1997, because of Big Rock Point's small size and the need for some equipment upgrades, it was determined to be uneconomic and was shut down. Decommissioning began almost immediately. The employees of the plant were retrained to work with the contractor responsible for the decommissioning. Because of its small size, the reactor vessel was removed and shipped to a low-level radioactive waste facility and the remainder of the plant was decontaminated and demolished. In 2007, the U.S. Nuclear Regulatory Commission released the cleaned-up site of 175 hectares (435 acres) with a mile of shoreline on Lake Michigan for unrestricted use. A small portion of the former nuclear plant site was reserved as storage for spent fuel rods. That function will cease when the Yucca Mountain facility begins accepting spent nuclear fuel. The decommissioning process took nearly nine years and $390 million.

Stage 2 leads to the final disposition of the facility. There are three options to this second stage of the decommissioning process.

1. Decontaminate and dismantle the plant as soon as it is shut down.
2. Secure the plant for many years to allow radioactive materials that have a short half-life to disintegrate and then dismantle the plant. (However, this process should be completed within 60 years.)
3. Entomb the contaminated portions of the plant by covering the reactor with reinforced concrete and placing a barrier around the plant. (Currently this option is only considered suitable for small research facilities.)

Today, over 90 nuclear power plants in the world have been shut down and are in various stages of being decommissioned.

Recent experience indicates that the cost for decommissioning a large plant will be between $200 million and $400 million, about 5 percent of the cost of generating electricity. Although the mechanisms vary among countries, the money for decommissioning is generally collected over the useful life of the plant.

THE FUTURE OF NUCLEAR POWER

SOCIAL FORCES

Many interacting forces will influence the future of nuclear power. Primary among them are environmental and economic issues, which become transformed into political stands. Plans to decommission plants, extend the life of plants, or build new plants are political decisions made by national governments in light of the opinions of their citizens. In many parts of the world there is a strong anti-nuclear sentiment. There are many reasons people take an anti-nuclear stance. Some oppose all things nuclear because they see it as a threat to world peace. Others are concerned about the environmental issues of potential nuclear contamination and the problem of waste disposal.

The acceptance of the threat of climate change has had major implications for nuclear power. Since nuclear power plants do not produce carbon dioxide, many people, including some environmental organizations, have reevaluated the value of nuclear power and see it as a continuing part of the energy equation.

Another strong social force is economics. As the cost of oil and natural gas has increased, electricity generation from these sources has become more costly. This has made nuclear power more attractive. Countries that have few fossil fuel reserves and those with developing economies are most likely to build nuclear power plants.

Democratic governments set policy based on how their citizens vote. Many governments have responded to these forces and declared themselves to be non-nuclear countries. For example, New Zealand, Denmark, Ireland, Portugal, Luxembourg, Greece, Austria, Italy, and the Netherlands have declared that they will not pursue nuclear energy as an option. Others have stated an intention to phase out nuclear power as a source of electricity. Even so, these policies can change as economic realities confront policy.

CAMPUS SUSTAINABILITY INITIATIVE

OREGON STATE UNIVERSITY AND PASSIVE NUCLEAR POWER PLANTS

In traditional nuclear power plants, the weakest links are all the moving parts and the pumps and pipes that move liquids. At Oregon State University, a team of researchers created a new reactor design that eliminates these weak links by using passive forces like gravity to supply emergency cooling water and natural convection to transfer heat. The result is a safer, smaller, more stream-lined reactor. The new reactor design allows the reactor to be contained within a single, 60-foot-long cylinder, which enables the reactor to fit on a single railcar. It is also designed to run for five years between refueling shutdowns and can be installed for a fraction of the cost of building a traditional nuclear power plant. Idaho National Engineering Laboratory and Bechtel Corporation are partnering with Oregon State on the project.

TECHNICAL TRENDS

Current Building Programs

Currently, there are 439 nuclear power reactors in 30 countries. In 2006, these provided 2658 billion kilowatt-hours, about 16 percent of the world's electricity. Nineteen countries, including the United States, depended on nuclear power for at least 20 percent of their electricity generation. The amount of electricity produced by nuclear power plants is expected to increase slightly in the next few years. Thirty-six nuclear power reactors are currently being constructed in 12 countries. Several countries have plans to build new power reactors beyond those now under construction. Most are expected to be in Asia, where China, India, Japan, and South Korea are projected to add about 50 new plants over the next 10 to 20 years. Russia has plans for an additional 10 plants, and electric utilities in the United States have requested approval for 12 new plants.

Plant Life Extensions

Most current nuclear power plants originally had a design lifetime of up to 40 years, but engineering assessments of many plants have established that many can operate safely and economically for much longer. In the United States, several reactors have been granted license renewals that extend their operating lives to 60 years. As additional plants reach the end of their 40-year licenses, it is likely that they will have their licenses extended as well. Extending the operating life of a plant reduces the cost of producing power because the decommissioning is delayed and no new power plant (nuclear or fossil-fuel) needs to be built to replace the lost capacity.

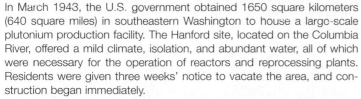

In March 1943, the U.S. government obtained 1650 square kilometers (640 square miles) in southeastern Washington to house a large-scale plutonium production facility. The Hanford site, located on the Columbia River, offered a mild climate, isolation, and abundant water, all of which were necessary for the operation of reactors and reprocessing plants. Residents were given three weeks' notice to vacate the area, and construction began immediately.

Because it was considered crucial to the development of weapons for World War II, the facility began industrial production of plutonium without first going through a small-scale development stage. The first nuclear reactor and chemical processing plants were completed in late 1944. Over the course of the following two decades, eight additional reactors and several processing plants were constructed.

Hanford's plutonium and uranium production facility is no longer in operation. However, it has left a legacy of contamination and waste that is one of the largest and most complex environmental cleanup sites in the United States. Early in the history of operations at Hanford, low-level waste was dumped onto the ground or pumped into wells. Other liquid wastes were stored in tanks. Over the years many tanks leaked before the waste was transferred to more secure containers. Consequently, soil and groundwater have been contaminated by about 4 million liters (1 million gallons) of waste. Burial of solid waste in unlined trenches also resulted in soil and groundwater contamination.

The cleanup task was huge:

- 202,000 cubic meters (53 million gallons) of radioactive waste
- 2100 metric tons (2300 U.S. tons) of spent nuclear fuel
- 8 metric tons (9 U.S tons) of plutonium
- 750,000 cubic meters (25 million cubic feet) of buried or stored waste
- 500 contaminated buildings
- 177 underground storage tanks

- 1700 individual waste sites
- 208 square kilometers (80 square miles) of contaminated groundwater

In 1989, the Department of Energy, the Environmental Protection Agency, and the Washington Department of Energy signed an agreement, known as the Tri-Party Agreement, calling for cleanup and better management of hazardous materials.

The following is a partial list of accomplishments:

- Contaminated groundwater is being pumped and treated at a rate of several hundred liters per minute.
- Soil contamination has been characterized.
- Liquid waste has been removed from tanks thought to be in danger of leaking.
- Six of nine reactors have had fuel removed and have been placed in long-term storage. Decommissioning of the remaining reactors is in progress.
- A lined landfill has been constructed on site to hold low-level radioactive and mixed waste.
- Over 400 shipments of transuranic waste have been sent to the Waste Isolation Pilot Plant.
- Uranium that had been in storage has been transferred off-site.
- Several buildings have been decontaminated and dismantled.
- Construction of a plant to convert low-level liquid waste to glass has begun, but it is not expected to begin operation until 2019.

However, there is still much to be done. The current goal for activities that will prevent contamination to the Columbia River is to have those activities completed by 2012. The more seriously contaminated sites at Hanford will require up to 40 years to clean up.

Yucca Mountain and Nuclear Waste Storage

Federal law requires the U.S. government to provide a solution to the storage of spent nuclear fuel. All nuclear power plants have been operating with the assumption that eventually their waste would be stored in a secure federal facility. In 2002, the Congress and the President selected Yucca Mountain as the site for long-term storage of spent nuclear fuel from nuclear power plants. The selection of the site and the construction of the storage facility had been delayed for about 20 years by legal action and political maneuvering. In 2008, the Department of Energy applied for a license to construct a repository to store nuclear waste at Yucca Mountain. The major arguments for and against constructing a nuclear waste repository at Yucca Mountain are as follows:

Pro

- Yucca Mountain is on federal land and is isolated. The nearest town is 65 kilometers (40 miles) away and the closest family is 22 kilometers (14 miles) away.
- Currently, spent nuclear fuel is stored at 131 temporary sites in 39 states. Many of the sites are aboveground. This situation increases the likelihood of radioactive releases as a result of accidents, natural disasters, or terrorist activity.
- The Yucca Mountain site is the best site for the purpose. It is very stable geologically and there is little chance of groundwater contamination.

Con

- Polls show 70 percent of Nevadans are opposed to the site.
- Critics doubt the Department of Energy's assurances of geologic stability and safety.
- The state of Nevada has actively opposed its designation as the host for the repository.

What Do You Think?

- Should Nevada be required to accept a nuclear repository for the good of the nuclear industry and the safety of the country?
- Should the state of Nevada be able to prevent the construction of the repository?
- If the site at Yucca Mountain is not built, what other solutions to the nuclear waste problem would you suggest?

SUMMARY

Nuclear fission is the splitting of the nucleus of the atom. The resulting energy can be used for a variety of purposes. The splitting of U-235 in a nuclear reactor can be used to heat water to produce steam that generates electricity. Various kinds of nuclear reactors have been constructed, including boiling-water reactors, pressurized-water reactors, heavy-water reactors, gas-cooled reactors, and experimental breeder reactors. Scientists are also conducting research on the possibilities of using fusion to generate electricity. All reactors contain a core with fuel, a moderator to control the rate of the reaction, and a cooling mechanism to prevent the reactor from overheating.

The nuclear fuel cycle involves mining and enriching the original uranium ore, fabricating it into fuel rods, using the fuel in reactors, and reprocessing or storing the spent fuel rods. The fuel and wastes must also be transported. At each step in the cycle, there is danger of exposure. During the entire cycle, great care must be taken to prevent accidental releases of nuclear material.

The use of nuclear materials for military purposes has resulted in the same kinds of problems caused by nuclear power generation. The reduced risk of nuclear warfare has led to a need to destroy nuclear weapons and clean up the sites contaminated by their construction.

Although the accidents at Three Mile Island in the United States and Chernobyl in Ukraine raised concerns about the safety of nuclear power plants for a time, rising energy prices have stimulated increased building of nuclear power plant in many countries and many more are in the planning stages. The disposal of nuclear waste is expensive and controversial. Long-term storage in geologically stable regions is the most commonly supported option. Russia, Japan, and the United Kingdom operate nuclear reprocessing facilities as an option for reducing the amount of nuclear waste that must ultimately be placed in long-term storage.

THINKING GREEN

1. Contact your local electric utility or visit its website and determine what percent of the electricity produced comes from nuclear power.
2. The United States has over 100 electricity-producing nuclear power plants. Identify the one that is closest to you geographically. An Internet search should allow you to find this information.
3. Call your local hospital and ask a representative how the hospital disposes of its nuclear waste.

WHAT'S YOUR TAKE?

The United States has over 100 nuclear power plants and generates about 20 percent of its electricity from nuclear power. Yet it seeks to prevent some other countries, such as North Korea and Iran, from developing nuclear power capabilities. Should the United States be able to prevent other countries from developing nuclear power?

Choose one side and develop arguments to support your point of view.

REVIEW QUESTIONS

1. How does a nuclear power plant generate electricity?
2. Name the steps in the nuclear fuel cycle.
3. What is a rem?
4. What is a nuclear chain reaction?
5. How are rising energy costs affecting the nuclear power industry?
6. What happened at Chernobyl, and why did it happen?
7. Describe a boiling-water reactor, and explain how it works.
8. How is plutonium-239 produced in a breeder reactor?
9. Why is plutonium-239 considered dangerous?
10. Why is fusion not being used as a source of energy?
11. What are the major environmental problems associated with the use of nuclear power?
12. List three environmental problems associated with the construction and subsequent de-emphasis on nuclear weapons.

CRITICAL THINKING QUESTIONS

1. Recent concerns about global warming have begun to revive the nuclear industry in the United States. Do you think nuclear power should be used instead of coal for generating electricity? Why?
2. The disposal of radioactive wastes is a big problem for the nuclear energy industry. What are some of the things that need to be evaluated when considering nuclear waste disposal? What criteria would you use to judge whether a storage proposal were adequate or not?
3. Nuclear weapons testing has released nuclear radiation into the environment. These tests have always been justified as necessary for national security. Do you agree or not? What are the risks? What are the benefits?
4. Some states allow consumers to choose an electric supplier. Would you choose an alternative to nuclear or coal even if it cost more?

CHAPTER 11

BIODIVERSITY ISSUES

One of the measures of biodiversity is the number of different species in an area. Coral reefs have great biodiversity. This photograph shows many different species of fish, corals, algae, and other organisms.

CHAPTER OUTLINE

Biodiversity Loss and Extinction
  Kinds of Organisms Prone to Extinction
  Extinction as a Result of Human Activity
Describing Biodiversity
  Genetic Diversity
  Species Diversity
  Ecosystem Diversity
The Value of Biodiversity
  Biological and Ecosystem Services Values
  Direct Economic Values
  Ethical Values
Threats to Biodiversity
  Habitat Loss
  Overexploitation
  Introduction of Exotic Species
  Control of Predator and Pest Organisms
  Climate Change
What Is Being Done to Preserve Biodiversity?
  Legal Protection
  Sustainable Management of Wildlife Populations
  Sustainable Management of Fish Populations

ISSUES & ANALYSIS
The Problem of Image 262

CASE STUDIES
Millennium Ecosystem Assessment Report and the Millennium Declaration 256
The California Condor 261

CAMPUS SUSTAINABILITY INITIATIVE
University of Kansas Biodiversity Partnership 254

GOING GREEN
Consumer Choices Related to Biodiversity 257

WATER CONNECTIONS
Freshwater Biodiversity 244

OBJECTIVES

After reading this chapter, you should be able to:

• Recognize that humans significantly modify natural ecosystems.

• State the major causes of biodiversity loss.

• Give examples of genetic diversity, species diversity, and ecosystem diversity.

• Describe the values of biodiversity.

• Appreciate the ways humans modify forests.

• Identify causes of desertification.

• Describe the role of endangered species legislation and the biodiversity treaty.

• Describe techniques that foster the sustainable use of wildlife and fisheries resources.

Global Perspectives on "Biodiversity Hot Spots," "The History of the Bison," and a Case Study on "The Northern Spotted Owl"
can be found on the book's website at www.mhhe.com/enger12e, along with other interesting readings.

BIODIVERSITY LOSS AND EXTINCTION

Biodiversity is a broad term that is used to describe the diversity of genes, species, and ecosystems in a region. Biodiversity is lost when populations are greatly reduced in size, when a species becomes extinct, or when ecosystems are destroyed or greatly modified. **Extinction** is the death of a species—the elimination of all the individuals of a particular kind. Extinction is a natural and common event in the long history of biological evolution. However, as we will see later in this section, extinction and loss of biodiversity are a major consequence of human domination of the Earth. Over the past few hundred years, humans are estimated to have increased the extinction rate by a factor of 1000 to 10,000 times above rates typical over the planet's history. About one-eighth of bird species, one-fourth of mammal species, one-third of amphibian species, and one-half of turtle species are threatened. Over 10 percent of the world's coral reefs have been lost. Nearly 60 percent of the remaining reefs are threatened by human activity. Mangrove forests, which are found in swampy areas near the ocean, are being reduced by over 1 percent per year. About 25 percent of the global land surface has been converted to raising crops. Approximately 60 percent of the ecosystem services (closely linked to biodiversity) are being degraded or used unsustainably. These include the maintenance of freshwater; the survival of fishery stocks; air and water purification; and the regulation of regional and local climate, natural hazards, and pests.

Currently, the most rapid changes affecting biodiversity are taking place in developing countries. This means that the harmful effects of biodiversity loss and the degradation of ecosystem services are borne disproportionately by the poor. Climate change and excessive nutrient loading are two major factors affecting biodiversity that are expected to become more severe in the future. Better protection of biodiversity and natural assets will require coordinated efforts across all levels of government, business, and international institutions. The productivity of ecosystems depends on policy choices in investment, trade, subsidy, taxation, and regulation, among others.

You may want to review chapter 5 before you begin this chapter so that you can apply ecological concepts to the issues regarding human impact on biodiversity.

KINDS OF ORGANISMS PRONE TO EXTINCTION

Complete extinction occurs when all the individuals of a species are eliminated. Besides complete extinction, we commonly observe *local extinctions* of populations. Although not as final, a local extinction indicates that the future of the species is not encouraging.

Furthermore, as a population is reduced in size, some of the genetic diversity in the population is likely to be lost.

Studies of modern local extinctions suggest that certain kinds of species are more likely than others to become extinct. Table 11.1 lists the major factors that affect extinction. First, species that have small populations of dispersed individuals (low population density) are more prone to extinction because successful breeding is more difficult for them than it is for species that have large populations of relatively high density. Second, organisms in small, restricted areas, such as islands, are also prone to extinction because an environmental change in their locale can eliminate the entire species at once. Organisms scattered over large areas are much less likely to be negatively affected by one event. Third, specialized organisms are more likely to become extinct than are generalized ones. Since specialized organisms rely on a few key factors in the environment, anything that negatively affects these factors could result in their extinction, whereas generalists can use alternate resources. Finally, some kinds of organisms, such as carnivores at higher trophic levels in food chains, typically have low populations but also have low rates of reproduction compared to their prey species.

Rabbits, raccoons, and rats are good examples of animals that are not likely to become extinct soon. They have high population densities and a wide geographic distribution. In addition, they have high reproductive rates and are generalists that can live under a variety of conditions and use a variety of items as food.

The cheetah is much more likely to become extinct because it has a low population density, is restricted to certain parts of Africa, has low reproductive rates, and has very specialized food habits. It must run down small antelope, in the open, during daylight, by itself. Similarly, the entire wild whooping crane species consists of about 375 individuals restricted to small winter and summer ranges that must have isolated marshes. (Captive and experimental

TABLE 11.1 Probability of Becoming Extinct

Most Likely to Become Extinct	Least Likely to Become Extinct
Low population density	High population density
Found in small area	Found over large area
Specialized niche	Generalized niche
Low reproductive rates	High reproductive rates

Most likely to become extinct

Least likely to become extinct

(a)

(b)

(c)

FIGURE 11.1 **Changes in the Ability of Humans to Modify Their World** As technology has advanced, the ability of people to modify their surroundings has increased significantly. (a) When humans lacked technology, they had only minor impacts on the natural world. (b) The agricultural revolution resulted in many of the suitable parts of the Earth being converted to agriculture. (c) Modern agricultural technology allows major portions of the Earth to be transformed into agricultural land.

populations bring the total number to slightly over 500 individuals.) In addition, their rate of reproduction is low. About one young bird is successfully raised by every two mated pairs.

EXTINCTION AS A RESULT OF HUMAN ACTIVITY

At one time, a human was just another consumer somewhere in the food chain. Humans fell prey to predators and died as a result of disease and accidents just like other animals. The simple tools they used would not allow major changes in their surroundings, so these people did not have a long-term effect on their surroundings.

As human populations grew, and as their tools and methods of using them became more advanced, the impact that a single human could have on his or her surroundings increased tremendously. Although in many ecosystems fires were natural events, the use of fire by humans to capture game and to clear land for gardens could destroy climax communities and return them to earlier successional stages more frequently than normal.

As technology advanced, wood was harvested for fuel and building materials, land was cleared for farming, streams were dammed to provide water power, and various mineral resources were exploited to provide energy and build machines. These modifications allowed larger human populations to survive, but always at the expense of previously existing ecosystems. (See figure 11.1.)

Today, with about 6.7 billion people on Earth, nearly all of the Earth's surface has been affected in some way by human activity. One of the major impacts of human activity has been to reduce biodiversity.

DESCRIBING BIODIVERSITY

Biodiversity is a broad term used to describe the diversity of genes, species, and ecosystems in a region. Loss of biodiversity is recognized by many as a major consequence of human domination of the Earth. Furthermore, a loss of biodiversity is a threat to the human occupants of the planet. To understand the implications of the threat, we must have a clear understanding of what the concept of biodiversity includes. Biodiversity can be examined at the genetic, species, and ecosystem levels.

GENETIC DIVERSITY

Genetic diversity is a term used to describe the number of different kinds of genes present in a population or a species. High genetic diversity indicates that there are many different kinds of genes present and that individuals within the population will have different structures and abilities. Low genetic diversity indicates that nearly all the individuals in the population have the same characteristics. Several things influence the genetic diversity of a population.

1. *Mutations* are changes in the genetic information of an organism. Mutations introduce new genetic information into a population by modifying genes that are already present. Most of the mutations we observe are harmful, but occasionally a mutation results in a new valuable characteristic. For example, at some time in the past, mutations occurred in the DNA of certain insect species that made some individuals tolerant to the insecticide DDT, even though the chemical had not yet been invented. These mutant characteristics remained very rare in these insect populations until DDT was used. Then, this characteristic became very valuable to the insects that carried it. Because insects that lacked the ability to tolerate DDT died when they came in contact with DDT, more of the DDT-tolerant individuals were left to reproduce the species, and therefore, the DDT-tolerance became much more common in these populations.

   If we think about the evolution of organisms, it is clear that incredible numbers of mutations were required for the evolution of complex plants, animals, and fungi from their single-celled ancestors. In the final analysis, all the characteristics shown by a species are the result of mutations.

2. *Migration* of individuals of a species from one population to another is also an important way to alter the genetic diversity of a population. Most species consist of many separate populations that are adapted to local environmental conditions. Whenever an organism leaves one population and

enters another, it subtracts its genetic information from the population it left and adds it to the population it joins. If the migrating individual contains rare characteristics, it may significantly affect the genetic diversity of both populations. The extent of migration need not be great. As long as genes are entering or leaving a population, genetic diversity will change.

3. *Sexual reproduction* is another process that influences genetic diversity. Although the process of sexual reproduction does not create new genetic information, it tends to generate new *genetic combinations* when the genetic information from two individuals mixes during fertilization, generating a unique individual. This doesn't directly change the genetic diversity of the population, but the new member may have a unique combination of characteristics so superior to those of other members of the population that the new member will be much more successful in producing offspring and will influence the genetic diversity of future generations.

4. *Population size* is a very important factor related to genetic diversity. The smaller the population, the less genetic diversity it can contain and the fewer the variations in the genes for specific characteristics. In addition, random events often can significantly alter the genetic diversity in small populations, with rare characteristics being lost from the population. For example, consider a population of a species of rose that consists of 100 individuals in which 95 of them have red blossoms and five have white blossoms. The death of five plants with red blossoms would not change genetic diversity very much. However, the death of the five plants with white blossoms could eliminate the characteristic from the population.

5. *Selective breeding* also can affect the genetic diversity of a species. Domesticated plants and animals have been modified over many generations by our choosing certain desired characteristics. Undesirable characteristics were eliminated, and desirable ones selected for. One of the consequences of this process is a loss of genetic diversity. (See figure 11.2.) Many modern domesticated organisms are very different from their wild ancestors and could not survive without our help. Often when genetic diversity is reduced, whether by reduced population size or selective breeding, deleterious genes for a characteristic may be present in both parents. When this occurs and the deleterious genes from both parents are passed to the offspring, the survival of the offspring may be jeopardized.

SPECIES DIVERSITY

Species diversity is a measure of the number of different species present in an area. Some localities naturally have high species diversity, while others have low species diversity. (See figure 11.3.) For example, it is well known that tropical rainforests have very high species diversity, while Arctic regions have low species diversity. Some people find it useful to distinguish two kinds of species diversity. One approach is to simply count the number of different kinds of species in an area. This is often referred to as *species richness*. Another way to look at species diversity is to take into account the number of different

(a) Genetically diverse Indian corn

(b) Genetically uniform field corn

FIGURE 11.2 **Genetic Diversity** Genetic diversity is a measure of the number of kinds of genes in a population. There is more genetic variety displayed in the multicolored and various shaped seeds of Indian corn than the yellow, uniform seeds of field corn.

taxonomic categories of the species present. This can be called *taxonomic richness*. A region with many different taxonomic categories of organisms (frogs, birds, mammals, insects, pine trees, mosses) would have a higher species diversity than one that had fewer taxonomic categories (grasses, insects, birds, mammals).

Measuring species diversity is not easy. For one thing, we do not know how many species there are. Estimates of the actual number of species range from a few million to 100 million. About 1.4 million species have been described, but each year, new species are discovered, and in some groups, such as the bacteria, there may be millions of species yet unnamed. Another problem associated with measuring species diversity is the difficulty of discovering all the species in an area. Many species are naturally rare, and others live in difficult-to-reach places or are active only at certain times. For example, an examination of the kinds of organisms

(a) Pine plantation—low species diversity

(b) Desert ecosystem—high species diversity

FIGURE 11.3 **Species Diversity** (a) Forest plantations have low species diversity—the dominant vegetation consists of a single species of tree. (b) This desert has several species making up the dominant vegetation. How many different species can you count?

	Approximate Number of Described
TABLE 11.2 Numbers of Described Species by Taxonomic Group	
Taxonomic Group	**Species**
Insects	950,000
Plants	270,000
Non-insect arthropods (spiders, mites, crustaceans, etc.)	113,000
Fungi	100,000
Roundworms	80,000
Mollusks	70,000
Protozoa	31,000
Fish	29,300
Algae	27,000
Flatworms	25,000
Earthworms and related organisms	12,000
Sponges	10,000
Birds	9900
Bacteria, blue-green algae	9000
Jellyfish, corals, comb jellies	9000
Reptiles	8240
Starfish, sea urchins, sand dollars, etc.	7000
Amphibians	5900
Mammals	5400
Rotifers	1800
Archaea	260
Other	27,000
Total	**1,800,800**

in the tops of trees in tropical forests led to the discovery of many species that were previously unknown. Table 11.2 gives current data about the number of different kinds of species in specific taxonomic groups.

Several factors are known to influence the species diversity of a particular location.

1. *The geologic and evolutionary history* of a region impacts its species diversity. As mentioned earlier, tropical rainforests naturally have greater species diversity than polar regions. Perhaps this is due to relatively recent climatic events. Twenty thousand years ago, glaciers covered much of northern Europe, the more northerly parts of North America, and parts of Asia. The glaciers would have drastically reduced the number of species present, and species would have migrated into the region as the glaciers receded. This same kind of drastic change did not take place in tropical regions; therefore, they retain more species.

2. *Migration* can introduce new species to an area where they were not present previously. While it is easy to see how such introductions can increase species diversity, some invading species actually result in a reduction in species diversity because the species originally present were unable to compete with the invaders. For example, the introduction of the zebra mussel to North America has had the effect of reducing the population size of many native mussels. Many of these native mussels are endangered.

3. *The size of the area* being considered also affects species diversity. In general, the larger the area being considered, the larger the species diversity. This is a natural consequence of including more kinds of spaces and the organisms that are adapted to them. For example, if species diversity were being measured in a desert, the diversity would increase greatly if a stream or water hole were included in the area under investigation.

4. *Human activity* has a great effect on the species diversity of a region. When humans exploit an area, they convert natural ecosystems to human-managed agricultural, forest, aquaculture, and urban ecosystems. They harvest certain species for

their use. They specifically eliminate species that compete with more desirable species. They introduce species that are not native to the area. These topics will be discussed in greater depth later in the chapter.

ECOSYSTEM DIVERSITY

Ecosystem diversity is a measure of the number of kinds of ecosystems present in an area. Many regions of the world appear to be quite uniform in terms of the kinds of ecosystems present. For example, large parts of central Australia, North Africa, and southwestern United States and adjacent Mexico are deserts. While there are general similarities (low rainfall, thorny woody plants, and animals that can survive on little water), each of these deserts is different and has specific organisms typical to the region. Furthermore, within each of these deserts are local regions where water is available. These areas may include rivers, springs, or rocky outcrops that collect and hold water. These locations support species of organisms not found elsewhere in the desert. In many other cases, natural events such as hurricanes, fires, floods, or volcanic eruptions may have destroyed the original vegetation, resulting in patches of early successional stages that contribute much to the diversity of organisms present.

In addition, local topographic conditions may create patches of the landscape that differ from the prevailing type. Ridges typically have different vegetation on north- and south-facing slopes. (See figure 11.4.) Differences in soil type will support unique vegetation. Rolling terrain may provide pockets that contain temporary ponds that support different vegetation types. Each of these unique localities also can have animals that are specialized to the locality. Since many microhabitats exist within a region, the larger the area studied, the more the kinds of ecosystems present.

THE VALUE OF BIODIVERSITY

Although most would agree that maintaining biodiversity is a good thing, a need exists to place a value on preserving biodiversity so that governments and individuals can apply logic to the daily decisions that affect biodiversity. There are many different ways to assign value to biodiversity. Some involve understanding the ecological roles played by organisms, others involve cold economic analysis, and still others stem from ethical considerations.

BIOLOGICAL AND ECOSYSTEM SERVICES VALUES

Our species is totally dependent on the diversity of organisms on Earth. It is important to recognize that each organism is involved in a vast network of relationships with other organisms. Symbiotic nitrogen-fixing bacteria live in the roots of certain plants. Soil-building organisms live on the dead organic matter provided by plants and animals. Animals eat plants or other animals. It is impossible to have an organism function optimally unless it has its supporting cast

FIGURE 11.4 **Ecosystem Diversity** This location shows high ecosystem diversity. There are at least four different ecosystem types visible; the lake, the deciduous trees surrounding the lake, the conifer trees in the middle, and low-growing vegetation on the mountainside.

of players that are part of the ecosystem. Although it is impossible to think of every possible role for each species on Earth, panels of experts have identified some broad categories of services provided by ecosystems and the organisms that make them up. This list describes ecosystem services that are not part of our economic system—we do not directly pay for them. Examining this list will help us recognize how important ecosystem services are to our lives.

Nutrient Cycling

Carbon, nitrogen, phosphorus, and many other chemical elements are cycled through ecosystems by a complex array of bacteria and other organisms. Nitrogen-fixing bacteria are particularly valuable for providing nitrogen in a usable form to plants.

Cultural Uses

People use the natural world for many kinds of nonconsumptive uses. Enjoyment of landscapes and individual organisms, scientific study and other educational activities, and the spiritual significance of specific places and things are all examples of cultural uses of ecosystems.

Water Regulation and Supply

Intact soil and vegetation slow the flow of water and allow water to penetrate the soil and recharge aquifers. These processes make water available for agricultural, industrial, and domestic uses.

In areas with intact ecosystems, water is released slowly so that intact ecosystems are less prone to damage from drought. New York City found that it could provide water to its residents less expensively by protecting the watershed from which the water comes, rather than building expensive water purification plants to clean water from local rivers.

Disturbance Regulation and Erosion Control

Colonization of disturbed sites—caused by fires, floods, windstorms, landslides, or human actions—by plants and animals heals the scars and prevents continued damage. Furthermore, intact ecosystems provide flood and erosion control. Mangrove forests, marshes, and other wetlands protect shorelands from erosion. The network of roots in forested areas and grasslands ties the soil together and protects watersheds.

Waste Treatment

Decomposer organisms recycle both natural and human-produced organic wastes. Excess nutrients are removed by organisms and pollutants are removed from air, soil, and water and converted to less harmful materials.

Food and Raw Materials

In many parts of the world, people are involved in a subsistence economy and rely directly on ecosystems for food and raw materials. They harvest wild plants and animals as food and medicine and use plants to provide food for livestock, building materials, and firewood. The UN Food and Agriculture Organization estimates that over 60 percent of all wood harvested worldwide is burned as fuel. Much of this is collected by individuals and used by them for heating homes and cooking.

Atmospheric and Climate Services

Many atmospheric gases are cycled between organisms and the atmosphere. In particular, removal of carbon dioxide during photosynthesis is important in controlling the warming of the planet. Many countries have planted trees to help remove carbon dioxide from the air as a response to concerns about global warming. Other pollutants such as nitrogen and sulfur compounds are also modified by organisms, particularly when they come in contact with soil. Ozone in the upper atmosphere protects from ultraviolet light.

Recreation

Natural areas provide important recreational opportunities for an increasingly urban population. Camping, hiking, kayaking, fishing, hunting, ecotourism, and sight-seeing provide exercise and enjoyment.

Biological Control Services

All organisms are involved in a complex set of interrelationships with other kinds of organisms. Some of these relationships are harmful. We label organisms that cause harm to us or our domesticated plants and animals as *pests*. However, every pest also has organisms that cause it harm—dragonflies, swallows, and bats eat mosquitoes that carry disease and are annoying; ladybird beetles eat aphids; and cats are frequently kept to control rodent populations on farms.

Pollination Services

Many different kinds of insects are pollinators that are extremely important to the successful fruiting of plants. The careless elimination of these beneficial insects by the broad use of insecticides can negatively affect agricultural production. Recent declines in honeybee populations have highlighted the value of this pollination service.

Habitat/Refuges

Refuges and other protected areas provide places that protect species of concern, serve as nursery sites for specific species, or provide temporary stopping places for migratory species. Ducks Unlimited, an organization that supports waterfowl hunting, uses money provided by its members to protect wetland nesting habitats for ducks and geese.

Genetic Resources

If we cause the extinction of a potentially useful organism, we have lost the opportunity to use it for our own ends. Most of the wild ancestors of our most important food grains, such as maize (corn), wheat, and rice, are thought to be extinct. Over 50 percent of the most common drugs used to control and cure disease are derived from plants and animals and new ones are discovered each year.

Soil Formation

The weathering of rock provides new mineral material for the building of soil. Bacteria, fungi, tiny animals, and the roots of plants are involved in building soil by breaking down organic matter, incorporating it into the mineral part of the soil, and creating a loose texture to the soil. All terrestrial ecosystems, including agricultural and commercial forest lands, rely on the soil-forming services provided by these organisms. Our food supply depends on the protection and management of the soil and the organisms that assist in soil building.

Assigning Value to Ecosystem Services

All of these services can be converted into monetary terms, since it takes money to purify water, purchase land, and buy and plant trees. Since choices between competing uses for ecosystems often are determined by financial values assignable to ecosystems, many environmental thinkers have begun to try to put a value on the many services provided by intact, functioning ecosystems. Obviously, this is not an easy task, and many will belittle these initial attempts to put monetary values on ecosystem services, but it is an important first step in forcing people to consider the importance of ecosystem services when making economic decisions about how ecosystems should be used. Table 11.3 presents approximate values for ecosystem services assigned by a

TABLE 11.3 Estimated Values for Ecosystem Services (1997)

Categories of Services	Estimated Yearly Value (Trillion $US)
Nutrient cycling	17.1
Cultural uses	3.0
Water regulation and supply	2.8
Disturbance regulation and erosion control	2.4
Waste treatment	2.3
Food and raw materials	2.1
Atmospheric gas and climate regulation	2.0
Recreation	0.8
Biological control	0.4
Pollination	0.1
Habitat/refuges	0.1
Genetic resources	0.08
Soil formation	0.05
Total	**33.0**

FIGURE 11.5 **Informal Economy** It is difficult to measure the real economic value of the agricultural, forestry, and fisheries industries because many products, like these fish, are sold or bartered in local markets.

panel of experts including ecologists, geographers, and economists. The study published in 1997 assigned an estimated value of $33 trillion per year, which many consider to be low. The current world gross national product is about $50 trillion per year. Therefore, the "free" services of ecosystems must not be overlooked when decisions are made about land use and how natural resources should be managed.

DIRECT ECONOMIC VALUES

There are three major economic sectors that are highly dependent on the ecological health of the planet: agriculture, forestry, and fisheries. Each contributes many billions of dollars to the global economy. The Food and Agriculture Organization of the United Nations estimates, based on exports, that agriculture contributes about US $670 billion to the world economy, while the harvest and trade in forest products is responsible for about US $200 billion and fisheries resources are responsible for about US $80 billion yearly. However, since much of this production is consumed in the country where it is produced—the products are not exported—the real economic value is much greater than the value of exports. For example, less than 40 percent of fisheries products are traded internationally, so the real value is greater than US $80 billion.

It is also important to understand that in many parts of the world, many agricultural, forestry, and fisheries products never enter the formal economy of a region, so it is difficult to assign specific monetary value to them. For example, firewood, fish, game meat, and other wild foods are often consumed directly by the people who harvest them, sold at local markets, or bartered for other goods and services. (See figure 11.5.)

There are also markets for specialty products such as medicinal plants, ivory, skins, and many other natural products. There are many historical examples of valued species that were or are being harvested beyond their capacity to reproduce. The harvesting of ivory was steadily reducing the number of elephants, and the harvesting of alligators for their hides was reducing alligator numbers. In both cases, regulation of the harvest has preserved the resource. Alligator populations have rebounded and they are again being harvested in the United States. Many countries in Africa maintain that elephant populations are so large that harvesting of elephants for meat and ivory should be allowed as well.

When looking at the economic value of biodiversity it is also important to consider the potential value of biodiversity. For example, many current drugs are derived from organisms. How many potential drugs are yet undiscovered? Protecting biodiversity can be thought of as protecting future economically important resources.

ETHICAL VALUES

Many would argue that a case can be made that all species have an intrinsic value and a fundamental right to exist without being needlessly eliminated by the unthinking activity of the human species. This is an ethical position that is unrelated to social or economic considerations. According to those who support this philosophical position, extinction by itself is not bad, but human-initiated extinction is. This contrasts with the philosophical position that humans are simply organisms that have achieved a preeminent position on Earth, and therefore, extinctions we cause are no different from extinctions caused by other forces.

Others would argue that experiencing natural landscapes and processes is an important human right. The beauty of a forested hillside in full autumn color, the graceful movements of deer, the raw power of a grizzly bear, and the wonder shown by a child when it discovers a garter snake or a butterfly are all emotional events. There is a value in their just being, so that they can be observed.

The ethical positions held by a person are typically shaped by experience. In 1800, 94 percent of North Americans lived in rural settings and had daily encounters with plants and animals in a natural setting. Furthermore, there was a basic understanding of how natural ecosystems operated. They milked their own cows, slaughtered animals for meat, and observed the effect of weather, seasons, and other cycles of change on the natural systems of which they were a part. Although they often changed ecosystems to provide for agriculture, they had a basic understanding of how ecosystems function. Today, about 80 percent of North Americans live in urban settings and have limited contact and experience with these natural processes. This same dynamic is at work throughout the world. Today, over 50 percent of the world population is urban.

Because urban dwellers have little personal experience with naturally functioning ecosystems, they have a different set of ethical values about what natural ecosystems provide. The values people place on biodiversity are important, since they shape the way political and social institutions respond to threats to biodiversity.

THREATS TO BIODIVERSITY

Efforts to preserve biodiversity involve a tension between the desire to use biotic resources and a wish to maintain biodiversity. The value of exploiting a resource can be measured in economic terms. Farmland, lumber, or animal products can be given a measurable monetary value by the economic marketplace. On the other hand, it is often difficult to put an economic value on the preservation of biodiversity and the environmental services provided by organisms. Therefore, environmentalists often must rely on ethical or biological arguments to make their points. Often the decisions that are made involve a compromise that allows some utilization of a resource while preserving some of the biodiversity.

Five major human impacts threaten to reduce biodiversity: habitat loss, overexploitation, introduction of exotic species, predator and pest control activities, and climate change.

HABITAT LOSS

Habitat loss occurs when human activities result in the conversion of natural ecosystems to human-dominated systems. The resulting changes eliminate or reduce the numbers of species that were a part of the original ecosystem.

The International Union for Conservation of Nature and Natural Resources (IUCN) estimates that about 80 percent to 90 percent of threatened species are under threat because of habitat loss or fragmentation. Similarly, habitat loss and fragmentation are thought to be a major cause of past extinctions. The primary activities that result in habitat loss are farming, forestry, grazing by livestock, modification of aquatic habitats, and conversion to urban and industrial landscapes.

Often the destruction of natural habitats is not total but leaves patches of relatively unaltered habitat interspersed with human-modified landscapes. However, scientists studying the effects of this activity on forest birds have noticed that as the forest is reduced to small patches, many species of birds disappear from the area.

Conversion to Agriculture

About 40 percent of the world's land surface has been converted to cropland and permanent pasture. Typically, the most productive natural ecosystems (forests and grasslands) are the first to be modified by human use and the most intensely managed. Since nearly all agricultural practices involve the removal of the original vegetation and substitution of exotic domesticated crops and animals, the loss of biodiversity is significant. For example, much of the Mediterranean and Middle East once supported extensive forests that were converted to agriculture but now consist of dry scrubland. In Europe, little of the original forest is left. In North America, the eastern deciduous forests were reduced, and almost all the original prairie in the United States has been converted to agricultural land, resulting in the loss of some species.

As the human population grows, it needs more space to grow food. Current pressures to modify the environment are greatest in areas that have high population density. (See figure 11.6.) Today, agricultural land is being pushed to feed more people, and its wise use is essential to the health and welfare of the people of the world. Chapters 13 and 14 take a close look at patterns of use of agricultural land.

Forestry Practices

The history and current status of the world's forests are well known. The economic worth of the standing timber can be assessed, and the importance of forests for wildlife and watershed protection can be given a value. Originally, almost half of the United States, three-fourths of Canada, almost all of Europe, and significant portions of the rest of the world were forested. The forests were removed for fuel, for building materials, to clear land for farming, and just because they were in the way. These forest-destroying activities are often called **deforestation.** This activity destroyed much of the forested land and returned other forests to an earlier successional stage, which resulted in the loss of certain animal and plant species that required mature forests for their habitat. In general, all continents still contain significant amounts

FIGURE 11.6 **Conversion of Forest to Agriculture** This farm in Uganda occupies a hillside that was once covered by forest.

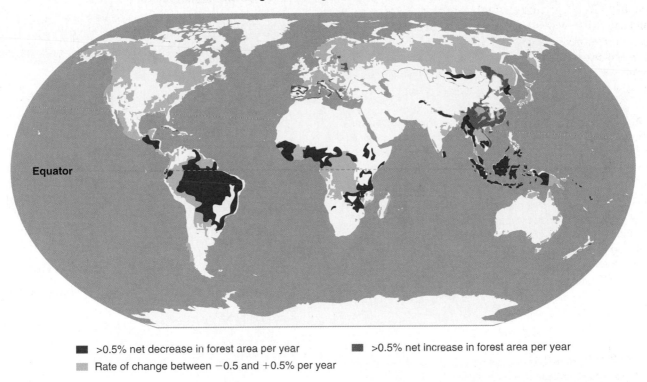

Countries with Large Net Changes in Forest Area 2000–2005

Equator

■ >0.5% net decrease in forest area per year ■ >0.5% net increase in forest area per year

▨ Rate of change between −0.5 and +0.5% per year

FIGURE 11.7 **Changes in Forest Area** Areas shown in red are losing forests at a rate of greater than 0.5 percent per year. Areas shown in brown are increasing at a rate of greater than 0.5 percent per year. Areas shown in green are maintaining their forests. It is obvious that most tropical forests in Central America, Africa, and much of Asia are being lost, while nontropical areas are holding their own or growing.

Source: Food and Agriculture Organization of the United Nations.

of forested land, although most of these forest ecosystems have been extensively modified by human activity. Figure 11.7 shows that temperate parts of the world (particularly North America, Europe, and China, which have large forested areas) are maintaining or increasing their forests, while tropical forests are being lost. The changes in China are recent and involve extensive reforestation projects.

Modern forest-management practices involve a compromise that allows economic exploitation while maintaining some of the biodiversity of the forest. Logging activities are often selective and preferentially remove certain species of trees. This causes a shift in the species diversity of plants. Logging also destroys the habitat for many kinds of animals that require mature stands of timber. The pine martin, grizzly bear, and cougar all require forested habitat that is relatively untouched by human activity. The removal of mature forests reduces their population size and consequently the genetic diversity of their populations.

For example, today in Australia, a squirrel-sized marsupial known as a numbat is totally dependent on old forests that have termite- and ant-infested trees. Termites are numbats' major source of food, and the hollow trees and limbs caused by the insects' activities provide numbats with places to hide. Clearly, tree harvesting will have a negative impact on this species, which is already in danger of extinction. (See figure 11.8.)

FIGURE 11.8 **A Specialized Marsupial—the Numbat** This small marsupial mammal requires termite- and ant-infested trees for its survival. Termites serve as food, and the hollow limbs and logs provide hiding places. Loss of old-growth forests with diseased trees will lead to the numbat's extinction.

Biodiversity Issues 243

Freshwater ecosystems are especially rich in species. The IUCN estimates that there are about 45,000 species that rely on freshwater ecosystems for their survival. River systems are linear and provide a variety of different habitats from their headwaters to their mouths. Different species are found in the cooler headwaters than at the warmer mouth of a river. Furthermore, each river system is isolated from others. Therefore, species have evolved to match the peculiar characteristics of each river system. Consequently, there is a high degree of *endemism* (local species found nowhere else) in river systems. Similarly, many lakes are isolated and show a high degree of endemism. Lake Victoria at one time perhaps had 300 species of fish found nowhere else in the world.

Biodiversity in freshwater ecosystems is under great threat from human actions.

- Rivers have been channelized to facilitate navigation, which increases the rate of flow in the river and reduces the amount of shallow water habitat.

- Dams have been built to provide for flood control and provide power, which changes the flow of the river and the temperature of the water.

- Pollution from industries, cities, and farms enters lakes and streams making the water unsuitable for sensitive species.

- Exotic species have been purposely and accidentally introduced. For example, the introduction of the Nile perch into Lake Victoria led to the extinction of about 200 species of native fish and the zebra mussel in the United States is responsible for the endangerment of many species of native mussels.

- Fish and other biotic resources have been overexploited.

- Withdrawal of water for municipal, industrial, and irrigation purposes has resulted in rivers drying up. The Aral Sea has shrunk in size and increased in salinity as a result of water withdrawals to irrigate farmland. Rivers such as the Rio Grande and Colorado in the United States and the Yellow River in China regularly dry up before they reach the sea as a result of water withdrawals for municipal and irrigation uses.

- Deforestation and agricultural activities lead to siltation and warming of waters.

The magnitude of the problem is illustrated in the following table:

Species Threatened with Extinction

Group of Organisms	Percent of Species
Freshwater turtles (worldwide)	50%
Amphibians (worldwide)	32
Fish (worldwide)	20
Fish (Europe)	38
Fish (Madagascar)	54
Fish (East Africa)	54
Fish (United States)	37
Mussels (United States)	69
Crayfish (United States)	51
Stoneflies (United States)	43
Dragonflies and damselflies (United States)	18

Data Sources: IUCN; U.S. data from http://www.natureserve.org.

In addition to serving as habitats for many species of plants and animals, forests provide many other ecosystem services. Forested areas modify the climate, reduce the rate of water runoff, protect soil from erosion, and provide recreational opportunities. Because trees transpire large quantities of water and shade the soil, their removal often leads to a hotter, drier climate. Trees and other plants hold water on their surfaces, thus reducing the rate of runoff. Slowing runoff also allows more water to sink into the soil and recharge groundwater resources. Therefore, removal of trees results in more rapid runoff so that flooding and soil erosion are more common. Soil particles wash into streams, where they cause siltation. The loss of soil particles reduces the soil's fertility. The particles that enter streams may cover spawning sites and reduce the biodiversity of fish populations. If the trees along the stream are removed, the water in the stream will be warmed by increased exposure to sunlight. This may also negatively affect fish populations.

In many parts of the world, the construction of logging roads increases access to the forest and results in colonization by peasant "squatters" who seek to clear the forest for agriculture. The roads also permit poachers to have greater access to wildlife in the forest. Finally, the "wilderness" nature of the area is destroyed, which results in a loss of value for many who like to visit mature forests for recreation. An area that has recently been logged is not very scenic, and the roads and other changes often irreversibly alter the area's wilderness character.

Environmental Implications of Various Forest Harvesting Methods

One of the most controversial logging practices is **clear-cutting.** (See figure 11.9.) As the name implies, all of the trees in a large area are removed. This is a very economical method of harvesting, but it exposes the soil to significant erosive forces. If large blocks of land are cut at one time, it may slow the reestablishment of forest and have significant effects on wildlife. On some sites with gentle slopes, clear-cutting is a reasonable method of harvesting trees, and environmental

FIGURE 11.9 Clear-cutting removes all of the trees in a large area and exposes the soil to erosive forces. This shows a clearcut in British Columbia, Canada.

FIGURE 11.10 **Tropical Deforestation** The deforestation of tropical regions leads to great reduction in biodiversity.

damage is limited. This is especially true if a border of undisturbed forest is left along the banks of any streams in the area. The roots of the trees help to stabilize the stream banks and retard siltation. Also, the shade provided by the trees helps to prevent warming of the water, which might be detrimental to some fish species.

Clear-cutting can be very destructive on sites with steep slopes or where regrowth is slow. Under these circumstances, it may be possible to use **patchwork clear-cutting.** With this method, smaller areas are clear-cut among patches of untouched forest. This reduces many of the problems associated with clear-cutting and can also improve conditions for species of game animals that flourish in successional forests but not in mature forests. For example, deer, grouse, and rabbits benefit from a mixture of mature forest and early-stage successional forest.

Clear-cut sites where natural reseeding or regrowth is slow may need to be replanted with trees, a process called **reforestation.** Reforestation is especially important for many of the conifer species, which often require bare soil to become established. Many of the deciduous trees will resprout from stumps or grow quickly from the seeds that litter the forest floor, so reforestation is not as important in deciduous forests.

Selective harvesting of some species of trees is also possible but is not as efficient or as economical as other methods, from the point of view of the harvesters. It allows them, however, to take individual, mature, high-value trees without completely disrupting the forest ecosystem. In many tropical forests, high-value trees, such as mahogany, are often harvested selectively. However, there may still be extensive damage to the forest by the construction of roads and to noncommercial trees by the felling of the selected species.

Special Concerns About Tropical Deforestation

Tropical forests have greater species diversity than any other terrestrial ecosystem. The diverse mixture of tree species requires harvesting techniques different from those traditionally used in northern temperate forests. In addition, because tropical soils have low fertility and are highly erodible, tropical forests are not as likely to regenerate after logging as are temperate forests. Currently, few tropical forests are being managed for long-term productivity; they are being harvested on a short-term economic basis only, as if they were nonrenewable resources. Worldwide tropical forests are being lost at a rate of about 0.6 percent per year. (See figure 11.10.)

Several concerns are raised by tropical deforestation. First, the deforestation of large tracts of tropical forest is significantly reducing the species diversity of the world. Second, because tropical forests very effectively trap rainfall and prevent rapid runoff and the large amount of water transpired from the leaves of trees tends to increase the humidity of the air, the destruction of these forests also can significantly alter climate, generally resulting in a hotter, more arid climate. Third, high rainfall coupled with the nature of the soils results in the deforested lands being easily eroded. Finally, people have become concerned about preserving the potential of forests to trap carbon dioxide. As they carry on photosynthesis, trees trap large amounts of carbon dioxide. This may help to prevent increased carbon dioxide levels that contribute to global warming. (See chapter 16 for a discussion of global warming and climate change.)

Another complicating factor is that the human population is growing rapidly in tropical regions of the world. More people need more food, which means that forestland will be converted to agriculture and the value of forest for timber, fuel, watershed protection, wildlife habitat, biodiversity, and carbon dioxide storage will be lost.

Plantation Forestry

Many forest products companies manage forest plantations in the same way farmers manage crops. They plant single-species, even-aged forests of fast-growing hybrid trees that have been developed in the same way as high-yielding agricultural crops. Fertilizer is applied if needed, and weeds and pests are controlled. Often controlled fires and aerial application of pesticides are used to control competing species and pests. Such forests

FIGURE 11.11 **Forest Plantation** This *Eucalyptus* forest plantation in South Africa shows both mature trees and young trees in the foreground. *Eucalyptus* species are not native to South Africa.

have low species diversity and are not as valuable for wildlife and other uses as are more natural, mixed species forests. Furthermore, the trees planted in many managed forests may be exotic species. *Eucalyptus* trees from Australia have been planted in South America, Africa, and other parts of the world; and most of the forests in northern England and Scotland have a mixture of native pines and imported species from the European mainland and North America.

In these intensively managed forests, some single-species plantations mature to harvestable size in 20 years, rather than in the approximately 100 years typical for naturally reproducing mixed forests. However, the quality of the lumber products is reduced. In many of these forests, clear-cutting is the typical method of harvest, and the cutover area is immediately replanted. (See figure 11.11.)

Rangeland and Grazing Practices

Rangelands consist of the many arid and semiarid lands of the world that support grasses or a mixture of grasses and drought-resistant shrubs. These lands are too dry to support crops but are often used to raise low-density populations of domesticated or semidomesticated animals. In some cases, the animals are maintained on permanent open ranges, while in others, nomadic herds are moved from place to place in search of suitable grazing. Usually, the animals are introduced species not native to the region. Sheep, cattle, and goats that are native to Europe and Asia have been introduced into the Americas, Australia, New Zealand, and many areas of Africa.

The conversion of rangelands to grazing by domesticated animals has major impacts on biodiversity. In an effort to increase the productivity of rangelands, management techniques may specifically eliminate certain species of plants that are poisonous or not useful as food for the grazing animals, or specific grasses may be planted that are not native to the area. In some cases, native animals are reduced if they are a threat to livestock because they are predators or because they may spread disease to the livestock. In addition, the selective eating habits of livestock tend to reduce certain species of native plants and encourage others.

Because rainfall is low and often unpredictable, it is important to regulate the number of livestock on the range. In areas where the animals are on permanent pastures, the numbers of animals can be adjusted to meet the capacity of the range to provide forage. In many parts of the world, nomadic herders simply move their animals from areas where forage is poor to areas that have better forage. This is often a seasonal activity that involves the movement of animals to higher elevations in the summer or to areas where rain has fallen recently. (See figure 11.12.)

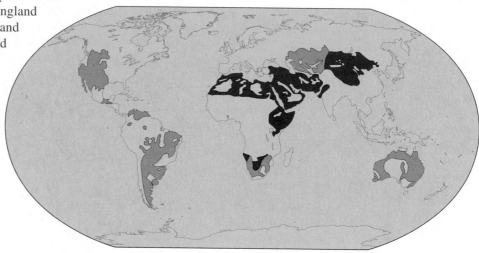

 **Nomadic herding**   **Stock raising on ranges**

FIGURE 11.12   **Use of Rangelands** The arid and semiarid regions of the world will not support farming without irrigation. In many of these areas, livestock can be raised. Permanent ranges occur where rainfall is low but regular. Nomadic herders can utilize areas that have irregular, sparse rainfall.

In many parts of the world where human population pressures are great, overgrazing is a severe problem. As populations increase, desperate people attempt to graze too many animals on the land. They also cut down the trees for firewood. If overgrazed, many plants die, and the loss of plant cover subjects the soil to wind erosion, resulting in a loss of fertility, which further reduces the land's ability to support vegetation. The cutting of trees for firewood has a similar effect, but it is especially damaging because many of these trees are legumes, which are important in nitrogen fixation. Their removal further reduces soil fertility. This severe overuse of the land results in conversion of the land to a more desertlike ecosystem. This process of converting arid and semiarid land to desert because of improper use by humans is called **desertification.** Desertification can be found throughout the world but is particularly prevalent in northern Africa and parts of Asia, where rainfall is irregular and unpredictable, and where many people are subsistence farmers or nomadic herders who are under considerable pressure to provide food for their families. (See figure 11.13.)

Habitat Loss in Aquatic Ecosystems

Habitat loss is also a problem in aquatic systems. In marine ecosystems, much of the harvest is restricted to shallow parts of the ocean where bottom dwelling fish can be easily harvested. The typical method used to harvest bottom-dwelling fish and shellfish involves the use of trawls which are nets that are dragged along the bottom. These nets capture various species, many of which are not commercially valuable. The trawls disturb the seafloor and create conditions that make it more difficult for the fish populations to recover. In addition, 25 percent of the catch typically consists of species that have no commercial value. These are discarded by being thrown overboard. However, they are usually dead, and their removal further alters the ecological nature of the seafloor. Some people have even advocated that the trawl should be banned as a fishing technique because of the damage done to the ocean bottom.

Freshwater lakes, streams, and rivers are modified for navigation, irrigation, flood control, or power production purposes, all of which may alter the natural ecosystem and change the numbers or

Risk of Human-Induced Desertification

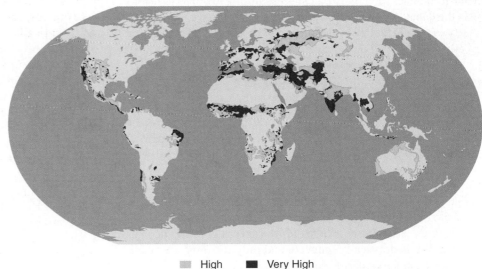

▨ High   ■ Very High

(a)

(b) Overgrazed landscape

FIGURE 11.13   Desertification (a) Areas of the world where desertification is prevalent. (b) Arid and semiarid areas can be converted to deserts by overgrazing or unsuccessful farming practices. The loss of vegetation increases erosion by wind and water, increases the evaporation rate, and reduces the amount of water that infiltrates the soil. All of these conditions encourage the development of desertlike areas.
Source: U.S. Department of Agriculture.

kinds of aquatic organisms present. These topics are discussed in greater detail in chapter 15.

In the Pacific Northwest, the extensive development of dams to provide power and aid navigation has made it nearly impossible for adult salmon to migrate upstream to spawn and difficult for young fish to migrate downstream to the ocean. Fish ladders and other techniques have not been successful in allowing the fish to pass the dams. As a result, many populations of Pacific salmon are nearly extinct. The only solution to the problem is to remove or greatly modify several of the dams.

Conversion to Urban and Industrial Uses

About 4.3 percent of U.S. land is developed as urban centers, industrial sites, and the transportation infrastructure that allows for the movement of people and products about the country. Although this is a relatively small percentage of the total land area of the United States, urban areas are the most heavily affected by human activity. Many such areas are covered with impermeable surfaces that prevent plant growth and divert rainfall to local streams and rivers. In addition, streams and other natural features are altered to serve the needs of the people. Biodiversity is drastically reduced, and only the most adaptable organisms can survive in such settings.

Many industrial sites are associated with urban centers, although some industries such as mining and oil and natural gas production may be located far from urban centers. Their impact, however, is similar to that of an urban center. The land is altered in such a way that the natural ecosystems are destroyed.

OVEREXPLOITATION

Overexploitation occurs when humans harvest organisms faster than the organisms are able to reproduce. Overexploitation has driven some organisms to extinction and threatens many others. According to the IUCN, overexploitation is responsible for over 30 percent of endangered species of animals and about 8 percent of plants.

Organisms are harvested for a wide variety of purposes. Animals of all kinds are killed and eaten as a source of protein. We use organisms for a variety of purposes in addition to food. Many plants and animals are used as ornaments. Flowers are picked, animal skins are worn, and animal parts are used for their purported aphrodisiac qualities. In the United States, many species of cactus are being severely reduced because people like to have them in their front yards. In other parts of the world, rhinoceros horn is used to make dagger handles or is powdered and sold as an aphrodisiac. Because some people are willing to pay huge amounts of money for these products, unscrupulous people are willing to break the law and poach these animals for the quick profit they can realize.

Overfishing of Marine Fisheries

The United Nations estimates that 70 percent of the world's marine fisheries are being overexploited or are being fully exploited and are in danger of overexploitation as the number of fishers increases. Figure 11.14 shows both the amount of fish captured and the amount produced by aquaculture. While the amount of fish captured by fishers has remained essentially constant since 1987, the amount produced by fish farming has increased. It is important to note that more effective capture methods and increased fishing

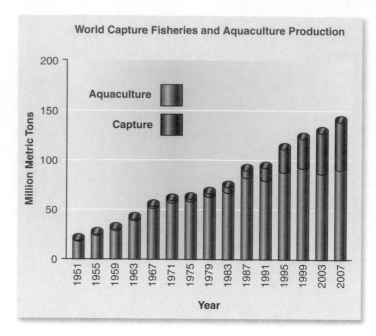

FIGURE 11.14 **Trends in World Fish Production** The amount of fish captured increased steadily until about 1987. Since then, the amount has remained essentially constant. This indicates that the world fisheries are being exploited to their capacity. Aquaculture has continued to increase so the total production continues to increase. These data have been called into question by many because it is thought that China has grossly overreported its fish production statistics. This would mean that the capture fisheries may actually have been declining in recent years.

Source: Food and Agriculture Organization of the United Nations.

effort were required to maintain the production levels for the capture of marine fish.

Another indication that marine fishery resources are being overexploited is the change in the kinds of fish being caught. The commercial fishing industry has been attempting to market fish species that previously were regarded as unacceptable to the consumer. These activities are the result of reduced catches of desired species. Examples of "newly discovered" fish in this category are monkfish and orange roughy.

Aquaculture

Fish farming (aquaculture) is becoming increasingly important as a source of fish production. (See figure 11.15.) Salmon farming has been particularly successful. This involves raising various species of salmon in "pens" in the ocean, which allows the introduction of food and other management techniques to achieve rapid growth of the fish. Norway, Chile, Canada, and Scotland are the leading countries in salmon production. The production of salmon from fish farms has increased rapidly. In 1988, less than 20 percent of the salmon sold were from fish farms, compared to over 65 percent in 2004. During the same period of time, the production of wild-caught salmon has been relatively constant. However, the farming method of producing fish is not without its environmental effects.

Raising fish in such concentrated settings results in increased nutrients in the surrounding water from uneaten food and the wastes the fish release. This can cause local algal blooms that

(a) Aquaculture in ocean

(b) Aquaculture in fresh water

FIGURE 11.15   Aquaculture Photo *a* shows aquaculture pens in the ocean. Fish reared in these pens are fed and managed like livestock. Photo *b* shows a freshwater fish pond. In many parts of the world such ponds provide important sources of protein for the local populace.

negatively affect other fisheries, such as shellfish. Many of the salmon species raised in farms are not native to the waters in which they are raised. Inevitably, some of these fish escape. The introduction of exotic species can have a negative impact on native species. Some people even worry about the genetic stocks of fish that are raised. Many wild salmon populations are in danger of extinction. The introduction of new genetic stocks that escape and interbreed with wild fish can alter the genetic makeup of the wild populations. On an economic front, the increase in the amount of farmed salmon is reducing the prices salmon fishers receive for their catch.

Currently, about 60 percent of all aquaculture production is from freshwater systems, and production is growing rapidly. Aquaculture of freshwater species typically involves the construction of ponds, which allow the close management of the fish, shrimp, or other species. This can be done with a very low level of technology and is easily accomplished in underdeveloped areas of the world.

The environmental impacts of freshwater aquaculture are similar to those of aquaculture in marine systems. Nutrient overloads from concentrations of fish can pollute local bodies of water, and the escape of exotic species may harm native species. In addition, freshwater aquaculture involves the conversion of land to a new use. Often the lands involved are mangrove swamps or other wetlands that many people feel should be protected. Regardless of environmental concerns, the productivity of freshwater aquaculture has the potential to provide for the protein needs of a growing population, so it is likely to continue to increase at a rapid rate.

Unsustainable Harvest of Wildlife and Plants

Wildlife may be harvested for a variety of reasons. Primary among them is the need for food in much of the world. Any animal that can be captured or killed is used for food. Meat from wild animals is often referred to as **bush meat,** since it is gathered from the "bush," a common term for the wild. Estimates about the extent of the problem vary, but there is wide agreement that the harvest is unsustainable. One estimate by the Wildlife Conservation Society is that about 70 percent of wildlife species in Asia and Africa and about 40 percent of species in Latin America are being hunted unsustainably. It is clear that this practice is causing local extinctions of certain species of wildlife. Furthermore, endangered species such as chimpanzees and gorillas are often harvested.

Trade in bush meat and other animal products has become a major problem for several reasons. Hunting of wildlife is a part of all subsistence cultures, so it is considered a normal activity. Poor people can earn money by harvesting and selling wild animals for meat or other purposes. Many kinds of wildlife are considered delicacies and are highly prized for the home and restaurant trade. In many parts of the developing world, regulations regarding the preservation of wildlife are poorly developed or widely ignored. The roads associated with logging operations in tropical forests give access to a greater part of the forest. Modern technology (guns and artificial lights) allows more efficient location and killing of animals.

The harvest of living animals for the pet and aquarium trade is a significant problem. Many kinds of birds, reptiles, and fish are threatened because they are desired as pets. Generally, it is not the isolated taking of an individual organism that is the problem but, rather, the method of capture, which may have far-reaching effects. For example, many nests are destroyed when nestling birds are taken. In the case of tropical marine fishes, toxins are often used to stun the fish, and the few valuable individuals are recovered while many of the others die. Furthermore, since much of the market in exotic pets involves transporting animals long distances, the mortality rate among those captured is high.

Wildlife species also are hunted because parts of the animal may have particular value. For example, ivory and animal skins are highly valued in many cultures as art objects or clothing. Parts of some animals are thought to have particular medicinal properties, and others are hunted because of their use in traditional cultural practices. For example, rhinoceros horn is thought to be an aphrodisiac, and some cultures prize rhinoceros horn as handles for knives. These beliefs lead to a brisk trade in the desired species.

FIGURE 11.16 **Traditional Medicine** This woman in Iquitos, Peru, is selling traditional medicine derived from local plants and animals.

Traditional medicine makes use of a great number of different plants and animals. (See figure 11.16.) Since about 80 percent of the world's population uses traditional cultural medicine as opposed to modern Western-style medical treatments, harvesting of medicinal plants and animals is a significant problem in much of the world. Many of the medicines used in these practices are derived from endangered plants and animals. Harvest of some species of plants has led to reductions in populations to the point that the species are endangered. For example, both the Asian and American ginseng plants have been greatly reduced because of their use in traditional medicine. Tiger bone and rhinoceros horn are important for a variety of traditional medicinal uses. The trade in these products puts a significant pressure on populations of these animals.

INTRODUCTION OF EXOTIC SPECIES

Introduction of exotic species can also have a significant effect on biodiversity because exotic species often kill or directly compete with native species and drive them to extinction.

The introduction of exotic species occurs for a variety of reasons. Some introductions are purposeful, while others are accidental. The reasons for purposeful introductions are quite varied. Most agriculturally important plants and animals are introductions. Other organisms were introduced because people had a fondness for a particular plant or animal. Many plants were introduced for horticultural purposes. Some introduced species are easily controlled. For example, most agriculturally important plants do not grow in the wild. Other introductions are accidental and arrived as stowaways on imported materials. Not all exotic species become problems; however, some organisms disperse broadly and displace native species.

The IUCN estimates that about 30 percent of birds and 15 percent of plants are threatened because they are unable to successfully compete against invasive exotic species. On many islands, the introduction of rats has had a major impact on nesting birds because the rats eat the eggs and kill the nestlings. The introduction of cats and foxes into Australia has resulted in the reduction of populations of many species of native marsupials. Cats and foxes are efficient predators of the native wildlife and caused the extinction of many kinds of native mammals on the Australian mainland. Some species of native marsupials are confined to isolated islands off the coast where cats and foxes are not found.

The introduction of diseases has also had considerable impact on American forests. Two fungal diseases of trees have significantly changed the nature of American forests. Chestnut blight essentially eliminated the American chestnut, and Dutch elm disease has greatly reduced the numbers of American elms.

Various kinds of insects have also had an effect on the structure of ecosystems. The gypsy moth has spread throughout North America. Its larvae eat the leaves of forest trees and have significantly altered forest ecosystems because mature oaks are more likely to die as a result of defoliation than some other forest trees. The Asian longhorned beetle is a recent arrival to North America. It was discovered in New York in 1996. Its larvae feed on the wood of deciduous trees and often cause their death. Asian longhorned beetles probably enter North America inside solid-wood packing material from China. Although currently confined to a few locations in the eastern United States and Canada, the beetle has the potential to do great damage to the deciduous forest of the region, and both the U.S. and Canadian governments are taking action to control the problem. The only method of controlling the infestations involves the destruction of infected trees.

Freshwater ecosystems also have been greatly affected by accidental and purposeful introductions. The zebra mussel and Eurasian milfoil are two accidental introductions that have caused great problems. Zebra mussels were first discovered in Lake St. Clair near Detroit about 1985. They were most likely released from ballast water from European ships. The zebra mussel has spread through much of the eastern United States and adjacent Canada in the Great Lakes region and the Mississippi River watershed. It has three major impacts. It clogs intake pipes for water treatment plants and other industrial users. It establishes colonies on the surface of native mussels, often resulting in their death. And it has altered freshwater ecosystems by filtering much plankton from the water and allowing more aquatic plants to grow.

Eurasian milfoil was first discovered in Washington, D.C., in the 1940s but has since spread to much of the United States and southern Canada. Although its introduction may have been on purpose, its spread has occurred by accidental transfer by boats, wildlife, and drifting. It forms thick mats in many bodies of water that can interfere with boat traffic. Figure 11.17 shows examples of these exotic organisms.

The introduction of exotic fish species also has greatly affected naturally occurring freshwater ecosystems. The Great Lakes, for example, have been altered considerably by the accidental and purposeful introduction of fish species. The sea lamprey, smelt, carp, alewife, brown trout, and several species of

(a) Adult gypsy moths

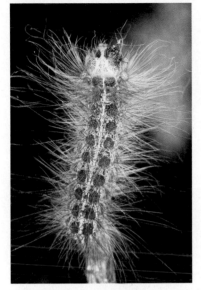

(b) Gypsy moth caterpillar

(c) Asian longhorned beetle

(d) Zebra mussel

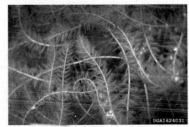

(e) Eurasian milfoil

FIGURE 11.17 **Invasive Species** The routes by which each of these species entered the United States is different. (a, b) The gypsy moth was purposely brought to the United States and accidentally escaped. (c) The Asian longhorned beetle appeared to have arrived in packing crates. (d) The zebra mussel arrived in ballast water from Europe, and (e) the Eurasian milfoil arrived as an aquarium plant.

salmon are all new to this ecosystem. (See figure 11.18.) The sea lamprey is parasitic on lake trout and other species and nearly eliminated the native lake trout population. Controlling the lamprey problem requires the use of a very specific larvicide that kills the immature lamprey in the streams. This technique works because mature lamprey migrate upstream to spawn, and the larvae spend several years in the stream before migrating downstream into the lake. With the partial control of lamprey, lake trout populations have been increasing. (See figure 11.19.) Recovery of the lake trout is particularly desirable, since at one time it was an important commercial species.

Another accidental introduction to the Great Lakes, the alewife, a small fish of little commercial or sport value, became a problem during the 1960s, when alewife populations were so great that they died in large numbers and littered beaches. Various species of salmon were introduced about this time in an attempt to control the alewife and replace the lake trout population, which had been depressed by the lamprey. While this salmon introduction has been an economic success and has generated millions of dollars for the sport fishing industry, it may have had a negative impact on some native fish such as the lake trout, with which the salmon probably compete. Salmon also migrate upstream, where they disrupt the spawning of native fish. Most species of salmon die after spawning, which causes a local odor problem.

Other more recent arrivals in the Great Lakes include a clam, the zebra mussel, and two fishes, the round goby and river ruffe. All of these probably arrived in ballast water from fresh or brackish water bodies in Europe. These exotic introductions are increasing rapidly and changing the mix of species present in Great Lakes waters.

CONTROL OF PREDATOR AND PEST ORGANISMS

The systematic killing of certain organisms because they interfere with human activities also results in reduced biodiversity. Many large predators have been locally exterminated because they preyed on the domestic animals that humans use for food. Mountain lions and grizzly bears in North America have been reduced to small, isolated populations, in part because they were hunted to reduce livestock loss. Tigers in Asia and the lion and wolf in Europe were reduced or eliminated for similar reasons.

Some extinctions of pest species are considered desirable. Most people would not be disturbed by the extinction of malaria parasite, mosquitoes, rats, or fleas. In fact, people work hard to drive some species to extinction. For example, in the *Morbidity and Mortality Weekly Report* (October 26, 1979), the U.S. Centers for Disease Control triumphantly announced that the virus that causes smallpox was extinct in the human population after many years of continuous effort to eliminate it.

At one time, it was thought that populations of game species could be increased substantially if predators were controlled. In Alaska, for example, the salmon-canning industry claimed that bald eagles were reducing the salmon population, and a bounty was placed on eagles. From 1917 to 1952 in Alaska, 128,000 eagles were killed for bounty money. This theory of predator control to increase populations of game species has not proven to be valid in most cases, however, since the predators do not normally take

Lake trout

Brown trout

Native	Introduced
Brook lamprey	Sea lamprey
Lake trout Brook trout Whitefish Herring Ciscoes	Brown trout Rainbow trout Pink salmon Coho salmon Chinook salmon Atlantic salmon
Suckers	
Chubs Shiners	Carp
Catfish Bullheads	
	Smelt Alewife
White bass	White perch
Smallmouth bass Rock bass	Sunfish
Yellow perch Walleye	Ruffe

Smelt

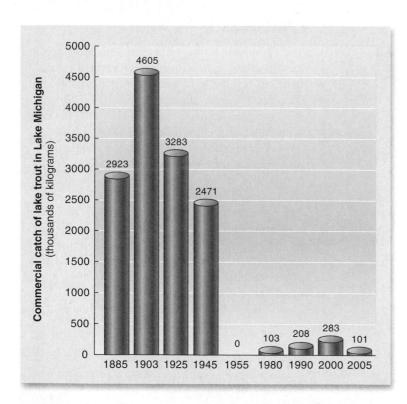
Yellow perch

FIGURE 11.18   Native and Introduced Fish Species in the Great Lakes The Great Lakes have been altered considerably by the introduction of many non-native fish species. Some were introduced accidentally (lamprey, alewife, and carp), and others were introduced on purpose (salmon, brown trout, rainbow trout).

FIGURE 11.19   The Impact of the Lamprey on Commercial Fishing The Lamprey entered the Great Lakes in 1932. Because it is an external parasite on lake trout, it had a drastic effect on the population of lake trout in the Great Lakes. As a result of programs to prevent the lamprey from reproducing, the number of lamprey has been reduced somewhat, and the lake trout population is recovering with the aid of stocking programs.

the prime animals anyway. They are more likely to capture sick or injured individuals not suitable for game hunting.

Although at one time they were thought to be helpful, bounties and other forms of predator management have been largely eliminated in North America. In fact, the pendulum has swung the other way, and predator control is not considered to be cost-effective in most cases. One exception to this general trend is the hunting and trapping of wolves in Alaska and Canada. One rationale for the taking of wolves is that they kill moose and caribou. Since many of the people who live in these areas rely on wild game as a significant part of their food source, controlling wolf populations is politically popular. However, the number of wolves taken is regulated.

The control of cowbird populations has been used to enhance the breeding success of the Kirtland's warbler. Cowbirds lay their eggs in the nests of other birds, including Kirtland's warblers. When the cowbird egg hatches, the hatchling pushes the warbler chicks out of the nest. Thus, trapping and killing of cowbirds in the vicinity of nesting Kirtland's warblers has been used to enhance the reproduction of this endangered species.

Many species may require refuges where they are protected from competing introduced species or human interference. Many native rangeland species of wildlife benefit from the exclusion of introduced grazing animals such as cattle and sheep because the absence of grazing livestock allows a more natural grassland community to be reestablished. (See figure 11.20.)

FIGURE 11.20 **Habitat Protection** The area on the left side of the fence has been protected from cattle grazing. This area provides a haven for many native species of plants and animals that cannot survive in heavily grazed areas.

FIGURE 11.21 **Climate Change Affects Arctic Animals** The polar bear was listed as a threatened species as a result of global warming. Warmer waters result in less sea ice and bears have greater difficulty capturing their primary prey—seals. Note the remains of a kill on the ice behind the bear.

CLIMATE CHANGE

As scientists and the public have come to recognize that the climate is changing, the role of climate change on the survival of species has become an issue. Many species exist near the limit of their physiological tolerance. A slight change in the temperature may push them over the brink. Three groups of organisms appear to be greatly affected by climate change: amphibians, corals, and Arctic species. The International Union for Conservation of Nature and Natural Resources (IUCN) estimates that about 30 percent of amphibian species are threatened because of the effects of climate change. Many species of frogs are threatened by a fungal disease that has become common in the last few decades. Many people think there is a link between a warming planet and the prevalence of the disease. Corals are declining throughout the world. Many scientists suspect that warming water is encouraging diseases that are causing the death of corals.

Arctic and Antarctic species are adapted to cold temperatures and often rely on ice cover. The melting of sea ice in both the Arctic and Antarctic is changing migration patterns and food availability. The listing of polar bears as vulnerable by the IUCN and as a threatened species by the U.S. Fish and Wildlife Service is recognition that breakup of ice cover in the Arctic oceans is affecting the ability of polar bears to obtain food. The bears use seals as a primary food source and wait at breathing holes or at the edge of ice flows to capture seals when they come up to breath. They also stalk seals that are resting on ice flows. With more open water, seals do not need to rely as much on breathing holes and bears are less successful at obtaining food. Furthermore, bears must swim greater distances to reach ice flows. (See figure 11.21.)

In the Antarctic, loss of sea ice has resulted in some populations of penguins migrating southward and appears to be related to a reduction in the amount of krill (small crustaceans), which are at the base of the Antarctic food web.

WHAT IS BEING DONE TO PRESERVE BIODIVERSITY?

Efforts to preserve biodiversity involve a variety of approaches. Foremost among them is the need to understand the life history of organisms so that effective measures can be taken to protect species from extinction. Several international organizations work to prevent the extinction of organisms. The IUCN lists over 16,000 species as threatened with extinction. The IUCN classifies species in danger of extinction into four categories: endangered, vulnerable, rare, and indeterminate. Endangered species are those whose survival is unlikely if the conditions threatening their extinction continue. These organisms need action by people to preserve them, or they will become extinct. Vulnerable species are those that have decreasing populations and will become endangered unless causal factors, such as habitat destruction, are eliminated. Rare species are primarily those that have small worldwide populations and that could be at risk in the future. Indeterminate species are those that are thought to be extinct, vulnerable, or rare, but so little is known about them that they are impossible to classify.

Although the IUCN is a highly visible international conservation organization, it has very little power to effect change. It generally seeks to protect species in danger by encouraging countries to complete inventories of plants and animals within their borders. It also encourages the training of plant and animal biologists within the countries involved. (There is currently a critical shortage of plant and animal biologists who are familiar

with the organisms of the tropics.) The IUCN also encourages the establishment of preserves to protect species in danger of extinction.

LEGAL PROTECTION

International efforts to preserve biodiversity have involved various activities of the United Nations. Most countries of the world have ratified the Convention on Biological Diversity, commonly known as the international biodiversity treaty. The United States is one of five countries that have not ratified the convention. The United States signed the treaty but it was not ratified by the U.S. Senate. Some of the key components of the Convention on Biological Diversity are that the signatory countries will:

- Develop national strategies for the conservation and sustainable use of biological diversity.

- Identify components of biological diversity important for its conservation.

- Monitor biological diversity.

- Identify activities that have adverse impacts on the conservation and sustainable use of biological diversity.

- Establish a system of protected areas.

- Rehabilitate and restore degraded ecosystems and promote the recovery of threatened species.

- Develop or maintain necessary legislation for the protection of threatened species and populations.

- Integrate consideration of the conservation and sustainable use of biological resources into national decision making.

Awareness and concern about loss of biodiversity are high in many of the developed countries of the world. However, these parts of the world have few undisturbed areas and many vulnerable species have already been eliminated. The potential for the loss of biodiversity is greatest in tropical, developing countries. Many biologists estimate that there may be as many species in the tropical rainforests of the world as in the rest of the world combined. Unfortunately, loss of biodiversity is not a high priority for the general public in developing countries, even though their national governments have ratified the biodiversity treaty. This difference in level of concern is understandable since the developed world has surplus food, higher disposable income, and higher education levels, while people in many of the less-developed countries, where population growth is high, are most concerned with immediate needs for food and shelter, not with long-term issues such as species extinction. The developed world should not be held blameless, however, because our activities may indirectly be involved in extinction. When we purchase lumber, agricultural products, and fish from the less-developed world at low prices, we often are indirectly encouraging people in the developing world to overexploit their resources and are thus affecting the rates of extinction.

In the United States, the primary action related to the preservation of biodiversity involved the passage of the U.S. Endangered Species Act in 1973. This legislation designates species as endangered or threatened and gives the federal government jurisdiction over any species designated as endangered. **Endangered species** are those that have such small numbers that they are in immediate jeopardy of becoming extinct. (See figure 11.22.) **Threatened species** could become extinct if a critical factor in their environment were changed. About 1,350 U.S. species and subspecies have been designated as endangered or threatened by the Office of Endangered Species of the Department of the Interior. The Endangered Species Act directs that

Blackfooted ferret

Nene (Hawaiian goose)

Mission blue butterfly

Galápagos tortoise

Giant panda

Green pitcher plant

FIGURE 11.22 **Endangered Species** Endangered species are those present in such low numbers that they are in immediate danger of becoming extinct. These organisms are examples of endangered species.

In 2005 a major study called the Millennium Ecosystem Assessment Synthesis Report was published. It involved input from about 1300 of the world's leading experts and a partnership among several international organizations including the United Nations, the World Bank, and the International Union for the Conservation of Nature (IUCN). The study was conducted over four years and involved experts from 95 countries. The report is recognized by governments as a mechanism to meet part of the assessment needs of four international environmental treaties—the UN Convention on Biological Diversity, the Ramscar Convention on Wetlands, the UN Convention to Combat Desertification, and the Convention on Migratory Species. Following are highlights of the main findings of the report.

- Humans have changed ecosystems more rapidly and extensively in the last 50 years than in any other period. This was done largely to meet rapidly growing demands for food, freshwater, timber, fiber, and fuel. More land was converted to agriculture since 1945 than in the eighteenth and nineteenth centuries combined. Experts say that this resulted in a substantial and largely irreversible loss in diversity of life on Earth, with 10 to 30 percent of the mammal, bird, and amphibian species currently threatened with extinction.

- Ecosystem changes that have contributed substantial net gains in human well-being and economic development have been achieved at growing costs in the form of degradation of other services. Only three ecosystem services have been enhanced in the last 50 years—increases in crop, livestock, and aquaculture production. Two services—capture fisheries and freshwater—are now well beyond levels that can sustain current, much less future, demands.

- The degradation of ecosystem services could grow significantly worse during the first half of this century. In all the plausible futures explored by the scientists, they project progress in eliminating hunger, but at a slower rate than needed to halve the number of people suffering from hunger by 2015. Experts warn that changes in ecosystems such as deforestation influence the abundance of human pathogens such as malaria and cholera, as well as increasing the risk from emerging new diseases.

- The challenge of reversing the degradation of ecosystems while meeting increasing demands can be met under some scenarios with significant policy and institutional changes. However, these changes will be large and are not currently under way. The report mentions options that exist to conserve or enhance ecosystem services that reduce negative trade-offs or that will positively affect other services. Protection of natural forests, for example, not only conserves wildlife but also supplies freshwater and reduces carbon emissions.

The major conclusion of this assessment is that it lies within the power of human societies to ease the strains we are putting on the nature services of the planet, while continuing to use them to bring better living standards to all. Achieving this, however, will require major changes in the way nature is treated at every level of decision-making and new forms of cooperation between government, business, and civil society.

VALUES UNDERLYING THE MILLENNIUM DECLARATION

The Millennium Declaration—which outlines 60 goals for peace; development; the environment; human rights; the vulnerable, hungry, and poor; Africa; and the United Nations—is founded on a core set of values described as follows:

"We consider certain fundamental values to be essential to international relations in the twenty-first century. These include:

- **Freedom.** Men and women have the right to live their lives and raise their children in dignity, free from hunger and from the fear of violence, oppression or injustice. Democratic and participatory governance based on the will of the people best assures these rights.

- **Equality.** No individual and no nation must be denied the opportunity to benefit from development. The equal rights and opportunities of women and men must be assured.

- **Solidarity.** Global challenges must be managed in a way that distributes the costs and burdens fairly in accordance with basic principles of equity and social justice. Those who suffer or who benefit least deserve help from those who benefit most.

- **Tolerance.** Human beings must respect one other, in all their diversity of belief, culture and language. Differences within and between societies should be neither feared nor repressed, but cherished as a precious asset of humanity. A culture of peace and dialogue among all civilizations should be actively promoted.

- **Respect for nature.** Prudence must be shown in the management of all living species and natural resources, in accordance with the precepts of sustainable development. Only in this way can the immeasurable riches provided to us by nature be preserved and passed on to our descendants. The current unsustainable patterns of production and consumption must be changed in the interest of our future welfare and that of our descendants.

- **Shared responsibility.** Responsibility for managing worldwide economic and social development, as well as threats to international peace and security, must be shared among the nations of the world and should be exercised multi-laterally. As the most universal and most representative organization in the world, the United Nations must play the central role."[1]

The Millennium Ecosystem Assessment Report is available at http://www.maweb.org/en/Article.aspx?id=58.

---

1. United Nations General Assembly, "United Nations Millennium Declaration," Resolution 55/2, United Nations A/RES/55/2, 8 September 2000.

no activity by a governmental agency should lead to the extinction of an endangered species and that all governmental agencies must use whatever measures are necessary to preserve these species.

Since one key to preventing extinctions is preservation of the habitat required by the endangered species, many U.S. governmental agencies and private organizations have purchased sensitive habitats or have managed areas to preserve suitable habitats for endangered species. Since the protection of a species requires the preservation of its habitat, inevitably landowners face losing the use of their land. Many lawsuits have been filed by landowners claiming that the federal government has taken their land from them by preventing them from using it for development purposes.

As a result of these pressures, the rules were changed to allow some of the land to be used if a species conservation plan is in place. Another modification to the Endangered Species Act occurred in 1978, when Congress amended it so that exemptions to the act could be granted for federally declared major disaster areas or for national defense, or by a seven-member Endangered Species Review Committee. Because this group has the power to sanction the extinction of an organism, it has been nicknamed the "God Squad." If the committee found that the economic benefits of a project outweighed the harmful ecological effects, it would exempt a project from the Endangered Species Act.

TABLE 11.4 Endangered and Threatened Species by Taxonomic Group

Taxonomic Group	Approximate Number of Described Species in the World	Number of U.S. Species Endangered or Threatened	IUCN Threatened List	Percent of Total Species on IUCN Threatened List
Insects	950,000	57	623	0.07%
Plants	270,000	742	8408	3.1%
Invertebrates other than insects or mollusks	257,800	36	507	0.2%
Mollusks	81,000	145	979	1.2%
Fishes	29,300	138	1201	4.0%
Birds	9900	90	1217	12.0%
Reptiles	8240	40	422	5.0%
Amphibians	5900	23	1808	29.0%
Mammals	5400	84	1094	20.0%
Bacteria, algae, fungi and protozoa	163,000	2	18	0.0%

Sources: Data from the IUCN Red List of Threatened Species and The U.S. Fish and Wildlife Service website.

The designation of a species as endangered or threatened is biased toward large organisms that are relatively easy to identify. Many groups of organisms are very poorly known, so it is difficult to assess whether specific species are endangered or not. Table 11.4 compares the number of existing species in different taxonomic categories and the numbers and percentages considered endangered. Most endangered or threatened species are birds, mammals, certain categories of plants, some insects (particularly butterflies), a few mollusks, and fish. Bacteria, fungi, most insects, and many other inconspicuous organisms rarely show up on endangered species lists, even though they play vital roles in the nitrogen and carbon cycles and as decomposers.

The Convention on International Trade in Endangered Species of Wild Fauna and Flora (CITES) is an international agreement among governments. Its aim is to ensure that international trade in specimens of wild animals and plants does not threaten their survival. Currently, there are over 170 member countries who agree to limit export and import of certain species in order to protect the populations in their countries of origin. There are about 5000 species of animals and 28,000 species of plants that are protected by CITES against overexploitation through international trade.

SUSTAINABLE MANAGEMENT OF WILDLIFE POPULATIONS

Many kinds of terrestrial wild animals are managed for their value as game animals, to protect them from extinction, or for other purposes. Several techniques, some of which result in ecosystem modifications, are used to enhance certain wildlife populations. These techniques include habitat analysis and management, population assessment and management, and establishing refuges.

Habitat Analysis and Management

Managing a particular species requires an understanding of the habitat needs of that species. An animal's habitat must provide the following: food, water, and cover. **Cover** refers to any set of physical features that conceals or protects animals from the elements or enemies. Several kinds of cover are important and include places where the animal can rest and raise young, where it can escape from enemies, and where it is protected from the elements.

Once the critical habitat requirements of a species are understood, steps can be taken to alter the habitat to improve the success of the species. The habitat modifications made to enhance the success of a species are known as **habitat management.** The endangered Kirtland's warbler builds nests near the ground in dense stands of young jack pine. The density is important since their nests are vulnerable to predators and are better hidden in dense cover. Jack pine is a fire-adapted species that releases seeds following forest fires and naturally reestablishes dense stands following fire. Therefore, planned fires to regenerate new dense jack pine stands is a technique that has been used to provide appropriate habitat for this endangered bird.

Habitat management also may take the form of encouraging some species of plants that are the preferred food of the game species. For example, habitat management for deer may involve encouraging the growth of many young trees, saplings, and low-growing shrubs by cutting the timber in the area and allowing the natural regrowth to supply the food and cover the deer need. Both forest management and deer management may have to be integrated in this case because some other species of animals, such as squirrels, will be excluded if mature trees are cut, since they rely on the seeds of trees as a major food source.

Population Assessment and Management

Population management is another important activity that requires planning. Species of game animals are often managed so that they do not exceed the carrying capacity of their habitat. Wildlife managers use several techniques to establish and maintain populations at an appropriate level. Population censuses are used to check the population regularly to see if it is within acceptable limits. These census activities include keeping records of the number of animals killed by hunters, recording the number of singing birds during the breeding season, counting the number of fecal pellets, direct counting of large animals from aircraft, and a variety of other techniques.

Since wildlife management often involves the harvesting of animals by hunting for sport and for meat, regulation of hunting activity is an important population management technique. Seasons are usually regulated to ensure adequate reproduction and provide the largest possible healthy population during the hunting season. Hunting seasons usually occur in the fall so that surplus animals are taken before the challenges of winter. Winter taxes the animal's ability to stay warm and is also a time of low food supplies in most temperate regions. A well-managed wildlife resource allows for a large number of animals to be harvested in the fall and still leaves a healthy population to survive the winter and reproduce during the following spring. (See figure 11.23.)

In some cases, when a population of a species is below the desired number, organisms may be artificially introduced. Introductions are most often made when a species is being reintroduced to an area where it had become extinct. Wolves have been successfully reintroduced to Yellowstone National Park and large tracts of wooded land in the northern Midwest (Michigan, Wisconsin, Minnesota). Similarly, an exchange between the state of Michigan and the province of Ontario reintroduced moose to Michigan and turkeys to parts of Ontario.

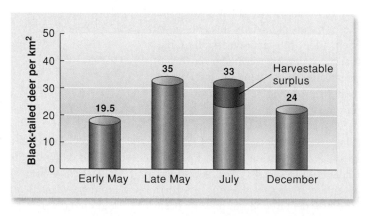

FIGURE 11.23 **Managing a Wildlife Population** The seasonal changes in this population of black-tailed deer are typical of many game species. The hunting season is usually timed to occur in the fall so that surplus animals will be harvested before winter, when the carrying capacity is lower.

Source: Data from R. D. Taber and R. F. Dasmann, "The Dynamics of Three Natural Populations of the Deer *Odocoileus hemionus columbianus*," *Ecology* 38, no. 2 (1957): 233–46.

Some game species are non-native species that have been introduced for sport hunting purposes. The ring-necked pheasant was originally from Asia but has been introduced into Europe, Great Britain, and North America. All of the large game animals of New Zealand are introduced species, since there was none in the original biota of these islands. Several species of deer have been introduced into Europe from Asia. Many of these are raised in deer parks, where the animals are similar to free-ranging domestic cattle. They may be hunted for sport or slaughtered to provide food. Introductions of exotic species were once common. However, today it is recognized that these introductions often result in the decreased viability of native species or in the introduced species becoming pests.

Special Issues with Migratory Animals

Waterfowl (ducks, geese, swans, rails, etc.) present some special management problems because they are migratory. **Migratory birds** can fly thousands of kilometers and can travel north in the spring to reproduce during the summer months and return to the south when cold weather freezes the ponds, lakes, and streams that serve as their summer homes. (See figure 11.24.)

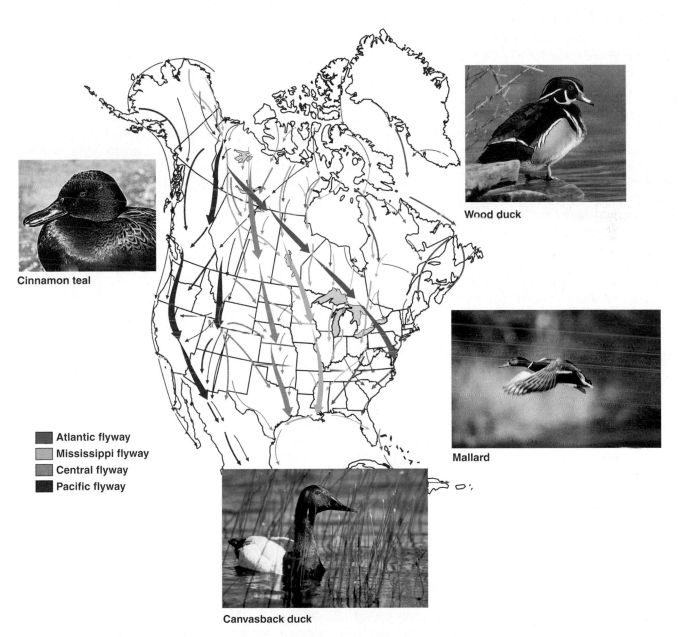

Cinnamon teal

Wood duck

Mallard

Canvasback duck

Atlantic flyway
Mississippi flyway
Central flyway
Pacific flyway

FIGURE 11.24   Migration Routes for North American Waterfowl Migratory waterfowl follow traditional routes when they migrate. These have become known as the Atlantic, Mississippi, Central, and Pacific flyways. Many of these waterfowl are hatched in Canada, migrate through the United States, and winter in the southern United States or Mexico.

Because many waterfowl nest in Canada and the northern United States and winter in the southern United States and Central America, an international agreement among Canada, the United States, and Mexico is necessary to manage and prevent the destruction of this wildlife resource. Habitat management has taken several forms. In Canada, where much of the breeding occurs, government and private organizations such as Ducks Unlimited have worked to prevent the draining of small ponds and lakes that provide nesting areas for the birds. In addition, new impoundments have been created where it is practical. Because birds migrate southward during the fall hunting season, a series of wildlife refuges provide resting places, food, and protection from hunting. In addition, these refuges may be used to raise local populations of birds. During the winter, many of these birds congregate in the southern United States. Refuges in these areas are important over-wintering areas where the waterfowl can find food and shelter.

A similar management problem exists in Africa and Eurasia. Many birds that rely on wetlands migrate between Eurasia and Africa. In 1995, an international Agreement on the Conservation of African-Eurasian Waterbirds was developed. Its development was a complicated process, since it involves about 40 percent of the world's land surface and includes 117 countries. Several important countries have signed the agreement, but many have yet to do so.

In sub-Saharan Africa, many species of wildlife migrate across large expanses of land that may involve several countries. Often different countries have differing rules for harvest, and often there are barriers to the movement of migratory animals. For example, fences are commonly used to designate the boundaries between nations. A relatively new concept of transboundary parks has been developed to accommodate the movements of these animals. One typical action that occurs with the formation of transboundary parks across national boundaries is the removal of structures that interfere with the movement of the animals.

SUSTAINABLE MANAGEMENT OF FISH POPULATIONS

One of the major problems associated with the management of marine fisheries resources is the difficulty in achieving agreement on limits to the harvest. The oceans do not belong to any country; therefore, each country feels it has the right to exploit the resource wherever it wants. Since each country seeks to exploit the resource for its advantage without regard for the sustainable use of the resource, international agreements are required to manage the resource. Such agreements are difficult to achieve because of political differences and disregard for the nature of the resource. (See Common Property Resource Problems—The Tragedy of the Commons in chapter 3.)

For several reasons, the most productive areas of the ocean are those close to land. In shallow water, the entire depth (water column) is exposed to sunlight. Plants and algae can carry on photosynthesis, and biological productivity is high. The nutrients washed from the land also tend to make these waters more fertile. Furthermore, land masses modify currents that bring nutrients up from the ocean bottom. Many of the commercially important fish and other seafood species are bottom dwellers, but fishing for them at great depths is not practical. Therefore, fishing pressure is concentrated on areas where the water is shallow and relatively nutrient rich. Countries have taken control over the waters near them by establishing a 200-nautical-mile limit (approximately 300 kilometers) within which they control the fisheries. This has not solved conflicts, however, as neighboring countries dispute the fishing practices used in waters they both claim. Canada and the United States have had continuing arguments over the management of the lobster fishery and argued as well over the North Atlantic cod fishery until its collapse in the early 1990s.

In recent years, several countries have designated areas within their territorial waters where fishing is not allowed. Several studies suggest that these reserves are effective ways to protect a portion of the fishery resource and serve to repopulate surrounding areas that are open to fishing. This approach is particularly effective for fish species that live on the bottom or are associated with specific structures on the bottom.

Although freshwater fisheries face some of the same problems as marine fisheries, they are typically easier to regulate because rivers, lakes, and streams are often contained within a particular country. Although people in many places in the world harvest freshwater fish for food, the quantities are relatively small compared to those produced by marine fisheries. Furthermore, the management of the freshwater fish resource is much more intense since the bodies of water are smaller and human populations have greater access to the resource. Freshwater fish management in much of the world includes managing for both recreational and commercial food purposes.

The fisheries resource manager must try to satisfy two interest groups. Both sports fishers and those who harvest for commercial purposes must adhere to regulations. However, the regulations usually allow the commercial fisher to use different harvesting methods, such as nets. In much of North America, freshwater fisheries are primarily managed for sport fishing. The management of freshwater fish populations is similar to that of other wild animals. Fish require cover, such as logs, stumps, rocks, and weed beds, so they can escape from predators. They also need special areas for spawning and raising young. These might be a gravel bed in a stream for salmon, a sandy area in a lake for bluegills or bass, or a marshy area for pike. A freshwater fisheries biologist tries to manipulate some of the features of the habitat to enhance them for the desired species of game fish. This might take the form of providing artificial spawning areas or cover. Regulation of the fishing season so that the fish have an opportunity to breed is also important.

In addition to these basic concerns, the fisheries biologist pays special attention to water quality. Whenever people use water or disturb land near the water, water quality is affected. For example, toxic substances kill fish directly, and organic matter in the water may reduce the oxygen. The use of water by industry or the removal of trees lining a stream warms the water and makes it unsuitable for certain species. Poor watershed management results in siltation, which covers spawning areas, clogs the gills of young fish, and changes the bottom so food organisms cannot live there. The fisheries biologist is probably as concerned about what happens outside the lake or stream as what happens within it.

THE CALIFORNIA CONDOR

The California condor *(Gymnogyps californianus)* is thought to have been adapted to feed on the carcasses of large mammals found in North America during the Ice Age. With the extinction of the large, Ice Age mammals, the condors' major food source disappeared. By the 1940s, their range had shrunk to a small area near Los Angeles, California. Further fragmentation of their habitat caused by human activity, death by shooting, and death by eating animals containing lead shot reduced the wild population to about 17 animals by 1986.

A low reproductive potential makes it difficult for the species to increase in numbers. They do not become sexually mature until 6 years of age, and females typically lay one egg every two years. Because of concerns about the survival of the species, in 1987, all the remaining wild condors were captured to serve as breeding populations. The total population was 27 individuals. The plan was to raise young condors in captivity, leading to a large population that would ultimately allow for the release of animals back into the wild. Captive breeding is a very involved activity. Extra eggs can be obtained by removing the first egg laid and incubating it artificially. The female will lay a second egg if the first is removed. Raising the young requires careful planning. Although the young are fed by humans, they must not associate food

with humans. This would result in inappropriate behaviors in animals eventually released to the wild. Therefore, puppets that resemble the parent birds are used to feed the young birds.

These efforts increased the number of offspring produced per female and resulted in a captive population of 54 individuals by 1991, when two condors were released into the wild north of Los Angeles. Six more were released in 1992. In 1996, a second population of condors was introduced in Arizona north of the Grand Canyon near the Utah border. The goal is to have two populations of up to 150 individuals each. In addition to natural mortality from predation, there has been unanticipated mortality from contact with power lines and lead poisoning from ingesting lead shot with food. To help reduce these mortality problems, the birds that are currently being released go through a period of training in which they are taught to avoid power lines. In addition, several condors have been recaptured to be treated for lead poisoning. These will eventually be released back to the wild.

By 2008, the population had increased to 332 birds, with 152 living in the wild. Since 2002, condors have been released in Baja, California, in conjunction with the Mexican government. There are 11 condors in Mexico. In 2007, for the first time, a pair laid an egg. However, the egg disappeared from the nest.

The Problem of Image

When we think of endangered species, we almost always visualize a mammal or a bird. In North America, we identify wolves, grizzly bears, various whales, bald eagles, whooping cranes, and similar species as endangered. We rarely think about clams, fish, plants, or insects. Because certain species are able to grab the attention of the public, they are called charismatic species. In addition, most of the charismatic species are carnivores at high trophic levels. Groups of people will organize to save the whales, whooping cranes, elephants, or osprey, but little interest is generated to save the Tar River spiny mussel, cave crayfish, or San Diego fairy shrimp. The graph shows the percentage of species in selected groups that are in various categories of concern in the United States. By far the most vulnerable category of organisms consists of freshwater species of mussels (clams), crayfish, fish, and stoneflies. The least vulnerable are birds and mammals, yet they capture most of the public's interest and are highlighted by the U.S. Fish and Wildlife Service on its endangered species website.

- What factors cause us to rank birds and mammals higher than clams and crayfish?
- Should we spend as much money to save the Tar River spiny mussel as we are spending on wolves or California condors?
- Since money is limited, how would you decide which species to spend limited resources on?

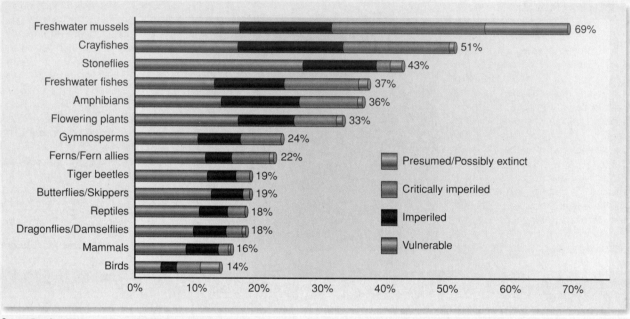

Source: Data from Natureserve, Precious Heritage.

SUMMARY

The loss of biodiversity has become a major concern. Biodiversity can be measured in several ways. Genetic diversity is reduced when populations are reduced in size. Species diversity is reduced when a species becomes extinct from a local population or worldwide. Ecosystems involve the interactions of organisms and their physical environment.

Concern about the loss of biodiversity is based on several ways of looking at the value of a species. Functioning ecosystems and their component organisms provide many valuable services that are often overlooked because they are not easily measured in economic terms. For example, nitrogen-fixing bacteria provide nitrogen for crops, photosynthesis is responsible for food production, and decomposers recycle organic materials. Other biotic resources can be measured in economic terms. Meat and fish harvested from the wild have great economic value, many kinds of drugs are derived from plants and animals, and ivory or animal hides are frequently traded. Many people also consider the loss of biodiversity to be an ethical problem. They ask the question, Do humans have the right to drive other species to extinction?

The primary causes of biodiversity loss are habitat loss by humans converting ecosystems to agriculture and grazing, overexploitation by harvesting species at unsustainable levels, the introduction of exotic species that disrupt ecosystems and compete with or prey on native organisms, purposeful killing of pest organisms and large predators, and climate change.

Protection of biodiversity typically involves two kinds of activities: legal protections by national laws and international agreements, and management of the use of species and ecosystems at sustainable levels. These management activities typically include the establishment of regulations for hunting, fishing, and other uses.

THINKING GREEN

1. Do not plant exotic species as landscape plants. They may become pests or carry diseases that harm native plant species.
2. Never release plants, fish, or other animals into a body of water unless they came out of that body of water.
3. Do not purchase exotic pets. Some become pests while others are harvested in unsustainable ways.
4. Participate in work parties to eliminate exotic, invasive species. Call your local nature center, Audubon club, Nature Conservancy, or similar organization.
5. Look for labels on products that indicate they were produced in a sustainable manner.

WHAT'S YOUR TAKE?

The United Nations stated in 2006 that the target of significantly reducing the rate of loss in biodiversity by 2010 would not be met without strong science and effective governance mechanisms. Some scientists estimate that the current rate of extinction is a thousand times greater than at any other time in the course of humanity's development. The director of the UN Environment Program stated that, "if we fail to demonstrate measurable success by 2010, political commitment will be undermined, public interest will be lost, investment in biodiversity research and management will be reduced, environmental institutions will be further weakened." Do you feel this is an accurate statement, or do you feel it is "alarmist" in tone? Defend your position.

REVIEW QUESTIONS

1. Name three ways humans directly alter ecosystems.
2. Why is the impact of humans greater today than at any time in the past?
3. Describe three factors that influence the genetic diversity of a population.
4. Describe three major causes of the loss of biodiversity.
5. What are the major causes of biodiversity loss in marine ecosystems?
6. List three problems associated with forest exploitation.
7. What is desertification? What causes it?
8. List three key components of the Convention on Biological Diversity (biodiversity treaty).
9. List six techniques utilized by wildlife managers.
10. What special problems are associated with waterfowl management?
11. What is extinction, and why does it occur?
12. List three examples of ecosystem services provided by biological resources.
13. List three actions that can be taken to prevent extinctions.

CRITICAL THINKING QUESTIONS

1. Perhaps 98 percent to 99 percent of all species that have ever existed are extinct. Nearly all went extinct long before humans arrived on the scene. Why should we be concerned about the extinction of organisms today?
2. Would you support clearing of forests and plowing of grasslands that are ecologically important in order to support agriculture in countries that have significant hunger? Where do you draw the line between preserving ecosystems and human interest?
3. A Pacific Northwest Native American tribe has been permitted to hunt an endangered whale species to preserve its traditional culture and uphold its treaty rights. Now there is fear that other countries will hunt the whales, too. Do you think the tribe should be denied its rights? Why or why not?
4. Pharmaceutical companies are helping some developing countries to preserve their rainforests so these companies can look for organisms with possible pharmacological value. How do you feel about these arrangements? What limits, if any, would you place on the pharmaceutical companies? Why?

CHAPTER 12

LAND-USE PLANNING

Land-use planning involves understanding the biological nature of the landscape and the values certain landscapes provide. This view of Yorkshire Dales National Park in England shows that human alterations such as roads, agriculture, and the use of water resources can be developed so that scenic and biological values are preserved.

CHAPTER OUTLINE

The Need for Planning
Historical Forces That Shaped Land Use
  The Rural-to-Urban Shift
  Urbanization in the Developing World
Migration from the Central City to the Suburbs
Factors That Contribute to Sprawl
  Lifestyle Factors
  Economic Factors
  Planning and Policy Factors
Problems Associated with Unplanned Urban Growth
  Transportation Problems
  Air Pollution
  Low Energy Efficiency
  Loss of Sense of Community
  Death of the Central City
  Higher Infrastructure Costs
  Loss of Open Space
  Loss of Farmland
  Water Pollution Problems
  Floodplain Problems
  Wetlands Misuse
  Other Land-Use Considerations
  Land-Use Planning and Aesthetic Pollution
Land-Use Planning Principles
Mechanisms for Implementing Land-Use Plans
  Establishing State or Regional Planning Agencies
  Purchasing Land or Use Rights
  Regulating Use
Special Urban Planning Issues
  Urban Transportation Planning
  Urban Recreation Planning
  Redevelopment of Inner-City Areas
  Smart Growth
Federal Government Land-Use Issues

ISSUES & ANALYSIS
Smart Communities' Success Stories 285

CAMPUS SUSTAINABILITY INITIATIVE
Green Building on Campus 278

GOING GREEN
Green Building 282

WATER CONNECTIONS
Waterways and Development 267

OBJECTIVES

After reading this chapter, you should be able to:

- Explain why most major cities are located on rivers, lakes, or the ocean.
- Describe the forces that result in farmland adjacent to cities being converted to urban uses.
- Explain why floodplains and wetlands are often mismanaged.
- Describe the economic and social values involved in planning for outdoor recreation opportunities.
- Explain why some land must be designated for particular recreational uses, such as wilderness areas, and why that decision sometimes invites disagreement from those who do not wish to use the land in the designated way.
- List the steps in the development and implementation of a land-use plan.
- Describe methods of enforcing compliance with land-use plans.
- Describe the advantages and disadvantages of both local and regional land-use planning.
- Describe the concept of smart growth.

The following additional Case Studies can be found on the book's website at www.mhhe.com/enger12e along with other interesting readings: "Wetlands Loss in Louisiana" and "Farmland Preservation in Pennsylvania."

THE NEED FOR PLANNING

A large proportion of the land surface of the world (about one-third to one-half) has been changed by human activity. Most of this change occurred as people converted the land to agriculture and grazing, but, in our modern world, significant amounts have been covered with buildings, streets, highways, and other products of society. In many cases, cities grew without evaluating and determining the most logical use for the land. Los Angeles and Mexico City have severe air pollution problems because of their geographic location, Venice and New Orleans are threatened by high sea levels, and San Francisco and Tokyo are subject to earthquakes. Currently, most land-use decisions are still based primarily on economic considerations or the short-term needs of a growing population rather than on careful analysis of the capabilities and unique values of the land and landscape. Each piece of land has specific qualities based on its location and physical makeup. Some is valued for the unique species that inhabit it, some is valued for its scenic beauty, and some has outstanding potential for agriculture or urban uses. Since land and the resources it supports (soil, vegetation, elevation, nearness to water, watersheds) are not being created today (except by such phenomena as volcanoes and river deltas), land should be considered a nonrenewable resource.

Once land has been converted from natural ecosystems or agriculture to intensive human use, it is generally unavailable for other purposes. As the population of the world grows, competition for the use of the land will increase, and systematic land-use planning will become more important. Furthermore, as the population of the world becomes more urbanized and cities grow, urban planning becomes critical.

Almost all population growth in the next 30 years will be concentrated in urban areas, and most of this growth will occur in less-developed countries.

HISTORICAL FORCES THAT SHAPED LAND USE

Today, most of the North American continent has been significantly modified by human activity. In the United States, about 47 percent of the land is used for crops and livestock, about 45 percent is forests and natural areas (primarily west of the Mississippi River), and nearly 5 percent is used intensively by people in urban centers and as transportation corridors. Canada is 54 percent forested and wooded and uses only 8 percent of its land for crops and livestock. Less than 1 percent of the land is in urban centers and transportation corridors. A large percentage of its remaining land is wilderness in the north.

This pattern of land use differs greatly from the original conditions experienced by the early European colonists who immigrated to the New World. The first colonists converted only small portions of the original landscape to farming, manufacturing, and housing, but as the population increased, more land was converted to agriculture, and settlements and villages developed into towns and cities. Although most of this early development was not consciously planned, it was not haphazard.

THE RURAL-TO-URBAN SHIFT

North America remained essentially rural until industrial growth began in the last third of the 1800s and the population began a trend toward greater urbanization. (See figure 12.1.) Several forces led to this rural-to-urban transformation. First, the Industrial Revolution led to improvements in agriculture that required less farm labor at the same time industrial jobs became available in the city. Thus, people migrated from the farm to the city. The average person was no longer a farmer but, rather, a factory worker, shopkeeper, or clerk (with a regular paycheck) living in a tenement or tiny apartment near where he or she worked. This pattern of rural-to-urban migration occurred throughout North America, and it is still occurring in developing countries today. A second factor that affected the growth of cities was the influx of immigrants from Europe. Although some became farmers, many of these new citizens settled in towns and cities, where jobs were available. A third reason for the growth of cities was that they offered a greater variety of cultural, social, and artistic opportunities than did rural communities. Thus, cities were attractive for cultural as well as economic reasons. San Antonio, Texas, and Las Vegas, Nevada, are examples of cities that developed around unique cultural attractions.

URBANIZATION IN THE DEVELOPING WORLD

Traditionally, most of the population of the developing world has been rural. However, in recent years, the number of people migrating to the cities has grown rapidly. By 2025, about 5 billion people are expected to be living in urban centers. Most of the increase in urban areas will be in the developing world. Many migrate to the cities because they feel they will have greater access to social services and other cultural benefits than are available in rural areas. Many also feel there are more employment opportunities. However, the increase in the urban population is occurring so rapidly that it is very difficult to provide the services needed by the population, and jobs are not being created as fast as the urban population is growing. Thus, many of the people live in poverty on the fringes of the city in shantytowns that lack water, sewer, and other services. Often these shantytowns are constructed without permission only a short distance from affluent urban dwellers. Because the poor lack safe drinking water and sewer services, they pollute the local water sources, and disease is common. Because they burn wood and other poor-quality fuels in inefficient stoves, air pollution is common. The additional people also create traffic problems of staggering proportions. Table 12.1 shows the seven cities with the largest rate of population increase. All are in the developing world.

MIGRATION FROM THE CENTRAL CITY TO THE SUBURBS

During the early stages of industrial development, there was little control of industry activities, so the waterfront typically became a polluted, unhealthy, undesirable place to live. As roads and rail

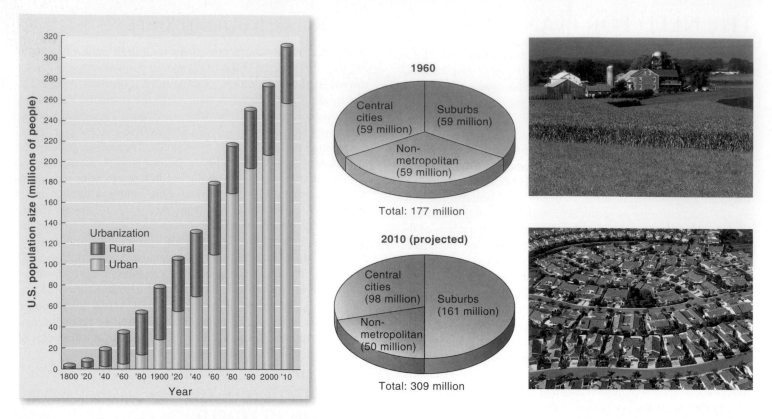

FIGURE 12.1 **Rural-to-Urban Population Shift** In 1800, the United States was essentially a rural country. Industrialization in the late 1800s began the shift to an increasing urban population. In 2010, 84 percent of the U.S. population is projected to be urban or suburban.
Sources: Data from Global Environmental Outlook. UNEP. 2007.

TABLE 12.1 Cities Expected to Grow by More Than 50 Percent Between 1995 and 2015

City	1995 Population (Millions)	Projected 2015 Population (Millions)
Bombay, India	15.14	26.22
Lagos, Nigeria	10.29	24.61
Delhi, India	9.95	16.86
Karachi, Pakistan	9.73	19.38
Metro Manila, Philippines	9.29	14.66
Jakarta, Indonesia	8.62	13.92
Dhaka, Bangladesh	8.55	19.49

Source: Data from *World Resources* 2004–2005.

transport became available, anyone who could afford to do so moved away from the original, industrial city center. The more affluent moved to the outskirts of the city, and the development of suburban metropolitan regions began. Thus, the agricultural land surrounding the towns was converted to housing. Most cities originally had good farmland near them, since the floodplain near rivers typically has a deep, rich soil and agricultural land adjacent to the city was one of the factors that determined whether the city grew or

not. This was true because until land transportation systems became well developed, farms needed to be close to the city so that farmers could transport their produce to the markets in the city. This rich farmland adjacent to the city was ideal for the expansion of the city. As the population of the city grew, demand for land increased. As the price of land in the city rose, people and businesses began to look for cheaper land farther away from the city. Developers and real estate agents were quick to respond and to help people acquire and convert agricultural land to residential or commercial uses. Land was viewed as a commodity to be bought and sold for a profit, rather than as a nonrenewable resource to be managed. As long as money could be made by converting agricultural land to other purposes, it was impossible to prevent such conversion. There were no counteracting forces strong enough to prevent it.

The conversion of land around cities in North America to urban uses destroyed many natural areas that people had long enjoyed. The Sunday drive from the city to the countryside became more difficult as people had to drive farther to escape the ever-growing suburbs. The unique character of neighborhoods and communities was changed by the erection of shopping malls, apartment complexes, and expressways. Most of these alterations occurred without considering how they would affect the biological community or the lives of the people who lived in the area.

As cities continued to grow, certain sections within each city began to deteriorate. Industrial activity continued to be concentrated near water in the city's center. Industrial pollution and urban crowding turned the cores of many cities into undesirable living

Water Connections

Waterways were the primary method of transportation, which allowed exploration and the development of commerce in the early European settlement of North America. Thus, early towns were usually built near rivers, lakes, and oceans. Typically, cities developed as far inland as rivers were navigable. Where abrupt changes in elevation caused waterfalls or rapids, goods being transported by boat or barge needed to be offloaded, transported around the obstruction, and loaded onto other boats. Cities often developed at these points. Buffalo, New York, and Sault Sainte Marie, Ontario, are examples of such cities. In addition to transportation, bodies of water provided drinking water, power, and waste disposal for growing villages and towns. Those towns and villages with access to waterways that provided easy transportation could readily receive raw materials and distribute manufactured goods. Some of these grew into major industrial or trade centers. Without access to water, St. Louis, Montreal, Chicago, Detroit, Vancouver, and other cities would not have developed. The availability of other natural resources, such as minerals, good farmland, or forests was also important in determining where villages and towns were established. Industrial development began on the waterfront, since water supplied transportation, waste disposal, and power. As villages grew into towns and cities, large factories replaced small gristmills, sawmills, and blacksmith shops. The waterfront became a center of intense industrial activity. As industrial activity increased in the cities, people began to move from rural to urban centers for the job opportunities these centers presented.

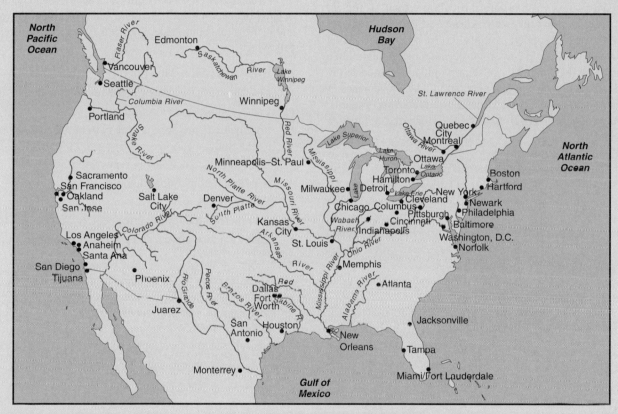

Water and Urban Centers Note that most of the large urban centers are located on water. Water is an important means of transportation and was a major determining factor in the growth of cities. The urban centers shown have populations of 1 million or more (except Edmonton, Winnipeg, and Quebec City).

areas. In the early 1900s, people who could afford to leave began to move to the outskirts. This trend continued after World War II, in the 1940s, 1950s, and 1960s, as a strong economy and government policies that favored new home purchases (tax deductions and low-interest loans) allowed more people to buy homes.

In 1947, William Levitt, a developer from the east coast of the United States, put forth a new plan for building affordable homes away from congested cities. Levitt bought thousands of hectares of potato fields on New York's Long Island. For no money down and $65 a month, a $6900 Cape Cod home could be bought—and many were! Levitt's dream was so successful that he built and sold thousands of houses, completing as many as 36 in a day. Levittown, New York, and the suburban tract home and lifestyle were born—and a new form of "the American dream" took hold.

At first, this form of land planning and building worked. Large numbers of people found themselves homeowners, with a yard where they could plant a garden reminiscent of the farms and country villages from which many of them had come. Land was plentiful, and with cheap oil and labor, it appeared the answer had been found: A person could work in the city and "live in the country." Cars were also cheap, allowing workers to drive to town and be home by sundown.

In 1950, about 60 percent of the urban population lived in the central city; by 1990, this number was reduced to about 30 percent. Most of these new single-family homes were in attractive suburbs, away from the pollution and congestion of the central city. These collections of houses were built on large lots that provided outdoor space for family activities. The blocks of homes were added to the periphery of the city in a decentralized pattern in which single-family homes were separated from multifamily structures and both were some distance from places of work, shopping, and other service needs. This pattern also made it very difficult to establish efficient public transportation networks, which decreased energy efficiency and increased the cost of supplying utility services. However, public transport was not considered important because rising automobile ownership and improved highway systems would accommodate the transportation needs of the suburban population. The convenience of a personal automobile escalated decentralized housing patterns, which, in turn, required better highways, which led to further decentralization.

By 1960, unplanned suburban growth had become known as urban sprawl. **Urban sprawl** is a pattern of unplanned, low-density housing and commercial development outside of cities that usually takes place on previously undeveloped land. In addition, blocks of housing are separated from commercial development, and the streets typically form branching patterns and often include cul-de-sacs. These large housing tracts surrounded cities, which made it difficult for people to find open space. A city dweller could no longer take a bus to the city limits and enjoy the open space of the countryside.

The way we plan the physical layout, or land use, of our communities is fundamental to sustainability. Two main features of our land-use practices over the past several decades have converged to generate haphazard, often inefficient, and unsustainable urban sprawl:

- Zoning ordinances that isolate employment locations, shopping, services, and housing locations from each other.
- Low-density growth planning aimed at creating automobile access to increasing expanses of land.

The complex problems shared by many cities are evidence of the impacts of urban sprawl, such as increasing traffic congestion and commute times, air pollution, inefficient energy consumption and greater reliance on oil, loss of open space and habitat, inequitable distribution of economic resources, and loss of sense of community.

Community sustainability requires a transition from poorly managed sprawl to land-use planning practices that create and maintain efficient infrastructure, ensure close-knit neighborhoods and a sense of community, and preserve natural systems.

Urban sprawl has been defined as auto-dependent development outside of the compact urban and village center, along highways, and in the rural countryside. Sprawl is typically characterized by the following:

- Excessive land consumption
- Low densities in comparison with older centers
- Lack of choice in ways to travel
- Fragmented open space, wide gaps between development, and a scattered appearance
- Lack of choice in housing types and prices
- Separation of uses into distinct areas
- Repetitive one-story development
- Commercial buildings surrounded by large areas for parking
- Lack of public spaces and community centers

Urban sprawl occurs in three ways. (See figure 12.2.) One type of growth involves the development of exclusive, wealthy suburbs adjacent to the city. These homes are usually on large individual lots in the more pleasing geographic areas surrounding the city. Often they are located along water, on elevated sites, or in wooded settings. A second development pattern is tract development.

(a)

(b)

(c)

FIGURE 12.2 **Types of Urban Sprawl** Note the three different types of growth depicted in these photos. (a) The wealthy suburbs with large lots are adjacent to the city. (b) Ribbon sprawl develops as a commercial strip along highways. (c) Tract development results in neighborhoods consisting of large numbers of similar houses on small lots.

Tract development is the construction of similar residential units over large areas. Initially, these tracts are often separated from each other by farmland. New roads are constructed to link new housing to the central city and other suburbs, which stimulates the development of a third form of urban sprawl along transportation routes. This is referred to as **ribbon sprawl** and usually consists of commercial and industrial buildings that line each side of the highway that connects housing areas to the central city and shopping and service areas. Ribbon sprawl results in high costs for the extension of utilities and other public services. It also makes the extent of urbanization seem much larger than it actually is, since the undeveloped land is hidden by the storefronts that face the highway.

As suburbs continued to grow, cities (once separated by farmland) began to merge, and it became difficult to tell where one city ended and another began. This type of growth led to the development of regional cities. Although these cities maintain their individual names, they are really just part of one large urban area called a **megalopolis**. (See figure 12.3.) The eastern seaboard of the United States, from Boston, Massachusetts, to Washington, D.C., is an example of a continuous city. In the midwestern United States, the area from Milwaukee, Wisconsin, to Chicago, Illinois, is a further example. Other examples are London to Dover in England, the Toronto-Mississauga region of Canada, and the southern Florida coast from Miami northward.

In some areas throughout North America, the growth of suburbs has been slowed due to the increased cost of housing and transportation. People have migrated back to some cities on a limited scale because of the lower cost of urban houses and the fact that public transportation is generally more efficient in the city than in the suburbs, thus freeing urban residents from the cost of daily commuting. This reverse migration, however, is still greatly offset by the continual growth of the suburban communities.

FACTORS THAT CONTRIBUTE TO SPRAWL

As we enter the twenty-first century, many, including government policy makers, are beginning to question the prevailing pattern of sprawling urban development. However, powerful social, economic, and policy forces have been behind the development of the current decentralized urban centers typical of North America. If this pattern is to change, these underlying factors must be understood.

Among the causes of sprawl are:

- Public investments in roads, public buildings, water, sewer, and other infrastructure in peripheral areas; disinvestments in existing centers
- Land-use regulations that promote spread-out, land-consumptive development
- Consumer desire for a rural lifestyle with large homes and large yards, a safe environment, and less traffic congestion
- Preference of business and industry for easy highway access, plenty of free parking
- Demands of commercial tenants for particular locations and designs for buildings and sites
- Other public policies, including tax and utility rate policies
- Higher costs of development in older, traditional centers
- Lower land prices in peripheral areas
- Telecommunications advances
- Commercial lending practices that favor suburban development

LIFESTYLE FACTORS

One of the factors that has supported urban sprawl is the relative wealth of the population. This wealth is reflected in material possessions, two of which are automobiles and homes. Based on 2007 census data, in the United States, there are 80 motor vehicles for

FIGURE 12.3   Regional Cities in the United States and Canada The lights in this satellite image show population concentrations. More than 30 major regional cities have developed, with each having more than 1 million people. Many of these cities merge with their neighbors to form huge regional cities. Major urban regions are the northeast coast of the United States (Boston to Washington, D.C.), the region south of the Great Lakes (Chicago to Pittsburgh), south Florida (Jacksonville to Miami), the Toronto and Montreal regions of Canada, and the west coast of California (San Francisco to San Diego).

every 100 people. Since over 20 percent of the population is too young to drive, this means that there is essentially one motor vehicle for every licensed driver in the United States. In addition, about two-thirds of the population live in family-owned homes. With this level of wealth, people are free to choose where and how they want to live. Many are attracted to a lifestyle that includes low-density residential settings, with easy access to open space, isolated from the problems of the city. They are also willing to drive considerable distances to live in these settings. Thus, a decentralized housing pattern is possible because of the high rate of automobile ownership, which allows for ease of movement.

Sprawl affects our quality of life in several ways, including:

- Increased dependency on the auto, fuel consumption, and air pollution
- Increased commuting times and costs
- Reduced opportunity for public transportation services
- Increased health problems in children and adults due to a sedentary lifestyle
- More time in cars and less time for family, friends, and recreation
- Loss of a sense of place and community decline, resulting in fragmented and dispersed communities, and a decline in social interaction and the isolation of some populations, such as the poor and the elderly, in urban areas.

ECONOMIC FACTORS

Several economic forces operate to encourage sprawl development. First of all, it is less expensive to build on agricultural and other nonurban land than it is to build within established cities. The land is less expensive, the regulations and permit requirements are generally less stringent, and there are fewer legal issues to deal with. An analysis of building costs in the San Francisco Bay area of California determined that it costs between 25 percent and 60 percent more to build in the city than it does to build in the suburbs.

Several tax laws also contributed to encouraging home ownership. The interest on home loans is deductible from income taxes, and, in the past, people could avoid paying capital gains taxes on homes they sold if they bought another home of equal or higher value. Since homes usually increased in value, this created a market for increasingly expensive homes.

Sprawl impacts the economy in several ways, including:

- Excessive public costs for roads, utility line extensions, and service delivery to dispersed development
- Decline in economic opportunity in traditional centers
- Premature disinvestments in existing buildings, facilities, and services in urban and village centers
- Relocation of jobs to peripheral areas at some distance from population centers
- Decline in the number of jobs in some sectors, such as retail
- Isolation of employees from activity centers, homes, daycare facilities, and schools
- Reduced ability to finance public services in urban centers

PLANNING AND POLICY FACTORS

Many planning and policy issues have contributed to sprawl development. First of all, until recently, little coordinated effort has been given to planning how development should occur in metropolitan areas. There are several reasons for this. First of all, most metropolitan areas include hundreds of different political jurisdictions (the New York City metropolitan area includes about 700 separate government units and involves the states of New York, New Jersey, and Connecticut). It is very difficult to integrate the activities of these separate jurisdictions in order to achieve coordinated city planning. In addition, it is very difficult for a small local unit of government to see the "big picture," and many are unwilling to give up their autonomy to a regional governmental body.

Local zoning ordinances have often fostered sprawl by prohibiting the mixing of different kinds of land use. Single-family housing, multiple-family housing, commercial, and light and heavy industry were restricted to specific parts of the community. In addition, many ordinances specify minimum lot sizes and house sizes. This tends to result in a decentralized pattern of development that is supported by the heavy use of automobiles and the roads and parking facilities necessary to support them.

In addition, many government policies actually subsidize the development of decentralized cities. For example, developers and the people who buy the homes and businesses they build are able to avoid paying for the full cost of extending services to new areas. The local unit of government picks up the cost of these improvements, and the cost is divided among all the taxpayers rather than just those who will benefit. The roads that are needed to support new developments are usually paid for with federal and state monies, so again the cost is not borne by those who benefit most but by the taxpayers of the state or nation. Furthermore, federal and state governments have not supported public transport in an equivalent way.

Sprawl impacts the environment in several ways, including:

- Fragmentation of open space and wildlife habitat
- Loss of productive farmland and forestland
- Decline in water quality from increased urban runoff, shoreline development, and loss of wetlands
- Inability to capitalize on unique cultural, historic, and public space resources (such as waterfronts) in urban and village centers

PROBLEMS ASSOCIATED WITH UNPLANNED URBAN GROWTH

As the population increased and metropolitan areas grew, several kinds of problems were recognized. They fall into several broad categories.

TRANSPORTATION PROBLEMS

Most cities experience continual problems with transportation. This is primarily because, as cities grew, little thought was given to how people were going to move around and through the city. Furthermore, when housing patterns and commercial sectors changed, transportation mechanisms had to be changed to meet the shifting needs of the public. This often involved the abandonment of old transportation corridors and the establishment of new ones. Paradoxically, the establishment of new transportation corridors stimulates growth in the areas served, and the transportation corridors soon become inadequate. The reliance on the automobile as the primary method of transportation has required the constant building of new highways. A strong automobile-based bias exists in the culture in the United States. In Los Angeles, for example, 70 percent of surface area in the city is dedicated in some way to the automobile (roads, parking garages, etc.), compared with only 5 percent devoted to parks and open space.

According to the U.S. Department of Transportation, the average person in the United States spends nine hours per week traveling in a car. In many metropolitan areas, the amount of time is much greater, and a significant amount of that time involves traffic jams. A study of traffic in the Washington, D.C., area concluded that the average commuter spent 80 hours per year stuck in traffic, in addition to the time normally needed to make the commute. (See figure 12.4.)

The local transportation commission in Ventura County, California (a semirural county with a population of approximately 650,000 located just north and west of Los Angeles County), found that the county will need to spend $1.35 billion over the next eight years to widen roads and build interchanges just to maintain and slightly improve the current level of service. That equals over $168 million per year. This example is not unique to southern California but rather is the norm now for most urban and semiurban areas.

AIR POLLUTION

Reliance on the automobile as the primary method of transportation has resulted in significant air pollution problems in many cities. Most of the large industrial sources of air pollution have been contained. However, the individual car with its single occupant going to work, to shop, or to eat a meal is a constant source of air pollution. A simple solution to this problem is a centralized, efficient public transportation system. However, this is difficult to achieve with a highly dispersed population.

LOW ENERGY EFFICIENCY

Energy efficiency is low for several reasons. First of all, automobiles are the least energy-efficient means of transporting people from one place to another. Second, the separation of blocks of homes from places of business and shopping requires that additional distances be driven. Third, congested traffic routes result in hours being spent in stop-and-go traffic, which wastes much fuel.

FIGURE 12.4 **Traffic and Suburbia** Traffic congestion is a common experience for people who work in cities but live in the suburbs. The popularity of automobiles and the desire to live in the suburbs are closely tied.

Finally, single-family homes require more energy for heating and cooling than multifamily dwellings.

LOSS OF SENSE OF COMMUNITY

Although it is difficult to measure, general agreement exists that in dispersed suburban developments, there is a loss of a sense of community. In many places, people stay within their homes and yards and do not routinely walk through the neighborhood. When they leave the confines of their home, they get in a car and go somewhere. This pattern of behavior reduces human interaction, isolates people from their neighbors, and greatly reduces the sense of community.

DEATH OF THE CENTRAL CITY

Currently, less than 10 percent of people work in the central city. When people leave the city and move to the suburbs, they take their purchasing power and tax payments with them. Therefore, the city has less income to support the services needed by the public. When the quality of services in urban centers drops, the quality of life declines, the flight from the city increases, and a downward spiral of decay begins. An additional problem is the decline of the downtown business district. When shopping malls are built to accommodate the people in the suburbs, the downtown business district declines. Because people no longer need to come to the city to shop, businesses in the city center fail or leave, which deprives the remaining residents of basic services. They must now travel greater distances to satisfy their basic needs.

HIGHER INFRASTRUCTURE COSTS

Infrastructure includes all the physical, social, and economic elements needed to support the population. Whenever a new housing or commercial development occurs on the outskirts of the city, municipal services must be extended to the area. Sewer and water services, natural gas and electric services, schools and police stations, roads and airports—all are needed to support this new population. Extending services to these new areas is much more costly than supplying services to areas already in the city because most of the basic infrastructure is already present in the city.

LOSS OF OPEN SPACE

One of the important features of a pleasing urban landscape is the presence of open space. Open fields, parks, boulevards, and similar land uses allow people to visually escape from the congestion of the city. Unplanned urban growth does not take this important factor into account. Consequently, the buildings must be torn down, and disused spaces must be renovated into parks and other open space at great expense to provide green space in the urban landscape.

LOSS OF FARMLAND

Most of the land that has recently been urbanized was previously used for high-value crops. Land that is flat, well drained, accessible to transportation, and close to cities is ideal farmland. However, it is also prime development land. Areas that once supported crops now support housing developments, shopping centers, and parking lots. (See figure 12.5.) For example, a once agricultural San Jose, California, is now known as a high-tech area. The orange groves that were characteristic of Miami and Los Angeles have long been transformed into suburbs, office complexes, and shopping malls.

Urban development of farmland is proceeding at a rapid pace. Currently, land is being converted to urban uses at a rate of over 400,000 hectares (1 million acres) per year. About one-third of this land is prime agricultural land.

Several states have established programs that provide protection to farmers who do not want to sell their land to developers. The programs may require farmers, in return for lower taxes on the land, to put their land in a conservation easement that prevents them or future owners from using the land for anything other than farming.

WATER POLLUTION PROBLEMS

A large impervious surface area results in high runoff and flash flood potential. A typical shopping mall has a paved parking lot

that is four times larger than the space taken up by the building. The runoff from paved parking lots carries pollutants (oil, coolant, pieces of rubber) into local streams and ponds.

FLOODPLAIN PROBLEMS

Because most cities were established along water, many cities are located in areas called **floodplains.** Floodplains are the low areas near rivers and, thus, are subject to periodic flooding. Some floodplains may flood annually, while others flood less regularly. They are generally flat and so are inviting areas for residential development even though they suffer periodic flooding. A better use of these areas is for open space or recreation, yet developers continue to build houses and light industry in them.

Usually, when a floodplain is developed for residential or commercial use, a retaining wall is built to prevent the periodic flooding

FIGURE 12.5 **Loss of Farmland** As cities grow outward, they eventually grow together to form a regional city. The land between them, once used for farming, becomes developed for residential and commercial purposes. Improved transportation routes and joint facilities (such as airports, shopping centers, and community colleges) hasten this loss of farmland.

272   CHAPTER 12

natural to the area. This boosts the cost of the development, increases the cost of insurance protection, and creates high-water problems downstream. Frequently, tax monies are used to repair the damage that results from the unwise use of floodplains. Floodplain development is one of the insidious problems associated with a rapidly expanding population. As long as the population continues to increase, these less desirable areas are likely to be used for housing, whether they are subject to annual flooding or less frequent damage.

Floods are natural phenomena. Contrary to popular impressions, no evidence supports the premise that floods are worse today than they were 100 or 200 years ago, except perhaps on small, isolated watersheds. There is little doubt, however, that urban flood control methods (walls, barriers, and levees) have had detrimental downstream effects during high-water periods, because preventing the water from spreading into the floodplain forces increased amounts of water to be channeled to areas downstream. What has increased is the economic loss from the flooding. In 1993, during an extensive flooding event that involved both the Mississippi and Missouri Rivers, the U.S. Army Corps of Engineers estimated over $1.5 billion in damage to residential and commercial property. (See figure 12.6.) This loss reflects the fact that more cities, industries, railroads, and highways have been constructed on floodplains. Sometimes there are no alternatives to floodplain development, but too often risks are simply ignored. Because flooding causes loss of life and property, floodplains should no longer be developed for uses other than agriculture and recreation.

Many communities have enacted **floodplain zoning ordinances** to restrict future building in floodplains. Although such ordinances may prevent further economic losses, what happens to individuals who already live in floodplains? Floodplain building ordinances usually allow current residents to remain. Relocation, usually at a financial loss, is the only alternative. Such situations are unfortunate; perhaps proper planning in the future will prevent these problems.

FIGURE 12.6 **Flooding in Floodplain** Urban flood control methods (walls, barriers, and levees) often channel flood waters downstream with catastrophic results. This photo shows some of the flooding that occurred in New Orleans, Louisiana, as a result of Hurricane Katrina.

WETLANDS MISUSE

Since access to water was and continues to be important to industrial development, many cities are located in areas with extensive wetlands. **Wetlands** are areas that periodically are covered with water. They include swamps, tidal marshes, coastal areas, and estuaries. Some wetlands, such as estuaries and marshes, are permanently wet, while others, such as many swamps, have standing water during only part of the year. Many wetlands may have standing water for only a few weeks a year, often in the spring of the year when the snow melts. Because wetlands breed mosquitoes and are sometimes barriers to the free movement of people, they have often been considered useless or harmful. Most of them have been drained, filled, or used as dumps. Many modern cities have completely covered over extensive wetland areas and may even have small streams running under streets, completely enclosed

in concrete. Not including Alaska, the United States has lost about 53 percent of its wetlands from pre-European settlement to the present. That equates to a loss of 89 million to 42 million hectares (220–100 million acres). The current loss rate is about 50,000 hectares (124,000 acres) per year. There are also 68 million hectares (170 million acres) of wetlands in Alaska. These are mostly peatlands, with almost no loss in the past 200 years.

Each kind of wetland has unique qualities and serves as a home to many kinds of plants and animals. Because most wetlands receive constant inputs of nutrients from the water that drains from the surrounding land, they are highly productive and excellent places for aquatic species to grow rapidly. Wetlands are frequently critical to the reproduction of many kinds of animals. Many fish use estuaries and marshes for spawning. Wetlands also provide nesting sites for many kinds of birds and serve as critical habitats for many other species. Waterfowl hunters and commercial and sport fisheries depend on these habitats to produce and protect the young of the species they harvest. Human impact on wetlands has severely degraded or eliminated these spawning and nursery habitats.

Besides providing a necessary habitat for fish and other organisms, wetlands provide natural filters for sediments and runoff. This filtration process allows time for water to be biologically cleaned before it enters larger bodies of water, such as lakes and oceans, and reduces the sediment load carried by runoff. Wetlands also protect shorelines from erosion. When they are destroyed, the natural erosion protection provided by wetlands must be replaced by costly artificial measures, such as breakwalls.

OTHER LAND-USE CONSIDERATIONS

The geologic status of an area must also be considered in land-use decisions. Building cities on the sides of volcanoes or on major earthquake-prone faults has led to much loss of life and property.

Building homes and villages on unstable hillsides or in areas subject to periodic fires is also unwise. Yet, every year more houses slide down California hillsides and wildfires consume homes throughout the dry West.

Another problem in some locations is lack of water. Southern California and metropolitan areas in Arizona must import water to sustain their communities. Wise planning would limit growth to whatever could be sustained by available resources. The risk of serious water shortages in Beijing, China, has forced the government to consider a massive 1200-kilometer (745-mile) diversion of water from the Yangtze River to the city. As the population of cities in such locations continues to grow, the strain on regional water resources will increase.

Water-starved cities often cause land-use dilemmas far from the city boundaries. Supplying water and power to cities often involves the construction of dams that flood valleys that may have significant agricultural, scenic, or cultural value. See chapter 15 on water use for a more extensive development of this topic.

LAND-USE PLANNING AND AESTHETIC POLLUTION

Unpleasant odors, disagreeable tastes, annoying sounds, and offensive sights can be aggravating. Yet it is difficult to get complete agreement on what is acceptable and when some aesthetic boundary has been crossed that is unacceptable. Furthermore, many useful activities generate stimuli that are offensive while the activity itself may be essential or at least very useful. Many of these do not harm us physically but may be harmful from an aesthetic point of view.

Odors are caused by various airborne chemicals. People who live near dairy farms, livestock-raising operations, paper mills, chemical plants, steel mills, and other industries may be offended by the odors originating from these sources, but the products of these activities are needed. Many of these industries discharge wastewater that contains materials that decompose or evaporate and cause odor pollution. People who are constantly exposed to an odor are usually not as offended by it as are people who are newly exposed to it. When an odor is constantly received, the brain ceases to respond to the stimulus. In other words, the person is not aware of the odor.

Chemicals can also affect the taste of things we eat or drink. Some naturally produced chemicals, as well as those discharged into waterways, can affect the taste of food or drinking water. Minute quantities of certain chemicals can affect the taste of our drinking water and our food. Algae in water produce flavors that are offensive to many. Some groundwater sources have sulfur or salts that cause unwanted flavors. In fish, a concentration of 100 ppm of phenol can be tasted. Although this small amount of chemical is not harmful biologically, it does make the fish very unappetizing.

Noise is unwanted sound. It is produced as an incidental by-product of industry, traffic, and other human activity. People vary considerably in their tolerance of unwanted sound. Many in cities adapt to the constant noise of the city, while those who live in more rural areas would find city noises annoying.

Visual pollution is a sight that offends us. This type of pollution is highly subjective and is, therefore, difficult to define or control. To most people, a dilapidated home or building is offensive, especially

FIGURE 12.7 **Aesthetic Pollution** Billboards line the landscape along Interstate Highway 40 in New Mexico.

if located in an area of higher-priced homes. A heavily littered highway or street is aesthetically offensive to most people, and litter along a wilderness trail is even more unacceptable. Some sources of visual pollution are not so clear-cut, however. To many people, roadside billboards are offensive, but they can be helpful to advertisers and to travelers looking for information. (See figure 12.7.)

Since aesthetic pollutants—odors, tastes, sounds, and sights—are extremely difficult to define, it is difficult to establish aesthetic pollution standards. One of the simplest ways to eliminate many of these annoyances is to separate the generator of the offensive stimulus from the general public. Proper land-use planning can greatly reduce the amount of annoying aesthetic pollution. If uses that produce annoying stimuli can be clustered together rather than dispersed, the effect on the public is reduced. It does not make sense to allow new home construction in the vicinity of farming operations that will produce odors or industrial operations or airports that will generate noise. Similarly, allowing homes to be built in areas that have poor-quality groundwater that will be used for drinking is a problem—unless an additional decision is made to provide a different source of drinking water. The roadside advertising that many find offensive can be regulated by allowing such signs only in specific areas rather than allowing them to be built anywhere along the roadway. None of these land-use remedies eliminates the aesthetic pollutant, but by segregating the public from the source of the annoyance, we reduce the impact.

LAND-USE PLANNING PRINCIPLES

Land-use planning is a process of evaluating the needs and wants of the population, the characteristics and values of the land, and various alternative solutions to the use of a particular land surface before changes are made. Planning land use brings with it the need to examine the desires of many competing interests. The economic and personal needs of the population are a central driving force that requires land-use decisions to be made. However, the unique qualities of particular portions of the land surface may prevent some uses, poorly accommodate others, but be highly suitable for still others. For example, the floodplain beside a river is unsuitable for building permanent structures; it can easily accommodate recreational uses such as parks, but it may be most useful as a nature preserve. Agricultural land near cities can be easily converted to housing but may be more valuable for growing fruits and vegetables that are needed by the people of the city. This is particularly true when agricultural land is in short supply near urban centers. To make good land-use decisions, each piece of land must be evaluated and one of several competing uses assigned to it. When land-use decisions are made, the decision process usually involves the public, private landowners, developers, government, and special interest groups. Each interest has special wants and will argue that its desires are most important. Since this is the case, people must have principles and processes to guide land-use decisions, irrespective of their personal wants and interests.

A basic rule should be to make as few changes as possible, but when changes are suggested or required, several things should be considered.

1. *Evaluate and record any unique geologic, geographic, or biologic features of the land.*

   Some land has unique features that should be preserved because of their special value to society. The Grand Canyon, Yellowstone National Park, and many wilderness areas have been set aside to preserve unique physical structures, scenic characteristics, special ecosystems, or unusual organisms. On a more local level, a stream may provide fishing opportunities near a city, or land may have excellent agricultural potential that should take precedence over other uses. New York City purchased and protects a watershed that provides water for the city at a much lower cost than treating water in the Hudson River.

2. *Preserve unique cultural or historic features.*

   Some portions of the landscape, areas within cities, and structures have important cultural, historic, or religious importance that should not be compromised by land-use decisions. In many cities, historic buildings have been preserved and particular sections set aside as historic districts. Sacred sites, many battlefields, and places of unique historic importance are usually protected from development.

3. *Conserve open space and environmental features.*

   It must be recognized that open space and natural areas are not unused, low-value areas. Many studies of human behavior have shown that when people are given choices, they will choose settings that provide a view of nature or that allow one to see into the distance. Some have argued that this is a deep-seated biological need, while others suggest that it is a culturally derived trait. Regardless of its origin, urban planners know that access to open space and natural areas is an important consideration when determining how to use land. Therefore, it makes sense to protect open space within and near centers of population.

4. *Recognize and calculate the cost of additional changes that will be required to accommodate altered land use.*

   Whenever land use is altered, additional modifications will be required to accommodate the change. For example, when a new housing development is constructed, schools and other municipal services will be required, roads will need to be improved, and the former use of the land is lost. Frequently, the cost of these changes is not borne by the developer or the homeowner but becomes the responsibility of the entire community; everyone pays for the additional cost through tax increases. In many areas, basic services, such as provision of water, are severe problems. It makes no sense to build new housing when there is not enough water to support the existing population.

5. *Plan for mixed housing and commercial uses of land in proximity to one another.*

   One of the major problems associated with development in North America is segregation of different kinds of housing from one another and from shopping and other service necessities. Mixing various kinds of uses together (single-family housing, apartments, shopping and other service areas, and offices) allows easier connection between uses without reliance on the automobile. Walking and biking become possible when these different uses are within a short distance of one another.

6. *Plan for a variety of transportation options.*

   Plan for transportation options other than the automobile. Currently, most urban and rural areas do not accommodate bicycles. Although bicycles can legally be ridden on streets, it is generally unsafe because of the high speeds of vehicular traffic and poor road surface conditions near the edge of the pavement. Special bike lanes are available in some areas, but this is rare in much of North America. Walking is a healthy, pleasant way to get from one task to another. However, crossing wide, busy streets is difficult, many areas lack sidewalks, and related service areas are often far apart. This tends to discourage this mode of transport. Clustering housing and service areas allows for easier planning of bus and rail routes to allow people to get from one place to another without relying on automobiles.

7. *Set limits and require managed growth with compact development patterns.*

   Much of unplanned growth occurs because there is no plan or the land-use plan is not enforced. One very effective tool that promotes efficient use of the land is to establish an urban growth limit for a municipality. An **urban growth limit** establishes a boundary within which development can occur. Development outside the boundary is severely restricted. One of the most important outcomes of setting urban growth boundaries is that a great deal of planning must precede the establishment of the limit. This lets all in the community know what is going on and can allow development to occur in logical stages that do not stress the community's ability to supply services. This mechanism also stimulates higher density uses of the urban land.

8. *Encourage development within areas that already have a supportive infrastructure so that duplication of resources is not needed.*

   Because all development of land for human activity requires that services be provided, it makes sense that housing and commercial development occur where the infrastructure is already present. This includes electric, phone, sewer, water, and transportation systems. It includes service industries, such as shopping, banking, restaurants, hotels, and entertainment. It includes schools, hospitals, and police protection. If development occurs far from these services, it is very costly to extend or duplicate those services in the new location. Furthermore, all large cities and most smaller ones have vacant lots or abandoned buildings that have outlived their usefulness or are vacant because of changing business or housing needs. This land is already urban and close to municipal services, and the buildings can be readily renovated or demolished and replaced. Many will argue that this is too expensive, but if there are no options, these spaces will be used, and their redevelopment can be important in revitalizing inner-city spaces.

MECHANISMS FOR IMPLEMENTING LAND-USE PLANS

Land-use planning is the construction of an orderly list of priorities for the use of available land. Developing a plan involves gathering data on current use and geological, biological, and sociological information. From these data, projections are made about what human needs will be. All of the data collected are integrated with the projections, and each parcel of land is evaluated and assigned a best use under the circumstances. There are basically three components that contribute to the successful implementation of a land-use development plan: land-use decisions can be assigned to a regional governmental body, the land or its development rights can be purchased, and laws or ordinances can be used to regulate land use. (See table 12.2.)

ESTABLISHING STATE OR REGIONAL PLANNING AGENCIES

National and regional planning is often more effective than local land-use planning because political boundaries seldom reflect the geological and biological database used in planning. Larger units contain more diverse collections of landscape resources and can afford to hire professional planners. A regional approach is also likely to prevent duplication of facilities and lead to greater efficiency. For example, airport locations should be based on a regional plan that incorporates all local jurisdictions. Three cities

---

TABLE 12.2 Land-Use Planning Strategies

Land-use planning for sustainability requires consideration of a wide spectrum of factors, including transportation, development density, energy efficiency, natural corridors and open space, and growth management. The following strategies are critical components of comprehensive planning to address the complex land-use issues facing our communities.

Transit-Oriented Design

Planning and design strategies for the development of mixed-use, walkable communities sited adjacent to transit access.

Mixed-Use Strategies

Development that promotes the coexistence of many community locales and services within close proximity, to reduce automobile dependency.

Urban Growth Boundaries

A regulatory strategy for limiting urban sprawl by creating a geographical boundary for new development over a period of time.

Infill Development

A strategy to promote greater development density and efficiency within existing urban boundaries.

Greenways

A strategy to preserve open spaces and natural systems, and provide recreation opportunities, by connecting cities, suburbs, and rural areas through linear corridors such as parks and trails.

Brownfield Redevelopment

A strategy for returning idle and often contaminated urban lands referred to as "brownfields" into productive use.

Transfer of Development Rights

A method of exchanging development rights among property developers to increase development density and protect open space and existing land uses.

Open Space Protection

Ways to protect a community's urban open space, farmland, wetlands, riparian lands, rangeland, forests and woodlands, and coastal lands.

Urban Forestry

Planting and maintenance of trees within a city or community as a strategy for reducing both carbon emissions and energy expenditures for heating and cooling.

Land Trusts

Local, regional, or statewide nonprofit organizations directly involved in protecting important land resources over the long term.

Agricultural Land Protection

Strategies for preserving the land that feeds and clothes us, provides open space, food, and habitat for diverse wildlife, and maintains a link to our nation's agricultural heritage.

Solar Access Protection

Regulatory measures to provide legal protection to property owners investing in solar energy systems through solar access ordinances.

Source: U.S. Department of Energy.

only 30 kilometers (20 miles) apart should not build three separate airports when one regional airport could serve their needs better and at a lower cost to the taxpayers.

In North America, the majority of regional governmental bodies are presently voluntary and lack any power to implement programs. Their only role is to advise the member governments. Members of local governments often are unwilling to give up power. They may view policy from a narrow perspective and put their own interests above the goals of the region.

One way to encourage regional planning is to develop policies at the state, provincial, or national level. The first state to develop a comprehensive statewide land-use program was Hawaii. During the early 1960s, much of Hawaii's natural beauty was being destroyed to build houses and apartments for the increasing population. The same land that attracted tourists was being destroyed to provide hotels and supermarkets for them. Local governments had failed to establish and enforce land-use controls. Consequently, in 1961, the Hawaii State Land-Use Commission was founded. This commission designated all land as urban, agricultural, or conservational. Each parcel of land could be used only for its designated purpose. Other uses were allowed only by special permit. To date, the record for Hawaii's action shows that it has been successful in controlling urban growth and preserving the islands' natural beauty, even though the population continues to grow. (See figure 12.8.)

Several states and cities are attempting to follow Hawaii's lead in land-use regulation. (See tables 12.3 and 12.4.) Some have passed legislation dealing with special types of land use. Examples include wetland preservation, floodplain protection, and scenic and historic site preservation. Although direct state involvement in land-use regulation is relatively new, it is expected to grow. Only large, well-financed levels of government can afford to pay for the growing cost of adequate land-use planning. State, provincial, and regional governments are also more likely to have the power to counter the political and economic influences of land developers, lobbyists, and other special-interest groups when conflicts over specific land-use policies arise.

National governments also have a role to play. Since national governments own and administer the use of large amounts of land, national policy will determine how those lands are used. The designation of lands as wilderness, forests, rangelands, or parks at the federal level often involves a balancing of national priorities with local desires. As with local land-use issues, the conflicts over the use of federal lands are often economic and involve compromise.

PURCHASING LAND OR USE RIGHTS

Probably the simplest way to protect desirable lands is to purchase them. When privately owned land is desired for

special purposes, it must be purchased from the owner, and the owner has a right to expect to get a fair price for the property. When it is determined that land has a high public value, then either the land

FIGURE 12.8 **Hawaii Land-Use Plan** Hawaii was the first state to develop a comprehensive land-use plan. This development in an agricultural area was stopped when the plan was implemented in the 1960s.

TABLE 12.3 **Examples of State Land-Use Planning Legislation**

State	Legislation
California	Coastal Act (1976)
	Coastal Zone Conservation Act (1972)
	Tahoe Regional Planning Compact (1969)
Florida	Omnibus Growth Management Act (1985)
	State Comprehensive Plan (1985)
	State and Regional Planning Act (1984)
	Environmental Land and Water Management Act (1972)
Georgia	Coordinated Planning Legislation (1989)
Hawaii	Hawaii State Plan (1978)
	Hawaiian Land Use Law (1961)
Maine	Comprehensive Planning and Land-Use Regulation Act (1988)
Maryland	Economic Growth, Resource Protection and Planning Act (1992)
Massachusetts	Cape Cod Commission Act (1989)
	Martha's Vineyard Commission Act (1974)
New Jersey	State Planning Act (1985)
	State Pinelands Protection Act (1974)
New York	Adirondack Park Agency Act (1971)
Oregon	Land Conservation and Development Act (1973)
Rhode Island	Comprehensive Planning and Land-Use Regulation Act (1988)
Vermont	Growth Management Act (1988)
Washington	Growth Management Act (1990)
	Amendments to the Growth Management Act (1991)

CAMPUS SUSTAINABILITY INITIATIVE

GREEN BUILDING ON CAMPUS

An increasing number of buildings on college and university campuses are being "built green." One example of a new campus green building is the Davis Student Center at the University of Vermont in Burlington, Vermont. The Davis Center is a 186,000 square foot facility and is located on 4.3 acres of space. In its construction the building used 1500 tons of structural steel, 280,000 bricks, and 23,000 tons of concrete. The environmental footprint of the new building was greatly reduced through efforts to obtain Leadership in Energy and Environmental Design (LEED) certification in green design.

During the construction of the Davis Center, efforts were made to use recycled, and locally produced materials whenever possible. For example the building's 62,000 slate shingles were made locally and 65 percent of all materials were manufactured within 500 miles. Ninety-two percent of all construction waste, including 94 percent of the weight of the building that was deconstructed to make way for the new building, was reused or recycled.

Features like the building's green roof, automatic faucets, and computerized energy management and lighting control system are anticipated to result in a 52 percent reduction in electricity usage and 40 percent estimated reduction in water cost. The green roof and radiant heating system will reduce stormwater runoff and help keep road salt pollution of the remaining runoff low. Efforts to reduce the building's environmental footprint will become educational experiences themselves as the monitoring system begins generating and displaying data about environmental technologies at work in the building.

Other features, including ample bicycle racks; showers on every floor for alternative transportation commuters; many south facing windows for maximization of daylight; and benches located throughout the building made from trees harvested on university grounds, promote a green lifestyle.

A similar green building campus plan is under way at the University of Tennessee, Knoxville. Under a policy enacted in 2007 all new buildings constructed on the campus at a cost of more than $5 million and any major renovations would require LEED standards. Policies similar to that of the University of Tennessee are being adopted by many colleges and universities in North America and globally.

TABLE 12.4 Examples of Local Growth Management Actions

Locality	Action
Livermore, California	Established limits on the number of building permits issued
Petaluma, California	Limited the number of building permits to 500 per year and established criteria for awarding permits that emphasized planning
San Diego, California	Adopted a growth management plan
Boulder, Colorado	Established a growth management plan and limited the number of new buildings
Larimer County, Colorado	Established regional urban growth areas
Westminster, Colorado	Limited the number of building permits issued
Boca Raton, Florida	Established a cap on growth at 40,000 dwellings in 1972. The ordinance was eventually overturned by the U.S. Supreme Court.
Broward County, Florida	Public facilities must be developed along with new housing.
Palm Beach County, Florida	Growth control to provide affordable housing
Sanibel, Florida	Established growth limits based on environmental carrying capacity. The city is adjacent to an important wildlife refuge.
Hardin County, Kentucky	Eliminated rural zoning through growth management techniques
Montgomery County, Maryland	Established growth management controls
Minneapolis, Minnesota	Containment of development within city boundaries
Ramapo, New York	Does not permit development until municipal services are established in an area
Austin, Texas	Neighborhood-based conservation overlay districts

or the rights to use it must be purchased. Many environmental organizations purchase lands of special historic, scenic, or environmental value.

In many cases, the owners may not be willing to sell the land but are willing to limit the uses to which the land can be put in the future. Therefore, landowners may sell the right to develop the land or may agree to place restrictions on the uses any future owners might consider.

REGULATING USE

Many communities are not in a financial position to purchase lands; therefore, they attempt to regulate land use by zoning laws.

Zoning is a common type of land-use regulation that restricts the kinds of uses to which land in a specific region can be put. When land is zoned, it is designated for specific potential uses. Common designations are agricultural, commercial, residential, recreational, and industrial. (See figure 12.9.)

Most local zoning boards are elected or appointed and often lack specific training in land-use planning. As a result, zoning regulations are frequently made by people who see only the short-term

FIGURE 12.9 **Zoning** Most communities have a zoning authority that designates areas for particular use. This sign indicates that decisions have been made about the "best" use for the land.

gain and not the possible long-term loss. Often the land is simply zoned so that its current use is sanctioned and is rezoned when another use appears to have a higher short-term value to the community. Even when well-designed land-use plans exist, they are usually modified to encourage local short-term growth rather than to provide for the long-range needs of the community. The public needs to be alert to variances from established land-use plans, because once the plan is compromised, it becomes easier to accept future deviations that may not be in the best interests of the community. Many times, individuals who make zoning decisions are real-estate agents, developers, or local businesspeople. These individuals wield significant local political power and are not always unbiased in their decisions. Concerned citizens must try to combat special interests by attending zoning commission meetings and by participating in the planning process.

SPECIAL URBAN PLANNING ISSUES

Urban areas present a large number of planning issues. Transportation, open space, and improving the quality of life in the inner city are significant problems.

URBAN TRANSPORTATION PLANNING

A growing concern of city governments is to develop comprehensive urban transportation plans. While the specifics of such plans might vary from region to region, urban transportation planning usually involves four major goals:

1. Conserve energy and land resources.
2. Provide efficient and inexpensive transportation within the city, with special attention to people who are unable to drive, such as many elderly, young, handicapped, and financially disadvantaged persons.

3. Provide suburban people opportunities to commute efficiently.
4. Reduce urban pollution.

Any successful urban transportation plan should integrate all of these goals, but funding and intergovernmental cooperation are needed to achieve this. The problems associated with current urban transportation will certainly not disappear overnight, but comprehensive planning is the first step toward solving them.

Since automobiles are heavily used, transportation corridors and parking facilities must be included in any urban transportation plan.

However, many urban planners recognize that the automobile's disadvantages may outweigh its advantages, so some cities, such as Toronto, London, San Francisco, and New York, have attempted to dissuade automobile use by developing mass transit systems and by allowing automobile parking costs to increase substantially.

The major urban mass transit systems are railroads, subways, trolleys, and buses. In many parts of the world, mass transportation is extremely efficient and effective. However, in the United States, where the automobile is the primary method of transportation, mass transportation systems are often underfunded and difficult to establish because mass transit is:

1. Economically feasible only along heavily populated routes
2. Less convenient than the automobile
3. Extremely expensive to build and operate
4. Often crowded and uncomfortable

Although mass transit meets a substantial portion of the urban transportation needs in some parts of the world, such as in Europe, its use in North America has grown slightly, while automobile and air travel have grown tremendously. One reason for this difference is the size of the region. Unlike North America, Japan and countries in Europe that have extensive public transportation are much smaller geographically. In addition, they have more uniformly dense populations. Thus, it is easier to make public transportation work in those settings. (See figure 12.10.) A variety of forces has contributed to this situation. As people became more affluent, they could afford to

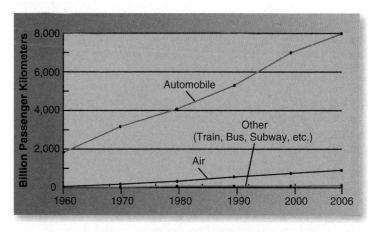

FIGURE 12.10 **Decline of Mass Transportation** Automobile use in the United States has increased consistently since 1960, while rail and bus transport use has remained low.

Source: Data from U.S. Department of Transportation, *National Transportation Statistics* 2008.

own automobiles, which are a convenient, individualized method of transportation. Governments in North America encourage automobile use by financing highways and expressways, by maintaining a cheap energy policy, and by withdrawing support for most forms of mass transportation. Thus, they encourage automobile transportation with hidden subsidies (highway construction and cheap gasoline) but maintain that rail and bus transportation should not be subsidized. North Americans will seek alternatives to private automobile use only when the cost of fuel, the cost of parking, or the inconvenience of driving becomes too high.

URBAN RECREATION PLANNING

About three-fourths of the population of North America live in urban areas. These urban dwellers value open space because it breaks up the sights and sounds of the city and provides a place for recreation. Inadequate land-use planning in the past has rapidly converted urban open spaces to other uses. Until recently, creating a new park within a city was considered an uneconomical use of the land, but people are now beginning to realize the need for parks and open spaces.

Some cities recognized the need for open space a long time ago and allocated land for parks. (See figure 12.11.) London, Toronto, and Perth, Australia, have centrally located and well-used parks. New York City set aside approximately 200 hectares (500 acres) for Central Park in the late 1800s. Boston has developed a park system that provides a variety of urban open spaces. Other cities have not dealt with this need for open space because they have lacked either the foresight or the funding. Recreation is a basic human need. The most primitive tribes and cultures all engaged in games or recreational activities. New forms of recreation are continually being developed. In the congested urban center, cities often must construct special areas where recreation can take place.

A major problem with urban recreation is locating recreational facilities near residential areas. Facilities that are not conveniently located may be infrequently used. For example, the hundreds of thousands of square kilometers of national parks in Alaska and the Yukon will be visited every year by a tiny proportion of the population of North America. Large urban centers are discovering that they must provide adequate, low-cost recreational opportunities within their jurisdiction. Some of these opportunities take the form of commercial establishments, such as bowling centers, amusement parks, and theaters. Others must be subsidized by the community. (See figure 12.12.) The responsibility for playgrounds, organized recreational activities, and open space typically resides in an arm of the municipal government known as the parks and recreation department. Cities spend millions of dollars to develop and maintain recreation programs. Often, conflict arises over the allocation of financial and land resources. These are closely tied because open land is scarce in urban areas, and it is expensive. Riverfront property is ideal for park and recreational use, but it is also prime land for industry, commerce, or high-rise residential buildings. Although conflict is inevitable, many metropolitan areas are beginning to see that recreational resources may be as important as economic growth for maintaining a healthy community.

(a)

(b)

FIGURE 12.11 **Urban Open Space** These photos show open space in two urban areas: (a) Stanley Park in Vancouver, British Columbia, and (b) Grant Park in Chicago, Illinois. If the land had not been set aside, it would have been developed.

An outgrowth of the trend toward urbanization is the development of **nature centers.** In many urban areas, so little natural area is left that the people who live there need to be given opportunities to learn about nature. Nature centers are basically teaching institutions that provide a variety of ways for people to learn about and appreciate the natural world. Zoos, botanical gardens, and some urban parks, combined with interpretative centers, also provide recreational experiences. Nature centers are usually located near urban centers, in places where some appreciation of the natural processes and phenomena can be developed. They may be operated by municipal governments, school systems, or other nonprofit organizations.

REDEVELOPMENT OF INNER-CITY AREAS

As people moved to the suburbs during the past 50 years, the inner city was abandoned. Many old industrial sites sit vacant. Businesses have moved to the suburban malls. The quality of housing has declined. Services have been reduced. To improve the quality of life of the residents of the city, special efforts must be made to revitalize the city. Although activities that will improve the quality of life in the inner city vary from city to city, several land-use processes can help. One problem that has plagued industrial cities is vacant industrial and commercial sites. Many of these buildings have remained vacant because the cost of cleanup and renovation is expensive. Such sites have been called **brownfields.** Many of these sites involved environmental contamination, and since the EPA required that they be cleaned up to a pristine condition, no one was willing to do so.

A new approach to utilizing these sites is called **brownfields development.** This involves a more realistic approach to dealing with the contamination at these sites. Instead of requiring complete cleanup, the degree of cleanup required is matched to the intended use of the site. Although an old industrial site with specific contamination problems may not be suitable for housing, it may be redeveloped as a new industrial site, since access to the contamination can be controlled. An old industrial site that has soil contamination may be paved to provide parking.

Another important focus in urban redevelopment is the remodeling of abandoned commercial buildings for shopping centers, cultural facilities, or high-density housing. Chattanooga, Tennessee, has achieved a reputation for revitalizing its inner-city area. The process of revitalization involved extensive planning activities that included the public, public and private funding of redevelopment activities, establishment of an electric bus system to alleviate air pollution, renovation of existing housing, redevelopment of old warehouses into a shopping center, and incorporation of a condemned bridge over the Tennessee River into a portion of a park that is also an important pedestrian connection between a residential area and the downtown business district.

SMART GROWTH

The current pattern of growth throughout most of North America is commonly referred to as "sprawl." Concern about sprawl is growing because of the sheer pace of land development, which, according to the U.S. Department of Agriculture, is roughly double what it was only a decade ago. Such development has had a number of negative cultural, economic, environmental, and social consequences. In central cities and older suburbs, these include deteriorating infrastructure, poor schools, and a shortage of affordable, quality housing. In newer suburban areas, these problems may include increased traffic congestion, declining air quality, and the loss of open space. Smart growth is an approach that argues that these problems are two sides

FIGURE 12.12   **Urban Recreation** In urban areas, recreation often takes the form of sports programs, playgrounds, and walking. Most cities recognize the need for such activities and facilities and develop extensive recreation programs for their citizens.

of the same coin and that the neglect of our central cities is fueling the growth and related problems of the suburbs.

To address these problems, smart growth advocates emphasize the concept of developing "livable" cities and towns. Livability suggests, among other things, that the quality of our built environment and how well we preserve the natural environment both directly affect our quality of life.

One aspect of smart growth is the building of "green buildings." Green buildings have been built over the past few years using a standard called Leadership in Energy and Environmental Design (LEED). Introduced in 2000 by the U.S. Green Building Council, the guidelines call for using recycled materials, ensuring better ventilation in buildings, and reducing water and energy use, among other goals.

Though they support growth, communities are questioning the economic costs of abandoning infrastructure in the city and rebuilding it farther out. They are questioning the necessity of spending increasing time in cars, locked in traffic or traveling miles to the nearest store. They are questioning the practice of abandoning older communities while developing open space and prime agricultural lands at the suburban fringe.

Smart growth recognizes the benefits of growth. It invests time, attention, and resources in restoring a sense of community and vitality to center cities and older suburbs. In new developments, smart growth practices create communities that are more town-centered, are transit- and pedestrian-oriented, and have a greater mix of housing, commercial, and retail uses. These practices also preserve open space and other environmental amenities. Smart growth recognizes connections between development and quality of life that can be presented as a list of principles.

Smart growth principles

1. Mix land uses.
2. Take advantage of compact building design.
3. Create a range of housing opportunities and choices.

Land-use conflicts also arise between business interests and recreational users of public lands. Federal and state governments give special use permits to certain users of public lands. Many ski resorts in the West make use of public lands. Grazing is also an important use of federal land. Based on "Animal Use Months" established by the Bureau of Land Management or the Forest Service, ranchers are allowed to graze cattle on certain public lands. Technically, failure to comply can mean a loss of grazing rights. However, since the establishment of regulations is highly political, many maintain that the political influence of ranchers allows them to use a public resource without adequately compensating the government. In addition, the regulatory agencies are understaffed and find it difficult to adequately regulate the actions of individual ranchers. As a result, some lands are overgrazed. Many people who want to use the publicly owned rangelands for outdoor recreation resent the control exercised by grazing interests. On the other hand, ranchers resent the intrusion of hikers and campers on land they have traditionally controlled.

An obvious solution to this problem is to allocate land to specific uses and to regulate the use once allocations have been made. Several U.S. governmental agencies, such as the National Park Service, the Bureau of Land Management, the Forest Service, and the Fish and Wildlife Service, allocate and regulate the lands they control. However, these agencies have conflicting roles. The Forest Service has a mandate to manage forested public lands for timber production. This mandate often comes into conflict with recreational uses. Similarly, the Bureau of Land Management has huge tracts of land that can be used for recreation, but it traditionally has been mandated to manage grazing rights.

A particularly sensitive issue is the designation of certain lands as wilderness areas. Obviously, if an area is to be wilderness, human activity must be severely restricted. This means that the vast majority of Americans will never see or make use of it. Many people argue that this is unfair because they are paying taxes to provide recreation for a select few. Others argue that if everyone were to use these areas, their charm and unique character would be destroyed and that, therefore, the cost of preserving wilderness is justifiable.

Areas designated as wilderness make up a very small proportion of the total public land available for recreation. (See figure 12.14.) Such areas, however, are becoming increasingly visible due to their unique character.

FIGURE 12.13   Conflict over Recreational Use of Land The people who use land for motorized and nonmotorized activities are often antagonists over the allocation of land for recreational use.

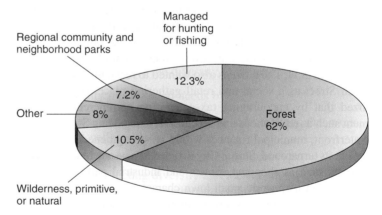

FIGURE 12.14   U.S. Federal Recreational Lands Of the approximately 108 million hectares (267 million acres) of federal recreational lands in the United States, approximately 10 percent is designated as wilderness, primitive, or natural.

Smart Communities' Success Stories

Cool Cities—Michigan

The Cool Cities initiative launched by the Michigan governor is an urban redevelopment endeavor aimed at attracting creative young workers to the state's cities. A series of forums throughout the state—and an on-line survey—collected input on ways to revitalize cities, strengthen the economy, and make Michigan a magnet state for new job creation and business expansion.

Learning Corridor—Hartford, Connecticut

Trinity College helped turn around the decline of its neighborhood by building a $175 million "Learning Corridor," according to a piece for the Elm Street Writers Group. The development combines a neighborhood school, job training facility, playing fields, and open space to revitalize what had been an urban area in serious decline. The "Learning Corridor" has become a model for other communities that want to avoid "school sprawl" that pushes neighborhood schools to suburbs accessible only by car.

Civano—Tucson, Arizona

The 700-hectare traditional neighborhood development is designed to promote economic growth while maintaining important social values and ecological harmony. Civano buildings use the best available technology to reduce energy and water demands. The project also focuses on land-use planning and sustainable transportation design.

Stapleton Development Corporation—Boulder, Colorado

This is the redevelopment of the former Stapleton International Airport site, near Denver. The project involves constructing a community of urban villages, employment centers, and greenways that focus on sustainability, environmental preservation, and economic and social development.

Jackson Meadow—Marine on St. Croix, Minnesota

This new 64-home residential development focuses on ecological land use. Jackson Meadow has a "commitment to create a sustainable environment that respects the unique nature of this special place." The development is combined with more than 100 hectares of open space for people and wildlife and includes a communally constructed wetlands and natural ponds for water filtration and runoff.

Grand Valley Metropolitan Council—Grand Rapids, Michigan

Grand Valley Metropolitan Council (GVMC) is an alliance of governmental units in the Grand Rapids, Michigan, metropolitan area that are appointed to plan for growth and development, improve the quality of community life, and coordinate governmental services. Through joint planning they coordinate land-use planning with transportation and other systems.

The Vail Environmental Strategic plan—Vail, Colorado

This plan describes a program that was adopted to maintain and improve environmental quality in the Vail Valley and to ensure the prolonged economic health of the region. Efforts include monitoring and improving air and water quality, preserving open space, and protecting the area's natural wildlife.

Community Greens—Arlington, Virginia

This concept promotes more urbane, livable cities through the creation of parks that are collectively owned and managed by the neighbors whose homes and backyards, decks, patios, and balconies enclose the green. According to Community Greens, such parks have multiple benefits: "accessible and safe play spaces for children, community building, increased safety and security and property values, and a prime antidote to sprawl by making city living more attractive, especially to families with children."

Would your community be considered a "smart community"?

Can you think of innovative ways to help your community grow smarter?

Would you be willing to submit your ideas to your local planning agency?

SUMMARY

Historically, waterways served as transportation corridors that allowed for the exploration of new land and for the transport of goods. Therefore, most large urban centers began as small towns located near water. Water served the needs of the towns in many ways, especially as transportation. Several factors resulted in the shift of the population from rural to urban. These included the Industrial Revolution, which provided jobs in cities, and the addition of foreign immigrants to the cities. As towns became larger, the farmland surrounding them became suburbs surrounding industrial centers. Unregulated industrial development in cities resulted in the degradation of the waterfront and stimulated the development of suburbs around the city as people sought better places to live and had the money to purchase new homes. The rise in automobile ownership further stimulated the movement of people from the cities to the suburbs.

Many problems have resulted from unplanned growth. Current taxation policies encourage the residential development of farmland, which results in a loss of valuable agricultural land. Floodplains and wetlands are often mismanaged. Loss of property and life results when people build on floodplains. Wetlands protect our shorelines and provide a natural habitat for fish and wildlife. Transportation problems and lack of open space are also typical in many large metropolitan areas.

Land-use planning involves gathering data, projecting needs, and developing mechanisms for implementing the plan. Good land-use planning should include assessment of the unique geologic, geographic, biological, and historic and cultural features of the land; the costs of providing additional infrastructure; preservation of open space; provision for a variety of transportation options; a mixture of housing and service establishments; redevelopment of disused urban land; and establishment of urban growth limits. Establishing regional planning agencies, purchasing land or its development rights, and enacting zoning ordinances are ways to implement land-use planning. The scale of local planning is often not large enough to be effective because problems may not be confined to political boundaries. Regional planning units can afford professional planners and are better able to withstand political and economic pressures. A growing concern of urban governments is to develop comprehensive urban transportation plans that seek to conserve energy and land resources, provide efficient and inexpensive transportation and commuting, and help to reduce urban pollution. Urban areas must also provide recreational opportunities for their residents and seek ways to rebuild decaying inner cities.

Federal governments own and manage large amounts of land; therefore, national policy must be developed. This usually involves designating land for particular purposes, such as timber production, grazing land, parks, or wilderness. The recreational use of public land often requires the establishment of rules that prevent conflict between potential users who have different ideas about what appropriate uses should be. Often federal policy is a compromise between competing uses and land is managed for multiple uses.

THINKING GREEN

1. Visit your city's website and locate its land-use planning page. What are the city's major rules associated with land use?
2. As you travel to class, observe changes that are occurring to land use. What is being lost with the change? What is being gained from the change?
3. Attend a local meeting of the planning or zoning board.
4. Be an advocate and user of public transportation in your community.
5. Become active in an organization that is promoting sustainability in your community.

WHAT'S YOUR TAKE?

In a 2006 article in the *Washington Post,* the architect and professor of architecture Roger Lewis addressed the rebuilding of New Orleans following Hurricane Katrina. "Why," he asks, "do we stubbornly refuse to acknowledge that there are places on the earth's surface—wetlands and floodplains, seismically active regions, arid deserts, steep hillsides and cliffs—where erecting cities endangers not only humans, but also the natural environment?" Develop a position paper that either supports or refutes Lewis's question as it pertains to the rebuilding of New Orleans.

REVIEW QUESTIONS

1. Why did urban centers develop near waterways? Are they still located near water?
2. Describe the typical changes that have occurred in cities from the time they were first founded until now.
3. Why do people move to the suburbs?
4. Why do some farmers near urban areas sell their land for residential or commercial development? If you were in this position, would you sell?
5. What is a megalopolis?
6. What land uses are suitable on floodplains?
7. What is multiple land use? Can land be used for multiple purposes?
8. Why is it important to provide recreational space in urban planning?
9. How can recreational activities damage the environment? Do you engage in any of those activities?
10. What is the monetary impact of recreational activities?
11. What are some strictly urban-related recreational activities?
12. List some conflicts that arise when an area is designated strictly as wilderness.
13. Describe the steps necessary to develop a land-use plan.
14. What are the advantages of regional or state planning?
15. List three benefits of land-use planning.
16. Describe the concept of smart growth.

1. Choose the city where you live. Interview local residents and look at old city maps. What did the city look like 75 years ago? What were the city's boundaries? Where did people do their shopping? How did they get around? How does this compare with the current situation in the city?

2. What historical factors brought members of your family to the city? How does this compare to the factors that are currently contributing to the growth of cities in the developing world?

3. Consider the outer rim of the city closest to you. Which, if any, of the problems associated with unplanned growth are associated with your city? What factors make them a problem? What do you think can be done about them?

4. There has been tremendous development in the arid West of the United States over the past few decades, creating demands for water. How should these demands be met? Should there be limits to this type of development? What kinds of limits, if any?

5. Imagine you are a U.S. Forest Service supervisor who is creating a 10-year plan that is in the public comment stage. What interests would be contacting you? What power would each interest have? How would you manage the competing interests of timber, mining, grazing, and recreation or those between motorized and nonmotorized recreation? What values, beliefs, and perspectives help you to form your recommendations?

6. Imagine that you lived in an area of the country that has the potential to be named a wilderness area. What conflicts do you think would arise from such a declaration? Who might be some of the antagonists? Which perspective do you think is most persuasive? How would you answer the objections of the other perspective?

7. After reading the Case Study in this chapter concerning wetland loss in Louisiana, what kinds of recommendations would you make to help preserve wetlands? What do you suppose might happen if nothing is done? What resistance might wetland preservation generate?

CHAPTER 13

SOIL AND ITS USES

Soil is a resource that supports plants, which are the base of most food chains. We should think of soil as a valuable resource that should be preserved.

CHAPTER OUTLINE

Geologic Processes
Soil and Land
Soil Formation
Soil Properties
Soil Profile
Soil Erosion
Soil Conservation Practices
    Soil Quality Management Components
    Contour Farming
    Strip Farming
    Terracing
    Waterways
    Windbreaks
Conventional Versus Conservation Tillage
Protecting Soil on Nonfarm Land

ISSUES & ANALYSIS
Soil Fertility and Hunger in Africa 309

CASE STUDY
Land Capability Classes 308

CAMPUS SUSTAINABILITY INITIATIVE
Composting on Campus 295

GOING GREEN
Green Landscaping 304

WATER CONNECTIONS
Water and Erosion 301

OBJECTIVES

After reading this chapter, you should be able to:

- Describe the geologic processes that build and erode the Earth's surface.
- List the physical, chemical, and biological factors involved in soil formation.
- Explain the importance of humus to soil fertility.
- Differentiate between soil texture and soil structure.
- Explain how texture and structure influence soil atmosphere and soil water.
- Explain the role of living organisms in soil formation and fertility.
- Describe the various layers in a soil profile.
- Describe the processes of soil erosion by water and wind.
- Explain how contour farming, strip farming, terracing, waterways, windbreaks, and conservation tillage reduce soil erosion.
- Understand that the misuse of soil reduces soil fertility, pollutes streams, and requires expensive remedial measures.
- Explain how land not suited for cultivation may still be productively used for other purposes.

A Global Perspective on "Worldwide Soil Degradation" and Case Studies on "Desertification and Global Security" and "The Conservation Security Program" can be found on the book's website at www.mhhe.com/enger12e along with other interesting readings.

GEOLOGIC PROCESSES

We tend to think of the Earth as being stable and unchanging until we recognize such events as earthquakes, volcanic eruptions, floods, and windstorms changing the surface of the places we live. There are forces that build new land and opposing forces that tear it down. Much of the building process involves shifting of large portions of the Earth's surface known as plates.

The Earth is composed of an outer crust, a thick layer of plastic mantle, and a central core. The **crust** is an extremely thin, less dense, solid covering over the underlying mantle. The **mantle** is a layer that makes up the majority of the Earth and surrounds a small core made up primarily of iron. The outermost portion of the mantle, adjacent to the crust, is solid. Collectively, the crust and solid outer mantle are known as the **lithosphere.** Just below the outer mantle is a thin layer known as the **asthenosphere,** which is capable of plastic flow. Below the asthenosphere, the mantle is more solid. The core consists primarily of iron and nickel and has a solid center and a liquid outer region. (See figure 13.1.)

Plate tectonics is the concept that the outer surface of the Earth consists of large plates composed of the crust and the outer portion of the mantle and that these plates are slowly moving over the surface of the liquid outer mantle. The heat from the Earth causes slow movements of the outer layer of the mantle similar to what happens when you heat a liquid on the stove, only much slower. The movements of the plates on this plastic outer layer of the mantle are independent of each other. Therefore, some of the plates are pulling apart from one another, while others are colliding.

Where the plates are pulling apart from one another, the liquid mantle moves upward to fill the gap and solidifies. Thus, new crust is formed from the liquid mantle. Approximately half of the surface of the Earth has been formed in this way in the past 200 million years. The bottom of the Atlantic and Pacific Oceans and the Rift Valley and Red Sea area of Africa are areas where this is occurring.

If plates are pulling apart on one portion of the Earth, they must be colliding elsewhere. Where plates collide, several things can happen. (See figure 13.2.) Often, one of the plates slides under the other and is melted. Often, when this occurs, some of the liquid mantle makes its way to the surface and volcanoes are formed, which results in the formation of mountains. The west coasts of North and South America have many volcanoes and mountain ranges where the two plates are colliding. The volcanic activity adds new material to the crust. When a collision occurs between

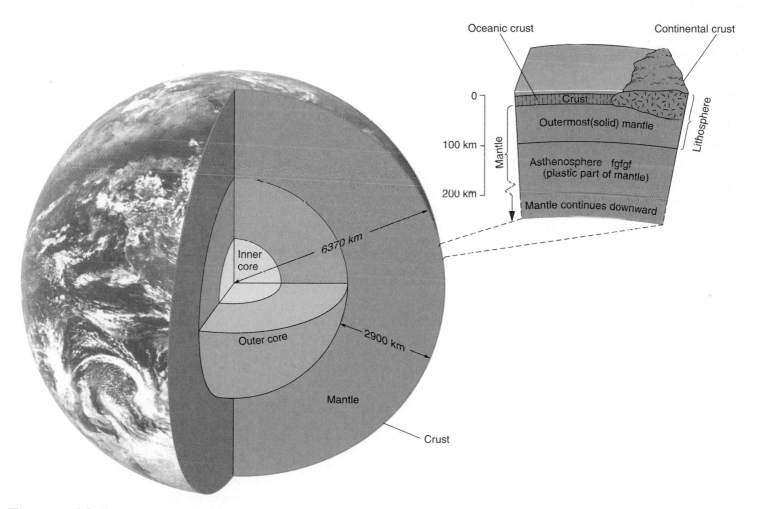

FIGURE 13.1 **Structure of the Earth** The Earth has a solid outer lithosphere that floats on a plastic mantle.

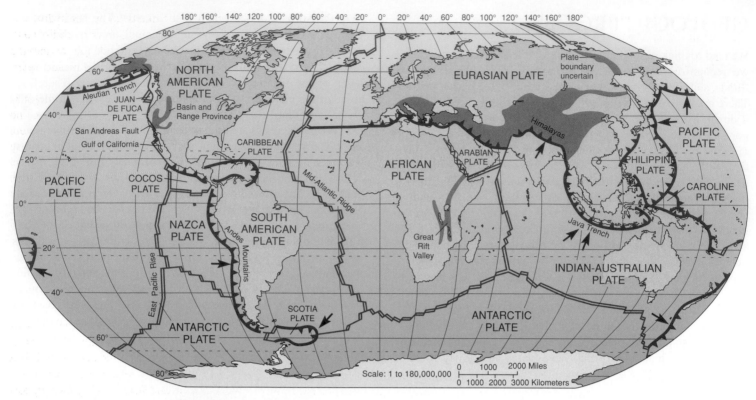

FIGURE 13.2 **Tectonic Plates** The plates that make up the outer surface of the Earth move with respect to one another. They pull apart in some parts of the world and collide in other parts of the world.

two plates under the ocean, the volcanoes may eventually reach the surface and form a chain of volcanic islands, such as can be seen in the Aleutian Islands and many of the Caribbean Islands. When two continental plates collide, neither plate slides under the other and the crust buckles to form mountains. The Himalayan, Alp, and Appalachian mountain ranges are thought to have formed from the collision of two continental plates. All of these movements of the Earth's surface are associated with earthquakes. The movements of the plates are not slow and steady sliding movements but tend to occur in small jumps. However, what is a small movement between two plates on the Earth is a huge movement for the relatively small structures and buildings produced by humans, so these small movements can cause tremendous amounts of damage.

These building processes are counteracted by processes that tend to make the elevated surfaces lower. Gravity provides a force that tends to wear down the high places. Moving water and ice (glaciers) and wind assist in the process; however, their effectiveness is related to the size of the rock particles. Several kinds of **weathering** processes are important in reducing the size of particles that can then be dislodged by moving water and air. **Mechanical weathering** results from physical forces that reduce the size of rock particles without changing the chemical nature of the rock. Common causes of mechanical weathering are changes in temperature that tend to result in fractures in rock, the freezing of water into ice that expands and tends to split larger pieces of rock into smaller ones, and the actions of plants and animals.

Because rock does not expand evenly, heating a large rock can cause it to fracture, so that pieces of the rock flake off. These pieces can be further reduced in size by other processes, such as the repeated freezing of water and thawing of ice. Water that has seeped into rock cracks and crevices expands as it freezes, causing the cracks to widen. Subsequent thawing allows more water to fill the widened cracks, which are enlarged further by another period of freezing. Alternating freezing and thawing breaks large rock pieces into smaller ones. (See figure 13.3.) The roots of plants growing in cracks can also exert enough force to break rock.

The physical breakdown of rock is also caused by forces that move and rub rock particles against each other (abrasion). For example, a glacier causes rock particles to grind against one another, resulting in smaller fragments and smoother surfaces. These particles are deposited by the glacier when the ice melts. In many parts of the world, the parent material from which soil is formed consists of glacial deposits. Wind and moving water also cause small particles to collide, resulting in further weathering. The smoothness of rocks and pebbles in a stream or on the shore is evidence that moving water has caused them to rub together, removing their sharp edges. Similarly, particles carried by wind collide with objects, fragmenting both the objects and the wind-driven particles.

Wind and moving water also remove small particles and deposit them at new locations, exposing new surfaces to the weathering process. For example, the landscape of the Grand Canyon in the southwest United States was created by a combination of wind and moving water that removed easily transported particles, while rocks more resistant to weathering remained. (See figure 13.4.)

The activities of organisms can also assist mechanical weathering. The roots of plants can exert considerable force and move particles apart from one another. The burrows of animals expose new surfaces that can be altered by freezing and thawing.

Chemical weathering involves the chemical alteration of the rock in such a manner that it is more likely to fragment or to be dissolved. Some small rock fragments exposed to the atmosphere may be oxidized; that is, they combine with oxygen from the air and chemically change to different compounds. Other kinds of rock may combine with water molecules in a process known as hydrolysis. Often, the oxidized or hydrolyzed molecules are more readily soluble in water and, therefore, may be removed by rain or moving water. Rain is normally slightly acidic, and the acid content helps dissolve rocks.

Because of gravity, the prevailing movement of particles is from high elevations to lower ones. This process of loosening and redistributing particles is known as **erosion.** Wind can move sand and dust and can cause the wearing away of rocky surfaces by sandblasting their surfaces. Glaciers can move large rocks and cause their surfaces to be rounded by being rubbed against each other and the surface of the Earth. Moving water transports much material in streams and rivers. In addition, wave action along the shores of lakes and the coasts of oceans constantly wears away and transports particles.

SOIL AND LAND

The geologic processes just discussed are involved in the development of both soil and land; however, soil and land are not the same. **Land** is the part of the world not covered by the oceans. **Soil** is a thin covering over the land consisting of a mixture of minerals, organic material, living organisms, air, and water that together support the growth of plant life. The proportions of the soil components vary with different types of soils, but a typical, "good" agricultural soil is about 45 percent mineral, 25 percent air, 25 percent water, and 5 percent organic matter. (See figure 13.5.) This combination provides good drainage, aeration, and organic matter. Farmers are

FIGURE 13.3 **Physical Fragmentation by Freezing and Thawing** The crack in the rock fills with water. As the water freezes and becomes ice, it expands. The pressure of the ice enlarges the crack. The ice melts, and water again fills the crack. The water freezes again and widens the crack. Alternate freezing and thawing splits the rock into smaller fragments.

FIGURE 13.4 **An Eroded Landscape** This landscape in Grand Canyon National Park was created by the action of wind and moving water. The particles removed by these forces were deposited elsewhere and may have become part of the soil in that new location.

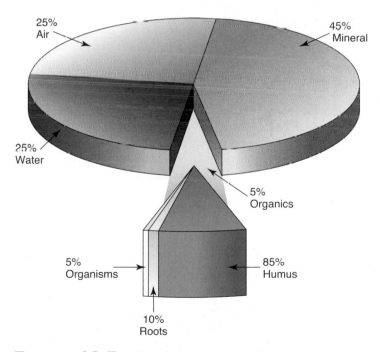

FIGURE 13.5 **The Components of Soil** Although soils vary considerably in composition, they all contain the same basic components: mineral material, air, water, and organic material. The organic material can be further subdivided into humus, roots, and other living organisms. The percentages shown are those that would be present in a good soil.

particularly concerned with soil because the nature of the soil determines the kinds of crops that can be grown and which farming methods must be employed. Urban dwellers should also be concerned about soil because its health determines the quality and quantity of food they will eat. If the soil is so abused that it can no longer grow crops or if it is allowed to erode, degrading air and water quality, both urban and rural residents suffer. To understand how soil can be protected, we must first understand its properties and how it is formed.

SOIL FORMATION

A combination of physical, chemical, and biological events acting over time is responsible for the formation of soil. Soil building begins with the fragmentation of the **parent material,** which consists of ancient layers of rock or more recent geologic deposits from lava flows or glacial activity. The kind and amount of soil developed depend on the kind of parent material present, the plants and animals present, the climate, the time involved, and the slope of the land. As was discussed earlier, the breakdown of parent material is known as weathering. The climate and chemical nature of the rock material greatly influence the rate of weathering. Similarly, the size and chemical nature of the particles have a great impact on the nature of the soil that will develop in an area.

The role of organisms in the development of soil is also very important. The first organisms to gain a foothold in this modified parent material also contribute to soil formation. Lichens often form a pioneer community that grows on the surface of rocks and traps small particles. The decomposition of dead lichens and other organic matter releases acids that chemically alter the underlying rock, causing further fragmentation. The release of chemicals from the roots of plants causes further chemical breakdown of rock particles. As other kinds of organisms, such as plants and small animals, become established, they contribute, through their death and decay, increasing amounts of organic matter, which are incorporated with the small rock fragments.

The organic material resulting from the decay of plant and animal remains is known as **humus.** It is a very important soil component that accumulates on the surface and ultimately becomes mixed with the top layers of mineral particles. This material contains nutrients that are taken up by plants from the soil. Humus also increases the water-holding capacity and the acidity of the soil so that inorganic nutrients, which are more soluble under acidic conditions, become available to plants. Humus also tends to stick other soil particles together and helps to create a loose, crumbly soil that allows water to soak in and permits air to be incorporated into the soil. Compact soils have few pore spaces, so they are poorly aerated, and water has difficulty penetrating, so it runs off.

Burrowing animals, soil bacteria, fungi, and the roots of plants are also part of the biological process of soil formation. One of the most important burrowing animals is the earthworm. One hectare (2.47 acres) of soil may support a population of 500,000 earthworms that can process as much as 9 metric tons of soil a year. These animals literally eat their way through the soil, resulting in further mixing of organic and inorganic material, which increases the amount of nutrients available for plant use. They often bring nutrients from the deeper layers of the soil up into the area where plant roots are more concentrated, thus improving the soil's fertility. Soil aeration and drainage are also improved by the burrowing of earthworms and other small soil animals, such as nematodes, mites, pill bugs, and tiny insects. Small soil animals also help to incorporate organic matter into the soil by collecting dead organic material from the surface and transporting it into burrows and tunnels. When the roots of plants die and decay, they release organic matter and nutrients into the soil and provide channels for water and air.

Fungi and bacteria are decomposers and serve as important links in many mineral cycles. (See chapter 5.) They, along with animals, improve the quality of the soil by breaking down organic material to smaller particles and releasing nutrients.

The position on the slope also influences soil development. Soil formation on steep slopes is very slow because materials tend to be moved downslope with wind and water. Conversely, river valleys often have deep soils because they receive materials from elsewhere by these same erosive forces.

Climate and time are also important in the development of soils. In general, extremely dry or cold climates develop soils very slowly, while humid and warm climates develop them more rapidly. Cold and dry climates have slow rates of accumulation of organic matter needed to form soil. Furthermore, chemical weathering proceeds more slowly at lower temperatures and in the absence of water. Under ideal climatic conditions, soft parent material may develop into a centimeter (less than ½ inch) of soil within 15 years. Under poor climatic conditions, a hard parent material may require hundreds of years to develop into that much soil. In any case, soil formation is a slow process. (See table 13.1.)

The amount of rainfall and the amount of organic matter influence the pH of the soil. In regions of high rainfall, basic ions such as calcium, magnesium, and potassium are leached from the soils, and more acid materials are left behind. In addition, the decomposition of organic matter tends to increase the soil's acidity. Soil pH is important, since it influences the availability of nutrients, which affects the kinds of plants that will grow, which affects the amount of organic matter added to the soil. Since calcium, magnesium, and potassium are important plant nutrients, their loss by leaching reduces the fertility of the soil. Excessively acidic soils also cause aluminum ions to become soluble, which in high amounts are toxic to many plants. (See the discussion of acid rain in chapter 16.) Most plants grow well in soils with a pH between 6 and 7, although some plants such as blueberries and potatoes grow well in acidic soils. In most agricultural situations, the pH of the soil is adjusted by adding chemicals to the soil. Lime can be added to make soils less acid, and acid-forming materials such as sulfates can be added to increase acidity.

SOIL PROPERTIES

Soil properties include soil texture, structure, atmosphere, moisture, biotic content, and chemical composition. **Soil texture** is determined by the size of the mineral particles within the soil.

TABLE 13.1 In the Time It Took to Form 1 Inch of Soil

2006	World population reaches 6.5 billion
2005	Hurricane Katrina devastates U.S. Gulf Coast region
2001	Terrorist attacks in U.S. kill thousands
1989	Berlin Wall torn down signaling an end to the Cold War
1975	U.S. Armed Forces withdraw from Vietnam
1973	Congress passes Endangered Species Act
1970	First "Earth Day" observed
1969	American astronaut Neil Armstrong walks on moon
1964	Rev. Martin Luther King, Jr., receives Nobel Peace Prize in Oslo, Norway
1963	President John F. Kennedy assassinated in Dallas, Texas
1962	British pop group The Beatles makes its first recordings
1961	Soviet cosmonaut Yuri Gagarin becomes first human to orbit Earth
1956	Elvis Presley's first film "Love Me Tender" premiers in New York City
1953	Korean War ends
1952	Dr. Jonas Salk successfully tests polio vaccine
1945	World War II ends
1935	U.S. Soil Conservation Service established
1934	Drought leads to severe dust storms in "Dust Bowl" region of Great Plains
1918	World War I ends
1908	First Model "T" Ford rolls off assembly line
1903	Orville Wright makes first successful flight in self-propelled airplane
1877	Thomas Edison invents phonograph
1875	Alexander Graham Bell invents telephone
1838	Samuel F. B. Morse develops code for electric telegraph systems
1829	Louis Braille publishes system of writing for the blind
1804	Lewis and Clark begin "Corps of Discovery" expedition to find a water route across North America
1803	U.S. purchases Louisiana Territory for $15 million
1789	French Revolution begins with attack on the Bastille
1776	Continental Congress adopts Declaration of Independence
1741	Astronomer Anders Celcius introduces Centigrade temperature scale
1732	George Washington born
1724	Daniel Gabriel Fahrenheit, inventor of mercury thermometer, devises Fahrenheit temperature scale
1689	Peter the Great becomes tsar of Russia
1682	Edmund Halley observes Great Comet, which is later named for him
1650	World population estimated at 500 million
1631	First newspaper, *Gazette de France,* published in Paris
1620	Pilgrim leaders at Plymouth Colony establish a governing authority through Mayflower Compact
1565	Spain settles colony in St. Augustine, Florida
1564	Shakespeare and Galileo born
1513	Juan Ponce de Leon becomes first European to set foot in Florida
1512	Michelangelo finishes painting Sistine Chapel ceiling
1507	Martin Waldseemuller, German geographer, first to call the New World "America"

Adapted from the NRCS poster PI-06, 2006.

The largest soil particles are gravel, which consists of fragments larger than 2.0 millimeters in diameter. Particles between 0.05 and 2.0 millimeters are classified as sand. Silt particles range from 0.002 to 0.05 millimeter in diameter, and the smallest particles are clay particles, which are less than 0.002 millimeter in diameter.

Large particles, such as sand and gravel, have many tiny spaces between them, which allow both air and water to flow through the soil. Water drains from this kind of soil very rapidly, often carrying valuable nutrients to lower soil layers, where they are beyond the reach of plant roots. Clay particles tend to be flat and are easily packed together to form layers that greatly reduce the movement of water through them. Soils with a lot of clay do not drain well and are poorly aerated. Because water does not flow through clay very well, clay soils tend to stay moist for longer periods of time and do not easily lose minerals to percolating water.

However, rarely does a soil consist of a single size of particle. Various particles are mixed in many different combinations, resulting in many different soil classifications. (See figure 13.6.) An ideal soil for agricultural use is a **loam,** which combines the good aeration and drainage properties of large particles with the nutrient-retention and water-holding ability of clay particles.

Soil structure is different from its texture. **Soil structure** refers to the way various soil particles clump together. The particles in sandy soils do not attach to one another; therefore, sandy soils have a granular structure. The particles in clay soils tend to stick to one another to form large aggregates. Other soils that have a mixture of particle sizes tend to form smaller aggregates. A good soil is **friable,** which means that it crumbles easily. The soil structure and its moisture content determine how friable a soil is. Sandy soils are very friable, while clay soils are not. If clay soil is worked when it is too wet, it can stick together in massive blocks that will be difficult to break up.

A good soil for agricultural use will crumble and has spaces for air and water. In fact, the air and water content depends on the presence of these pore spaces. (See figure 13.7.) In good soil, about one-half to two-thirds of the spaces contain air after the excess water has drained. The air provides a source of oxygen for plant root cells and all the other soil organisms. The relationship between the amount of air and water is not fixed. After a heavy rain, most of the spaces may be filled with water and less oxygen is available to plant roots and other organisms. If some of the excess water does not drain from the soil, the plant roots may die from lack of oxygen. They are literally drowned. On the other hand, if there is not enough soil moisture, the plants wilt from lack of water. Soil moisture and air are also important in determining the numbers and kinds of soil organisms.

Protozoa, nematodes, earthworms, insects, algae, bacteria, and fungi are typical inhabitants of soil. (See figure 13.8.) The role of protozoa in the soil is not firmly established, but they seem to act as parasites and predators on other forms of soil organisms and, therefore, help to regulate the populations of those organisms. Nematodes, which are often called wireworms or roundworms, may aid in the breakdown of dead organic matter. Some nematodes

Since the roots of the plants rot in place when the grasses die, a deep layer of topsoil develops. This lack of leaching also results in a thin layer of subsoil, which is low in mineral and organic content and supports little root growth.

Forest soils develop in areas of more abundant rainfall. Water moves down through the soil so that deeper layers of the soil have a great deal of moisture. The roots of the trees penetrate to this layer and extract the water they need. The leaves and other plant parts that fall to the soil surface form a thin layer of organic matter on the surface. This organic matter decomposes and mixes with the mineral material of the top layers of the soil. The water that moves through the soil tends to carry material from the topsoil to the subsoil, where many of the roots of the plants are located. One of the materials that accumulates in the *B* horizon is clay. In some soils, particularly forest soils, clay or other minerals may accumulate and form a relatively impermeable "hardpan" layer that limits the growth of roots and may prevent water from reaching the soil's deeper layers.

Desert soils have very poorly developed horizons. Since there is little rainfall, deserts do not support a large amount of plant growth, and much of the soil is exposed. Therefore, little organic matter is added to the soil, and little leaching of materials occurs from upper layers to lower layers. Since much of the soil is exposed to wind and water erosion, much of the organic material and smaller particles are carried away by wind or by flash floods when it rains.

In cold, wet climates, typical of the northern parts of Europe, Russia, and Canada, there may be considerable accumulations of organic matter, since the rate of decomposition is reduced. The extreme acidity of these soils also reduces the rate of decomposition. Hot, humid climates also tend to have poorly developed soil horizons because the organic matter decays very rapidly, and soluble materials are carried away by the abundant rainfall.

Because tropical rainforests support such a vigorous growth of plants and an incredible variety of plant and animal species, it is often assumed that tropical soils must be very fertile. Consequently, many people have tried to raise crops on tropical soils. It is possible to grow certain kinds of crops that are specially adapted to tropical soils, but raising of most traditional crop species is not successful. To understand why this is so, it is important to understand the nature of tropical rainforest soils. Two features of the tropical rainforest climate have a great influence on the nature of the soil. High temperature results in rapid decomposition of organic matter, so that the soils have very little litter and humus. High rainfall tends to leach nutrients from the upper layers of the soil, leaving behind a soil that is rich in iron and aluminum. The high iron content results in a reddish color for most of these soils.

Because the nutrients are quickly removed, these soils are very infertile. Furthermore, when the vegetation is removed, the soil is quickly eroded.

In addition to the differences caused by the kind of vegetation and rainfall, topography influences the soil profile. (See figure 13.11.) On a relatively flat area, the topsoil formed by soil-building processes will collect in place and gradually increase in depth. The topsoil formed on rolling hills or steep slopes is often transported down the slope as fast as it is produced. On such slopes, the accumulation of topsoil may not be sufficient to support a cultivated crop. The topsoil removed from these slopes is eventually deposited in the flat floodplains. These regions serve as collection points for topsoil that was produced over extensive areas. As a result, these river-bottom and delta regions have a very deep topsoil layer and are highly productive agricultural land.

SOIL EROSION

Erosion is the wearing away and transportation of soil by water, wind, or ice. Concerns about soil degradation and increased runoff from loss of tree cover data as far back as Plato writing about Attica in the fourth century B.C.:

> . . . there are remaining only the bones of the wasted body, as they may be called, as in the case of small islands, all the richer and softer parts of the soil having fallen away, and the mere skeleton of the land being left. But in the primitive state of the country, its mountains were high hills covered with soil, and the plains, as they are termed by us, of Phelleus were full of rich earth, and there was abundance of wood in the mountains. Of this last the traces still remain, for although some of the mountains now only afford sustenance to bees, not so very long ago, there were

FIGURE 13.11   **The Effect of Slope on a Soil Profile** The topsoil formed on a large area of the hillside is continuously transported down the slope by the flow of water. It accumulates at the bottom of the slope and results in a thicker *A* horizon. The resulting "bottomland" is highly productive because it has a deep, fertile layer of topsoil, while the soil on the slope is less productive.

A horizon } 15 cm (6 inches)
B horizon } 25 cm (10 inches)
A horizon } 150 cm (60 inches)
B horizon } 15 cm (6 inches)

still to be seen roofs of timber cut from trees growing there, which were of a size sufficient to cover the largest houses; and there were many other high trees, cultivated by man and bearing abundance of food for cattle. Moreover, the land reaped the benefit of the annual rainfall, not as now losing the water which flows off the bare earth into the sea, but, having an abundant supply in all places, and receiving it into herself and treasuring it up in the close clay soil, it let off into the hollows the streams which it absorbed from the heights, providing everywhere abundant fountains and rivers, of which there may still be observed sacred memorials in place where fountains once existed;
and this proves the truth of what I am saying.

Soil erosion takes place everywhere in the world, but some areas are more exposed than others. Erosion occurs wherever grass, bushes, and trees are disappearing. Deforestation and desertification both leave land open to erosion. In deforested areas, water washes down steep, exposed slopes, taking the soil with it. In desertified regions, exposed soils, cleared for farming, building, or mining, or overgrazed by livestock, simply blow away. Wind erosion is most extensive in Africa and Asia. Blowing soil not only leaves a degraded area behind but can bury and kill vegetation where it settles. It will also fill drainage and irrigation ditches. When high-tech farm practices are applied to poor lands, soil is washed away and chemical pesticides and fertilizers pollute the runoff. Every year, erosion carries away far more topsoil than is created, primarily because of agricultural practices that leave the soil exposed. (See figure 13.12.)

Worldwide, erosion removes about 25.4 billion metric tons of soil each year. In Africa, soil erosion has reached critical levels, with farmers pushing farther onto deforested hillsides. In Ethiopia, for example, soil loss occurs at a rate of between 1.5 billion and 2 billion cubic meters (53–70 billion cubic feet) a year,

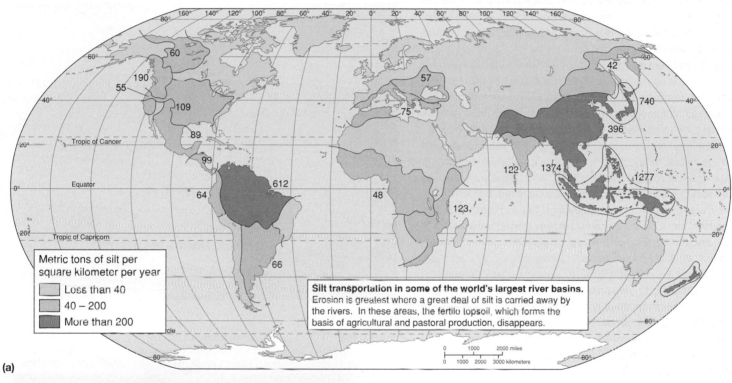

(a)

Metric tons of silt per square kilometer per year

Less than 40

40 – 200

More than 200

Silt transportation in some of the world's largest river basins. Erosion is greatest where a great deal of silt is carried away by the rivers. In these areas, the fertile topsoil, which forms the basis of agricultural and pastoral production, disappears.

(b)

FIGURE 13.12 **Worldwide Soil Erosion** (a) Soil erosion is widespread throughout the world. (b) Silty rivers are evidence of poor soil conservation practices upstream.

Source: (a) U.S. Soil Conservation Service.

Soil and Its Uses 299

with some 4 million hectares (about 10 million acres) of highlands considered irreversibly degraded. In Asia, in the eastern hills of Nepal, 38 percent of the land area is fields that have been abandoned because the topsoil has washed away. In the Western Hemisphere, Ecuador is losing soil at 20 times the acceptable rate.

According to the International Fund for Agricultural Development (IFAD), traditional labor-intensive, small-scale soil conservation efforts that combine maintenance of shrubs and trees with crop growing and cattle grazing work best at controlling erosion. In parts of Pakistan, a program begun by IFAD in 1980 to control rainfall runoff, erosion, and damage to rivers from siltation has increased crop yields and livestock productivity by 20 percent to 30 percent.

Badly eroded soil has lost all of the topsoil and some of the subsoil and is no longer productive farmland. Most current agricultural practices lose soil faster than it is replaced. Farming practices that reduce erosion, such as contour farming and terracing, are discussed in the Soil Conservation Practices section of this chapter.

Wind is also an important mover of soil. Under certain conditions, it can move large amounts. (See figure 13.13a.) Wind erosion may not be as evident as water erosion, since it does not leave gullies. Nevertheless, it can be a serious problem. Wind erosion is most common in dry, treeless areas where the soil is exposed. In the Sahel region of Africa, much of the land has been denuded of vegetation because of drought, overgrazing, and improper farming practices. This has resulted in extensive wind erosion of the soil. (See figure 13.13b.) In the Great Plains region of North America, there have been four serious periods of wind erosion since European settlement in the 1800s. If this area receives less than 30 centimeters (12 inches) of rain per year, there is not enough moisture to support crops. When this occurs for several years in a row, it is called a drought. Farmers plant crops, hoping for rain. When the rain does not come, they plow their fields again to prepare them for another crop. Thus, the loose, dry soil is left exposed, and wind erosion results. Because of the large amounts of dust in the air during those times, the region is known as the Dust Bowl. During the 1930s, wind destroyed 3.5 million hectares (over 8.5 million acres) of farmland and seriously damaged an additional 30 million hectares (75 million acres) in the Dust Bowl.

In the United States in the late 1920s, good crop yields and high prices for wheat encouraged a rapid increase in the cropped area. When drought hit in the following decade, there was catastrophic soil erosion that drove many farmers from the land. By 1940, 2.5 million people had left the Great Plains.

During the 1930s, the U.S. government responded with a comprehensive package of measures, both to give short-term economic relief and provide for long-term agricultural research and development. Examples of these initiatives include:

- The Emergency Farm Mortgage Act—to prevent farm closures by helping farmers who could not pay their mortgages.

(a)

(b)

FIGURE 13.13 **Wind Erosion** (a) The dry, unprotected topsoil from this field is being blown away. The force of the wind is capable of removing all the topsoil and transporting it several thousand kilometers. (b) The semiarid region just south of the Sahara Desert is in an especially vulnerable position. The rainfall is unpredictable, which often leads to crop failure. In addition, population pressure forces people in this region to try to raise crops in marginal areas. This often results in increased wind erosion.

- The Farm Bankruptcy Act—restricting banks from dispossessing farmers in times of crisis.
- The Drought Relief Service—buying cattle in emergency areas at reasonable prices.
- The Resettlement Administration—buying land that could be set aside from agriculture.
- The Soil Conservation Service was created within the Department of Agriculture to develop and implement new soil conservation programs. It also undertook a national soil survey.

Fortunately, many soil conservation practices have been instituted that protect soil. However, much soil is being lost to erosion, and more protective measures should be taken.

The Grand Canyon of the Colorado River, the floodplains of the Nile in Egypt, the little gullies on hillsides, and the deltas that develop at the mouths of rivers all attest to the ability of water to move soil. Anyone who has seen muddy water after a rainstorm has observed soil being moved by water. The force of moving water allows it to carry large amounts of soil. While erosion is a natural process, it is greatly accelerated by agricultural practices that leave the soil exposed. Each year, the Mississippi River transports over 325 million metric tons of soil from the central regions of North America to the Gulf of Mexico. This is equal to the removal of a layer of topsoil approximately 1 millimeter (0.04 inch) thick from the entire region. Although the rate of erosion varies from place to place, movement of soil by water occurs in every stream and river in the world. Dry Creek, a small stream in California, has only 500 kilometers (310 miles) of mainstream and tributaries; however, each year, it removes 180,000 metric tons of soil from a 340-square-kilometer (130-square-mile) area.

Water Erosion The force of moving water is able to pick up soil particles and remove them. In cases of prolonged erosion, gullies (such as the one shown here) are likely to form.

SOIL CONSERVATION PRACTICES

The kinds of agricultural activities that land can be used for are determined by soil structure, texture, drainage, fertility, rockiness, slope of the land, amount and nature of rainfall, and other climatic conditions. A relatively large proportion—about 20 percent—of U.S. land is suitable for raising crops. However, only 2 percent of that land does not require some form of soil conservation practice. (See figure 13.14.) This means that nearly all of the soil in the United States must be managed in some way to reduce the effects of soil erosion by wind or water.

Not all parts of the world are as well supplied as the United States with land that has agricultural potential. (See table 13.2.) For example, worldwide, approximately 11 percent of the land surface is suitable for crops, and an additional 24 percent is in permanent pasture. In the United States, about 20 percent is cropland, and 25 percent is in permanent pasture. Contrast this with the continent of Africa, in which only 6 percent is suitable for crops and 29 percent can be used for pasture. Canada has only 5 percent suitable for crops and 3 percent for pasture. Europe has the highest percentage of cropland with 30 percent, but it has only 17 percent in permanent pasture.

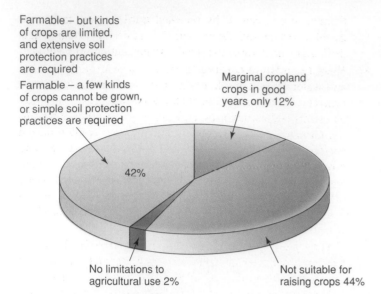

Farmable – but kinds of crops are limited, and extensive soil protection practices are required

Farmable – a few kinds of crops cannot be grown, or simple soil protection practices are required

Marginal cropland crops in good years only 12%

42%

No limitations to agricultural use 2%

Not suitable for raising crops 44%

FIGURE 13.14   **U.S. Land Used for Agricultural Purposes**
Only 2 percent of the land in the United States can be cultivated without some soil conservation practices. This 2 percent is primarily flatland, which is not subject to wind erosion. On 42 percent of the remaining land, some special considerations for protecting the soil are required, or the kinds of crops are limited. Twelve percent of the land is marginal cropland that can provide crops only in years when rainfall and other conditions are ideal. Forty-four percent is not suitable for cultivating crops but may be used for other purposes, such as cattle grazing or forests.

TABLE 13.2   Percentage of Land Suitable for Agriculture

Country	Percent Cropland	Percent Pasture
World	11.0	26.0
Africa	6.3	28.8
Egypt	2.8	5.0
Ethiopia	12.7	40.7
Kenya	7.9	37.4
South Africa	10.8	66.6
North America	13.0	16.8
Canada	4.9	3.0
United States	19.6	25.0
South America	6.0	28.3
Argentina	9.9	51.9
Venezuela	4.4	20.2
Asia	15.2	25.9
China	10.3	30.6
Japan	12.0	1.7
Europe	29.9	17.1

Source: Data from *World Resources*, 2007.

Since very little land is left that can be converted to agriculture, we must use what we have wisely. A study by the Food and Agricultural Organization states that up to 40 percent of the world's agricultural lands are seriously affected by soil degradation. The World Resources Institute in 2007 stated that land degradation affects around 70 percent of the world's rangelands, 40 percent of rain-fed agricultural lands, and 30 percent of irrigated lands. According to the United Nations Environment Programme, an estimated 500 million hectares (1.24 billion acres) of land in Africa have been affected by soil degradation since about 1950, including as much as 65 percent of agricultural land.

While the global expansion of agricultural area has been modest in recent decades, intensification has been rapid, as irrigated area increased and fallow time decreased to produce more output per hectare.

Many techniques can protect soil from erosion while allowing agriculture. Some of the more common methods are discussed here. Whenever soil is lost by water or wind erosion, the topsoil, the most productive layer, is the first to be removed. When the topsoil is lost, the soil's fertility decreases, and larger amounts of expensive fertilizers must be used to restore the fertility that was lost. This raises the cost of the food we buy. In addition, the movement of excessive amounts of soil from farmland into streams has several undesirable effects. First, a dirty stream is less aesthetically pleasing than a clear stream. Second, too much sediment in a stream affects the fish population by reducing visibility, covering spawning sites, and clogging the gills of the fish. Fishing may be poor because of unwise farming practices hundreds of kilometers upstream. Third, the soil carried by a river is eventually deposited somewhere. In many cases, this soil must be removed by dredging to clear shipping channels. We pay for dredging with our tax money, and it is a very expensive operation.

For all of these reasons, proper soil conservation measures should be employed to minimize the loss of topsoil. Figure 13.15 contrasts poor soil conservation practices with proper soil protection. When soil is not protected from the effects of running water, the topsoil is removed and gullies result. This can be prevented by slowing the flow of water over sloping land.

SOIL QUALITY MANAGEMENT COMPONENTS

- **Enhance organic matter:** Whether soil is naturally high or low in organic matter, adding new organic matter every year is perhaps the most important way to improve and maintain soil quality. Regular additions of organic matter improve soil structure, enhance water- and nutrient-holding capacity, protect soil from erosion and compaction, and support a healthy community of soil organisms. Practices that increase organic matter include leaving crop residues in the field, choosing crop rotations that include high-residue plants, using optimal nutrient and water management practices to grow healthy plants with large amounts of roots and residue, growing cover crops, applying manure or compost, using low- or no-tillage systems, and mulching.

- **Avoid excessive tillage:** Reducing tillage minimizes the loss of organic matter and protects the soil surface with plant residue. Tillage is used to loosen surface soil, prepare the seedbed, and control weeds and pests. But tillage can also break up soil structure, speed the decomposition and loss of organic matter, increase the threat of erosion, destroy the habitat

(a)

(b)

FIGURE 13.15 **Poor and Proper Soil Conservation Practices** (a) This land is no longer productive farmland since erosion has removed the topsoil. (b) This rolling farmland shows strip contour farming to minimize soil erosion by running water. It should continue indefinitely to be productive farmland.

of helpful organisms, and cause compaction. New equipment allows crop production with minimal disturbance of the soil.

- **Manage pests and nutrients efficiently:** An important function of soil is to buffer and detoxify chemicals, but soil's capacity for detoxification is limited. Pesticides and chemical fertilizers have valuable benefits, but they also can harm nontarget organisms and pollute water and air if they are mismanaged. Nutrients from organic sources also can pollute when misapplied or overapplied. Efficient pest and nutrient management means testing and monitoring soil and pests; applying only the necessary chemicals, at the right time and place to get the job done; and taking advantage of nonchemical approaches to pest and nutrient management such as crop rotations, cover crops, and manure management.

- **Prevent soil compaction:** Compaction reduces the amount of air, water, and space available to roots and soil organisms.

Compaction is caused by repeated traffic, heavy traffic, or traveling on wet soil. Deep compaction by heavy equipment is difficult or impossible to remedy, so prevention is essential.

- **Keep the ground covered:** Bare soil is susceptible to wind and water erosion and to drying and crusting. Ground cover protects soil, provides habitats for larger soil organisms, such as insects and earthworms, and can improve water availability. Ground can be covered by leaving crop residue on the surface or by planting cover crops. In addition, to ground cover, living cover crops provide additional organic matter, and continuous cover and food for soil organisms. Ground cover must be managed to prevent problems with delayed soil warming in spring, diseases, and excessive buildup of phosphorus at the surface.

- **Diversify cropping systems:** Diversity is beneficial for several reasons. Each plant contributes a unique root structure and type of residue to the soil. A diversity of soil organisms can help control pest populations, and a diversity of cultural practices can reduce weed and disease pressures. Diversity across the landscape can be increased by using buffer strips, small fields, or contour strip cropping. Diversity over time can be increased by using long crop rotations. Changing vegetation across the landscape or over time increases not only plant diversity but also the types of insects, microorganisms, and wildlife that live on a farm.

CONTOUR FARMING

Contour farming, which is tilling at right angles to the slope of the land, is one of the simplest methods for preventing soil erosion. This practice is useful on gentle slopes and produces a series of small ridges at right angles to the slope. (See figure 13.16.) Each ridge acts as a dam to prevent water from running down the incline. This allows more of the water to soak into the soil. Contour farming reduces soil erosion by as much as 50 percent and, in drier regions, increases crop yields by conserving water.

FIGURE 13.16 **Contour Farming** Tilling at right angles to the slope creates a series of ridges that slows the flow of the water and prevents soil erosion. This soil conservation practice is useful on gentle slopes.

Traditional landscaping and current landscape maintenance practices, while frequently meeting human needs and aesthetics, often have harmful environmental impacts. These impacts include:

- Gasoline-powered landscape equipment (mowers, trimmers, blowers) account for over 5 percent of urban air pollution in North America.

- Residential application of pesticides is typically at a rate 20 times that of farm applications per acre.

- Yard wastes (mostly grass clippings) comprise 20 percent of municipal solid waste collected and most still ends up in landfills.

- A lawn has less than 10 percent of the water absorption capacity of a natural woodland—often a reason for suburban flooding.

Natural or native landscaping attempts to balance our needs and aesthetics with those of the environment. The principles of natural landscaping include:

- Protect existing natural areas to the greatest extent possible (woodlands and wetlands, stream corridors, and meadows).

- Select regionally native plants to form the backbone of the landscape.

- Reduce use of turf; instead, install woodland, meadow, or other natural plantings.

- Reduce use of pesticides.

- Reduce use of power landscape equipment. Shrinking the size of the lawn and planting appropriate native species in less formal arrangements will reduce the need for extensive use of power equipment.

- Avoid use of invasive exotics that outcompete native plants and result in declines in biodiversity. Examples include: Norway maples, kudzu, purple loosestrife, and multiflora rose.

- Practice soil and water conservation. Stabilize slopes with natural plantings, mulch around plants, and install drought tolerant species.

- Compost and mulch on site to eliminate solid waste. Generate your own mulch—a soil additive can replace the need for most fertilizers.

A growing number of communities are developing municipal compost facilities. These communities offer curbside pickup of "organic green waste," including yard waste, food scraps, and biodegradable products. Often the resulting compost is available to community members for their garden use.

Does your community offer curbside pickup of organic green waste and make the compost available for your use?

FIGURE 13.17   **Strip Farming**   On rolling land, a combination of contour and strip farming prevents excessive soil erosion. The strips are planted at right angles to the slope, with bands of closely sown crops, such as wheat or hay, alternating with bands of row crops, such as corn or soybeans.

STRIP FARMING

When a slope is too steep or too long, contour farming alone may not prevent soil erosion. However, a combination of contour and strip farming may work. **Strip farming** is alternating strips of closely sown crops such as hay, wheat, or other small grains with strips of row crops such as corn, soybeans, cotton, or sugar beets. (See figure 13.17.) The closely sown crops retard the flow of water, which reduces soil erosion and allows more water to be absorbed into the ground. The type of soil, steepness, and length of slope dictate the width of the strips and determine whether strip or contour farming is practical.

TERRACING

On very steep land, the only practical method of preventing soil erosion is to construct terraces. **Terraces** are level areas constructed at right angles to the slope to retain water and greatly reduce the amount of erosion. (See figure 13.18.) Terracing has been used for centuries in nations with a shortage of level farmland. The type of terracing seen in figure 13.18a requires the use of small machines and considerable hand labor and is not suitable for the mechanized farming typical in much of the world. The modification shown in figure 13.18b allows for the use of large farm machines. Terracing is an expensive method of controlling erosion, since it requires moving soil to construct the level areas, protecting the steep areas between terraces, and constant repair and maintenance. Many factors, such as length and steepness of slope, type of soil, and amount of precipitation, determine whether terracing is feasible.

WATERWAYS

Even with such soil conservation practices as contour farming, strip farming, and terracing, farmers must often provide protected channels for the movement of water. **Waterways** are depressions on sloping land where water collects and flows off the land. When not properly maintained, these areas are highly susceptible to erosion.

(a)

(b)

FIGURE 13.18 **Terraces** Since the construction of terraces requires the movement of soil and the protection of the steep slope between levels, terraces are expensive to build. (a) The terraces seen here are extremely important for people who live in countries that have little flatland available. They require much energy and hand labor to maintain but make effective agricultural use of the land without serious erosion. (b) This modification of the terracing concept allows the use of the large farm machines typical of farming practices in Canada, Europe, and the United States.

(See figure 13.19.) If a waterway is maintained with a permanent sod covering, the speed of the water is reduced, the roots tend to hold the soil particles in place, and soil erosion is decreased.

WINDBREAKS

Contour farming, strip farming, terracing, and maintaining waterways are all important ways of reducing water erosion, but wind is also a problem with certain soils, particularly in dry areas of the world. Wind erosion can be reduced if the soil is protected. The best protection is a layer of vegetation over the surface. However, the process of preparing soil for planting and the method of planting often leave the soil exposed to wind. **Windbreaks** are plantings of trees or other plants that protect bare soil from the full force of the wind. Windbreaks reduce the velocity of the wind, thereby decreasing the amount of soil that it can carry away. (See figure 13.20.) In some

(a)

(b)

FIGURE 13.19 **Protection of Waterways Prevents Erosion** (a) An unprotected waterway has been converted into a gully. (b) A well-maintained waterway is not cultivated; a strip of grass retards the flow of water and protects the underlying soil from erosion.

cases, rows of trees are planted at right angles to the prevailing winds to reduce their force, while in other cases, a kind of strip farming is practiced in which strips of hay or grains are alternated with row crops that leave large amounts of the soil exposed. In some areas of the world, the only way to protect the soil is to not cultivate it at all but to leave it in a permanent cover of grasses.

CONVENTIONAL VERSUS CONSERVATION TILLAGE

Conventional tillage methods in much of the world require the extensive use of farm machinery to prepare the soil for planting and to control weeds. Typically, a field is plowed and then disked or harrowed one to three times before the crop is planted. The plowing,

Soil and Its Uses 305

(a)

(b)

FIGURE 13.20 **Windbreaks** *(a)* In sections of the Great Plains, trees provide protection from wind erosion. The trees along the road protect the land from the prevailing winds. *(b)* In this field, temporary strips of vegetation serve as windbreaks.

which turns the soil over, has several desirable effects: any weeds or weed seeds are buried, thus reducing the weed problem in the field. Crop residue from previous crops is incorporated into the soil, where it will decay faster and contribute to soil structure. Nutrients that had been leached to deeper layers of the soil are brought near the surface. And the dark soil is exposed to the sun so that it warms up faster. This last effect is most critical in areas with short growing seasons. In many areas, fields are plowed in the fall, after the crop has been harvested, and the soil is left exposed all winter.

After plowing, the soil is worked by disks or harrows to break up any clods of earth, kill remaining weeds, and prepare the soil to receive the seeds. After the seeds are planted, there may still be weed problems. Farmers often must cultivate row crops to kill the weeds that begin to grow between the rows.

Each trip over the field costs the farmer money, while at the same time increasing the amount of time the soil is exposed to wind or water erosion.

In recent years, several new systems of tillage have developed, as innovations in chemical herbicides and farm equipment have taken place. These tilling practices protect the soil by leaving the crop residue on the soil surface, thus reducing the amount of time it is exposed to erosion forces. **Reduced tillage** is a method that uses less cultivation to control weeds and to prepare the soil to receive seeds but generally leaves 15 percent to 30 percent of the soil surface covered with crop residue after planting. **Conservation tillage** methods further reduce the amount of disturbance to the soil and leave 30 percent or more of the soil surface covered with crop residue following planting. Selective herbicides are used to kill unwanted vegetation before planting the new crop and to control weeds afterward. Several variations of conservation tillage are used:

1. Mulch tillage involves tilling the entire surface just before planting or as planting is occurring.
2. Strip tillage is a method that involves tilling only in the narrow strip that is to receive the seeds. The rest of the soil and the crop residue from the previous crop are left undisturbed.
3. Ridge tillage involves leaving a ridge with the last cultivation of the previous year and planting the crop on the ridge with residue left between the ridges. The crop may be cultivated during the year to reduce weeds.
4. No-till farming involves special planters that place the seeds in slits cut in the soil that still has on its surface the residue from the previous crop.

Both reduced tillage and conservation tillage methods reduce the amount of time and fuel needed by the farmer to produce the crop and, therefore, represent an economic savings. By 2000, over half of the cropland in the United States was being farmed using reduced or conservation tillage methods. (See table 13.3.) For many kinds of crops, yields are comparable to those produced by conventional tillage methods.

Other positive effects of reduced tillage, in addition to reducing erosion, are:

1. The amount of winter food and cover available for wildlife increases, which can lead to increased wildlife populations.

TABLE 13.3 Comparison of Various Tillage Methods

Tillage Method	Fuel Use (Liters per Hectare)	Time Involved (Hours per Hectare)
Conventional plowing	8.08	3.00
Reduced tillage	5.11	2.20
Mulch tillage	4.64	2.07
Ridge tillage	4.12	2.25
No tillage	2.19	1.21

Source: Data from University of Nebraska, Institute of Agriculture and Natural Resources, 2004.

2. Since there is less runoff, siltation in streams and rivers is reduced. This results in clearer water for recreation and less dredging to keep waterways open for shipping.

3. Row crops can be planted on hilly land that cannot be converted to such crops under conventional tilling methods. This allows a farmer to convert low-value pastureland into cropland that gives a greater economic yield.

4. Since fewer trips are made over the field, petroleum is saved, even when the petrochemical feedstocks necessary to produce the herbicides are taken into account.

5. Two crops may be grown on a field in areas that had been restricted to growing one crop per field per year. In some areas, immediately after harvesting wheat, farmers have planted soybeans directly in the wheat stubble.

6. Because conservation tillage reduces the number of trips made over the field by farm machinery, the soil does not become compacted as quickly.

However, there are also some drawbacks to conservation tillage methods:

1. The residue from previous vegetation may delay the warming of the soil, which may, in turn, delay planting some crops for several days.

2. The crop residue reduces evaporation from the soil and the upward movement of water and soil nutrients from deeper layers of the soil, which may retard the growth of plants.

3. The accumulation of plant residue can harbor plant pests and diseases that will require more insecticides and fungicides. This is particularly true if the same crop is planted repeatedly in the same location.

Conservation tillage is not the complete answer to soil erosion problems but may be useful in reducing soil erosion on well-drained soils. It also requires that farmers pay close attention to the condition of the soil and the pests to be dealt with.

In 1972, 12 million hectares (30 million acres) in the United States were under some form of conservation tillage. About 1.3 million hectares (3.2 million acres) were being farmed using no-till methods. By 1992, this had risen to 60 million hectares (150 million acres) of conservation-tilled land, of which 6.2 million hectares (15 million acres) were being farmed using no-till methods. It is estimated that by the year 2012, 95 percent of U.S. cropland will be under some form of reduced-tillage practice.

PROTECTING SOIL ON NONFARM LAND

Each piece of land has characteristics, such as soil characteristics, climate, and degree of slope, that influence the way it can be used. When all these factors are taken into consideration, a proper use can be determined for each portion of the planet. Wise planning and careful husbandry of the soil are necessary if the land and its soil are to provide food and other necessities of life. Not all land is suitable for crops or continuous cultivation. Some has highly erodible soil and must never be plowed for use as cropland, but it can still serve other useful functions. With the use of appropriate soil conservation practices, much of the land not usable for crops can be used for grazing, wood production, wildlife production, or scenic and recreational purposes. Figure 13.21 shows land that is not suitable for cultivation because it is too arid but that, if it is used properly, can provide grazing for cattle or sheep.

The land shown in figure 13.22 is not suitable for either crops or grazing because of the steep slope and thin soil. However, it is still a valuable and productive piece of land, since it can be used to furnish lumber, wildlife habitats, and recreational opportunities.

FIGURE 13.21   **Noncrop Use of Land to Raise Food** As long as it is properly protected and managed, this land can produce food through grazing, but it should never be plowed to plant crops because the topsoil is too shallow and the rainfall is too low.

FIGURE 13.22   **Forest and Recreational Use** Although this land is not capable of producing crops or supporting cattle, it furnishes lumber, a habitat for wildlife, and recreational opportunities.

CASE STUDY 13.1

LAND CAPABILITY CLASSES

Not all land is suitable for raising crops or urban building. Such factors as the degree of slope, soil characteristics, rockiness, erodibility, and other characteristics determine the best use for a parcel of land. In an attempt to encourage people to use land wisely, the U.S. Soil Conservation Service has established a system to classify land-use possibilities. The table shows the eight classes of land and lists the characteristics and capabilities of each. The photo shows how these classes can coexist.

Unfortunately, many of our homes and industries are located on type I and II land, which has the least restrictions on agricultural use. This does not make the best use of the land. Zoning laws and land-use management plans should consider the land-use capabilities and institute measures to ensure that land will be used to its best potential.

- Can you provide examples from your community where you feel land has not been used properly?

- In your opinion, should there be more or less land-use planning?
- What are some of the consequences of building or farming on land that is not suitable?

Land Class		Characteristics	Capability	Special Conservation Measures
Land suitable for cultivation	I	Excellent, flat, well-drained land	Cropland	Normal good practices adequate
	II	Good land; has minor limitations, such as slope, sandy soil, or poor drainage	Cropland Pasture	Strip cropping Contour farming
	III	Moderately good land with important limitations of soil, slope, or drainage	Cropland Pasture Watershed	Contour farming Strip cropping Terraces Waterways
	IV	Fair land with severe limitations of soil, slope, or drainage	Pasture Orchards Urban Industry Limited cropland	Crops on a limited basis Contour farming Strip cropping Terraces Waterways
Land not suitable for cultivation	V	Use for grazing and forestry; slightly limited by rockiness, shallow soil, or wetness	Grazing Forestry Watershed Urban Industry	No special precautions if properly grazed or logged; must not be plowed
	VI	Moderate limitations for grazing and forestry because of moderately steep slopes		Grazing or logging may be limited at times.
	VII	Severe limitations for grazing and forestry because of very steep slopes vulnerable to erosion		Careful management is required when used for grazing or logging.
	VIII	Unsuitable for grazing and forestry because of steep slope, shallow soil, lack of water, or too much water		Not to be used for grazing or logging; steep slope and lack of soil present problems

Soil Fertility and Hunger in Africa

Africa south of the Sahara is the only remaining region of the world where per capita food production has remained stagnant over the past 40 years. About 180 million Africans—up 100 percent since 1970—do not have access to sufficient food to lead healthy and productive lives.

Africa's food insecurity is directly related to insufficient total food production, in contrast to South Asia and other regions, where food insecurity is primarily due to poor distribution and lack of purchasing power. Depletion of soil fertility is a major cause of low per capita food production in Africa. Over decades, small-scale farmers have removed large quantities of nutrients from their soils without using sufficient quantities of manure or fertilizer to replenish the soil.

The traditional way to overcome nutrient depletion is the use of mineral fertilizers. But fertilizers in Africa cost two to six times as much as those in Europe, North America, or Asia.

A soil fertility replenishment approach has been developed during the past decade using resources naturally available in Africa. Participatory research trials were conducted in which farmers tested and adapted several new practices. These included the use of nitrogen-fixing leguminous trees and biomass transfer of the leaves of nutrient-accumulating shrubs.

Leguminous trees are interplanted into a young maize (corn) crop and allowed to grow as fallows during dry seasons. The quantities of nitrogen captured are similar to those applied as fertilizers by commercial farmers to grow maize in developed countries. After the wood is harvested from the trees, nitrogen-rich leaves, pods, and green branches are hoed into the soil before maize is planted at the start of a subsequent season. This litter decomposes with the tree roots, releasing nitrogen and other nutrients to the soil.

The transfer of the leaf biomass of nutrient-accumulating shrubs from roadsides and hedges into cropped fields also adds nutrients and can double maize yields without fertilizer additions. This organic source of nutrients is effective because it also adds other plant nutrients, particularly potassium and micronutrients. Because of the high labor requirements for cutting and carrying the biomass to fields, this is profitable with high-value crops such as vegetables but not with relatively low-valued maize.

Tens of thousands of farm families in Kenya, Tanzania, Malawi, Zambia, Zimbabwe, and Mozambique use various combinations of tree fallows and biomass transfers with good and consistent results. Adoption of this approach is taking place by the transfer of knowledge from farmer to farmer and village to village by community-based organizations and by a multitude of national research and extension institutes, universities, nongovernmental organizations, and development projects. The challenge, now, is to accelerate the adoption rate to reach tens of millions of farm families.

- Do you feel that such low-tech solutions as those discussed in this essay have a role in solving the challenge of feeding the world's growing population?
- What are the barriers to expanding the programs discussed?
- Are low-tech solutions equally applicable to the developed nations in the world? Why or why not?

SUMMARY

The surface of the Earth is in constant flux. The movement of tectonic plates results in the formation of new land as old land is worn down by erosive activity. Soil is an organized mixture of minerals, organic material, living organisms, air, and water. Soil formation begins with the breakdown of the parent material by such physical processes as changes in temperature, freezing and thawing, and movement of particles by glaciers, flowing water, or wind. Oxidation and hydrolysis can chemically alter the parent material. Organisms also affect soil building by burrowing into and mixing the soil, releasing nutrients, and decomposing.

Topsoil contains a mixture of humus and inorganic material, both of which supply soil nutrients. The ability of soil to grow crops is determined by the inorganic matter, organic matter, water, and air spaces in the soil. The mineral portion of the soil consists of various mixtures of sand, silt, and clay particles.

A soil profile typically consists of the O horizon of litter; the A horizon, which is rich in organic matter; an E horizon from which materials have been leached; the B horizon, which accumulates materials leached from above; and the C horizon, which consists of slightly altered parent material. Forest soils typically have a shallow A horizon and an E horizon and a deep, nutrient-rich B horizon with much root development. Grassland soils usually have a thick A horizon containing most of the roots of the grasses. They lack an E horizon and have few nutrients in the thin B horizon.

Soil erosion is the removal and transportation of soil by water or wind. Proper use of such conservation practices as contour farming, strip farming, terracing, waterways, windbreaks, and conservation tillage can reduce soil erosion. Misuse reduces the soil's fertility and causes air- and water-quality problems. Land unsuitable for crops may be used for grazing, lumber, wildlife habitats, or recreation.

THINKING GREEN

1. Work with a local group to plant trees and native vegetation along rivers and streams.
2. Compost your biodegradable home and yard waste.
3. Support natural or native landscaping in your community and college.
4. Plant a garden—get some soil under your fingernails!
5. Find out what kind of soil is present where you live. Dig a small hole. Is the soil clay, sand, or rubble from a previous building?

WHAT'S YOUR TAKE?

It has been stated by some soil scientists that the only way to have successful action against soil erosion on a large scale is to implement land management practices that increase production capacity over a large area. This belief is counter to another school of thought that soil protection is best served by smaller organic farms. Develop an argument in support of one of these contradictory attitudes on soil conservation and farming practices.

REVIEW QUESTIONS

1. How are soil and land different?
2. Name the five major components of soil.
3. Describe the process of soil formation.
4. Name five physical and chemical processes that break parent material into smaller pieces.
5. In addition to fertility, what other characteristics determine the usefulness of soil?
6. How does soil particle size affect texture and drainage?
7. Describe a soil profile.
8. Define erosion.
9. Describe three soil conservation practices that help to reduce soil erosion.
10. Besides cropland, what are other possible uses of soil?

CRITICAL THINKING QUESTIONS

1. Minimum tillage soil conservation often uses greater amounts of herbicides to control weeds. What do you think about this practice? Why?
2. As populations grow, should we try to bring more land into food production, or should we use technology to aid in producing more food on the land we already have in production? What are the trade-offs?
3. Given what you know about soil formation, how might you explain the presence of a thick *A* horizon in soils in the North American Midwest?
4. Why should nonfarmers be interested in soil conservation?
5. Imagine that you are a scientist hired to consult on a project to evaluate land-use practices at the edge of a small city. The area in question has deep ravines and hills. What kinds of agricultural, commercial, and logging practices would you recommend in this area to help preserve the environment?
6. Look at your own community. Can you see examples of improper land use (urban or rural)? What are the consequences of these land-use practices? What recommendations would you make to improve land use?

CHAPTER 14

AGRICULTURAL METHODS AND PEST MANAGEMENT

Agricultural use of land results in the modification of the natural world to allow for the growth of crops. Often pest species are controlled with the use of pesticides.

CHAPTER OUTLINE

The Development of Agriculture
    Shifting Agriculture
    Labor-Intensive Agriculture
    Mechanized Agriculture
Fossil Fuel Versus Muscle Power
The Impact of Fertilizer
Agricultural Chemical Use
    Insecticides
    Herbicides
    Fungicides and Rodenticides
    Other Agricultural Chemicals
Problems with Pesticide Use
    Persistence
    Bioaccumulation and Biomagnification
    Pesticide Resistance
    Effects on Nontarget Organisms
    Human Health Concerns
Why Are Pesticides So Widely Used?
Alternatives to Conventional Agriculture
    Sustainable Agriculture
    Techniques for Protecting Soil and Water Resources
    Integrated Pest Management
    Genetically Modified Crops

ISSUES & ANALYSIS
What Does "Certified Organic" Food Mean? 332

CASE STUDIES
DDT—A Historical Perspective 316
Economic Development and Food Production in China 322

CAMPUS SUSTAINABILITY INITIATIVE
Integrated Pest Management at Seattle University 320

GOING GREEN
Organic Farming: Helping to Promote Sustainable Agriculture 325

WATER CONNECTIONS
The Dead Zone of the Gulf of Mexico 319

OBJECTIVES

After reading this chapter, you should be able to:

- Explain how mechanization encouraged monoculture farming.
- List the advantages and disadvantages of monoculture farming.
- Explain why chemical fertilizers are used.
- Understand how fertilizers alter soil characteristics.
- Explain why modern agriculture makes extensive use of pesticides.
- Differentiate between persistent pesticides and nonpersistent pesticides.
- List four problems associated with pesticide use.
- Define biomagnification.
- Define organic farming.
- Explain why integrated pest management depends on a complete knowledge of the pest's life history.
- Recognize that genetically modified crops are created by using biotechnological techniques to insert genes from one species into another.

The following additional Case Studies can be found on the book's website at www.mhhe.com/enger12e along with other interesting readings:
"A New Generation of Insecticides," "Industrial Production of Livestock," and "Promoting Environmental Conservation in the Specialty Coffee Industry."

FIGURE 14.1 **Shifting Agriculture** In many areas of the world where the soils are poor and human populations are low, crops can be raised by disturbing small parts of the ecosystem followed by several years of recovery. The burning of vegetation releases nutrients that can be used by crops for one or two years before the soil is exhausted. The return of the natural vegetation prevents erosion and repairs the damage done by temporary agricultural use.

THE DEVELOPMENT OF AGRICULTURE

Our early ancestors obtained food from nature by hunting and gathering. The development of agriculture involved manipulating the natural environment to produce the kinds of foods humans want and allowed for an increase in the size of the human population. The history of the development of agriculture has involved various kinds of innovations. One of the simplest is shifting agriculture, also known as "slash and burn" agriculture.

SHIFTING AGRICULTURE

Shifting agriculture involves cutting down the trees and burning the trees and other vegetation in a small area of the forest. (See figure 14.1.) Burning releases nutrients that were tied up in the biomass and allows a few crops to be raised before the soil is exhausted. Once the soil is no longer suitable for raising crops (within two or three years), the site is abandoned. The surrounding forest recolonizes the area, which will return over time to the original forest through the process of succession.

In some parts of the world with poor soil and low populations, shifting agriculture is still used successfully. This method is particularly useful on thin, nutrient-poor tropical soils and on steep slopes. The small size of the openings in the forest and their temporary existence prevent widespread damage to the soil, and erosion is minimized. While this system of agriculture is successful when human population densities are low, it is not suitable for large, densely populated areas. When populations become too large, the size and number of the garden plots increase and the time between successive uses of the same plot of land decreases. When a large amount of the forest is disturbed and the time between successive uses is decreased, the forest cannot return and repair the damage done by the previous use of the land, and the nature of the forest is changed.

The traditional practices of the people who engage in small-scale, shifting agriculture have been developed over hundreds of years and often are more effective for their local conditions than other methods of gardening. Typically, these gardens are planted with a mixture of plants, a system known as **polyculture.** Mixing plants together in a garden often is beneficial, since shade-requiring species may be helped by taller plants, or nitrogen-fixing legumes may provide nitrogen for species that require it. In addition, mixing species may reduce insect pest problems because some plants produce molecules that are natural insect repellents. The small, isolated, temporary nature of the gardens also reduces the likelihood of insect infestations. While today we see this form of agriculture practiced most commonly in tropical areas, it is important to note that many Native American cultures used shifting agriculture and polyculture in temperate areas.

LABOR-INTENSIVE AGRICULTURE

In many areas of the world with better soils, more intense forms of agriculture developed that involved a great deal of manual labor to till, plant, and harvest the crop. This style of agriculture is still practiced in much of the world today. Three situations favor this kind of farming: (1) when the growing site does not allow for mechanization, (2) when the kind of crop does not allow it, and (3) when the economic condition of the people does not allow them to purchase the tools and machines used for mechanized agriculture. Crops or terrain that requires that fields be small discourages mechanization, since large tractors and other machines cannot be used efficiently on small, oddly shaped fields. Many mountainous areas of the world fit into this category. In addition, some crops require such careful handling in planting, weeding, or harvesting that large amounts of hand labor are required. The planting of paddy rice and the harvesting of many fruits and vegetables are examples. However, the primary reason for labor-intensive farming is economic. Many densely populated countries have numerous small farms of 1 to 2 hectares (a hectare is about the size of a soccer field) that can be effectively managed with human labor, supplemented by that of draft animals and a few small gasoline-powered engines. (See figure 14.2.) In addition, in the less-developed regions of the world, the cost of labor is low, which encourages the use of hand labor rather than relatively expensive machines to do planting, weeding, and other activities.

Mechanization requires large tracts of land that could be accumulated only by the expenditure of large amounts of money or the development of larger cooperative farms from many small units. Even if social and political obstacles to such large land-holdings could be overcome, there is still the problem of obtaining the necessary capital to purchase the machines. Large parts of the developing world fit into this category, including much of Africa, many areas in Central and South America, and many areas in Asia. In countries such as China and India, which have a combined population of over 2 billion people, about 70 percent of the population is rural. Many of these people are engaged in non-mechanized agriculture.

MECHANIZED AGRICULTURE

The development of various kinds of machines following the Industrial Revolution resulted in changes in agriculture. Horse-drawn farm implements and the subsequent development of

FIGURE 14.2 **Labor-Intensive Agriculture** In many of the less-developed countries of the world, the extensive use of hand labor allows for impressive rates of production with a minimal input of fossil fuels and fertilizers. This kind of agriculture is also necessary in areas that have only small patches of land suitable for farming.

FIGURE 14.3 **Mechanized Monoculture** This wheat field is an example of monoculture, a kind of agriculture that is highly mechanized and requires large fields for the efficient use of machinery. In this kind of agriculture, machines and fossil-fuel energy have replaced the energy of humans and draft animals.

tractors led to the modern mechanized agriculture typical of North America, much of Europe, and other parts of the world where money and land are available to support this form of agriculture. In large measure, machines and fossil-fuel energy replace the energy formerly supplied by human and animal muscles. Mechanization requires large expanses of fairly level land for the machines to operate effectively. In addition, large tracts of land must be planted in the same crop for efficient planting, cultivating, and harvesting, a practice known as **monoculture.** (See figure 14.3.) Small sections of land with many kinds of crops require many changes of farm machinery, which takes time. Also, many crops interspersed with each other reduces the efficiency of farming operations because farmers must skip parts of the field, which increases travel time and uses expensive fuel.

Even though mechanized monoculture is an efficient method of producing food, it is not without serious drawbacks. When large tracts of land are prepared for planting, they are often left uncovered by vegetation, and soil erosion increases. Because of problems with erosion, many farmers are now using methods that reduce the time the fields are left bare. (See chapter 13.)

Traditionally, mechanized farming has removed much of the organic matter each year when the crop was harvested. This tended to reduce the soil organic matter. As agricultural scientists and farmers have recognized the need to improve the organic-matter content of soils, many farmers have been leaving increased amounts of organic matter after harvest, or they specifically plant a crop that is later plowed under to increase the soil's organic content.

To ensure that a crop can be planted, tended, and harvested efficiently by machines, farmers rely on seeds that are identical genetically and thus provide uniform plants with characteristics suitable for mechanized farming. These special seeds can ensure that all the plants germinate at the same time, resist the same pests, ripen at the same time, and grow to the same height. These are valuable characteristics to the farmer, but these plants have little genetic variety. When all the farmers in an area plant the same

genetic varieties, pest control becomes a serious problem. If diseases or pests begin to spread, the magnitude of the problem becomes devastating because all the plants have the same characteristics and, thus, are susceptible to the same diseases. If genetically diverse crops are planted or crops are rotated from year to year, this problem is not as great.

Because farm equipment is expensive, farmers tend to specialize in a few crops. This means that the same crop may be planted in the same field several years in a row. This lack of crop rotation may deplete certain essential soil nutrients, thereby requiring special attention be paid to soil chemistry. In addition, planting the same crop repeatedly encourages the growth of insect and fungus pest populations because they have a huge food supply at their disposal. This requires the frequent use of insecticides and fungicides or other methods of pest control.

Even though there are problems associated with mechanized, monoculture agriculture, it has greatly increased the amount of food available to the world over the past 100 years. Yields per hectare of land being farmed have increased over much of the world, particularly in the developed world, which includes the United States. (See figure 14.4.) This increase has come about because of improved varieties of crops, irrigation, better farming methods, the use of agricultural chemicals, more efficient machines, and the use of energy-intensive as opposed to labor-intensive technology.

Throughout the 1950s, 1960s, and 1970s, the introduction of new plant varieties and farming methods resulted in increased agricultural production worldwide. This has been called the **Green Revolution.** Both the developed world, which uses highly mechanized farming methods, and the developing world, where labor-intensive farming is typical, have benefited from these advances, and food production has increased significantly. The Green Revolution has also had some negative effects. For example, many modern varieties of plants require fertilizer and pesticides that the traditional varieties they replaced did not need. In addition, many of the crops require higher amounts of water and increase the

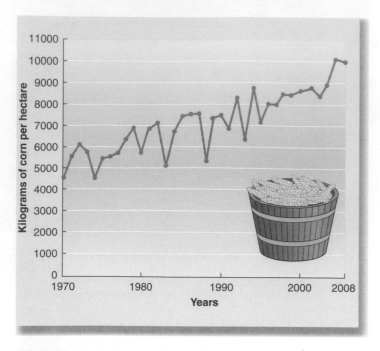

FIGURE 14.4 **Increased Yields Resulting from Modern Technology and the Total Amount of Land Available for Cultivation** The increased crop yields in the United States and many other parts of the world are the result of a combination of factors, including the development of high-yielding varieties, changed agricultural methods, the application of fertilizers and pesticides, and more efficient machinery.

Sources: Data from U.S. Department of Agriculture, *Agriculture Statistics,* and USDA.

demand for irrigation. Farmers also have become more dependent on the industries that provide the specialized seeds. When the positives and negatives are balanced, however, the end result is that food production per hectare has increased. This has not solved the world's food problem, however, because the population of the world continues to increase and more food is needed.

Long-term solutions to feeding the world's hungry are complex and are unlikely to be found without global answers to problems that have been difficult for decades. These issues include the following:

- Food subsidies and trade barriers—Governments like to protect their farmers with subsidies and trade barriers. Such barriers hinder the development of farming in poorer nations.

- Higher fuel costs—The increase in the price of fuel adds to the cost of food by making it more expensive to produce and transport.

- Alternative fuels—Increased energy prices divert crops to alternative fuels. Growing corn for fuel (and, in many cases, cutting back on wheat production) means less grain for food.

- Rising demand—The rise in global wealth has meant more demand for higher-quality food that is more expensive to produce. In India and China, people are eating more meat. That is a sign of prosperity, but it requires greater supply to sustain. That problem, which also is the major driver behind rising oil prices, is only going to increase.

FOSSIL FUEL VERSUS MUSCLE POWER

Mechanized agriculture has substituted the energy stored in petroleum products for the labor of humans. For example, in the United States in 1913, it required 135 hours of labor to produce 2500 kilograms (5500 pounds) of corn. In 1980, it required only about 15 hours of labor to produce the same amount of corn. The energy supplied by petroleum products replaced the equivalent of 120 hours of labor. Energy is needed for tilling, planting, harvesting, and pumping irrigation water. The manufacture of fertilizer and pesticides also requires the input of large amounts of fossil fuels, both as a source of energy for the industrial process and as raw material from which these materials are made. For example, about 5 metric tons of fossil fuel are required to produce about 1 metric ton of fertilizer. In addition, the pesticides used in mechanized agriculture require the use of oil as a raw material. Since the developed world depends on oil to run machines and manufacture fertilizer and pesticides to support its agriculture, any change in the availability or cost of oil will have a major impact on the world's ability to feed itself.

THE IMPACT OF FERTILIZER

Various experts estimate that approximately 25 percent of the world's crop yield can be directly attributed to the use of chemical fertilizers. The use of fertilizer has increased significantly over the last few decades and is projected to increase even more. (See figure 14.5.) However, since fertilizer production relies on energy from fossil fuels, the price and availability of chemical fertilizers are strongly influenced by world energy prices. If the price of oil increases, the price of fertilizer goes up, as does the cost of food. This is felt most acutely in parts of the world where money is in short supply, since the farmers are unable to buy fertilizer and crop yields fall accordingly.

Fertilizers are valuable because they replace the soil nutrients removed by plants. Some of the chemical building blocks of plants, such as carbon, hydrogen, and oxygen, are easily replaced by carbon dioxide from the air and water from the soil, but others are less easily replaced. The three primary soil nutrients often in short supply are nitrogen, phosphorus, and potassium compounds. They are often referred to as **macronutrients** and are the common ingredient of chemical fertilizers. Their replacement is important because when the crop is harvested, the chemical elements that are a part of the crop are removed from the field. Since many of those elements originated from the soil, they need to be replaced if another crop is to be grown. Certain other elements are necessary in extremely small amounts and are known as **micronutrients.** Examples are boron, zinc, and manganese. As an example of the difference between macronutrients and micronutrients, harvesting 1 metric ton of potatoes removes 10 kilograms (22 pounds) of nitrogen (a macronutrient) but only 13 grams (0.03 pound) of boron (a micronutrient). When the same crop is grown repeatedly in the same field, certain micronutrients may be depleted, resulting in reduced yields. These necessary elements can be returned to the soil in sufficient amounts by incorporating them into the fertilizer the farmer applies to the field.

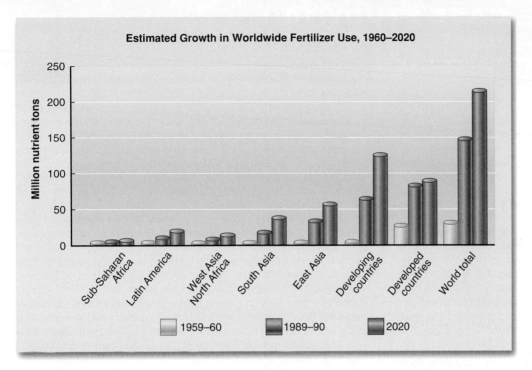

FIGURE 14.5   Increasing Fertilizer Use The use of fertilizer is increasing rapidly. Growth in use will be most rapid in the developing countries over the next 20 years.

Although chemical fertilizers replace inorganic nutrients, they do not replace soil organic matter. Organic material is important because it modifies the structure of the soil, preventing compaction and maintaining pore space, which allows water and air to move to the roots. The decomposition of organic matter produces humus, which helps to maintain proper soil chemistry because it tends to loosely bind many soil nutrients and other molecules and modifies the pH so that nutrients are not released too rapidly. Soil bacteria and other organisms use organic matter as a source of energy. Since these organisms serve as important links in the carbon and nitrogen cycles, the presence of organic matter is important to their function. Thus, total dependency on chemical fertilizers usually reduces the amount of organic matter and can change the physical, chemical, and biological properties of the soil.

As water moves through the soil, it dissolves soil nutrients (particularly nitrogen compounds) and carries them into streams and lakes, where they may encourage the growth of unwanted plants and algae. This is particularly true when fertilizers are applied at the wrong time of the year, just before a heavy rain, or in such large amounts that the plants cannot efficiently remove them from the soil before they are lost. These ideas are covered in greater detail in chapter 15 on water management.

AGRICULTURAL CHEMICAL USE

In addition to chemical fertilizers, mechanized monoculture requires large amounts of other agricultural chemicals, such as pesticides, growth regulators, and preservatives. These chemicals have specific scientific names but are usually categorized into broad groups based on their effects. A **pesticide** is any chemical used to kill or control populations of unwanted fungi, animals, or plants, often called **pests.** The term *pest* is not scientific but refers to any organism that is unwanted. Insects that feed on crops are pests, while others, such as bees, are beneficial for pollinating plants. Unwanted plants are generally referred to as **weeds.**

Pesticides can be subdivided into several categories based on the kinds of organisms they are used to control. **Insecticides** are used to control insect populations by killing them. Unwanted fungal pests that can weaken plants or destroy fruits are controlled by **fungicides.** Mice and rats are killed by **rodenticides,** and plant pests (weeds) are controlled by **herbicides.** Since pesticides do not kill just pests but can kill a large variety of living things, including humans, these chemicals might be more appropriately called **biocides.** They kill many kinds of living things. A perfect pesticide is one that kills or inhibits the growth of only the specific pest organism causing a problem. The pest is often referred to as the **target organism.** However, most pesticides are not very specific and kill many **nontarget organisms** as well. For example, most insecticides kill both beneficial and pest species, rodenticides kill other animals as well as rodents, and most herbicides kill a variety of plants, both pests and nonpests.

Many of the older pesticides were very stable and remained active for long periods of time. These are called **persistent pesticides.** Pesticides that break down quickly are called **nonpersistent pesticides.**

INSECTICIDES

If insects are not controlled, they consume a large proportion of the crops produced by farmers. In small garden plots, insects can be controlled by manually removing them and killing them. However, in large fields, this is not practical, so people have sought other ways to control pest insects.

Nearly 3000 years ago, the Greek poet Homer mentioned the use of sulfur to control insects. For centuries, it was known that natural plant products could repel or kill insect pests. Plants with insect-repelling abilities were interplanted with crops to help control the pests. Nicotine from tobacco, rotenone from tropical legumes, and pyrethrum from chrysanthemums were extracted and used to control insects. In fact, these compounds are still used today. However, because plant products are difficult to extract and apply and have short-lived effects, other compounds were sought. In 1867, the first synthetic inorganic insecticide, Paris green, was

The first synthetic organic insecticide to be used was DDT [1,1,1-trichloro-2,2-*bis*-(p-chlorophenyl)ethane] (see the figure). DDT was originally thought to be the perfect insecticide. It was inexpensive, long-lasting, relatively harmless to humans, and very deadly to insects. After its discovery, it was widely used in agriculture and to control disease-carrying insects. During the first 10 years of its use (1942–52), DDT is estimated to have saved 5 million lives, primarily because of its use in controlling disease-carrying mosquitoes.

However, DDT had several drawbacks. Although most of these drawbacks are not unique to DDT, they were first recognized with DDT because it was the first widely used insecticide. One problem involved the development of resistance in insect populations that were repeatedly subjected to spraying by DDT. Another was the fact that it affected many nontarget organisms, not just the disease-carrying insects that were the original targets. This was a particular problem because DDT is a persistent chemical, which is another major drawback. In temperate regions of the world, DDT has a half-life (the amount of time required for half of the chemical to decompose) of 10 to 15 years. This means that if 1000 kilograms (2200 pounds) of DDT were sprayed over an area, 500 kilograms (1100 pounds) would still be present in the area 10 to 15 years later; 30 years from the date of application, 250 kilograms (550 pounds) would still be present. The half-life of DDT varies depending on soil type, temperature, the kinds of soil organisms present, and other factors. In tropical parts of the world, the half-life may be as short as six months. An additional complication is that persistent pesticides may break down into products that are still harmful. Furthermore, because it is persistent, DDT tends to accumulate and reach higher concentrations in older animals and in animals at higher trophic levels.

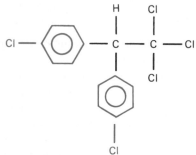

The chemical structure of DDT.

When these effects were recognized, the United States and many other developed countries banned the use of DDT. A key force in bringing about this change in thinking was the book *Silent Spring,* by Rachel Carson. (See Case Study: Early Philosophers of Nature in chapter 2.) The results of this ban are clearly seen in the reduced levels of DDT in the environment and in the recovery of populations of such organisms as eagles, cormorants, and pelicans.

Because of these problems, the World Wildlife Fund asked delegates at a United Nations Environment Programme conference on persistent organic pollutants in 1999 to ban DDT from use throughout the world because it was showing up worldwide in the bodies of many kinds of animals that were great distances from sources of DDT and was present in the breast milk of women where DDT is used. This resulted in a great deal of protest from public health people concerned about controlling malaria, since DDT is still widely used in many parts of the world, where it is sprayed on the walls of houses to kill mosquitoes that spread malaria. The use of DDT is still effective and is much less expensive than materials that would be substituted for it. They feared that the total ban of DDT would result in less effective control of mosquitoes and increased deaths from malaria. Where the risk of dying from malaria is high and the risk of possible health effects from exposure to DDT is low, it makes sense to use an inexpensive, effective material such as DDT. In countries where malaria and other serious insect-transmitted diseases are rare (most of the developed world), it is possible to eliminate the use of DDT. Ultimately, continued use of DDT was sanctioned for the control of malaria and other diseases transmitted by insects, and it continues to be used for these purposes today.

formulated. It was a mixture of acetate and arsenide of copper and was used to control Colorado potato beetles.

In addition, many insects harm humans because they spread diseases, such as sleeping sickness, bubonic plague, and malaria. Mosquitoes are known to carry over 30 diseases harmful to humans. Currently, the World Health Organization estimates that between 300 million and 500 million people have malaria and 1.1 million people die of the disease each year. Most of these deaths are children. Malaria is one of the top five causes of death in children. The discovery of chemicals that could kill insects was celebrated as a major advance in the control of disease and the protection of crops.

DDT was the first synthetic organic insecticide produced. (See Case Study: DDT—A Historical Perspective.) Since then, over 60,000 different compounds that have potential as insecticides have been synthesized. However, most of these have never been put into production because cost, human health effects, or other drawbacks make them unusable. Several categories of these compounds have been developed. Three that are currently used are chlorinated hydrocarbons, organophosphates, and carbamates.

Chlorinated Hydrocarbons

Chlorinated hydrocarbons are a group of pesticides of complex, stable structure that contain carbon, hydrogen, and chlorine. DDT was the first such pesticide manufactured, but several others have been developed. Other chlorinated hydrocarbons are chlordane, aldrin, heptachlor, dieldrin, and endrin. It is not fully understood how these compounds work, but they are believed to affect the nervous systems of insects, resulting in their death.

One of the major characteristics of these pesticides is that they are very stable chemical compounds. This is both an advantage and a disadvantage. They can be applied once and be effective for a long time. However, since they do not break down easily, they tend to accumulate in the soil and in the bodies of animals in the food chain. Thus, they affect many nontarget organisms, not just the original target insects.

Because of their negative effects, most of the chlorinated hydrocarbons are no longer used in many parts of the world. DDT, aldrin, dieldrin, toxaphene, chlordane, and heptachlor have been banned in the United States and many other developed countries. However, many developing countries still use chlorinated hydrocarbons for insect control to protect crops and public health. Because of their persistence and continued use in many parts of the world, chlorinated hydrocarbons are still present in the food chain, although the level of contamination has dropped. These molecules continue to enter parts of the world where their use has been banned through the atmosphere and as trace contaminants of imported products.

Organophosphates and Carbamates

Because of the problems associated with persistent insecticides, nonpersistent insecticides that decompose to harmless products in a few hours or days were developed. However, like other insecticides, these are not species-specific; they kill beneficial insects as well as harmful ones. Although the short half-life prevents the accumulation of toxic material in the environment, it is a disadvantage for farmers, since more frequent applications are required to control pests. This requires more labor and fuel and, therefore, is more expensive.

Both **organophosphates** and **carbamates** work by interfering with the ability of the nervous system to conduct impulses normally. Under normal conditions, a nerve impulse is conducted from one nerve cell to another by means of a chemical known as a neurotransmitter. One of the most common neurotransmitters is acetylcholine. When this chemical is produced at the end of one nerve cell, it causes an impulse to be passed to the next cell, thereby transferring the nerve message. As soon as this transfer is completed, an enzyme known as cholinesterase destroys acetylcholine, so the second nerve cell in the chain is stimulated for only a short time. Organophosphates and carbamates interfere with cholinesterase, preventing it from destroying acetylcholine. This results in nerve cells being continuously stimulated, causing uncontrolled spasms of nervous activity and uncoordination that result in death.

Although these pesticides are less persistent in the environment than are chlorinated hydrocarbons, they are generally much more toxic to humans and other vertebrates because these insecticides affect their nerve cells as well. Persons who apply such pesticides must use special equipment and should receive special training because improper use can result in death. Since organophosphates interfere with cholinesterase more strongly than do carbamates, they are considered more dangerous and, for many applications, have been replaced by carbamates.

Common organophosphates are malathion, parathion, and diazinon. Malathion is widely used for such projects as mosquito control, but parathion is a restricted organophosphate because of its high toxicity to humans. Diazinon is widely used in gardens. Carbaryl, propoxur, and aldicarb are examples of carbamates. Carbaryl (Sevin) is widely used in home gardens and in agriculture to control many kinds of insects. Propoxur is not used on crops but is used to control insects around homes and farms. Aldicarb is a restricted-use insecticide that is used primarily on cotton, soybeans, and peanuts. It has been associated with groundwater contamination and has been discontinued for some uses, such as control of insects in potatoes.

HERBICIDES

Herbicides are another major class of chemical control agents. In fact, about 60 percent of the approximately 440 million kilograms (about 1 billion pounds) of pesticides used in U.S. agriculture are herbicides. They are widely used to control unwanted vegetation along power-line rights-of-way, railroad rights-of-way, and highways, as well as on lawns and cropland, where they are commonly referred to as weed killers.

Weeds are plants we do not want to have growing in a particular place. Weed control is extremely important for agriculture because weeds take nutrients and water from the soil, making them unavailable to the crop species. In addition, weeds may shade the crop species and prevent it from getting the sunlight it needs for rapid growth. At harvest time, weeds reduce the efficiency of harvesting machines. Also, weeds generally must be sorted from the crop before it can be sold, which adds to the time and expense of harvesting.

Traditionally, farmers have expended much energy trying to control weeds. Initially, weeds were eliminated with manual labor and the hoe. Tilling the soil also helps to control weeds. Once the crop is planted, row crops such as corn or sugar beets may be cultivated to remove weeds from between the rows. All of these activities are expensive in terms of time and fuel. Selective use of herbicides can have a tremendous impact on a farmer's profits.

Many of the recently developed herbicides can be very selective if used appropriately. Some are used to kill weed seeds in the soil before the crop is planted, while others are used after the weeds and the crop begin to grow. In some cases, a mixture of herbicides can be used to control several weed species simultaneously. Figure 14.6 shows the effects of using a herbicide that kills grasses but not other kinds of plants.

Several major types of herbicides are in current use. One type is synthetic plant-growth regulators that mimic natural-growth regulators known as **auxins**. Two of the earliest herbicides were of this

FIGURE 14.6 **The Effect of Herbicides** The grasses in this photograph have been treated with herbicides. The soybeans are unaffected and grow better without competition from grasses.

type: 2,4-dichlorophenoxyacetic acid (2,4-D) and 2,4,5-trichloro-phenoxyacetic acid (2,4,5-T). When applied to broadleaf plants, these chemicals disrupt normal growth, causing the death of the plant. 2,4-D has been in use for about 50 years and is still one of the most widely used herbicides. Many newer herbicides have other methods of action. Some disrupt the photosynthetic activity of plants, causing their death. Others inhibit enzymes, precipitate proteins, stop cell division, or destroy cells directly. Depending on the concentration of the herbicide used, some are toxic to all plants, while others are very selective as to which species of plants they affect. One such herbicide is diuron. In proper concentrations and when applied at the appropriate time, it can be used to control annual grasses and broadleaf weeds in over 20 different crops. However, at higher concentrations, it kills all vegetation in an area. Fenuron is an herbicide that kills woody plants. In low concentration, it is used to control woody weed plants in cropland. In high concentrations, it is used on noncroplands, such as power-line rights-of-way. (See figure 14.7.)

Atrazine is widely used to control broad-leaf and grassy weeds in corn, sorghum, sugarcane, pineapple, Christmas trees, and other crops, and in conifer reforestation plantings. Glyphosate is a broad-spectrum, nonselective, systemic herbicide used to control annual and perennial plants, including grasses, sedges, broad-leaved weeds, and woody plants. It can be used on noncropland as well as on a great variety of crops.

FUNGICIDES AND RODENTICIDES

Fungus pests can be divided into two categories. Some are natural decomposers of organic material, but when the organic material being destroyed happens to be a crop or other product useful to humans, the fungus is considered a pest. Other fungi are parasites on crop plants; they weaken or kill the plants, thereby reducing the yield. Fungicides are used as fumigants (gases) to protect agricultural products from spoilage, as sprays and dusts to prevent the spread of diseases among plants, and as seed treatments to protect seeds from rotting in the soil before they have a chance to germinate. Methylmercury is often used on seeds to protect them from spoilage before germination. However, since methylmercury is extremely toxic to humans, these seeds should never be used for food. To reduce the chance of a mix-up, treated seeds are usually dyed a bright color.

Like fungi, rodents are harmful because they destroy food supplies. In addition, they can carry disease and damage crops in the field. In many parts of the world, such as India, the government pays a bounty to people who kill rats, because this is an inexpensive way to protect the food supply. Several kinds of rodenticides have been developed to control rodents. One of the most widely used is warfarin, a chemical that causes internal bleeding in animals that consume it. It is usually incorporated into a food substance so that rodents eat warfarin along with the bait. Because it is effective in all mammals, including humans, it must be used with care to prevent nontarget animals from having access to the chemical. As with many kinds of pesticides, some populations of rodents have become tolerant of warfarin, while others avoid baited areas. In many cases, rodent problems can be minimized by building storage buildings that are rodent-proof, rather than relying on rodenticides.

OTHER AGRICULTURAL CHEMICALS

In addition to herbicides, other agrochemicals are used for special applications. For example, a synthetic auxin sprayed on cotton plants before harvest causes the leaves to drop off, which facilitates the harvesting process by reducing clogging of the mechanical cotton picker.

NAA (naphthaleneacetic acid) is used by fruit growers to prevent apples from dropping from the trees and being damaged. This chemical can keep the apples on the trees for up to 10 extra days, which allows for a longer harvest period and fewer lost apples.

Under other conditions, it may be valuable to get fruit to fall more easily. Cherry growers use ethephon to promote loosening of the fruit so that the cherries will fall more easily from the tree when shaken by the mechanical harvester. This method lowers the cost of harvesting the fruit. (See figure 14.8.)

FIGURE 14.7   Herbicide Use to Maintain Rights-of-Way Power-line rights-of-way are commonly maintained by herbicides that kill the woody vegetation, which might grow so tall that it interferes with the power lines.

FIGURE 14.8   Chemical Loosening of Cherries to Allow Mechanical Harvesting By using chemicals and machinery, this farmer can rapidly harvest the cherry crop. This practice reduces the amount of labor required to pick the cherries but requires the application of chemicals to loosen the fruit.

Each summer, a major "dead zone" of about 18,000 square kilometers (7000 square miles) develops in the Gulf of Mexico off the mouth of the Mississippi River. This dead zone contains few fish and bottom-dwelling organisms. It is caused by low oxygen levels (hypoxia) brought about by the rapid growth of algae and bacteria in the nutrient-rich waters. The nutrients can be traced to the extensive use of fertilizer in the major farming areas of the central United States, farming areas drained by the Mississippi River and its tributaries. About 1.6 million metric tons of nitrogen—mostly fertilizer runoff from midwestern farms—flows out of the Mississippi River every year.

The hypoxia problem begins when nitrogen and other nutrients wash down the Mississippi River to the Gulf of Mexico, where they trigger a bloom of microscopic plants and animals. The dead cells and fecal matter of the organisms then fall to the seafloor. As growing colonies of bacteria digest this waste, they consume dissolved oxygen faster than it can be replenished. The flow of oxygen-rich water from the Mississippi cannot rectify the problem, because differences in temperature and density cause the warm freshwater to float above the colder, salty ocean water. Crustaceans, worms, and any other animals that cannot swim out of the hypoxic zone die.

According to the EPA, nutrient pollution has degraded more than half of U.S. estuaries. In 2002, the National Research Council named nutrient pollution and the sustainability of fisheries as the most important problems facing the U.S. coastal waters in the next decade.

Gulf of Mexico fisheries, which generate some $2.8 billion a year in revenue, are one potential casualty of hypoxia. Hypoxia can block crucial migration of shrimp, which must move from inland nurseries to feed and spawn offshore. In other places in the world, such as the Black and Baltic Seas, hypoxia has been responsible for the collapse of some commercial fisheries.

Several approaches could reduce the amount of hypoxia-causing nitrogen released into the Mississippi River Basin. These include:

- Reduce use of nitrogen-based fertilizers and improve storage of manure. Reduce runoff from feedlots.

- Plant perennial crops instead of fertilizer-intensive corn and soybeans on 10 percent of the acreage.

- Remove nitrogen and phosphorus from domestic wastewater.

- Restore 2 million to 4 million hectares (5–10 million acres) of wetlands, which absorb nitrogen runoff.

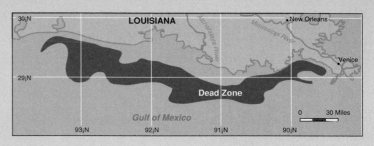

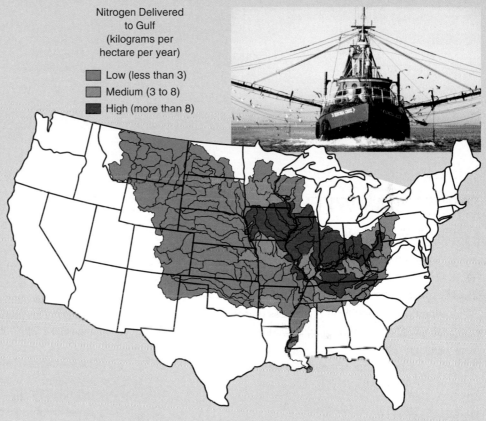

Nitrogen Delivered to Gulf (kilograms per hectare per year)

- Low (less than 3)
- Medium (3 to 8)
- High (more than 8)

Source: Reprinted by permission from Macmillan Publishers Ltd: *Nature*, 403; 761, Effect of stream channel size on the delivery of nitrogen to the Gulf of Mexico, copyright © 2000.

Source: Dead Zone map modified from R. B. Alexander, R. A. Smith, and G. E. Schwartz, "Effect of Stream Channel Size on the Delivery of Nitrogen to the Gulf of Mexico," *Nature*, 17 (February 2000): 761.

INTEGRATED PEST MANAGEMENT AT SEATTLE UNIVERSITY

Seattle University differs from many other campuses by rejecting the notion that pesticide application is a viable last resort in dealing with harmful insects and disease infestation on university grounds. A shift to sustainable landscape practices began in 1979 with the adoption of an Integrated Pest Management (IPM) program. The Grounds Department has successfully and beautifully maintained the campus since 1986 without the use of pesticides. The main components of Seattle University's IPM program include the following:

- Mechanical—removal or disruption of pests using various traps and tools
- Cultural—planting plants of different species that are naturally disease or insect resistant

- Biological—release of beneficial insects that prey on harmful pests
- Chemical—use of compounds (i.e., insecticidal soaps, compost tea, vinegar, and citric acid)

In addition, plants for use on the campus are chosen that need little water, are well suited to the northwest climate, resist insect and disease infestation, and are not invasive species that out-compete native species. Selecting combinations of plants that meet these criteria and are also aesthetically pleasing, low maintenance, affordable, and contribute to the creation of wildlife habitat is a subjective and sometimes difficult process that the university gardeners work hard to achieve.

PROBLEMS WITH PESTICIDE USE

A perfect pesticide would have the following characteristics:

1. It would be inexpensive.
2. It would affect only the target organism.
3. It would have a short half-life.
4. It would break down into harmless materials.

However, the perfect pesticide has not been invented. Many of the more recently developed pesticides have fewer drawbacks than the early pesticides, but none is without problems.

PERSISTENCE

Although the trend has been away from using persistent pesticides in North America and much of the developed world, some are still allowed for special purposes, and they are still in common use in other parts of the world. Because of their stability, these chemicals have become a long-term problem. Persistent pesticides become attached to small soil particles, which are easily moved by wind or water to any part of the world. Persistent pesticides and other pollutants have been discovered in the ice of the poles and are present in detectable amounts in the body tissues of animals, including humans, throughout the world. Thus, chemicals originally sprayed to control mosquitoes in Africa or to protect a sugarcane field in Brazil may be distributed throughout the world.

BIOACCUMULATION AND BIOMAGNIFICATION

A problem associated with persistent chemicals is that they may accumulate in the bodies of animals. If an animal receives small quantities of persistent pesticides or other persistent pollutants in its food and is unable to eliminate them, the concentration within the animal increases. This process of accumulating higher and higher amounts of material within the body of an animal is called **bioaccumulation.** Many of the persistent pesticides and their breakdown products are fat soluble and build up in the fat of animals. When affected animals are eaten by a carnivore, these toxins are further concentrated in the body of the carnivore, causing disease or death, even though lower-trophic-level organisms are not injured. This phenomenon of acquiring increasing levels of a substance in the bodies of higher-trophic-level organisms is known as **biomagnification.**

The well-documented case of DDT is an example of how biomagnification occurs. DDT is not very soluble in water but dissolves in oil or fatty compounds. When DDT falls on an insect or is consumed by the insect, the DDT is accumulated in the insect's fatty tissue. Large doses kill insects but small doses do not, and their bodies may contain as much as one part per billion of DDT. This is not very much, but it can have a tremendous effect on the animals that feed on the insects.

If an aquatic habitat is sprayed with a small concentration of DDT or receives DDT from runoff, small aquatic organisms may accumulate a concentration that is up to 250 times greater than the concentration of DDT in the surrounding water. These organisms

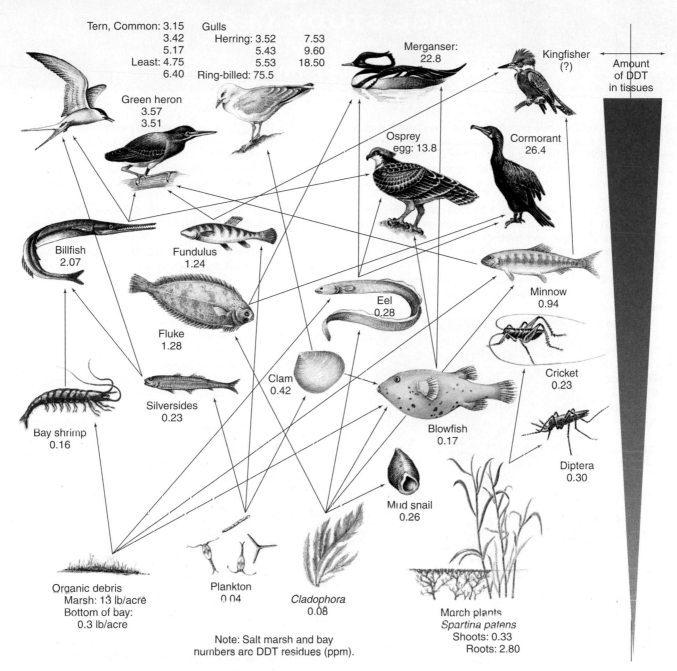

Tern, Common: 3.15
3.42
5.17
Least: 4.75
6.40

Gulls
Herring: 3.52   7.53
5.43   9.60
5.53   18.50
Ring-billed: 75.5

Merganser: 22.8

Kingfisher (?)

Amount of DDT in tissues

Green heron 3.57 3.51

Osprey egg: 13.8

Cormorant 26.4

Billfish 2.07

Fundulus 1.24

Minnow 0.94

Fluke 1.28

Eel 0.28

Clam 0.42

Cricket 0.23

Silversides 0.23

Blowfish 0.17

Diptera 0.30

Bay shrimp 0.16

Mud snail 0.26

Organic debris
Marsh: 13 lb/acre
Bottom of bay:
0.3 lb/acre

Plankton 0.04

Cladophora 0.08

Marsh plants
Spartina patens
Shoots: 0.33
Roots: 2.80

Note: Salt marsh and bay numbers are DDT residues (ppm).

FIGURE 14.9   The Biomagnification of DDT All the numbers shown are in parts per million (ppm). A concentration of one part per million means that in a million equal parts of the organism, one of the parts would be DDT. Notice how the amount of DDT in the bodies of the organisms increases as we go from producers to herbivores to carnivores. Because DDT is persistent, it builds up in the top trophic levels of the food chain.

are eaten by shrimp, clams, and small fish, which are, in turn, eaten by larger fish. DDT concentrations of large fish can be as much as 2000 times the original concentration sprayed on the area. What was a very small initial concentration has now become so high that it could be fatal to animals at higher trophic levels. This has been of particular concern for birds, since DDT interferes with the production of eggshells, making them much more fragile. This problem is more common in carnivorous birds because they are at the top of the food chain. Although all birds of prey have probably been affected to some degree, those that rely on fish for food seem to have been affected most severely. Eagles, osprey, cormorants, and pelicans are particularly susceptible species. (See figure 14.9.)

Other persistent molecules are known to behave in similar fashion. Mercury, aldrin, chlordane, and other chlorinated hydrocarbons, such as polychlorinated biphenyls (PCBs) used as insulators in electric transformers, are all known to accumulate in ecosystems. PCBs have been strongly implicated in the decline of cormorants in the Great Lakes. As PCB levels have declined, the cormorant population has returned to former levels.

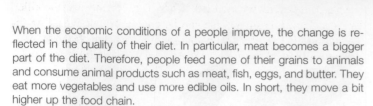
When the economic conditions of a people improve, the change is reflected in the quality of their diet. In particular, meat becomes a bigger part of the diet. Therefore, people feed some of their grains to animals and consume animal products such as meat, fish, eggs, and butter. They eat more vegetables and use more edible oils. In short, they move a bit higher up the food chain.

A sizable percentage of China's nearly 1.3 billion citizens are moving up the food chain. In the early 1980s, the typical urban Chinese diet consisted of rice, porridge, and cabbage. By the middle 1990s, the diet had dramatically changed to include meat, eggs, or fish at least once a day.

China has about 21 percent of the world's population but only 7 percent of the arable land; in other words, it has less arable land than the United States but nearly five times the population. China's farm sector is simply unable to keep up with the surging demand for what people regard as better food.

Total meat consumption in China is growing by 10 percent a year; feed for animal consumption is growing by 15 percent. Demand for poultry, which requires 2 to 3 kilograms of feed per kilogram of bird, has doubled in five years. China's new diet will make it more dependent on the United States, Canada, and Australia for feed

KFC and McDonald's in Beijing. Demand for poultry and beef is soaring.

grains. By 1997, China had gone from being a net exporter of grain to importing 16 million metric tons. The switch in corn is even more dramatic. As recently as 1995, China was the second-largest corn exporter in the world. But with the chickens and pigs eating so much corn, in one year, China moved from exporting 12 million metric tons of corn to importing 4 million metric tons.

To support its increased agricultural needs, China has become the world's largest importer of fertilizer. In addition, the government is converting some marginal lands in the north for agriculture. Even so, its farmland is declining by at least 0.5 percent per year because of the loss of fertile, multiple-cropped farmland in the southern coastal provinces.

As in much of the world, irrigation is responsible for a major portion of agricultural productivity in China. In the Yellow River Valley's large grain belt, irrigation projects tripled crop yields during the 1950s and 1960s, resulting in severe overpumping of groundwater. Today, the water table is falling, aquifers are vanishing, and farmers now have to compete with industry and households for water.

China is only a part—but certainly the biggest part—of a larger story. What's true for China is also happening throughout much of Asia. Large, populous countries such as Indonesia, Thailand, and the Philippines are rapidly urbanizing. They are gaining purchasing power and losing farmland, and the people are adding more animal proteins and processed foods to their diets. Given the magnitude of the changes, the world's ability to feed itself in the future is difficult to predict.

Because of their persistence, their effects on organisms at higher trophic levels, and concerns about long-term human health problems, most chlorinated hydrocarbon pesticides have been banned from use in the United States and some other countries. The use of DDT was prohibited in the United States in the early 1970s. Aldrin, dieldrin, heptachlor and chlordane have also been prohibited from use on crops, although heptachlor and chlordane were still used for termite control until recently. In 1987, Velsicol Chemical Corporation agreed to stop selling chlordane in the United States.

The populations of several species of birds, including the brown pelican, bald eagle, osprey, and cormorant, all of which feed primarily on fish, were severely affected because of biomagnification of persistent organic chemicals. With the control or restriction of most persistent pesticides and several other chemicals, the levels of these chemicals in their body tissues have declined, and their populations have rebounded. The bald eagle has been removed from the endangered species list, and the pelican has been removed from the endangered species list in part of its range.

PESTICIDE RESISTANCE

Another problem associated with pesticides is the ability of pest populations (insects, weeds, rodents, fungi) to become resistant to them. Pesticide resistance develops because not all organisms within a given species are identical. Each individual has a slightly different genetic composition and slightly different characteristics. If an insecticide is used for the first time on the population of a particular insect pest, it kills all the individuals that are susceptible. Individuals with characteristics that allow them to tolerate the insecticides may live to reproduce.

If only 5 percent of the individuals possess genes that make them resistant to an insecticide, the first application of the insecticide will kill 95 percent of the exposed population and so will be of great benefit in controlling the size of the insect population. However, the surviving individuals that are tolerant of the insecticide will constitute the majority of the breeding populations. Since these individuals possess genetic characteristics for tolerating the insecticide, so will many of their offspring. Therefore, in the next generation, the number of individuals able to tolerate the insecticide will increase, and the second use of the insecticide will not be as effective as the first. Since some species of insect pests can produce a new generation each month, this process of selecting individuals capable of tolerating the insecticide can result in resistant populations in which 99 percent of the individuals are able to tolerate the insecticide within five years. As a result, that particular insecticide is no longer as effective in controlling insect pests, and increased dosages or more frequent spraying may be necessary. Over 500 species of insects have populations resistant to insecticides.

For example, a study of insecticide resistance of houseflies in poultry facilities used for egg production shows that these populations of houseflies are resistant to a wide variety of insecticides compared to a nonresistant test population. Poultry facilities are an excellent situation for developing resistance, since the flies spend their entire lives within the facility and are subjected to repeated applications of insecticide. Table 14.1 shows the effect of the insecticide Cyfluthrin on resistance in a particular housefly population. The standard dose that killed susceptible houseflies did not kill the resistant population. Even at a dosage of 100 times the standard dose, nearly 40 percent of the houseflies in the resistant population survived

EFFECTS ON NONTARGET ORGANISMS

Most pesticides are not specific and kill beneficial species as well as pest species. With herbicides, this is usually not a problem because an herbicide is chosen that does not harm the desired crop plant, and generally all other plants are competing pests. However, with insecticides, several problems are associated with the effects on nontarget organisms. The use of insecticides can harm populations of birds, mammals, and harmless or beneficial insects. In general, insecticides that harm vertebrates are restricted in their use. However, it is difficult to apply insecticides in such a way that only the harmful species are affected. When beneficial insects are killed by insecticides, pesticide use can be counterproductive. If an insecticide kills predator and parasitic insects that normally control the pest insects, there are no natural checks to control the population growth of the pest species. Additional applications of insecticides are necessary to prevent the pest population from rebounding to levels even higher than the initial one. Once the decision is made to use pesticides, it often becomes an irreversible tactic, because stopping their use would result in rapid increases in pest populations and extensive crop damage.

An associated problem is that the use of insecticides may change the population structures of the species present so that a species that was not previously a problem becomes a serious pest. For example, when synthetic organic insecticides came into common use with cotton in the 1940s, the insect parasites and predators were eliminated, and the bollworm and tobacco budworm became major pests. In the mid-1990s, a similar situation developed when repeated use of malathion removed predator insects and allowed beet army worms to become a major pest. The repeated use of insecticides caused a different pest problem to develop.

HUMAN HEALTH CONCERNS

Short-term and long-term health effects to persons applying pesticides and the public that consumes pesticide residues in food are also concerns. If properly applied, most pesticides can be used with little danger to the applicator. However, in many cases, people applying pesticides are unaware of how they work and what precautions should be used in their application. In many parts of the world, farmers may not be able to read the caution labels on the packages or do not have access to the protective gear specified for use with the pesticide. Therefore, many incidences of acute poisoning occur each year. In most cases, the symptoms disappear after a period free from exposure. Estimates of the number of poisonings are very difficult to obtain, since many go unreported, but in the United States, pesticide poisonings requiring medical treatment are in the thousands per year. The World Health Organization estimates that each year, there are between 1 million and 5 million acute pesticide poisonings and that about 20,000 deaths occur from pesticide poisonings.

For most people, however, the most critical health problem related to pesticide use is inadvertent exposure to very small quantities of pesticides. Many pesticides have been proven to cause mutations, produce cancers, or cause abnormal births in experimental animals. Studies of farmers who were occupationally exposed to pesticides over many years show that they have higher levels of certain kinds of cancers than the general public.

TABLE 14.1 Insecticide Resistance of Houseflies to the Insecticide Cyfluthrin

	Houseflies from a Susceptible Population	Houseflies from a Poultry Facility Population		
Concentration (ng/cm²)	8.3	8.3 (1×)	83 (10×)	830 (100×)
Percent survival	0	100	90	38

Source: Data from Scott, Jeffrey G., et al., "Insecticide Resistance in Houseflies from Caged-Layer Poultry Facilities," *Pest Management Science* 56 (2000): 147–153.

There also are questions about the effects of chronic, minute exposures to pesticide residues in food or through contamination of the environment.

Following the rapid advance of modern agriculture in developed countries, developing countries have, over the past half century, increasingly adopted a pest management approach that centers on the use of chemical pesticides. The pesticide world market's total value surpasses US $26 billion. Developing countries' share of this is approximately one-third. It is reasonable to expect that developing countries will experience the same kinds of health problems and environmental side effects that led industrialized countries to develop pesticide-regulating laws in the 1970s.

A variety of factors is likely to make people in developing countries—especially farmers and agricultural workers—more vulnerable to the toxicological effects of pesticides. These include:

- Low literacy and education levels
- Weak or absent legislative and regulatory frameworks
- Climate factors (which make the use of protective clothing while spraying pesticides uncomfortable)
- Inappropriate or faulty spraying technology
- Lower nutritional status (weaker physiological defenses against toxic substances)

The first four of these factors increase the likelihood of higher exposure; the last factor increases the toxic effects from that exposure on the human body. In addition, it has been found that when pesticides are banned or restricted, people in developing countries have often turned to substances that exhibit even higher toxicity.

Although the risks of endangering health by consuming tiny amounts of pesticides are very small compared to other risk factors, such as automobile accidents, smoking, or poor eating habits, many people find pesticides unacceptable and seek to prohibit their sale and use. The U.S. Environmental Protection Agency requires careful studies of the effectiveness and possible side effects of each pesticide licensed for use. It has banned several pesticides from further use because new information suggests that they are not as safe as originally thought. For example, it banned the use of dinoseb, an herbicide, because tests by the German chemical company Hoechst AG indicated that dinoseb causes birth defects in rabbits. Other studies indicate that dinoseb causes sterility in rats. Similarly, in 1987, Velsicol Chemical Corporation signed an agreement with the EPA to stop producing and distributing chlordane in the United States. Chlordane had been banned previously for all applications except for termite control. After it was shown that harmful levels of chlordane could exist in treated homes, its use was finally curtailed. Also in 1987, the Dow Chemical Company announced that it would stop producing the controversial herbicide 2,4,5-T. In both of these cases, new pesticides were developed to replace the older types, so that discontinuing them did not result in an economic hardship for farmers or other consumers.

Finally, during the 1990s, the EPA required that many pesticides be reregistered. During this process, manufacturers needed to justify the continued production of the product. In many cases, companies simply did not want to spend the money to go through the registration process; therefore, the product was withdrawn from the market. In other cases, the variety of uses was modified. For example, chlorpyrifos, one of the most widely used insecticides, had its use altered. Through negotiations with the producers, its use in homes and schools will be eliminated, while it will still be used in most agricultural settings.

WHY ARE PESTICIDES SO WIDELY USED?

Figure 14.10 shows changes in the use of pesticides throughout the world. Use is projected to continue to increase significantly. If pesticides have so many drawbacks, why are they used so extensively? There are three primary reasons. First, the use of pesticides has increased, at least in the short term, the amount of food that can be grown in many parts of the world. In the United States, pests are estimated to consume 33 percent of the crops grown. On a worldwide basis, pests consume approximately 35 percent of crops. This represents an annual loss of $18.2 billion in the United States alone. Farmers, grain-storage operators, and the food industry continually seek to reduce this loss. A retreat from dependence on pesticides would certainly reduce the amount of food produced. Agricultural planners in most countries are not likely to suggest changes in pesticide use that would result in malnutrition and starvation for many of their inhabitants.

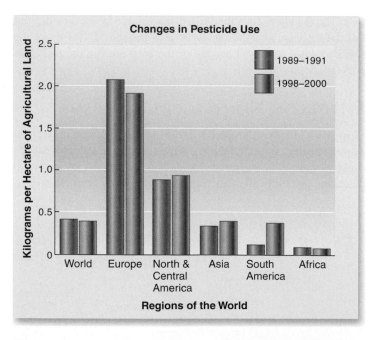

FIGURE 14.10   **Changes in Pesticide Use** The use of pesticides is greatest in the developed world, comprising about 67 percent of the total. However, all parts of the world except Europe are experiencing increased use of pesticides. The use in Africa is low and relatively stable.
Source: Data from Food and Agriculture Organization of the United Nations.

Organic food is more than a trendy industry that provides healthful produce to co-ops and upscale markets. Sustainable agriculture can help fight world hunger and some of its related challenges, including a shrinking safe water supply and pesticide contamination. This news comes at a critical time for reasons such as the following:

- The majority of the people who are chronically hungry in the world are small-scale farmers and their families who depend on the land to give them the food they need to survive.

- In many regions where hunger is a major problem, water for agriculture is also scarce.

- Pesticides used in conventional farming get into bodies of water, suffocate aquatic plants and animals, and accumulate in the food chain, eventually contaminating plants and animals that are consumed by humans and throwing off the environmental balance.

Sustainable agriculture methods, which include organic farming and crop rotation, are providing satisfying results for many small-scale farmers around the world. According to the United Nations, in 57 of the world's poorest regions, small-scale farmers who use sustainable agriculture techniques have increased their yields by 71 percent. These farmers are experiencing other benefits:

- Because farmers who practice sustainable agriculture methods use few or no pesticides, they can avoid that expense.

- Organic farming also requires less water, because soils that are high in organic material are much more efficient at holding water than is poorer soil. High-quality soil is better able to deal with drought conditions, a major consideration in many impoverished nations, especially in Africa.

- Organic farming is often more labor intensive than conventional farming, especially in the beginning stages. However, because organic farming typically involves crop rotation throughout the year, farmers have more evenly distributed planting and harvesting schedules and better distribution of labor.

- Crop rotation spreads out the risk of crop failure; it can also make farmers more competitive in the marketplace if they sell some of their produce. It also provides farmers and their families with a more nutritious and varied diet.

The economic value of pesticides is the second reason they are used so extensively. The cost of pesticides is more than offset by increased yields and profits for the farmer. In addition, the production and distribution of pesticides is big business. Companies that have spent millions of dollars developing a pesticide are going to argue very strongly for its continued use. Since farmers and agrochemical interests have a powerful voice in government, they have successfully lobbied for continued use of pesticides.

A third reason for extensive pesticide use is that many health problems are currently impossible to control without insecticides. This is particularly true in areas of the world where insect-borne diseases would cause widespread public health consequences if insecticides were not used.

ALTERNATIVES TO CONVENTIONAL AGRICULTURE

Before the invention of synthetic fertilizers, herbicides, fungicides, and other agrochemicals, animal manure and crop rotation provided soil nutrients; a mixture of crops prevented regular pest problems; and manual labor killed insects and weeds. With the development of mechanization, larger areas could be farmed, draft animals were no longer needed, and many farmers changed from mixed agriculture, in which animals were an important ingredient, to monoculture. Chemical fertilizers replaced manure as a source of soil nutrients, and crop rotation was no longer as important, since hay and grain were no longer grown for draft animals and cattle. The larger fields of crops such as corn, wheat, and cotton presented opportunities for pest problems to develop, and chemical pesticides were used to "solve" this problem.

Today, many people feel current agricultural practices are not sustainable and that we should look for ways to reduce reliance on fertilizer and pesticides, while ensuring good yields and controlling the pests that compete with us for the food that we raise. Several alternative approaches have somewhat different but overlapping goals. Some of the terms used to describe these approaches are: *alternative agriculture, sustainable agriculture,* and *organic agriculture.* **Alternative agriculture** is the broadest term. It includes all nontraditional agricultural methods and encompasses sustainable agriculture, organic agriculture, alternative uses of traditional crops, alternative methods for raising crops, and producing crops for industrial use. The **sustainable agriculture** movement maintains that current practices are degrading natural resources and seeks methods to produce adequate, safe food in an economically viable manner while enhancing the health of agricultural land and related ecosystems.

SUSTAINABLE AGRICULTURE

There has been a growing interest in sustainable agriculture. There has also been a growing amount of misleading information surrounding sustainable agriculture. In recent decades, sustainable and researchers around the world have responded to the ~~~ive industrial model with ecology-based approaches, variously ~~~ed natural, organic, low-input, alternative, regenerative, holis-~~c, biodynamic, biointensive, and biological farming systems. All of them, representing thousands of farms, have contributed to our understanding of what sustainable systems are, and all share a vision of "farming with nature" and agro-ecology that promotes biodiversity, recycles plant nutrients, protects soil erosion, conserves and protects water, uses minimum tillage, and integrates crop and livestock enterprises on the farm.

But no matter how elegant the system or how accomplished the farmer, no agriculture is sustainable it it's not also profitable and able to provide a healthy family income and a good quality of life. Sustainable practices lend themselves to smaller, family-scale farms. These farms, in turn, tend to find their best niches in local markets, within local food systems, often selling directly to consumers. As alternatives to industrial agriculture evolve, so must their markets and the farmers who serve them. Creating and serving new markets remains one of the key challenges for sustainable agriculture.

Sustainable agriculture is a method of agriculture that does not deplete soil, water, air, wildlife, or human community resources. Sustainable agriculture is a term used globally to refer to farming practices that strive for this ideal as opposed to methods that rely heavily on products such as gasoline, chemical fertilizers, and pesticides. A growing number of farmers throughout the world are growing all of the major food groups (grains, dairy, meat, vegetables, fruit) using sustainable methods. Though yields differ by crop and growing region, in general, sustainable methods do achieve average yields.

There are many different types of sustainable pest and weed control practices. They are, however, guided by some general principles. These include:

- Keeping insecticide, herbicide, fungicide, and fertilizer to a minimum.
- Biological diversity (many different types of plants and animals) should be encouraged.
- Healthy, biologically active soils lead to healthier, more insect- and disease-resistant plants and animals.
- Natural or supplemented populations of beneficial insects (good bugs) will keep pests (bad bugs) below economically damaging levels. Many sustainable farmers purposely grow plants that will attract beneficial insects.

Fertile soils have a balanced mix of minerals, organic matter, microorganisms, and macroorganisms (like earthworms). Sustainable farmers try to keep these components in balance by adding compost, minerals, and naturally occurring fertilizers (like bloodmeal or bat guano) and by plowing back into the soil crop residues or crops grown specifically for fertility.

Organic Food

Organically grown is a legally defined term in the United States that tells you how a food or fiber crop was grown. Thirty states also have their own legal definitions. The U.S. Federal Organic Foods Production Act of 1990 defines national organic standards, which generally:

- Require organic farms and organic handlers to be "certified," that is, inspected by a disinterested third party
- Require that an organic farm be increasing its soil fertility through sustainable soil-building techniques
- Prohibit synthetic pesticides and fertilizers

One concern about organically grown food is its cost to the consumer. While the cost of organic products can be more expensive, they can also be thought of as less expensive in the long run. The organic market, while growing, is still limited by low supply, so prices are higher than they might otherwise be. In addition, most organic food is produced by smaller farms that do not have the economics of large-scale industrial agribusiness. You can support sustainable agriculture by growing your own food in whatever space you have and by buying organic foods at food co-ops, natural food stores, or farmer markets. You can also ask your local supermarket to carry organic foods.

TECHNIQUES FOR PROTECTING SOIL AND WATER RESOURCES

In order to reduce the negative impacts of conventional agriculture while making a profit, farmers must deal with the twin problems of protecting the quality and fertility of the soil while controlling pests of their crops. A wide variety of techniques can be used to reduce the negative impact of conventional agricultural activity and maintain economic viability for the farmer.

Conventional farming practices have several negative effects on soil and water. Soil erosion is a problem throughout the world. (See chapter 13.) However, two other problems are also important: compaction of the soil and reduction in soil organic matter. Several changes in agricultural production methods can help to reduce these problems. Reducing the number of times farm equipment travels over the soil will reduce the degree of compaction. Leaving crop residue on the soil and incorporating it into the soil reduces erosion and increases soil organic matter. In addition, introducing organic matter into the soil makes compaction less likely.

Conventional agriculture has several negative impacts on watersheds: fertilizer runoff stimulates aquatic growth and degrades water resources; pesticides can accumulate in food chains; and groundwater resources can be contaminated by fertilizer, pesticides, or animal waste. Reducing or eliminating these sources of contamination would enhance ecological harmony and reduce a threat to human health. Fertilizer runoff can be lessened by reducing the amount of fertilizer applied and the conditions under which it is applied. Applying fertilizer as plants need it will ensure that more of it is taken up by plants and less runs off. Increased organic matter in the soil also tends to reduce runoff. More careful selection, timing, and use of pesticides would decrease the extent to which these materials become environmental contaminants.

Precision agriculture is a new technique that addresses many of these concerns. With modern computer technology and geographic information systems, it is now possible, based on the soil and topography, to automatically vary the chemicals applied to the crop at different places within a field. Thus, less fertilizer is used, and it is used more effectively.

True organic agriculture that uses neither chemical fertilizers nor pesticides is the most effective in protecting soil and water resources from these forms of pollution, but it requires several adjustments in the way in which farming is done. Crop rotation is an effective way to enhance soil fertility, reduce erosion, and control pests. The use of nitrogen-fixing legumes, such as clover, alfalfa, beans, or soybeans, in crop rotation increases soil nitrogen but places other demands on the farmer. For example, it typically requires that cattle be a part of the farmer's operation in order to make use of forage crops and provide organic fertilizer for subsequent crops. Crop rotation also requires a greater investment in farm machinery, since certain crops require specialized equipment. Also, the raising of cattle requires additional expenditures for feed supplements and veterinary care. Critics say that organic farming cannot produce the amount of food required for today's population and that it can be economically successful only in specific cases. Proponents disagree and stress that when the hidden costs of soil erosion and pollution are included, organic agriculture or some modification of it is a viable alternative approach to conventional means of food production. Furthermore, organic farmers are willing to accept lower yields because they do not have to pay for expensive chemical fertilizers and pesticides. In addition, organic farmers often receive premium prices for products that are organically grown. Thus, even with lower yields, they can still make a profit.

INTEGRATED PEST MANAGEMENT

Integrated pest management uses a variety of methods to control pests rather than relying on pesticides alone. Integrated pest management is a technique that depends on a complete understanding of all ecological aspects of the crop and the particular pests to which it is susceptible to establish pest control strategies. It requires information about the metabolism of the crop plant, the biological interactions between pests and their predators or parasites, the climatic conditions that favor certain pests, and techniques for encouraging beneficial insects. It may involve the selective use of pesticides. Much of the information necessary to make integrated pest management work goes beyond the knowledge of the typical farmer. The metabolic and ecological studies necessary to pinpoint weak points in the life cycles of pests can usually be carried out only at universities or government research institutions. These studies are expensive and must be completed for each kind of pest, since each pest has a unique biology. Once a viable technique has been developed, an educational program is necessary to provide farmers with the information they need in order to use integrated pest management rather than the "spray and save" techniques that they used previously and that pesticide salespeople continually encourage.

Several methods are employed in integrated pest management. These include disrupting reproduction, using beneficial organisms to control pests, developing resistant crops, modifying farming practices, and selectively using pesticides.

Disrupting Reproduction

In some species of insects, a chemical called a **pheromone** is released by females to attract males. Males of some species of moths can detect the presence of a female from a distance of up to 3 kilometers (nearly 2 miles). Since many moths are pests, synthetic odors can be used to control them. Spraying an area with the pheromone confuses the males and prevents them from finding females, which results in a reduced moth population the following year. In a similar way, a synthetic sex attractant molecule known as Gyplure is used to lure gypsy moths into traps, where they become stuck. Since the females cannot fly and the males are trapped, the reproductive rate drops, and the insect population may be controlled.

Another technique that reduces reproduction is male sterilization. In the southern United States and Central America, the screwworm fly weakens or kills large grazing animals, such as cattle, goats, and deer. The female screwworm fly lays eggs in open wounds on these animals, where the larvae feed. However, it was discovered that the female mates only once in her lifetime. Therefore, the fly population can be controlled by raising and releasing large numbers of sterilized male screwworm flies. Any female that mates with a sterile male fails to produce fertilized eggs and cannot reproduce. In Curaçao, an island 65 kilometers (40 miles) north of Venezuela, a program of introducing sterile male screwworm flies eliminated this disease from the 25,000 goats on the island. In parts of the southwestern United States, the sterile male technique has also been very effective. The screwworm fly has been eliminated from the United States and northern Mexico, and much of Central America may become free of them as well. In 1990, sterile males were released in Libya to begin eliminating screwworm flies that had been introduced with a South American cattle shipment.

During an epidemic of Mediterranean fruitflies in southern California and northern Mexico in the early 1980s, a similar technique was employed. Unfortunately, the X-ray technique used to sterilize the males was ineffective, and most of the flies released were not sterile, which made the problem worse rather than better. Pesticides were eventually used to control the fruitflies. Recent concern about the Mediterranean fruitfly in California has resulted in the controversial aerial spraying of malathion.

Using Beneficial Organisms to Control Pests

The manipulation of predator-prey relationships can also be used to control pest populations. For instance, the ladybird beetle, commonly called a ladybug, is a natural predator of aphids and scale insects. (See figure 14.11a.) Artificially increasing the population of ladybird beetles reduces aphid and scale populations. In California during the late 1800s, scale insects on orange trees damaged the trees and reduced crop yields. The introduction of a species of ladybird beetle from Australia quickly brought the pests under control. Years later, when chemical pesticides were first used in the area, so many ladybird beetles were accidentally killed that scale insects once again became a serious problem. When pesticide use was discontinued, ladybird beetle populations rebounded, and the scale insects were once again brought under control. (See figure 14.11b.)

(a)

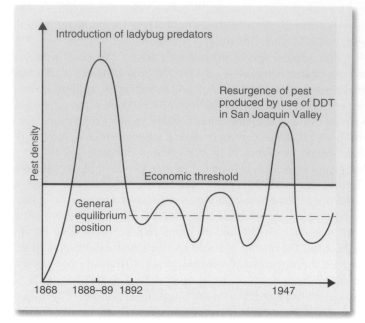

(b)

FIGURE 14.11 Insect Control with Natural Predators
(a) The ladybird beetle is a predator of many kinds of pest insects, including aphids. (b) In 1889, the introduction of ladybird beetles (ladybugs) brought the cottony cushion scale under control in the orange groves of the San Joaquin Valley. In the 1940s, DDT reduced the ladybird beetle population, and the cottony cushion scale population increased. Stopping the use of DDT allowed the ladybird population to increase, reducing the pest population and allowing the orange growers to make a profit.

Source: (b) V. M.Stern, et al., *Hilgardia*, 29: 93, 1959.

(a) Purple loosestrife

(b) Galerucella calmariensis

(c) Galerucella pusilla

(d) Hylobius transversovittatus

FIGURE 14.12 Biological Control of Purple Loosestrife (a) Purple loosestrife is a European plant that invades wetlands and prevents the growth and reproduction of native species of plants. Several kinds of European beetles that feed on purple loosestrife have been released as biological control agents. (b, c) Two species (*Galerucella calmariensis* and *Galerucella pusilla*) feed on the leaves, shoots, and flowers of the newly growing purple loosestrife. (d) Another species (*Hylobius transversovittatus*) has larvae that feed on the roots. It appears that they are actually slowing the spread of purple loosestrife.

Herbivorous insects can also be used to control weeds. Purple loosestrife (*Lythrum salicaria*) is a wetlands plant accidentally introduced from Europe in the mid-1800s. It takes over sunny wetlands and eliminates native vegetation such as cattails, thereby eliminating many species that rely on cattails for food, nesting places, or hiding places. (See figure 14.12.) The plant exists in most states and provinces in the United States and Canada. In Europe, the plant is not a pest because there are several insect species that attack it in various stages of the plant's life cycle. Since purple loosestrife is a European plant, a search was made to identify candidate species of insects that would control it. The criteria were that the insects must live only on purple loosestrife and not infest other plants and must have the capacity to do major damage to purple loosestrife. Five species of beetles have been identified that will attack the plant in various ways. After extensive studies to determine if introductions of these insects from Europe would be likely to cause other problems, several were selected as candidates to help control purple loosestrife in the United States and Canada. Some combination of these beetles has been released in a large proportion of the states and provinces infested with purple loosestrife. Two species (*Galerucella calmariensis* and *Galerucella pusilla*) feed on the leaves, shoots, and flowers of the newly growing purple loosestrife; one (*Hylobius transversovittatus*) has larvae that feed on the roots. Two other species (*Nanophyes brevis* and *Nanophyes marmoratus*) feed on the flowers. This multipronged attack by several species of beetles has been effective in reducing populations of purple loosestrife.

The use of specific strains of the bacterium *Bacillus thuringiensis* to control mosquitoes and moths is another example of the use of one organism to control another. A crystalline toxin produced by the bacterium destroys the lining of the gut of the feeding insect, resulting in its death. One strain of *B. thuringiensis* is used to control mosquitoes, while another is primarily effective against the caterpillars of leaf-eating moths, including the gypsy moth.

Naturally occurring pesticides found in plants also can be used to control pests. For example, marigolds are planted to reduce the number of soil nematodes, and garlic plants are used to check the spread of Japanese beetles.

Developing Resistant Crops

One outcome of the major advances in molecular genetics is the development of specialized organisms that have had specific, valuable genes inserted into their genetic makeup. Although this is a relatively new technology, it is an extension of an ancient art. Farmers have been involved in manipulating the genetic makeup of their plants and animals since these organisms were first domesticated. Initially, farmers either consciously or accidentally chose to plant specific seeds or breed certain animals that had specific desirable characteristics. This resulted in local varieties with particular characteristics. When the laws of genetics began to be understood in the early 1900s, scientists began to make precise crosses between carefully selected individuals to enhance the likelihood that their offspring would have certain highly desirable characteristics. This led to the development of hybrid seeds and specific breeds of domesticated animals. Controlled plant and animal breeding resulted in increased yields and better disease resistance in domesticated plants and animals. These activities are still the major driving force for developing improved varieties of domesticated organisms.

When the structure of DNA was discovered and it was determined that the DNA of organisms could be manipulated, an entirely new field of plant and animal breeding arose. **Genetic engineering** or **biotechnology** involves inserting specific pieces of DNA into the genetic makeup of organisms. The organism with the altered genetic makeup is known as a **genetically modified organism.** The DNA inserted could be from any source, even an entirely different organism. In agricultural practice, two kinds of genetically modified organisms have received particular attention. One involves the insertion of genes from a specific kind of

bacterium called *Bacillus thuringiensis israeliensis* (Bti). Bti produces a material that causes the destruction of the lining of the gut of insects that eat it. It is a natural insecticide. To date, the gene has been inserted into the genetic makeup of several crop plants, including corn. In field tests, the genetically engineered corn was protected against some of its insect pests, but there was some concern that pollen grains from the corn might be blown to neighboring areas and affect nontarget insect populations. In particular, a study of monarch butterflies indicated that populations of butterflies adjacent to fields of this genetically engineered corn were negatively affected. One could argue that since the use of Bti corn results in less spraying of insecticides in cornfields, this is just a trade-off.

A second kind of genetically engineered plant involves inserting a gene for herbicide resistance into the genome of certain crop plants. The value of this to farmers is significant. For example, a farmer could plant cotton with very little preparation of the field to rid it of weeds. When both the cotton and the weeds began to grow, the field could be sprayed with a specific herbicide that would kill the weeds but not harm the herbicide-resistant cotton. This has been field-tested and it works. Critics have warned that the genes possibly could escape from the crop plants and become part of the genome of the weeds that we are trying to control, thus creating superweeds.

Many groups oppose the use of genetically modified organisms. They argue that this technology is going a step too far, that no long-term studies have been done to ensure their safety, that there are dangers we cannot anticipate, and that if such crops are grown, they should be labeled so that the public knows when they are consuming products from genetically modified organisms. The European Union (EU) will not buy genetically modified grains from the United States. This has resulted in farmers having to segregate their stores of grains so that they can guarantee to an EU buyer that a crop is not genetically modified and at the same time sell a genetically modified crop to other buyers.

Supporters argue that all plant and animal breeding involves genetic manipulation and that this is just a new kind of genetic manipulation. A great deal of evidence exists that genes travel between species in nature and that genetic engineering simply makes a common, natural process more frequent. Over the next 20 to 30 years, scientists hope to use biotechnology to produce high-yield plant strains that are more resistant to insects and disease, thrive on less fertilizer, make their own nitrogen fertilizer, do well in slightly salty soils, withstand drought, and use solar energy more efficiently during photosynthesis. These new kinds of genetically modified organisms will continue to be developed and tested, and the political arguments about their appropriateness will continue as well.

Modifying Farming Practices

Modification of farming practices can often reduce the impact of pests. In some cases, all crop residues are destroyed to prevent insect pests from finding overwintering sites. For example, shredding and plowing under the stalks of cotton in the fall reduces overwintering sites for boll weevils and reduces their numbers significantly, thereby reducing the need for expensive insecticide applications. Many farmers are also returning to crop rotation, which tends to prevent the buildup of specific pests that typically occurs when the same crop is raised in a field year after year.

Selective Use of Pesticides

Pesticides can also play a part in integrated pest management. Identifying the precise time when the pesticide will have the greatest effect at the lowest possible dose has these advantages: It reduces the amount of pesticide used and may still allow the parasites and predators of pests to survive. Such precise applications often require the assistance of a trained professional who can correctly identify the pests, measure the size of the population, and time pesticide applications for maximum effect. In several instances, pheromone-baited traps capture insect pests from fields, and an assessment of the number of insects caught can be a guide to when insecticides should be applied.

Integrated pest management will become increasingly popular as the cost of pesticides rises and knowledge about the biology of specific pests becomes available. However, as long as humans raise crops, there will be pests that will outwit the defenses we develop. Integrated pest management is just another approach to a problem that began with the dawn of agriculture.

GENETICALLY MODIFIED CROPS

Genetically modified (GM) foods have attracted intense media coverage in recent years. Depending on one's point of view, they have been hailed as "super foods" or reviled as "frankenfoods." Yet, despite the controversy, GM foods have found their way onto dining tables. In the United States, an estimated 60 percent of processed foods in supermarkets—from breakfast cereals to softdrinks—contain a GM ingredient, such as soy, corn, or canola.

Selective plant breeding is not a new concept. Casual selection of observed desirable traits by our ancestors essentially tamed wild plants and made them suitable for agriculture. In the past, if pests devastated a field of crops and a few plants stayed alive and healthy, the seeds from these healthy plants were used to generate the next crop. Thus the beneficial factors that made the plants resistant were transferred to the next generation, making the new generation of crops slightly more resistant to the same pests. Such selections have been used for over 10,000 years, since the beginning of agriculture, and have resulted in significant advances for humanity with increased yields, disease resistance, and, overall, greater productivity. A good example is corn; the original crop was Teosinte with very small seeds and very few seeds in each cob. Over the centuries people selected for various traits, thus improving the size of the corn kernel, the number of kernels in each cob, and the stronger attachment of the kernels to the cob so that it could be harvested, resulting in the corn that we are familiar with today.

The advent of genetic engineering greatly enhanced this process of transferring a beneficial trait into plants by directly

TABLE 14.2 List of Genetically Modified Crops and Their Altered Traits

Modified Trait	Crop
Input Traits	
Herbicide resistance	Sugar beet, soybean, corn, canola, cotton, flax
Insect/herbicide resistance	Corn
Insect resistance	Tomato, corn, potato, cotton
Virus resistance	Squash, papaya
Male sterile	Corn
Output Traits	
Modified oil	Soybean, canola
Modified fruit ripening	Tomato
Provitamin A enriched	Rice
Iron fortification	Rice
Beta-carotene, lycopene enriched	Tomato
Detoxification of mycotoxins	Corn
Detoxification of cyanogens	Cassava
Caffeine-free	Coffee beans
Vitamin E enriched	Canola

Source: USDA.

transferring the gene or genes responsible for the beneficial attribute. (See table 14.2.) So in one generation, or one planting season, a plant can be created that is the same in all respects except for the addition of the beneficial trait. Since genes in all living organisms code for similar proteins and properties, it is possible to transfer a gene from one good corn variety to another corn variety or from fish to strawberry plants.

We can now define a genetically modified organism: It is any organism that has been modified by altering one or more genes by recombinant techniques. A recombinant technique is the method used to transfer a gene of interest from one organism to another. A genetically modified food therefore is any food that is produced from plants or animals that have been genetically altered using this method.

The first genetically altered plant created was a tobacco plant with resistance to antibiotics in 1983. It was almost 10 years later when the first commercial genetically altered crop, a delayed-ripening tomato, "Flavr-savr," was commercially released.

Genetically altered foods are prevalent in the United States and the Western world. More than 60 percent of the foods we purchase from the supermarket today have ingredients derived from genetically modified crops. Most of these are either from corn or soybeans, which are the bases for numerous ingredients manufactured for the food industry, including starch, oils, proteins, and other ingredients. Despite this prevalence, a U.S. Department of Agriculture consumer focus group survey revealed that most consumers were unaware of the use of biotechnology in foods. Furthermore, the benefits of biotechnology were viewed as skewed toward producers and manufacturers, with little benefit to the consumer. There was also skepticism about the long-term health effects and the environmental impact of genetically altered foods. It is therefore essential that we disseminate the information about genetic engineering so that we can have an informed debate on the merits and shortcomings of this technology.

Like all new technologies, the potential for applications is tremendous. Many of them must wait until they can be done more efficiently or more economically, or until their effects are better understood. Recently scientists created a genetically altered variety of rice, termed "golden rice," that contains higher levels of beta-carotene, a precursor for vitamin A. This is promising, considering that rice is the staple food for more than half the population of the world, many of whom are under- or malnourished. The future of the technology holds promise in a number of different ways:

1. Produce more food economically by improving yields and agricultural practices associated with farming.

2. Improve the nutritional quality of foods and enhance the levels of compounds that confer health benefits.

3. Improve the shelf life and quality of fresh fruits and vegetables.

4. Decrease allergenic compounds in foods such as peanuts and wheat.

5. Create crops that can be used to provide vaccines and other medical benefits.

6. Convert toxic soils into more productive land for agriculture.

The scope of the technology at the present time is only limited by our imagination, and its adoption must be a result of acceptance by the public at large.

The United States accounts for over two-thirds of all genetically modified crops planted globally. (See figure 14.13.) GM food crops grown by U.S. farmers include corn, cotton, soybeans, canola, squash, and papaya. Other major producers of GM crops are Argentina, which plants primarily genetically modified soybeans, and Canada, whose principal genetically modified crop is canola.

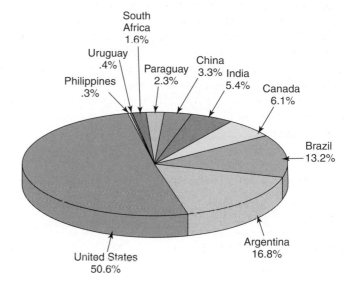

FIGURE 14.13   Global distribution of genetically modified crop land area.

SUMMARY

Although small slash-and-burn garden plots are common in some parts of the world, most of the food in the world is raised on more permanent farms. In countries where population size is high and money is in short supply, much of the farming is labor-intensive, using human labor for many of the operations necessary to raise crops. However, much of the world's food is grown on large, mechanized farms that use energy rather than human muscle for tilling, planting, and harvesting crops, and for producing and applying fertilizers and pesticides.

Monoculture involves planting large areas of the same crop year after year. This causes problems with plant diseases, pests, and soil depletion. Although chemical fertilizers can replace soil nutrients that are removed when the crop is harvested, they do not replace the organic matter necessary to maintain soil texture, pH, and biotic richness.

Mechanized monoculture depends heavily on the control of pests by chemical means. Persistent pesticides are stable and persist in the environment, where they may biomagnify in ecosystems.

Consequently, many of the older persistent pesticides have been quickly replaced by nonpersistent pesticides that decompose much more quickly and present less of an environmental hazard. However, most nonpersistent pesticides are more toxic to humans and must be handled with greater care than the older persistent pesticides.

Pesticides can be divided into several categories based on the organism they are used to control. Insecticides are used to control insects, herbicides are used for plants, fungicides for fungi, and rodenticides for rodents. Because of the problems of persistence, biomagnification, resistance of pests to pesticides, and effects on human health, many people are seeking pesticide-free alternatives to raising food. Several different philosophies that seek the same ends—less use of chemicals and better stewardship of soil—are alternative agriculture, sustainable agriculture, and organic agriculture. One ingredient in all of these approaches is the use of integrated pest management, which employs a complete understanding of an organism's ecology to develop pest-control strategies that use no or few pesticides.

What Does "Certified Organic" Food Mean?

"Certified organic" refers to agricultural products that have been grown and processed according to uniform standards and verified by independent state or private organizations accredited by the USDA. All products sold as "organic" must be certified. Certification includes annual submission of an organic system plan and inspection of farm fields and processing facilities. Inspectors verify that organic practices such as long-term soil management, buffering between organic farms and neighboring conventional farms, and recordkeeping are being followed. Processing inspections include review of the facility's cleaning and pest control methods, ingredient transportation and storage, and recordkeeping and audit control. Organic foods are minimally processed to maintain the integrity of food without artificial ingredients or preservatives. Certified organic requires the rejection of synthetic agrochemicals, irradiation, and genetically engineered foods or ingredients. Since 2002, organic certification in the United States has taken place under the authority of the USDA National Organic Program, which accredits organic certifying agencies and oversees the regulatory process. To find out more about the national organic certification requirements and organic program, please go to the USDA National Organic Program website at www.arms.usda .gov/nop.

Organic Crops	Organic Livestock	Organic Labels
Land must not have prohibited substances applied to it for at least three years.	Animals for slaughter must be raised under organic management from the last third of gestation, or no later than the second day of life for poultry.	Foods labeled "100 percent organic" and "organic" must contain (excluding water and salt) only organically produced ingredients.
Use of genetic engineering, ionizing radiation, and sewage sludge is prohibited.	Livestock must be fed products that are 100 percent organic. Vitamin and mineral supplements are permitted.	Products labeled "organic" must consist of at least 95 percent organically produced ingredients (excluding water and salt).
No chemical fertilizer can be used. Only animal and crop wastes and tillage methods may be used to provide fertility.	No antibiotics or hormones may be used.	Processed products that contain at least 70 percent organic ingredients can use the phrase "made with organic ingredients."
Synthetic pesticides must not be used, but pesticides made from plants and other natural sources may be used.	Animal welfare conditions require that: 1. Animals have access to the outdoors. (They may be temporarily confined only for reasons of health, safety, the animal's stage of production, or protection of soil or water quality.) 2. Preventive health management practices must be employed, including the use of vaccines to keep animals healthy. 3. Sick or injured animals must be treated; however, animals treated with a prohibited medication may not be sold as organic.	Processed products that contain less than 70 percent organic ingredients cannot use the term *organic* anywhere on the label but may list the ingredients that are organic.
Farms and handling operations that sell less than $5,000 annually of organic agricultural products are exempt from certification. But if they comply with all other national standards for organic products, they may label their products as organic.	A dairy herd can be converted to organic production if 80 percent organically produced feed is used for nine months, followed by three months of 100 percent organically produced feed.	Products with pesticide residues up to 5 percent of the EPA pesticide tolerance may be sold as organic.

- Many people feel there is no difference in food quality between foods that are labeled organic and those that are not. Do you think the nutritional quality of organic foods differs from that of conventionally produced foods?

- The rules permit small amounts of pesticide residues. Why do you think these are allowed?

- The rule that animals must have access to the outdoors has been challenged. Do you think being outdoors changes the quality of the meat, milk, or eggs produced?

THINKING GREEN

1. Reduce or eliminate the use of pesticides and herbicides on your yard or garden.
2. Support local or regional producers of sustainable agricultural products.
3. Visit your local "farmers market."
4. Visit a grocery store and identify the origins of three fruits. How far did they travel to get to the store?
5. Plant your own organic garden even if it is only a small container in your apartment.

WHAT'S YOUR TAKE?

In 2006, a U.S. court of appeals upheld a ban on phosphorus use in lawn fertilizers in Madison, Wisconsin. The ban was originally enacted in 2004 to reduce the amount of phosphorus runoff into Madison's lakes and diminish the algae blooms that plague the lakes each summer. Fertilizer manufacturers appealed the ban, claiming that local government could not regulate the fertilizers, since state law controls the use of pesticides and local law cannot supersede state law. The court of appeals rejected this argument, stating that the "weed and feed" products are a fertilizer-pesticide mixture, and that since local governments can regulate fertilizers, the combination can be regulated by local ordinances. Develop a position paper that supports or refutes the court of appeals ruling.

REVIEW QUESTIONS

1. What is monoculture?
2. List three reasons why fossil fuels are essential for mechanized agriculture.
3. Describe why pesticides are commonly used in mechanized agriculture.
4. Why are fertilizers used? What problems are caused by fertilizer use?
5. How do persistent and nonpersistent pesticides differ?
6. What is biomagnification? What problems does it cause?
7. How do organic farms differ from conventional farms?
8. Name three nonchemical methods of controlling pest populations.
9. What are the advantages and disadvantages of integrated pest management?
10. List three uses of food additives.
11. List three actions farmers could take to reduce the effect of pesticides on the environment.

CRITICAL THINKING QUESTIONS

1. If you were a public health official in a developing country, would you authorize the spraying of DDT to control mosquitoes that spread malaria? What would be your reasons?
2. Look at table 14.1. What caused the changes in the effectiveness of the insecticide? If you were an agricultural extension agent, what alternatives to pesticides might you recommend?
3. Imagine that you are a scientist examining fish in Lake Superior and you find toxaphene in the fish you are studying. Toxaphene was used primarily in cotton farming and has been banned since 1982. How can you explain its presence in these fish?
4. Are the risks of pesticide use worth the benefits? What values, beliefs, and perspectives lead you to this conclusion?
5. Do you think that current agricultural practices are sustainable? Why or why not? What changes in agriculture do you think will need to happen in the next 50 years?
6. Imagine you are an EPA official who is going to make a recommendation about whether an agricultural pesticide can remain on the market or should be banned. What are some of the facts you would need to make your recommendation? Who are some of the interest groups interested in the outcome of your decision? What arguments might they present regarding their positions? What political pressures might they be able to bring to bear on you?
7. Why are few consumers demanding alternative methods of crop production, and why are farmers not using those methods?

CHAPTER 15

WATER MANAGEMENT

Freshwater is essential to maintain life for most terrestrial plants and animals. Human use of water for domestic, industrial, and agricultural use often requires diversion of water. While we must manage water use, we should also preserve the scenic and other aesthetic values of water as well.

CHAPTER OUTLINE

The Water Issue

The Hydrologic Cycle

Human Influences on the Hydrologic Cycle

Kinds of Water Use
  Domestic Use of Water
  Agricultural Use of Water
  Industrial Use of Water
  In-Stream Use of Water

Kinds and Sources of Water Pollution
  Municipal Water Pollution
  Agricultural Water Pollution
  Industrial Water Pollution
  Thermal Pollution
  Marine Oil Pollution
  Groundwater Pollution

Water-Use Planning Issues
  Water Diversion
  Wastewater Treatment
  Salinization
  Groundwater Mining
  Preserving Scenic Water Areas and Wildlife Habitats

ISSUES & ANALYSIS
Is There Lead in Our Drinking Water? 362

CASE STUDIES
Growing Demands for a Limited Supply of Water in the West 345
Restoring the Everglades 354

CAMPUS SUSTAINABILITY INITIATIVE
Conserving Water on Campus 361

GOING GREEN
Water Reuse 349

WATER CONNECTIONS
The Bottled Water Boom 340

OBJECTIVES

After reading this chapter, you should be able to:

- Explain how water is cycled through the hydrologic cycle.
- Explain the significance of groundwater, aquifers, and runoff.
- Explain how land use affects infiltration and surface runoff.
- List the various kinds of water use and the problems associated with each.
- List the problems associated with water impoundment.
- List the major sources of water pollution.
- Define biochemical oxygen demand (BOD).
- Differentiate between point and nonpoint sources of pollution.
- Explain how heat can be a form of pollution.
- Differentiate among primary, secondary, and tertiary sewage treatments.
- Describe some of the problems associated with stormwater runoff.
- List sources of groundwater pollution.
- Explain how various federal laws control water use and prevent misuse.
- List the problems associated with water-use planning.
- Explain the rationale behind the federal laws that attempt to preserve certain water areas and habitats.
- List the problems associated with groundwater mining.
- Explain the problem of salinization associated with large-scale irrigation in arid areas.
- List the water-related services provided by local governments.

Global Perspectives on "Comparing Water Use and Pollution in Industrialized and Developing Countries," "The Cleanup of the Holy Ganges," and "Death of a Sea" can be found on the book's website at www.mhhe.com/enger12e along with other interesting readings.

THE WATER ISSUE

Water in its liquid form is the material that makes life possible on Earth. All living organisms are composed of cells that contain at least 60 percent water. Furthermore, their metabolic activities take place in a water solution. Organisms can exist only where they have access to adequate supplies of water. Water is also unique because it has remarkable physical properties. Water molecules are polar; that is, one part of the molecule is slightly positive and the other is slightly negative. Because of this, the water molecules tend to stick together, and they also have a great ability to separate other molecules from each other. Water's ability to act as a solvent and its capacity to store heat are a direct consequence of its polar nature. These abilities make water extremely valuable for societal and industrial activities. Water dissolves and carries substances ranging from nutrients to industrial and domestic wastes. A glance at any urban sewer will quickly point out the importance of water in dissolving and transporting wastes. Because water heats and cools more slowly than most substances, it is used in large quantities for cooling in electric power generation plants and in other industrial processes. Water's ability to retain heat also modifies local climatic conditions in areas near large bodies of water. These areas do not have the wide temperature-change characteristic of other areas.

For most human as well as some commercial and industrial uses, the quality of the water is as important as its quantity. Water must be substantially free of dissolved salts, plant and animal waste, and bacterial contamination to be suitable for human consumption. The oceans, which cover approximately 70 percent of the Earth's surface, contain over 97 percent of its water. However, saltwater cannot be consumed by humans or used for many industrial processes. Freshwater is free of the salt found in ocean waters. Of the freshwater found on Earth, only a tiny fraction is available for use. (See figure 15.1.) Unpolluted freshwater that is suitable for drinking is known as **potable water.** Early human migration routes and settlement sites were influenced by the availability of drinking water. At one time, clean freshwater supplies were considered inexhaustible. Today, despite advances in drilling, irrigation, and purification, the location, quality, quantity, ownership, and control of potable waters remain important concerns.

Some areas of the world have abundant freshwater resources, while others have few. In addition, demand is increasing for freshwater for industrial, agricultural, and personal needs.

Shortages of potable freshwater throughout the world can also be directly attributed to human abuse in the form of pollution. Water pollution has negatively affected water supplies throughout the world. In many parts of the developing world, people do not have access to safe drinking water. The World Health Organization estimates that about 25 percent of the world's people do not have access to safe drinking water. Even in the economically advanced regions of the world, water quality is a major issue. According to the United Nations Environment Programme, 5 million to 10 million deaths occur each year from water-related diseases. The illnesses include cholera, malaria, dengue fever, and dysentery. The United Nations also reports that these illnesses have been increasing over the past decade and that without large economic investments in safe drinking-water supplies, the rate of increase will continue.

Unfortunately, the outlook for the world's freshwater supply is not very promising. According to studies by the United Nations

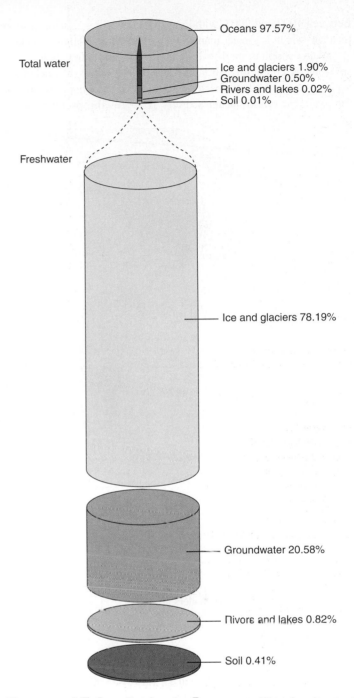

FIGURE 15.1 **Freshwater Resources** Although water covers about 70 percent of the Earth's surface, over 97 percent is saltwater. Of the less than 3 percent that is freshwater, only a tiny fraction is available for human use.

and the International Joint Commission (a joint U.S.-Canadian organization that studies common water bodies between the two nations), many sections of the world are currently experiencing a shortage of freshwater, and the problem will only intensify.

Water, used by households, agriculture, and industry, is clearly the most important good provided by freshwater systems. Humans now withdraw about one-fifth of the world's rivers' base flow (the dry-weather flow or the amount of available water in rivers most of the time), but in river basins in arid or populous regions the proportion can be much higher. This has implications for the species living in or dependent on these systems, as well as for human water supplies.

Between 1900 and 2007, withdrawals increased by a factor of more than six, which is greater than twice the rate of population growth.

Water supplies are distributed unevenly around the world, with some areas containing abundant water and others a much more limited supply. In water basins with high water demand relative to the available runoff, water scarcity is a growing problem. Many experts, governments, and international organizations are predicting that water availability will be one of the major challenges facing human society in the twenty-first century and that the lack of water will be one of the key factors limiting development.

Figure 15.2 shows areas of the world experiencing water stress. Water experts define areas where per capita water supply drops below

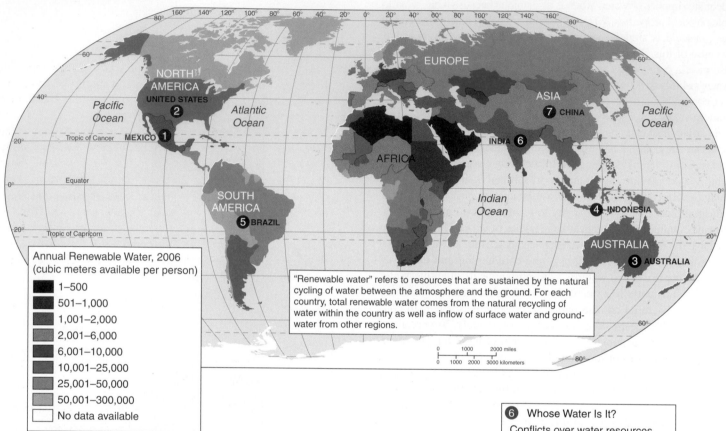

Annual Renewable Water, 2006
(cubic meters available per person)

- 1–500
- 501–1,000
- 1,001–2,000
- 2,001–6,000
- 6,001–10,000
- 10,001–25,000
- 25,001–50,000
- 50,001–300,000
- No data available

"Renewable water" refers to resources that are sustained by the natural cycling of water between the atmosphere and the ground. For each country, total renewable water comes from the natural recycling of water within the country as well as inflow of surface water and groundwater from other regions.

① Antiquated Systems

Mexico City, with a population of 20 million people, suffers from water shortages. Aging water pipes also waste millions of liters of water every year.

② Depletion of Resources

Ninety five percent of the freshwater in the United States is underground in aquifers. Today many aquifers are being depleted faster than they can be replenished.

③ Mismanagement

Australia is the world's driest inhabited continent. Lack of proper management of water and the effects of global climate change have caused critical water supply problems in parts of Australia.

④ Industrial Needs

Nearly half of Indonesia's 232 million citizens lack sufficient water because of pollution and poor water storage. Competition for available water is severe between domestic and industrial users.

⑤ Scarcity Where Water Seems Abundant

While parts of Brazil have abundant freshwater resources, because of the Amazon River, other parts of the country suffer water shortages. Pollution also severely impacts the country's freshwater supplies.

⑥ Whose Water Is It?

Conflicts over water resources impact many countries. For example India and Bangladesh have bitterly disagreed over the water resources of the Ganges River for decades.

⑦ Population Pressures

Inadequate freshwater supply is a major problem facing China. While the south of China has adequate supplies the north suffers from frequent shortages. It is the northern part of China that has many of its agricultural and population centers. It is also estimated that 300 million Chinese do not have adequate clean drinking water.

FIGURE 15.2   Areas of the World Experiencing Water Stress Today, a person whose annual water resources fall below 2,000 cubic meters runs the risk of joining the 700 million people in 43 countries living under the critical threshold of 1,700 cubic meters per person. At this level; water-stressed countries have difficulty meeting water requirements for agriculture, industry, energy, and the environment. By 2025, 3 billion people could be living in water-stressed countries—and 14 countries will slip from being water-stressed to being water-scarce (under 1,000 cubic meters per person).

Source: United Nations, *Population Division, UNDP,* and *AQUASTAT-FAQ;* Population Reference Bureau.

1700 m³/year as experiencing water stress—a situation in which disruptive water shortages can frequently occur. The map shows that by 2025, assuming current consumption patterns continue, at least 3.5 billion people—or 48 percent of the world's projected population—will live in water-stressed river basins. Of these, 2.4 billion will live under high water stress conditions. This per capita water supply calculation, however, does not take into account the coping capabilities of different countries to deal with water shortages. For example, high-income countries that are water scarce may be able to cope to some degree with water shortages by investing in desalination or reclaimed wastewater.

In the map, a selected number of river basins have been outlined. These watersheds represent basins that are in or approaching water scarcity and where the projected population for 2025 is expected to be higher than 10 million. Six of these basins, including the Volta in Ghana, Nile in Sudan and Egypt, Tigris and Euphrates in the Middle East, Narmada in India, and the Colorado River basin in the United States, will go from having more than 1700 m³ to less than 1700 m³ of water per capita per year. Another 29 basins will descend further into scarcity by 2025, including the Indus in Pakistan, Huang He in China, Seine in France, Balsas in Mexico, and Rio Grande in the United States.

Increasing scarcity, competition, and arguments over water in the first quarter of the twenty-first century could dramatically change the way we value and use water and the way we mobilize and manage water resources. Furthermore, changes in the amount of rain from year to year result in periodic droughts for some areas and devastating floods for others. However, rainfall is needed to regenerate freshwater and, therefore, is an important link in the cycling of water.

THE HYDROLOGIC CYCLE

All water is locked into a constant recycling process called the **hydrologic cycle.** (See figure 15.3.) Two important processes involved in the cycle are the evaporation and condensation of water. Evaporation involves adding energy to molecules of a liquid so that it becomes a gas in which the molecules are farther apart. Condensation is the reverse process in which molecules of a gas give up energy, get closer together, and become a liquid. Solar energy provides the energy that causes water to evaporate from the ocean surface, the soil, bodies of freshwater, and the surfaces of plants. The water evaporated from plants comes from two different sources. Some is water that has fallen on plants as rain, dew, or snow. In addition, plants take up water from the soil and transport it to the leaves, where it evaporates. This process is known as **evapotranspiration.**

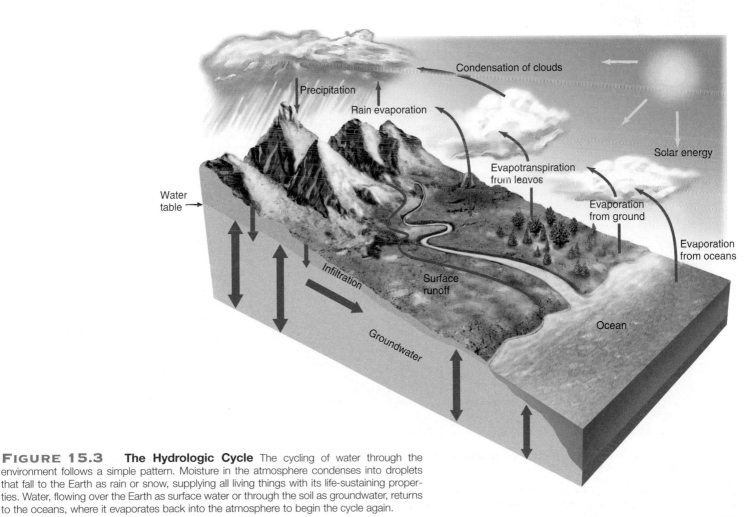

FIGURE 15.3 **The Hydrologic Cycle** The cycling of water through the environment follows a simple pattern. Moisture in the atmosphere condenses into droplets that fall to the Earth as rain or snow, supplying all living things with its life-sustaining properties. Water, flowing over the Earth as surface water or through the soil as groundwater, returns to the oceans, where it evaporates back into the atmosphere to begin the cycle again.

The water vapor in the air moves across the surface of the Earth as the atmosphere circulates. As warm, moist air cools, water droplets form and fall to the land as precipitation. Although some precipitation may simply stay on the surface until it evaporates, most will either sink into the soil or flow downhill and enter streams and rivers, which eventually return the water to the ocean. Surface water that moves across the surface of the land and enters streams and rivers is known as **runoff.** Water that enters the soil and is not picked up by plant roots moves slowly downward through the spaces in the soil and subsurface material until it reaches an impervious layer of rock. The water that fills the spaces in the substrate is called **groundwater.** It may be stored for long periods in underground reservoirs.

The porous layer that becomes saturated with water is called an **aquifer.** There are three basic kinds of aquifers: unconfined, semiconfined, and confined. (See figure 15.4.) An **unconfined aquifer** usually occurs near the land's surface where water enters the aquifer from the land above it. The top of the layer saturated with water is called the **water table.** The lower boundary of the aquifer is an impervious layer of clay or rock that does not allow water to pass through it. Unconfined aquifers are replenished (recharged) primarily by rain that falls on the ground directly above the aquifer and infiltrates the layers below. The water in such aquifers is at atmospheric pressure and flows in the direction of the water table's slope, which may or may not be similar to the surface of the land above it. Above the water table and below the land surface is a layer

known as the **vadose zone** (also known as the unsaturated zone or zone of aeration) that is not saturated with water.

A **confined aquifer** is bounded on both the top and bottom by layers that are impervious to water and is saturated with water under greater-than-atmospheric pressure. An impervious confining layer is called an **aquiclude.** If water can pass in and out of the confining layer, the layer is called an **aquitard.** A confined aquifer is primarily replenished by rain and surface water from a recharge zone (the area where water is added to the aquifer) that may be many kilometers from where the aquifer is tapped for use. If the recharge area is at a higher elevation than the place where an aquifer is tapped, water will flow up the pipe until it reaches the same elevation as the recharge area. Such wells are called **artesian wells.** If the recharge zone is above the elevation of the top of the well pipe, it is called a flowing artesian well because water will flow from the pipe.

The nature of the substrate in the aquifer influences the amount of water the aquifer can hold and the rate at which water moves through it. **Porosity** is a measure of the size and number of the spaces in the substrate. The greater the porosity, the more water it can contain. The rate at which water moves through an aquifer is determined by the size of the pores, the degree to which they are connected, and any cracks or channels present in the substrate. The rate at which the water moves through the aquifer determines how rapidly water can be pumped from a well per minute.

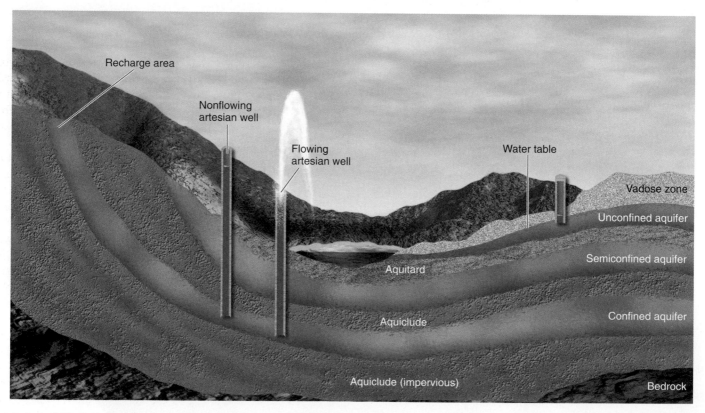

FIGURE 15.4 **Aquifers and Groundwater** Groundwater is found in the pores in layers of sediment or rock. The various layers of sediment and rock determine the nature of the aquifer and how it can be used.

HUMAN INFLUENCES ON THE HYDROLOGIC CYCLE

The world's oceans are the primary regulator of global climate and an important sink for greenhouse gases. At continental, regional, and ocean basin scales, the water cycle is being affected by long-term changes in climate, threatening human security. These changes are affecting Arctic temperatures, sea and land ice, including mountain glaciers. They also affect ocean salinity and acidification, sea levels, precipitation patterns, extreme weather events, and, possibly, the ocean's circulatory regime. The trend to increasing urbanization and tourism development has considerable impacts on coastal ecosystems. The socio-economic consequences of all these changes are potentially immense.

Human activities can significantly impact evaporation, runoff, and infiltration. When water is used for cooling in power plants or to irrigate crops, the rate of evaporation is increased. Water impounded in reservoirs also evaporates rapidly. This rapid evaporation can affect local atmospheric conditions. Runoff and the rate of infiltration also are greatly influenced by human activity. Removing the vegetation by logging or agriculture increases runoff and decreases infiltration. Because there is more runoff, there is more erosion of soil. Urban complexes with a high percentage of impervious, paved surfaces have increased runoff and reduced infiltration. A major concern in urban areas is providing ways to carry stormwater away rapidly. This involves designing and constructing surface waterways and storm sewers. Often cities have significant flooding problems when heavy rains overtax the ability of their stormwater management systems to remove excess water. Many cities combine their storm-sewer water with wastewater at their treatment plants, which can cause serious pollution problems after heavy rains. The treatment plants cannot process the increased volume of water and must discharge the combined untreated wastewater and sewer water into the receiving body of water.

Cities also have problems supplying water for industrial and domestic use. In many cases, cities rely on surface water sources for their drinking water. The source may be a lake or river, or an impoundment that stores water. Over one-third of the world's largest cities rely on large areas of protected forests to maintain a steady flow of clean drinking water. For example, New York City, with a population of over 8 million people, relies on a 998 square kilometer (385 square mile) watershed protected by the Catskill State Park for 90 percent of its drinking water. In addition, aquifers are extremely important in supplying water. Many large urban areas in the United States depend on underground water for their water supply. This groundwater supply can be tapped as long as it is not used faster than it can be replaced. Determining how much groundwater or surface water can be used and what the uses should be is a major concern, especially in water-poor areas of the world.

There are several ways to monitor water use from surface and groundwater sources. Water withdrawals are measurements of the amount of water taken from a source. This water may be used temporarily and then returned to its source and used again.

For example, when a factory withdraws water from a river for cooling purposes, it returns most of the water to the river; thus, the water can be used later. Water that is incorporated into a product or lost to the atmosphere through evaporation or evapotranspiration cannot be reused in the same geographic area and is said to be consumed. Much of the water used for irrigation is lost to evaporation and evapotranspiration or is removed with the crop when it is harvested. Therefore, much of the water withdrawn for irrigation is consumed.

KINDS OF WATER USE

Water use varies considerably around the world, depending on availability of water and degree of industrialization. However, use can be classified into four broad categories: (1) domestic use, (2) agricultural use, (3) industrial use, and (4) in-stream use. It is important to remember that some uses of water are consumptive, while others are nonconsumptive. Freshwater availability and use, as well as the conservation of aquatic resources, are key to human well-being. The quality and quantity of surface and groundwater resources, and life-supporting ecosystem services, are being jeopardized by the impacts of population growth, rural-to-urban migration, and rising wealth and resource consumption and by climate change. If present trends continue, 1.8 billion people will be living in countries or regions with absolute water scarcity by 2025, and two-thirds of the world population could be subject to water stress.

DOMESTIC USE OF WATER

Over 90 percent of the water used for domestic purposes in North America is supplied by municipal water systems, which typically include complex, costly storage, purification, and distribution facilities. Many rural residents, however, can obtain safe water from untreated private wells. Nearly 37 percent of municipal water supplies come from wells.

Regardless of the water source (surface or groundwater), water supplied to cities in the developed world is treated to ensure its safety. Treatment of raw water before distribution usually involves some combination of the following processes. The raw water is filtered through sand or other substrates to remove particles. Chemicals may be added to the water that will cause some dissolved materials to be removed. Then, before the water is released for public use, it is disinfected with chlorine, ozone, or ultraviolet light to remove any organisms that might still be present. Where no freshwater is available, expensive desalinization of saltwater may be the only option available.

Domestic activities in highly developed nations require a great deal of water. This domestic use includes drinking, air conditioning, bathing, washing clothes, washing dishes, flushing toilets, and watering lawns and gardens. On average, each person in a North American home uses about 400 liters (about 100 gallons) of **domestic water** each day. About 69 percent is used as a solvent to carry away wastes (nonconsumptive use; laundry, bathing,

Consumers spend billions of dollars on bottled water, making it the fastest-growing drink of choice in many parts of the world. Some people drink bottled water as an alternative to other beverages; others drink it because they prefer its taste or because their tap water is not always safe to drink.

In the United States, the Environmental Protection Agency (EPA) and the Food and Drug Administration (FDA) set drinking water standards. EPA sets standards for tap water provided by public water suppliers; FDA sets standards for bottled water based on EPA standards. FDA regulates bottled water as a packaged food under the Federal Food, Drug and Cosmetic Act.

Any bottled water sold in interstate commerce in the United States, including products that originate overseas, must meet federal standards. Bottled water must meet FDA standards for physical, chemical, microbial, and radiological contaminants.

Bottlers use standard identifiers, prescribed by FDA regulations, to describe their water but the meanings may be different than you expect. The terms refer to both the geological sources of the water and the treatment methods applied to the water. The terms do not necessarily describe the geographic location of the source or determine its quality. For instance, *spring water* can be collected at the point where water flows naturally to the Earth's surface or from a borehole that taps into the underground source. Other terms used on the label about the source, such as *glacial water* and *mountain water,* are not regulated standards of identity and may not indicate that the water is necessarily from a pristine area. Likewise, the term *purified* refers to processes that remove chemicals and pathogens. Purified water is not necessarily free of microbes, though it might be. The following are the most commonly used bottled water terms:

Artesian water, groundwater, spring water, well water—water from an underground aquifer that may or may not be treated. Well water and artesian water are tapped through a well. Spring water is collected as it flows to the surface or via a borehole. Groundwater can be either.

Distilled water—steam from boiling water that is recondensed and bottled. Distilling water kills microbes and removes water's natural minerals, giving it a flat taste.

Drinking water—water intended for human consumption that is sealed in bottles or other containers with no ingredients except that it may optionally contain safe and suitable disinfectants. Fluoride may be added within limitations set in the bottled water quality standards.

Mineral water—groundwater that naturally contains 250 or more parts per million of total dissolved solids.

Purified water—water that originates from any source but has been treated to meet the U.S. *Pharmacopeia* definition of purified water. Purified water is essentially free of all chemicals (it must not contain more than 10 parts per million of total dissolved solids) and may also be free of microbes if treated by reverse osmosis. Purified water may alternately be labeled according to how it is treated.

Carbonated water, soda water, seltzer water, sparkling water, and tonic water are considered soft drinks and are not regulated as bottled water.

Bottled Water—Did You Know?

- Tap water is much less expensive than bottled water—in some cases, you are paying for little more than the bottle itself. At least 25 percent (some experts say as much as 40 percent) of bottled water is nothing more than processed tap water.

- There is an environmental cost—it takes three to four times the amount of water in the bottle just to make the plastic for the bottle, and that's not including how much oil is used and how much carbon dioxide is created when the water is shipped to the store.

  - Buy domestic—if you cannot break the bottled water habit, look for a brand that has not been shipped across the world. The less distance the water has to travel, the fewer greenhouse gases are produced.

  - 28 billion—the number of plastic water bottles purchased in the United States annually, of which only 16 percent are recycled. 1.5 million barrels—the amount of oil used to make those bottles. $11 billion—the amount that people in the United States spend on bottled water annually.

  - 2.5 million—the number of plastic water bottles that people in the United States discard hourly.

In an effort to encourage recycling and discourage consumption, the city of Chicago became the first major U.S. city to put a 5 cent tax on bottled water. Many other cities are debating similar measures.

---

toilets, and dishes), about 29 percent is used for lawns and gardens (consumptive use), and only a tiny amount (about 2 percent) is actually consumed by drinking or in cooking. (See figure 15.5.) Yet all water that enters the house has been purified and treated to make it safe for drinking. Natural processes cannot cope with the highly concentrated wastes typical of a large urban area. The unsightly and smelly wastewater also presents a potential health problem, so cities and towns must treat it before returning it to a local water source. Until recently, the cost of water in almost every community has been so low that there was very little incentive to conserve, but shortages of water and increasing purification costs have raised the price of domestic water in many parts

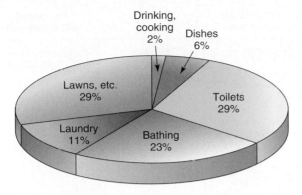

Water Use by a Typical North American Family of Four

Drinking, cooking 2%
Dishes 6%
Lawns, etc. 29%
Toilets 29%
Laundry 11%
Bathing 23%

FIGURE 15.5 **Urban Domestic Water Uses** Over 150 billion liters (40 billion gallons) of water are used each day for urban domestic purposes in North America. (Nonconsumptive uses are shown in blue, and consumptive uses are shown in green.)

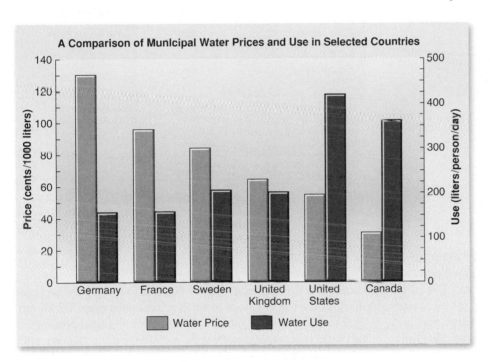

A Comparison of Municipal Water Prices and Use in Selected Countries

Germany · France · Sweden · United Kingdom · United States · Canada

■ Water Price   ■ Water Use

FIGURE 15.6 **Water Use Decreases as Water Price Increases** A general correlation exists between the amount of water that is used and its price.

of the world, and it is becoming evident that increased costs do tend to reduce use. (See figure 15.6.)

Although domestic use of water is a relatively small component of the total water-use picture (see figure 15.7), urban growth has created problems in the development, transportation, and maintenance of quality water supplies. In regions experiencing rapid population growth, such as Asia, domestic use is expected to increase sharply. Many cities in China are setting quotas on water use that are enforced by higher prices for larger users. In the coastal city of Dalian, a family that uses more than

8000 liters (2113 gallons) of water per month will pay four times more than a family that conserves water. In North America, more than 36 states expect a water crisis in the next 10 years. Fast-growing cities in the West, especially those in arid areas, are experiencing water shortages. Demand for water in urban areas sometimes exceeds the immediate supply, particularly when the supply is local surface water. This is especially true during the summer, when water demand is high and precipitation is often low. Many communities have begun public education campaigns designed to help reduce water usage. Figure 15.8 shows some ways that people could reduce their water use. In addition, at the federal level, the U.S. Environmental Protection Agency (EPA) is in the process of developing a national consumer labeling program of water-efficient products modeled after its highly successful Energy STAR program. Mexico City, with a population of nearly 20 million and minimal access to surface water, has one of the most serious water management problems in the world. The water supply for the Mexico Basin is severely stressed by water management practices that do not ensure a sustainable supply of clean water. Since extractions from the Valley of Mexico aquifer began nearly 100 years ago, groundwater levels have dropped significantly. As the city's population continues to grow, managers have been forced to look at alternative sources of water from more than 160 kilometers (100 miles) away.

In addition to encouraging the public to conserve water, municipalities need to pay attention to losses that occur within the distribution system. Leaking water pipes and mains account for significant losses of water. Even in the developed world, such losses may be as high as 20 percent. Poorer countries typically exceed this, and some may lose over 50 percent of the water to leaks. Another major cause of water loss has been public attitudes. As long as water is considered a limitless, inexpensive resource, people will make little effort to conserve it. In the United States, water rates have been increasing in the recent past at about 4 percent a year. The World Water Council based in France, has ranked 147 countries according to the efficiency of their water use. It ranked the United States last—as the most wasteful and least efficient. One explanation for this is that, historically, water in the United States has been very cheap. For example, the Germans pay $1.82 for a cubic meter of water, the French $1.12, and the British $1.34. In the United States, the cost is 55 cents, or roughly $15 a month for the average family of four. As the cost of water rises and attitudes toward water change, so will usage and efforts to conserve. For example, the California Urban Conservation Council, together with the EPA, has launched a highly innovative virtual home tour that allows Internet users to click on different areas of the H_2O use floor-plan for facts and advice on saving water.

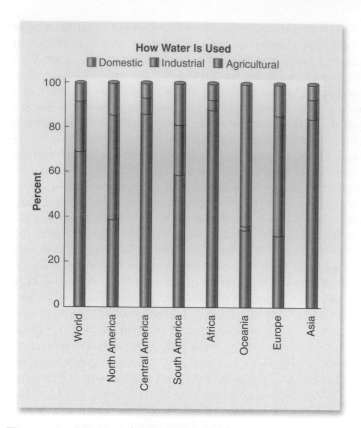

How Water Is Used
■ Domestic ■ Industrial ■ Agricultural

FIGURE 15.7 **World Uses of Water** Domestic, industrial, and agricultural uses dominate the allocation of water resources. However, there is considerable variety in different parts of the world in how these resources are used.

Source: World Resources Institute, 2007.

Water Savings Guide

Conservative use will save water		Normal use will waste water
Wet down, soap up, rinse off 15 liters (4 gal)	**Shower**	Regular shower 95 liters (25 gal)
May we suggest a shower?	**Tub bath**	Full tub 135 liters (36 gal)
Minimize flushing Each use consumes 20-25 liters (5-7 gal) New toilets use 6 liters (1.6 gal)	**Toilet**	Frequent flushing is very wasteful
Fill basin 4 liters (1 gal)	**Washing hands**	Tap running 8 liters (2 gal)
Fill basin 4 liters (1 gal)	**Shaving**	Tap running 75 liters (20 gal)
Wet brush, rinse briefly 2 liters (1/2 gal)	**Brushing teeth**	Tap running 38 liters (10 gal)
Take only as much as you require	**Ice**	Unused ice goes down drain
Please report immediately	**Leaks**	A small drip wastes 95 liters (25 gal) per week
Turn off light, TV, heaters, and air conditioning when not in room	**Energy**	Wasting energy also wastes water

Thank you for using
this column...and not this one

FIGURE 15.8 **Ways to Conserve Water** Minor changes in the way people use water could significantly reduce domestic water use. Note: 1 gallon equals approximately 3.785 liters.

Source: Reprinted by permission of the San Francisco Convention & Visitors Bureau.

AGRICULTURAL USE OF WATER

In North America, groundwater accounts for about 37 percent of the water used in agriculture and surface water about 63 percent. **Irrigation** is the major consumptive use of water in most parts of the world and accounts for about 80 percent of all the water consumed in North America. About 500 billion liters per day (134 billion gallons per day or 150 million acre-feet per year) are used in irrigation in the United States. The amount of water used for irrigation and livestock continues to increase throughout the world. Future agricultural demand for water will depend on the cost of water for irrigation; the demand for agricultural products, food, and fiber; governmental policies; the development of new technology; and competition for water from a growing human population.

Since irrigation is common in arid and semiarid areas, local water supplies are often lacking, and it is often necessary to transport water great distances to water crops. This is particularly true in the western United States, where about 14 million hectares (35 million acres) of land are irrigated.

Four methods of irrigation are commonly used. Surface or flood irrigation involves supplying the water to crops by having the water flow over the field or in furrows. This requires extensive canals and is not suitable for all kinds of crops. Spray irrigation involves the use of pumps to spray water on the crop. Trickle irrigation uses a series of pipes with strategically placed openings so that water is delivered directly to the roots of the plants. Subirrigation involves supplying water to plants through underground pipes. Often this method is used where soils require draining at certain times of the year. The underground pipes can be used to drain excess water at one time of the year and supply water at others. Each of these methods has its drawbacks and advantages as well as conditions under which it works well.

Construction and maintenance of irrigation structures, such as dams, canals, pipes, and pumps, are expensive. Costs for irrigation water have traditionally been low, since many of the dams and canals were constructed with federal assistance, and farmers have often used water wastefully. As competition has grown between urban areas and agriculture for scarce water resources, there has been pressure to raise the cost of water used for irrigation. Increasing the cost of water will stimulate farmers to conserve, just as it does homeowners. Another way to reduce the demand for irrigation water is to reduce the quantity of water-demanding crops grown in dry areas or change from high water-demanding to lower water-demanding crops. For example, wheat or soybeans require less water than do potatoes or sugar beets. It is also becoming increasingly important to modify irrigation practices to use less water. (See figure 15.9.) For example, the use of trickle irrigation and some variations of spray irrigation use water more efficiently than the more traditional flood irrigation methods.

Many forms of irrigation require a great deal of energy. This is particularly true when pumps are used to deliver the water to the crop. It is estimated that 40 percent of the energy devoted to agriculture in Nebraska is used for irrigation. Increasing energy costs may force some farmers to reduce or discontinue irrigation. In addition, much of western Nebraska relies on groundwater for irrigation, and the water table is dropping rapidly. If a water shortage develops, land values will decline. Land use and water use are interrelated and cannot be viewed independently. In the Salinas Valley, along the central coast of California, farmers are irrigating their fields with recycled water from the Monterey Regional Water Pollution Control Agency's (MRWPCA) local wastewater treatment plant. Because seawater had entered the groundwater as far as 10 kilometers (6 miles) inland, well water was becoming too salty for agricultural use. MRWPCA, aware that this problem may be caused by overpumping of groundwater, in the 1980s began studying the possibility of irrigating with treated municipal wastewater. Although California regulations allow the use of treated water for irrigation, health officials expressed concerns that its use on cold-season crops such as broccoli—many of which are consumed raw—may contaminate the crops with pathogens. The MRWPCA conducted a study that showed that crops irrigated with recycled water were not contaminated with pathogens. California then approved the plan, which required an upgrade of the local wastewater treatment plant. Sixty kilometers (40 miles) of pipeline were constructed to deliver the water to some 75 users throughout the system, which has the capacity to provide 24 billion liters (19,500 acre-feet) of water per year and irrigate 4900 hectares (12,000 acres) of coastal farmland. Such systems may become a common solution to the problem of deteriorating and diminishing groundwater supplies in California and perhaps throughout the world.

INDUSTRIAL USE OF WATER

Industrial water use accounts for nearly half of total water withdrawals in the United States, about 70 percent in Canada, and about 23 percent worldwide. Since most industrial processes involve heat exchanges, 90 percent of the water used by industry is for cooling and is returned to the source, so only a small amount is actually consumed. Industrial use accounts for less than 20 percent of the water consumed in the United States. For example, electric-power generating

FIGURE 15.9 **Types of Irrigation** Many arid areas require irrigation to be farmed economically. (a) Surface or flood irrigation uses irrigation canals and ditches to deliver water to the crops. The land is graded so that water flows from the source into the fields. Water is siphoned from a canal into ditches between rows of crops. (b) Spray irrigation uses a pump to spray water into the air above the plants. (c) Trickle irrigation conserves water by delivering water directly to the roots of the plants but requires an extensive network of pipes.

plants use water to cool steam so that it changes back into water. Many industries, especially power plants, actually can use saltwater for cooling purposes. About 30 percent of the cooling water used by power plants is saltwater. If the water heated in an industrial process is dumped directly into a watercourse, it significantly changes the water temperature. This affects the aquatic ecosystem by increasing the metabolism of the organisms and reducing the water's ability to hold dissolved oxygen.

Industry also uses water to dissipate and transport waste materials. In fact, many streams are now overused for this purpose, especially in urban centers. The use of watercourses for waste dispersal degrades the quality of the water and may reduce its usefulness for other purposes, as well as directly harming fish and wildlife. This is especially true if the industrial wastes are toxic.

Historically, industrial waste and heat were major causes of pollution. However, most industrialized nations have passed laws that severely restrict industrial discharges of wastes or heated water into watercourses. In the United States, the federal government's role in maintaining water quality began in 1972, with the passage of the Federal Water Pollution Control Act (PL 92-500). This act provided federal funds and technical assistance to strengthen local, state, and interstate water-quality programs. The act (and subsequent amendments in 1977, 1981, 1987, and 1993) is commonly referred to as the "Clean Water Act." The Clean Water Act is a comprehensive and technically rigorous piece of legislation. It seeks to protect the waters of the United States from pollution. To do this, the act specifically regulates pollutant discharges into "navigable waters" by implementing two concepts: setting water-quality standards for surface water and limiting effluent discharges into the water. The policy objectives of the Clean Water Act are to restore and maintain the "chemical, physical and biological integrity of the nation's waters." Enforcement of this act has been extremely effective in improving surface-water quality. However, many countries in the developing world have done little to control industrial pollution, and water quality is significantly reduced by careless use.

IN-STREAM USE OF WATER

In-stream water use does not remove water but makes use of it in its channels and basins. Therefore, all in-stream uses are nonconsumptive. Major in-stream uses of water are for hydroelectric power, recreation, and navigation. Although in-stream uses do not remove water, they may require modification of the direction, time, or volume of flow and can negatively affect the watercourse.

Electricity from hydroelectric power plants is an important energy resource. Presently, hydroelectric power plants produce about 13 percent of the total electricity generated in the United States. (See figure 15.10.) They do not consume water and do not add waste products to it. However, the dams needed for the plants have definite disadvantages, including the high cost of construction and the resulting destruction of the natural habitat in streams and surrounding lands. The sudden discharge of impounded water from a dam can seriously alter the downstream environment. If the discharge is from the top of the reservoir, the stream temperature rapidly increases. Discharging the colder water at the bottom of the reservoir causes a sudden decrease in the stream's water temperature. Either of these changes is harmful to aquatic life. The impoundment of water also reduces the natural scouring action of a flowing stream. If water is allowed to flow freely, the silt accumulated in the river is carried downstream during times of high water. This maintains the river channel and carries nutrient materials to the river's mouth. But if a dam is constructed, the silt is

(a)

(b)

FIGURE 15.10 **Dams Interrupt the Flow of Water** The flow of water in most large rivers is controlled by dams. Most of these dams provide electricity. In addition, they prevent flooding and provide recreational areas. (a) Large dams, however, destroy the natural river system, while smaller dams (b), do not always have a significant impact.

As one of the fastest-growing population centers in the United States, Las Vegas has a daily growing demand for water. Every month 5000 to 7000 newcomers arrive to retire or find jobs, meaning the already swollen population could double in 20 to 30 years. This is forcing Las Vegas to look for water as far away as the Snake Valley (see map).

In 2005 the city started to file for groundwater rights in counties hundreds of kilometers away, setting off a water war that could be repeated across the parched but popular Southwest United States. It is argued that if Las Vegas is successful in its claims, it could upset a complex web of aquifers that run as far away as California's Death Valley and western Utah.

That Las Vegas has real water concerns cannot be denied. The city exceeded the capacity of its own groundwater field several decades ago and currently is 90 percent dependent on a limited allotment from the Colorado River—an allotment it is fast outgrowing. That is what has driven the city to petition for water rights in the outlying counties, but if the history of Western development has shown one thing, it's that this kind of water shopping can go seriously wrong. In the early part of the twentieth century, Los Angeles famously—and secretly—bought up thousands of hectares of land in California's Owens Valley and then began to drain the surface and subsurface water for the city's use. After decades of pumping, a dozen Owens Valley springs have dried up, and water tables in places are too low to support once-abundant native grasses and shrubs.

Equally controversial plans such as Las Vegas's are developing in other Western cities. A multiyear drought, which eased only in 2006, dropped water levels in the Colorado River's vast reservoirs to historic lows, raising the specter of involuntary water rationing. This raised concerns among water managers in numerous states, causing Denver, for example, to be looking at the headwaters of the Gunnison River, which is across the Continental Divide, and Los Angeles to consider exploiting a groundwater field in the Mojave Desert.

A major study of the Western water resources will be completed in 2010 by federal and state hydrologists and geologists. By itself, collecting better data will not resolve the current conflict, but it is a start. The more scientists learn about Nevada's aquifer system, the more accurately they can model potential effects, and the more confidence the public in that state and elsewhere will have in the decisions the government agencies make.

Few urban areas are more vulnerable to water shortages than Las Vegas, which has been referred to as the dryest big city in the United States. Las Vegas takes 90 percent of its water from Lake Mead, although Nevada gets by far the smallest share of water among the seven states that border the Colorado—just 2 percent of the total. Las Vegas has worked to conserve water, paying residents to replace lawns with desert-appropriate landscaping. The city's overall water use has dropped since 2002, even as population and visitor numbers have continued to rise.

If the rest of the Southwest United States can use its water more efficiently, it should have enough for decades. One solution could involve diverting more of the Colorado River's water away from agriculture—which claims 85 percent of the supply—in favor of the region's thirsty cities. That would be challenging politically, but something has to give. While Lake Mead has shrunk to just 52 percent of capacity, the immense reservoir still contains 35 trillion cu L of water (9 trillion gallons). But the dry sky above and the rock all around reinforce the inescapable fact that this land was a desert, is a desert, and always will be a desert. When the explorer J.C. Ives visited the present location of the Hoover Dam in 1857, he declared the land "worthless," adding, "There is nothing there to do but leave." Today's residents are hoping there is another choice.

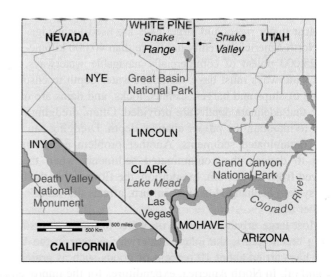

dissolved oxygen is removed, aquatic organisms die. The amount of oxygen required to decay a certain amount of organic matter is called the **biochemical oxygen demand (BOD).** (See figure 15.12.) Measuring the BOD of a body of water is one way to determine how polluted it is. If too much organic matter is added to the water, all of the available oxygen will be used up. Then, anaerobic (not requiring oxygen) bacteria begin to break down wastes. Anaerobic respiration produces chemicals that have a foul odor and an unpleasant taste and generally interfere with the well-being of humans.

Disease-causing organisms are a very important pollution problem in most of the world. Untreated or inadequately treated human or domesticated animal waste is most often the source of these organisms. In the developed world, sewage treatment and drinking-water treatment plants greatly reduce this public health problem.

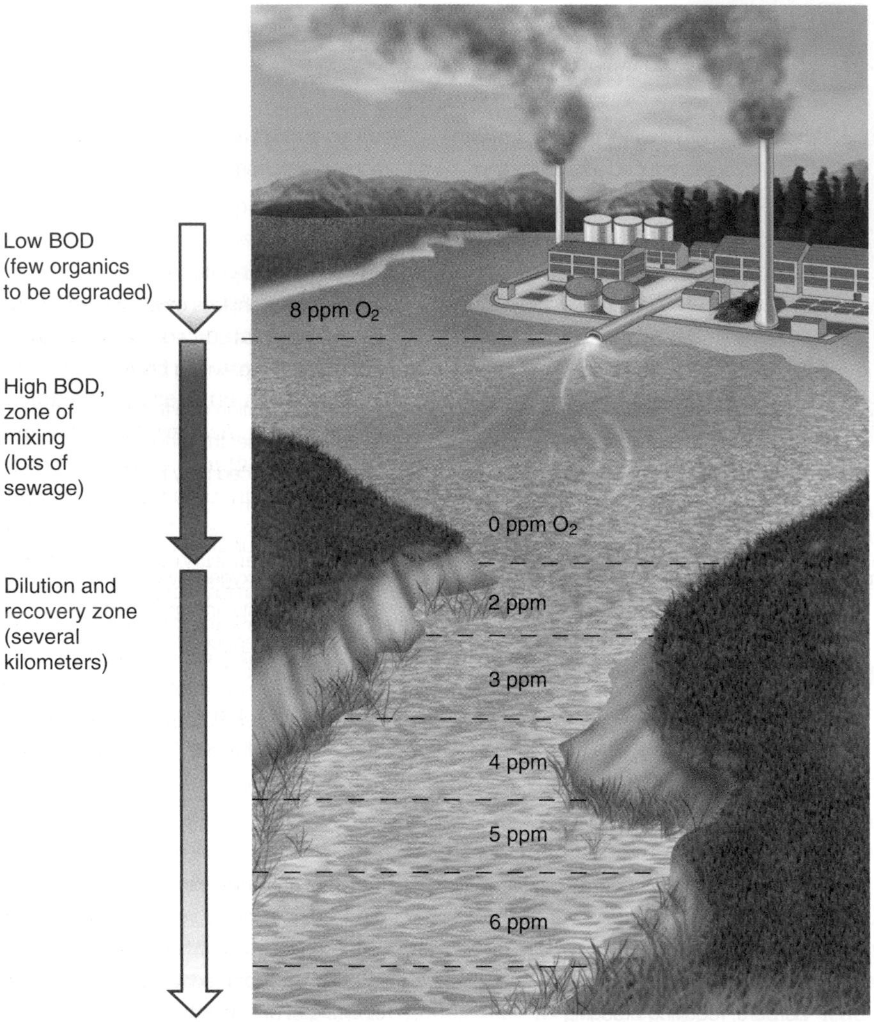

Nutrients are also a pollution problem. Additional nutrients in the form of nitrogen and phosphorus compounds from fertilizer, sewage, detergents, and animal waste increase the rate of growth of aquatic plants and algae. However, phosphates and nitrates are generally present in very limited amounts in unpolluted freshwater and, therefore, are a limiting factor on the growth of aquatic plants and algae. (A **limiting factor** is a necessary material that is in short supply, and because of the lack of it, an organism cannot reach its full potential growth. See chapter 5.) Thus, when phosphates or nitrates are added to the surface water, they act as a fertilizer and promote the growth of undesirable algae populations. The excessive growth of algae and aquatic plants due to added nutrients is called **eutrophication.** Algae and larger aquatic plants may interfere with the use of the water by fouling boat propellers, clogging water-intake pipes, changing the taste and odor of water, and causing the buildup of organic matter on the bottom. As this organic matter decays, oxygen levels decrease, and fish and other aquatic species die.

Physical particles also can negatively affect water quality. Particles alter the clarity of the water, can cover spawning sites, act as abrasives that injure organisms, and carry toxic materials.

Ideally, we should think in terms of eliminating all pollution, but any human use is going to have at least some minor negative impact. Even activities such as swimming and boating add particles and chemicals to the water. So, determining acceptable water quality involves economic considerations. Removing the last few parts per million of some materials from the water may not significantly improve water quality and may not be economically justifiable. This is certainly true of organic matter, which is biodegradable. However, radioactive wastes and toxins that may accumulate in living tissue are a different matter. Vigorous attempts to remove these materials are often justified because of the materials' potential harm to humans and other organisms.

Sources of pollution are classified as either point sources or nonpoint sources. When a source of pollution can be readily identified because it has a definite source and place where it enters the water, it is said to come from a **point source.** Municipal and industrial discharge pipes are good examples of point sources. Diffuse pollutants, such as from agricultural land and urban paved surfaces, acid rain, and runoff, are said to come from **nonpoint sources** and are much more difficult to identify and control. Initial attempts to control water pollution were focused on point sources of pollution, since these were readily identifiable and economic pressure and adverse publicity could be brought to bear on companies that continued to pollute from point sources. In North America, most point sources of water pollution have been identified and are regulated. In the United States, the EPA is responsible for identifying point sources of pollution, negotiating the permissible levels of pollution allowed from each source, and enforcing the terms of the permits.

Low BOD (few organics to be degraded)

High BOD, zone of mixing (lots of sewage)

Dilution and recovery zone (several kilometers)

8 ppm O_2

0 ppm O_2

2 ppm

3 ppm

4 ppm

5 ppm

6 ppm

FIGURE 15.12   Effect of Organic Wastes on Dissolved Oxygen
Sewage contains a high concentration of organic materials. When these are degraded by organisms, oxygen is removed from the water. This is called the biochemical oxygen demand (BOD). An inverse relationship exists between the amount of organic matter and oxygen in the water. The greater the BOD, the more difficult it is for aquatic animals to survive and the less desirable the water is for human use. The more the organic pollution, the greater the BOD.

Water reuse is the use of surface water that has been used beneficially once under a water right or the use of groundwater that has been used. For example, treating wastewater and piping it to a golf course for irrigation is water reuse. There are two types of water reuse: direct reuse and indirect reuse. Direct reuse is the use of effluent from a wastewater treatment plant that is piped directly from the plant to the place where it is used, such as the golf course example just mentioned. Indirect reuse is the use of water, usually treated effluent, that is placed back into a river or stream and then diverted further downstream to be used again. An example of indirect reuse is treating wastewater, discharging the effluent into a river, and using the river to transport it downstream where a golf course diverts the water from the river for irrigation.

Reuse is a promising source of additional water in the future. As municipal water supply and wastewater facilities expand to support a growing customer base, the volume of treated wastewater can be expected to expand as well. Because treated wastewater is often located close to its customers, the cost of conveying this water supply source can be much less than that of other water supply options. A number of communities and water providers use treated wastewater for direct and indirect reuse. Although wastewater can be treated to achieve compliance with federal and state drinking water standards, treated wastewater is not currently used for drinking water purposes. Municipal and other water utilities do, however, distribute treated wastewater for a variety of other purposes. For example, El Paso, Texas, Northwest Reclaimed Water Project provides more than a million liters of reclaimed water per year to schools, parks, a golf course, and multifamily housing developments and residential customers for irrigation.

The amount of existing supply related to water reuse is based on the amount of water that can be produced with current permits and existing infrastructure. As the amount of effluent and the need for additional supplies of water increase, water reuse will have an important role in meeting future water supply needs. However, policy issues related to permitting and environmental flows will have to be addressed before the full potential of this water management strategy is realized.

Nonpoint sources of water pollution are being addressed, but this is much more difficult to do, since regulating many small, individual human acts is necessary.

There are many things that you can do to protect surface and ground waters from nonpoint source pollution. The following are some specific actions you can take:

- Be aware that many chemicals commonly used around the home are toxic. Select less toxic alternatives. Use nontoxic substitutes wherever possible.

- Buy chemicals only in the amount you expect to use, and apply them only as directed. More is not better.

- Take unwanted household chemicals to hazardous waste collection centers; do not pour them down the drain.

- Never pour unwanted chemicals on the ground. Soil cannot destroy most chemicals, and they may eventually contaminate runoff.

- Use water-based products whenever possible.

- When landscaping your yard, select native plants that have low requirements for water, fertilizers, and pesticides.

- Test your soil before applying fertilizers. Overfertilization is a common problem, and the excess can leach into groundwater or contaminate rivers or lakes.

MUNICIPAL WATER POLLUTION

Municipalities are faced with the double-edged problem of providing suitable drinking water and disposing of wastes. These wastes consist of stormwater runoff, wastes from industry, and wastes from homes and commercial establishments. Wastes from homes consist primarily of organic matter from garbage, food preparation, cleaning of clothes and dishes, and human wastes. Human wastes are mostly undigested food material and a concentrated population of bacteria, such as *Escherichia coli* and *Streptococcus faecalis*. These particular bacteria normally grow in the large intestine (colon) of humans and are present in high numbers in the feces of humans; therefore, they are commonly called **fecal coliform bacteria.** Fecal coliform bacteria are also present in the wastes of other warm-blooded animals, such as birds and mammals. Low numbers of these bacteria in water are not harmful to healthy people. However, because they can be easily identified, their presence in the water is used to indicate the amount of pollution from the fecal wastes of humans and other warm-blooded animals. The numbers of these types of bacteria present in water are directly related to the amount of fecal waste entering the water.

When human wastes are disposed of in water systems, potentially harmful bacteria from humans may be present in amounts too small to detect by sampling. Even in small numbers, these harmful bacteria may cause disease epidemics. It is estimated that some 1.5 million people in the United States become ill each year from infections caused by fecal contamination. In 1993, for example, a protozoan pathogen called *Cryptosporidium* was identified in the Milwaukee, Wisconsin, public water system. This resulted in over 400,000 people becoming ill and at least 100 deaths. The total costs of such diseases amount to billions of dollars per year in the United States alone. The greater the amount of wastes deposited in the water, the more likely it is that there will be populations of disease-causing bacteria. Therefore, the presence of fecal coliform bacteria is used as an indication that other, more harmful organisms may be present as well.

Wastewater from cleaning dishes and clothing contains some organic material along with the soap or detergent, which helps to separate the contaminant from the dishes or clothes. Soaps and detergents are useful because one end of the molecule dissolves in dirt or grease and the other end dissolves in water. When the soap or detergent molecules are rinsed away by the water, the dirt or grease goes with them.

At one time, many detergents contained phosphates as a part of their chemical makeup, which contributed to eutrophication. However, because of the environmental effects of phosphate on aquatic environments, since 1994 most major detergent manufacturers in North America and other developed countries have eliminated phosphates from most of their formulations. Today, the majority of phosphate entering water in North America is from human waste and runoff from farm fields and livestock operations. A study conducted by the Toxic Substances Hydrology Program of the U.S. Geological Survey (USGS) revealed that a broad range of chemicals found in residential, industrial, and agricultural wastewaters commonly occurs at low concentrations downstream from areas of intense urbanization and animal production. The chemicals include human and veterinary drugs (including antibiotics), natural and synthetic hormones, detergent metabolites, plasticizers, insecticides, and fire retardants. The report found that in addition to caffeine, the most frequently detected compounds were cholesterol and coprostanol, which is a by-product of DEET, a common insect repellent. The compounds found in the water are sold on supermarket shelves and are in virtually every medicine cabinet and broom closet, as well as farms and factories. Although they are flushed or rinsed down the drain every day, they do not disappear.

In a study of 139 streams throughout the United States, the USGS found that one or more of these chemicals were in 80 percent of the streams sampled. Half of the streams contained seven or more of these chemicals, and about one-third contained 10 or more. This was the first national-scale examination of streams for these organic wastewater contaminants. The chemicals identified largely escape regulation and are not removed by municipal wastewater treatment. At this point, the long-term effects of exposure to such chemicals are not clear; however, further study is anticipated.

AGRICULTURAL WATER POLLUTION

Agricultural activities are the primary cause of water pollution problems. Excessive use of fertilizer results in eutrophication in many aquatic habitats because precipitation carries dissolved nutrients (nitrogen and phosphorus compounds) into streams and lakes. In addition, groundwater may become contaminated with fertilizer and pesticides. The exposure of land to erosion results in increased amounts of sediment being added to water courses. Runoff from animal feedlots carries nutrients, organic matter, and bacteria. Water used to flush irrigated land to get rid of excess salt in the soil carries a heavy load of salt that degrades the water body. And the use of agricultural chemicals results in contamination of

sediments and aquatic organisms. One of the largest water pollution problems is agricultural runoff from large expanses of open fields. See chapter 13 for a general discussion of methods for reducing runoff and soil erosion.

Farmers can reduce runoff in several ways. One is to leave a zone of undisturbed, permanently vegetated land, called a conservation buffer, near drains or stream banks. This retards surface runoff because soil covered with vegetation tends to slow the movement of water and allows the silt to be deposited on the surface of the land rather than in the streams. This can be costly because farmers may need to remove valuable cropland from cultivation. One goal of the Clean Water Action Plan is to establish 3.2 million kilometers (2 million miles) of conservation buffer strips. By 2004, the plan was about halfway to its goal. Another way to retard runoff is keep the soil covered with a crop as long as possible. Careful control of the amount and the timing of fertilizer application can also reduce the amount of nutrients lost to streams. This makes good economic sense because any fertilizer that runs off or leaches out of the soil is unavailable to crop plants and results in less productivity.

INDUSTRIAL WATER POLLUTION

Factories and industrial complexes frequently dispose of some or all of their wastes into municipal sewage systems. Depending on the type of industry involved, these wastes contain organic materials, petroleum products, metals, acids, toxic materials, organisms, nutrients, or particulates. Organic materials and oil add to the BOD of the water. The metals, acids, and specific toxic materials need special treatment, depending on their nature and concentration. In these cases, a municipal wastewater treatment plant will require that the industry pretreat the waste before sending it to the wastewater treatment plant. If this is not done, the municipal sewage treatment plants must be designed with their industrial customers in mind. In most cases, cities prefer that industries take care of their own wastes. This allows industries to segregate and control toxic wastes and design wastewater facilities that meet their specific needs.

Since industries are point sources of pollution, they have been relatively easy to identify as pollution sources, have been vigorously regulated, and have responded to mandates that they clean up their effluent. Most companies, when they remodel their facilities, include wastewater treatment as a necessary part of an industrial complex. However, some older facilities continue to pollute. These companies discharge acids, particulates, heated water, and noxious gases into the water. While industrial water pollution in the industrialized world has been significantly regulated, in much of the developing world this is not the case, and many lakes, streams, and harbors are severely polluted with heavy metals and other toxic materials, organic matter, and human and animal waste.

A special source of industrial water pollution is mining. By its very nature, mining disturbs the surface of the Earth and increases the chances that sediment and other materials will pollute surface waters. Hydraulic mining is practiced in some countries and involves spraying hillsides with high-pressure water jets to dislodge valuable ores. Often, chemicals are used to separate the valuable metals from the ores, and the waste from these processes is released into streams as well. Water that drains from current or abandoned coal mines is often very acidic. Pyrite is a mineral associated with many coal deposits. It contains sulfur, and when exposed to weathering, the sulfur reacts with oxygen, resulting in the formation of sulfuric acid. In addition, fine coal-dust particles are suspended in the water, which makes the water chemically and physically less valuable as a habitat. Dissolved ions of iron, sulfur, zinc, and copper also are present in mine drainage. Control involves containing mine drainage and treating it before it is released to surface water. Although federal legislation was passed in the 1990s requiring backfilling and land restoration after a mine is closed down, issues of responsibility and compliance with the law persist.

THERMAL POLLUTION

Amendments to the Federal Water Pollution Control Act of 1972 mandated changes in how industry treats water. Industries were no longer allowed to use water and return it to its source in poor condition. One of the standards regulates the temperature of the water that is returned to its source. Because many industries use water for cooling, thermal pollution can be a problem. **Thermal pollution** occurs when an industry removes water from a source, uses the water for cooling purposes, and then returns the heated water to its source.

Power plants heat water to convert it into steam, which drives the turbines that generate electricity. For steam turbines to function efficiently, the steam must be condensed into water after it leaves the turbine. This condensation is usually accomplished by taking water from a lake, stream, or ocean to absorb the heat. This heated water is then discharged. The least expensive and easiest method of discharging heated water is to return the water to the aquatic environment, but this can create problems for aquatic organisms. Although an increase in temperature of only a few degrees may not seem significant, some aquatic organisms are very sensitive to minor temperature changes. Many fish are triggered to spawn by increases in temperature, while others may be inhibited from spawning if the temperature rises. For example, lake trout will not spawn in water above 10°C (50°F). If a lake has a temperature of 8°C (46°F), the lake trout will reproduce, but an increase of 3°C (5°F) would prevent spawning and result in this species' eventual elimination from that lake. Another problem associated with elevated water temperature is that it results in a decrease in the amount of oxygen dissolved in the water.

Ocean estuaries are very fragile. The discharge of heated water into an estuary may alter the type of plants present. As a result, animals with specific food habits may be eliminated because the warm water supports different food organisms. The entire food web in the estuary may be altered by only slight temperature increases.

Cooling water used by industry does not have to be released into aquatic ecosystems. Today in the industrialized world, most cooling water is not released in such a way that aquatic ecosystems are endangered. Three other methods of discharging the heat are commonly used. One method is to construct a large shallow pond. Hot water is pumped into one end of the pond, and cooler water is removed from the other end. The heat is dissipated from the pond into the atmosphere and substrate.

A second method is to use a cooling tower. In a cooling tower, the heated water is sprayed into the air and cooled by evaporation. The disadvantage of cooling towers and shallow ponds is that large amounts of water are lost by evaporation. The release of this water into the air can also produce localized fogs.

The third method of cooling, the dry tower, does not release water into the atmosphere. In this method, the heated water is pumped through tubes, and the heat is released into the air. This is the same principle used in an automobile radiator. The dry tower is the most expensive to construct and operate.

MARINE OIL POLLUTION

Marine oil pollution has many sources. One source is accidents, such as oil-drilling blowouts or oil tanker accidents. The *Exxon Valdez,* which ran aground in Prince William Sound, Alaska, in 1989, released over 42 million liters (11 million gallons) of oil and affected nearly 1500 kilometers (930 miles) of Alaskan coastline. The event had a great effect on the algae and animal populations of the sound, and the economic impact on the local economy was severe. A U.S. National Oceanic and Atmospheric Administration study estimates that 50 percent of the oil biodegraded on beaches or in the water; 20 percent evaporated; 14 percent was recovered; 12 percent is at the bottom of the sea, mostly in the Gulf of Alaska; 3 percent lies on shorelines; and less than 1 percent still drifts in the water column. River otter, seabird, and bald eagle populations recovered to prespill numbers by 1992. Long-term impacts on reproduction and susceptibility to disease for many marine species are still under study. This points out the tremendous resilience of natural ecosystems to respond to and recover (within limits) from disastrous events.

Although accidents such as the *Exxon Valdez* are spectacular events, much more oil is released as a result of small, regular discharges from other, less-visible sources. Nearly two-thirds of all human-caused marine oil pollution comes from three sources: (1) runoff from streets, (2) improper disposal of lubricating oil from machines or automobile crankcases, and (3) intentional oil discharges that occur during the loading and unloading of tankers. With regard to the latter, pollution occurs when the tanks are cleaned or oil-contaminated ballast water is released. Oil tankers use seawater as ballast to stabilize the craft after they have discharged their oil. This oil-contaminated water is then discharged back into the ocean when the tanker is refilled. In addition to human-caused oil pollution, oil naturally seeps into the water from underlying oil deposits in many places.

As the number of offshore oil wells and the number and size of oil tankers have grown, the potential for increased oil pollution has also grown. Many methods for controlling marine oil pollution have been tried. Some of the more promising methods are recycling and reprocessing used oil and grease from automobile service stations and from industries, and enforcing stricter regulations on the offshore drilling, refining, and shipping of oil. As a result of oil spills from shipping tankers, an international agreement was reached in 1992 that required that all new oil tankers be constructed with two hulls—one inside the other. Such double-hulled vessels would be much less likely to rupture and spill their contents. Today, approximately 25 percent of oil tankers are double hulled.

GROUNDWATER POLLUTION

A wide variety of activities, some once thought harmless, have been identified as potential sources of groundwater contamination. In fact, possible sources of human-induced groundwater contamination span every facet of social, agricultural, and industrial activities. (See figure 15.13.) Once groundwater pollution has occurred, it is extremely difficult to remedy. Pumping groundwater and treating it is very slow and costly, and it is difficult to know when all of the contaminated water has been removed. A much better way to deal with the issue of groundwater pollution is to work very hard to prevent the pollution from occurring in the first place.

Major sources of groundwater contamination include:

1. Agricultural products. Pesticides contribute to unsafe levels of organic contaminants in groundwater. Seventy-three different pesticides have been detected in the groundwater in Canada and the United States. Accidental spills or leaks of pesticides pollute groundwater sources with 10 to 20 additional pesticides. Other agricultural practices contributing to groundwater pollution include animal-feeding operations, fertilizer applications, and irrigation practices.

2. Underground storage tanks. For many years in North America, a large number of underground storage tanks containing gasoline and other hazardous substances have leaked. Four liters (1 gallon) of gasoline can contaminate the water supply of a community of 50,000 people. A major program of replacing leaking underground storage tanks recently was completed in the United States. However, the effects of past leaks and abandoned tanks will continue to be a problem for many years.

3. Landfills. Even though recently constructed landfills have special liners and water collection systems, approximately 90 percent of the landfills in North America have no liners to stop leaks to underlying groundwater, and 96 percent have no system to collect the leachate that seeps from the landfill. Sixty percent of landfills place no restrictions on the waste accepted, and many landfills are not inspected even once a year.

4. Septic tanks. Poorly designed and inadequately maintained septic systems have contaminated groundwater with nitrates, bacteria, and toxic cleaning agents. Over 20 million septic tanks are in use in the United States, and up to a third have been found to be operating improperly.

5. Surface impoundments. Over 225,000 pits, ponds, and lagoons are used in North America to store or treat wastes. Seventy-one percent are unlined, and only 1 percent use a plastic or other synthetic, nonsoil liner. Ninety-nine percent of these impoundments have no leak-detection systems. Seventy-three percent have no restriction on the waste placed in the impoundment. Sixty percent are not even inspected annually. Many of these ponds are located near groundwater supplies.

Other sources of groundwater contamination include mining wastes, salting for controlling road ice, land application of treated wastewater, open dumps, cemeteries, radioactive disposal sites, urban runoff, construction excavation, fallout from the atmosphere, and animal feedlots.

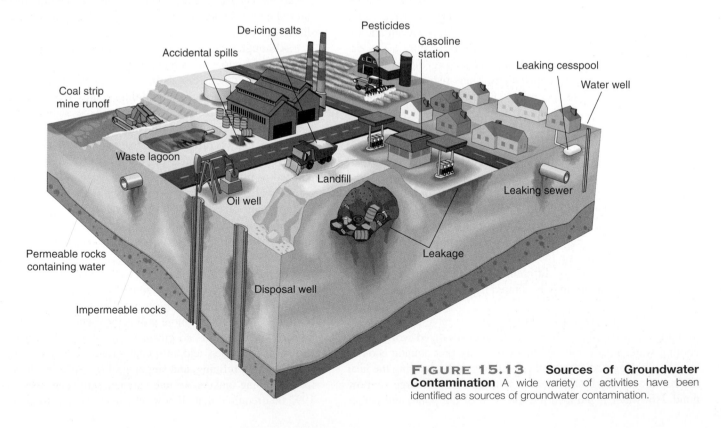

FIGURE 15.13 **Sources of Groundwater Contamination** A wide variety of activities have been identified as sources of groundwater contamination.

WATER-USE PLANNING ISSUES

In the past, wastes were discharged into waterways with little regard for the costs imposed on other users by the resulting decrease in water quality. Furthermore, as the population has grown and the need for irrigation and domestic water has intensified, in many parts of the world, there has not been enough water to satisfy everyone's needs. With today's increasing demands for high-quality water, unrestrained waste disposal and unlimited withdrawal of water could lead to serious conflicts about water uses, causing social, economic, and environmental losses at both local and international levels. (Table 15.2 summarizes some of the areas of controversy involving water throughout the world.)

Metropolitan areas must deal with a variety of issues and maintain an extensive infrastructure to provide three basic water services:

1. Water supply for human and industrial needs
2. Wastewater collection and treatment
3. Stormwater collection and management

Water sources must be identified and preserved for use. Some cities obtain all their municipal water from groundwater and must have a thorough understanding of the size and characteristics of the aquifer they use. Some cities, such as New York City, obtain potable water by preserving a watershed that supplies the water needed by the population. Other cities have abundant water in the large rivers that flow by them but must deal with pollution problems caused by upstream users. In many places where water is in short supply, municipal and industrial-agricultural needs for water conflict.

Water for human and industrial use must be properly treated and purified. It is then pumped through a series of pipes to consumers. After the water is used, it flows through a network of sewers to a wastewater treatment plant, where it is treated before it is released. Maintaining the infrastructure of pipes, pumps, and treatment plants is expensive.

Metropolitan areas must also deal with great volumes of excess water during storms. This water is known as **stormwater runoff.** Because urban areas are paved and little rainwater can be absorbed into the ground, managing stormwater is a significant problem. Cities often have severe local flooding because the water is channeled along streets to storm sewers. If these sewers are overloaded or blocked with debris, the water cannot escape and flooding occurs.

The Water Quality Act of 1987 requires that municipalities obtain permits for discharges of stormwater runoff so that nonpoint sources of pollution are controlled. In the past, many cities had a single system to handle both sewage and stormwater runoff. During heavy precipitation or spring thaws, the runoff from streets could be so large that the wastewater treatment plant could not handle the volume. The wastewater was then diverted directly into the receiving body of water without being treated. Because of these new requirements, some cities have created areas in which to store this excess water until it can be treated. This is expensive and, therefore, is done only if federal or state funding is available. Many cities have also gone through the expensive process of separating their storm sewers from their sanitary sewers. A good example of the long and costly process of separation of sewers is Portland, Oregon. By 2006, Portland had completed only half of a 20-year project to separate its storm and sanitary sewer systems. The final costs are estimated to exceed the project's $1 billion budget.

Providing water services is expensive. We must understand that water supplies are limited. We must also understand that water's ability to dilute and degrade pollutants is limited and that proper land-use planning is essential if metropolitan areas are to provide services and limit pollution.

TABLE 15.2 International Water Disputes

River/Lakes	Countries Involved	Issues
Asia		
Brahmaputra, Ganges, Farakka	Bangladesh, India, Nepal	Alluvial deposits, dams, floods, irrigation, international quotas
Mekong	Cambodia, Laos, Thailand, Vietnam	Floods, international quotas
Salween	Tibet, China (Yunan), Burma	Alluvial deposits, floods
Middle East		
Euphrates, Tigris	Iraq, Syria, Turkey	International quotas, salinity levels
West Bank Aquifer, Jordan, Litani, Yarmuk	Israel, Jordan, Lebanon, Syria	Water diversion, International quotas
Africa		
Nile	Mainly Egypt, Ethiopia, Sudan	Alluvial deposits, water diversion, floods, irrigation, international quotas
Lake Chad	Nigeria, Chad	Dam
Okavango	Namibia, Angola, Botswana	Water diversion
Europe		
Danube	Hungary, Slovak Republic	Industrial pollution
Elbe	Germany, Czech Republic	Industrial pollution, salinity levels
Meuse, Escaut	Belgium, Netherlands	Industrial pollution
Szamos (Somes)	Hungary, Romania	Water allocation
Tagus	Spain, Portugal	Water allocation
Americas		
Colorado, Rio Grande	United States, Mexico	Chemical pollution, international quotas, salinity levels
Great Lakes	Canada, United States	Pollution
Lauca	Bolivia, Chile	Dams, salinity
Paran	Argentina, Brazil	Dams, flooding of land
Cenepa	Ecuador, Peru	Water allocation

Everglades National Park is a unique, subtropical, freshwater wetland visited by about a million people per year. This unique ecosystem exists because of an unusual set of conditions. South Florida is a nearly flat landscape with a slight decrease in elevation from the north to the south. In addition, the substrate is a porous limestone that allows water to flow through rather easily. Originally, water drained in a broad sheet from Lake Okeechobee southward to Florida Bay. This constant flow of water sustained a vast, grassy wetland interspersed with patches of trees.

After Everglades Park was established in 1947, about 800,000 hectares (2 million acres) of wetlands to the north were converted to farms and urban development. South Florida boomed. Some 4.5 million people currently live in the horseshoe crescent around the Everglades region, and new residents arrive each day. Conversion of land to agriculture and urban development required changes in the natural flow of water. Dams, drainage canals, and water diversion supported and protected the human uses of the area but cut off the essential, natural flow of freshwater to the Everglades. Changes in the normal pattern of water flow to the Everglades resulted in periods of drought and a general reduction in the size of this unique wetland region. Populations of wading birds in the southern Everglades fell drastically as their former breeding and nesting areas dried up. Other wildlife, such as alligators, Florida panthers, snail kites, and wood storks, also were negatively affected because drought reduces suitable habitat during parts of the year and the animals are forced to congregate around the remaining sources of water. The quality of the water is also important. The original wetland ecosystem was a nutrient-poor system. The introduction of nutrients into the water from agricultural activities encouraged the growth of exotic plants that replaced natural vegetation.

As people recognized that the key element necessary to preserve the Everglades was a constant, reliable source of clean freshwater, several steps were taken to modify water use to preserve the Everglades. The Kissimmee River, which flows into Lake Okeechobee, was channelized into an arrow-straight river in the late 1960s. This project destroyed the marshes and allowed nutrients from dairy farms and other agricultural activities to pollute the lake and Everglades Park, to which the water from Lake Okeechobee eventually flowed. To help alleviate this problem, in 1990, the U.S. Army Corps of Engineers began to return the Kissimmee to its natural state, with twisting oxbow curves and extensive wetlands. This allows the plants in the natural wetlands to remove much of the nutrient load before it enters the lake. To reduce the likelihood of further development near the park, in 1989, Congress approved the purchase of 43,000 hectares (100,000 acres) for an addition to the east section of the park. Florida obtained an additional 60,000 hectares (150,000 acres) as an additional buffer zone for the park.

For several years, the U.S. Army Corps of Engineers and the South Florida Water Management District cooperated in the development of a comprehensive restoration plan that was finished in 2000. The development of the plan involved the input of scientists, politicians, various business interests, and environmentalists. Key components of the plan are:

1. Developing facilities to store surface water and pump water into aquifers so that it can be released when needed

2. Developing wetlands to treat municipal and agricultural runoff so that nutrient loads are reduced

3. Using clean wastewater to recharge aquifers and supply water to wetlands in the Miami area

4. Reducing the amount of water lost through levees and redirecting water to the Everglades

5. Removing barriers to the natural flow of water through the Everglades

The plan received strong support from Congress in the fall of 2001, when it allocated $1.4 billion to begin implementing the plan. It will require many more billions of dollars and up to 30 years to accomplish all aspects of the plan, but if the plan continues to be implemented over the

In pursuing these objectives, city planners encounter many obstacles. Large metropolitan areas often have hundreds of local jurisdictions (governmental and bureaucratic areas) that divide responsibility for management of basic water services. The Chicago metropolitan area is a good example. This area is composed of six counties and approximately 2000 local units of government. It has 349 separate water-supply systems and 135 separate wastewater disposal systems. Efforts to implement a water-management plan, when so many layers of government are involved, can be complicated and frustrating.

To meet future needs, urban, agricultural, and national interests will need to deal with a number of issues, such as the following:

- Increased demand for water will generate pressure to divert water to highly populated areas or areas capable of irrigated agriculture.

- Increased demand for water will force increased treatment of wastewater and reuse of existing water supplies.

- In many areas where water is used for irrigation, evaporation of water from the soil over many years results in a buildup of salt in the soil. When the water used to flush the salt from the soil is returned to a stream, the quality of the water is lowered.

- In some areas, wells provide water for all categories of use. If the groundwater is pumped out faster than it is replaced, the water table is lowered.

- In coastal areas, seawater may intrude into the aquifers and ruin the water supply.

- The demand for water-based recreation is increasing dramatically and requires high-quality water, especially for activities involving total body contact, such as swimming.

next few decades, the Everglades ecosystem will be restored to a more stable condition and will have a more hopeful future.

In 2008, an agreement was entered into between the state of Florida and U.S. Sugar, the largest producer of cane sugar in the United States. Under the agreement, the state of Florida will purchase U.S. Sugar for $1.7 billion. U.S. Sugar will continue operations for

six years and will then transfer to Florida 187,000 acres (760 km²). The newly acquired land will remain undeveloped to allow it to be restored to its pre-drainage state. The land will be used to reestablish a part of the connection between Lake Okeechobee and the Everglades through a managed system of water storage and treatment and, at the same time, safeguard the St. Lucie and Caloosahatchee rivers and estuaries. The land acquisition would also prevent massive amounts of nutrient phosphorus from entering the Everglades every year. Phosphorus is used as a fertilizer for sugar production. Phosphorus runoff pollutes the water to 20 times the tolerable level, endangering native wildlife.

Phosphorus changes the chemistry of the water and destroys the microbial populations, an essential source of food for many aquatic organisms, which then do not flourish. As one result, 90 percent of the wading birds in the Everglades have disappeared and 68 species of plants and animals are either endangered or threatened.

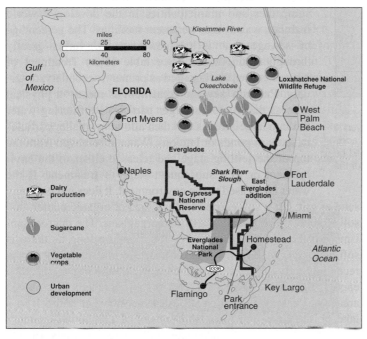

Source for Everglades map: Copyright 1990 U.S. News and World Report, L.P. Reprinted with permission.

Everglades wading birds once numbered more than a million, but pollution and drought brought on by water diversion have decimated their number.

Construction threatens to destroy the fragile ecosystem of the Everglades.

WATER DIVERSION

Water diversion is the physical process of transferring water from one area to another. The aqueducts of ancient Rome are early examples of water diversion. Thousands of diversion projects have been constructed since then. New York City, for example, receives 90 percent of its water supply from the Catskill Mountains (Schoharie and Ashokan Reservoirs) and from four reservoirs collecting water from tributaries to the Delaware River west of the Catskills. About 10 percent of its supply comes from the Croton watershed east of the Hudson River. The most distant watershed serving New York City is some 200 kilometers (125 miles) away. Los Angeles is another example. It began importing water from the Owens Valley, 400 kilometers (250 miles) to the north, as early as 1913.

While diversion is seen as a necessity in many parts of the world, it often generates controversy. An example of this is the Garrison Diversion Unit in North Dakota, which was originally

envisioned as a way to divert water from the Missouri River system for irrigation. (See figure 15.14.) The initial plan to irrigate portions of the Great Plains was developed during the Dust Bowl era of the 1930s. The Federal Flood Control Act of 1944 authorized the construction of the Garrison Diversion Unit. However, one of the original intents of the plan—to divert water from the Missouri River to the Red River—has not been met. Since the Red River flows north into Canada, the project requires international cooperation. The Canadian government has concerns about the effects on water quality of releasing additional water into the Red River. There are also concerns about environmental consequences. Portions of wildlife refuges, native grasslands, and waterfowl breeding marshes would be damaged or destroyed. Some states also have expressed concern about diverting water from the Missouri River, since any water diverted is not available for those downstream on the Missouri River.

While the plan has been modified several times, two sections of canal and a pumping station at Lake Sakakawea have been

A growing number of larger sewage treatment plants use additional processes called tertiary sewage treatment. **Tertiary sewage treatment** involves a variety of techniques to remove inorganic nutrients left after primary and secondary treatments. (See table 15.4.) The tertiary treatment of municipal wastewater is often used to remove phosphorus and nitrogen that could increase aquatic plant growth. Some municipalities are using natural or constructed wetlands to serve as tertiary sewage treatment systems. In other cases, the effluent from the treatment facility is used to irrigate golf courses, roadside vegetation, or cropland. The vegetation removes excess nutrients and prevents them from entering streams and lakes where they would present a pollution problem. Tertiary treatment of industrial and other specialized wastewater streams is very costly because it requires specific chemical treatment of the water to eliminate specific problem materials. Many industries maintain their own wastewater facilities and design specific tertiary treatment processes to match the specific nature of their waste products.

As water has become scarce in many parts of the world, people have looked at wastewater as a source of water for other purposes. Water from wastewater plants can be used for irrigation, industrial purposes, cooling, and many other activities. Ultimately, it is possible to have a closed loop system for domestic water in which the output of the wastewater plant becomes the input for the drinking water supply.

TABLE 15.4 Tertiary Treatment Methods

Kind of Tertiary Treatment	Problem Chemicals	Methods
Biological	Phosphorus and nitrogen compounds	1. Large ponds are used to allow aquatic plants to assimilate the nitrogen and phosphorus compounds from the water before the water is released.
		2. Columns containing denitrifying bacteria are used to convert nitrogen compounds into atmospheric nitrogen.
Chemical	Phosphates and industrial pollutants	1. Water can be filtered through calcium carbonate. The phosphate substitutes for the carbonate ion, and the calcium phosphate can be removed.
		2. Specific industrial pollutants, which are nonbiodegradable, may be removed by a variety of specific chemical processes.
Physical	Primarily industrial pollutants	1. Distillation
		2. Water can be passed between electrically charged plates to remove ions.
		3. High-pressure filtration through small-pored filters
		4. Ion-exchange columns

SALINIZATION

Another water-use problem results from **salinization,** an increase in salinity caused by growing salt concentrations in soil. This is primarily a problem in areas where irrigation has been practiced for several decades. As water evaporates from soil or plants extract the water they need, the salts present in all natural waters become concentrated. Since irrigation is most common in hot, dry areas that have high rates of evaporation, there is generally an increase in the concentration of salts in the soil and in the water that runs off the land. (See figure 15.16.) Every river increases in salinity as it flows to the ocean. The salinity of the Colorado River water increases 20 times as it passes through irrigated cropland between Grand Lake in north-central Colorado and the Imperial Dam in southwest Arizona. The problem of salinity will continue to increase as irrigation increases.

GROUNDWATER MINING

Groundwater mining means that water is removed from an aquifer faster than it is replaced. When this practice continues for a long time, the water table eventually declines. Groundwater mining is common in areas of the western United States and throughout the world. In North America, it is a particular problem due to growing cities and increasing irrigation. In aquifers with little or no recharge, virtually any withdrawal constitutes mining, and sustained withdrawals will eventually exhaust the supply. This problem is particularly serious in communities that depend heavily on groundwater for their domestic needs.

FIGURE 15.16 **Salinization** As water evaporates from the surface of the soil, the salts it was carrying are left behind. In some areas of the world, this has permanently damaged cropland.

Groundwater mining can also lead to problems of settling or subsidence of the ground surface. Removal of the water allows the ground to compact, and large depressions may result. For example, in the San Joaquin Valley of California, groundwater has been withdrawn for irrigation and cultivation since the 1850s, and groundwater levels have fallen over 100 meters (300 feet). More than 1000 hectares (approximately 2500 acres) of ground have subsided, some as much as 6 meters (20 feet). Currently, the ground surface in that area is sinking 30 centimeters (12 inches) per year. London, Mexico City, Venice, Houston, and Las Vegas are some other cities that have experienced subsidence as a result of groundwater withdrawal. Table 15.5 lists some estimated amounts of groundwater depletion in certain areas of the United States and the world. As people recognize the severity of the problem, public officials are beginning to develop water conservation plans for their cities. Albuquerque, New Mexico, which relies on groundwater for its water supply, has an extensive public education program to encourage people to reduce their water consumption. Since grass demands water, people are encouraged to use desert plants for landscaping or collect rainwater to water their lawns. Finding and correcting leaks, reducing the amount of water used in bathing, and recycling water from swimming pools are other conservation strategies.

An average of 85 billion gallons (320 billion liters) of groundwater are withdrawn daily in the United States. More than 90 percent of these withdrawals are used for irrigation, public supply (deliveries to homes, businesses, industry), and self-supplied industrial uses. Irrigation is the largest use, accounting for about two-thirds of the amount. The percentage of total irrigation withdrawals provided by groundwater increased from 23 percent in 1950 to 42 percent in 2007. Groundwater provides about half of the drinking water in the United States with nearly all those in rural areas reliant upon groundwater.

The importance of groundwater withdrawals in the United States is similar to that in the rest of the world, with some variations from country to country. Rapid expansion in groundwater use occurred between 1950 and 1975 in many industrial nations and subsequently in much of the developing world. The intensive use of groundwater for irrigation in arid and semi-arid countries has been called a "silent revolution" as millions of independent farmers worldwide have chosen to become increasingly dependent on the reliability of groundwater resources, reaping abundant social and economic benefits but with limited management controls by government water agencies. Perhaps as many as 2 billion people worldwide depend directly upon groundwater for drinking water. The dependence on groundwater for drinking water is particularly high in Europe, where about 75 percent of the drinking-water supply is obtained from groundwater.

Groundwater mining poses a special problem in coastal areas. As the fresh groundwater is pumped from wells along the coast, the saline groundwater moves inland, replacing fresh groundwater with unusable saltwater. This process, called **saltwater intrusion,** is shown in figure 15.17. Saltwater intrusion is a serious problem in heavily populated coastal areas throughout the world.

Achieving an acceptable trade-off between groundwater use and the long-term effects of that use is a central theme in the evolving concept of groundwater sustainability. Initially, groundwater was viewed as a convenient resource for general use, and attention was focused on the economic aspects of groundwater development. Sustainability concerns emerging in the early 1980s have brought environmental viewpoints to the forefront in discussions about groundwater availability.

Groundwater sustainability is commonly defined in a broad context as the development and use of groundwater resources in a manner that can be maintained for an indefinite amount of time without causing unacceptable environmental, economic, or social

TABLE 15.5 Groundwater Depletion in Major Regions of the World

Region/Aquifer	Estimates of Depletion
California	Groundwater overdraft exceeds 1.7 billion cubic meters (60 billion cubic feet) per year. The majority of the depletion occurs in the Central Valley, which is referred to as the vegetable basket of the United States.
Southwestern United States	In parts of Arizona, water tables have dropped more than 120 meters (400 feet). Projections for parts of New Mexico indicate that water tables will drop an additional 22 meters (70 feet) by 2020.
High Plains aquifer system, United States	The Ogallala aquifer underlies nearly 20 percent of all the irrigated land in the United States. To date, the net depletion of the aquifer is in excess of 350 billion cubic meters (12 trillion cubic feet), or roughly 15 times the average annual flow of the Colorado River. Most of the depletion has been in the Texas high plains, which witnessed a 26 percent decline in irrigated land from 1979 to 1989. Current depletion is estimated to be in excess of 13 billion cubic meters (450 billion cubic feet) a year.
Mexico City and Valley of Mexico	Use exceeds natural recharge by 60 to 85 percent, causing land subsidence and falling water tables.
African Sahara	North Africa has vast nonrecharging aquifers where current depletion exceeds 12 billion cubic meters (425 billion cubic feet) a year.
India	Water tables are declining throughout much of the most productive agricultural land in India. In parts of the country, groundwater levels have declined 90 percent during the past two decades.
North China	The water table underneath portions of Beijing has dropped 40 meters (130 feet) during the past 40 years. A large portion of northern China has significant groundwater overdraft.
Arabian peninsula	Groundwater use is nearly three times greater than recharge. At projected depletion rates, exploitable groundwater reserves could be exhausted within the next 50 years. Saudi Arabia depends on nonrenewable groundwater for roughly 75 percent of its water. This includes irrigation of 2 million to 4 million metric tons of wheat per year.

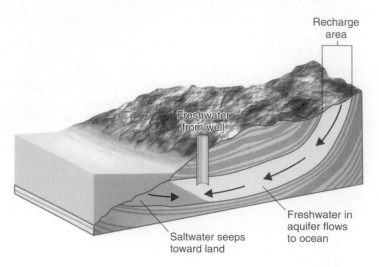

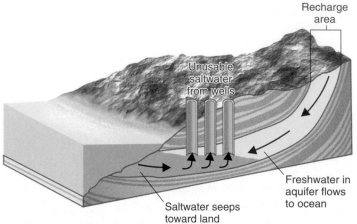

FIGURE 15.17 **Saltwater Intrusion** When saltwater intrudes on fresh groundwater, the groundwater becomes unusable for human consumption and for many industrial processes.

consequences. Groundwater sustainability management strategies are composed of a small number of general approaches, including:

• Use of sources of water other than local groundwater, by shifting the local source of water (either completely or in part) from groundwater to surface water or importing water from outside the local water-system boundaries (the California Central Valley and Houston have implemented these approaches).

• Control or regulation of groundwater pumping through implementation of guidelines, policies, taxes, or regulations by water management authorities.

• Use of groundwater and surface water through the coordinated and integrated use of the two sources to ensure optimum, long-term economic and social benefits.

• Conservation practices, techniques, and technologies that improve the efficiency of water use, often accompanied by public education programs on water conservation.

• Reuse of wastewater (grey water) and treated wastewater (reclaimed water) for nonpotable purposes such as irrigation of crops, lawns, and golf courses.

• Desalinization of brackish groundwater or treatment of otherwise impaired groundwater to reduce dependency on fresh groundwater sources.

PRESERVING SCENIC WATER AREAS AND WILDLIFE HABITATS

Some bodies of water have unique scenic value. To protect these resources, the way in which the land adjacent to the water is used must be consistent with preserving these scenic areas.

The U.S. Federal Wild and Scenic Rivers Act of 1968 established a system to protect wild and scenic rivers from development. All federal agencies must consider the wild, scenic, or recreational values of certain rivers in planning for the use and development of the rivers and adjacent land. The process of designating a river or part of a river as wild or scenic is complicated. It often encounters local opposition from businesses dependent on growth. Following reviews by state and federal agencies, rivers may be designated as wild and scenic by action of either Congress or the secretary of the interior. Sections of over 150 streams comprising about 12,000 kilometers (7700 miles) in the United States have been designated as wild or scenic.

Many unique and scenic shorelands have also been protected from future development. Until recently, estuaries and shorelands have been subjected to significant physical modifications, such as dredging and filling, which may improve conditions for navigation and construction but destroy fish and wildlife habitats. Recent actions throughout North America have attempted to restrict the development of shorelands. Development has been restricted in some particularly scenic areas, such as Cape Cod National Seashore in Massachusetts and the Bay of Fundy in the Atlantic provinces of Canada.

Historically, poorly drained areas were considered worthless. Subsequently, many of these wetlands were filled or drained and used for building sites. At the time of European settlement, the area that is now the conterminous United States contained an estimated 89 million hectares (221 million acres) of wetlands. Over time, wetlands have been drained, dredged, filled, leveled, and flooded to the extent that less than half of the original wetland acreage remains. Figure 15.18 shows the most recent causes of wetlands loss.

Today, 95 percent of the remaining 20 million hectares (50 million acres) of wetlands in the United States are inland freshwater wetlands. The remaining 5 percent are in saltwater estuarine

Causes of Wetlands Loss (1986–2008)

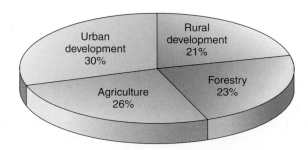

FIGURE 15.18 **Wetlands Conversion** The loss of wetlands occurs because people convert wetlands to other uses. Conversion to urban and rural housing and other infrastructure accounts for just over 50 percent of the wetlands loss. Draining wetlands for agriculture accounts for an additional 26 percent.

Source: U.S. Geological Survey.

CAMPUS SUSTAINABILITY INITIATIVE

CONSERVING WATER ON CAMPUS

Efforts to reduce water consumption are expanding rapidly on university and college campuses. Three such programs are underway at Duke University in North Carolina, Princeton University in New Jersey, and Cuyamaca College in California.

A Duke initiative included the dispensing of 5000 low-water-flow showerheads to faculty, staff, and off-campus students. It is estimated that each 1.5-gallons-per-minute showerhead will save an estimated 7300 gallons of water annually, compared to a standard 2.5 gallons per minute fixture. In addition, Duke's Residence Life and Housing Services replaced more than 1000 showerheads in campus residence halls with the 1.5-gallons-per-minute fixtures.

Residence halls at Princeton are having older toilets replaced with new dual flushing systems designed to save water. The new toilets allow users to push the flush handle one way to use less water for liquid waste and another way to release more water for solid waste. By 2020, Princeton aims to decrease per capita water usage to 25 percent of current levels.

Cuyamaca College is promoting water conservation in the southern California landscape through exhibits and programs that educate and inspire the public. One project is the Water Conservation Garden. The 5-acre garden has displays that showcase water conservation in themed gardens, such as a native plant garden and a vegetable garden, as well as how-to displays, such as mulch and irrigation exhibits.

---

environments. Freshwater forested wetlands make up the single largest category of all wetlands in the conterminous United States. The natural and economic importance of wetlands has been recognized only recently. In addition to providing spawning and breeding habitats for many species of wildlife, wetlands act as natural filtration systems by trapping nutrients and pollutants and preventing them from entering adjoining lakes, streams, or estuaries. Wetlands also slow down floodwaters and permit nutrient-rich particles to settle out. In addition, wetlands can act as reservoirs and release water slowly into lakes, streams, or aquifers, thereby preventing floods. (See figure 15.19.) Coastal estuarine zones and adjoining sand dunes also provide significant natural flood control. Sand dunes act as barriers and absorb damaging waves caused by severe storms. In recent years, public appreciation of the ecological, social, and economic values of wetlands has increased substantially. The increased awareness of how much wetland acreage has been lost or damaged since the time of European settlement and the consequences of those losses has led to the development of many federal, state, and local wetland protection programs and laws.

- Filter toxic wastes, excess nutrients, sediments, and other pollutants
- Help prevent erosion
- Reduce flooding by storing stormwater

- Reduce storm damage by absorbing waves
- Provide feeding and resting spots for migratory waterfowl

- Provide food and habitat for other aquatic species

- Provide nursery sites for the young of a number of species, including oysters, clams, crabs, and shrimp

FIGURE 15.19 **The Value of Wetlands** Wetlands are areas covered with water for most of the year that support aquatic plant and animal life. Wetlands can be either fresh or saltwater and may be isolated potholes or extensive areas along rivers, lakes, and oceans. We once thought of wetlands as only a breeding site for mosquitoes. Today, we are beginning to appreciate their true value.

Is There Lead in Our Drinking Water?

A reader of a major metropolitan newspaper wrote a letter to the editor outlining her family's decision to move from the suburbs to the city to experience arts, culture, and greater diversity and to help prevent urban sprawl. Her family was very happy with their decision until they received a letter from the city water department informing them that everyone in their neighborhood had lead in their drinking water, and it would cost each homeowner thousands of dollars to replace old pipes—a measure that might or might not fix the problem. The lead level was 400 parts per billion (ppb), which is more than 20 times higher than safe levels. The Environmental Protection Agency takes action if the level exceeds 15 ppb for 90 percent of homes tested.

What is the health risk from lead exposure? Children, babies, and pregnant women face the highest risk because children absorb lead at a faster rate than adults, and lead has a greater impact on developing bodies. Studies from the Centers for Disease Control and Prevention (CDC) have established that high lead exposure levels are likely to result in learning disorders in babies and children under age 6, and may even lead to behavioral problems. Adults are not exempt from effects, either. Exposure to high lead levels is associated with high blood pressure, infertility, decreased kidney and muscular function, and possibly even cancer. Lead mimics calcium in the body, moving through the liver, kidneys, and brain. It can be stored in the bones and teeth over long periods of time. In the brain, lead disrupts the electrical messages sent to cells throughout the body.

How widespread is the problem? Lead paint in older buildings is the leading cause of lead exposure; however, drinking water accounts for 20 percent of exposure. The EPA estimates that 98 percent of homes in the country have plumbing that can slowly leach lead into tap water as the pipes and fixtures deteriorate over time. Heat and acidity can accelerate the process, as can changes in the chemicals used to treat drinking water. Lead exposure levels from water increase to 50 percent for babies who are fed formula. While residential drinking water is routinely tested, there is no federal mandate to test for lead at schools and daycare centers. Concern by parents and school officials has led to voluntary testing for lead in selected areas of the country.

What has been done? The federal government has long recognized the severe health problems for children and pregnant women associated with lead. Leaded gasoline and house paint were banned in the 1970s. The Safe Drinking Water Act amendments of 1986 banned lead pipes and solder in new construction and repairs. However, there are miles of old lead pipes in drinking water systems throughout the country. For example, in Washington, D.C., the city is trying to replace 23,000 lead service lines. The good news is that these actions by the federal government have resulted in a significant decrease in lead exposure levels. In 2001, 2.2 percent of children ages 1 to 5 had blood lead levels that exceeded CDC standards, compared to 90 percent in 1976. The bad news is that urban and suburban areas in Maryland, Pennsylvania, Washington, New York, and the District of Columbia, to name a few, have recently found high lead levels in drinking water in homes, schools, and daycare centers. In 2004, Senator Jim Jeffords of Vermont introduced the Lead-Free Drinking Water Act, which would ban all lead plumbing fixtures and grant $200 million to the District of Columbia to replace lead plumbing throughout the city.

What do you think should be done?

- Should schools and daycare centers be required to monitor drinking water for lead?
- If high lead levels are found, who should pay for short-term solutions such as bottled water and additional testing or for permanent solutions such as replacing all plumbing?
- Who should pay to replace lead pipes in homes and cities throughout the nation?

SUMMARY

Water is a renewable resource that circulates continually between the atmosphere and the Earth's surface. The energy for the hydrologic cycle is provided by the sun. Water loss from plants is called evapotranspiration. Water that infiltrates the soil and is stored underground in the tiny spaces between rock particles is called groundwater, as opposed to surface water that enters a river system as runoff. There are two basic kinds of aquifers. Unconfined aquifers have an impervious layer at the bottom and receive water that infiltrates from above. The top of the layer of water is called the water table. A confined aquifer is sandwiched between two impervious layers and is often under pressure. The recharge area may be a great distance from where the aquifer is tapped for use. The way in which land is used has a significant impact on rates of evaporation, runoff, and infiltration.

The four human uses of water are domestic, agricultural, instream, and industrial. Water use is measured by either the amount withdrawn or the amount consumed. Domestic water is in short supply in many metropolitan areas. Most domestic water is used for waste disposal and washing, with only a small amount used for drinking. The largest consumptive use of water is for agricultural irrigation. Major in-stream uses of water are for hydroelectric power, recreation, and navigation. Most industrial uses of water are for cooling and for dissipating and transporting waste materials.

Major sources of water pollution are municipal sewage, industrial wastes, and agricultural runoff. Nutrients, such as nitrates and phosphates from wastewater treatment plants and agricultural runoff, enrich water and stimulate algae and aquatic plant growth. Organic matter in water requires oxygen for its decomposition and, therefore, has a large biochemical oxygen demand (BOD). Oxygen depletion can result in fish death and changes in the normal algal community, which leads to visual and odor problems.

Point sources of pollution are easy to identify and resolve. Nonpoint sources of pollution, such as agricultural runoff and mine drainage, are more difficult to detect and control than those from municipalities or industries.

Thermal pollution occurs when an industry returns heated water to its source. Temperature changes in water can alter the kinds and numbers of plants and animals that live in it. The methods of controlling thermal pollution include cooling ponds, cooling towers, and dry cooling towers.

Wastewater treatment consists of primary treatment, a physical settling process; secondary treatment, biological degradation of the wastes; and tertiary treatment, chemical treatment to remove specific components. Two major types of secondary wastewater treatments are the trickling filter and the activated-sludge sewage methods.

Groundwater pollution comes from a variety of sources, including agriculture, landfills, and septic tanks. Marine oil pollution results from oil drilling and oil-tanker accidents, runoff from streets, improper disposal of lubricating oil from machines and car crankcases, and intentional discharges from oil tankers during loading or unloading.

Reduced water quality can seriously threaten land use and in-place water use. In the United States and other nations, legislation helps to preserve certain scenic water areas and wildlife habitats. Shorelands and wetlands provide valuable services as buffers, filters, reservoirs, and wildlife areas. Water management concerns of growing importance are groundwater mining, increasing salinity, water diversion, and managing urban water use. Urban areas face several problems, such as providing water suitable for human use, collecting and treating wastewater, and handling stormwater runoff in an environmentally sound manner. Water planning involves many governmental layers, which makes effective planning difficult.

THINKING GREEN

1. Plant native and/or drought-tolerant grasses, ground covers, shrubs, and trees. Once established, they do not need water as frequently and usually will survive a dry period without watering. Group plants together based on similar water needs.

2. Install a new water-saving toilet.

3. Get involved in water management issues. Voice your questions and concerns at public meetings conducted by your local government or water management district.

4. Conserve water because it is the right thing to do. Don't waste water just because someone else is paying for it, such as when you are staying at a hotel.

5. Support projects that will lead to an increased use of reclaimed wastewater for irrigation and other uses.

6. Encourage your friends and neighbors to be part of a water-conscious community. Promote water conservation in community newsletters, on bulletin boards, and by example. Encourage your friends, neighbors, and co-workers to "do their part."

WHAT'S YOUR TAKE?

It has been stated that while water resources have rarely, if ever, been the sole source of violent conflict or war, the next major war could be a water war. This is in response to the growing pressure on natural resources that is being experienced throughout the world in the context of increasing demand. With the very high numbers of international watercourses that are shared between countries, water and its use are undoubtedly a cause of tension and often strains relations between countries. Water is a security concern for many countries. Develop a position paper on the likelihood of a future "water war" and where you feel it could develop.

REVIEW QUESTIONS

1. Describe the hydrologic cycle.
2. Distinguish between withdrawal and consumption of water.
3. What are the similarities between domestic and industrial water use? How are they different from in-stream use?
4. How is land use related to water quality and quantity? Can you provide local examples?
5. What is biochemical oxygen demand? How is it related to water quality?
6. How can the addition of nutrients such as nitrates and phosphates result in a reduction of the amount of dissolved oxygen in the water?
7. Differentiate between point and nonpoint sources of water pollution.
8. How are most industrial wastes disposed of? How has this changed over the past 25 years?
9. What is thermal pollution? How can it be controlled?
10. Describe primary, secondary, and tertiary sewage treatment.
11. What are the types of wastes associated with agriculture?
12. Why is stormwater management more of a problem in an urban area than in a rural area?
13. Define groundwater mining.
14. How does irrigation increase salinity?
15. What are the three major water services provided by metropolitan areas?

1. Leakage from freshwater distribution systems accounts for significant losses. Is water so valuable that governments should require systems that minimize leakage to preserve the resource? Under what conditions would you change your evaluation?

2. Do nonfarmers have an interest in how water is used for irrigation? Under what conditions should the general public be involved in making these decisions along with the farmers who are directly involved?

3. Should the United States allow Mexico to have water from the Rio Grande and the Colorado River, both of which originate in the United States and flow to Mexico?

4. Do you believe that large-scale hydroelectric power plants should be promoted as a renewable alternative to power plants that burn fossil fuels? What criteria do you use for this decision?

5. What are the costs and the benefits of the proposed Garrison Diversion Unit? What do you think should happen with this project?

6. How might you be able to help save freshwater in your daily life? Would the savings be worth the costs?

7. Look at the hydrologic cycle in figure 15.3. If global warming increases the worldwide temperature, how should increased temperature directly affect the hydrologic cycle?

CHAPTER 16

AIR QUALITY ISSUES

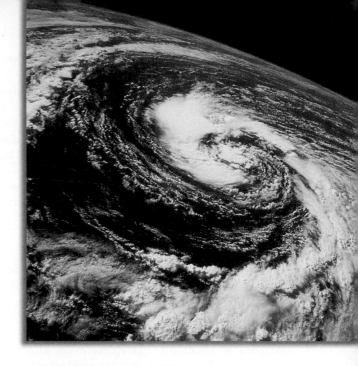

The atmosphere contains gases that are constantly mixed by wind caused by uneven heating of the Earth's surface by the sun.
One of the ways we recognize this mixing process is in changes in weather brought about by the movement of air masses.

CHAPTER OUTLINE

The Atmosphere
Pollution of the Atmosphere
Categories of Air Pollutants
    Carbon Monoxide
    Particulate Matter
    Sulfur Dioxide
    Nitrogen Dioxide
    Lead
    Volatile Organic Compounds
    Ground Level Ozone and Photochemical Smog
    Hazardous Air Pollutants
Control of Air Pollution
    The Clean Air Act
    Actions That Have Reduced Air Pollution
Acid Deposition
    Causes of Acid Precipitation
    Effects on Structures
    Effects on Terrestrial Ecosystems
    Effects on Aquatic Ecosystems
Ozone Depletion
    Why Stratospheric Ozone Is Important
    Ozone Destruction
    Actions to Protect the Ozone Layer
Global Warming and Climate Change
    Causes of Global Warming and Climate Change
    Potential Consequences of Global Warming and Climate Change
Addressing Climate Change
    Energy Efficiency
    The Role of Biomass
    Technological Approaches
    Political and Economic Forces
Indoor Air Pollution
    Sources of Indoor Air Pollutants
    Significance of Weatherizing Buildings
    Secondhand Smoke
    Radon
Noise Pollution

ISSUES & ANALYSIS
Pollution, Policy, and Personal Choice 389
CASE STUDY
Air Pollution in Mexico City 369
CAMPUS SUSTAINABILITY INITIATIVE
New York University's Co-Generation Plant 376
GOING GREEN
Germany's Energy Policy 384
WATER CONNECTIONS
Decline in Arctic Sea Ice 386

OBJECTIVES

After reading this chapter, you should be able to:

- Recognize that air can accept and disperse significant amounts of pollutants.
- List the major sources and effects of the six criteria air pollutants.
- Describe how photochemical smog is formed and how it affects humans.
- Explain how acid rain is formed.
- Understand that human activities can alter the atmosphere in such a way that they can change climate.
- Describe the kinds of changes that could occur as a result of global warming.
- Describe the link between chlorofluorocarbon use and ozone depletion.
- Recognize that there are many positive actions that have improved air quality.
- Recognize that enclosed areas can trap air pollutants that are normally diluted in the atmosphere.

A Global Perspective feature on "The Kyoto Protocol on Greenhouse Gases" can be found on the book's website
at www.mhhe.com/enger12e along with other interesting readings.

THE ATMOSPHERE

The atmosphere (air) is composed of 78.1 percent nitrogen, 20.9 percent oxygen, and a number of other gases such as argon, carbon dioxide, methane, and water vapor that total about 1 percent. Most of the atmosphere is held close to the Earth by the pull of gravitational force, so it gets less dense with increasing distance from the Earth. Throughout the various layers of the atmosphere, nitrogen and oxygen are the most common gases present, although the molecules are farther apart at higher altitudes.

The atmosphere is composed of four layers. (See figure 16.1.)

- The *troposphere* extends from the Earth's surface to about 10 kilometers (about 6.2 miles) above the Earth. It actually varies from about 8 to 18 kilometers (5–11 miles), depending on the position of the Earth and the season of the year. The temperature of the troposphere declines by about 6°C (11°F) for every kilometer above the surface. The troposphere contains most of the water vapor of the atmosphere and is the layer in which weather takes place.

- The *stratosphere* extends from the top of the troposphere to about 50 kilometers (about 31 miles) and contains most of the ozone. The ozone is in a band between 15 and 30 kilometers (9–19 miles) above the Earth's surface. Because the ozone layer absorbs sunlight, the upper layers of the stratosphere are warmer than the lower layers.

- The *mesosphere* is a layer with decreasing temperature from 50 to 80 kilometers (31–50 miles) above the Earth.

- The *thermosphere* is a layer with increasing temperature that extends to about 300 kilometers (186 miles) above the Earth's surface.

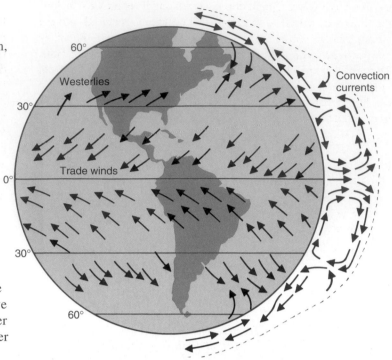

FIGURE 16.2 **Global Wind Patterns** Wind is the movement of air caused by the rotation of the Earth and atmospheric pressure changes brought about by temperature differences. Both of these contribute to the patterns of world air movement. In North America, most of the winds are westerlies (from the west to the east).

Even though gravitational force keeps the majority of the air near the Earth, the air is not static. As it absorbs heat from the Earth, it expands and rises. When its heat content is radiated into space, the air cools, becomes more dense, and flows toward the Earth. As the air circulates vertically due to heating and cooling, it also moves horizontally over the surface of the Earth because the Earth rotates on its axis. The combination of all air movements creates the wind and weather patterns characteristic of different regions of the world. (See figure 16.2.)

POLLUTION OF THE ATMOSPHERE

Pollution is any addition of matter or energy that degrades the environment for humans and other organisms. Because human actions are the major cause of pollution we can do something to prevent it. There are several natural sources of gases and particles that degrade the quality of the air, including material emitted from volcanoes, dust from wind erosion, and gases from the decomposition of dead plants and animals. Since these events are not controlled by humans, we cannot do much to control them. However, automobile emissions, chemical odors, factory smoke, and similar materials are considered air pollution and will be the focus of this chapter.

The problem of air pollution is directly related to the number of people living in an area and the kinds of activities in which they

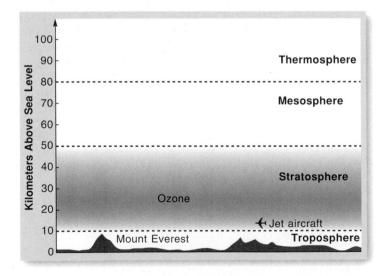

FIGURE 16.1 **The Atmosphere** The atmosphere is divided into the troposphere, the relatively dense layer of gases close to the surface of the Earth; the stratosphere, more distant with similar gases but less dense; the mesosphere; and the thermosphere. Weather takes place in the troposphere, and the important ozone layer is present in the stratosphere.

are involved. When a population is small and its energy use is low, the impact of people is minimal. The pollutants released into the air are diluted, carried away by the wind, washed from the air by rain, or react with oxygen in the air to form harmless materials. Thus, the overall negative effect is slight. However, our urbanized, industrialized civilization has dense concentrations of people that use large quantities of fossil fuels for manufacturing, transportation, and domestic purposes. These activities release large quantities of polluting by-products into our environment.

Gases or small particles released into the atmosphere are likely to stay near the Earth due to gravity. We do not get rid of them; they are just diluted and moved out of the immediate area. In industrialized urban areas, pollutants cannot always be diluted sufficiently before the air reaches another city. The polluted air from Chicago is further polluted when it reaches Gary, Indiana, supplemented by the wastes of Detroit and Cleveland, and finally moves over southeastern Canada and New England to the ocean. While not every population center adds the same kind or amount of waste, each adds to the total load carried.

A good example to illustrate the effects of population centers on pollution levels involves the production of ground-level ozone. Ground-level ozone is a by-product of automobile usage. (This topic will be dealt with in great detail later in the chapter.) Although ozone in the upper atmosphere is valuable in protecting the Earth from ultraviolet light, ground-level ozone can severely damage lung tissue. The map in figure 16.3 shows the peak values for ground-level ozone for June 25, 2003. This was a particularly bad day, but it points out the regional nature of air pollution problems.

Air pollution is not just an aesthetic problem. It also causes health problems. Thousands of deaths have been directly related to poor air quality in cities. Many of the megacities of the developing world have extremely poor air quality. The causes of this air pollution are open fires, large numbers of poorly maintained motor vehicles, and poorly regulated industrial plants. The World Health Organization estimates that urban air pollution accounts for 2 million deaths annually. Not only does poor air quality in such cities increase the death rate, but the general health of the populace is lowered. Approximately 20 to 30 percent of all respiratory diseases appear to be caused by air pollution. Chronic coughing and susceptibility to infections are common in these cities. Deaths from air pollution occur primarily among the elderly, the infirm, and the very young. Bronchial inflammations, allergic reactions, and irritation of the mucous membranes of the eyes and nose all indicate that air pollution must be reduced.

CATEGORIES OF AIR POLLUTANTS

Around the world, five major types of substances are released directly into the atmosphere in their unmodified forms in sufficient quantities to pose a health risk and are called **primary air pollutants.** They are carbon monoxide, volatile organic compounds (hydrocarbons), particulate matter, sulfur dioxide, and oxides of nitrogen. Primary air pollutants may interact with one another in the presence of sunlight to form new compounds such as ozone that are known as **secondary air pollutants.** Secondary air pollutants also form from reactions with substances that occur naturally in the atmosphere. In addition, the U.S. Environmental Protection Agency (EPA) has established air quality standards for six principal air pollutants, which are called the **criteria air pollutants.**

The criteria air pollutants are carbon monoxide (CO), particulate matter (PM), sulfur dioxide (SO_2), nitrogen dioxide (NO_2), lead (Pb), and ozone (O_3). Four of these pollutants (CO, SO_2, NO_2, and Pb) are emitted directly from a variety of sources. Ozone is not directly emitted but is formed when nitrogen dioxide, other oxides of nitrogen, and volatile organic compounds (VOCs) react in the presence of sunlight. Particulate matter can be directly emitted, or it can be formed when emissions of nitrogen oxides, sulfur oxides (SO_x)—primarily SO_2 and SO_3—, ammonia, organic compounds, and other gases react in the atmosphere. (See table 16.1). In addition, certain compounds with high toxicity are known as **hazardous air pollutants** or **air toxics.**

CARBON MONOXIDE

Carbon monoxide (CO) is produced when organic materials such as gasoline, coal, wood, and trash are burned with insufficient oxygen. When carbon-containing compounds are burned with abundant oxygen present, carbon dioxide is formed ($C + O_2 \rightarrow CO_2$). When the amount of oxygen is restricted, carbon monoxide is formed instead of carbon dioxide ($2C + O_2 \rightarrow 2CO$). Any process that involves the burning of fossil fuels has the potential to produce carbon monoxide. The single largest source of carbon monoxide is the automobile. (See figure 16.4.) About 60 percent of CO comes from vehicles driven on roads and 20 percent comes from vehicles not used on roads. Most of the remainder comes from other

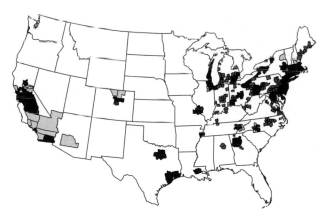

■ Nonattainment areas (263 entire counties)
☐ Nonattainment areas (30 partial counties)
■ Maintenance areas (141 entire or partial counties)
Partial counties, those with part of the county designated nonattainment and part attainment, are shown as full counties on the map.

FIGURE 16.3   Air Pollution and Population Centers
The population concentration in eastern North America creates the conditions that lead to regional air pollution problems. Since the prevailing winds are from west to east, each city adds its pollutants, and the air quality deteriorates
Source: U.S. Environmental Protection Agency.

TABLE 16.1 Air Pollutant Standards

Air Pollutant	Measurement Period	U.S. National Ambient Air Quality Standards	European Union Air Quality Standards
EPA Criteria Air Pollutants			
Carbon monoxide (CO)	8-hour average	(9 ppm) (10 mg/m³)	10 mg/m³
	1-hour average	(35 ppm) (40 mg/m³)	
Nitrogen dioxide (NO₂)	Annual mean	(0.053 ppm) (100 µg/m³)	40 µg/m³
	1-hour		200 µg/m³
Ozone (O₃)	8-hour average	(0.075 ppm) (147 µg/m³)	120 µg/m³
	1-hour average	(0.12 ppm) (235 µg/m³)	
Lead (Pb)	3-month average	(1.5 µg/m³)	0.5 µg/m³
Particulate matter (PM₁₀)	Annual mean		20 µg/m³
	24-hour average	(150 µg/m³)	50 µg/m³
Particulate matter (PM₂.₅)	Annual mean	(15 µg/m³)	
	24-hour average	(35 µg/m³)	(25 µg/m³)
Sulfur dioxide (SO₂)	Annual mean	(0.03 ppm) (78 µg/m³)	
	24-hour average	(0.14 ppm) (365 µg/m³)	125 µg/m³
	3-hour average	(0.50 ppm) (1300 µg/m³)	
	1-hour average		350 µg/m³
Other Common Air Pollutants			
Benzene	Annual mean	No standards set, but current levels are below 2.5 µg/m³	5 µg/m³
Volatile organic compounds		No standards set, but reductions needed to reduce ground-level ozone	

Source: U.S. Environmental Protection Agency and European Commission.

FIGURE 16.4 **Carbon Monoxide** The major source of carbon monoxide is the internal combustion engine, which is used to provide most of our transportation. The more concentrated the number of automobiles, the more concentrated the pollutants. Carbon monoxide concentrations of a hundred parts per million are not unusual in rush-hour traffic in large metropolitan areas. These concentrations are high enough to cause fatigue, dizziness, and headaches.

processes that involve burning (power plants, industry, burning leaves, etc.). Although increased fuel efficiency and the use of catalytic converters have reduced carbon monoxide emissions per kilometer driven, carbon monoxide remains a problem because the number of automobiles on the road and the number of kilometers driven have risen. In urban areas, as much as 90 percent of carbon monoxide is from motor vehicles. In many parts of the world, automobiles are poorly maintained and may have inoperable pollution control equipment, resulting in even greater amounts of carbon monoxide.

Carbon monoxide is dangerous because it binds to the hemoglobin in blood and makes the hemoglobin less able to carry oxygen. Because carbon monoxide remains attached to hemoglobin for a long time, even small amounts tend to accumulate and reduce the blood's oxygen-carrying capacity. Several hours of exposure to air containing only 0.001 percent of carbon monoxide can cause death. Carbon monoxide is most dangerous in enclosed spaces, where it is not diluted by fresh air entering the space. The amount of carbon monoxide produced in heavy traffic can cause headaches, drowsiness, and blurred vision. Cigarette smoking is also an important source of carbon monoxide because the smoker is inhaling it directly. A heavy smoker in congested traffic is doubly exposed and may experience severely impaired reaction time compared to a nonsmoking driver.

Fortunately, carbon monoxide is not a persistent pollutant. It readily combines with oxygen in the air to form carbon dioxide ($2CO + O_2 \rightarrow 2CO_2$). Therefore, the air can be cleared of its carbon monoxide if no new carbon monoxide is introduced into it. Control of carbon monoxide in the United States has been very good. The U.S. EPA reports that carbon monoxide levels decreased by about 67 percent between 1990 and 2007, and nearly all communities now meet the standards set by the EPA. This was accomplished with a variety of controls on industry and, in particular, on motor vehicles. Catalytic converters reduce the amount of carbon monoxide released by vehicle engines, and specially formulated fuels that produce less carbon monoxide are used in many cities that have a carbon monoxide problem. Often these special fuels are only used in winter, when car engines run less efficiently and produce more carbon monoxide.

PARTICULATE MATTER

Particulate matter consists of minute (10 microns and smaller) solid particles and liquid droplets dispersed into the atmosphere. A micron is one millionth of a meter. Many bacteria are about 1 micron in diameter. The Environmental Protection Agency has set standards

In the 1990s, Mexico City was labeled the city with the worst air pollution ever recorded. The air over Mexico City exceeded ozone limits set by the World Health Organization on more than 300 days in one year. The city had about 35,000 factories and 3.6 million vehicles. A large proportion of the air pollution was the result of automobiles. Most of these vehicles were older models that were poorly tuned and, therefore, polluted the air with a mixture of hydrocarbons, carbon monoxide, and nitrogen oxides. The city's altitude of over 2000 meters (6500 feet) results in even greater air pollution from automobiles because automobile engines do not burn fuel efficiently at such high altitudes. Mexico City's location in a valley also allows for conditions suitable for thermal inversions during the winter. This combination of hydrocarbons, nitrogen oxides, and thermal inversions results in the production of ground-level ozone (smog).

Improvement in air quality has required major changes in industry and transportation in Mexico City.

- A polluting, government-owned oil refinery was shut down.
- Power plants and many industries have switched from oil to natural gas, which pollutes less.
- Some polluting industries have been relocated to areas outside Mexico City.
- The government is improving public transportation to make it more attractive for people to switch from private automobiles to public transport.
- Public information campaigns encourage people to keep their automobiles tuned.
- Lead-free gasoline is being used.
- The use of cleaner-burning gasoline is required.

- Catalytic converters are now required on all automobiles.
- All pre-1991 taxis have been replaced.
- Vapor recovery mechanisms have been installed at gasoline stations and distribution centers.

These actions have had significant effects on the quality of the air. Although ozone in the air continues to be a major problem, as it is in metropolitan areas around the world, lead, carbon monoxide, nitrogen dioxide, and sulfur dioxide concentrations have been reduced to levels below current air quality standards. To control ozone, increasingly strong restrictions on polluting industries and the use of private automobiles will be necessary.

Changes in Air Quality of the Mexico City Metropolitan Area

Pollutant	Number of Days Concentrations Exceeded Air Quality Standards	
	1990	2006
Ozone	325	209
Particulates	58	43
Carbon monoxide	43	0
Nitrogen dioxide	31	1
Sulfur dioxide	11	0
Lead	4	0

for particles smaller than 10 microns (PM_{10}) and 2.5 microns ($PM_{2.5}$). Most of the coarse particles (between 10 and 2.5 microns) are primary pollutants such as dust and carbon particles that are released directly into the air. Dust from travel on roads accounts for about 50 percent of PM_{10} particles. Dust from agricultural activities, construction sites, industrial processes, and smoke particles from fires are the other primary sources of coarse particles. Fine particles (2.5 microns or less) are mostly secondary pollutants that form in the atmosphere from interactions of primary air pollutants. Sulfates and nitrates formed from sulfur dioxide and nitrogen oxides are examples. Other sources of fine particulates are fires and road dust.

Particulates cause problems ranging from the annoyance of reduced visibility caused by fine particulates to soot settling on a backyard picnic table to the **carcinogenic** (cancer-causing) effects of asbestos. Particulates frequently get attention because the coarse particles are so readily detected by the public. Heavy black smoke from a factory or a view obscured by dust can be observed without expensive monitoring equipment and generally causes an outcry.

Particles can accumulate in the lungs and interfere with their ability to exchange gases. Such lung damage usually occurs in people who are repeatedly exposed to large amounts of particulate matter on the job. Miners and others who work in dusty conditions are most likely to be affected. Droplets and solid particles can also serve as centers for the deposition of other substances from the atmosphere. As we breathe air containing particulates, we come in contact with concentrations of other potentially more harmful materials that have accumulated on the particulates. Sulfuric, nitric, and carbonic acids, which irritate the lining of our respiratory system, frequently are associated with particulates.

According to the U.S. EPA, the amount of PM_{10} pollution decreased by about 28 percent between 1990 and 2007. The EPA has been setting standards for $PM_{2.5}$ particles for a shorter period, and pollution by these particles decreased by about 11 percent between 2000 and 2007. Most communities now meet the standards set for PM_{10} but some still exceed the standards set for $PM_{2.5}$.

Source	Necessary Conditions	Reactions Take Place in Atmosphere		Products
Primarily automobiles	Volatile organic compounds (VOC) present	$VOC + O^* \text{ or } O_3 \rightarrow$ Highly reactive organic radicals $+ NO_2$		Peroxyacetyl nitrates
				Aldehydes
Primarily automobiles	Nitrogen monoxide (NO) present	$NO +$ Organic radicals $\rightarrow NO_2$		
From automobiles and formed from NO	Nitrogen dioxide (NO_2) present	$O^* + O_2 \rightarrow O_3$ (ozone) $\rightarrow$		Ozone
		$NO_2 \rightarrow NO + O^*$ (atomic oxygen)		
		Sunlight		
Sun	Sunlight			
Sun (summer temperatures)	Heat	Reactions take place more rapidly at higher temperatures.		

FIGURE 16.6 **Major Steps in the Development of Photochemical Smog** Photochemical smog develops when specific reactants and conditions are present. The necessary reactants are volatile organic compounds, nitrogen monoxide, and nitrogen dioxide. The conditions that cause these compounds to react are warm temperatures and the presence of sunlight. When these conditions exist, ozone and other components of smog are formed as secondary pollutants.

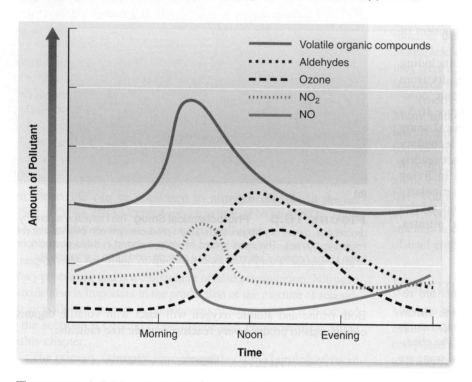

FIGURE 16.7 **Daily Changes in Pollutants During a Photochemical Smog Incident** The development of photochemical smog begins with a release of nitrogen oxides and volatile organic compounds associated with automobile use during morning traffic. As the sun rises and the day warms up, these reactants interact to form ozone and other secondary pollutants. These peak during the early afternoon and decline as the sun sets.

The development of photochemical smog in an area involves the interaction of climate, time of day, and motor vehicle emissions in the following manner:

1. During morning rush-hour traffic, the amounts of nitrogen monoxide and volatile organic compounds increase.

2. The presence of NO and VOCs leads to an increase in the amount of NO_2. NO levels fall because NO is converted to NO_2.

3. Ozone, peroxyacetyl nitrates, and aldehyde levels rise and remain high throughout the middle of the day.

4. Since sunlight and warm temperatures support the production of photochemical smog, in the evening as the sun sets and temperatures fall, the production of ozone lessens. In addition, ozone and other smog components react with their surroundings and are destroyed, so the destructive components of smog decline in the evening. Figure 16.7 shows how the concentration of these various molecules changes during the day.

The Role of Climate and Geography

While smog can develop in any area, some cities have greater problems because of their climate, traffic, and geographic features. Cities with warm climates and those that have lots of sunlight are more prone to develop photochemical smog because the chemical reactions reponsible for smog are supported by warm temperatures and sunlight. Similarly, smog is more likely to be a problem during summer months because of the higher temperatures and longer days.

Cities that are located adjacent to mountain ranges or in valleys have greater problems because the pollutants can be trapped by thermal inversions. Normally, air is warmer at the surface of the Earth and gets cooler at higher altitudes. (See figure 16.8a.) In some instances, a layer of warmer air may be above a layer of cooler air at the Earth's surface. This condition is known as a **thermal inversion.**

In cities located in valleys, as the surface of the Earth cools at night, the cooler air on the sides of the valley can flow down into the valley and create a thermal inversion. (See figure 16.8b.) As cool air flows into these valleys, it pushes the warm air upward. Simiarly, in cities such as Los Angeles that have mountains to the east and the ocean to the west, cool air from the ocean may push in under a layer of warm air to create a thermal inversion. (See figure 16.8c.) In either case, the warm air becomes sandwiched between two layers of cold air and acts like a lid on the valley. The lid of warm air cannot rise because it is covered by a layer of cooler, denser air pushing down on it. It cannot move out of the area because of the mountains. Without normal air circulation, smog accumulates. Harmful chemicals continue to increase in concentration until a major weather change causes the lid of warm air to rise and move over the mountains. Then the underlying cool air can begin to circulate, and the polluted air can be diluted.

Normal Situation

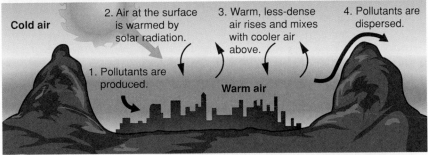

Cold air

2. Air at the surface is warmed by solar radiation.

3. Warm, less-dense air rises and mixes with cooler air above.

4. Pollutants are dispersed.

1. Pollutants are produced.

Warm air

(a)

Thermal Inversion

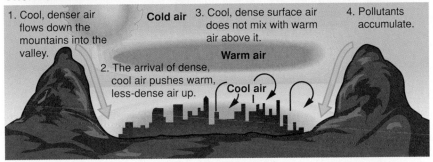

1. Cool, denser air flows down the mountains into the valley.

Cold air

3. Cool, dense surface air does not mix with warm air above it.

4. Pollutants accumulate.

Warm air

2. The arrival of dense, cool air pushes warm, less-dense air up.

Cool air

(b)

Thermal Inversion

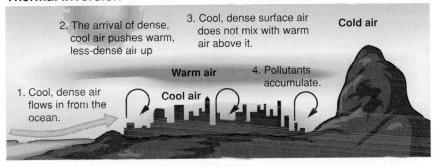

2. The arrival of dense, cool air pushes warm, less-dense air up

3. Cool, dense surface air does not mix with warm air above it.

Cold air

Warm air

4. Pollutants accumulate.

1. Cool, dense air flows in from the ocean.

Cool air

(c)

FIGURE 16.8 **Thermal Inversion** Under normal conditions, (a) the air at the Earth's surface is heated by the sun and rises to mix with the cooler air above it. When a thermal inversion occurs (b and c), a layer of heavier cool air flows into a valley and pushes the warmer air up. The heavy cooler air is then unable to mix with the less-dense warm air above and cannot escape because of surrounding mountains. The cool air is trapped, sometimes for several days, and accumulates pollutants. If the thermal inversion continues, the levels of pollution can become dangerously high.

Smog problems could be substantially decreased by reducing the NO_X and VOCs associated with the use of internal combustion engines (perhaps eliminating them completely) or by moving population centers away from the valleys where thermal inversions occur. It is highly unlikely that population centers can be moved; however, it is possible to reduce the molecules that lead to the production of smog. Reformulating gasoline and installing devices on automobiles that reduce the amount of NO_X and VOCs released have been beneficial. Although the emission of VOCs has been reduced substantially and NO_X levels have been reduced a small amount, ozone levels have not been decreasing. Ozone levels fell by only about

9 percent from 1990 to 2007 and are still a problem, particularly in Southern California and the U.S. Northeast. Since ozone is produced as a result of interactions between VOCs and NO_X, it will be necessary to further reduce the levels of these two components to decrease the production of ozone. In particular, it will be necessary to reduce NO_X levels, and this will require changes in how automobiles are designed or fueled.

HAZARDOUS AIR POLLUTANTS

While all of the major air pollutants already discussed are environmental health concerns, hundreds of other dangerous chemical compounds are purposely or accidentally released into the air that can cause harm to human health or damage the environment. These compounds are collectively known as hazardous air pollutants (HAP) or air toxics. Pesticides are toxic materials that are purposely released to kill insects or other pests. Other dangerous materials are released as a result of consumer activities. Benzene in gasoline escapes when gasoline is put into the tank, and the use of some consumer products such as glues and cleaners releases toxic materials into the air. The majority of air toxics, however, are released as a result of manufacturing processes. Perchloroethylene is released from dry cleaning establishments, and toxic metals are released from smelters. The chemical and petroleum industries are the primary sources of hazardous air pollutants. Although air toxics are harmful to the entire public, their presence is most serious for people who are exposed on the job, since they are likely to be exposed to higher concentrations of hazardous substances over longer time periods. Chapter 19 deals with hazardous materials and the issues related to regulating their production, use, and disposal.

CONTROL OF AIR POLLUTION

All of the air pollutants we have examined thus far are produced by humans. That means their release into the atmosphere can be controlled. In the United States, implementation of the requirements of the Clean Air Act has been primary means of controlling air pollution.

THE CLEAN AIR ACT

Under the Clean Air Act, EPA has a number of responsibilities, including:

- Establishing air quality standards, developing strategies for meeting the standards, and ensuring that the standards are met.
- Conducting periodic reviews of the six criteria air pollutants that are considered harmful to public health and the environment.
- Reducing emissions of SO_2 and NO_X that cause acid rain.

- Reducing air pollutants such as PM, SO_x, and NO_x, which can reduce visibility across large regional areas, including many of the nation's most treasured parks and wilderness areas.

- Ensuring that sources of toxic air pollutants that may cause cancer and other adverse human health and environmental effects are well controlled and that the risks to public health and the environment are substantially reduced.

- Limiting the use of chemicals that damage the stratospheric ozone layer in order to prevent increased levels of harmful ultraviolet radiation.

The current version of the Clean Air Act established a series of detailed control requirements to meet the goals of improving air quality:

1. All industries are required to obtain permits to release materials into the air.

2. All new and existing sources of air pollution are subject to national ambient air quality standards established for sulfur dioxide (SO_2), nitrogen dioxide (NO_2), particulate matter, carbon monoxide (CO), ozone, and lead.

3. Newly constructed facilities are subject to more stringent control technology and permitting requirements than are pre-existing facilities.

4. Hazardous air pollutants are specifically identified (there are 188 substances identified) and regulated. Any source emitting 9.1 metric tons per year of any listed substance, or 22.7 metric tons of combined substances, is considered a major source and is subject to strict regulations.

5. Power plants are allowed to sell their sulfur dioxide release permits to other companies. This program encourages power plants that can easily reduce their emissions to do so. They can then sell their permits to other power plants that are having a more difficult time reducing emissions. The net result of this program has been a rapid reduction in sulfur dioxide emissions.

6. A program for the phaseout of ozone-depleting substances (CFCs, halons, carbon tetrachloride, and methyl chloroform) was established.

ACTIONS THAT HAVE REDUCED AIR POLLUTION

As a result of the implementation of the requirements of the Clean Air Act, a variety of pollution control mechanisms have been effectively employed. As a result of many different actions, air quality has improved significantly in the past 27 years. (See figure 16.9.)

Motor Vehicle Emissions

Motor vehicles are the primary source of several important air pollutants: carbon monoxide, volatile organic compounds, and nitrogen oxides. In addition, ozone is a secondary pollutant of automobile use.

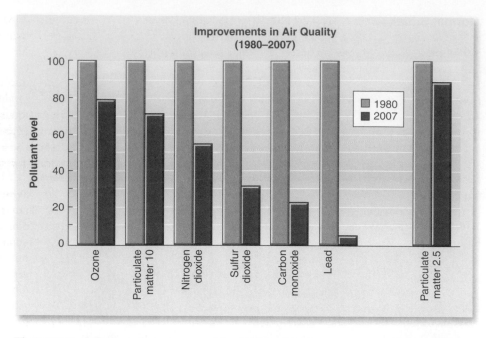

FIGURE 16.9 **Improvement in Air Quality** Many initiatives to improve air quality have been effective. Between 1980 and 2007, all of the major air pollutants have decreased significantly. The data for $PM_{2.5}$ are for the years 2000 and 2007, since the EPA did not begin collecting data on $PM_{2.5}$ until 2000.
Source: U.S. Environmental Protection Agency.

Placing controls on the emissions from motor vehicles has resulted in a significant improvement in the air quality in North America. Many engineering changes in automobiles have reduced the amount of VOCs that escape from the gas tank and crankcase. Modifications to the pumps at gas stations and to the filler pipes of cars have also been beneficial. Better fuel efficiency and specially blended fuels that produce less carbon monoxide and unburned organic compounds have also improved air quality. Catalytic converters reduced carbon monoxide, oxides of nitrogen, and volatile organic compounds in emissions and necessitated the use of lead-free fuel. This lead-free fuel requirement, in turn, greatly reduced the amount of lead (and other metal additives) in the atmosphere. Figure 16.9 shows that levels of carbon monoxide, nitrogen dioxide, and lead are down significantly. Ozone levels, which are dependent on VOC and NO_2 levels, have also fallen somewhat but still need improvement in some areas of the country—particularly California and the Northeast part of the United States.

Particulate Matter Emissions

Particulate matter comes from a variety of sources. Road dust is the major source of PM_{10} particles. In addition, many industrial activities involve processes that produce dusts. Mining and other earth-moving activities, farming operations, and the transfer of grain or coal from one container to another all produce dust.

Fires (forest fires, grass fires, leaf burning, and fires from fireplaces and woodstoves, etc.) are a significant source of particulate matter—particularly $PM_{2.5}$. Burning of fossil fuels is another major source of particulate matter. Because of air quality regulations, industries have done much to reduce the amount of particulate matter released from the burning of fossil fuels. Various kinds of devices can be used to trap particles so they do not escape from the smokestack. These devices are very effective, and particulate matter from the burning of fossil fuels is greatly reduced. However, the smaller particles that form from gaseous emissions (SO_2, NO_X) are still a problem. Diesel engines are a significant source of particulate matter, and the gaseous emissions from motor vehicles contribute to the formation of smaller particles in the same way that industrial sources do.

While industrial activities, motor vehicle use, and land-use practices are major sources of particulate matter, many individual personal activities are also important. Many people in the world use wood as their primary source of fuel for cooking and heating. In developed countries such as the United States and Canada, some people use fireplaces and wood-burning stoves as a primary source of heat, but most use them for supplemental heat or for aesthetic purposes. However, the use of large numbers of wood-burning stoves and fireplaces can generate a significant air-pollution problem. Some municipalities with air-pollution problems regulate or ban the use of wood-burning stoves and fireplaces. High-efficiency wood-burning stoves significantly reduce the amount of particulate emissions.

Power Plant Emissions

The two primary pollutants associated with electric power plants are particulates and sulfur dioxide (SO_2). Most PM_{10} emissions have been controlled with filters and other mechanical means. The control of sulfur dioxide requires more fundamental changes to the way electricity is produced. The general approach of the EPA has been to set limits and allow the electric utilities to decide which options are the best for them. Switching from a high-sulfur to a low-sulfur coal reduces the amount of sulfur dioxide released into the atmosphere by 66 percent. Switching to oil, natural gas, or nuclear fuels would reduce sulfur dioxide emissions even more.

A second alternative is to remove the sulfur from the fuel before the fuel is used. Chemical or physical treatment of coal before it is burned can remove nearly 40 percent of the sulfur. This is technically possible, but it increases the cost of electricity to the rate payer.

Scrubbing the gases emitted from a smokestack is a third alternative. The technology is available, but, of course, these control devices are costly to install, maintain, and operate.

ACID DEPOSITION

Acid deposition is the accumulation of potential acid-forming particles on a surface. The acid-forming particles can be dissolved in rain, snow, or fog or can be deposited as dry particles. When dry particles are deposited, an acid does not actually form until these materials mix with water. Even though the acids are

formed and deposited in different ways, all of these sources of acid-forming particles are commonly referred to as **acid rain.**

CAUSES OF ACID PRECIPITATION

Acids result from natural causes, such as vegetation, volcanoes, and lightning, and from human activities, such as the burning of coal and use of the internal combustion engine. (See figure 16.10.) These combustion processes produce sulfur dioxide (SO_2) and oxides of nitrogen (NO_X). Oxidizing agents, such as ozone, hydroxide ions, or hydrogen peroxide, along with water, are necessary to convert the sulfur dioxide or nitrogen oxides to sulfuric or nitric acid.

Acid rain is a worldwide problem. Reports of high acid-rain damage have come from Canada, England, Germany, France, Scandinavia, and the United States. Rain is normally slightly acidic (pH between 5.6 and 5.7), since atmospheric carbon dioxide dissolves in water to produce carbonic acid. But acid rains sometimes have a concentration of acid a thousand times higher than normal. In 1969, New Hampshire had a rain with a pH of 2.1. In 1974, Scotland had a rain with a pH of 2.4. Currently, rain in much of the northeastern part of the United States and parts of Ontario has a pH between 4.4 and 4.8. This compares with levels in 1994, when pH readings were between 4.2 and 4.6. This is a substantial improvement brought about primarily by reductions in SO_2 emissions and, to a lesser extent, a reduction in NO_2 emissions.

EFFECTS ON STRUCTURES

Acid rain can cause damage in several ways. Buildings and monuments are often made from materials that contain limestone (calcium carbonate, $CaCO_3$), because limestone is relatively soft and easy to work. Sulfuric acid (H_2SO_4), a major component of acid rain, converts

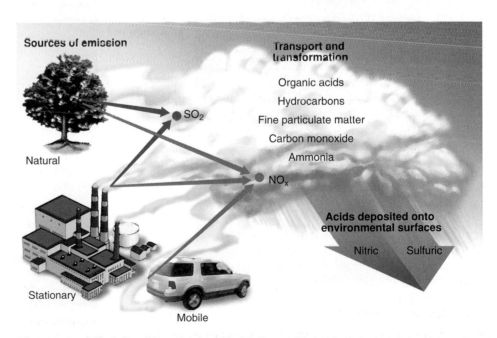

FIGURE 16.10   Sources of Acid Deposition Molecules from natural sources, power plants, and internal combustion engines react to produce the chemicals that are the source of acid deposition.

New York University's Co-Generation Plant

As part of New York University's "Green Action Plan," it began building a new co-generation plant in 2008. Co-generation refers to the simultaneous production of electricity and thermal energy. Electric power plants release a great deal of heat. By using the "waste" heat from the power plant to heat buildings on campus, the new co-generation facility at New York University will lead to a 75 percent reduction in regulated pollutants and an annual decrease of greenhouse gas pollutants of over 5000 tons.

The new facility will triple the university's capacity to provide power to its buildings. By producing its own electricity, the university will provide the energy in a more environmentally benign way, and remove its buildings from the local utility grid, which is overburdened.

Another advantage of the facility is its location—it will be located underground. The land over the facility will be landscaped as new open space. New York University worked with the New York City Parks Department and a local community advisory committee to design the new park. Perhaps this is the first park that can claim to be green both above and below ground.

limestone to gypsum ($CaSO_4$), which is more soluble than calcium carbonate and is eroded over many years of contact with acid rain. (See figure 16.11.) Metal surfaces can also be attacked by acid rain.

Effects on Terrestrial Ecosystems

The effects of acid rain on ecosystems are often subtle and difficult to quantify. However, in many parts of the world, acid rain is suspected of causing the death of many forests and reducing the vigor and rate of growth of others. (See figure 16.12.) In central Europe, many forests have declined significantly, resulting in the death of about 6 million hectares (14.8 million acres) of trees. Northeastern North America has been affected with significant tree death and reduction in vigor, particularly at higher elevations. Some areas have had 50 percent mortality of red spruce trees.

A strong link can be established between the decline of the forests and acid rain. Sulfur dioxide and oxides of nitrogen are the primary molecules that contribute to acid rain. The deposition of acids causes major changes to the soils in areas where the soils are not able to buffer the additional acid. As soil becomes acidic, aluminum is released from binding sites and becomes part of the soil water, where it interferes with the ability of plant roots to absorb nutrients. A recent long-term study in New Hampshire strongly

FIGURE 16.11 **Damage Due to Acid Deposition** Sulfuric acid (H_2SO_4), which is a major component of acid deposition, reacts with limestone ($CaCO_3$) to form gypsum ($CaSO_4$). Since gypsum is water soluble, it washes away with rain. The damage to this tombstone is the result of such acid reacting with the stone.

FIGURE 16.12 **Forest Decline** Many forests, particularly at high elevations in northeastern North America, have shown significant decline, and dead trees are common.

suggests that the many years of acid precipitation have reduced the amount of calcium and magnesium in the soil, which are essential for plant growth. Because there are no easy ways to replace the calcium even if acid rain were to stop, it would still take many years for the forests to return to health. Reduction in the pH of the soil may also change the kinds of bacteria in the soil and reduce the availability of nutrients for plants. While none of these factors alone would necessarily result in tree death, each could add to the stresses on the plant and may allow other factors, such as insect infestations, extreme weather conditions (particularly at high elevations), or drought, to further weaken trees and ultimately cause their death.

EFFECTS ON AQUATIC ECOSYSTEMS

The effects of acid rain on aquatic ecosystems are much more clear-cut. In several experiments, lakes were purposely made acidic and the changes in the ecosystems recorded. The experiments showed that as lakes became more acidic, there was a progressive loss of many kinds of organisms. The food web becomes less complicated, many organisms fail to reproduce, and many others die. Most healthy lakes have a pH above 6. At a pH of 5.5, many desirable species of fish are eliminated; at a pH of 5, only a few starving fish may be found, and none is able to reproduce. Lakes with a pH of 4.5 are nearly sterile.

Several reasons account for these changes. Many of the early developmental stages of insects and fish are more sensitive to acid conditions than are the adults. In addition, the young often live in shallow water, which is most affected by a flood of acid into lakes and rivers during the spring snowmelt. The snow and its acids have accumulated over the winter, and the snowmelt releases large amounts of acid all at once. Crayfish and other crustaceans need calcium to form their external skeletons. As the pH of the water decreases, the crayfish are unable to form new exoskeletons and so they die. Reduced calcium availability also results in the development of some fish with malformed skeletons.

As mentioned earlier, increased acidity also results in the release of aluminum, which impairs the function of a fish's gills. About 14,000 lakes in Canada and 11,000 in the United States have been seriously altered by becoming acidic. Many lakes in Scandinavia are similarly affected. The extent to which acid deposition affects an ecosystem depends on the nature of the bedrock in the area and the ecosystem's proximity to acid-forming pollution sources. (See figure 16.13.) Soils derived from igneous rock are not capable of buffering the effects of acid deposition, while soils derived from sedimentary rocks such as limestone release bases that neutralize the effects of acids. Because of this, eastern Canada and the U.S. Northeast are particularly susceptible to acid rain. These areas have a high proportion of granite rock and are downwind from the major air-pollution sources of North America. Scandinavian countries have a similar geology and receive pollution from industrial areas in the United Kingdom and Europe. Thousands of kilometers of streams and up to 200,000 lakes in

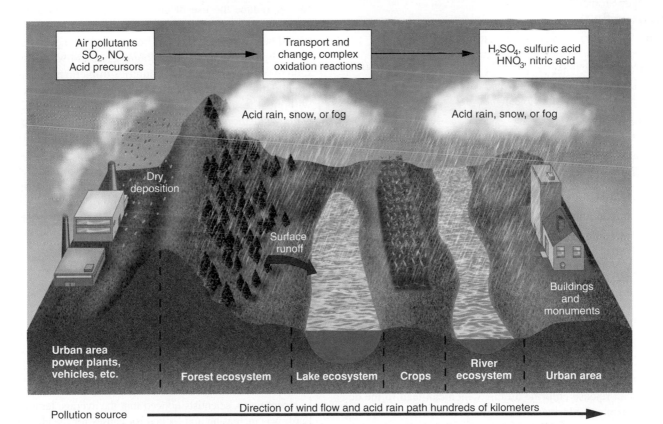

FIGURE 16.13 **Factors That Contribute to Acid Rain Damage** In any aquatic ecosystem, the following factors increase the risk of damage from acid deposition: (1) location downwind from a major source of pollution; (2) hard, insoluble bedrock with a thin layer of infertile soil in the watershed; (3) low buffering capacity in the soil of the watershed; (4) a low lake surface area-to-watershed ratio.

Source: Data from U.S. Environmental Protection Agency, *Acid Rain*.

eastern Canada and the northeastern United States are thought to be in danger of becoming acidified because of their location and geology.

OZONE DEPLETION

Ozone is a molecule of three atoms of oxygen bonded together (O_3). In the 1970s, various sectors of the scientific community became concerned about the possibility that the ozone layer in the Earth's upper atmosphere (the stratosphere) was being reduced. In 1985, it was discovered that a significant thinning of the ozone layer over the Antarctic occurred during the Southern Hemisphere spring (September–November); this area became known as the "ozone hole." Some regions of the ozone layer showed 95 percent depletion. Ozone depletion also was found to be occurring farther north than previously. Measurements in Arctic regions suggest a thinning of the ozone layer there also.

WHY STRATOSPHERIC OZONE IS IMPORTANT

The ozone in the outer layers of the atmosphere, approximately 15 to 35 kilometers (9–21 miles) from the Earth's surface, shields the Earth from the harmful effects of ultraviolet light radiation. Ozone absorbs ultraviolet light and is split into an oxygen molecule and an oxygen atom:

$$O_3 \xrightarrow{\text{Ultraviolet light}} O_2 + O$$

Oxygen molecules are also split by ultraviolet light to form oxygen atoms:

$$O_2 \xrightarrow{\text{Ultraviolet light}} 2O$$

Recombination of oxygen atoms and oxygen molecules allows ozone to be formed again and to be available to absorb more ultraviolet light.

$$O_2 + O \rightarrow O_3$$

This series of reactions results in the absorption of 99 percent of the ultraviolet light energy that comes from the sun and prevents it from reaching the Earth's surface. Less ozone in the upper atmosphere results in more ultraviolet light reaching the Earth's surface. Ultraviolet light is strongly linked to skin cancers and cataracts in humans and increased mutations in all living things.

OZONE DESTRUCTION

Chlorofluorocarbons are strongly implicated in the ozone reduction in the upper atmosphere. Chlorofluorocarbons and similar compounds can release chlorine atoms, which can lead to the destruction of ozone. Chlorine reacts with ozone in the following way to reduce the quantity of ozone present:

$$Cl + O_3 \rightarrow ClO + O_2$$
$$ClO + O \rightarrow Cl + O_2$$

These reactions both destroy ozone and reduce the likelihood that it will be formed because atomic oxygen (O) is removed as well. It is also important to note that it can take 10 to 20 years for chlorofluorocarbon molecules to get into the stratosphere, and then they can react with the ozone for up to 120 years.

ACTIONS TO PROTECT THE OZONE LAYER

Since the 1970s, when chlorofluorocarbons were linked to the depletion of the ozone layer in the upper atmosphere, their use as propellants in aerosol cans has been banned in the United States, Canada, Norway, and Sweden, and the European Union agreed to reduce use of chlorofluorocarbons in aerosol cans.

In 1987, several industrialized countries, including Canada, the United States, the United Kingdom, Sweden, Norway, Netherlands, the Soviet Union, and West Germany, agreed to freeze chlorofluorocarbon and halon (used in fire extinguishers) production at current levels and reduce production by 50 percent by the year 2000. This document, known as the Montreal Protocol, was ratified by the U.S. Senate in 1988. Although the initial concerns related to the problem of ozone depletion, efforts to reduce chlorofluorocarbons have been effective at removing a greenhouse gas as well. As a result of the 1987 Montreal Protocol, chlorofluorocarbon emissions dropped dramatically from their peak in 1988.

In 1990, in London, international agreements were reached to further reduce the use of chlorofluorocarbons and added carbon tetrachloride and methyl chloroform as chemicals whose use would be eliminated. A major barrier to these negotiations was the reluctance of the developed countries of the world to establish a fund to help less-developed countries implement technologies that would allow them to obtain refrigeration and air conditioning without the use of chlorofluorocarbons. In 1991, DuPont announced the development of new refrigerants that would not harm the ozone layer. These and other alternative refrigerants are now used in refrigerators and air conditioners in many nations, including the United States. In 1996, the United States stopped producing chlorofluorocarbons. As a result of these international efforts and rapid changes in technology, the use of chlorofluorocarbons has dropped rapidly, and concentrations of chlorofluorocarbons in the atmosphere will slowly fall over the next few decades. Although the size of the ozone hole has fluctuated in recent years, current trends suggest that the size may be stabilizing. (See figure 16.14.)

GLOBAL WARMING AND CLIMATE CHANGE

In recent years, scientists noticed that the average temperature of the Earth was increasing and looked for causes for the change. It is clear that the Earth has had changes in its average temperature many times in the geologic past before humans were present. So, scientists initially tried to determine if the warming was a natural phenomenon or the

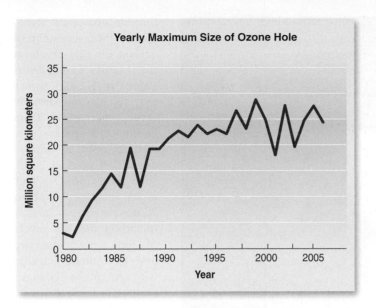

FIGURE 16.14 **Size of Ozone Hole** The graph shows a steady increase in the size of the ozone hole from 1980 to 2000. The ban on the use of chlorofluorocarbons and other ozone-destroying chemicals appears to have had a positive effect. The trend since 2000 appears to be stable or declining.

result of human activity. Several gases such as carbon dioxide, chlorofluorocarbons, methane, and nitrous oxide are known as **greenhouse gases** because they let sunlight enter the atmosphere but slow the loss of heat from the Earth's surface. Evidence of past climate change going back as far as 160,000 years indicates a close correlation between the concentration of greenhouse gases in the atmosphere and global temperatures. Computer simulations of climate indicate that global temperatures will rise as atmospheric concentrations of greenhouse gases increase, and there are many other effects predicted by an increase in temperature. Since these predictions are based on computer models of climate, some scientists criticized them as being inaccurate and constructed from sketchy data.

Because major disagreements arose over the significance of global warming, the United Nations Environment Programme established an Intergovernmental Panel on Climate Change (IPCC) to study the issue and make recommendations. A main activity of the IPCC is to provide at regular intervals an assessment of the state of knowledge about climate change.

The Fourth Assessment Report, *Climate Change 2007*, presented several important conclusions:

1. Evidence of increased temperature is clear.
   - The average temperature of the Earth has increased 0.56–0.92°C (1.0–1.7°F) in the past 100 years. 1998 was the warmest year on record and 2005 was the second warmest. Eight of the warmest years on record occurred in the 10-year period between 1998 and 2007. (See figure 16.15.)
   - Sea level is rising about 1.8 mm/year. This equates to a rise of 18 cm (7 inches) over 100 years.
   - Sea ice has decreased.
   - The arrival of spring is earlier in many parts of the world.

2. A strong correlation exists between the increase in temperature and the concentration of greenhouse gases in the atmosphere.

3. Human activity has greatly increased the amounts of these greenhouse gases. Greenhouse gases increased 70 percent between 1970 and 2004.

In 2007, the Nobel Peace Prize Committee recognized the IPCC and former U.S. Vice President Al Gore for their efforts to educate people about the causes and dangers of global warming and climate change.

CAUSES OF GLOBAL WARMING AND CLIMATE CHANGE

Several gases in the atmosphere are transparent to ultraviolet and visible light but absorb infrared radiation. These gases allow sunlight to penetrate the atmosphere and be absorbed by the Earth's surface. This sunlight energy is reradiated as infrared

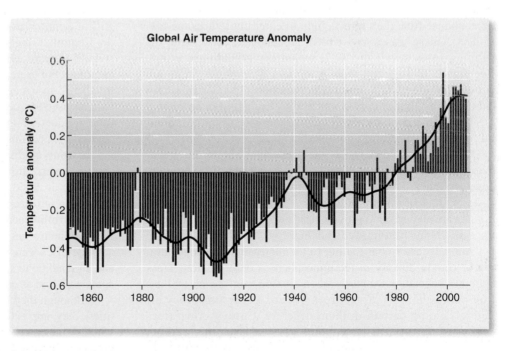

FIGURE 16.15 **Changes in Average Global Temperature** The graph shows the changes in average surface temperature of the Earth. The warmest year was 1998. Eight of the warmest years on record occurred within the past 10 years.

Sources: Data from Climatic Research Unit, University of East Anglia and Hadley Centre.

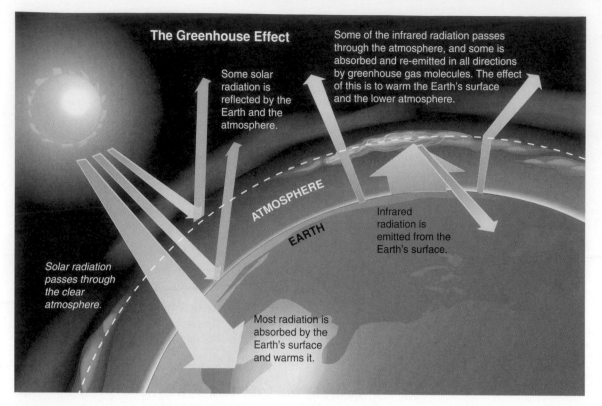

FIGURE 16.16 **Greenhouse Effect** The greenhouse effect naturally warms the Earth's surface. Without it, Earth would be 33°C (60°F) cooler than it is today—uninhabitable for life as we know it.

Source: Data from *Climate Change—State of Knowledge*, October 1997, Office of Science and Technology Policy, Washington, D.C.

radiation (heat), which is absorbed by the greenhouse gases in the atmosphere. Because the effect is similar to what happens in a greenhouse (the glass allows light to enter but retards the loss of heat), these gases are called greenhouse gases, and the warming from their increase is called the **greenhouse effect.** (See figure 16.16.) The most important greenhouse gases are carbon dioxide (CO_2), chlorofluorocarbons (primarily CCl_3F and CCl_2F_2), methane (CH4), and nitrous oxide (N_2O). Table 16.2 lists the relative contribution of each of these gases to the potential for global warming.

Carbon dioxide (CO_2) is the most abundant of the greenhouse gases. It occurs as a natural consequence of respiration. However, much larger quantities are put into the atmosphere as a waste product of energy production. Coal, oil, natural gas, and biomass are all burned to provide heat and electricity for industrial processes, home heating, and cooking.

Another factor contributing to the increase in the concentration of carbon dioxide in the atmosphere is deforestation. Trees and other vegetation remove carbon dioxide from the air and use it for photosynthesis. Since trees live for a long time, they effectively tie up carbon in their structure. Cutting down trees to convert forested land to other uses releases this carbon, and a reduction in the amount of forest lessens its ability to remove carbon dioxide from the atmosphere. The combination of these factors (fossil-fuel burning and deforestation) has resulted in an increase in the concentration of carbon dioxide in the atmosphere. Measurement of carbon dioxide levels at the Mauna Loa Observatory in Hawaii shows that the carbon dioxide level increased from about 315 parts Permillion (ppm) in 1958 to about 384 ppm in 2007. (See figure 16.17.) This is an increase of about 22 percent. Since changes in carbon dioxide levels in the atmosphere are due to human activity, we can make changes that will stabilize or reduce atmospheric carbon dioxide. The actions required will be discussed later.

Methane comes primarily from biological sources, although some enters the atmosphere from fossil-fuel sources. (See figure 16.17.) Several kinds of bacteria that are particularly abundant in wetlands and rice paddies release methane into the atmosphere. Methane-releasing bacteria are also found in large numbers in the guts of termites and various kinds of ruminant animals such as cattle. Control of methane sources is unlikely, since the primary sources involve agricultural practices that would be very difficult to change. For example, nations would have to convert rice paddies to other forms of agriculture and drastically reduce the number of animals used for meat production. Neither is likely to occur, since food production in most parts of the world needs to be increased, not decreased.

Nitrous oxide, a minor component of the greenhouse gas picture, enters the atmosphere primarily from fossil fuels and fertilizers. It could be reduced by more careful use of nitrogen-containing fertilizers.

Chlorofluorocarbons (CFCs) are also a minor component of the greenhouse gas picture and are entirely the result of human activity. CFCs were widely used as refrigerant gases in refrigerators and air conditioners, as cleaning solvents, as propellants in aerosol containers, and as expanders in foam products.

Although they are present in the atmosphere in minute quantities, they are extremely efficient as greenhouse gases (about 15,000 times more efficient at retarding heat loss than is carbon dioxide). Because chlorofluorocarbons are a major cause of ozone destruction, production of chlorofluorocarbons has been sharply reduced and will be eliminated in the future. Atmospheric concentrations have begun to decline.

TABLE 16.2 Principal Greenhouse Gases

Greenhouse Gas	Pre-1750 Concentration (ppm)	2007 Concentration (ppm)	Contribution to Global Warming (percent)	Principal Sources
Carbon dioxide (CO_2)	280	382	60	• Burning of fossil fuels • Deforestation
Methane (CH_4)	0.608	1.78	20	• Produced by bacteria in wetlands, rice fields, and guts of livestock • Release of fossil fuels
Chlorofluorocarbons (CFCs)	0	0.00088	14	• Release from foams, aerosols, refrigerants, and solvents
Nitrous oxide (N_2O)	0.270	0.321	6	• Burning of fossil fuels • Fertilizers • Deforestation

Source: Data from Intergovernmental Panel on Climate Change, with updates from Oak Ridge National Laboratory.

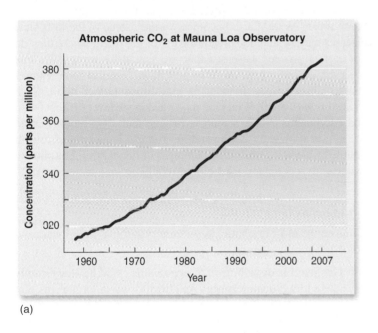

(a)

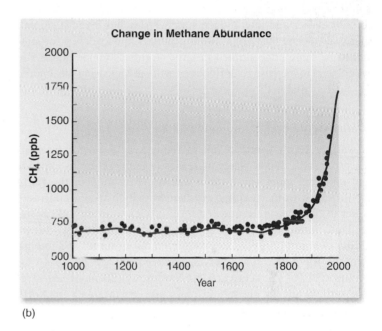

(b)

FIGURE 16.17   **Change in Atmospheric Carbon Dioxide and Methane** (a) Since the establishment of a carbon dioxide monitoring station at Mauna Loa Observatory in Hawaii, a steady increase in carbon dioxide levels has been observed. (b) Methane comes primarily from biological sources, although some enters the atmosphere from fossil-fuel sources.

Sources: (a) Data from Scripps Institution of Oceanography. (b) Data from U.N. Intergovernmental Panel on Climate Change, 2005.

POTENTIAL CONSEQUENCES OF GLOBAL WARMING AND CLIMATE CHANGE

It is important to recognize that although a small increase in the average temperature of the Earth may seem trivial, such an increase could set in motion changes that could significantly alter the climate of major regions of the world. Computer models suggest that rising temperature will lead to a cascade of consequences that affect the hydrologic cycle, sea level, human health,

the survival and distribution of organisms, and the use of natural resources by people. Furthermore, some natural ecosystems or human settlements will be able to withstand or adapt to the changes, while others will not.

Poorer nations are generally more vulnerable to the consequences of global warming. These nations tend to be more dependent on climate-sensitive sectors, such as subsistence agriculture, and lack the economic resources to buffer themselves against the changes that global warming may bring. The Intergovernmental

Panel on Climate Change has identified Africa as "the continent most vulnerable to the impacts of projected changes because widespread poverty limits adaptation capabilities."

Disruption of the Hydrologic Cycle

Among the most fundamental effects of climate change is disruption of the hydrologic cycle. Rising temperatures are expected to result in increased evaporation, which will cause some areas to become drier, while the increased moisture in the air will result in greater rainfall in other areas. This is expected to cause droughts in some areas and flooding in others. In those areas where evaporation increases more than precipitation, soil will become drier, lake levels will drop, and rivers will carry less water. Lower river flows and lake levels could impair navigation, hydroelectric power generation, and water quality and reduce the supplies of water available for agricultural, residential, and industrial uses.

Some areas may experience increased flooding during winter and spring, as well as lower supplies during summer. In California's Central Valley, for example, melting snow provides much of the summer water supply; warmer temperatures would cause the snow to melt earlier and thus reduce summer supplies, even if rainfall increased during the spring. More generally, the tendency for rainfall to be more concentrated in large storms as temperatures rise would tend to increase river flooding, without increasing the amount of water available.

Rising Sea Level

A warmer Earth would result in rising sea levels for two different reasons. When water increases in temperature, it expands and takes up more space. In addition, a warming of the Earth would result in the melting of glaciers, which would add more water to the oceans. Rising sea level erodes beaches and coastal wetland, inundates low-lying areas, and increases the vulnerability of coastal areas to flooding from storm surges and intense rainfall. By 2100, sea level is expected to rise by 15 to 90 centimeters (6–35 inches). A 50-centimeter (20-inch) sea-level rise will result in substantial loss of coastal land in North America, especially along the southern Atlantic and Gulf coasts, which are subsiding and are particularly vulnerable. Many coastal cities would be significantly affected by an increase in sea level. The land area of some island nations and countries such as Bangladesh would change dramatically as flooding occurred. The oceans will continue to expand for several centuries after temperatures stabilize.

Health Effects

Climate change will impact human health in a variety of ways.

Heat Affects Health   The most direct effect of climate change would be the impacts of hotter temperatures. Extremely hot temperatures increase the number of people who die (of various causes) on a given day. For example, people with heart problems are vulnerable because the cardiovascular system must work harder to keep the body cool during hot weather. Heat exhaustion and some respiratory problems increase. In August 2003, Europe experienced a prolonged heat wave. France recorded its highest temperatures on record. Thousands of people (primarily the elderly) died in France and throughout southern Europe as a result of the heat. Carbon dioxide concentrations of 550 ppm (double preindustrial levels) could cause such heat wave events to occur six times more frequently.

Heat Affects Air Pollution   Climate change will also aggravate air quality problems. Higher air temperature increases the concentration of ozone at ground level, which leads to injury of lung tissue and intensifies the effects of airborne pollen and spores that cause respiratory disease, asthma, and allergic disorders. Because children and the elderly are the most vulnerable, they are likely to suffer disproportionately with both warmer temperatures and poorer air quality.

Tropical Diseases Could Migrate to Former Temperate Regions
Throughout the world, the prevalence of particular diseases depends largely on local climate. Several serious diseases appear only in warm areas. As the Earth becomes warmer, some of these tropical diseases may be able to spread to parts of the world where they do not currently occur. Diseases that are spread by mosquitoes and other insects could become more prevalent if warmer temperatures enabled those insects to become established farther north. Such "vector-borne" diseases include malaria, dengue fever, yellow fever, and encephalitis. Some scientists believe that algal blooms could occur more frequently as temperatures rise, particularly in areas with polluted waters, in which case outbreaks of diseases such as cholera that tend to accompany algal blooms could become more frequent.

Changes to Ecosystems

Some of the most dramatic projections regarding global warming involve natural systems:

- **Geographic distribution of organisms** could be significantly altered by climate change. As climates warm, organisms that were formerly restricted to warmer regions will become more common toward the poles. The tundra biomes of the world will be greatly affected because of the thawing of the permafrost, which will allow the northward migration of boreal species. Similarly, mountainous areas will have less snow and earlier melting of the snow that does accumulate during the winter.

- **Coral reefs are especially challenged** because they are affected both by an increase in water temperature and by an increase in the acidity of the ocean. When carbon dioxide dissolves in water, it forms an acid. An increase in acidity would cause the skeletons of corals and the shells of many other organisms to tend to dissolve. This would make it more difficult for these organisms to precipitate calcium salts from the ocean to construct their skeletons and shells.

- **Low-lying islands and shorelines** will be especially impacted by rising sea level. Mangrove forests and marshes will be inundated and subjected to violent weather and storm surges.

TABLE 16.3 Consequences of Increases in Greenhouse Gases

Change	Climate Consequences	Ecosystem Consequences	Cultural Consequences
Warmer air	Permafrost melts	Tundra biome changed	Arctic native communities affected
	Less snowfall and accumulation	Changes in flow of rivers	Less regular release of water from snowmelt
	Glaciers melt	Changes in flow of rivers	Glaciers less reliable as a source of water
Warmer oceans		Coral reefs threatened by warmer water	Loss of biodiversity and marine resources
	Less sea ice	Arctic and Antarctic food chains altered	Impact on fishing industry
	Sea water expands	Coastal flooding affects mangroves and salt marshes	Cost of combating flooding
Changes in weather patterns	Warmer weather	Northward shift of plant and animal distributions	Tropical disease may migrate to temperate regions
			Air pollution problems become more intense
			Patterns of agriculture will change
	Less rainfall in mid latitudes and some subtropical regions	Drier deserts and droughts in some regions	Water shortages in arid climates
	More rainfall in high latitudes and parts of the tropics	Increased runoff causing erosion and increased flow in rivers	Flooding of cities and agricultural land
	More heat waves and severe storms	Ecosystems altered by wind, flooding, and erosion	Heat effects of health Destruction of buildings and loss of agricultural productivity
	Carbon dioxide dissolved in water acidifies the ocean	Corals and organisms that make shells threatened	Loss of biodiversity and marine resources

Challenges to Agriculture and the Food Supply

Climate strongly affects crop yields. Yields will fall in regions where drought and heat stress will increase. In regions that will receive increased rainfall and warming temperatures yields should increase. However, episodes of severe weather will cause crop damage that will affect yields. A warmer climate would reduce flexibility in crop distribution and increase irrigation demands. Expansion of the ranges of pests could also increase vulnerability and result in greater use of pesticides. Despite these effects, total global food production is not expected to be altered substantially by climate change, but negative regional impacts are likely. Agricultural systems in the developed countries are highly adaptable and can probably cope with the expected range of climate changes without dramatic reductions in yields. It is the poorest countries, where many already are subject to hunger, that are the most likely to suffer significant decreases in agricultural productivity. Table 16.3 summarizes some of the main points related to the consequences of global warming and climate change.

ADDRESSING CLIMATE CHANGE

Approaches to dealing with climate change involve technological change coupled with political will and economic realities.

ENERGY EFFICIENCY

A major step toward slowing global warming would be to increase the efficiency of energy utilization. More efficient use of fuels conserves the shrinking supplies of energy resources. It makes sense to increase energy efficiency, thus reducing carbon dioxide production, even if global warming is not a concern. One way to stimulate a move toward greater efficiency would be to place a tax on the amount of carbon individuals and corporations release into the atmosphere. This would increase the cost of fuels and stimulate a demand for fuel efficient products because the cost of fuel would rise. It would also stimulate the development of alternative fuels with a lower carbon content and generate funds for research in many aspects of fuel efficiency and alternative fuel technologies.

Increases in energy efficiency and reductions in greenhouse gas emissions are likely to have important related benefits that could offset the costs. For example, the World Bank estimated that in 2003, air pollution in China, primarily attributable to poorly controlled burning of coal, caused 200,000 premature deaths, 1.8 million cases of chronic bronchitis, 1.7 billion restricted-activity days, and more than 5 billion cases of respiratory illness. Each of these statistics can be converted into monetary terms so that costs of improving fuel efficiency and reducing pollution can be offset by lower health care costs and higher worker productivity.

Improved energy efficiency also reduces the need for new power plants and related energy infrastructure, now estimated at about $100 billion annually in developing nations.

The European Union ratified the Kyoto Treaty in 2002. Since that time, many European countries have made major commitments to reduce their production of greenhouse gases—primarily carbon dioxide. Germany, in particular, developed an energy policy that has resulted in lower carbon dioxide emissions.

The following are some of the significant policy initiatives of the German government:

- The Renewable Energy Resources Act set a goal to provide 20 percent of electricity from renewable sources by 2020.
- Low-interest loans were provided to upgrade the energy efficiency of buildings.
- Energy use certificates were required for buildings.
- A tax was imposed on the use of coal, coke, and lignite.
- A tax was imposed on motor freight that varied with the distance traveled.

- Blue Angel eco-labeling of appliances was established.
- A carbon dioxide emissions trading law was enacted.
- Low-interest loans were provided for the installation of solar energy technology.

These policies have led to significant changes in energy use and carbon dioxide emissions. Although Germany is not in the most geographically advantageous location for using solar energy, it is second only to Japan in installed capacity. Most of the solar capacity is used for heating purposes. The growth in solar capacity is driven in large part by subsidies that are part of the government's energy policy. By 2006, about 12 percent of electricity was produced from renewable sources. Wind provides about 7 percent of its electricity needs and hydroelectricity provides about 4 percent. Biomass—wood, landfill gas, and biogas—also provides significant amounts of energy. As a result of these policies, energy efficiency has improved and carbon dioxide emissions have declined.

THE ROLE OF BIOMASS

Another approach to the problem is to increase the amount of carbon dioxide removed from the atmosphere. If enough biomass is present, the excess carbon dioxide can be used by vegetation during photosynthesis, thereby reducing the impact of carbon dioxide released by fossil-fuel burning. Australia, the United States, and several other countries have instituted plans to plant billions of trees to help remove carbon dioxide from the atmosphere. Many critics argue that this approach will provide only a short-term benefit since, eventually, the trees will mature and die, and their decay will release carbon dioxide into the atmosphere at some later time.

An associated concern is the destruction of vast areas of rainforest in tropical regions of the world. These ecosystems are extremely efficient at removing carbon dioxide and storing the carbon atoms in plant structures. The burning of tropical rainforests to provide farm or grazing land not only adds carbon dioxide to the atmosphere, but it also reduces their ability to remove carbon dioxide from the atmosphere, since the grasslands or farms created do not remove carbon dioxide as efficiently as do the rainforests. Furthermore, the grazing lands and farms in such regions of the world are often abandoned after a few years and do not return to their original forest condition.

TECHNOLOGICAL APPROACHES

Many candidates for this longer-term technological transition have been identified, and some have already begun to penetrate the market. For example, wind energy systems are improving

rapidly and, in ideal conditions, already compare favorably with conventional coal-burning power plants. Direct conversion of sunlight to electricity is now possible with photovoltaic and solar thermal technologies. While relatively expensive today, they are already competitive in areas remote from electric utility grids. The costs of these technologies are likely to decline significantly over time.

The U.S. Department of Energy has concluded that, relying primarily on already proven technology, the United States could reduce its carbon emissions by almost 400 million metric tons in 2010, or enough to stabilize U.S. emissions in that year at 1990 levels, with savings from reduced energy costs roughly equal to the added cost of investment.

Resources and policies to increase investment in renewables and other longer-term technologies will be needed. Good examples from the industrialized nations include wind power purchase programs in Denmark and Germany, the "10,000 rooftops" photovoltaic program in Japan, and the evolution of "green marketing" campaigns in the United States and Europe to capture consumer willingness to pay modest premiums for electricity from clean energy technologies.

POLITICAL AND ECONOMIC FORCES

International Agreements

With the recognition that chlorofluorocarbons were destroying the stratospheric ozone layer that protects us from ultraviolet radiation, international agreements were reached that have led to a sharp reduction in the amount of chlorofluorocarbons being released. The phasing out of chlorofluorocarbons required significant technological changes, but the changes were very rapid once a broad consensus was reached and a plan established. Changes made to protect the ozone layer, therefore, have had the side benefit of reducing the release of a potent greenhouse gas. See the section on ozone depletion in this chapter.

In 1997, the development of an international agreement known as the Kyoto Protocol on greenhouse gas emissions was a first step toward a worldwide approach to alleviating the problem. Most countries (except the United States and a few small countries) of the world have ratified the Treaty and have begun to address the causes of global climate change. Progress is slow but the European Union and its member countries have made considerable progress in reducing their carbon emissions. There is still a long way to go and some countries have greater resolve than others, but there is a commitment and policy changes have been made that back up that commitment. The Kyoto Protocol expires in 2012, so there is a need to decide what to do next.

In December 2007, a United Nations conference on climate change was held in Bali, Indonesia. While it did not result in any commitments to curtail or reduce greenhouse gas emissions, it did include several important steps:

- Countries agreed to hold further talks to arrive at a new accord by 2009 that would set guidelines for reducing greenhouse gases.
- Attendees committed to adding rapidly developing countries such as India and China to the list of those that need to come up with measurable, verifiable evidence of progress in reducing greenhouse gas emissions.
- Industrialized countries committed to provide technical and financial aid to assist developing countries in dealing with the problem.
- Countries committed to work to reduce carbon emissions attributable to deforestation.

The Global Economy

It will be more difficult to achieve a similar consensus to reduce carbon dioxide emissions because carbon dioxide is released as a result of energy consumption, and energy consumption affects all parts of the economy. However, the fundamental benefit of cleaner and more energy-efficient production potentially could encourage changes to industry, energy production, and transportation that would result in lower carbon dioxide releases.

Incentive for change is reinforced by the globalization of the economy, which encourages manufacturers to make products that have the widest possible market acceptance. Thus, many developing nations that were allowed to continue using chlorofluorocarbons beyond the time limits placed on more-developed nations quickly discovered that this grace period was a mixed blessing. If they continued to manufacture products that contained chlorofluorocarbons, they were excluded from the largest export markets. Multinational firms reinforced the trend by requiring that products they purchased be "chlorofluorocarbon-free." Financial assistance to countries through the Ozone Multilateral Fund further encouraged a more rapid transition.

INDOOR AIR POLLUTION

Even though we spend on average almost 90 percent of our time indoors, the movements to reduce indoor air pollution lag behind regulations governing outdoor air pollution. In the United States, the Environmental Protection Agency is conducting research to identify and rank the human health risks that result from exposure to individual indoor pollutants or mixtures of multiple indoor pollutants.

A growing body of scientific evidence indicates that the air within homes and other buildings can be more seriously polluted than outdoor air in even the largest and most industrialized cities. Many indoor air pollutants and pollutant sources are thought to have an adverse effect on human health.

SOURCES OF INDOOR AIR POLLUTANTS

Indoor air pollutants come from a variety of sources:

- Asbestos fibers from older ceiling and floor tiles and insulation;
- Formaldehyde, which is associated with many consumer products, including certain wood products and aerosols;
- Radon in certain parts of the world with the appropriate geology;
- Lead from lead-based paints in older homes;
- Volatile organic compounds such as airborne pesticide residues, perchloroethylene (associated particularly with dry cleaning), paradichlorobenzene (from mothballs and air fresheners), and gases from cleaning agents, polishes, and paints;
- Carbon monoxide and nitrogen dioxide from stoves, furnaces, kerosene heaters, fireplaces and tobacco smoke;
- Particulate matter from burning materials;
- Mold spores from molds growing in damp places; and
- Other disease-causing or allergy-producing organisms.

SIGNIFICANCE OF WEATHERIZING BUILDINGS

A recent contributing factor to the concern about indoor air pollution is the weatherizing of buildings to reduce heat loss and save on fuel costs. In most older homes, a complete exchange of air occurs every hour. This means that fresh air leaks in around doors and windows and through cracks and holes in the building. In a weatherized home, a complete air exchange may occur only once every five hours. Such a home is more energy efficient, but it also tends to trap air pollutants.

SECONDHAND SMOKE

Smoking is the most important air pollutant source in the United States in terms of human health. The surgeon general estimates that 350,000 people in this country die each year from emphysema, heart attacks, strokes, lung cancer, or other diseases caused by tobacco smoking. Banning smoking probably would save more lives than would any other pollution-control measure.

Many nonsmoking people, are exposed to environmental tobacco smoke (secondhand smoke) because they live and work in

Since 1978, satellites have made continual observations on the area of Arctic Ocean covered by ice. There are several factors that influence the amount of ice present. During the summer months, increased air and water temperatures cause melting of the ice; and the area covered by ice shrinks to a minimum during September. The general trend since 1978 has been that the area covered by sea ice at the end of the Arctic summer has been declining. In September 2007, the Arctic sea ice reached its smallest extent since records have been kept—4.13 million km^2. The previous record was 5.32 million km^2 in 2005.

There are several factors that affect the rate at which the ice melts:

- Long-term trends show that the melting of ice begins earlier in the spring than it had previously. Thus, there is a longer period of time during which melting occurs, resulting in a smaller extent of sea ice at the end of the summer.

- The nature of the ice is also important. Ice that is formed during the winter is thinner than ice that has remained unmelted during the summer. This thinner ice tends to break up more easily than perennial ice. This is important because intact ice reflects sunlight, reduces warming, and slows the rate at which the ice melts. On the other hand, open water tends to absorb sunlight and is warmed, resulting in further melting of the ice.

- There is evidence that there have been changes in ocean currents that have introduced warmer water from the both the Atlantic and Pacific Oceans into the Arctic Ocean, which could be contributing to the more rapid melting seen in recent years.

The extent of sea ice is important to the people and wildlife that live in the Arctic. Reduced sea ice appears to be reducing the success of polar bears in obtaining food. Similarly, native peoples typically hunt from the ice and rely on ice cover to allow them to travel from place to place. The red lines on the map shows the long-term average extent of sea ice at the summer minimum. The white area shows the extent of sea ice during September 2007.

Sea Ice Extent

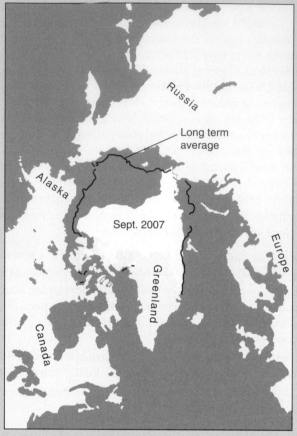

Total extent = 4.3 million sq km

■ Median ice edge

Source: Data the National Snow and Ice Data Center, Boulder, Co.

---

spaces where people smoke. The U.S. Environmental Protection Agency estimates that approximately 3000 nonsmokers die each year of lung cancer as a result of breathing air that contains secondhand smoke. Young children who are exposed to secondhand smoke are much more likely to have respiratory infections.

In July 1993, the EPA recommended several actions to prevent people from being exposed to secondhand indoor smoke, including that:

- People not smoke in their homes or permit others to do so

- All organizations that deal with children have policies that protect children from secondhand smoke

- Every company have a policy that protects employees from secondhand smoke

- Smoking areas in restaurants and bars be placed so that the smoke will have little chance of coming into contact with nonsmokers

California has some of the most restrictive smoking laws in the United States. In 1998, California passed a no-smoking law that effectively banned all smoking in indoor public places, including restaurants and bars. Today only 16 states do not have some kind of a statewide ban on smoking ban in public places. Even in states that do not have legislation to protect people from secondhand smoke, many cities and other jurisdictions within those states do.

RADON

Radon-222, an inert radioactive gas with a half-life of 3.8 days, is one of the products formed during the breakdown of uranium-238. Uranium-238 is a naturally occuring element that makes up about 3 parts per million of the Earth's crust. Uranium-238 goes through 14 steps of decay before it becomes stable, nonradioactive lead-206. One of the intermediate steps is radon-222.

As the radon gas is formed in the rocks, it usually diffuses up through the rocks and soil and escapes harmlessly into the atmosphere. It can also diffuse into groundwater. Radon usually enters a home through an open space in the foundation. A crack in the basement floor or the foundation, the gap around a water or sewer pipe, or a crawl space allows the radon to enter the home. It may also enter in the water supply from wells.

Since radon is an inert gas, it does not enter into any chemical reactions within the body, but it can be inhaled. Once in the lungs, it may undergo radioactive decay, producing other kinds of atoms called "daughters" of radon. These decay products (daughters) of radon—plutonium 218, which has a three-minute half-life; lead 214, which has a 27-minute half-life; bismuth 214, which has a 20-minute half-life; and polonium 214, which has a millisecond half-life—are solid materials that remain in the lungs and are chemically active.

Increased incidence of lung cancer is the only known health effect associated with radon decay products. It is estimated that the decay products of radon are responsible for about 20,000 lung cancer deaths annually in the United States. This is about 10 percent of lung cancer deaths. The radon risk evaluation chart (table 16.4) indicates that individuals exposed to greater than 4 picocuries per liter are at increased risk of developing lung cancer.

About 10 percent of the homes in the United States have a potential radon problem. Figure 16.18 shows those regions of the United States that are likely to have elevated levels of radon. In addition, the Environmental Protection Agency and the U.S. surgeon general recommend that all Americans (other than those living in apartment buildings above the second floor) test their homes for radon. If the tests indicate radon levels at or above 4 picocuries per liter, the EPA recommends that the homeowner take action to lower the level. This is usually not expensive and consists of blocking the places where radon is entering or venting radon sources to the outside. People who are concerned about radon should contact their state's public health department or environmental protection agency.

NOISE POLLUTION

Noise is referred to as unwanted sound. However, noise can be more than just an unpleasant sensation. Research has shown that exposure to noise can cause physical, as well as mental, harm. (See

TABLE 16.4 Radon Risk Evaluation Chart

pCi/L	WL	Estimated Lung-Cancer Deaths Due to Radon Exposure (out of 1000)	Comparable Exposure Levels	Comparable Risk
200	1.0	440–770	One thousand times average outdoor level	More than 60 times nonsmoker risk / Four-pack-a-day smoker
100	0.5	270–630	One hundred times average indoor level	Twenty thousand chest X-rays per year
40	0.2	120–380		Two-pack-a-day smoker
20	0.1	60–120	One hundred times average outdoor level	One-pack-a-day smoker
10	0.05	30–120	Ten times average indoor level	Five times nonsmoker risk
4	0.02	13–50	Level at which EPA suggests remedial action	Two hundred chest X-rays per year
2	0.01	7–30	Ten times average outdoor level	Nonsmoker risk of dying from lung cancer
1	0.005	3–13	Average indoor level	
0.2	0.001	1–3	Average outdoor level	Twenty chest X-rays per year

Note: Measurement results are reported in one of two ways: (1) pCi/L (picocuries per liter)—measurement of *radon gas*, or (2) WL (working levels)—measurement of *radon decay products*.

Source: Data from U.S. Environmental Protection Agency, Office of Air and Radiation Programs.

figure 16.19.) The loudness of the noise is measured by **decibels** (db). Decibel scales are logarithmic rather than linear. Thus, the change from 40 db (a library) to 80 db (a dishwasher or garbage disposal) represents a ten-thousandfold increase in sound loudness.

The frequency or pitch of a sound is also a factor in determining its degree of harm. High-pitched sounds are the most annoying. The most common sound pressure scale for high-pitched sounds is the A scale, whose units are written "dbA." Hearing loss begins with prolonged exposure (eight hours or more per day) to 80 or 90 dbA levels of sound pressure. Sound pressure becomes painful at around 140 dbA and can kill at 180 dbA. (See table 16.5.)

In addition to hearing loss, noise pollution is linked to a variety of other ailments, ranging from nervous tension headaches to neuroses. Research has also shown that noise may cause blood vessels to constrict (which reduces the blood flow to key body

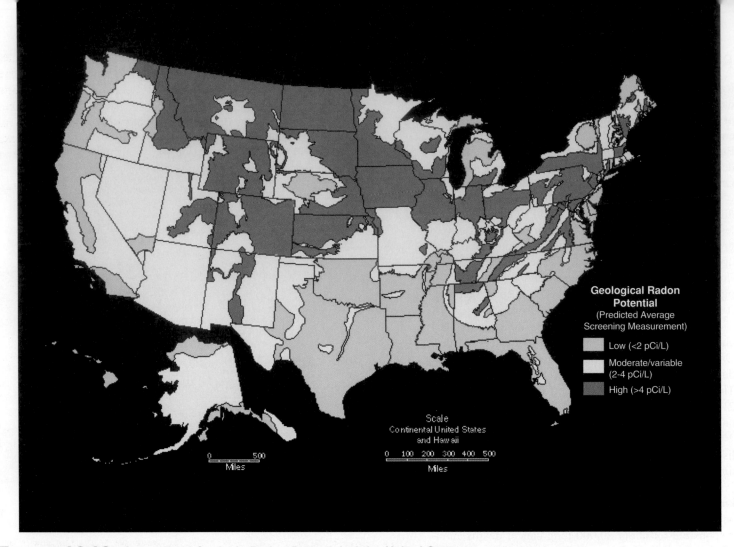

FIGURE 16.18 Generalized Geologic Radon Potential of the United States The EPA recommends that people check their homes for radon. Those regions of the country shown in pink have the highest likelihood of elevated radon levels.
Source: U.S. Geological Survey.

Geological Radon Potential
(Predicted Average Screening Measurement)

- Low (<2 pCi/L)
- Moderate/variable (2-4 pCi/L)
- High (>4 pCi/L)

Scale
Continental United States and Hawaii

0 500
Miles

0 100 200 300 400 500
Miles

FIGURE 16.19 People whose jobs require them to work with noisy machinery often wear devices to prevent hearing loss.

TABLE 16.5 Intensity of Noise

Source of Sound	Intensity in Decibels
Jet aircraft at takeoff	145
Pain occurs	140
Hydraulic press	130
Jet airplane (160 meters overhead)	120
Unmuffled motorcycle	110
Subway train	100
Farm tractor	98
Gasoline lawnmower	96
Food blender	93
Heavy truck (15 meters away)	90
Heavy city traffic	90
Vacuum cleaner	85
Hearing loss after long exposure	85
Garbage disposal unit	80
Dishwasher	65
Window air conditioner	60
Normal speech	60

parts), disturbs unborn children, and sometimes causes seizures in epileptics. The U.S. Environmental Protection Agency has estimated that noise causes about 40 million U.S. citizens to suffer hearing damage or other mental or physical effects. Up to 64 million people are estimated to live in homes affected by aircraft, traffic, or construction noise.

The Noise Control Act of 1972 was the first major attempt in the United States to protect the public health and welfare from detrimental noise. This act also attempted to coordinate federal research and activities in noise control, to set federal noise emission standards for commercial products, and to provide information to the public. After the passage of the Noise Control Act, many local communities in the United States enacted their own noise ordinances. While such efforts are a step in the right direction, the United States is still controlling noise less than are many European countries. Several European countries have developed quiet construction equipment in conjunction with strongly enforced noise ordinances. The Germans and Swiss have established maximum day and night noise levels for certain areas. Regarding noise-pollution abatement, North America has much to learn from European countries.

ISSUES & ANALYSIS

Pollution, Policy, and Personal Choice

The air pollution problem that has shown the least improvement in North America is the control of ground-level ozone. Some geographic regions are more prone to developing conditions that generate ground-level ozone, and we know that the presence of nitrogen oxides and volatile organic compounds triggers the events leading to the development of unhealthy levels of ground-level ozone. Furthermore, it is clear that automobiles are the major source of the volatile organic compounds and nitrogen oxides.

It is also clear that a large proportion of the population is regularly exposed to unhealthy levels of ozone and that many millions of people become ill or have their activities restricted as a direct result of poor air quality.

There are several ways to attack this problem, but all of them involve restricting the freedom of people to use automobiles or restricting the kinds of automobiles they can drive. The following list describes some of the approaches used and their effects on the driving public.

1. *Restrict the kinds of cars that people can buy to less-polluting types.* Requirements that automobiles obtain better mileage with reduced emissions have successfully produced less pollution per mile driven. However, the number of miles driven is increasing, which negates some of the gains attributable to higher-mileage vehicles.

2. *Provide incentives to encourage fewer people to drive.* Many metropolitan areas have carpool lanes restricted to cars with two or more passengers. These lanes are largely unused. People choose to drive separately.

3. *Provide economic penalties for the use of automobiles.* In many cities of the world, people pay a fee to drive in parts of the city. In addition, toll roads and bridges are an economic incentive not to drive. However, this only reduces traffic if the fees are high and there are no "free" roads that people can use.

4. *Provide affordable, safe, and convenient alternative means of transport.* For every dollar spent on public transport in the United States, about $6 are spent on roads for cars.

These approaches to dealing with ground-level ozone raise several issues that affect personal choice.

- Should the use of automobiles be restricted to protect the health of people who are highly sensitive to ground-level ozone?
- Should governments determine what kinds of automobiles are allowed?
- Should fees, taxes, or assessments be used to increase the cost of using a car?
- A typical driver spends $1000 or more per year for gasoline. Would you pay an annual fee of $1000 to use public transport if it were available?

SUMMARY

The atmosphere has a tremendous ability to disperse pollutants. Carbon monoxide, hydrocarbons, particulate matter, sulfur dioxide, and oxides of nitrogen compounds are the primary air pollutants. They can cause a variety of health problems. The U.S. Environmental Protection Agency establishes standards for six pollutants known as criteria air pollutants. They are carbon monoxide, nitrogen dioxide, sulfur dioxide, volatile organic compounds (hydrocarbons), ozone, and lead. The EPA also regulates hazardous air pollutants.

Photochemical smog is a secondary pollutant, formed when hydrocarbons and oxides of nitrogen are trapped by thermal inversions and react with each other in the presence of sunlight to form peroxyacetyl nitrates and ozone. Elimination of photochemical smog requires changes in technology, such as more fuel-efficient automobiles, special devices to prevent the loss of hydrocarbons, and catalytic converters to more completely burn hydrocarbons in exhaust gases.

Acid rain is caused by emissions of sulfur dioxide and oxides of nitrogen in the upper atmosphere, which form acids that are washed from the air when it rains or snows or settle as particles on surfaces. Direct effects of acid rain on terrestrial ecosystems are difficult to prove, but changes in many forested areas are suspected of being partly the result of additional stresses caused by acid rain. Recent evidence suggests that loss of calcium from the soil may be a major problem associated with acid rain. The effect of acid rain on aquatic ecosystems is easy to quantify. As waters become more acidic, the complexity of the ecosystem decreases, and many species fail to reproduce. The control of acid rain requires the use of scrubbers, precipitators, and filters—or the removal of sulfur from fuels. However, oxides of nitrogen are still a problem.

Currently, many are concerned about the damaging effects of greenhouse gases: carbon dioxide, nitrous oxide, methane, and chlorofluorocarbons. These gases are likely to be causing an increase in the average temperature of the Earth and, consequently, are leading to major changes in the climate. Human and ecological systems are already vulnerable to a range of environmental pressures, including climate extremes and variability. Global warming is likely to amplify the effects of other pressures and disrupt our lives

in numerous ways. Significant impacts on our health, the vitality of forests and other natural areas, the distribution of freshwater supplies, and the productivity of agriculture are among the probable consequences of climate change. Chlorofluorocarbons also lead to the destruction of ozone in the upper atmosphere, which results in increased amounts of ultraviolet light reaching the Earth. Concern about the effects of chlorofluorocarbons has led to international efforts that have resulted in significant reductions in the amount of these substances reaching the atmosphere. Many commonly used materials release gases into closed spaces (indoor air pollution), where they cause health problems. The most important of these health problems are associated with tobacco smoking.

THINKING GREEN

1. You can reduce automotive emissions by driving less or joining a ride pool.
2. You can reduce emissions from power plants by reducing your electrical energy use
3. Eat less meat—cows produce methane.
4. Don't burn leaves and debris outdoors—let them decay naturally.
5. Use only water-based paints, which do not release VOCs.
6. Purchase green energy from your electric utility.
7. Dispose of old CFC-containing air conditioners, refrigerators, freezers, and dehumidifiers appropriately.

WHAT'S YOUR TAKE?

China's economic boom has an environmental dark side. While China's economy continued to grow at a rate of more than 11 percent in 2008, the country's environment and the Chinese people are paying a steep price. China is home to 16 of the world's 20 most polluted cities. Air pollution is blamed for the premature death of some 450,000 Chinese annually, and trends indicate that such levels of pollution and environmental degradation will worsen. Over the next decade Chinese consumers are expected to purchase hundreds of millions of automobiles that will add to the current air-pollution problems. Chinese consumers argue that they want to enjoy the same standard of living as those in the West, and owning their own cars is part of that goal. Develop a position paper on car ownership from a Chinese citizen's perspective and another paper from the perspective of a citizen in Japan (which is downwind of China) or a citizen in North America.

REVIEW QUESTIONS

1. List the five primary air pollutants commonly released into the atmosphere and their sources.
2. List the six criteria air pollutants and their sources.
3. Define *secondary air pollutants* and give an example.
4. List three health effects of air pollution.
5. Why is air pollution such a large problem in urban areas?
6. What is photochemical smog? What causes it?
7. Describe three actions that can be taken to control air pollution.

8. What causes acid rain? List three probable detrimental consequences of acid rain.
9. Why is carbon dioxide (a nontoxic normal component of the atmosphere) called a "greenhouse gas"?
10. What would the consequences be if the ozone layer surrounding the Earth were destroyed?
11. How does energy conservation influence air quality?

1. What could you do to limit the air pollution you create?

2. Do you agree with a ban on smoking, as in California, that includes all indoor public places, even privately owned restaurants and bars? Why or why not?

3. Some developing countries argue that they should be exempt from limits on the production of greenhouse gases and that developed countries should bear the brunt of the changes that appear to be necessary to curb global climate change. What values, beliefs, and perspectives underlie this argument? What do you think about this argument?

4. As a nation, the United States provides many subsidies to make energy cheap because policy makers feel that economic development depends on cheap energy. If these subsidies were withdrawn or taxes on energy were added, what effect would this have on your own energy consumption? Would you be willing to support high gasoline prices, in the $7 to $10-a-gallon range as in many European countries, if it would cut greenhouse gas emissions?

5. Why do you think air pollution is so much worse in developing countries than in developed countries? What should developed countries do about this, if anything?

6. What common indoor air pollutants are you exposed to? What can you do to limit this exposure?

7. What kinds of noise pollution do you encounter? How important is noise pollution to you? What can you do to reduce noise pollution? How can we make cities quieter?

8. Is it possible to have zero emissions of pollutants? What level of risk are you willing to live with?

SOLID WASTE MANAGEMENT AND DISPOSAL

As we use materials in our daily lives we generate waste. Much of that waste is sent to landfills. However, many materials that are no longer useful in their current form can be recycled for use in a different way. Most communities have recycling programs that reduce the amount of waste sent to landfills.

CHAPTER OUTLINE

Kinds of Solid Waste
Municipal Solid Waste
Methods of Waste Disposal
    Landfills
    Incineration
    Producing Mulch and Compost
    Source Reduction
    Recycling

ISSUES & ANALYSIS
Paper or Plastic or Plastax? 406

CASE STUDIES
Resins Used in Consumer Packaging 400
Beverage Container Deposit-Refund Programs 403

CAMPUS SUSTAINABILITY INITIATIVE
Recycling Partnership at Northern Arizona University 404

GOING GREEN
Garbage Goes Green 395

WATER CONNECTIONS
Landfills' Impact on Water 397

OBJECTIVES

After reading this chapter, you should be able to:

- Explain why solid waste is a problem throughout the world.
- Understand that the management of municipal solid waste is directly affected by economics, changes in technology, and citizen awareness and involvement.
- Describe the various methods of waste disposal and the problems associated with each method.
- Understand the difficulties in developing new municipal landfills.
- Define the problems associated with incineration as a method of waste disposal.
- Describe some methods of source reduction.
- Describe composting and how it fits into solid waste disposal.
- List some benefits and drawbacks of recycling.

The following additional Case Study can be found on the book's website at www.mhhe.com/enger12e along with other interesting readings: "Recycling Is Big Business."

KINDS OF SOLID WASTE

The several kinds of waste produced by a technological society can be categorized in many ways. Some kinds of wastes are released into the air and water. Some are purposely released, while others are released accidentally. Many wastes that are purposely released are treated before their release. Wastes that are released into the water and air are discussed in chapters 15 and 16, respectively. There are wastes with particularly dangerous characteristics, such as nuclear wastes, medical wastes, industrial hazardous wastes, and household hazardous wastes. Chapter 10 deals with nuclear wastes, and chapter 18 considers hazardous waste issues. The focus of this chapter is solid waste.

Solid waste is generally made up of objects or particles that accumulate on the site where they are produced, as opposed to water- and airborne wastes that are carried away from the site of production. Solid wastes are typically categorized by the sector of the economy responsible for producing them, such as mining, agriculture, manufacturing, and municipalities.

We have very good information about those waste streams that are tightly regulated (hazardous wastes, municipal solid waste, medical wastes, nuclear wastes) but only general estimates for many of the other kinds of wastes, such as mining and agricultural waste.

Mining waste is generated in three primary ways. First, in most mining operations, large amounts of rock and soil need to be removed to get to the valuable ore. This waste material is generally left on the surface at the mine site. Second, milling operations use various technologies to extract the valuable material from the ore. These techniques vary from relatively simple grinding and sorting to sophisticated chemical separation processes. Regardless of the technique involved, once the valuable material is recovered, the remaining waste material, commonly known as tailings, must be disposed of. Solid materials are typically dumped on the land near the milling site, and liquid wastes are typically stored in ponds. It is difficult to get vegetation to grow on these piles of waste rock and tailings, so they are unsightly and remain exposed to rain and wind. Finally, the water that drains or is pumped from mines or that flows from piles of waste rock or tailings often contains hazardous materials (such as asbestos, arsenic, lead, and radioactive materials) or high amounts of acid that must be contained or treated—but often are not. In addition, failures of the earthen dams used to form waste ponds result in the release of contaminated water into local streams. Between 1993 and 2005, about 35 such failures occurred worldwide, seven of them in the United States.

It is difficult to get more than a rough estimate of the amount of mining waste produced, but the U.S. Environmental Protection Agency estimates that between 1 billion and 2 billion metric tons of mining waste are produced each year in the United States. Of the total waste produced, about 700 million to 800 million metric tons are considered hazardous.

Many types of mining operations require vast quantities of water for the extraction process. The quality of this water is degraded, so it is unsuitable for drinking, irrigation, or recreation. Since mining disturbs the natural vegetation in an area, water may carry soil particles into streams and cause erosion and siltation. Some mining operations, such as strip mining, rearrange the top layers of the soil, which lessens or eliminates its productivity for a long time. Strip mining has disturbed approximately 75,000 square kilometers (30,000 square miles) of U.S. land, an area equivalent to the state of Maine.

Agricultural waste is the second most common form of waste and includes waste from the raising of animals and the harvesting and processing of crops and trees. The amount of animal manure produced annually is estimated at about 1240 million metric tons. Other wastes associated with agriculture, such as waste from processing operations (peelings, seeds, straw, stems, sludge, and similar materials), might bring the total agricultural waste to about 1.5 billion metric tons per year. Since most agricultural waste is organic, approximately 90 percent is used as fertilizer or for other soil-enhancement activities. Other materials are burned as a source of energy, so little of this waste needs to be placed in landfills. However, when too much waste is produced in one place, there may not be enough farmland available to accept the agricultural waste without causing water pollution problems associated with runoff or groundwater contamination due to infiltration.

Industrial solid waste from sources other than mining is variously estimated to be between 200 million and 600 million metric tons of solid waste per year. It includes a wide variety of materials such as demolition waste, foundry sand, scraps from manufacturing processes, sludge, ash from combustion, and other similar materials. These materials are tested to determine if they are hazardous. If they are classified as hazardous waste, their disposal requires that they be placed in special hazardous waste landfills. Hazardous wastes are discussed in chapter 18. In addition to solid wastes, industries produce several billion metric tons of aquatic waste. See chapter 15 for a discussion of industrial use of water.

Municipal solid waste (MSW) consists of all the materials that people in a region no longer want because they are broken, spoiled, or have no further use. It includes waste from households, commercial establishments, institutions, and some industrial sources and amounts to about 210 million metric tons per year. Table 17.1 summarizes estimates of the quantity of various kinds of solid waste produced in the United States. The remainder of this chapter will focus on the generation and disposal of municipal solid waste.

TABLE 17.1 Estimates of Solid Waste Produced per Year in the United States

Category of Waste	Amount of Waste (million metric tons)
Mining waste	1000–2000
Agricultural waste	1500
Industrial waste	200–600
Municipal solid waste	210

Source: Estimates from U.S. Environmental Protection Agency, U.S. Geological Survey, and U.S. Department of Agriculture.

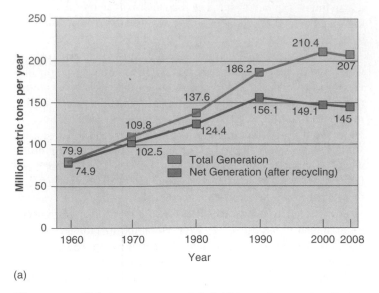

Total and Net Municipal Solid Waste Generation—1960 to 2008

(a)

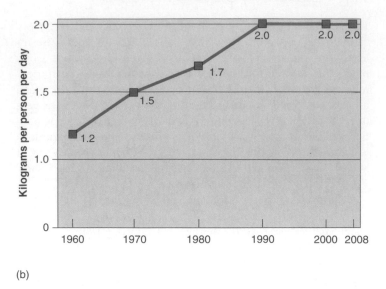

Per Capita Municipal Solid Waste Generation

(b)

FIGURE 17.1 **Municipal Solid Waste Generation Rates** The generation of municipal solid waste in the United States has increased steadily. However, because of increased recycling rates, the net production rates (after recyclables have been removed) has actually fallen since 1990 (a), and the per capita rate has stabilized (b).
Source: U.S. Environmental Protection Agency, Washington, D.C.

MUNICIPAL SOLID WASTE

Wherever people exist, waste disposal is a problem. Archaeologists are eager to find the middens, or waste heaps, of ancient civilizations. The kinds of articles discarded can tell us a great deal about the nature of the society that produced them. In modern society, many products are discarded when they are broken or worn out, while others have only a temporary use. Those that have only temporary uses make up the majority of solid waste.

The United States produces about 210 million metric tons of municipal solid waste each year. This equates to about 2 kilograms (4.4 pounds) of trash per person per day, or 0.73 metric tons per person per year. The amount of municipal solid waste has more than doubled since 1960, and the per capita rate has increased by nearly 70 percent in that same time, although per capita rates began to stabilize about 1990. When recycling is included, the net waste produced has actually fallen since 1990. (See figure 17.1.)

Nations with high standards of living and productivity tend to have more municipal solid waste per person than less-developed countries. (See figure 17.2.) The United States and Canada, therefore, are world leaders in waste production. Large metropolitan areas have the greatest difficulty dealing with their solid waste because of the large volume and the challenge of finding suitable landfill sites near the city. For example, Toronto, Canada, has no local landfill for its waste. As a result, Toronto developed an ambitious plan to divert 100 percent of its solid waste from landfills by 2011. By 2004, it had met its intermediate target and was diverting about 40 percent. Toronto has a

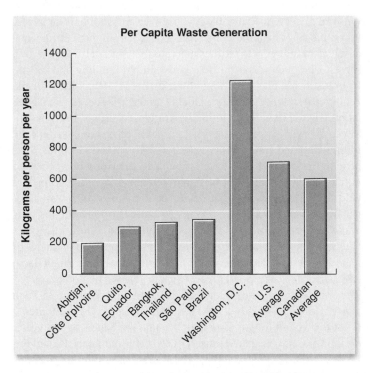

FIGURE 17.2 **Waste Generation and Lifestyle** The waste generation rates of people are directly related to their economic condition. People in richer countries produce more garbage than those in poorer countries.
Source: Data from *World Resources* 1996–2006, and U.S. Environmental Protection Agency.

long-term contract with a landfill site in Michigan to dispose of its remaining solid waste (about 4000 metric tons per day) at a cost of about US $40 per metric ton. In response to public

Nearly 400 municipal solid waste (MSW) landfills in the United States recover and combust landfill gas to generate heat or electricity. Landfill gas is generated during the natural process of bacterial decomposition of organic material contained in MSW landfills. A number of factors influence the quantity of gas that a MSW landfill generates and the components of that gas. These factors include, but are not limited to, the types and age of the waste buried in the landfill, the quantity and types of organic compounds in the waste, and the moisture content and temperature of the waste. Temperature and moisture levels are influenced by the surrounding climate.

Landfill gas can be an asset when it is used as a source of energy to create electricity or heat. It is classified as a medium-Btu gas with a heating value of 350 to 600 Btu per cubic foot—approximately one-half that of natural gas. Landfill gas can often be used in place of conventional fossil fuels in certain applications. It is a reliable source of energy because it is generated 24 hours a day, seven days a week. By using landfill gas to produce energy, landfills can significantly reduce their emissions of methane and avoid the need to generate energy from fossil fuels, thus reducing emissions of carbon dioxide, sulfur dioxide, nitrogen oxides, and other pollutants from fossil fuel combustion.

Landfill gas recovery projects provide a highly effective means of reducing overall greenhouse gas emissions from landfills. By using the otherwise wasted methane contained in the collected landfill gas to generate electricity or directly as a fuel, fossil fuels such as oil and coal are displaced. This displacement of fossil fuels is an environmental benefit, the magnitude of which would depend on the actual amount of electricity generated or landfill gas used. Some of the places making the best use of landfill gas include:

Jackson County, NC—Gas from a closed landfill warms a greenhouse where the county grows plants for public landscaping.

Sonoma County, CA—Landfill gas is compressed to natural gas and used to fuel the county's 45 public buses.

Columbus, OH—Compressed natural gas runs the landfill's 26 trash trucks.

Greer, SC—The plant that makes the BMW Roadster powers its paint shop with landfill gas.

---

concern, the Michigan legislature has enacted several laws that restrict the nature of the waste that can be imported into the state.

In March 2001, New York City closed its Fresh Kills Landfill on Staten Island. (It was reopened for a time following the September 11, 2001, terrorist attack to serve as a place to take the debris from the World Trade Center so that it could be processed.) Before its closure, it was the largest landfill in the world and received about 12,600 metric tons of trash each day. Today, New York City is exporting all of its solid waste to landfills in other parts of New York State, Pennsylvania, Virginia, and other states. In contrast to Toronto, which diverts about 40 percent of its waste, New York City diverts less than 20 percent of its waste.

Archaeologists rely on the waste of past societies to tell them about the nature of the culture and lifestyle of ancient civilizations. In the same way today, our municipal solid waste is a reflection of our society. Figure 17.3 shows how the composition of our trash has changed since 1960. Notice particularly the increase in the amount of paper and plastic and the effect that recycling has had on the amount of glass in the trash. In the United States, the two most common items in the waste stream are paper products and yard waste, and other significant segments are wood, metal, glass, plastics, and food waste. (See figure 17.4.) An analysis of the composition of our waste will present us with possible approaches to reducing the amount of waste we generate.

METHODS OF WASTE DISPOSAL

From prehistory through the present day, the favored means of disposal was simply to dump solid wastes outside of the city or village limits. Frequently, these dumps were in wetlands adjacent to a river or lake. To minimize the volume of the waste, the dump was often burned. Unfortunately, this method is still being used in remote or sparsely populated areas in the world. (See figure 17.5.)

As better waste-disposal technologies were developed and as values changed, more emphasis was placed on the environment and quality of life. Dumping and open burning of wastes is no longer an acceptable practice from an environmental or health perspective. While the technology of waste disposal has evolved during the past several decades, our options are still limited. Realistically, there are no ways of dealing with waste that have not been known for many thousands of years. Essentially, five techniques are used: (1) landfills, (2) incineration, (3) composting, (4) source reduction, and (5) recycling.

LANDFILLS

Landfills have historically been the primary method of waste disposal because this method is the cheapest and most convenient, and because the threat of groundwater contamination was not initially recognized. As we have recognized some of the problems associated

FIGURE 17.3 The Changing Nature of Trash

Paper products are the largest component of the waste stream. Changes in lifestyle and packaging have led to a change in the nature of trash. Note the increase in the amount of plastics in the waste stream. Most of what is currently disposed of could be recycled.

Source: Data from the U.S. Environmental Protection Agency.

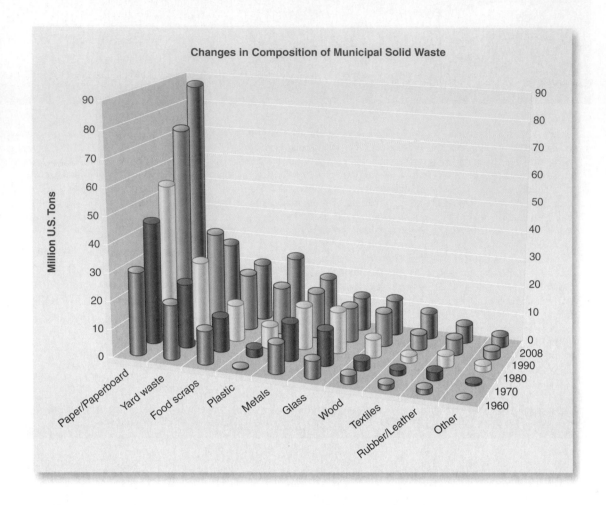

Changes in Composition of Municipal Solid Waste

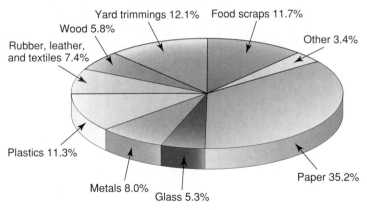

FIGURE 17.4 **Composition of Trash in the United States (2008)** Paper and yard waste are the most common materials disposed of, accounting for about 50 percent of the waste stream.

Source: Data from the U.S. Environmental Protection Agency.

with poorly designed landfills, efforts to reduce the amount of material placed in landfills have been substantial. Although the amount of waste has increased, composting and recycling have removed significant amounts of materials from the waste stream, and the amount of material entering landfills has declined. (See figure 17.6.) However, the landfill of today is far different from a simple hole in the ground into which garbage is dumped.

A modern **municipal solid waste landfill** is typically constructed above an impermeable clay layer that is lined with an impermeable membrane and includes mechanisms for dealing with liquid and gas materials generated by the contents of the landfill. Each day's deposit of fresh garbage is covered with a layer of soil to prevent it from blowing around and to discourage animals from scavenging for food. The selection of landfill sites is based on an understanding of local geologic conditions such as the presence of a suitable clay base, groundwater geology, and soil type. In addition, it is important to address local citizens' concerns. Once the site is selected, extensive construction activities are necessary to prepare it for use. New landfills have complex bottom layers to trap contaminant-laden water, called **leachate,** leaking through the buried trash. The water that leaches through the site must be collected and treated. In addition, monitoring systems are necessary to detect methane gas production and groundwater contamination. In some cases, methane produced by decomposing waste is collected and used to produce heat or to generate electricity. As a result of the technology involved, new landfills are becoming increasingly more complex and expensive. They currently cost up to $1 million per hectare ($400,000 per acre) to prepare. (See figure 17.7.)

Today, about 55 percent of the municipal solid waste from the United States and about 80 percent of Canadian municipal solid waste goes into landfills. The number of landfills is declining. In 1988, there were about 8000 landfills, and in 2008, there were

Although landfills are an indispensable part of our society, they may present long-term threats to groundwater and surface waters that are hydrologically connected. In the United States, federal standards to protect groundwater quality were implemented in 1991 and required some landfills to use plastic liners and collect and treat leachate. However, many disposal sites were either exempted from these rules or grandfathered (excused from the rules owing to previous usage).

There is an increasing belief among solid waste experts that unless further steps are taken to detoxify landfilled materials, today's society will be placing a burden on upcoming generations to address future landfill impacts.

The precipitation that falls into a landfill, coupled with any disposed liquid waste, results in the extraction of the water-soluble compounds and particulate matter of the waste, and the subsequent formation of leachate. The creation of leachate, sometimes called "garbage soup," presents a major threat to the current and future quality of groundwater.

It is for this reason that new landfills are required to have two or more liners and leachate collection systems above and between liners. Liners can be made of highly compacted clay and/or plastic that is nonpermeable. Landfills should also be strategically located well above the water table in a soil with a low permeability so as to help prevent further seepage of the leachate. These soils usually have a very high clay content.

FIGURE 17.5 **Burning Landfills** In the past, it was common practice to burn the waste in landfills to reduce the volume. Waste is still being burned in sparsely populated areas in North America and other parts of the world.

about 1625 active sanitary landfills. (See figure 17.8.) The number of landfills has decreased for two reasons. Many small, poorly run landfills have been closed because they were not meeting regulations. Others have closed because they reached their capacity. The overall capacity, however, has remained relatively constant because new landfills are much larger than the old ones.

A prolonged public debate over how to replace lost landfill capacity is developing where population density is high and available land is scarce. Selecting sites for new landfills in locations such as Toronto, New York, and Los Angeles is extremely difficult because of (1) the difficulty in finding a geologically suitable site and (2) local opposition, which is commonly referred to as the

Solid Waste Management and Disposal   397

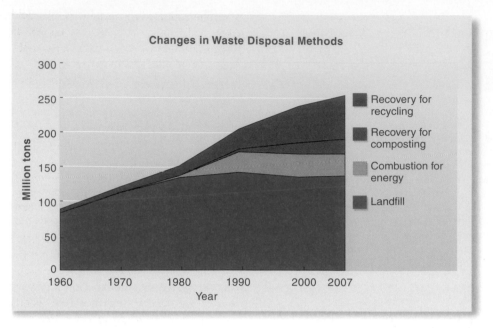

FIGURE 17.6 **Changes in Waste Disposal Methods** The landfill is still the primary method of waste disposal. Historically, landfills have been the cheapest means of disposal, but this may not be the case in the future. Notice that recycling and composting have grown over the past decade, while the amount of waste going to landfills has declined somewhat.
Source: Data from the U.S. Environmental Protection Agency.

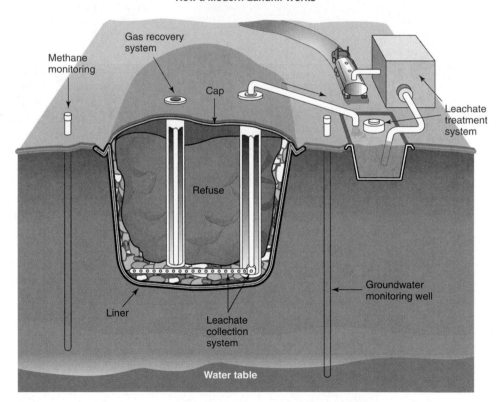

FIGURE 17.7 **A Well-Designed Modern Landfill** A modern sanitary landfill is far different from a simple hole in the ground filled with trash. A modern landfill is a self-contained unit that is separated from the soil by impermeable membranes and sealed when filled. Methane gas and groundwater are continuously monitored to ensure that wastes are not escaping to the air or the groundwater.
Source: National Solid Waste Management Association.

NIMBY, or "not-in-my-backyard," syndrome. Resistance by the public comes from concern over groundwater contamination, rodents and other vectors of disease, odors, and truck traffic. Public officials look for alternatives to landfills to avoid controversy over landfill site selection. Although sites must be chosen for new landfills, politicians are often unwilling to take strong positions that might alienate their constituents.

Japan and many Western European countries have already moved away from landfills as the primary method of waste disposal because of land scarcities and related environmental concerns. Switzerland and Japan dispose of less than 15 percent of their waste in landfills, compared to 55 percent in the United States. Instead, recycling and incineration are the primary methods. (See figure 17.9.) In addition, the energy produced by incineration can be used for electric generation or heating.

INCINERATION

Incineration is the process of burning refuse in a controlled manner. Today, about 15 percent of the municipal solid waste in the United States is incinerated; Canada incinerates about 8 percent. While some incinerators are used just to burn trash, most are designed to capture the heat, which is then used to make steam to produce electricity. The production of electricity partially offsets the cost of disposal. There are about 125 combustors with energy recovery in the United States, with the capacity to burn up to 96,000 metric tons of MSW per day. Most incineration facilities burn unprocessed municipal solid waste. This is often referred to as **mass burn** technology. About one-fourth of the incinerators use refuse-derived fuel—collected refuse that has been processed into pellets prior to combustion. This is particularly useful with certain kinds of materials, such as tires.

Incinerators drastically reduce the amount of municipal solid waste—up to 90 percent by volume and 75 percent by weight. Primary risks of incineration, however, involve air quality problems and the toxicity and disposal of the ash.

Modern incinerators have many pollution control devices that trap nearly all of the pollutants produced. However, tiny amounts of pollutants are released into the

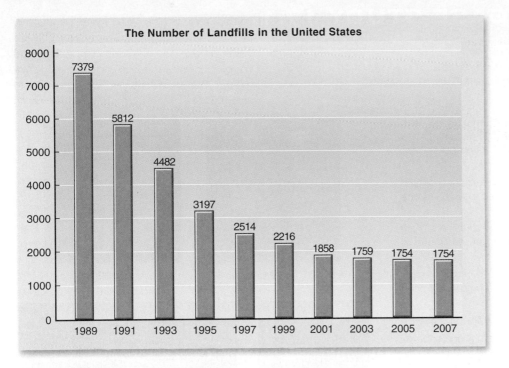

FIGURE 17.8 **Reducing the Number of Landfills** The number of landfills in the United States is declining because they are filling up or because some have been closed because their design and operation do not meet environmental standards.
Source: Data from the U.S. Environmental Protection Agency.

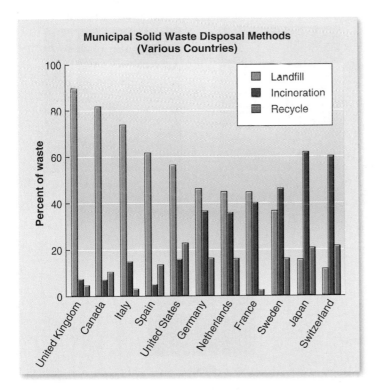

FIGURE 17.9 **Disposal Methods Used in Various Countries** Many countries have difficulty finding adequate space for landfills. Therefore, they rely on other technologies, such as incineration and recycling, to reduce the amount of waste that must be placed in a landfill.
Source: Data from the U.S. Environmental Protection Agency.

atmosphere, including certain metals, acid gases, and classes of chemicals known as dioxins and furans, which have been implicated in birth defects and several kinds of cancer. The long-term risks from the emissions are still a subject of debate.

Ash from incineration is also an important issue. Small concentrations of heavy metals are present in both the fly ash captured from exhaust stacks and the bottom ash collected from these facilities. Because the ash contains lead, cadmium, mercury, and arsenic in varying concentrations from such items as batteries, lighting fixtures, and pigments, the ash is tested to determine if it should be designated as a hazardous waste. This is a concern because the toxic substances are more concentrated in the ash than in the original garbage and can seep into groundwater from poorly sealed landfills. In nearly all cases, the ash is not designated as hazardous and can be placed in a landfill or used as aggregate for roads and other purposes.

The cost of the land and construction for new incinerators are also major concerns facing many communities. Incinerator construction is often a municipality's single largest bond issue. Incinerator construction costs in North America in 2000 ranged from $45 million to $350 million, and the costs are not likely to decline.

Incineration is also more costly than landfills in most situations. As long as landfills are available, they will have a cost advantage. When cities are unable to dispose of their trash locally in a landfill and must begin to transport the trash to distant sites, incinerators become more cost effective. The U.S. Environmental Protection Agency has not looked favorably on the construction of new waste-to-energy facilities and has encouraged recycling and source reduction as more effective ways to reduce the solid waste problem. Critics have argued that cities and towns have impeded waste reduction and recycling efforts by putting a priority on incinerators and committing resources to them. Proponents of incineration have been known to oppose source reduction. They argue that incinerators need large amounts of municipal solid waste to operate and that reducing the amount of waste generated makes incineration impractical. Many communities that have opposed incineration say that they support a vigorous waste-reduction and recycling effort.

PRODUCING MULCH AND COMPOST

Mulch is organic material that is used to cover the soil. It is often used to protect areas where the soil is disturbed or to control the growth of unwanted vegetation in certain kinds of plantings. Typically, large branches, bark, and other organic materials are chopped or shredded into smaller pieces. Mulches can be sorted by the size of the pieces and can be colored for decorative purposes. Since mulch is organic matter, it eventually breaks down and becomes part of the soil.

Solid Waste Management and Disposal   399

Thermoplastics are plastics that can be remelted and reprocessed, usually with only minor changes in their properties. About 30 percent of thermoplastic resins produced are used in consumer packaging applications. The most commonly used resins are the following:

1. Polyethylene terephthalate (PET) is used extensively in rigid containers, particularly beverage bottles for carbonated beverages and medicine containers.

2. High-density polyethylene (HDPE) is used for rigid containers, such as milk and water jugs, household-product containers, and motor oil bottles.

3. Polyvinyl chloride (PVC) is a tough plastic often used in construction and plumbing. It is also used in some food, shampoo, oil, and household-product containers.

4. Low-density polyethylene (LDPE) is often used in films and bags.

5. Polypropylene (PP) is used in a variety of areas, from yogurt containers to battery cases to disposable diaper linings. It is frequently interchanged for polyethylene or polystyrene.

6. Polystyrene (PS) is used in foam cups, trays, and food containers. In its rigid form, it is used in plastic cutlery.

7. Other. These usually contain layers of different kinds of resins and are most commonly used for squeezable bottles (for example, for ketchup).

Currently, HDPE and PET are the two most commonly recycled resins because containers made of these resins are typically recovered by municipal recycling programs. Other plastics are less frequently accepted. Most of the recycled LDPE is from commercial establishments that receive large numbers of shipments wrapped in LDPE.

PET

HDPE

PVC

LDPE

PS

PP

Other

Recycling Rates for Plastic Resins Used in Packaging

Type of Plastic Resin	Percent Recycled (2007)
High-density polyethylene (HDPE)	25.6%
Polyethylene terephthalate (PET)	20.7%
Polypropylene (PP)	3.9%
Low-density polyethylene (LDPE) films	3–4%
Polyvinyl chloride (PVC)	Less than 1%
Polystyrene (PS)	Less than 1%

Source: Data from American Plastics Council.

Composting is the process of allowing the natural process of decomposition to transform organic materials—anything from manure and corncobs to grass and soiled paper—into compost, a humuslike material with many environmental benefits. In nature, leaves and branches that fall to the forest floor form a rich, moist layer that protects the roots of plants and provides a habitat for worms, insects, and a host of bacteria, fungi, and other microorganisms.

In composting operations, proper management of air and moisture provides ideal conditions for these organisms to transform large quantities of organic material into compost in a few weeks. A good small-scale example is a backyard compost pile. Green materials (grass, kitchen vegetable scraps, and flower clippings) mixed with brown materials (twigs, dry leaves, and soiled paper towels) at a ratio of 1:3 provide a balance of nitrogen and carbon that helps microbes efficiently decompose these materials.

Large-scale municipal composting uses the same principles of organic decomposition to process large volumes of organic materials. Composting facilities of various sizes and technological sophistication accept materials such as yard trimmings, food scraps, biosolids from sewage treatment plants, wood shavings, unrecyclable paper, and other organic materials. These materials undergo processing—shredding, turning, and mixing—and, depending on the materials, can be turned into compost in a period ranging from eight to 24 weeks. About 3800 composting facilities are in use in the United States. Roughly 62 percent of yard trimmings is converted into mulch or composted in the United States through municipal programs. (See figure 17.10.) Most municipal programs involve one of three composting methods: windrows, aerated piles, or enclosed vessels.

- *Windrow* systems involve placing the compostable materials into long piles or rows called windrows. Tractors with front-end loaders or other kinds of specialized machinery are used to turn the piles periodically. Turning mixes the different kinds of materials and aerates the mixture.

- *Aerated piles* are large piles of material that have air pumped through them (aeration) so that no mechanical turning or agitation is necessary. They also typically are covered with a layer of mature compost or other material to insulate the pile to keep it at an optimal temperature.

- *Enclosed vessels* can also be used to compost materials very rapidly (within days). However, these systems are much more technologically complex. In such systems, compostable material is fed into a drum, silo, or other structure where the environmental conditions are closely controlled, and the material is mechanically aerated and mixed.

In addition to keeping wastes from entering a landfill, composting has other significant benefits. The addition of compost to soil will improve it by making clay soils more porous or increasing the water-holding capacity of sandy soils. Nitrogen, potassium, iron, phosphorus, sulfur, and calcium are all common in compost and are beneficial to plant growth.

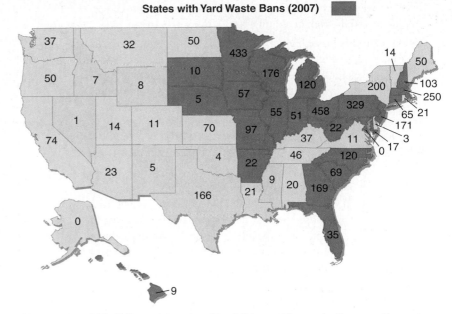

States with Yard Waste Bans (2007)

FIGURE 17.10 **Diverting Yard Waste Through Composting** Since yard waste is such an important segment of the solid waste stream, many states have passed laws that prohibit the deposition of yard waste in landfills. The states shown in green have laws that ban yard waste from going into a landfill. The states shaded blue have yard waste composting programs. This will extend the useful life of the landfill. To accommodate their citizens, many communities still collect yard waste but have instituted composting programs to deal with that waste. Even states that do not ban yard waste from landfills can extend the life of landfills by diverting yard waste to composting programs. The numbers on the map indicate the number of composting programs in each state.
Source: Data from the U.S. Environmental Protection Agency.

SOURCE REDUCTION

Throughout the life cycle of a product—from extraction of raw materials, to transport to processing and manufacturing facilities, to manufacture and use—waste is generated. The simplest way to reduce waste is to prevent it from ever becoming waste in the first place. **Source reduction** is the practice of designing, manufacturing, purchasing, using, and reusing materials so that the amount of waste or its toxicity is reduced.

Design changes to soft drink bottles and milk jugs are good examples of source reduction. Since 1977, the weight of a 2-liter plastic soft drink bottle has been reduced from 68 grams (2.4 ounces) to 51 grams (1.8 ounces). That translates to 114 million kilograms (250 million pounds) of plastic per year that has been kept out of the waste stream. The weight of a plastic milk jug has been reduced 30 percent.

Manufacturing processes have been changed in many industries to reduce the amount of waste produced. One of the simplest ways to reduce waste is to pay careful attention to leaks, spills, and accidents during the manufacturing process. All of these incidents generate waste, and their prevention reduces the amount of raw material needed and the amount of waste generated.

Purchasing decisions can significantly reduce the amount of waste produced. In many cases, consumers and businesses can choose to purchase things that have reduced packaging waste. You can choose to purchase products in larger sizes so that the amount of waste produced is reduced. In addition, careful planning of the quantities purchased can prevent unused surplus materials from becoming part of the waste stream.

Using materials in such a way that waste is not generated is an important means of curbing waste. Using less-hazardous alternatives for certain items (e.g., cleaning products and pesticides), sharing products that contain hazardous chemicals instead of throwing out leftovers, following label directions carefully, and using the smallest amount necessary are ways to reduce waste or its toxicity.

Reusing items is a way to reduce waste at the source because it delays or prevents the entry of reused items into the waste collection and disposal system. For example, many industries participate in waste exchanges that allow a waste product from one industry to be used as a raw material in another industry. In such cases, both industries benefit because the waste producer does not need to pay to dispose of the waste and the industry using the waste has an inexpensive source of a raw material.

Most businesses and manufacturers have a strong economic incentive to make sure that they get the most from all the materials used in their operations. As regulations on the production of waste become more stringent, industries will need to become even more efficient. Most businesses also see the value in reducing waste disposal costs. Any activities that reduce the amount of waste produced reduce the cost of waste disposal, the amount of raw materials needed, and the amount of pollution generated. This economic incentive also works at the consumer level. In the United States, over 4000 communities have instituted "pay-as-you-throw" programs in which citizens pay for each can or bag of trash they set out for disposal rather than through the tax base or a flat fee. When these households reduce waste at the source, they dispose of less trash and pay lower trash bills.

RECYCLING

Recycling is one of the best environmental success stories of the late twentieth century. (See figure 17.11.) In the United States, recycling, including composting, diverted about 30 percent of the solid waste stream from landfills and incinerators in 2007, up from about 16 percent in 1990. Several kinds of programs have contributed to the increase in the recycling rate. Some benefits of recycling are resource conservation, pollutant reduction, energy savings, job creation, and reduced need for landfills and incinerators. However, incentives are needed to encourage people to participate in recycling programs. Several kinds of programs have been successful.

Container laws provide an economic incentive to recycle. (See the case study on Beverage Container Deposit-Refund Programs.) In October 1972, Oregon became the first state to enact a "bottle bill." The law required a deposit of two to five cents on all beverage containers that could be reused. It banned the sale of one-time-use beverage bottles and cans. One of the primary goals of the law was to reduce the amount of litter, and it worked. Within two years of when it went into effect, beverage-container litter decreased by about 49 percent.

Many argue that a national bottle bill is long overdue. A national bottle bill would reduce litter, save energy and money, and

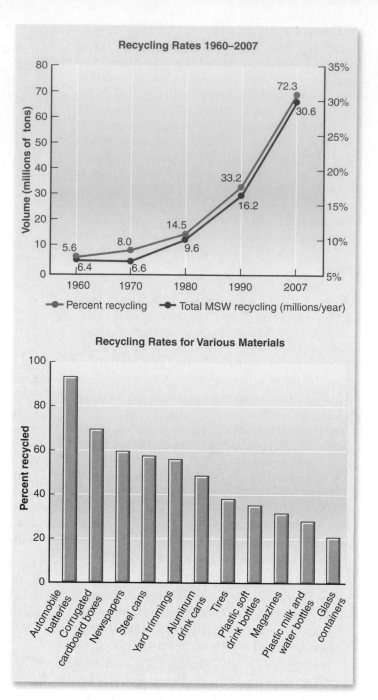

FIGURE 17.11 **Recycling Percentage for Selected Materials (2005) and Recycling Rates from 1960–2007** Recycling rates for materials that have high value such as automobile batteries are extremely high. Other materials are more difficult to market. But recycling rates today are much higher than in the past as technology and markets have found uses for materials that once were considered valueless.

Source: Data from the U.S. Environmental Protection Agency, *Characterization of Municipal Solid Waste in the United States*, 2007.

create jobs. It would also help to conserve natural resources. But the lobbying efforts of the soft drink and brewing industries are very strong, and Congress currently has failed to pass a national container law.

Mandatory recycling laws provide a statutory incentive to recycle. Many states and cities have passed mandatory recycling laws.

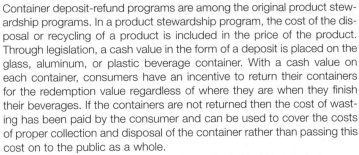

Container deposit-refund programs are among the original product stewardship programs. In a product stewardship program, the cost of the disposal or recycling of a product is included in the price of the product. Through legislation, a cash value in the form of a deposit is placed on the glass, aluminum, or plastic beverage container. With a cash value on each container, consumers have an incentive to return their containers for the redemption value regardless of where they are when they finish their beverages. If the containers are not returned then the cost of wasting has been paid by the consumer and can be used to cover the costs of proper collection and disposal of the container rather than passing this cost on to the public as a whole.

Deposit-refund laws are the most effective public or private recycling policies adopted over the past 30 years. The 11 states with bottle bills recycle more bottles and cans than the other 39 states combined and account for over 90 percent of the container recycling that occurs. Glass and other materials collected through deposit systems, unlike those collected through curbside recycling programs, are of a higher quality and are more marketable. The amount of the deposit also has an impact on the recovery rate. Michigan, the only state with a 10 cent deposit, also has the highest recovery rate: over 90 percent.

BEVERAGE CONTAINER WASTE IS A PROBLEM

In 2008, consumers in the United States failed to recycle an estimated 155 billion aluminum, glass, and plastic beverage containers. This is 33 percent more containers than were disposed of a decade ago. The problems associated with the disposal of these beverage containers are many:

- Glass, aluminum, and plastic containers disposed of in landfills and incinerators represent a loss of valuable resources and reduce employment opportunities in the domestic recycling industry.
- Disposal of these containers leads to increased greenhouse gas emissions and other forms of pollution when replacement containers are manufactured
- Manufacturing aluminum cans from used cans requires only 4 percent of the energy compared to making new cans from bauxite ore.
- Beverage containers account for 40 percent of the total volume of litter on our roads and highways.

WHAT ARE THE CURRENT RECOVERY RATES?

Container recovery is down in the United States for aluminum and plastic containers even though the number of households with access to curbside recycling has increased 600 percent since the early 1990s.

- Aluminum recovery is down from a high of 65 percent in 1992 to 48 percent in 2008.
- Plastic bottle recovery is down from a high of 40 percent in 1995 to less than 20 percent in 2008.

WHAT IS CAUSING THIS DECLINE?

- More beverages are being consumed away from home, where consumers are less likely and less able to recycle.
- New types of beverages, including bottled waters and designer drinks, have been introduced into the market and are not accepted in some curbside or bottle bill programs. Of the 11 states with container deposits, only three states cover these new types of drinks and containers.
- In 2008, Chicago became the first major U.S. city to put a 5-cent tax on bottled water in an effort to encourage recycling and discourage consumption.

COMPARING OREGON AND WASHINGTON STATE'S CONTAINER DISPOSAL TONNAGES

In 1971, Oregon was the first state in the United States to pass a bottle bill. Oregon's bill requires that all beer, carbonated soft drink, and malt-based glass and aluminum beverage containers be returnable and have a minimum refund value. Oregon's 5-cent deposit begins with the distributor and is refunded to the customer when the empty container is returned. A current effort is underway in Oregon to raise the deposit to a dime to provide a better incentive to return the containers and to add the new drinks in plastic bottles to the deposit system.

A comparison was made between Washington and Oregon to see if there was a difference in the quantities of containers appearing in the garbage. The following data clearly show the impact of container deposits. As a percentage of the waste stream, Washington disposed of four times more aluminum cans and 2.5 times more glass bottles than Oregon. The percentages of plastic bottles are almost the same. Since only plastic soda bottles are a part of the Oregon system, it makes sense that they are similar to Washington numbers. The data show that Oregon residents throw away far fewer glass bottles and aluminum cans than Washingtonians.

Do you support beverage container deposit-refund programs? Should all beverage containers, including bottled water, be included in such programs? Who would be opposed to such laws? Why?

(See figure 17.12.) Some of these laws simply require that residents separate their recyclables from other trash. Other laws are aimed at particular products such as beverage containers and require that they be recycled. Some are aimed at businesses and require them to recycle certain kinds of materials such as cardboard or batteries. Finally, some laws forbid the disposal of certain kinds of materials in landfills. Therefore, the materials must be recycled or dealt with in some other way. For example, banning yard waste from landfills has resulted in extensive composting programs.

Those states and cities with mandatory recycling laws understandably have high recycling rates. For example, California mandated 50 percent diversion of waste through recycling and other

ISSUES & ANALYSIS

Paper or Plastic or Plastax?

Before 2002 Ireland's 3.9 million people were using 1.2 billion plastic bags per year. These bags were generally nonrecyclable, took 20 to 1000 years to break down in the environment, were littering the countryside and clogging storm drains, and were adding to the burden on the country's landfill sites.

Worldwide some 100,000 birds, whales, seals and turtles are killed as a result of suffocating on plastic bags. Ireland's population is mostly rural, and its waste disposal system is poorly funded, making it an ideal place for this kind of initiative.

The idea of the "plastax" or tax on plastic bags was first announced in 1999 and in 2002 the environment minister launched the program, one of the first of its kind in the world. For every bag used at the checkout counter of the supermarket a 15 euro cents (about 25 U.S. cents) surcharge was added. The revenue raised from this tax would be put toward a "green fund" for environmental projects such as recycling refrigerators and other large appliances.

In 2008, China banned free plastic bags from shops and supermarkets. The bags are also banned from all public transportation, including buses, trains, and planes and from airports and scenic locations. The Chinese were using 3 billion plastic bags every day, which accounted for 3 to 5 percent of the total weight of landfills. It is estimated that it took 37 million barrels of crude oil a year to make all the bags needed for China.

In the United States, which has less than one-quarter of China's 1.3 billion people, almost 100 billion plastic bags are thrown out each year. Over 12 million barrels of oil are needed to manufacture the bags, of which only a fraction make it to the recycling bin. It is estimated that if every one of New York City's 8 million people used one less grocery bag per year, it would reduce waste by about 220,000 pounds. In 2008, San Francisco became the first major U.S. city to ban petroleum-based plastic bags in large grocery stores and drugstore chains. Plastic bags had previously been banned in at least 30 remote Alaskan villages.

Other countries have enacted various controls or bans on plastic bags, including the following:

Bangladesh: Banned the manufacturing and distribution of plastic bags in 2003 after it was found that they were blocking drainage systems and had been a major problem during the 1988 and 1998 floods that submerged two-thirds of the country.

Denmark: In 2004 Denmark introduced the Greentax. This tax was placed on plastic bags and paper bags and is included in the wholesale price of the bags and is not apparent to the consumer.

Hong Kong: Prohibits retailers over a certain size to provide plastic bags to customers free of charge.

South Africa: South Africa passed a tax on plastic bags. Plastic bags have been dubbed the "national flower" because so many can be seen flapping from fences and caught in bushes.

Australia: IKEA—the retailer introduced its own 10 cent plastic bag levy. Since its introduction, IKEA have reduced their plastic bag consumption from 8000 per week to 250 per week—a 97 percent reduction. Aldi supermarkets—this supermarket chain charges for plastic bags and provides four options for customers to carry their goods. These are: 15 cent plastic bag; 69 cent cotton bag; $1.49 cooler bag; reused boxes (free); or no bag or own bag. The most common option chosen is the reused boxes, or for small purchases no bags.

India: In the Delhi Capital Territory, legislation makes the use of non-biodegradable plastic bags a punishable offense.

Mauritius: This nation has banned the import or local manufacture of non-degradable plastic bags, and has specified that only oxo-biodegradable can be considered degradable.

Malta: Malta charges a lower tax on bags made from degradable plastic.

Barbados: Barbados charges 60 percent import surtax on non-degradable plastic bags but only 15 percent on oxo-biodegradable plastic bags.

Japan: Almost any store you visit in Japan, from convenient stores to street vendors, will also net you a free plastic bag for your purchase. Although there are some supermarkets (like Kyoto Co-op) which charge for plastic bags, this is by no means the norm. Many supermarkets (like Izumiya) will give you extra points on your point-card if you bring your own bag.

- Do you think your community would support such a tax?
- Would you favor a tax on plastic bags?
- How would a ban on plastic bags alter your lifestyle?

SUMMARY

Municipal solid waste is managed by landfills, incineration, composting, waste reduction, and recycling. Landfills are the primary means of disposal; however, a contemporary landfill is significantly more complex and expensive than the simple holes in the ground of the past. The availability of suitable landfill land is also a problem in large metropolitan areas.

About 15 percent of the municipal solid waste in the United States is incinerated. While incineration does reduce the volume of municipal solid waste, the problems of ash disposal and air quality continue to be major concerns. There are several forms of composting that can keep organic wastes from entering a landfill.

The most fundamental way to reduce waste is to prevent it from ever becoming waste in the first place. Using less material in packaging, reusing items, and composting yard waste are all examples of source reduction. On an individual level, we can all attempt to reduce the amount of waste we generate.

About 30 percent of the waste generated in North America is handled through recycling. Recycling initiatives have grown rapidly in North America during the past several years. As a result, the markets for some recycled materials have become very volatile. Recycling of municipal solid waste will be successful only if markets exist for the recycled materials. Another problem in recycling is the current inability to mix various plastics.

Future management of municipal solid waste will be an integrated approach involving landfills, incineration, composting, source reduction, and recycling. The degree to which any option will be used will depend on economics, changes in technology, and citizen awareness and involvement.

THINKING GREEN

1. Buy things that last, keep them as long as possible, and have them repaired, if possible.
2. Buy things that are reusable or recyclable, and be sure to reuse and recycle them.
3. Use plastic or metal lunch boxes and metal or plastic garbage containers without throwaway plastic liners.
4. Use rechargeable batteries.
5. Skip the bag when you buy only a quart of milk, a loaf of bread, or anything you can carry with your hands.
6. Recycle all newspaper, glass, and aluminum, and any other items accepted for recycling in your community.
7. Reduce the amount of junk mail you get. This can be accomplished by writing to Mail Preference Service, Attn: Dept: 13885751, Direct Marketing Association, P.O. Box 282, Carmel, NY 10512. Ask that your name not be sold to large mailing-list companies. You may also register on-line by going to the website. Type Mail Preference Service into your browser. Of the junk mail you do receive, recycle as much of the paper as possible.
8. Push for mandatory trash separation and recycling programs in your community and schools.
9. Compost your yard and food wastes, and pressure local officials to set up a community composting program.

WHAT'S YOUR TAKE?

The growth of recycling programs has been pushed in part by municipalities' need to reduce waste and by the satisfaction people feel in taking the responsibility to recycle. These two forces have driven recycling's rise in spite of the fact that it has often not been financially profitable. In fact, many of the increasingly popular municipal recycling programs are run at an economic loss. Formulate an argument that justifies the continuation of a community recycling program that currently operates at a loss.

REVIEW QUESTIONS

1. How is lifestyle related to the quantity of municipal solid waste generated?
2. What conditions favor incineration over landfills?
3. Describe some of the problems associated with modern landfills.
4. What are four concerns associated with incineration?
5. Describe examples of source reduction.
6. Describe the importance of recycling household solid wastes.
7. Name several strategies that would help to encourage the growth of recycling.
8. Describe the various types of composting and the role of composting in solid waste management.

CRITICAL THINKING QUESTIONS

1. How can you help solve the solid waste problem?

2. Given that you have only so much time, should you spend your time acting locally, as a recycling coordinator, for example, or advocating for larger political and economic changes at the national level, changes that would solve the waste problems? Why? Or should you do nothing? Why?

3. How does your school or city deal with solid waste? Can solid waste production be limited at your institution or city? How?

What barriers exist that might make it difficult to limit solid waste production?

4. It is possible to have a high standard of living, as in North America and Western Europe, and not produce large amounts of solid waste. How?

5. Incineration of solid waste is controversial. Do you support solid waste incineration in general? Would you support an incineration facility in your neighborhood?

CHAPTER 18

ENVIRONMENTAL REGULATIONS: HAZARDOUS SUBSTANCES AND WASTES

Many kinds of hazardous materials are a common part of our lives. As with most risks, the danger can be minimized if the use and disposal of hazardous materials are done properly. Proper labeling and storing of hazardous materials are essential to their management.

CHAPTER OUTLINE

Hazardous and Toxic Materials in Our Environment
Hazardous and Toxic Substances—Some Definitions
Defining Hazardous Waste
Determining Regulations
  Identification of Hazardous and Toxic Materials
  Setting Exposure Limits
  Acute and Chronic Toxicity
  Synergism
  Persistent and Nonpersistent Pollutants
Environmental Problems Caused by Hazardous Wastes
Health Risks Associated with Hazardous Wastes
Hazardous-Waste Dumps—A Legacy of Abuse
  Toxic Chemical Releases
Hazardous-Waste Management Choices
  Reducing the Amount of Waste at the Source
  Recycling Wastes
  Treating Wastes
  Disposal Methods
International Trade in Hazardous Wastes
Hazardous-Waste Management Program Evolution

ISSUES & ANALYSIS
Household Hazardous Waste 424

CASE STUDY
Determining Toxicity 414

CAMPUS SUSTAINABILITY INITIATIVE
Chemical Exchange at the University of British Columbia 421

GOING GREEN
Guide to Electronics Recycling 419

WATER CONNECTIONS
Dioxins in the Tittabawassee River Floodplain 415

OBJECTIVES

After reading this chapter, you should be able to:

- Distinguish between hazardous substances and hazardous wastes.
- Distinguish between hazardous and toxic substances.
- Describe the four characteristics used to identify hazardous substances.
- Describe the kinds of environmental problems caused by hazardous and toxic substances.
- Understand the difference between persistent and nonpersistent pollutants.
- Describe the difference between chronic and acute exposures to hazardous wastes.
- Describe why hazardous-waste dump sites developed
- Describe how hazardous wastjes are managed, and list five technologies used in their disposal.
- Describe the importance of source reduction with regard to hazardous wastes.
- Describe when and why an Environmental Site Assessment would be conducted.
- Describe the benefits from enacting the ISO 14000 Environmental Management System.
- Understand the difficulties associated with determining cleanup criteria for hazardous substances/wastes.

Global Perspectives on "Hazardous Wastes and Toxic Materials in China" and "Lead and Mercury Poisoning," and Case Studies on "Computers—A Hazardous Waste" and "Micro-Scale Chemistry" can be found on the book's website at www.mhhe.com/enger12e along with other interesting readings.

HAZARDOUS AND TOXIC MATERIALS IN OUR ENVIRONMENT

Our modern technological society makes use of a large number of substances that are hazardous or toxic. The benefits gained from using these materials must be weighed against the risks associated with their use. When these materials are released into the environment in an inappropriate way through accident or neglect, significant environmental and human health consequences result. The following items indicate the magnitude of the problem.

- Dioxins are found in Tittabawassee River floodplain sediments in Michigan from the historical releases of chemical production facilities.

- Pesticides thought to degrade in soils are found in rural drinking-water wells.

- There are elevated levels of mercury downwind of coal-fired power plants.

- Chemicals leaching from abandoned waste sites contaminate city water supplies.

- Pesticides spilled into the Rhine River from a warehouse near Basel, Switzerland, in 1988 destroyed a half million fish, disrupted water supplies, and caused considerable ecological damage.

- The collapse of an oil storage tank in Pennsylvania in 1988 spilled over 3 million liters (750,000 gallons) of oil into the Monongahela River and threatened the water supply of millions of residents.

- The release of polychlorinated biphenyls (PCBs) from electrical parts manufacturing plants before the 1970s contaminated hundreds of hectares of marine sediments in New Bedford Harbor, Massachusetts, resulting in marine life being reduced and the area being closed to fishing.

Toxic and hazardous products and by-products are becoming a major issue of our time. At sites around the world, accidental or purposeful releases of hazardous and toxic chemicals are contaminating the land, air, and water. The potential health effects of these chemicals range from minor, short-term discomforts, such as headaches and nausea, to serious health problems, such as cancers and birth defects (that may not manifest themselves for years), to major accidents that cause immediate injury or death. Today, such names as Love Canal, New York, and Times Beach, Missouri, in the United States; Lekkerkerk in the Netherlands; Vác, Hungary; and Minamata Bay, Japan, are synonymous with the problems associated with the release of hazardous and toxic wastes into the environment.

Increasingly, governments and international agencies are attempting to control the growing problem of hazardous substances

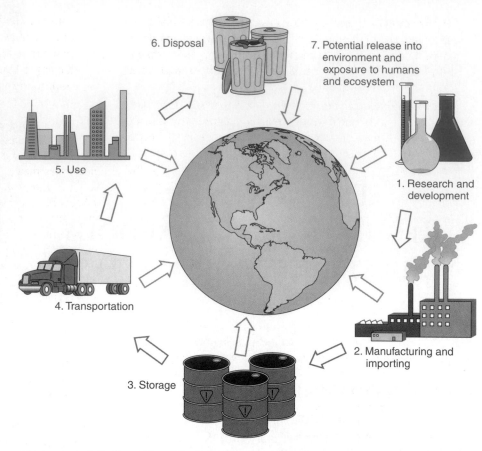

FIGURE 18.1 **The Life Cycle of Toxic Substances** Controlling the problems of hazardous substances is complicated because of the many steps involved in a substance's life cycle.

in our environment. Controlling the release of these substances is difficult since there are so many places in their cycles of use at which they may be released. (See figure 18.1.)

HAZARDOUS AND TOXIC SUBSTANCES—SOME DEFINITIONS

To begin, it is important to clarify various uses of the words *hazardous* and *toxic,* as well as to distinguish between things that are wastes and those that are not. **Hazardous substances** or **hazardous materials** are those that can cause harm to humans or the environment. Different government agencies have slightly different definitions for what constitutes a hazardous substance. The U.S. Environmental Protection Agency defines hazardous substances as having one or more of the following characteristics:

1. **Ignitability**—Describes materials that pose a fire hazard during routine management. Fires not only present immediate dangers of heat and smoke but also can spread harmful particles over wide areas. Common examples are gasoline, paint thinner, and alcohol.

2. **Corrosiveness**—Describes materials that require special containers because of their ability to corrode standard materials or that require segregation from other materials because of their ability to dissolve toxic contaminants. Common examples are strong acids and bases.

3. **Reactivity** (or explosiveness)—Describes materials that, during routine management, tend to react spontaneously, to react vigorously with air or water, to be unstable to shock or heat, to generate toxic gases, or to explode. Common examples are gunpowder, which will burn or explode; the metal sodium, which reacts violently with water; and nitroglycerine, which explodes under a variety of conditions.

4. **Toxicity**—Describes materials that, when improperly managed, may release toxicants (poisons) in sufficient quantities to pose a substantial hazard to human health or the environment. Almost everything that is hazardous is toxic in high enough quantities. For example, tiny amounts of carbon dioxide in the air are not toxic, but high levels are.

Some hazardous materials fall into several of these categories. Gasoline, for example, is ignitable, can explode, and is toxic. It is even corrosive to certain kinds of materials. Other hazardous materials meet only one of the criteria. PCBs are toxic but will not burn, explode, or corrode other materials. While the terms *toxic* and *hazardous* are often used interchangeably, there is a difference. **Toxic** commonly refers to a narrow group of substances that are poisonous and cause death or serious injury to humans and other organisms by interfering with normal body physiology. **Hazardous,** the broader term, refers to all dangerous materials, including toxic ones, that present an immediate or long-term human health risk or environmental risk.

DEFINING HAZARDOUS WASTE

It is important to distinguish between hazardous substances and hazardous wastes. Although the health and safety considerations regarding hazardous substances and hazardous wastes are similar, the legal and regulatory implications are quite different. Hazardous substances are materials that are used in business and industry for the production of goods and services. Typically, hazardous substances are consumed or modified in industrial processes. **Hazardous wastes** are by-products of industrial, business, or household activities for which there is no immediate use. These wastes must be disposed of in an appropriate manner, and there are stringent regulations pertaining to their production, storage, and disposal.

Although the definition of hazardous waste varies from one country to another, one of the most widely used definitions is contained in the U.S. **Resource Conservation and Recovery Act** of 1976 (RCRA). This act created the "cradle-to-grave" concept of hazardous waste management by regulating generators, transporters, and Treatment Storage and Disposal Facilities (TSDF) as well as underground storage tanks (USTs) and petroleum products. It also attempted to define toxic and/or hazardous waste by using the terms *listed* and *characteristic* waste. RCRA considers wastes toxic and/or hazardous if they:

cause or significantly contribute to an increase in mortality or an increase in serious irreversible, or incapacitating reversible, illness; or pose a substantial present or potential hazard to human health or the environment when improperly treated, stored, transported, disposed of, or otherwise managed.

This very complex definition gives one an appreciation for the complexity of hazardous-waste regulation.

Because of the difficulty in developing simple definitions of hazardous materials, listing is the most common method for defining hazardous waste in many countries. The U.S. Environmental Protection Agency has compiled such a list of hazardous wastes (the so-called listed wastes). The EPA also requires that a waste be tested to determine if it possesses any one of the four characteristics discussed earlier: ignitability, corrosiveness, reactivity, or toxicity. If it does, it is "characteristic waste," and is subject to regulation under RCRA. It should be noted that although EPA is discussed here, other U.S. agencies such as the Occupational Safety and Health Administration (OSHA) and the Department of Transportation (DOT) have compiled their own lists, and there is a comprehensive "list of lists" that documents many of these. If a waste appears on any other agency list, it is to be considered hazardous by the EPA.

There are many types of hazardous waste, ranging from waste contaminated with dioxins and heavy metals (such as mercury, cadmium, and lead) to organic wastes. These wastes can also take many forms, such as barrels of liquid waste or sludge, old computer parts, used batteries, and incinerator ash. In industrialized countries, industry and mining are the main sources of hazardous wastes, though small-scale industry, hospitals, military establishments, transport services, and small workshops also contribute to the generation of significant quantities of such wastes in both the industrialized and developing worlds.

Improper handling and disposal of hazardous wastes can affect human health and the environment by contaminating groundwater, soil, waterways, and the atmosphere. The environmental and health effects can be immediate (acute), as when exposure to toxins at a particular site causes sudden illness, or long-term (chronic), as when contaminated waste leaches into groundwater or soil and then works its way into the food chain. Although chronic exposures to low levels of toxic materials do not cause immediate health effects, they may cause serious health problems later in life. The damage caused by the release of hazardous wastes into the environment also takes an economic toll, since cleaning up contaminated sites can be costly for local authorities. The handling of hazardous waste requires special training to prevent worker exposure to the materials.

Exact figures regarding the amounts of hazardous waste generated worldwide are difficult to obtain. However, the United Nations Environment Programme estimates between 300 million and 500 million metric tons are generated annually. A large proportion comes from developed countries. Data from the EPA indicate that the United States generates about 36 million metric tons of hazardous waste each year. The European Union generates about 40 million metric tons. The rapidly industrializing countries of Asia are generating increasing amounts of hazardous waste, but it is difficult to obtain estimates of the amounts.

DETERMINING REGULATIONS

The United States has attempted to deal with hazardous substances and wastes by using "command-and-control" methods of governmental regulations. This attempt began relatively recently, starting with the development of EPA and OSHA in 1970. There were attempts to regulate certain substances prior to that date (see table 18.1), but the main focus of environmental regulation and enforcement began much later. This may have been mainly due to our lack of understanding.

Many states, as well as some countries, have tried to mirror these federal regulations by codifying their own statutes specific to their needs. State regulations can be equal to or more stringent than federal statutes. Creating these regulations is difficult due to the vast number of variables involved. Of course, the human factor that complicates the issue is the concept of NIMBY (Not In My Backyard). No one, whether you live in Peoria, Illinois, or Seoul South Korea, wants to have hazardous waste/substances located anywhere near their home.

Whether a hazardous substance is a raw material, an ingredient in a product, or a waste, there are problems in determining how to regulate it. Mining, electric generation, and chemical and petrochemical industries are the major producers of hazardous wastes. These industries produce many useful materials that are converted into the products we use every day, such as metals, energy sources, and many synthetic materials. (See figure 18.2.)

While determining how and what to regulate, governments must assess who will pay for cleanup if hazardous substances or wastes enter the environment. This liability issue is often the most contentious of all. Remember, the L and C in CERCLA (Superfund) stand for Liability and Compensation. The PRP (Potentially Responsible Party) will have to pay for dealing with the remediation of the site once hazardous substances/wastes are found in the environment. Regulatory agencies try to identify the PRP and force that party or parties to comply with remediation requirements. Regulation and enforcement is one thing, compliance is another. Most governments have a difficult time with all of these.

Some voluntary industry standards have been developed and have even been incorporated into federal acts (SBLRBRA, 2002). One example is the ASTM International (formerly American Society for Testing and Materials) Phase I Environmental Site Assessment (ESA) standard E-1527. Before financing a loan to acquire commercial real estate, or before receiving grant monies under the federal SBLRBRA program, a business must first conduct an ESA to determine potential environmental liabilities associated with the real estate. This will help to determine who will pay for potential cleanup activities or how much liability a PRP will have. This is the reality of property acquisition in the United States as well as in many other countries.

Other international voluntary consensus standards such as ISO 14000 (Environmental Management Systems) attempt to use voluntary actions rather than regulatory command and control to achieve compliance or to go beyond regulatory compliance. Some countries have adopted these standards in an attempt to avoid extra regulatory cost and to achieve better conformance to cleaner environmental goals.

IDENTIFICATION OF HAZARDOUS AND TOXIC MATERIALS

In attempting to regulate the use of toxic and hazardous substances and the generation of toxic and hazardous wastes, most countries simply draw up a list of specific substances that have been scientifically linked to adverse human health or environmental effects. However, since many potentially harmful chemical compounds have yet to be tested adequately, most lists include only the known offenders. Historically, we have often identified toxic materials only after their effects have shown up in humans or other animals. Asbestos was identified as a cause of lung cancer in humans who were exposed on the job, and DDT was identified as toxic to birds when robins began to die and eagles and other fish-eating birds failed to reproduce. Once these substances were identified as toxic, their use was regulated. Countries contemplating the regulation of hazardous and toxic materials and wastes must consider not only how toxic each one is but also how flammable, corrosive, or explosive it is, and whether it will produce mutations or cause cancer. Also, governments and their regulatory agencies must attempt to determine how to fairly enforce these measures to successfully control exposures to humans and the environment. Sometimes the reason for the regulation is lost in the effort to enforce it. The purpose of all environmental regulations is to control and/or stop pollution and environmental degradation. If an industry is put out of business due to environmental fines and cleanup costs, it will not be able to contribute to the control and/or remediation of the toxic material, the area will revert to an "orphaned" site, and other sources of money will have to be sought for its cleanup.

TABLE 18.1	Select Federal Environmental Statutes, by Year of Enactment	
Year	**Statute**	
1947	FIFRA:	Federal Insecticide, Fungicide, Rodenticide Act
1970	CAA:	Clean Air Act
1970	OSHA:	Occupational Safety and Health Act
1972	CWA:	Clean Water Act
1974	SDWA:	Safe Drinking Water Act
1975	HMTA:	Hazardous Materials Transportation Act
1976	TSCA:	Toxic Substance Control Act
1976	RCRA:	Resource Conservation and Recovery Act
1980	CERCLA:	Comprehensive Environmental Response Compensation and Liability Act (Superfund)
1986	SARA:	Superfund Amendments and Reauthorization Act
2002	SBLRBRA:	Small Business Liability Relief, Brownfield Revitalization Act

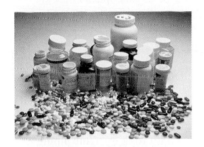

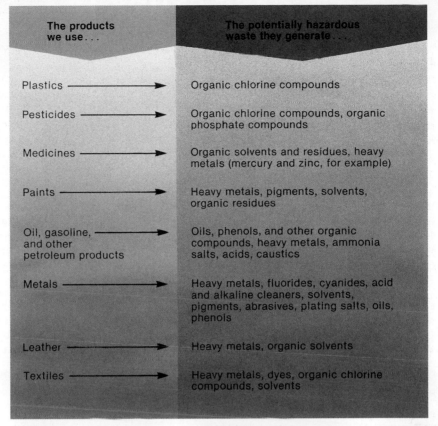

The products we use...	The potentially hazardous waste they generate...
Plastics	Organic chlorine compounds
Pesticides	Organic chlorine compounds, organic phosphate compounds
Medicines	Organic solvents and residues, heavy metals (mercury and zinc, for example)
Paints	Heavy metals, pigments, solvents, organic residues
Oil, gasoline, and other petroleum products	Oils, phenols, and other organic compounds, heavy metals, ammonia salts, acids, caustics
Metals	Heavy metals, fluorides, cyanides, acid and alkaline cleaners, solvents, pigments, abrasives, plating salts, oils, phenols
Leather	Heavy metals, organic solvents
Textiles	Heavy metals, dyes, organic chlorine compounds, solvents

FIGURE 18.2   Common Materials Can Produce Hazardous Wastes Many commonly used materials generate hazardous wastes during their manufacture. Some consumer products such as pesticides, paints, and solvents are also hazardous and should be disposed of in an approved manner.

Oftentimes these areas are cleaned up through the use of state and federal funds. In other words, we all pay for it.

SETTING EXPOSURE LIMITS

Even after a material is identified as hazardous or toxic, there are problems in determining appropriate exposure limits. Nearly all substances are toxic in sufficiently high doses. The question is, When does a chemical cross over from safe to toxic? There is no easy way to establish acceptable levels. Several government agencies set limits for different purposes. The Occupational Safety and Health Administration, the Food and Drug Administration, the U.S. Public Health Service, the Environmental Protection Agency, and others publish guidelines or set exposure limits for hazardous substances in the air, water, and soil. Organizations such as National Institute of Occupational Safety and Health (NIOSH) and American Conference of Governmental Industrial Hygienists (ACGIH), among others, also test and set exposure limits in a variety of units (PEL—Permissible Exposure Limits; STELs—Short Term Exposure Limits; TWA—Time Weighted Average; CL—Ceiling Limit; to name a few). Furthermore, it is important to recognize that people can be exposed in three

We are all exposed to potentially harmful toxins. The question is, at what levels is such exposure harmful? One measure of toxicity of a substance is its **LD$_{50}$**, the dosage of a substance that will kill (lethal dose) 50 percent of a test population (LD$_{50}$ = lethal dose 50 percent). Toxicity is measured in units of poisonous substance per kilogram of body weight. For example, the deadly chemical that causes botulism, a form of food poisoning, has an LD$_{50}$ in adult human males of 0.0014 milligram per kilogram. This means that if each of 100 human adult males weighing 100 kilograms consumed a dose of only 0.14 milligram—about the equivalent of a few grains of table salt—approximately 50 of them would die.

Lethal doses are not the only danger from toxic substances. During the past decade, concern has been growing over minimum harmful dosages, or threshold dosages, of poisons, as well as their sublethal effects.

The length of exposure further complicates the determination of toxicity values. Acute exposure refers to a single exposure lasting from a few seconds to a few days. Chronic exposure refers to continuous or repeated exposure for several days, months, or even years. Acute exposure usually is the result of a sudden accident, such as the tragedy at Bhopal, India, mentioned earlier in this text. Acute exposures often make disaster headlines in the press, but chronic exposure to sublethal quantities of toxic materials presents a much greater hazard to public health. For example, millions of urban residents are continually exposed to low levels of a wide variety of pollutants. Many deaths attributed to heart failure or such diseases as emphysema may actually be brought on by a lifetime of exposure to sublethal amounts of pollutants in the air.

primary ways (routes of entry): by breathing the material, by consuming the material through the mouth, or by absorbing the material through the skin. Each of these routes may require different exposure classifications. Regardless, for any new compounds that are to be brought on the market, extensive toxicology studies must be done to establish their ability to do harm. Usually these involve tests on animals. (See Case Study: Determining Toxicity.) Typically, the regulatory agency will determine the level of exposure at which none of the test animals is affected (**threshold level**) and then set the human exposure level lower to allow for a safety margin. This safety margin is important because it is known that threshold levels vary significantly among species, as well as among members of the same species. Even when concentrations are set, they may vary considerably from country to country. For example, in the Netherlands, 50 milligrams of cyanide per kilogram of waste is considered hazardous; in neighboring Belgium, the toxicity standard is fixed at 250 milligrams per kilogram.

ACUTE AND CHRONIC TOXICITY

Regulatory agencies must look at both the effects of one massive dose of a substance (**acute toxicity**) and the effects of exposure to small doses over long periods (**chronic toxicity**). Acute toxicity is readily apparent because organisms respond to the toxin shortly after being exposed. Chronic toxicity is much more difficult to determine because the effects may not be seen for years. Furthermore, an acute exposure may make an organism ill but not kill it, while chronic exposure to a toxic material may cause death.

A good example of this effect is alcohol toxicity. Consuming extremely high amounts of alcohol can result in death (acute toxicity and death). Consuming moderate amounts may result in illness (acute toxicity and full recovery). Consuming moderate amounts over a number of years may result in liver damage and death (chronic toxicity and death).

Another example of chronic toxicity involves lead. Lead was used in paints, gasoline, and pottery glazes for many years, but researchers discovered that it has harmful effects. The chronic effects on the nervous system are most noticeable in children, particularly when children eat paint chips.

SYNERGISM

Another problem in regulating hazardous materials is assessing the effects of mixtures of chemicals. Most toxicological studies focus on a single compound, even though industry workers may be exposed to a variety of chemicals, and hazardous wastes often consist of mixtures of compounds. Although the materials may be relatively harmless as separate compounds, once mixed, they may become highly toxic and cause more serious problems than do individual pollutants. This is referred to as **synergism.** For example, all uranium miners are exposed to radioactive gases, but those who smoke tobacco and thus are exposed to the toxins in tobacco smoke have unusually high incidences of lung cancer. Apparently, the radioactive gases found in uranium mines interact synergistically with the carcinogens found in tobacco smoke.

An example of how difficult it can be to determine exposure limits, toxicity, regulatory standards, and threshold planning quantities occurs in the case of the group of chemicals commonly called dioxins. Dioxin is the name given to a group of chemicals that are formed as unwanted by-products of industrial manufacturing and burning activities. Major sources of dioxins include chemical and pesticide manufacturing, the burning of household trash, forest fires, and the burning of industrial and medical waste products. There are approximately 210 chemicals with similar structures and properties included among the dioxins. In the environment, they usually appear in a mixture. The highest levels are usually found in soil, sediment, and animal fat. Much lower levels are found in air and water. The problem is that most dioxins are not toxic, or are less toxic than other forms of the same chemicals. As discussed previously, toxicity is the ability to cause illness and adverse health effects. The most toxic chemical in the dioxin group is 2,3,7,8-tetrachlorodibenzo-para-dioxin (2,3,7,8-TCDD). As the most toxic, 2,3,7,8-TCDD is the standard to which other dioxins are compared. Since the harmful effects of dioxins are not fully known, and only certain types or forms of dioxin are considered dangerous or more toxic than others, scientists have agreed to report dioxin levels by combining dioxin forms and converting them to an "equivalent" of that form considered most harmful, the "total toxic equivalent" (TEQ) concentration.

The Tittabawassee River floodplain in central Michigan contains higher-than-normal levels of dioxins in soil and sediment samples. The State of Michigan Department of Environmental Quality (MDEQ), the U.S. EPA, the Centers for Disease Control (CDC), the World Health Organization (WHO), and the U.S. Agency for Toxic Substances and Disease Registry (ATSDR) have all provided guidance and toxicity profiles for dioxins. The MDEQ uses a "90 parts per trillion (ppt) Direct Contact Clean-up Criteria" standard, while other agencies have provided interim policy guidelines containing other concentrations as their standards. The key is that no one can agree on what is safe. Other countries have set their own levels and continue to conduct research into how much is too much. However, the levels found in some sample locations along the floodplain exceeded the CDC and U.S. ATSDR action levels of 1000 ppt for dioxin. Soil samples taken from areas outside the floodplain were at normal background levels.

People living in the floodplain area could be exposed to dioxins in the soil and sediments. In some animal studies, low-level chronic exposure to dioxins caused cancer, liver damage, and hormone changes. In some other studies, dioxin exposure also decreased the ability to fight infection and caused reproductive damage, miscarriages, and birth defects. Skeleton and kidney defects, lowered immune responses, and effects on the development of the brain and nervous system were among the birth defects seen. However, other studies did not yield similar results. Industry, federal and state, and academic institutions continue to conduct toxicity testing on animals (humans included) in an effort to determine the truth. In the meantime, state and federal agencies have issued health advisories, cautions, fish and game consumption advisories, and exposure pathway interim mitigation strategies for people living in and/or using the floodplain. In particular, consumption advisories focus on fish intake for pregnant women and other at-risk individuals.

Find out more: For further study of this ongoing issue, go to www.michigan.gov/deq and use the "Dioxin Information" link located there to see specific data, maps, and health advisories for the Tittabawassee River and other information on dioxin.

PERSISTENT AND NONPERSISTENT POLLUTANTS

The regulation of hazardous and toxic materials is also influenced by the degree of persistence of the pollutant. **Persistent pollutants** are those that remain in the environment for many years in an unchanged condition. Most of the persistent pollutants are human-made materials. An estimated 30,000 synthetic chemicals are used in the United States. They are mixed in an endless variety of combinations to produce all types of products used in every aspect of daily life. They are part of our food, transportation, clothing, building materials, home appliances, medicine, recreational equipment, and many other items. Our way of life depends heavily on synthetic materials.

An example of a persistent pollutant is DDT. It was used as an effective pesticide worldwide and is still used in some countries because it is so inexpensive and very effective in killing pests. However, once released into the environment, it accumulates in the food chain and causes death when its concentration is high enough. (See chapter 14 for a discussion of DDT as a pesticide.)

Another widely used group of synthetic compounds of environmental concern is polychlorinated biphenyls. PCBs are highly stable compounds that resist changes from heat, acids, bases, and oxidation. These characteristics make PCBs desirable for industrial use but also make them persistent pollutants when they are released into the environment. At one time, these materials were commonly used in transformers and electrical capacitors. Other uses included inks, plastics, tapes, paints, glues, waxes, and polishes. Although the manufacture of PCBs in the United States stopped in 1977, these persistent chemicals are still present in the soil and sediments and continue to do harm. PCBs are harmful to fish and other aquatic forms of life because they interfere with reproduction. In humans, PCBs produce liver ailments and skin lesions. In high concentration, they can damage the nervous system, and they are suspected carcinogens.

In addition to synthetic compounds, our society uses heavy metals for many purposes. Mercury, beryllium, arsenic, lead, and cadmium are examples of heavy metals that are toxic. These metals are used as alloys with other metals, in batteries, and have many other special applications. In addition, these metals may be released as by-products of the extraction and use of other metals. When released into the environment, they enter the food chain and become concentrated. In humans, these metals can produce kidney and liver disorders, weaken the bone structure, damage the central nervous system, cause blindness, and lead to death. Because these materials are persistent, they can accumulate in the environment even though only small amounts might be released each year. When industries use these materials in a concentrated form, it presents a hazard not found naturally.

A **nonpersistent pollutant** does not remain in the environment for very long. Most nonpersistent pollutants are biodegradable. Others decompose as a result of inorganic chemical reactions. Still others quickly disperse to concentrations that are too low to cause harm. A biodegradable material is chemically changed by living organisms and often serves as a source of food and energy for decomposer organisms, such as bacteria and fungi. Phenol and many other kinds of toxic organic materials can be destroyed by decomposer organisms.

Other toxic materials, such as many insecticides, are destroyed by sunlight or reaction with oxygen or water in the atmosphere. These include the "soft biocides." For example, organophosphates usually decompose within several weeks. As a result, organophosphates do not accumulate in food chains because they are pollutants for only a short period of time.

Other toxic and hazardous materials such as carbon monoxide, ammonia, or hydrocarbons can be dispersed harmlessly into the atmosphere (as long as their concentration is not too great), where they eventually react with oxygen.

Because persistent materials can continue to do harm for a long time (chronic toxicity), their regulation is particularly important. Nonpersistent materials need to be kept below threshold levels to protect the public from acute toxicity. They are not likely to present a danger of chronic toxicity, since they either disperse or decompose.

ENVIRONMENTAL PROBLEMS CAUSED BY HAZARDOUS WASTES

Hazardous wastes enter the environment in several ways. Many molecules that evaporate readily are vented directly to the atmosphere. Many kinds of solvents used in paints and other industrial processes fall into this category. Other materials escape from faulty piping and valves. These materials are often not even thought of as hazardous waste but are called fugitive emissions. Uncontrolled or improper incineration of hazardous wastes, whether on land or at sea, can contaminate the atmosphere and the surrounding environment.

Other wastes are in solid or liquid form and are more easily contained. Some of these wastes can be treated to reduce or eliminate their hazardous nature, and these treated wastes can be released to the environment. Many kinds of liquid wastes, such as sewage and acid discharges, are handled in this manner. Other wastes cannot be treated and must be stored.

In the past, many kinds of hazardous wastes were improperly disposed of on land, in containers, storage lagoons, or landfills. Groundwater contamination has resulted from leaking land disposal facilities. Once groundwater is polluted with hazardous wastes, the cost of reversing the damage is prohibitive. In fact, if an aquifer is contaminated with organic chemicals, restoring the water to its original state is seldom physically or economically feasible. Chapter 15 deals with many other aspects of groundwater contamination.

Even improper labeling and recordkeeping procedures can result in releases. If workers are unable to distinguish hazardous waste from other kinds of waste materials, hazardous wastes are not properly disposed of. Especially in the past, wastes were not known to be hazardous to the environment or to people and were therefore released, buried, burned, and essentially "dumped" wherever it was easiest. This usually meant the local river, wetland, lowland, gully/ravine, or other area considered "useless." Out-of-sight, out-of-mind were the words of the day. This was not done to purposely harm, but through ignorance. Our understanding of hazardous wastes and their potential dangers continues to undergo changes today. We struggle to understand what the wastes generated today can do, let alone what we have done in the past.

HEALTH RISKS ASSOCIATED WITH HAZARDOUS WASTES

Because most hazardous wastes are chemical wastes, controlling chemicals and their waste products is a major issue in most developed countries. Every year, roughly 1000 new chemicals join the nearly 70,000 in daily use. Many of these chemicals are toxic, but they pose little threat to human health unless they are used or disposed of improperly. For example, many insecticides are extremely toxic to humans. However, if they are stored, used, and disposed of properly, they do not constitute a human health hazard. Unfortunately, at the center of the hazardous waste problem is the fact that the products and by-products of industry are often handled and disposed of improperly. Table 18.2 is a list of 10 top toxic materials as identified by the federal Agency for Toxic Substances and Disease Registry.

Establishing the medical consequences of exposure to toxic chemicals is extremely complicated. The problem of linking a particular chemical to specific injuries or diseases is further compounded by the lack of toxicity data on most hazardous substances.

Although assessing environmental contamination from toxic substances and determining health effects is extremely difficult, what little is known is cause for concern. Most older hazardous-waste dumpsites, for example, contain dangerous and toxic chemicals along with heavy-metal residues and other hazardous substances.

HAZARDOUS-WASTE DUMPS— A LEGACY OF ABUSE

In the United States prior to the passage of the Resource Conservation and Recovery Act (RCRA) in 1976, hazardous waste was essentially unregulated. Similar conditions existed throughout most of the industrial nations of the world. Hazardous wastes were simply buried or dumped without any concern for potential environmental or health risks. Such uncontrolled sites included open dumps, landfills, bulk storage containers, and

TABLE 18.2 Top 10 Hazardous Substances, 2008

Substance	Source	Toxic Effects
Arsenic	From elevated levels in soil or water	Multiple organ systems affected. Heart and blood vessel abnormalities, liver and kidney damage, impaired nervous system function
Lead	Lead-based paint Lead additives in gasoline	Neurological damage. Affects brain development in children Large doses affect brain and kidneys in adults and children.
Mercury	Air or water at contaminated sites Methylmercury in contaminated fish and shellfish	Permanent damage to brain, kidneys, developing fetus
Vinyl chloride	Plastics manufacturing Air or water at contaminated sites	Acute effects: dizziness, headache, unconsciousness, death Chronic effects: liver, lung, and circulatory damage
Polychlorinated biphenyls (PCBs)	Eating contaminated fish Industrial exposure	Probable carcinogens; acne and skin lesions
Benzene	Industrial exposure Glues, cleaning products, gasoline	Acute effects: drowsiness, headache, death at high levels Chronic effects: damage to blood-forming tissues and immune system; also carcinogenic
Cadmium	Released during combustion Living near a smelter or power plant Picked up in food	Probable carcinogen: kidney damage, lung damage, high blood pressure
Polycyclic aromatic hydrocarbons	Exposure to smoke from a variety of sources	Probable carcinogen; possible birth defects
Benzo[a]pyrene	Product of combustion of gasoline or other fuels In smoke and soot	Probable carcinogen; possible birth defects
Benzo[b]fluoranthene	Product of combustion of gasoline and other fuels Inhaled in smoke	Probable carcinogen

Source: Data from Agency for Toxic Substances and Disease Registry.

surface impoundments. As indicated earlier, these sites were typically located convenient to the industry and were often in environmentally sensitive areas, such as floodplains or wetlands. At the time, these land areas were considered unimportant and were frequently abandoned or filled in to make room for further development. Rain and melting snow soaked through the sites, carrying chemicals that contaminated underground waters. When these groundwaters reached streams and lakes, they were contaminated as well. When the sites became full or were abandoned, they were sometimes left uncovered, thus increasing the likelihood of water pollution from leaching or flooding and of people having direct contact with the wastes. At some sites, specifically the uncovered ones, the air was also contaminated by toxic vapors from evaporating liquid wastes or from uncontrolled chemical reactions. (See figure 18.3.) In North America alone, the number of abandoned or uncontrolled sites was over 25,000.

Europeans are also paying a heavy price for their negligence. The Netherlands is a good example. Authorities estimate that up to 8 million metric tons of hazardous chemical wastes may be buried in the Netherlands. Estimates for cleaning up those wastes run as high as $9 billion. In the republics of the former Soviet Union and Eastern Europe, many hazardous waste sites have been identified recently, and there is no money to pay for cleanup. It is assumed that there are many hazardous waste sites in countries of the developing world, but our knowledge about these is very limited. In the Czech Republic, Slovakia, Serbia, and other former Eastern Bloc countries, environmental concerns have grown steadily as the dangers associated with past practices are appreciated. In the past industrial production was the emphasis while ignoring environmental health

and safety issues was the norm. Therefore, environmental degradation was immense. Rivers and streams were used as industrial dumping conduits. Hazardous waste was routinely dumped in open wells and mine shafts. Unexploded munitions, chemical weapons, and other materials were left in open pits, fields, rivers, lakes, and marshes. Since the early 1990s, these countries have just begun to undertake measures to understand the extent of contamination and to undertake cleanup operations. Regulation of current environmental issues is ongoing.

In the United States, the federal government has become the principal participant in the cleanup of hazardous-waste sites. The **Comprehensive Environmental Response, Compensation, and Liability Act (CERCLA)** was enacted in 1980. This program deals with financing the cleanup of large, uncontrolled hazardous-waste sites and has popularly become known as **Superfund.** Superfund was established when Congress responded to public pressure to clean up hazardous-waste dumps and protect the public against the dangers of such wastes. CERCLA had several key objectives:

1. To develop a comprehensive program to set priorities for cleaning up the worst existing hazardous-waste sites.

2. To make responsible parties pay for those cleanups whenever possible.

3. To set up a $1.6 billion Hazardous Waste Trust Fund—popularly known as Superfund—to support the identification and cleanup of abandoned hazardous-waste sites.

4. To advance scientific and technological capabilities in all aspects of hazardous-waste management, treatment, and disposal.

FIGURE 18.3 **Improper Toxic Chemical Storage** This is a site where toxic wastes were improperly stored. Often such sites are abandoned, and it is difficult to assign responsibility for cleanup because the company that produced the waste is unknown or no longer exists.

A **National Priorities List** of hazardous-waste sites requiring urgent attention was drawn up for Superfund action. Under CERCLA, over 44,000 sites were evaluated, and about 11,000 were considered serious enough to warrant further investigation. The number of sites listed on the National Priorities List fluctuates as new sites are added and sites are removed because they are cleaned up or deleted from the list. Currently about 1200 sites are on the list.

Although the purpose of the Superfund program was clear, it became controversial during its early history. Millions of dollars were spent by both the federal government and industry, but most of the money went for litigation and technical studies to support or disprove the claims of the parties. Little of the money paid for cleanup. One of the primary reasons for this problem was the way CERCLA was written. It provided that anyone who contributed to a specific hazardous-waste site could be required to pay for the cleanup of the entire site, regardless of the degree to which they contributed to the problem. Since many industries that contributed to the problem had gone out of business or could not be identified, those that could be identified were asked to pay for the cleanup. Most businesses found it cost-effective to hire lawyers to fight their inclusion in a cleanup effort rather than to pay for the cleanup. Consequently, little cleanup occurred.

After a slow start, however, the Superfund program is showing significant results. By 2006, there were about 1200 active

sites on the National Priorities List and about 65 were being considered for inclusion. About 900 sites have been cleaned up. Most of the remaining sites were in the process of being cleaned up or were under study to determine the best way to proceed. This has been an expensive undertaking. Total expenditures for the Superfund program have been about $27 billion. In addition, the EPA estimates that the settlements it has reached with responsible parties amount to over $20 billion.

As discussed earlier in this chapter, the public had to determine who might be liable for cleanup costs under CERCLA. Superfund language indicated that prior to the purchase of commercial real estate, a person or entity must conduct "all appropriate inquiry" using "good and customary practices" to receive what is called the "ILD" or "innocent landowner defense." Further legislation enacted in 2002 under SBLRBRA also included new language for the defense of liability under CERCLA. This meant that most purchasers of commercial real estate must perform an ASTM Phase I Environmental Site Assessment (E-1527) prior to purchasing the land. In an attempt to get money from any and all PRPs, CERCLA uses "strict, joint and several" liability. In other words, you own the problem regardless of whether you are "innocent" (did not contribute to the contamination); that is called liability without fault. It also means third-party liability; any party may be held responsible for the entire cost of cleanup (especially those with "deep pockets"—the most money). This holds true even if it has only contributed a small amount to the problem. If the property is purchased without conducting due diligence and an environmental problem is encountered, the purchaser is responsible for damages and the cost of cleanup. It is truly a "buyer beware" situation.

TOXIC CHEMICAL RELEASES

In 1987, as the result of EPA requirements, certain industries in the United States had to begin reporting toxic chemicals released into the environment. Any industrial plant that released 23,000 kilograms (50,000 pounds) or more of toxic pollutants was required to file a report. These were primarily manufacturing industries. The information collected allowed the EPA to target specific industries for enforcement action. Since the information is public, many industries were encouraged to take action to reduce their emissions. These activities have been successful, and industrial emissions of toxic chemicals were reduced by about 60 percent from 1988 through 2004. In the intervening years, changes have been made in who must report. The quantity that required reporting was reduced to 11,400 kilograms (25,000 pounds). In addition, industries that were originally exempt from reporting now must report toxic releases. About 2.1 billion kilograms (4.5 billion pounds) of toxic chemicals were reported released into the environment by industry in 2008.

We all have old electronics (e-waste) stored in our basements or closets waiting for the day we can easily recycle them. It can be difficult to figure out where to take e-waste, and once we find a recycling center, there is no guarantee that our e-waste will be recycled safely. Up to 80 percent of North American e-waste is exported to developing countries where toxic components are burned, dumped, or smashed apart without proper protection for human health or the environment.

E-waste is more than just household waste. It contains large amounts of substances such as lead, cadmium, mercury, and chromium that can leach into soil and contaminate groundwater or, if incinerated, can be released into the air. Most e-waste is highly reusable, and up to 97 percent of the parts can be recycled—either to use as upgraded components in other computers or melted down as scrap. Responsible recycling of e-waste is important to protect workers' health and the environment. Unfortunately, only 10 percent of unwanted and obsolete e-waste is recycled responsibly. The following points should help you in properly recycling your e-waste:

- Send your old equipment back to the electronics companies. Because of consumer demand, electronics companies are beginning to recycle their old products. Dell, HP, and Apple all have take-back policies.

- Pick from a list of responsible recyclers. There are a growing number of recyclers who have signed a pledge to end the practice of shipping e-waste to developing countries. The Basel Action Network and Computer TakeBack Campaign have developed a list of responsible recyclers.

- Learn about the current or pending e-waste laws in your community. State e-waste and recycling laws are not all the same. State policies have developed that ban most electronic products from landfills; others have put a consumer tax on electronic products for the cost of their disposal and others require electronic companies to pay for recycling. Washington and Maine have the most extensive e-waste laws in the United States.

- Check to see if the company that produced the product uses extended producer responsibility (EPR). This is a policy that is being used in some parts of the world, such as Japan and Europe. EPR includes electronics, as well as many other products including cars. EPR requires companies to take responsibility for the effects of their products—from the materials used in the production to their disposal. EPR is a solution that encourages companies to reduce toxins from their products and to reuse the materials from their recycling.

- There are about 600 million obsolete cellphones in the United States. If you have one, you can recycle it by logging on to gooddeedfoundation.org/aarp for instructions. It is simple and free.

Today, the primary industries involved in toxic releases to the environment are the mining, power generation, chemical, and metal manufacturing industries. (See figure 18.4.) The large amount of toxic waste produced by the mining industry is also reflected in figure 18.5. Mining waste typically contains metals and is deposited on the surface of the land. In addition, you would expect those states with large mining operations to have high amounts of waste.

HAZARDOUS-WASTE MANAGEMENT CHOICES

In discussing the kinds of hazardous wastes released to the environment, it is important to point out that the amount released is less than one-quarter (about 23 percent) of the total amount of hazardous waste produced. In 2008, a total of about 29 million metric tons (30 million tons) of hazardous waste was produced by industries in the United States. About 77 percent of this material was recycled, burned for energy, or treated so that it was not released to the environment. What forces contribute to this trend toward reduced amounts of waste being released?

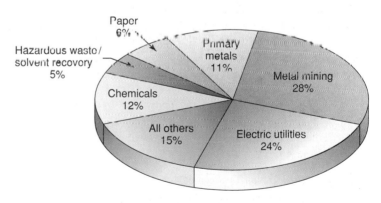

FIGURE 18.4 **Sources of Toxic Releases** Mining industries are responsible for about 28 percent of all toxic waste material released to the environment. These are primarily deposited on the surface of the land. Electric utilities are responsible for about 24 percent of releases. These are primarily releases to the atmosphere. The chemical industry and metal processing industries are also significant sources of toxic releases.
Source: Data from the U.S. Environmental Protection Agency.

The environmental costs of the uncontrolled release of toxic materials are astronomical. Recognition of these costs has led to laws and regulations that govern the release of hazardous wastes, require industries to report the hazardous waste they release, and

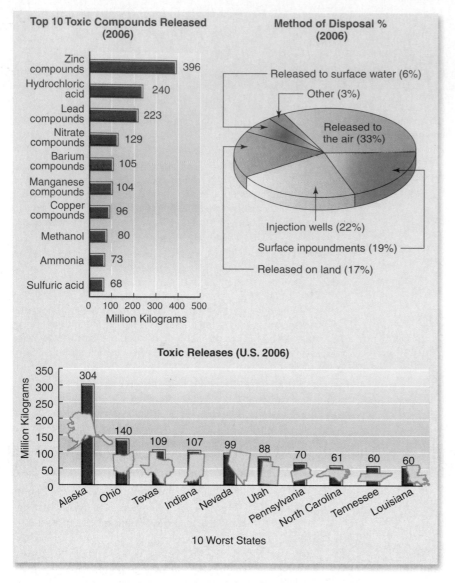

Top 10 Toxic Compounds Released (2006)

Compound	Million Kilograms
Zinc compounds	396
Hydrochloric acid	240
Lead compounds	223
Nitrate compounds	129
Barium compounds	105
Manganese compounds	104
Copper compounds	96
Methanol	80
Ammonia	73
Sulfuric acid	68

Method of Disposal % (2006)

Released to surface water (6%)
Other (3%)
Released to the air (33%)
Injection wells (22%)
Surface inpoundments (19%)
Released on land (17%)

Toxic Releases (U.S. 2006)

State	Million Kilograms
Alaska	304
Ohio	140
Texas	109
Indiana	107
Nevada	99
Utah	88
Pennsylvania	70
North Carolina	61
Tennessee	60
Louisiana	60

10 Worst States

FIGURE 18.5   Toxic Releases Dominated by Mining and Electric Generation The large amount of waste generated by the mining industry is reflected in the kinds of waste produced (toxic metals), the environmental compartment affected (land disposal), and the states with the most toxic releases. Electric power generation is responsible for large amounts of toxic materials released to the air.

Source: Data from the U.S. Environmental Protection Agency.

other countries have emphasized pollution prevention and waste minimization. This includes concepts such as Design for Environment (DfE), which incorporates the idea of taking into account environmental concerns and costs prior to production—engineering the product to be more "green" or environmentally friendly before it is produced. This also involves looking at things more holistically. The concepts of ISO 14000 and Environmental Management Systems are more frequently being implemented. Strong regulatory control requires that industries report the hazardous wastes they produce and that the wastes be stored, transported, and disposed of properly.

The EPA now promotes a **pollution-prevention hierarchy** (commonly referred to as "P2") that emphasizes reducing the amount of hazardous waste produced. This involves the following strategy:

1. Reduce the amount of pollution at the source.
2. Recycle wastes wherever possible.
3. Treat wastes to reduce their hazard or volume.
4. Dispose of wastes on land or incinerate them as a last resort. (See figure 18.6.)

REDUCING THE AMOUNT OF WASTE AT THE SOURCE

Pollution prevention (P2) encourages changes in the operations of business and industry that prevent hazardous wastes from being produced in the first place. Source reduction is the preferred starting point as indicated by EPA and other entities. The idea of "end of pipe" control through regulations has nearly reached its maximum effectiveness. If the waste is never created in the first place, the need for regulatory compliance is eliminated, along with the associated liabilities. Many of the actions are simple to perform and cost little. Primary among them are activities that result in fewer accidental spills, leaks from pipes and valves, loss from broken containers, and similar mishaps. These reductions often can be achieved at little cost through better housekeeping and awareness training for employees. Many industries actually save money because they need to buy less raw material because less is being lost. There are economic as well as environmental reasons for practicing P2. Incentives from regulatory agencies with programs like Michigan's Business Pollution Prevention Partnership (MBP3) or Clean Corporate Citizen programs, along with lower insurance rates, less regulatory scrutiny, and marketing advantages, have induced many businesses and industries to implement P2 programs. Some voluntary programs like ISO 14001 (EMS Registration) require not only internal P2 compliance with the standard, but also compliance from a company's suppliers.

force industries to pay for the measures needed to eliminate or reduce the waste they produce. Strong, enforceable laws have forced industries to absorb the cost of waste disposal and look for better and cheaper ways to treat and dispose of hazardous wastes. These legal and economic forces are driving industrial behavior toward pollution prevention and waste minimization.

In the past, the management of hazardous waste was always added on to the end of the industrial process. The effluents from pipes or smokestacks were treated to reduce their toxicity or concentration.

In recent years, it has become obvious that a better way to deal with the problem of hazardous waste is not to produce it in the first place. To this end, the EPA and regulatory agencies in

CHEMICAL EXCHANGE AT THE UNIVERSITY OF BRITISH COLUMBIA

A program at the University of British Columbia (UBC) in Vancouver, British Columbia, allows faculty and students to share research chemicals through the Chemical Exchange Database (CED). CED is an on-line tool that is helping UBC reduce lab waste and get more value for the ever-rising cost of chemicals and disposal. The site connects those looking for research chemicals with those who have too much of a given substance. The substances have already been paid for, so everything in the database is available for free.

Most universities have to handle surplus chemicals. For example, an experiment may only need 10 milligrams of a chemical, but the chemical may only come from suppliers in 4-liter quantities. Researchers often need to buy more of a substance than they require. This poses a problem because proper chemical disposal is very expensive.

Currently, the CED lists more than 200 available chemicals. To make a request, researchers place an order online. Within one to two days, the materials are delivered. To post a surplus chemical a researcher needs to enter the substance, amount, and producer. Within one to two days, the materials will be properly stored or delivered to the new owners if another lab has already made a request.

The database, launched in 2004, processes about 300 exchanges (1500 kilograms in chemicals) a year. Savings are estimated at about $75,000 in disposal and purchasing costs.

Reduce and Reuse

Stops leaks
Redesign processes
Reuse waste

Prevent spills
Concentrate wastes
Clean solvents

Educate employees
Substitute less–toxic raw materials

Recycle

Recycle solvents for other uses
Exchange waste with other industries
Burn wastes as fuel

Treatment

Neutralization of acids and bases
Filtration and separation
Biodegradation
Solidification

Disposal

Hazardous waste landfill
Incineration
Deep-well injection

FIGURE 18.6
Pollution-Prevention Hierarchy The simplest way to deal with hazardous wastes is not to produce them in the first place. The pollution-prevention hierarchy stresses reductions in the amount of hazardous waste produced by employing several different strategies.

contains a nonpolluting tungsten core and does not contain lead, which contaminates the soil and air around firing ranges. The U.S. military uses about 700 million rounds of small-caliber ammunition each year at about 3000 firing ranges. Lead contamination has closed hundreds of outdoor firing ranges on military bases across the United States. In 1998, when lead concentrated in firing berms was found to be leaching into Cape Cod's water supply, the Environmental Protection Agency ordered the Massachusetts Military Reservation to stop live-fire training.

Waste minimization involves changes that industries could make in the way they manufacture products, changes that would reduce the amount of waste produced. For example, it may be possible to change a process so that a solvent that is a hazardous material is replaced with water, which is not a hazardous material. This is an example of source reduction: any change or strategy that reduces the amount of waste produced.

Another strategy is to use the waste produced in a process in another aspect of the process, thus reducing the amount of waste produced. For example, water used to clean equipment might be included as a part of the product rather than being discarded as a contaminated waste.

Another technique that can be used to reduce the amount of waste produced is to clean solvents used in processes. Using a still to purify solvents results in a lower total volume of hazardous waste being produced because the same solvent can be used over and over again.

Pollution prevention can be applied in unusual ways. In 2000, the U.S. Army announced that it would begin issuing an environmentally friendly "green bullet" that contains no lead. The bullet

The simple process of allowing water to evaporate from waste can reduce the total amount of waste produced. Obviously, the hazardous components of the waste are concentrated by this process.

RECYCLING WASTES

Often it is possible to use a waste for another purpose and thus eliminate it as a waste. Many kinds of solvents can be burned as a fuel in other kinds of operations. For example, waste oils can be used as fuels for power plants, and other kinds of solvents can be burned as fuel in cement kilns. Care needs to be taken that the contaminants in the oils or solvents are not released into the environment during the burning process, but the burning of these wastes destroys them and serves a useful purpose at the same time.

Similarly, many kinds of acids and bases are produced as a result of industrial activity. Often these can be used by other industries. Ash or other solid wastes can often be incorporated into concrete or other building materials and therefore do not require disposal. Thus, the total amount of waste is reduced.

TREATING WASTES

Wastes can often be treated in such a way that their amount is reduced or their hazardous nature is modified.

Neutralization of dangerous acids and bases by reacting them with one another can convert hazardous substances to materials that are not hazardous.

Biodegradation of organic materials by the actions of microorganisms can convert hazardous chemicals to innocuous substances. Sewage treatment plants perform their function by biodegradation.

Air stripping is sometimes used to remove volatile chemicals from water. Volatile chemicals, which have a tendency to vaporize easily, can be forced out of liquid when air passes through it. Steam stripping works on the same principle, except that it uses heated air to raise the temperature of the liquid and force out volatile chemicals that ordinary air would not. The volatile compounds can be captured and reused or disposed of.

Carbon absorption tanks contain specifically activated particles of carbon to treat hazardous chemicals in gaseous and liquid waste. The carbon chemically combines with the waste or catches hazardous particles just as a fine wire mesh catches grains of sand. Contaminated carbon must then be disposed of or cleaned and reused.

Precipitation involves adding special materials to a liquid waste. These bind to hazardous chemicals and cause them to precipitate out of the liquid and form large particles called floc. Floc that settles can be separated as sludge; floc that remains suspended can be filtered, and the concentrated waste can be sent to a hazardous waste landfill.

DISPOSAL METHODS

Today, the two most common methods for disposing of hazardous wastes are incineration and land disposal. The primary factors that determine which method is used are economic concerns and acceptance by the public.

Incineration (thermal treatment) involves burning waste at high temperatures and can be used to destroy a variety of kinds of wastes.

A hazardous-waste incinerator can be used to burn organic wastes but is unable to destroy inorganic wastes. A well-designed and well-run incinerator can destroy 99.9999 percent of the hazardous materials that go through it. The relatively high costs of incineration (compared with landfills) and concerns about the emissions affecting surrounding areas have kept incineration from becoming a major method of treatment or disposal in North America. In North America, abundant land is available for land disposal, making it the most economical and most widely used method. Incineration accounts for the disposal of only about 4 percent of the hazardous wastes. In Europe and Japan, where land is in short supply and expensive, incineration is more cost-effective and is used to dispose of large amounts of hazardous wastes, but it still accounts for the disposal of less than 50 percent of the hazardous wastes produced. However, because of concerns about the emissions from incinerators, significant amounts of hazardous waste are incinerated at sea on specially designed ships.

Land disposal is still the primary method for the disposal of hazardous wastes when all other options have been exhausted. Land disposal can take several forms:

1. Deep-well injection into porous geological formations or salt caverns
2. Discharge of treated and untreated liquids into municipal sewers, rivers, and streams
3. Placement of liquid wastes or sludges in surface pits, ponds, or lagoons
4. Storage of solid wastes in specially designed hazardous-waste landfills

Of these methodologies, deep-well injection is the most important. Nearly 50 percent of all hazardous waste disposed of on land is injected into deep wells. About 20 percent is released to aquatic environments, and about 5 percent is stored in landfills and other surface sites.

There are techniques that reduce the chance that hazardous materials will escape from these locations and become a problem for the public. Immobilizing a waste puts it into a solid form that is easier to handle and less likely to enter the surrounding environment. Waste immobilization is useful for dealing with wastes, such as certain metals, that cannot be destroyed. Two popular methods of immobilizing waste are fixation and solidification. Engineers and scientists mix materials such as fly ash or cement with hazardous wastes. This either "fixes" hazardous particles, in the sense of immobilizing them or making them chemically inert, or "solidifies" them into a solid mass. Solidified waste is sometimes made into solid blocks that can be stored more easily than can a liquid.

INTERNATIONAL TRADE IN HAZARDOUS WASTES

The growth in the uncontrolled movement of hazardous wastes between countries has been one of the most contentious environmental issues on the international political agenda. There is particular concern about rich, industrialized countries exporting such wastes to poorer, developing countries lacking the administrative and technological resources to safely dispose of or recycle the waste. For example, in 1999, between 3000 and 4000 metric tons of mercury-contaminated concrete waste packed in plastic bags was found in an open dump in a small town in Cambodia. The waste, labeled as "construction waste" on import documents, came from a Taiwanese petrochemical company. In this case, the waste was tracked down and returned to its point of origin. Unfortunately, most such cases are not reported or detected. International awareness of the problems associated with the trade in hazardous wastes has increased noticeably owing to several factors: the growing amounts of such wastes being generated; closure of old waste disposal facilities and political opposition to the development of new ones; and the dramatically higher costs associated with the disposal of hazardous wastes in industrialized countries (and thus the potential to earn profits by exporting such wastes to developing countries with low disposal costs). The debate over controlling hazardous-waste movements between countries culminated in 1989 with the creation of the *Basel Convention.*

The Basel Convention was negotiated under the auspices of the United Nations Environmental Programme between 1987 and 1989. The objectives of the convention are to minimize the generation of hazardous wastes and to control and reduce their transboundary movements to protect human health and the environment. To achieve these objectives, the convention prohibits exports of hazardous waste to Antarctica, to countries that have banned such imports as a national policy, and to nonparties to the convention (unless those transactions are subject to an agreement that is as stringent as the Basel Convention). Though not part of the original agreement, there is now a broad ban on the export of hazardous wastes from the Northern to the Southern Hemisphere. The waste transfers that are permitted under the Basel regime are subject to the mechanism of prior notification and consent, which requires parties not to export hazardous wastes unless a "competent authority" in the importing country has been properly informed and has consented to the trade.

While the Basel regime may not be perceived as being as successful or significant as some other multilateral environmental agreements, it remains an important part of the international community's attempt to protect the global environment and human health from hazardous materials. Now that the convention has been in place for more than a decade, the focus is on assisting parties with the environmentally sound management of hazardous wastes and with reducing the amount of wastes generated.

The export of hazardous wastes for recycling from industrialized nations to developing nations raises other questions. Those who support the export of hazardous substances for recycling argue that this practice offers two major benefits: reducing the quantity of such substances that get into the environment through final disposal and slowing down the depletion of natural resources. This argument is undoubtedly correct, provided the receiving country has the proper recycling facilities and adequate environmental standards. Environmentally beneficial trade in hazardous wastes ordinarily requires that there be an established market for these wastes and that the trade be economically viable.

HAZARDOUS-WASTE MANAGEMENT PROGRAM EVOLUTION

Fundamentally, the goal of a hazardous-waste management program is to change the behavior of those who generate hazardous wastes so that they routinely store, transport, treat, and dispose of them in an environmentally safe manner. The focus on hazardous-waste management typically comes in the second phase of countries' environmental programs, after efforts to address more immediate threats to public health, such as safe drinking water. Economics play a major role in this because countries that are most concerned with simply feeding and providing drinking water and shelter for their citizens in reality don't have time to worry about the management of waste (or so might be their perception). In the early years of many countries' hazardous-waste management programs, uncontrolled disposal of hazardous waste was the norm. Few, if any, proper treatment and disposal facilities existed. Currently, information on who is generating waste, what types are being generated, and where it is being disposed of is meager or nonexistent.

The transition from an unregulated environment to a regulated one is complex, but hazardous-waste management programs typically evolve through the following major stages: identifying the problem and enacting legislation, designating a lead agency, establishing rules and regulations, developing treatment and disposal capacity, and creating a compliance and enforcement program. Each of these stages takes a number of years, and at each stage, there are many difficult issues to be resolved. Denmark, Germany, and the United States began this process by passing their first major hazardous-waste laws between 1972 and 1976, Canada did so in 1980. In the decade that followed, all four of these countries developed hazardous-waste regulations and requirements, so that by the end of the 1980s, their regulatory systems were largely operational. Laws and policies developed during the 1990s have focused mainly on waste minimization and recycling, as well as on harmonization with international standards (such as ISO and ASTM) and the cleanup of contaminated sites. Again, traditionally, command-and-control regulatory authority has been used to force entities to comply. More advanced international voluntary standards/systems (ISO, ASTM) have recently been employed to go beyond compliance to achieve higher levels of environmental "friendliness" yielding businesses and industries that produce less pollution.

The more recent evolution of regulatory programs in Hong Kong, Indonesia, Malaysia, and Thailand has followed a similar pattern. By the early 1980s, all of these countries had enacted some form of environmental legislation providing at least limited authority to regulate hazardous waste. However, hazardous-waste management received little attention until the late 1980s, after periods of rapid economic growth and the expansion of the countries' manufacturing sectors. From 1989 to 1998, all of these countries passed major legislation that addressed hazardous waste or developed regulations outlining comprehensive programs: Malaysia did so in 1989, Hong Kong in 1991, Thailand in 1992, and

Indonesia in 1995–98. All four countries now have at least one modern hazardous-waste treatment, storage, and disposal facility.

While all countries pass through the same stages of program development, no two countries follow precisely the same path. Differences in geography, demographics, industrial profile, politics, and culture lead countries to make different choices at each stage. Of course, the funding for such actions is always a problem. To create successful programs, countries must educate the regulated community and the public. All of this takes a great deal of investment in time and money. Priority must be placed on facing these issues and allocating resources to carry them out.

ISSUES & ANALYSIS

Household Hazardous Waste

Since the passage of the Resource Conservation and Recovery Act in 1978 and other environmental legislation, industries have been required to document the amounts of hazardous wastes they produce and to account for their appropriate disposal. As a result, the amount of industrial hazardous waste released to the environment has been significantly reduced. At the same time that citizens insist on business and industry compliance with hazardous-waste regulations, our personal behavior does not always reflect the same level of concern.

Americans generate about 1.45 million metric tons of household hazardous waste per year. Frequently, these materials become waste when they become contaminated or are unwanted. Common forms of household hazardous waste are unwanted paints, varnishes, and other finishes; contaminated solvents such as alcohol and paint thinner; automobile fluids such as waste oil, brake fluid, and antifreeze; appliance and automobile batteries containing heavy metals; caustic cleaning products; unused pesticides; and fluorescent light bulbs.

We have several options to help us properly handle household hazardous waste. Most containers of such materials carry instructions for the proper disposal of the unwanted material or the empty container. In addition, thousands of communities have regular household hazardous waste collection days staffed by people who are specially trained to assist the public in disposing of such wastes. However, these collection days are often infrequent, which requires that we store and label our waste, find out when and where the waste will be accepted, and deliver it to a particular place on a particular day. Often a reservation is required. This is obviously inconvenient.

We can reduce the amount of hazardous waste we produce by purchasing products that are not hazardous such as water-based paints instead of oil-based paints, carefully choosing the quantities we purchase so that we don't have leftover materials, and choosing not to do our own automobile maintenance.

Common toxics found in a home include the following:

Polybrominated diphenylethers (PBDEs)—used as flame-retardants, PBDEs are found in foam mattresses and pillows, carpet and padding, and electronics.

Phthalates—vinyl wallpaper, plastic containers and bottles, garden hoses, lawn furniture, extension cords.

Pesticides—pet flea collars, various garden products, products to kill pests such as roaches, rats, and mice.

Bisphenols—polycarbonate plastics, found in some rigid plastic bottles such as water bottles, plastic baby bottles, and plastic containers.

Metals—toxic metals are some of the most common industrial poisons in the home. Old paint contains lead; fish, such as tuna, contain mercury; and a pressure-treated wood deck can contain arsenic and chromium.

Therefore, we need to examine our behavior toward the use and disposal of hazardous materials. Some pertinent questions are:

- Do we think about the potential a product has to become hazardous waste when we purchase it?
- Do we read label directions pertaining to proper use and disposal of products and follow them?
- Do we dispose of these materials in our household trash or by dumping them down the drain, or do we look for a more proper method of disposal?

SUMMARY

Public awareness of the problems of hazardous substances and hazardous wastes is relatively recent. The industrialized countries of Europe and North America began major regulation of hazardous materials only during the past 30 years, and most developing countries exercise little or no control over such substances. As a result, many countries are living with serious problems from prior uncontrolled dumping practices, while current systems for management of hazardous and toxic waste remain incomplete and incapable of even identifying all hazardous waste.

A number of fundamental problems are involved in hazardous-waste management, some of which were discussed earlier. First, there is no agreement as to what constitutes a hazardous waste. Moreover, little is known about the amounts of hazardous wastes generated throughout the world. The issue is further complicated by our limited understanding of the health effects of most hazardous wastes and the fact that large numbers of potentially hazardous chemicals are being developed faster than their health risks can be evaluated.

Hazardous-waste management must move beyond burying and burning. Industries need to be encouraged to generate less hazardous waste in their manufacturing processes. Although toxic wastes cannot be entirely eliminated, technologies are available for minimizing, recycling, and treating wastes. It is possible to enjoy the benefits of modern technology while avoiding the consequences of a poisoned environment. The final outcome rests with governmental and agency policy makers, as well as with an educated public.

THINKING GREEN

1. Conduct a "chemical inventory" of your garage, bathroom, kitchen, or other area in your home where chemicals are stored. Properly dispose of products that are no longer of use or outdated.
2. Contact your local solid waste operator or local government solid waste office. Ask about hazardous waste collection programs, oil recycling, and programs for e-waste collection in your community. If there are no programs, ask why. Become a champion for the development of programs.
3. Participate in hazardous waste minimization programs on your campus.
4. Use rechargeable batteries and nontoxic cleaning products.

WHAT'S YOUR TAKE?

State and federal agencies have traditionally relied on command-and-control regulatory actions to control pollution. Recently, due to declining budgets and other factors, some states have turned to voluntary standards for dealing with current and future pollution problems. Many regulatory agencies believe this is the way of the future. They have even turned to trading air pollution permits on the stock exchange. Develop an argument for or against the concept of relying on business and industry to police itself through voluntary programs and standards.

REVIEW QUESTIONS

1. Explain the problems associated with hazardous-waste dump sites and how such sites developed.
2. Distinguish between acute and chronic toxicity.
3. Give two reasons why regulating hazardous wastes is difficult.
4. In what ways do hazardous wastes contaminate the environment?
5. Describe how hazardous wastes contaminate groundwater.
6. Why is it often a problem to link a particular chemical or hazardous waste to a particular human health problem?
7. Describe what is meant by the U.S. National Priorities List.
8. Describe five technologies for managing hazardous wastes.
9. What is meant by pollution prevention and waste minimization?
10. Describe the pollution-prevention hierarchy.
11. What are RCRA and CERCLA? Why is each important for managing hazardous wastes?
12. Explain why RCRA and CERCLA might be referred to as "companion" statutes. (How do they differ and how are they similar?)
13. Why might it be necessary for a company to do a Phase I Environmental Site Assessment prior to acquiring a piece of property for commercial development?
14. What is meant by "strict, joint and several" liability?
15. How might a company benefit from enacting an environmental management system?

CRITICAL THINKING QUESTIONS

1. Scientists at the EPA have to make decisions about thresholds in order to identify which materials are toxic. What thresholds would you establish for various toxic materials? What is your reasoning for establishing the limits you do? What, if any, type of testing might you conduct to arrive at these thresholds?
2. Go to the EPA's website (www.epa.gov/enviro/html/ef_overview .html) and identify the major releasers of toxic materials in your area. Were there any surprises? Are there other releasers of toxic materials that might not be required to list their releases?
3. In North America alone, there are over 25,000 abandoned and uncontrolled hazardous-waste dumps. Many were abandoned before the RCRA of 1976 was passed. Who should be responsible for cleaning up these dumps? How might you go about identifying where these abandoned sites might be located, and once located, how would you determine who is responsible for cleaning them up? What types of variables might be involved in determining how to conduct the remediation?
4. Look at this chapter's section called "Hazardous-Waste Dumps—A Legacy of Abuse." Do the authors present the information from a particular point of view? What other points of view might there be on this issue? What information do you think these other viewpoints would provide?

5. Many economically deprived areas, Native American reservations, and developing countries that need an influx of cash have agreed, over significant local opposition, to site hazardous-waste facilities in their areas. What do you think about this practice? Should outsiders have a say in what happens within these sovereign territories?

6. After reading about the problem with hazardous wastes and toxic materials in China, do you think that the United States or any other country should have the right to intervene if another country is creating significant environmental damage? Why or why not? How might this relate to what some countries are saying about the United States and its environmental policies?

7. Review the Case Study dealing with dioxins. How might the area be cleaned up? Who should be responsible for conducting the cleanup? To what levels would you suggest the area be remediated? Should the river water and sediments be treated as well? Should the residents in the area be consulted, and should they be compensated and given medical treatment options? Consider the plant and animal life in the floodplain—what, if anything, should be done about that?

8. Go to www.ASTM.org and find the ASTM standard for conducting a Phase I ESA. How might a commercial developer use this document when purchasing a piece of property? Could the developer conduct the Phase I ESA, or would they need to hire an environmental professional?

CHAPTER 19

ENVIRONMENTAL POLICY AND DECISION MAKING

Government regulation is often necessary to control the actions of uncaring individuals or corporations. The federal government has passed many environmental laws directed at forcing better behavior from individuals and corporations.

CHAPTER OUTLINE

New Challenges for a New Century
   Forces and Trends
   Kinds of Policy Responses
   Government and Governance
   Learning from the Past
   Thinking About the Future
   Defining the Future
Development of Environmental Policy in the United States
   Legislative Action
   The Role of Nongovernmental Organizations
   The Challenge for U.S. Environmental Policy
Environmental Policy and Regulation
   The Significance of Administrative Law
   National Environmental Policy Act—Landmark Legislation
   Other Important Environmental Legislation
   Role of the Environmental Protection Agency
The Greening of Geopolitics
   International Aspects of Environmental Problems
   National Security Issues
Terrorism and the Environment
International Environmental Policy
   The Role of the United Nations
   Earth Summit on Environment and Development
   Environmental Policy and the European Union
   New International Instruments
It All Comes Back to *You*

ISSUES & ANALYSIS
Gasoline, Taxes, and the Environment 448

CASE STUDY
The Environmental Effects of Hurricane Katrina 439

CAMPUS SUSTAINABILITY INITIATIVE
College and University Presidents' Climate Commitment 446

GOING GREEN
Investing in a Green Future 436

WATER CONNECTIONS
Shared Water Resources 431

OBJECTIVES

After reading this chapter, you should be able to:

• Explain how the executive, judicial, and legislative branches of the U.S. government interact in forming policy.

• Understand how environmental laws are enforced in the United States.

• Describe the forces that led to changes in environmental policy in the United States during the past three decades.

• Understand the history of the major U.S. environmental legislation.

• Understand why some individuals in the United States are concerned about environmental regulations.

• Understand what is meant by "green" politics.

• Describe the reasons environmentalism is a growing factor in international relations.

• Understand the factors that could result in "ecoconflicts."

• Understand why it is not possible to separate politics and the environment.

• Explain how citizen pressure can influence governmental environmental policies.

Global Perspective features on "ISO Standards for Environmental Management Systems," and "Copper River International Migratory Bird Initiative: A Case Study of International Cooperation," and the Case Study on "Can the Environment Become a Political Bridge Between Conservatives and Liberals?" can be found on the book's website at www.mhhe.com/enger12e along with other interesting readings.

NEW CHALLENGES FOR A NEW CENTURY

We live in remarkable times. This is an era of rapid and often bewildering alterations in the forces and conditions that shape human life.

FORCES AND TRENDS

The end of the Cold War has been accompanied by the swift advance of democracy in places where it was previously unknown and an even more rapid spread of market-based economies. The authority of central governments is eroding, and power has begun to shift to local governments and private institutions. In some countries, freedom and opportunity are flourishing, while in others, these changes have unleashed the violence of old conflicts and new ambitions. In countries like China and India, we have seen rapid economic growth.

Population Growth

Tomorrow's world will be shaped by the aspirations of a much larger global population. The number of people living on Earth has doubled in the last 50 years. Growing populations demand more food, goods, services, and space. Where there is scarcity, population increase aggravates it. Where there is conflict, rising demand for land and natural resources exacerbates it. Struggling to survive in places that can no longer sustain them, growing populations overfish, overharvest, and overgraze. (See figure 19.1.)

Globalization

Internationally, trade, investment, information, and even people flow across borders largely outside of governmental control. Domestically, deregulation and the shift of responsibilities from federal to state and local governments are changing the relationships among levels of government and between government and the private sector.

Communication Revolution

Communications technology has enhanced people's ability to receive information and influence events that affect them. This has sparked explosive growth in the number of organizations, associations, and networks formed by citizens, businesses, and communities seeking a greater voice for their interests. As a result, society outside of government—civil society—is demanding a greater role in governmental decisions, while at the same time impatiently seeking solutions outside government's power to decide.

Knowledge Economy

But technological innovation is changing much more than communication. It is changing the ways in which we live, work, produce, and consume. Knowledge has become the economy's most important and dynamic resource. It has rapidly improved efficiency as those who create and sell goods and services substitute information and innovation for raw materials. During the past 30 years, the

FIGURE 19.1   Population Increase Aggravates Scarcity
Struggling to survive in places that can no longer sustain them, growing populations overfish, overharvest, and overgraze.

amount of energy and natural resources the U.S. economy uses to produce each constant dollar of output has steadily declined, as have many forms of pollution. When U.S. laws first required industry to control pollution, the response was to install cleanup equipment. The shift to a knowledge-driven economy has emphasized the positive connection among efficiency, profits, and environmental protection and helped launch a trend in profitable pollution prevention. More and more people now understand that pollution is waste, waste is inefficient, and inefficiency is expensive.

KINDS OF POLICY RESPONSES

As we begin the new century, it is important that we recognize that economic, environmental, and social goals are integrally linked and that we develop policies that reflect that interrelationship. Thinking narrowly about jobs, energy, transportation, housing, or ecosystems—as if they were not connected—creates new problems even as it attempts to solve old ones. Asking the wrong questions is a sure way to get misleading answers that result in short-term remedies for symptoms, instead of cures for long-term problems.

All of this will require new modes of decision making, ranging from the local to the international level. While trend is not always destiny, the trend that has been evolving over the past several years has been toward more collaborative forms of decision making. Perhaps such collaborative structures will involve more people and a broader range of interests in shaping and making public policy. It is hoped that this will improve decisions, mitigate conflict, and begin to counteract the corrosive trends of cynicism and civic disengagement that seem to be growing.

More collaborative approaches to making decisions can be arduous and time-consuming, and all of the players must change

FIGURE 19.2 **Trend Is Not Destiny** Scenes such as these were very common in North America only a short time ago. Fortunately, for the most part, such photos are today only historic in nature. Positive change is possible.

their customary roles. For government, this means using its power to convene and facilitate, shifting gradually from prescribing behavior to supporting responsibility by setting goals, creating incentives, monitoring performance, and providing information.

For their part, businesses need to build the practice and skills of dialogue with communities and citizens, participating in community decision making and opening their own values, strategies, and performance to their community and the society.

Advocates, too, must accept the burdens and constraints of rational dialogue built on trust, and communities must create open and inclusive debate about their future.

Does all of this sound too idealistic? Perhaps it is; however, without a vision for the future, where would we be? As was stated previously, trend is not destiny. In other words, we are capable of change regardless of the status quo. This is perhaps nowhere more important than in the world of environmental decision making. (See figure 19.2.)

GOVERNMENT AND GOVERNANCE

"Who let this happen? Who's responsible for this problem?" These are typical questions people ask in reaction to a local environmental problem or to the steady deterioration of global environmental conditions. For most people, it is not obvious who is "in charge" of the environment or how decisions are made about developing, using, or managing ecosystems.

Government

Government is the set of institutions we normally associate with political authority. Clearly, governments are important players in how

ecosystems are managed and how natural resources are exploited or conserved. National laws and regulatory frameworks set the formal rules for managing natural resources by recognizing discrete property, mineral, or water rights. They also establish the legal mandates of government agencies with responsibility for environmental protection and resource management. It is these government institutions that we frequently associate with big environmental decisions and the responsibility to govern natural resources.

Governments also act internationally (often through the United Nations) to set ground rules about pollution, water use, fishing rights, and other activities that affect resources across political boundaries. One of the most visible aspects of this global environmental governance is a large set of international environmental treaties, such as the Convention on Biological Diversity, the Convention on the Law of the Sea, and the Montreal Protocol to protect the stratospheric ozone layer. Multinational bodies such as the World Bank and the World Trade Organization are also assuming greater environmental significance in an increasingly globalized and interdependent world economy.

Governance

Governance is about decisions and how we make them. It is about the exercise of authority, about being in charge. It relates to decision makers at all levels—government managers and ministers, businesspeople, property owners, farmers, and consumers. In short, governance deals with who is responsible, how they wield their power, and how they are held accountable.

Environmental governance is inevitably associated with institutions—the organizations in which official authority often resides. These commonly include government departments of environment, agriculture, and mining, and environmental regulatory agencies. But governance also encompasses oversight or advisory groups, corporate councils and trade groups, and even private think tanks and advocacy groups that help to formulate policy. Overall, then, environmental governance takes in the whole range of institutions and decision-making practices that communities use to manage their environment and control natural resources.

Sometimes we use the term *governance* very broadly to describe not just the process of decision making but the actual management actions—where and when to log or how to limit fishing or distribute grazing permits—that result. In other words, in our day-to-day experience, we intertwine environmental governance and ecosystem management, which is where the real impact of decisions becomes visible.

But environmental governance goes beyond the official actions of governments. Sometimes, corporations or individuals act in the state's place to harvest or manage resources. For instance, states may grant forest or mining concessions to companies for a fee, allowing them broad discretion to cut trees, build roads, or make other important land-use decisions. Or the state may privatize once-public functions such as the delivery of water, electricity, or wastewater treatment, again putting a host of environmental choices—from water pricing to power plant construction—into private hands.

Nongovernmental organizations (NGOs), such as environmental organizations, civic groups, land trusts, labor unions, and neighborhood groups, have become strong advocates for better and fairer environmental decisions in the past few decades. The actions of industry groups, trade associations, and shareholder groups also influence the way companies do business by promoting or obstructing cleaner processes, better environmental accounting practices, or pointing out the financial liabilities of business practices that harm the environment.

Governance includes our individual choices and actions when these influence larger public policies or affect corporate behavior. Voting, lobbying, participating in public hearings, and joining an environmental watchdog or monitoring group are typical ways that individuals can influence environmental decisions. Our actions as consumers are powerful governance forces. For example, the choice to purchase environmentally friendly products such as organic produce, certified lumber, or a fuel-efficient car influences the environmental behavior of businesses through the marketplace. Consumer choices can sometimes be as powerful as government regulations in tempering business decisions that affect the environment.

LEARNING FROM THE PAST

For the past quarter century, the basic pattern of environmental protection in economically developed nations has been to react to specific crises. Institutions have been established, laws passed, and regulations written in response to problems that already were posing substantial ecological and public health risks and costs or that already were causing deep-seated public concern.

The United States is no exception. The U.S. Environmental Protection Agency (EPA) has focused its attention almost exclusively on present and past problems. The political will to establish the agency grew out of a series of highly publicized, serious environmental problems, such as the fire on the Cuyahoga River in Ohio, smog in Los Angeles, and the near extinction of the bald eagle. During the 1970s and 1980s, Congress enacted a series of laws intended to solve these problems, and the EPA, which was created in 1970, was given the responsibility for enforcing most environmental laws.

Despite success in correcting a number of existing environmental problems, there has been a continuing pattern of not responding to environmental problems until they pose immediate and unambiguous risks. Such policies, however, will not adequately protect the environment in the future. People are recognizing that the agencies and organizations whose activities affect the environment must begin to anticipate future environmental problems and take steps to avoid them. One of the most important lessons learned during the past quarter century of environmental history is that the failure to think about the future environmental consequences of prospective social, economic, and technological changes may impose substantial and avoidable economic and environmental costs on future generations.

THINKING ABOUT THE FUTURE

Thinking about the future is more important today than ever before because the accelerating rate of change is shrinking the distance between the present and the future. Technological capabilities that seemed beyond the horizon just a few years ago are now outdated. Scientific developments and the flow of information are accelerating. For example, who would have envisioned cellular phones that take photos and connect to the Internet, voice-activated computers, and a map navigator in your automobile only 20 years ago?

Thinking about the future is valuable also because the cost of avoiding a problem is often far less than the cost of solving it later. The U.S. experience with hazardous-waste disposal provides a compelling example. Some private companies and federal facilities undoubtedly saved money in the short term by disposing of hazardous wastes inadequately, but those savings were dwarfed by the cost of cleaning up hazardous-waste sites years later. In that case, foresight could have saved private industry, insurance companies, and the federal government (i.e., taxpayers) billions of dollars, while reducing exposure to pollutants and public anxieties in the affected communities.

Environmental foresight can preserve the environment for future generations. When one generation's behavior necessitates environmental remediation in the future, an environmental debt is bequeathed to future generations just as surely as unbalanced government budgets bequeath a burden of financial debt. (See figure 19.3.)

By anticipating environmental problems and taking steps now to prevent them, the present generation can minimize the environmental and financial debts that its children will incur.

Today, we face new classes of environmental problems that are more diffuse than those of the past and thus demand different approaches. Since the first Earth Day in 1970, the vast majority of the significant "point" sources of air and water pollution—large industrial facilities and municipal sewage systems—that once spewed untreated wastes into the air, rivers, and lakes have been

Shared water resources are the rivers and estuarine regions that form borders or flow across borders, the lakes that span political boundaries, marine areas with multiple jurisdictions, and the groundwater aquifers that lie beneath political boundaries. Because water is essential for supporting all life processes, many nations view its adequate availability as a fundamental human right. Conflicts over water rights were recorded as early as 2500 B.C., and such conflicts are expected to arise more frequently in the future as human populations and economic development continue to grow and climate patterns change.

North America has extensive shared water resources, but there are vast differences in the quality and quantity of those resources across the continent. Canada and the United States share water along their almost 9000-kilometer border, from the relatively water-rich areas in the east to the more arid regions in the west. Likewise, Mexico and the United States share water along their 3000-kilometer border, which runs along the arid regions from Texas to California. Potential conflicts over shared water resources in North America are addressed through bilateral water treaties, agreements, and protocols.

Canada and the United States have negotiated agreements to resolve water issues. In 1909, the Boundary Waters Treaty established the International Joint Commission to prevent and resolve disputes. In 1972, Canada and the United States signed the first Great Lakes Water Quality Agreement (revised in 1978 and 1987) to control pollution in the Great Lakes and to clean up wastes from industries and communities.

Major shared resources between Mexico and the United States are the Colorado River and Rio Grande/Rio Bravo and the Gulf of Mexico. Ensuring that both countries have sufficient shared water resources has been a major point of contention along the border. The Convention of 1906 between Mexico and the United States addressed water distribution. The U.S.-Mexico Water Treaty of 1944 distributed waters in the lower Rio Grande, the Colorado River, and the Tijuana River and also created the U.S.-Mexico International Boundary Water Commission. In 1983, an agreement to prevent, reduce, and eliminate sources of pollution was enacted by both countries.

Many of the water quality issues along country borders are similar. DDT and other pesticides, as well as PCBs, have contaminated fish tissue from the Gulf of Maine to the Great Lakes to the Gulf of Mexico. Mercury contamination of fish tissue in top predator fish, such as walleye and king mackerel, has led to fish consumption advisories in the Great Lakes and throughout the Gulf of Mexico. Rivers and streams throughout North America also exhibit degraded water quality because of oxygen-consuming organic matter, sedimentation that decreases water clarity and water depth and volume, and nutrients that contribute to nuisance and harmful algal blooms.

---

controlled. The most important remaining sources of pollution are diffuse and widespread: sediment, pesticides, and fertilizers that run off farmland; oil and toxic heavy metals that wash off city streets and highways; and air pollutants from automobiles, outdoor grills, and woodstoves. Pollution from these sources cannot always be controlled with sewage treatment plants or the same regulatory techniques used to check emissions from large industries. Furthermore, we now have the global environmental problems of biodiversity loss, ozone depletion, and climate change. These problems will require cooperative international responses. We are also recognizing that controlling pollutants alone, no matter how successful, will not achieve an environmentally sustainable economy, since many global concerns are related to the size of the human population and the unequal distribution of resources.

For the most fortunate, the past 50 years have produced a quality of life unprecedented in human history. For another 3 billion people, it has brought about marked improvements in living standards, including significant increases in life expectancy, infant survival, literacy, and access to safe drinking water. This progress notwithstanding, however, more than 20 percent of today's

FIGURE 19.3 **Environmental Debt** The plight of these fishermen, resulting from the collapse of the cod fishery in the northeast United States and Canada and the salmon fishery in the northwest United States and Canada, is an environmental debt inherited from past generations of abuse and misuse.

human population still lives in poverty, 15 percent experience persistent hunger, and at least 10 percent is homeless. Moreover, the gap between the very rich and very poor is widening, with whole regions of the world clearly losing ground.

Through both its successes and failures, modern human development has transformed the planet. Human activities have doubled the planet's rate of nitrogen fixation, tripled the rate of invasion by exotic organisms, increased sediment loads in rivers fivefold, and vastly increased natural rates of species extinctions. Clearly, in the future, we will need a different vision from that which shaped our past.

DEFINING THE FUTURE

We are progressing from an environmental paradigm based on cleanup and control to one including assessment, anticipation, and avoidance.

Changing Technologies

Expenditures to develop technologies that prevent environmental harm are beginning to pay off. Agricultural practices are becoming less wasteful and more sustainable, manufacturing processes are becoming more efficient in the use of resources, and consumer products are being designed with the environment in mind. The infrastructures that supply energy, transportation services, and water supplies are becoming more resource efficient and environmentally benign. Remediation efforts are cleaning up a large portion of existing hazardous-waste sites. Our ability to respond to emerging problems is being aided by more advanced monitoring systems and data analysis tools that continually assess the state of the local, regional, and global environment. Finally, we are developing effective ways of restoring or re-creating severely damaged ecosystems to preserve the long-term health and productivity of our natural resource base.

Involved Public

In the long run, environmental quality is not determined solely by the actions of government, regulated industries, or nongovernmental organizations. It is largely a function of the decisions and behavior of individuals, families, businesses, and communities everywhere. Consequently, the extent of environmental awareness and the strength of environmental institutions will be two critical factors driving changes in environmental quality in the future.

Looking ahead, consensus is growing that the next 50 years will see a world in which people are more crowded, more connected, and more consuming than at any time in human history. The natural environment in which those people live will be stressed as never before. It will be almost certainly warmer, more polluted, and less species-rich. Many of these trends have generated a good deal of discussion recently, under headings ranging from "globalization" to "climate change." Less remarked upon is the fundamental transition under way in the growth of human populations: Rates of increase are now falling almost everywhere in the world, with the result that the number of people on the planet is expected to level off at 10 billion or 11 billion by the end

of the twenty-first century, reaching around 9 billion—still half again as many as today's count—by 2050.

Let's put all of this into another perspective that may ring closer to home. Current projections are that the U.S. population will grow by 44 percent between 2008 and 2050. This works out to about a 50 percent increase over 50 years (2008–2058). Potentially, this translates into a 50 percent increase in U.S. infrastructure as well, which means 50 percent more cars, trucks, planes, airports, parking lots, streets, and freeways; 50 percent more houses and apartment buildings; 50 percent more landfills, wastewater treatments plants, hazardous-waste treatment facilities, and chemicals (pesticides and herbicides) for agriculture. (See figure 19.4.) In short, this vision is not very appealing. What happens to wilderness areas, remote and quiet places, and habitat for songbirds, waterfowl, and other wild creatures? What happens to our quality of life?

Another way of looking at the future would be along the lines of what was previously mentioned about the analogy to the agricultural and industrial revolutions. This translates into a future of profound change—a future in which virtually everthing we do will change. This future of profound change will also be one of phenomenal opportunity and excitement. It is a future in which we will farm, build, and transport in entirely new ways. This vision of the future could well become known as the environmental revolution.

DEVELOPMENT OF ENVIRONMENTAL POLICY IN THE UNITED STATES

Public **policy** is the general principle by which government **branches**—the **legislative, executive,** and **judicial**—are guided in their management of public affairs. The legislature (Congress) is directed to declare and shape national policy by passing legislation, which is the same as enacting law. The executive (president) is directed to enforce the law, while the judiciary (the court system) interprets the law when a dispute arises. (See figure 19.5.)

LEGISLATIVE ACTION

When Congress considers certain conduct to be against public policy and against the public good, it passes legislation in the form of acts or statutes. Congress specifically regulates, controls, or prohibits activity that is in conflict with public policy and attempts to encourage desirable behavior.

Through legislation, Congress regulates behavior, selects agencies to implement new programs, and sets general procedural guidelines. When Congress passes environmental legislation, it also declares and shapes the national environmental policy, thus fulfilling its policy-making function. (See figure 19.6.)

Over 90 years ago, President Teddy Roosevelt declared that nothing short of defending your country in wartime "compares in leaving the land even better land for our descendants than it is for us." The environmental issues that Roosevelt strongly believed in, however, did not become major political issues until the early 1970s.

FIGURE 19.4 **Options and Trade-offs for the New Century** Unless we become more creative in areas such as transportation and land use, it will be necessary to develop in the United States alone, in the next 50 years, an amount of farmland and scenic countryside equal to the amount developed over the past 200 years.

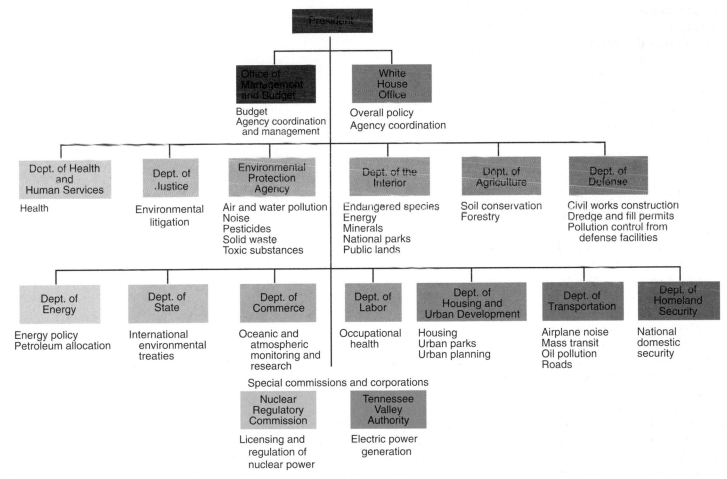

President

Office of Management and Budget	White House Office
Budget Agency coordination and management	Overall policy Agency coordination

Dept. of Health and Human Services	Dept. of Justice	Environmental Protection Agency	Dept. of the Interior	Dept. of Agriculture	Dept. of Defense
Health	Environmental litigation	Air and water pollution Noise Pesticides Solid waste Toxic substances	Endangered species Energy Minerals National parks Public lands	Soil conservation Forestry	Civil works construction Dredge and fill permits Pollution control from defense facilities

Dept. of Energy	Dept. of State	Dept. of Commerce	Dept. of Labor	Dept. of Housing and Urban Development	Dept. of Transportation	Dept. of Homeland Security
Energy policy Petroleum allocation	International environmental treaties	Oceanic and atmospheric monitoring and research	Occupational health	Housing Urban parks Urban planning	Airplane noise Mass transit Oil pollution Roads	National domestic security

Special commissions and corporations

Nuclear Regulatory Commission	Tennessee Valley Authority
Licensing and regulation of nuclear power	Electric power generation

FIGURE 19.5 **Major Agencies of the Executive Branch** Major agencies of the executive branch are shown with their environmental responsibility.

Environmental Policy and Decision Making

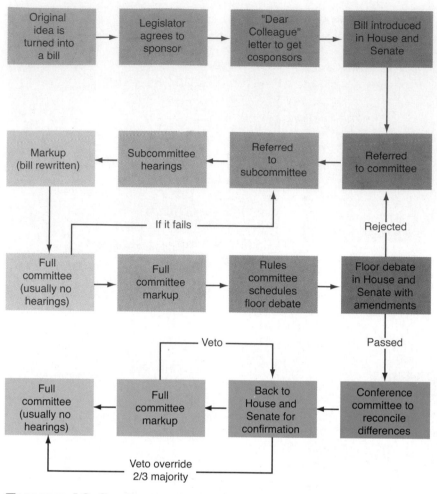

FIGURE 19.6 **Passage of a Law** This figure illustrates the path of a bill in the U.S. Congress from organization to becoming a law. As we can see, the process is not a quick one.

legislative focus during the beginning of the Obama administration. Initial policy directions included:

- Renewables like wind and solar beginning to displace coal from the electricity-generating market, with solar panels mounted on the roofs of nearly all big buildings.

- Electric cars becoming more widespread, and car batteries used to store excess electricity or as a source for more electricity when it is needed.

- Energy efficiency and conservation becoming the foundation of energy policy.

THE ROLE OF NONGOVERNMENTAL ORGANIZATIONS

The United States is commonly recognized as having been an environmental innovator in the 1970s. U.S. environmental laws and institutions became models for consideration by many other countries. Important to U.S. prominence as an environmental leader was the growth of environmental nongovernmental organizations. Environmental groups were aided by amendments to tax laws that made it relatively easy for groups to incorporate as not-for-profit organizations even while engaged in some lobbying activities. This provided them with important benefits, including the possibility of receiving contributions that are tax-deductible for the donor and favorable postage rates for reaching out to the public. Environmental groups could turn to both members and large, private financial institutions for donations; they were also eligible to receive governmental grants and contracts. U.S. environmental groups also won the important right to stand in court to sue on behalf of environmental interests. Using these new rights, environmental groups became increasingly professionalized and diversified.

While the publication of Rachel Carson's *Silent Spring* in 1962 is considered to be the beginning of the modern environmental movement, the first Earth Day on April 22, 1970, was perhaps the single event that put the movement into high gear. In 1970, as a result of mounting public concern over environmental deterioration—cities clouded by smog, rivers on fire, waterways choked by raw sewage—many nations, including the United States, began to address the most obvious, most acute environmental problems.

Public opinion polls indicate that a permanent change in national priorities followed Earth Day 1970. When polled in May 1971, 25 percent of the U.S. public declared protecting the environment to be an important goal—a 2500 percent increase over the proportion in 1969.

During the 1970s, many important pieces of environmental legislation were enacted in the United States. In addition, the percent of discretionary U.S. budget spent on natural resources and the environment grew significantly. Many of the identified environmental problems were so immediate and so obvious that it was relatively easy to see what had to be done and to summon the political will to do it. (See table 19.1.) Energy policy became a primary area of

THE CHALLENGE FOR U.S. ENVIRONMENTAL POLICY

The ongoing challenge for U.S. environmental policy is to build, maintain, and constantly renew public support for effective environmental governance, at home and worldwide. To meet that need, U.S. environmental policy today must recover an essential missing element: a broadly shared vision of the common environmental good. Such visions have emerged at several points in the past. Examples include the sanitation movement of the nineteenth century; the City Beautiful movement of the 1890s; and the Progressive civic reform and conservation movements that followed it the New Deal vision of the 1930s of combining ecological, social, and economic recovery; and the vision of a modern society in harmony with its natural environment that was widely voiced by the public on Earth Day in 1970.

TABLE 19.1 Major U.S. Environmental and Resource Conservation Legislation

Wildlife conservation

Anadromous Fish Conservation Act of 1965

Fur Seal Act of 1966

National Wildlife Refuge System Act of 1966, 1976, 1978

Species Conservation Act of 1966, 1969

Marine Mammal Protection Act of 1972

Marine Protection, Research, and Sanctuaries Act of 1972

Endangered Species Act of 1973, 1982, 1985, 1988, 1995

Fishery Conservation and Management Act of 1976, 1978, 1982, 1996

Whale Conservation and Protection Study Act of 1976

Fish and Wildlife Improvement Act of 1978

Fish and Wildlife Conservation Act of 1980 (Nongame Act)

Fur Seal Act Amendments of 1983

Land use and conservation

Taylor Grazing Act of 1934

Wilderness Act of 1964

Multiple Use Sustained Yield Act of 1968

Wild and Scenic Rivers Act of 1968

National Trails System Act of 1968

National Coastal Zone Management Act of 1972, 1980

Forest Reserves Management Act of 1974, 1976

Forest and Rangeland Renewable Resources Act of 1974, 1978

Federal Land Policy and Management Act of 1976

National Forest Management Act of 1976

Soil and Water Conservation Act of 1977

Surface Mining Control and Reclamation Act of 1977

Antarctic Conservation Act of 1978

Endangered American Wilderness Act of 1978

Alaskan National Interests Lands Conservation Act of 1980

Coastal Barrier Resources Act of 1982

Food Security Act of 1985

Emergency Wetlands Resources Act of 1986

North American Wetlands Conservation Act of 1989

Coastal Development Act of 1990

California Desert Protection Act of 1994

Federal Agriculture Improvement and Reform Act of 1996

General

National Environmental Policy Act (NEPA) of 1969

International Environmental Protection Act of 1983

Energy

Energy Policy and Conservation Act of 1975

National Energy Act of 1978, 1980

Northwest Power Act of 1980

National Appliance Energy Conservation Act of 1987

Energy Policy Act of 1992

Water quality

Refuse Act of 1899

Water Quality Act of 1965

Water Resources Planning Act of 1965

Federal Water Pollution Control Acts of 1965, 1972

Ocean Dumping Act of 1972

Safe Drinking Water Act of 1974, 1984, 1996

Clean Water Act of 1977, 1987

Great Lakes Toxic Substance Control Agreement of 1986

Great Lakes Critical Programs Act of 1990

Oil Spill Prevention and Liability Act of 1990

Air quality

Clean Air Act of 1963, 1965, 1970, 1977, 1990

Noise control

Noise Control Act of 1965

Quiet Communities Act of 1978

Resources and solid waste management

Solid Waste Disposal Act of 1965

Resources Recovery Act of 1970

Resource Conservation and Recovery Act of 1976

Waste Reduction Act of 1990

Toxic substances

Toxic Substances Control Act of 1976

Resource Conservation and Recovery Act of 1976

Comprehensive Environmental Response, Compensation, and Liability (Superfund) Act of 1980, 1986, 1990

Nuclear Waste Policy Act of 1982

Pesticides

Food, Drug, and Cosmetics Act of 1938

Federal Insecticide, Fungicide, and Rodenticide Control Act of 1972, 1988

Food Quality Protection Act of 1996

Such visions are largely absent in the United States today. The modern environmental movement has been far more effective as an opposition movement rallying the public to arms against corporate polluters and government despoilers than in articulating a positive vision.

Perhaps the closest such vision today is the idea of sustainable development as articulated by the United Nations World Commission for Environment and Development in 1987. The Commission envisioned sustainable development as a pattern of development that would meet the needs of human communities today without jeopardizing those of the future. Its vision specifically included economic development, ecological sustainability, and social equity as essential and interdependent elements.

ENVIRONMENTAL POLICY AND REGULATION

Environmental laws are not a recent phenomenon. As early as 1306, London adopted an ordinance limiting the burning of coal because of the degradation of local air quality. Such laws became more common as industrialization created many sources of air and water pollution throughout the world. In the United States, environmental laws often evolved from ordinances passed by local governments. Interested in protecting public health, officials of towns and cities enacted local laws to limit the activities of private citizens for the common good. For example, to have "healthy air,"

In 2006, after ringing the opening bell of the New York Stock Exchange, Secretary General of the United Nations, Kofi Annan, launched the Principles for Responsible Investment (PRI). Six months later, PRI had 94 institutional investors from 17 countries representing US $5 trillion in investments.

The launch of the principles created the first-ever global network of investors who addressed many of the same environmental, social, and governance issues that the United Nations is asked to address. One of the goals of the PRI community is to work with policy makers to address issues of long-term importance to both investors and society. Investors representing more than 10 percent of global capital market value have, therefore, sent the strongest of signals to the marketplace that environment, social, and good governance issues count in investment policy making and decision making. The PRI has evolved because investors have recognized that systemic issues of sustainability are material to long-term investment returns.

There is a range of areas where policy makers could create the necessary environment that would encourage investors to take longer-term views on environmental, social, and governance issues. Mandatory disclosure of environmental performance is one such area. Once investors are able to assess the risks involved in various activities, they are able to put pressure on companies to address those risks. But they are unable to do this if they are unaware of what the company is doing. Mandatory disclosure policies would allow investors to take action when required.

many communities enacted ordinances in the 1880s to regulate rubbish burning within city limits. Public health issues were the foundations on which the environmental laws of today were built.

THE SIGNIFICANCE OF ADMINISTRATIVE LAW

Environmental law in the United States is governed by administrative law. Administrative law is a relatively new concept, having been developed only during the twentieth century. In 1946, Congress passed the Federal Administrative Procedure Act. This act designated general procedures to be used by federal agencies when they exercised their rule-making, adjudicatory, and enforcement powers. This is a rapidly expanding area of law and defines how governmental organizations such as agencies, boards, and commissions develop and implement the regulatory programs they are legislatively authorized to create. Some of the many U.S. federal agencies that influence environmental issues include the Environmental Protection Agency, the Council on Environmental Quality, the National Forest Service, and the Bureau of Land Management.

Administrative law applies to government agencies and to those that are affected by agency actions. In the United States, many federal environmental programs are administered by the states under the authority of federal and related state laws. States often differ from both the federal government and each other in the way they interpret, implement, and enforce federal laws. In addition, each state has its own administrative guidelines that govern and define how state agencies act. All actions of federal agencies must comply with the 1946 Administrative Procedure Act.

NATIONAL ENVIRONMENTAL POLICY ACT—LANDMARK LEGISLATION

The National Environmental Policy Act (NEPA) was enacted in 1969 and signed into law by President Richard Nixon on New Year's Day, 1970. It is a short, general statute designed to institutionalize within the federal government a concern for the "quality of the environment." NEPA helps encourage environmental awareness among all federal agencies, not just those that prior to NEPA had to consider environmental factors in their planning and decision making. Until 1970, most federal agencies acted within their delegated authority without considering the environmental impacts of their actions. However, in the 1960s, Congress seriously began to study pollution problems. Because Congress has found that the federal government is both a major cause of environmental degradation and a major source of regulatory activity, all actions of the federal government now fall under NEPA.

NEPA forces federal agencies to consider the environmental consequences of their actions before implementing a proposal or recommendation. NEPA has two purposes: first, to advise the president on the state of the nation's environment; and second, to create an advisory council called the Council on Environmental Quality (CEQ). The CEQ outlines NEPA compliance guidelines. The CEQ also provides the president with consistent expert advice on national environmental policies and problems.

NEPA has been interpreted narrowly by the federal courts. As a result, many states have passed much stronger state environmental protection acts (SEPAs) as well. Today, NEPA analysis is undertaken as part of almost every recommendation or proposal for federal action. This includes not only actions by agencies of the federal government but also actions of states, local municipalities, and private corporations. NEPA is Congress's mission statement that mandates the means by which the federal government, through the guidance of the CEQ, will achieve its national environmental policy.

OTHER IMPORTANT ENVIRONMENTAL LEGISLATION

In addition to the passage of NEPA, the 1970s saw a series of new environmental laws passed, including the Resource Conservation

EPA Enforcement Options

1. Warning letter
- Describes alleged violation
- Outlines action that must be taken to remedy
  - Deadline for compliance or enforcement action
  - Allow for an opportunity to discuss the situation

2. Administrative order
- Requires certain activity be ceased, or
- Requires compliance by a certain date, or
- Requires certain tests be performed
- Provides rights
  - Respond by citing defenses or objections
  - Right to confer with EPA

3. Permit action
- Revoke or modify existing permit
- Seek to add conditions to permits that are being negotiated

4. Civil enforcement
- Injunctive
- Monetary relief varies by statute per violation, per day

5. Criminal proceedings
- Fines
- Incarceration

FIGURE 19.7 **Enforcement Options of the U.S. Environmental Protection Agency** The enforcement options of the U.S. Environmental Protection Agency range from a warning letter to a jail sentence.

Source: U.S. Environmental Protection Agency, Office of Enforcement, Washington, D.C.

and Recovery Act (RCRA); the Comprehensive Environmental Response, Compensation and Liability Act (CERCLA); and the Clean Air Act (CAA). All of these acts are broadly worded to identify existing problems that Congress believes can be corrected to protect human health, welfare, and the environment.

Protecting human health, welfare, and the environment is the national policy that Congress has chosen to encourage. For example, under NEPA, the national policy is to "promote efforts which will prevent or eliminate damage to the environment and biosphere and stimulate the health and welfare of [humans]." Similarly, under the Clean Air Act, the policy is to "protect and enhance the quality of the Nation's air resources so as to promote the public health and welfare and the productive capacity of its population."

Each of the preceding statutes declares national policy on environmental issues and addresses distinct problems. These statutes also authorize the use of some or all of the administrative functions discussed previously, such as rule making, adjudication, administrative law, civil and criminal enforcement, citizen suits, and judicial review. (See figure 19.7.)

ROLE OF THE ENVIRONMENTAL PROTECTION AGENCY

Congress established the Environmental Protection Agency in 1970 as the primary agency to implement the statutes. Administrative functions empower EPA, the states, and private citizens to take responsibility for enforcing the various authorized programs. These administrative functions not only shape environmental law but also control the daily operations of both the regulated industry and agencies authorized to protect the environment.

Command and Control Philosophy

To date, much environmental law has reflected the perception that environmental problems are localized in time, space, and media (i.e., air, water, soil). For example, many hazardous-waste sites in the United States have been "cleaned" by simply shipping the contaminated dirt someplace else, which not only does not solve the problem but creates the danger of incidents during removal and transportation. Environmental regulation has focused on specific phenomena and adopted the so-called command-and-control approach, in which restrictive and highly specific legislation and regulation are implemented by centralized authorities and used to achieve narrowly defined ends.

Such regulations generally have very rigid standards, often mandate the use of specific emission-control technologies, and generally define compliance in terms of "end-of-pipe" requirements. Examples in the United States include the Clean Water Act (which applied only to surface waters), the Clean Air Act (urban air quality), and the Comprehensive Environmental Response, Compensation, and Liability Act (Superfund), which applied to specific landfill sites.

If properly implemented, command-and-control methods can be effective in addressing specific environmental problems. For example, rivers such as the Potomac and the Hudson in the United States are much cleaner as a result of the Clean Water Act. (See figure 19.8.) Moreover, where applied against particular substances, such as the ban on tetraethyl lead in gasoline in the United States, the command-and-control approach has clearly worked well.

Regardless of how environmental decisions are made, most decisions are complex and involve compromise. Large-scale and complicated ecological policy problems, such as reversing the decline of salmon, deciding on the proper role of wild fires on public lands, the consequences of declining biological diversity, and making sense about the confusing policy choices surrounding interpretations of sustainability all share several qualities:

1. **Complexity**—innumerable options and trade-offs;
2. **Polarization**—clashes between competing values;
3. **Winners and losers**—for each policy choice, some will clearly benefit, some will be harmed, and the consequences for others are uncertain;
4. **Delayed consequences**—no immediate "fix" and the benefits, if any, of painful concessions will often not be evident for decades;
5. **National vs. regional conflict**—national (or international) priorities often differ substantially from those at the local or regional level; and
6. **Ambiguous role for science**—science is often not pivotal in evaluating policy options, but science often ends up serving inappropriately as a surrogate for debates over values and preferences.

(a)

(b)

FIGURE 19.8   Effect of the Command-and-Control Approach The environmental quality of rivers has dramatically improved over the past 30 years as a direct result of the Clean Water Act. (a) A visibly impaired river in New England in the 1970s. (b) An environmentally healthy Hudson River in the late 1990s.

THE GREENING OF GEOPOLITICS

Environmental or "green" politics have emerged from minority status and become a political movement in many nations. Issues such as transboundary water supply and pollution, acid precipitation, and global climate change have served to bolster the emergence of green politics.

INTERNATIONAL ASPECTS OF ENVIRONMENTAL PROBLEMS

Concern about the environment is not limited to developed nations. A 1989 treaty signed in Switzerland limits what poorer nations call toxic terrorism—use of their lands by richer countries as dumping grounds for industrial waste. In 1990, more than 100 developing nations called for a "productive dialogue with the developed world" on "protection of the environment." As was covered in chapter 2, the Earth Summit in Rio brought together nearly 180 governments to address world environmental concerns, and the 1997 conference on global warming in Kyoto brought together some 120 nations.

Environmental concern is also a growing factor in international relations. Many world leaders see the concern for environment, health, and natural resources as entering the policy mainstream. A sense of urgency and common cause about the environment is leading to cooperation in some areas. Ecological degradation in any nation is now understood almost inevitably to impinge on the quality of life in others. Drought in Africa and deforestation in Haiti have resulted in large numbers of refugees whose migrations generate tensions both within and between nations. From the Nile to the Rio Grande, conflicts flare over water rights. The growing megacities of the developing world are areas of potential civil unrest. Sheer numbers of people overwhelm social services and natural resources. The government of the Maldives has pleaded with the industrialized nations to reduce their production of greenhouse gases, fearing that the polar ice caps may melt and inundate the island nation.

Economic progress in developing nations could also bring the possibility for environmental peril and international tension. Energy consumption in China is having a major impact on world energy prices and efforts to control atmospheric carbon dioxide. China's energy consumption increased by 75 percent between 2002 and 2007. China now consumes about 16.8 percent of the world's energy. It is the second-largest energy consumer in the world, following the United States, which consumes about 21 percent of the world's energy. China's energy use is currently growing at nearly 8 percent per year (compared with growth of less than 1.7 percent for the United States).

In 2007, coal accounted for over 70 percent of China's primary energy production. Because of China's extensive domestic coal resources and its wish to minimize dependence on foreign energy sources, it is expected that coal will remain the main energy source for the foreseeable future. One of the biggest concerns regarding energy consumption in China focuses on carbon emissions and the threat of increasing global warming. China is putting a large number of new coal-fired power plants on line every year. Most of these will be coal-fired. Thus, China's carbon emissions are expected to increase by more than 10 percent per year.

More than half of China's oil imports now come from the Middle East. Consequently, China is now in the same position as

THE ENVIRONMENTAL EFFECTS OF HURRICANE KATRINA

Hurricane Katrina, which hit the Gulf of Mexico coast of the United States in the autumn of 2005, is perhaps the worst environmental catastrophe ever to befall the country as a result of a natural disaster. The scope and magnitude of the hurricane was massive, and the environmental impact will be felt for years to come.

According to U.S. Coast Guard and EPA data there were 575 Katrina-related spills of petroleum or hazardous chemicals reported. It is estimated that the oil spills totaled 20 million liters.

It is also estimated that there were 350,000 or more ruined automobiles and other vehicles damaged by the flooding. The amount of gasoline and hazardous fluids in these vehicles could add another 12 million liters to the figure mentioned previously. By comparison, some 40 million liters of oil were released in the Exxon Valdez disaster.

The scope of devastation caused by Hurricane Katrina is difficult to comprehend.

At least four Superfund hazardous-waste sites in the New Orleans area were hit by Katrina. Across the storm-ravaged areas of Louisiana, Mississippi, and Alabama, dozens of other toxic waste sites, major industrial facilities, ports, barges, and vessels that handle enormous quantities of oil and hazardous chemicals took a direct blow from the hurricane. In addition to oil and chemical spills, and potential releases from toxic waste or industrial facilities, one major source of toxins is the stirring up of the toxic sediment that has accumulated at the bottom of many of the lakes, rivers, and streams in industrialized areas over many decades due to Industrial spills. This toxic sediment could be redeposited in residential communities.

Katrina generated an estimated 100 million cubic yards of debris—enough to cover over 1000 football fields with 20 meters of waste. While some of this debris Is merely downed trees or vegetation, much of it is destroyed housing, commercial buildings, and a wide array of other detritus, much of which is intermixed with plastics and other materials that will become toxic if burned.

Katrina also had enormous ecological impacts. The associated spills, storm surge, and floodwaters carried saltwater and pollution into sensitive and ecologically important waters and marshes that serve as the nursery for many rare birds, as well as fish, shrimp, and other forms of life. The overwhelming human and financial impacts of Katrina are evidence that political and economic decisions have often failed to account for our dependence on a healthy resource base. The alteration of the Mississippi River and the destruction of wetlands at its mouth left the area around New Orleans vulnerable to the forces of nature. Many scientists also argue that global warming is causing levels of water in the Gulf of Mexico to rise, and rising sea levels may also have exacerbated the destructive power of Katrina.

According to the Worldwatch Institute, the long-term lessons of Katrina should include:

1. Maintaining the integrity of natural ecosystems should be a priority: Indiscriminate economic development and ecologically destructive policies have left many communities more vulnerable to disasters than they realize. This, together with rapid population growth in vulnerable areas, has contributed to worldwide economic losses from weather-related catastrophes totaling $567 billion over the last 10 years, exceeding the combined losses from 1950 through 1989. Losses in 2004 exceeded $100 billion for the second time ever, and a new record will certainly be set once Katrina's damages are totaled.

2. Short-term thinking is a dangerous approach to policy. During the past few years, the U.S. government has diverted funding from disaster preparedness and has reduced protections for wetlands in order to spur economic development. Both decisions are now exacting costs that far exceed the money saved. Natural ecosystems such as wetlands and forests are often more valuable when left intact so as to protect communities from floods, landslides, drought, and other natural occurrences. Failure to protect ecosystems contributed to the massive loss of life when the tsunamis swept across the Indian Ocean in 2004 and when Hurricane Mitch killed 10,000 people in Central America in 1998.

3. The links between climate change and weather-related catastrophes need to be addressed by decision makers: Although no specific storm can be definitively linked to climate change, scientists agree that warm water is the fuel that increases the intensity of such storms and that tropical seas have increased in temperature over the past century. (Katrina transformed rapidly from a Category 1 to a Category 5 hurricane when it passed from the Atlantic Ocean to the much warmer Gulf of Mexico.) In the next few decades, water temperatures and sea levels will continue to rise, greatly increasing the vulnerability of many communities. Global warming and its anticipated effects on the hydrological cycle will make some areas more vulnerable as storms, floods, and droughts increase in frequency and intensity.

4. There is an urgent need to diversify energy supplies: Decades of failure to invest in new energy options have left the world dependent on oil and natural gas that are concentrated in some of the world's most vulnerable regions—the U.S. Gulf Coast, the Persian Gulf, and the Niger Delta in Africa. Biofuels and other renewable resources now represent viable alternatives to fossil fuels, which are not only vulnerable to natural disasters but could have a big impact on the severity of future disasters.

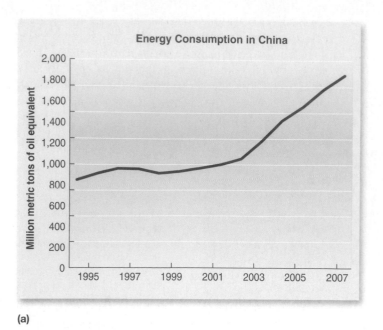

FIGURE 19.9 **The Growth of Energy Consumption in China** (a) Energy. (b) Oil.

FIGURE 19.10 **Developing Concerns** If China's current "modernization" continues, the boom will be fueled by coal to the possible detriment of the planet as a whole.

most of the industrialized world—its economic stability is dependent on access to oil supplies from the world's most volatile region. China's demand for oil has been an important contributing factor in the worldwide increase in oil prices.

Figure 19.9 shows the growth in total energy use and the growth in use of oil in China in recent years. Both have doubled over the time period. (See figure 19.10.)

Some experts estimate that the developing world, which today produces one-fourth of all greenhouse gas emissions, could be responsible for nearly two-thirds by the middle of the twenty-first century. Developing nations have repeatedly indicated that they are not prepared to slow down their own already weak economic growth to help compensate for decades of environmental problems caused largely by the industrialized world.

Some developing countries may resist environmental action because they see a chance to improve their bargaining leverage with foreign aid donors and international bankers. Where before the poor nations never had a strategic advantage, they now may have an ecological edge. Ecologically, there could be more parity than there ever was economically or militarily.

NATIONAL SECURITY ISSUES

National security may no longer be about fighting forces and weaponry alone. It also relates increasingly to watersheds, croplands, forests, climate, and other factors rarely considered by military experts and political leaders but that, when taken together, deserve to be viewed as equally crucial to a nation's security as are military factors. It is interesting to note that the North Atlantic Treaty Organization (NATO) has developed an Office for Scientific and Environmental Affairs, and the U.S. Department of Defense has created an Office of Environmental Security.

The Hungarian Academy of Sciences has pointed out that 90 percent of the water entering the tributaries of Hungary's rivers flows through Romania, Ukraine, Slovakia, and Austria. Numerous mines, chemical plants, oil refineries, and other sources of pollutants line those tributaries. The Hungarian government stated that the management of environmental security cannot be stopped at the borders. The government has filed lawsuits seeking monetary damage against the operators of the Baia-Mare gold extraction lagoon, and it has threatened to sue the Romanian government to help recover the cleanup costs. The accident stoked bilateral tensions: Romanian officials accused the Hungarian side of exaggerating the extent of the damage, while Hungarians asserted that the Romanians were downplaying the spill. The issue of environmental security continues to expand in the dialogue between the two nations.

The increased attention to the environment as a foreign policy and national security issue is only the beginning of what will be necessary to avert problems in the future. The most formidable obstacle may be the entrenched economic and political interests of the world's most advanced nations. If North Americans wish to stem the supply of hardwood from a fragile jungle or furs from endangered species, then they will have to stem the demand for fancy furniture and fur coats. If they wish to preserve wilderness from the intrusions of the oil industry, then they will have to find alternative sources of energy and use all fuels more efficiently.

TERRORISM AND THE ENVIRONMENT

A study of environmental terrorism requires understanding motivations, identifying vulnerabilities and risks, and working on effective solutions. At a time when world populations are increasing, the existing resource base (water, energy, soils) is being stretched to provide for more people and is being consumed at a faster rate. As the value and vulnerability of these resources increase, so does their attractiveness as terrorism targets. Terrorists often choose targets because of what they represent, for example, skyscrapers and government buildings. Rivaling both of those, however, for the amount of long-term damage that can be inflicted on a country, environmental resources should be considered as being at risk. **Environmental terrorism** is defined as the unlawful use of force against environmental resources so as to deprive populations of their benefits or destroy other property. Over the past several decades, several types of environmental terrorism have been waged, from acts of war to acts by individuals or small groups.

The effects of war on the natural environment are catastrophic. By its nature, war is destructive, and the environment is an innocent victim. While the effects of war on the environment have been with us over the centuries, the 1990–91 conflict in the Persian Gulf witnessed a new role for the environment in war: its degradation as a weapon.

The term *ecoterrorism* was applied to former Iraqi leader Saddam Hussein's use of oil as a weapon in the war. Iraq dumped an estimated 1.1 billion liters (290 million gallons) of crude oil into the Persian Gulf from Kuwait's Sea Island terminal. The oil spill that resulted from the dumping was the world's largest—over 30 times the size of the 1989 *Exxon Valdez* disaster in Alaska. (See figure 19.11a.)

Saddam Hussein may have engineered the spill to block allied plans for an amphibious invasion of Kuwait, but he was also probably trying to shut down seaside desalinization plants that provide much of the freshwater for Saudi Arabia's Eastern Province. Whatever the military or political effect, the environmental effect was much greater.

Another act of ecoterrorism that took place during the Iraq-Kuwait war was the intentional burning of hundreds of oil wells in Kuwait. (See figure 19.11b.) The air pollution resulting from the burning oil covered a large area. It took several months to extinguish

(b)

FIGURE 19.11 **Ecoterrorism** (a) The Persian Gulf was the victim of environmental terrorism during the 1991 Gulf War. (b) Oil well fires burning in Kuwait.
Source: (a) United Nations Development Program.

(a)

Environmental Policy and Decision Making 441

the fires following the war. Using environmental damage as a weapon of war is a frightening concept. The environmental consequences of such acts of ecoterrorism will not be fully understood for years.

Concerns over the use of biological and chemical weapons as a means of terrorism became, unfortunately, all too real following the events of September 11, 2001. The use of such weapons, however, is not new. During the Vietnam War, a chemical defoliant called Agent Orange was used extensively. During the First World War, several chemicals were used as poison gas. History is full of examples of both chemical and biological weapons being used. There is a difference, however, between biological and chemical weapons.

Biological weapons are naturally occurring organisms that cause disease. The two most common examples are the bacterium *Bacillus anthracis* (anthrax), which produces a toxin, and the virus that causes smallpox, a highly infectious disease. Anthrax bacteria produce spores that allow them to live in a dormant state in soil. When used as a weapon, the spores enter the lungs, where they are carried into the blood and the immune system. The spores become active, reproduce in large numbers, and release a devastating toxin that is lethal to cells. If enough spores are inhaled, anthrax can be fatal.

The history of biological weapons is a long one. Almost as soon as humans figured out how to make arrows, they were dipping them in animal feces to poison them. The Roman Empire used animal carcasses to contaminate their enemies' water wells. This practice was used again in Europe's many wars, in the American Civil War, and even into the twentieth century. In the fifteenth century, Spanish conquistador Francisco Pizarro gave clothing contaminated with the smallpox virus to natives in South America. Britain's Lord Amherst continued the practice into the late eighteenth century, spreading smallpox among Native Americans during the French-Indian War by giving them blankets that had been used at a hospital treating smallpox victims.

Chemical weapons are poisons such as mustard gas and nerve gases such as sarin. In the First World War, poison gas was used on both the Eastern and Western fronts from 1915 to 1918. Chlorine gas was the most widely used of the poison gases. Chlorine gas burns and destroys lung tissue. Chlorine is not an exotic chemical. Most municipal water systems use it to kill bacteria.

Modern chemical weapons tend to be made with agents having much greater power, meaning that it takes a lot less of the chemical to kill the same number of people. Many of them use the types of chemicals found in insecticides. When you spray your lawn or garden with a chemical to control aphids, you are, in essence, waging a chemical war on aphids.

In 1972, 103 countries signed the Biological Weapons Convention (BWC), which prohibited the development and use of biological and chemical weapons. Even so, several countries are known to have developed biological weapons since the convention was signed. The BWC still allows research for defense, such as vaccines, against biological weapons. In 2001, the United States announced it was developing a new form of anthrax for defensive research.

While war is obviously harmful to the environment, the reverse could also be argued in certain circumstances. Armed conflict—or the threat of it—can sometimes be good for the environment. The demilitarized zone between North and South Korea is a 250-kilometer (155-mile) long strip of mountains, jungle, and wetlands untouched by humans since 1953. It is now home to wildlife that is extinct elsewhere on the peninsula. Land mines laid during civil wars in Africa have discouraged hunters and allowed game to flourish in areas from which it had previously disappeared. In Congo, anarchy has prevented mining companies and timber firms from spreading into the country's remaining wild areas. Although many large animals have been killed by gun-toting soldiers, recent aerial surveys suggest that Congo's rhinos have survived the conflict well; some 6000 elephants remain as well. The people of Congo would doubtless prefer less anarchy, even if it meant fewer elephants. But the fact remains that when men are busy killing each other, nature sometimes gains.

INTERNATIONAL ENVIRONMENTAL POLICY

If there must be a war, let it be against environment contamination, nuclear contamination, chemical contamination; against the bankruptcy of soil and water systems; against the driving of people away from the lands as environmental refugees. If there must be war, let it be against those who assault people and other forms of life by profiteering at the expense of nature's capacity to support life. If there must be war, let the weapons be your healing hands, the hands of the world's youth in defense of the environment.

Mustafa Tolba
Former Secretary General
United Nations Environment
Programme

THE ROLE OF THE UNITED NATIONS

There are many institutions that address the global environment. In the United Nations system, 21 separate agencies deal with environmental issues. In addition to the United Nations, the World Bank and other institutions charged with economic development play an important role in the implementation of policies and projects that affect the global environment. Institutions that deal with trade such as the World Trade Organization (WTO) and the North American Free Trade Agreement (NAFTA) also affect the global environment.

Successes and Failures

While there have been examples of successful initiatives, global organizations have not been able to achieve significant progress in reversing global environmental degradation. There are several reasons for this. Some fail because they are controlled by a disparate membership with competing interests who are unable to reach consensus on complex and difficult issues. An example of this is the UN Food and Agriculture Organization (FAO), under whose policies only limited progress has been made in dealing with global fishing and agriculture issues. Although the FAO has pursued the collection of statistics on fish stocks and it is clear that most fisheries are being fished at or above capacity, it has not acted to develop systems that would conserve fish stocks, and individual governments continue to set catch quotas above sustainable levels.

Other institutions do not succeed because they are responsible only for specific functions or activities and are unable to address whole issues on their own. The World Bank, for example, can provide guidelines for dealing with air pollution or biodiversity preservation only for development projects that use World Bank funds. Still other institutions have failed to take even initial steps to weave the environment into their policies and programs as a permanent concern. For example, the WTO focuses on trade issues but has yet to confront the nexus of trade—sustainable economic growth and the environment.

International Coordination

International coordination and political resolve are necessary if the goals of preserving and protecting the global environment are to be realized. A step in that direction was the 1972 United Nations Conference in Stockholm, Sweden. This was the first international conference specifically dealing with global environmental concerns. The UN Environment Programme, a separate department of the United Nations that deals with environmental issues, developed out of that conference. The United Nations Conference on the Law of the Sea produced a comprehensive convention that addressed many of the issues concerning jurisdiction over ocean waters and the use of ocean resources. This treaty is viewed by many as a model for international environmental protection. Positive results are already evident from application of the agreements on pollution control, marine mammal protection, navigation safety, and other aspects of the marine environment. The issue of deep ocean mining, primarily of manganese nodules, has to date kept many industrial nations, such as Germany and the United States, from ratifying the treaty.

Over 150 global environmental treaties have been negotiated since the start of the twentieth century. In addition, at least 500 bilateral agreements are now in effect dealing with cross-border environmental issues. These agreements generally address single issues, such as the Framework Convention on Climate Change or the Convention on International Trade in Endangered Species of Wild Fauna and Flora. History has shown that most of the existing agreements are lacking in ambition or are framed in such a way that they result in little concrete action.

However, there have been several successful international conventions and treaties. The Antarctic Treaty of 1961 reserves the Antarctic continent for peaceful scientific research and bans all military activities in the region. The 1979 Convention on Long-Range Transboundary Air Pollution was the first multilateral agreement on air pollution and the first environmental accord involving all the nations of Eastern and Western Europe and North America. The 1987 Montreal Protocol on Substances That Deplete the Stratospheric Ozone Layer addresses the ozone protective shield problem.

Agreed to in 1987, the Montreal Protocol's objective was to phase out the manufacture and use of chemicals depleting the Earth's protective ozone layer (see chapter 16). Since its adoption over a decade ago, 160 countries are now parties, representing over 95 percent of the Earth's population. Production and consumption of chlorofluorocarbons (CFCs), carbon tetrachloride, halons, and methyl chloroform have been phased out in developed countries, with reduction schedules set for their use in developing countries. In 1999, developing countries ended the production and consumption of CFCs, with phaseout scheduled for 2010.

By the late 1990s, the concentration of some CFCs in the atmosphere had started to decline, and predictions are that the ozone layer could recover by the middle of this century. Among the many reasons for these achievements, three merit attention:

- *Global agreement on the nature and seriousness of the threat.* Even the strongest skeptics could not deny the Antarctic ozone hole, which was first brought to international attention by British scientists in 1985. It was understood for the first time that emissions of ozone-depleting substances were in reality putting our lives and the lives of future generations at risk. Decisive action was required.

- A *cooperative approach, especially between developed and developing countries.* In the developed countries, it was recognized that their industries had contributed significantly to this global problem and that they had to take the lead in stopping emissions and finding alternatives. It was also recognized that solving the ozone problem required a global solution, with all countries committed to eliminating ozone-depleting substances. Innovative partnerships, including early controls for developed countries and a grace period, funding, and technology transfer for developing countries, were set in place.

- *Policy based on expert and impartial advice.* The parties to the Montreal Protocol were able to receive impartial advice from their science, technical, and economic committees. These drew together experts from around the world to evaluate the need for further action and to propose options that were technically and economically feasible.

Put these three factors together—acknowledgment of the threat, agreement to cooperate, and commitment to take effective action based on expert advice—and you have a potentially strong recipe to solve global issues, such as climate change and biodiversity loss.

Even in cases where countries have ratified treaties that have entered into force, parties to the treaties do not always comply with their provisions. An example is Russia's noncompliance with the Montreal Protocol. There are few penalties for noncompliance other than public anger, in part because countries are unwilling to give up their sovereignty.

Barriers to the Implementation of International Agreements

Why is the implementation of global environmental agreements so difficult? One answer is that for most of these accords to function effectively, participation must be truly global. Yet because of the unique nature of individual nations and their differing economies, it is difficult to reach meaningful agreement among the necessary participants. Another obstacle to implementing global environmental agreements is poorer nations' lack of capacity to comply with international treaty requirements. Some treaties have addressed this problem by offering financial assistance to countries that are unable to comply: for example, the Multilateral Fund of the Montreal Protocol. Other agreements have established more relaxed timetables for compliance for these countries. In many cases, however, the financing has not been forthcoming, and differentiating among countries based on their ability to act has become politically controversial. The Kyoto Protocol, for example, deals with emissions reductions from developed countries only, which has opened the treaty to criticism in the United States. The challenge of developing accords that are both effective and fair has proved to be a major obstacle to progress.

Despite these problems, treaties are the best traditional tools of global environmental governance. Frameworks for action that are negotiated among different nations are key to leveling the playing field and establishing the rules under which governments, businesses, nongovernmental organizations, and citizens can work together toward a common goal.

In 1997, the General Assembly of the United Nations held a special session and adopted a comprehensive document entitled Programme for the Further Implementation of Agenda 21 prepared by the Commission on Sustainable Development. It also adopted the program of work of the commission for 1999–2003. The Commission on Sustainable Development (CSD) was created in 1992 to ensure effective follow-up of the United Nations Conference on Environment and Development held in Rio de Janeiro, Brazil.

The Commission on Sustainable Development consistently generates a high level of public interest. Over 50 national leaders attend the CSD each year, and more than 1000 nongovernmental organizations are accredited to participate in the commission's work. The commission ensures the high visibility of sustainable development issues within the UN system and helps to improve the UN's coordination of environment and development activities. The CSD also encourages governments and international organizations to host workshops and conferences on different environmental and cross-sectoral issues. The results of these expert-level meetings enhance the work of CSD and help it work better with national governments and various nongovernmental partners in promoting sustainable development worldwide.

There is no international legislature with authority to pass laws; nor are there international agencies with power to regulate resources on a global scale. An international court at The Hague in the Netherlands has no power to enforce its decisions. Nations can simply ignore the court if they wish. However, a network is growing of multilateral environmental organizations that have developed a greater sense of their roles and a greater incentive to work together. These include not only the United Nations Environment Programme but also the Environment Committee of the Organization for Economic Cooperation and Development and the Senior Advisors on Environmental Problems of the Economic Commission for Europe. Such institutions perform unique functions that cannot be carried out by governments acting alone or bilaterally.

This environmental "coming of age" is reflected in the broadening of intellectual perspective. Governments used to be preoccupied with domestic environmental affairs. Now, they are beginning to broaden their scope to confront problems that cross international borders, such as transboundary air and water pollution, and threats of a planetary nature, such as stratospheric ozone depletion and climatic warming. It is becoming increasingly evident that only decisive mutual action can secure the kind of world we seek.

EARTH SUMMIT ON ENVIRONMENT AND DEVELOPMENT

In June 1992, representatives from 178 countries, including 115 heads of state, met in Rio de Janeiro, Brazil, at the Earth Summit. Officially, the meeting was titled the United Nations Conference on Environment and Development (UNCED), and it was the largest gathering of world leaders ever held. The first Earth Summit had been held 20 years earlier in Stockholm, Sweden. At that time, the planet was divided into rival East and West blocs and was preoccupied with the perils of the nuclear arms race. With the collapse of the eastern bloc and the thawing of the Cold War, a fundamental shift in the global base of power had occurred.

The idea behind the Earth Summit was that the relaxation of Cold War tensions, combined with the growing awareness of ecological crises, offered a rare opportunity to persuade countries to look beyond their national interests and agree to some basic changes in the way they treat the environment. The major issues are clear: The developed countries of the North have grown accustomed to lifestyles that are consuming a disproportionate share of natural resources and generating the bulk of global pollution. Many of the developing countries of the South are consuming irreplaceable global resources to provide for their growing populations.

Although the hopes of some developing nations for large commitments of new foreign assistance did not fully materialize, much was accomplished during the summit.

- The *Rio Declaration on Environment and Development* sets out 27 principles to guide the behavior of nations toward more environmentally sustainable patterns of development. The declaration, a compromise between developing and industrialized countries that was crafted at preparatory meetings, was adopted in Rio without negotiation due to fears that further debate would jeopardize any agreement.

- States at UNCED also adopted a voluntary action plan called *Agenda 21,* named because it is intended to provide an agenda for local, national, regional, and global action into the twenty-first century. UNCED Secretary General Maurice Strong called Agenda 21 "the most comprehensive, the most far-reaching and, if implemented, the most effective program of international action ever sanctioned by the international community." Agenda 21 includes hundreds of pages of recommended actions to address environmental problems and promote sustainable development. It also represents a process of building consensus on a "global work-plan" for the economic, social, and environmental tasks of the United Nations as they evolve over time.

- The third official product of UNCED was a "*non-legally binding authoritative statement of principles for a global consensus on the management, conservation, and sustainable development of all types of forests."* Negotiations on the forest statement, begun as negotiations for a legally binding convention on forests, were among the most difficult of the UNCED process. Many states and experts, dissatisfied with the end result, came away from UNCED seeking further negotiations toward agreement on a framework convention on forests.

ENVIRONMENTAL POLICY AND THE EUROPEAN UNION

"The environment knows no frontiers" was the slogan of the 1970s, when the European Community—known today as the European Union—began to develop its first environmental legislation.

The early laws focused on the testing and labeling of dangerous chemicals, testing drinking water, and controlling air pollutants such as sulfur dioxide, oxides of nitrogen, and particulates from power plants and automobiles. Many of the directives from the 1970s and 1980s were linked to Europe's desire to improve the living and working conditions of its citizens.

In 1987, the Single European Act gave this growing body of environmental legislation a formal legal basis and set three objectives: protection of the environment, protection of human health, and prudent and rational use of natural resources.

The treaty reflected what many governments had already understood: that countries are part of an interconnected and interdependent world of people who are bound together by the air they breathe, the food they eat, the products they use, the wastes they throw away, and the energy they consume.

Similarly, a factory in one European nation may import supplies and raw material from several other neighboring nations; it may consume energy produced from imported gas, produce wastes that affect the air and water quality across the border or downstream, and export products whose wastes become the risk and responsibility of governments and peoples several hundred or many thousand kilometers away.

The 1992 Maastricht Treaty formally established the concept of sustainable development in European Union law. Then, in 1997, the Amsterdam Treaty made sustainable development one of the overriding objectives of the European Union. The treaty considerably strengthened the commitment to the principle that the European Union's future development must be based on the principle of sustainable development and a high level of protection of the environment. The environment must be integrated into the definition and implementation of all of the Union's other economic and social policies, including trade, industry, energy, agriculture, transport, and tourism.

NEW INTERNATIONAL INSTRUMENTS

Over the past few decades, the global community has responded to emerging environmental problems with unprecedented international agreements—notably the Montreal Protocol, the Framework Convention on Climate Change, the Convention on Biological Diversity, and the Convention to Combat Desertification, among others. (See table 19.2.)

In the course of crafting these global agreements, many important lessons have been learned. As the experience with the Montreal Protocol shows, the scientific community can play a crucial role in two ways: first, confirming the links between human activities and global environmental problems; and second, showing what could happen to human health and the global environment if nothing is done.

When the evidence is in hand, an international consensus to act can emerge quickly. The same process is underway in the current international debate on climate change, but the process has been more difficult because the linkages between human activities and global environmental impacts are more complex and still not completely understood. Nevertheless, a global consensus for action is emerging. Another important lesson learned has to do with the structure of international agreements and the elements that can contribute to an effective structure. In the case of the Montreal Protocol, the agreement was not punitive and favored

TABLE 19.2 Major Environmental Treaties and U.S. Status

Treaty	Status
Convention on International Trade in Endangered Species, in force since 1975	The U.S. is a party.
Geneva Convention on Long-Range Transboundary Air Pollution, in force since 1983	The U.S. is a party.
Bonn Convention on Conservation of Migratory Species, in force since 1983	The U.S. is not a party.
Vienna Convention for the Protection of the Ozone Layer, in force since 1988	The U.S. is a party.
Montreal Protocol to the Vienna Convention, in force since 1989. This protocol set the first explicit limits to emissions of gases that erode the ozone layer.	The U.S. is a party.
Basel Convention on Transboundary Movements of Hazardous Wastes and Their Disposal, in force since 1992. This convention was a response to the growing practice of dumping wastes in developing countries and Eastern Europe.	The U.S. is a party.
Convention on Biological Diversity, in force since 1993	The U.S. is not a party.
Framework Convention on Climate Change, in force since 1994	The U.S. is a party.
Convention on the Law of the Sea, in force since 1994	The U.S. is not a party.
Convention to Combat Desertification, in force since 1996	The U.S. is a party.
Kyoto Protocol to the Framework Convention. The negotiators adopted the text in 1997, but it has not yet collected enough ratifications to go into force.	The U.S. has said that it will not ratify.
Rotterdam Convention on the Prior Informed Consent Procedure for Certain Hazardous Chemicals and Pesticides in International Trade. The text was adopted in 1998 but is not yet in force.	The U.S. has not ratified.
Stockholm Convention on Persistent Organic Pollutants. The text was adopted in 2001 but is not yet in force.	The U.S. has not ratified.

CAMPUS SUSTAINABILITY INITIATIVE

COLLEGE AND UNIVERSITY PRESIDENTS' CLIMATE COMMITMENT

The American College and University Presidents' Climate Commitment is a high-visibility effort to address global warming by gaining institutional commitments to neutralize greenhouse gas emissions, and to accelerate the research and educational efforts of higher education to equip society to re-stabilize the Earth's climate.

Building on the growing momentum for leadership and action on climate change, the Presidents' Climate Commitment provides a framework and support for America's colleges and universities to go climate neutral. The Commitment recognizes the unique responsibility that institutions of higher education have as role models for their communities and in training the people who will develop the social, economic, and technological solutions to address global warming.

Presidents who sign the Commitment pledge to eliminate, over a period of time, greenhouse gas emissions on their campus. This involves:

- Completing an emissions inventory;

- Within two years, setting a target date and interim milestones for becoming climate neutral;

- Taking immediate steps to reduce greenhouse gas emissions by choosing from a list of short-term actions;

- Integrating sustainability into the curriculum and making it part of the educational experience; and

- Making the action plan, inventory, and progress reports available to the public.

The college and university presidents and chancellors who are joining the Commitment believe that exerting leadership in addressing climate change will help stabilize and reduce their long-term energy costs, attract excellent students and faculty, attract new sources of funding, and increase the support of alumni and local communities. Universities and colleges in all 50 states have signed the Commitment.

Did the president of your college/university sign the Commitment? You might want to contact the Office of President at your college or university and ask why the Commitment was signed or not signed.

---

incentives and results-oriented approaches. All nations participated in the agreement but at different levels of responsibility in recognition of their differing conditions. The Kyoto Protocol to the Climate Change Convention has benefited from the experience with the Montreal Protocol and included many of the same elements.

A third vital lesson is that, to the extent possible, every interested party must have an opportunity to participate as a full partner in the process and to voice concerns. It is particularly helpful for environmental advocacy groups and the business community to be part of this process. International agreements need to provide incentives to foster public-private partnerships and to provide a role for business leaders to seek innovative technical solutions.

A fourth lesson concerns the role of governments in implementing these conventions. Government actions need to be consistent and predictable; provide sufficient lead times; favor government-led incentives over direct industry subsidies; and use flexible, market-based solutions where they are appropriate.

Finally, a fifth lesson is that agreements mark the beginning of a process, not the end. Scientists and nongovernmental organizations must continue to further global understanding of environmental problems and communicate what they have learned to the public and policy makers. Policy makers, in turn, must be flexible and respond to changing circumstances with new or modified policy solutions.

One example of new policy solutions is grounded in consumerism. *Green consumerism* is the concept of rational consumption of our scarce resources for the benefit of the environment and future generations. The old saying "the world is enough for everyone's need but not for everyone's greed" calls for a change in our behavior and lifestyle in favor of a sustainable future. Ecolabels have been introduced in a number of countries to help consumers choose products with a proven environmental edge, determined by the product's choice of raw materials, production process, product life cycle, and associated disposal problems. Ecolabels provide evidence that products have met the safety, quality, and environmental protection requirements of the authority that issues the label.

The first ecolabel was introduced in Germany in 1978. This distinctive Blue Angel label was followed by other environmental certification, including the White Swan of Northern European countries, Environmental Choice of Canada, Green Label of Singapore, and Environmental Label of China. By 2000, over a dozen countries had adopted this system of providing consumers with information enabling them to become responsible green consumers. The ecolabel of the European Union is significant because it is the world's first regional scheme to apply the same minimum standards across national markets. (See figure 19.12.)

In recent years, a new class of green certification programs has emerged. Perhaps best described as a hybrid between an environmental management system (EMS) standard and an ecolabeling

1992
European Union—'Eco-label'

1992
Singapore—'Green Label'

1989
Nordic Council—'White Swan'

1993
China—'Environmental Label'

1978
Germany—'Blue Angel'

1988
Canada—'Environmental Choice'

1997
'Marine Stewardship Council'

FSC
1993
'Forest Stewardship Council'

The Marine Stewardship Council is the only generally applicable, multicriteria ecolabel for fisheries worldwide.

The FSC logo identifies products that contain wood from well-managed forests certified in accordance with the rules of the Forest Stewardship Council.

FIGURE 19.12 **Ecolabels** A sampling of ecolabels from around the world.

Sources: ECCO, *Bulletin of the Environmental Campaign Committee.* © 1996 Forest Stewardship Council; and A.C. Marine Stewardship Council.

FIGURE 19.13 **Individual Responsibility** While laws protecting the environment are important, it is clear that individuals do not always respect these laws. Individuals must become accountable for their actions.

program, this type of labeling is based on third-party verification of compliance with specific performance criteria for environmental management practices. Some of these programs are managed by government-approved certifiers and/or administered by state agencies, such as the Costa Rican government's Certification for Sustainable Tourism, the Energy Star program in the United States, the U.S. Department of Agriculture's Organic seal, and the U.S. Department of Commerce's Dolphin-Safe label (tuna products).

However, most of the programs have been set up by nongovernmental bodies and are inspected by certifiers accredited by those bodies. Examples include the Forest Stewardship Council, the Marine Stewardship Council, the Sustainable Tourism Stewardship Council, Green Seal (various products and services), and a wide variety of sustainable agriculture, organic agriculture, and food safety programs.

IT ALL COMES BACK TO *YOU*

No set of policies, no system of incentives, no amount of information can substitute for individual responsibility when it comes to ensuring that our grandchildren will enjoy a quality of life that comes from a quality environment. Information can provide a basis for action. Vision and ideas can influence perceptions and inspire change. New ways to make decisions can empower those who seek a role in shaping the future. However, all of this will be meaningless unless individuals acting as citizens, consumers, investors, managers, workers, and professionals decide that it is important to them to make choices on the basis of a broader, longer view of their self-interest; to get involved in turning those choices into action; and, most important, to be held accountable for their actions. (See figure 19.13.)

The combination of political will, technological innovation, and a very large investment of resources and human ingenuity in pursuit of environmental goals has produced enormous benefits over the past two decades. This is an achievement to celebrate, but in a world that steadily uses more materials to make more goods for more people, we must recognize that we will have to achieve more in the future for the sake of the future. Finally, we must recognize that the pursuit of one set of goals affects others and that we must pursue policies that integrate economic, environmental, and social goals.

Gasoline, Taxes, and the Environment

In 2008, the average price of a gallon of gasoline in the United States crossed the $4 line. Given the increase in the price of gasoline, the thought of higher taxes on each gallon is not a welcome one. However, the question remains: Should there be higher taxes on gasoline? Would higher taxes encourage conservation, stimulate mass transit use, and, hence, reduce local air pollution?

Gasoline taxes vary dramatically in different countries. The United Kingdom has a gasoline tax equivalent to $2.80 per gallon—the highest among industrial countries. The United States, on the other hand, has the lowest tax—40 cents per gallon (18 cents federal tax and, on average, 22 cents state tax). It is commonly thought that Europeans have a greater tolerance for high fuel taxes than Americans because they have shorter distances to travel and better access to public transportation.

A number of arguments are made for implementing high gasoline taxes. By discouraging driving and fuel consumption, gasoline taxes help reduce local air pollution, emissions of carbon dioxide (a greenhouse gas), traffic congestion, traffic accidents, and oil dependency. Taxing gasoline is one way of forcing people to take into account the social costs of these problems when deciding how much—and what type of vehicle—to drive.

Gasoline taxes also provide a source of government revenues. In Britain, gasoline tax revenues are several times the amount of highway spending, and the government has argued that if gasoline taxes are reduced, some schools and hospitals will have to close. This argument is somewhat misleading because the revenues could always be made up through other sources, such as income taxes. The real issue is what level of gasoline taxation might be justified when taking into account the full social costs of driving and the appropriate balance between gasoline taxes and other taxes in raising revenues for the government.

- Is an increase in gasoline taxes feasible in our current political climate?
- Would you vote for a candidate who advocated higher gasoline taxes?
- Would you support increasing the taxes on gasoline?
- Would higher taxes on gasoline encourage you to travel by mass transit more often? To purchase vehicles that are more energy-efficient?

SUMMARY

Politics and the environment cannot be separated. In the United States, the government is structured into three separate branches, each of which impacts environmental policy.

The increase in environmental regulations in the United States over the past 30 years has caused concerns in some sectors of society.

The late 1980s and early 1990s witnessed a new international concern about the environment in both the developed and developing nations of the world. Environmentalism is also seen as a growing factor in international relations. This concern is leading to international cooperation where only tension had existed before.

While there exists no world political body that can enforce international environmental protection, the list of multilateral environmental organizations is growing.

It remains too early to tell what the ultimate outcome will be, but progress is being made in protecting our common resources for future generations. Several international conventions and treaties have been successful. In the final analysis, however, each of us has to adjust our lifestyle to clean up our own small part of the world.

THINKING GREEN

1. Speak out. Tell your elected representatives and the businesses where you shop what you, as a voter and a consumer, want. Businesses are especially responsive to customers who have concerns. If you have compliments regarding a company's environmental performance, let the owner know.

2. Become active in an organization that works to influence environmental or sustainability policy.

3. Volunteer on a political campaign

4. Do not be apathetic. Do not wait for "someone else" to help bring about change. Be aware and active and be a leader. Great change always begins one step at a time.

WHAT'S YOUR TAKE?

Public opinion polls in the United States continue to indicate that the public is concerned about environmental issues. Many state and local governments have been very active in adopting environmental laws and policies ranging from sustainable building practices to vehicle emission standards. The federal government, however, has not followed suit with such laws.

Should the federal government play a more aggressive role in establishing environmental policy or should that role be delegated to the state and local levels of government? Develop a position paper that supports why environmental policy is best developed at either the federal or state/local level of government.

REVIEW QUESTIONS

1. What are the major responsibilities of each of the three branches of the U.S. government?
2. What are some of the enforcement options in U.S. environmental policy?
3. What role does administrative law play in U.S. environmental policy?
4. What are some of the criticisms of U.S. environmental policy?
5. In the past 10 years, how has public opinion in the United States changed concerning the protection of the environment?
6. Why is environmentalism a growing factor in international relations?
7. Give some examples of international environmental conventions and treaties.

CRITICAL THINKING QUESTIONS

1. Does chapter 19 have an overall point of view? If you were going to present the problems of environmental policy making and enforcement to others, what framework would you use?
2. The authors of this text say that "we are progressing from an environmental paradigm based on cleanup and control to one including assessment, anticipation, and avoidance." Do you agree with this assessment? Are there environmental problems that are harder to be proactive about than others?
3. Does a command-and-control approach to environmental problems, an approach that emphasizes regulation and remediation, make sense with global environmental problems such as global climate change, habitat destruction, and ozone depletion?
4. What values, perspectives, and beliefs does the Wise Use Movement exhibit in its response to environmental legislation? How are they similar to or different from your own?
5. How is it best, as a global society with many political demarcations, to preserve the resources that are held in common? What special problems does this kind of preservation entail?
6. Do you agree with William Ruckelshaus that current environmental problems require a change on the part of industrialized and developing countries that would be "a modification in society comparable in scale to the agricultural revolution . . . and Industrial Revolution"? What kinds of changes might that mean in your life? Would these be positive or negative changes?
7. New treaties regarding free trade might enable some nations to argue that other nations' environmental legislation is too restrictive, thereby imposing a barrier to trade that is subject to sanction. What special problems and possibilities might the new global economy present for environmental preservation? What do you think about that?

PERIODIC TABLE OF THE ELEMENTS

Traditionally, elements are represented in a shorthand form by letters. For example, the formula for water, H_2O, shows that a molecule of water consists of two atoms of hydrogen and one atom of oxygen. These chemical symbols for each of the atoms can be found on any periodic table of the elements. Using the periodic table, we can determine the number and position of the various parts of atoms.

Notice that atoms numbered 3, 11, 19, and so on are in column one. The atoms in this column act in a similar way, since they all have one electron in their outermost layer. In the next column, Be, Mg, Ca, and so on act alike because these metals all have two electrons in their outermost electron layer. Similarly, atoms numbered 9, 17, 35, and so on all have seven electrons in their outer layer.

Knowing how fluorine, chlorine, and bromine act, you can probably predict how iodine will act under similar conditions. At the far right in the last column, argon, neon, and so on all act alike. They all have eight electrons in their outer electron layer. Atoms with eight electrons in their outer electron layer seldom form bonds with other atoms.

1 1A																	18 8A
1 **H** Hydrogen 1.008	2 2A											13 3A	14 4A	15 5A	16 6A	17 7A	2 **He** Helium 4.003
3 **Li** Lithium 6.941	4 **Be** Beryllium 9.012											5 **B** Boron 10.81	6 **C** Carbon 12.01	7 **N** Nitrogen 14.01	8 **O** Oxygen 16.00	9 **F** Fluorine 19.00	10 **Ne** Neon 20.18
11 **Na** Sodium 22.99	12 **Mg** Magnesium 24.31	3 3B	4 4B	5 5B	6 6B	7 7B	8	9 8B	10	11 1B	12 2B	13 **Al** Aluminum 26.98	14 **Si** Silicon 28.09	15 **P** Phosphorus 30.97	16 **S** Sulfur 32.07	17 **Cl** Chlorine 35.45	18 **Ar** Argon 39.95
19 **K** Potassium 39.10	20 **Ca** Calcium 40.08	21 **Sc** Scandium 44.96	22 **Ti** Titanium 47.88	23 **V** Vanadium 50.94	24 **Cr** Chromium 52.00	25 **Mn** Manganese 54.94	26 **Fe** Iron 55.85	27 **Co** Cobalt 58.93	28 **Ni** Nickel 58.69	29 **Cu** Copper 63.55	30 **Zn** Zinc 65.39	31 **Ga** Gallium 69.72	32 **Ge** Germanium 72.59	33 **As** Arsenic 74.92	34 **Se** Selenium 78.96	35 **Br** Bromine 79.90	36 **Kr** Krypton 83.80
37 **Rb** Rubidium 85.47	38 **Sr** Strontium 87.62	39 **Y** Yttrium 88.91	40 **Zr** Zirconium 91.22	41 **Nb** Niobium 92.91	42 **Mo** Molybdenum 95.94	43 **Tc** Technetium (98)	44 **Ru** Ruthenium 101.1	45 **Rh** Rhodium 102.9	46 **Pd** Palladium 106.4	47 **Ag** Silver 107.9	48 **Cd** Cadmium 112.4	49 **In** Indium 114.8	50 **Sn** Tin 118.7	51 **Sb** Antimony 121.8	52 **Te** Tellurium 127.6	53 **I** Iodine 126.9	54 **Xe** Xenon 131.3
55 **Cs** Cesium 132.9	56 **Ba** Barium 137.3	57 **La** Lanthanum 138.9	72 **Hf** Hafnium 178.5	73 **Ta** Tantalum 180.9	74 **W** Tungsten 183.9	75 **Re** Rhenium 186.2	76 **Os** Osmium 190.2	77 **Ir** Iridium 192.2	78 **Pt** Platinum 195.1	79 **Au** Gold 197.0	80 **Hg** Mercury 200.6	81 **Tl** Thallium 204.4	82 **Pb** Lead 207.2	83 **Bi** Bismuth 209.0	84 **Po** Polonium (210)	85 **At** Astatine (210)	86 **Rn** Radon (222)
87 **Fr** Francium (223)	88 **Ra** Radium (226)	89 **Ac** Actinium (227)	104 **Rf** Rutherfordium (257)	105 **Db** Dubnium (260)	106 **Sg** Seaborgium (263)	107 **Bh** Bohrium (262)	108 **Hs** Hassium (265)	109 **Mt** Meitnerium (266)	110 **Ds** Darmstadtium (269)	111 **Rg** Roentgenium (272)	112	(113)	114	(115)	116	(117)	(118)

Atomic number — 9 **F** Fluorine 19.00 — Atomic mass

Metals
Metalloids
Nonmetals

58 **Ce** Cerium 140.1	59 **Pr** Praseodymium 140.9	60 **Nd** Neodymium 144.2	61 **Pm** Promethium (147)	62 **Sm** Samarium 150.4	63 **Eu** Europium 152.0	64 **Gd** Gadolinium 157.3	65 **Tb** Terbium 158.9	66 **Dy** Dysprosium 162.5	67 **Ho** Holmium 164.9	68 **Er** Erbium 167.3	69 **Tm** Thulium 168.9	70 **Yb** Ytterbium 173.0	71 **Lu** Lutetium 175.0
90 **Th** Thorium 232.0	91 **Pa** Protactinium (231)	92 **U** Uranium 238.0	93 **Np** Neptunium (237)	94 **Pu** Plutonium (242)	95 **Am** Americium (243)	96 **Cm** Curium (247)	97 **Bk** Berkelium (247)	98 **Cf** Californium (249)	99 **Es** Einsteinium (254)	100 **Fm** Fermium (253)	101 **Md** Mendelevium (256)	102 **No** Nobelium (254)	103 **Lr** Lawrencium (257)

The 1–18 group designation has been recommended by the International Union of Pure and Applied Chemistry (IUPAC) but is not yet in wide use. In this text we use the standard U.S. notation for group numbers (1A–8A and 1B–8B). No names have been assigned for elements 112, 114, and 116. Elements 113, 115, 117, and 118 have not yet been synthesized.

APPENDIX 2

METRIC UNIT CONVERSION TABLES

The Metric System				
Standard metric units				**Abbreviations**
Standard unit of mass	Gram			g
Standard unit of length	Meter			m
Standard unit of volume	Liter			L
Common prefixes				**Examples**
Tera (T)	one trillion	1,000,000,000,000	10^{12}	A terawatt is 10^{12} watts.
Giga (G)	one billion	1,000,000,000	10^{9}	A gigawatt is 10^{9} watts.
Mega (M)	one million	1,000,000	10^{6}	A megagram is 10^{6} grams.
Kilo (k)	one thousand	1000	10^{3}	A kilogram is 10^{3} grams.
Centi (c)	one-hundredth	0.01	10^{-2}	A centimeter is 10^{-2} meter.
Milli (m)	one-thousandth	0.001	10^{-3}	A milliliter is 10^{-3} liter.
Micro (μ)	one-millionth	0.000001	10^{-6}	A micrometer is 10^{-6} meter.
Nano (n)	one-billionth	0.000000001	10^{-9}	A nanogram is 10^{-9} gram.
Pico (p)	one-trillionth	0.000000000001	10^{-12}	A picogram is 10^{-12} gram.

Units of Length		
Unit	**Abbreviation**	**Equivalent**
Kilometer	km	1000 m
Meter	m	——
Centimeter	cm	10^{-2} m
Millimeter	mm	10^{-3} m
Micrometer	μm	10^{-6} m
Nanometer	nm	10^{-9} m
Angstrom	Å	10^{-10} m

Length conversions

1 in = 2.54 cm	1 mm = 0.0394 in
1 ft = 30.5 cm	1 cm = 0.394 in
1 ft = 0.305 m	1 m = 39.4 in
1 yd = 0.914 m	1 m = 3.28 ft
1 mi = 1.61 km	1 m = 1.094 yd
	1 km = 0.621 mi

Units of Area

Unit	Abbreviation	Equivalent
Square meter	m^2	$1\ m^2$
Square kilometer	km^2	$1{,}000{,}000\ m^2$
Hectare	ha	$10{,}000\ m^2$

Area conversions

$1\ ft^2 = 0.093\ m^2$	$1\ m^2 = 10.76\ ft^2$
$1\ yd^2 = 0.84\ m^2$	$1\ m^2 = 1.19\ yd^2$
$1\ acre = 0.4\ ha$	$1\ ha = 2.47\ acre$
$1\ mi^2 = 2.6\ km^2$	$1\ km^2 = 0.386\ mi^2$
$1\ mi^2 = 259\ ha$	$1\ km^2 = 247\ acres$

Units of Volume

Unit	Abbreviation	Equivalent
Liter	L*	1
Milliliter	mL	$10^{-3}\ L$ $(1\ mL = 1\ cm^3 = 1\ cc)$
Microliter	μL	$10^{-6}\ L$

Volume conversions

$1\ tsp = 5\ mL$	$1\ mL = 0.034\ fl\ oz$
$1\ tbsp = 15\ mL$	$1\ L = 2.11\ pt$
$1\ fl\ oz = 30\ mL$	$1\ L = 1.057\ qt$
$1\ cup = 0.24\ L$	$1\ L = 0.265\ gal$
$1\ pt = 0.474\ L$	$1\ L = 33.78\ fl\ oz$
$1\ qt = 0.946\ L$	$1\ cubic\ meter\ (m^3) = 61{,}000\ cubic\ inches$
$1\ gal = 3.77\ L$	$1\ cubic\ meter\ (m^3) = 35.3\ cubic\ feet$
	$1\ cubic\ meter\ (m^3) = 0.00973\ acre\text{-}inch$

*Note: Many people use an uppercase "L" as the symbol for liter to avoid confusion with the number one (1). Similarly, milliliter is written mL.

Units of Weight

Unit	Abbreviation	Equivalent
Tonne (metric ton)	t	$10^3\ kg$
Kilogram	kg	$10^3\ g$
Gram	g	1
Milligram	mg	$10^3\ g$
Microgram	μg	$10^6\ g$
Nanogram	ng	$10^{-9}\ g$
Picogram	pg	$10^{-12}\ g$

Weight conversions

$1\ oz = 28.4\ g$	$1\ g = 0.0352\ oz$
$1\ lb = 454\ g$	$1\ kg = 2.205\ lb$
$1\ lb = 0.454\ kg$	
$1\ U.S.\ ton = 0.91\ metric\ ton$	$1\ metric\ ton = 1.102\ U.S.\ tons$

Temperature conversions

$$°C = \frac{(°F - 32)}{9} \times 5$$

$$°F = \frac{(°C \times 9)}{5} + 32$$

Some equivalents

$0°C = 32°F$

$37°C = 98.6°F$

$100°C = 212°F$

GLOSSARY

A

abiotic factors   Nonliving factors that influence the life and activities of an organism.

absorbed dose   The amount of energy absorbed by matter, measured in grays or rads.

abyssal ecosystem   The collection of organisms and the conditions that exist in the deep portions of the ocean.

acid   Any substance that, when dissolved in water, releases hydrogen ions.

acid deposition   The accumulation of potential acid-forming particles on a surface.

acid mine drainage   A kind of pollution, associated with coal mines, in which bacteria convert the sulfur in coal into compounds that form sulfuric acid.

acid rain (acid precipitation)   The deposition of wet acidic solutions or dry acidic particles from air.

activated-sludge sewage treatment   Method of treating sewage in which some of the sludge is returned to aeration tanks, where it is mixed with incoming wastewater to encourage degradation of the wastes in the sewage.

activation energy   The initial energy input required to start a reaction.

active solar system   A system that traps sunlight energy as heat energy and uses mechanical means to move it to another location.

acute toxicity   A serious effect, such as a burn, illness, or death, that occurs shortly after exposure to a hazardous substance.

age distribution   The comparative percentages of different age groups within a population.

agricultural products   Any output from farming: milk, grain, meat, etc.

agricultural runoff   Surface water that carries soil particles, nutrients, such as phosphate, nitrates, and other agricultural chemicals as it runs off agricultural land into lakes and streams.

agricultural waste   Waste from the raising of animals and harvesting and processing of crops and trees.

air stripping   The process of pumping air through water to remove volatile materials dissolved in the water.

air toxics   See **hazardous air pollutants.**

alpha radiation   A type of radiation consisting of a particle with two neutrons and two protons.

alpine tundra   The biome that exists above the tree line in mountainous regions.

alternative agriculture   All nontraditional agricultural practices.

animal rights/welfare   A movement that makes an ethical commitment to the well-being of nonhuman animals.

anthropocentrism   A theory in ethics that views human values as primary and the environment as solely a resource for humankind.

aquiclude   An impervious confining layer of an aquifer.

aquifer   A porous layer of earth material that becomes saturated with water.

aquitard   A partially permeable layer in an aquifer.

artesian well   The result of a pressurized aquifer being penetrated by a pipe or conduit, within which water rises without being pumped.

asthenosphere   Part of Earth's mantle capable of plastic flow.

ASTM International (formerly American Society of Testing and Materials)   International voluntary consensus standard-setting organization, originally founded in the United States in 1898, that develops and maintains environmental standards and processes.

atom   The basic subunit of elements, composed of protons, neutrons, and electrons.

auxin   A plant hormone that stimulates growth.

B

base   Any substance that, when dissolved in water, removes hydrogen ions from solution; forms a salt when combined with an acid.

benthic   Describes organisms that live on the bottom of marine and freshwater ecosystems.

benthic ecosystem   A type of marine or freshwater ecosystem consisting of organisms that live on the bottom.

beta radiation   A type of radiation consisting of electrons released from the nuclei of many fissionable atoms.

bioaccumulation   The buildup of a material in the body of an organism.

biocentrism   A theory in ethics that acknowledges the value of all living organisms.

biogeochemical cycles   Movement of matter within or between ecosystems; caused by living organisms, geological forces, or chemical reactions. The cycling of nitrogen, carbon, sulfur, oxygen, phosphorus, and water are examples.

biochemical oxygen demand (BOD)   The amount of oxygen required by microbes to degrade organic molecules in aquatic ecosystems.

biocide   A kind of chemical that kills many different types of living things.

biodegradable   Able to be broken down by natural biological processes.

biodiversity   A measure of the variety of kinds of organisms present in an ecosystem.

biogeochemical cycles   The processes by which atoms are cycled in ecosystems.

biomagnification   The increases in the amount of a material in the bodies of organisms at successively higher trophic levels.

biomass   Any accumulation of organic material produced by living things.

biome   A kind of plant and animal community that covers large geographic areas. Climate is a major determiner of the biome found in a particular area.

biotechnology   Inserting specific pieces of DNA into the genetic makeup of organisms.

biotic factors   Living portions of the environment.

biotic potential   The inherent reproductive capacity.

birthrate   The number of individuals born per thousand individuals in the population per year.

black lung disease   A respiratory condition resulting from the accumulation of large amounts of fine coal dust particles in miners' lungs.

boiling-water reactor (BWR)   A type of light-water reactor in which steam is formed directly in the reactor and is used to generate electricity.

boreal forest A broad band of mixed coniferous and deciduous trees that stretches across northern North America (and also Europe and Asia); its northernmost edge is integrated with the Arctic tundra.

brownfields Buildings and land that have been abandoned because they are contaminated and the cost of cleaning up the site is high.

brownfields cleanup Cleaning a contaminated industrial site to the point that it is safe to use for specific purposes.

brownfields development The concept that abandoned contaminated sites can be cleaned up sufficiently to allow some specified uses without totally removing all of the contaminants.

bush meat Meat from wild animals.

C

carbamate A class of soft pesticides that work by interfering with normal nerve impulses.

carbon absorption The use of carbon particles to treat chemicals by having the chemicals attach to the carbon particles.

carbon cycle The cyclic flow of carbon from the atmosphere to living organisms and back to the atmospheric reservoir.

carbon dioxide (CO_2) A normal component of the Earth's atmosphere that in elevated concentrations may interfere with the Earth's heat budget.

carbon monoxide (CO) A primary air pollutant produced when organic materials, such as gasoline, coal, wood, and trash, are incompletely burned.

carcinogen A substance that causes cancer.

carcinogenic The ability of a substance to cause cancer.

carnivores Animals that eat other animals.

carrying capacity The optimum number of individuals of a species that can be supported in an area over an extended period of time.

catalyst A substance that alters the rate of a reaction but is not itself changed.

cause-and-effect relationship A relationship between two events or things in which a change in the first leads to a change in the second.

chemical bond The physical attraction between atoms that results from the interaction of their electrons.

chemical weathering Processes that involve the chemical alteration of rock in such a manner that it is more likely to fragment or to be dissolved.

chlorinated hydrocarbon A class of pesticide consisting of carbon, hydrogen, and chlorine; these pesticides are very stable.

chlorofluorocarbons (CFC) Stable compounds containing carbon, hydrogen, chlorine, and fluorine. They were formerly used as refrigerants, propellants in aerosol containers, and expanders in foam products. They are linked to the depletion of the ozone layer.

chronic toxicity A serious effect, such as an illness or death, that occurs after prolonged exposure to small doses of a toxic substance.

clear-cutting A forest harvesting method in which all the trees in a large area are cut and removed.

climax community Last stage of succession; a relatively stable, long-lasting, complex, and interrelated community of plants, animals, fungi, and bacteria.

coevolution Two or more species of organisms reciprocally influencing the evolutionary direction of the other.

combustion The process of releasing chemical bond energy from fuel.

commensalism The relationship between organisms in which one organism benefits while the other is not affected.

community Interacting groups of different species.

competition An interaction between two organisms in which both require the same limited resource, which results in harm to both.

competitive exclusion principle A theory that no two populations of different species will occupy the same niche and compete for exactly the same resources in the same habitat for very long.

composting The process of harnessing the natural process of decomposition to transform organic materials into compost, a humuslike material with many environmental benefits.

compound A kind of matter composed of two or more different kinds of atoms bonded together.

Comprehensive Environmental Response, Compensation, and Liability Act (CERCLA) The 1980 U.S. law that addressed the cleanup of hazardous-waste sites.

confined aquifer An aquifer that is bounded on the top and bottom by impermeable confining layers.

conservation To use in the best possible way so that the greatest long-term benefit is realized by society.

conservation approach An approach that seeks a balance between the development and preservation approaches.

conservation tillage A tillage method in which 30 percent or more of the soil surface is left covered with crop residue following planting.

consumers Organisms that use other organisms as food.

contour farming A method of tilling and planting at right angles to the slope, which reduces soil erosion by runoff.

controlled experiment An experiment in which two groups are compared. One, the control, is used as a basis of comparison and the other, the experimental, has one factor different from the control.

coral reef ecosystem A tropical, shallow-water, marine ecosystem dominated by coral organisms that produce external skeletons.

corporation A business structure that has a particular legal status.

corrosiveness Ability of a chemical to degrade standard materials.

cost-benefit analysis A method used to determine the feasibility of pursuing a particular project by balancing estimated costs against expected benefits.

cover A term used to refer to any set of physical features that conceals or protects animals from the elements or their enemies.

criteria air pollutants Those air pollutants for which specific air quality standards have been set by the U.S. Environmental Protection Agency.

crust The thin, outer, solid surface of the Earth.

cultural relativism The view that right and wrong are to be determined from within a particular society or cultural group.

D

death phase The portion of the population growth curve of some organisms that shows the population declining.

death rate The number of deaths per thousand individuals in the population per year.

debt-for-nature exchange The purchase of a nation's debt by a third party that requires conservation on the part of the debtor nation in exchange for relief from the debt.

deceleration phase A part of the population growth curve in which the rate of population increase begins to decline.

decibel A unit used to measure the loudness of sound.

decommissioning Decontaminating and disassembling a nuclear power plant and safely disposing of the radioactive materials.

decomposers Small organisms, such as bacteria and fungi, that cause the decay of dead organic matter and recycle nutrients.

deep ecology   The generally ecocentric view that a new spiritual sense of oneness with the Earth is the essential starting point for a more healthy relationship with the environment.

deforestation   Activities that destroy forests.

deferred cost   A cost that is not paid immediately when an economic decision is made but must be paid at a later date.

demand   Amount of a product that consumers are willing and able to buy at various prices.

demographic transition   The hypothesis that economies proceed through a series of stages, beginning with growing populations with high birth and death rates and low economic development and ending with stable populations with low birth and death rates and high economic development.

demography   The study of human populations, their characteristics, and their changes.

denitrifying bacteria   Bacteria that convert nitrogen compounds into nitrogen gas.

design for environment (DfE)   U.S. Environmental Protection Agency's (EPA) program for utilizing "green" (environmentally friendly) methods and processes in the development and design of products and services.

density-dependent limiting factors   Those limiting factors that become more severe as the size of the population increases.

density-independent limiting factors   Those limiting factors that are not affected by population size.

desert   A biome that receives less than 25 centimeters (10 inches) of precipitation per year.

desertification   The conversion of arid and semiarid lands into deserts by inappropriate farming practices or overgrazing.

detritus   Tiny particles of organic material that result from fecal waste material or the decomposition of plants and animals.

development approach   An approach that encourages humankind to transform nature as it pleases to satisfy human needs.

dioxins   A general term for a group or family of chemicals containing hundreds of members (some of which are toxic) that are persistent in the environment and that are unintentionally formed by-products of industrial processes that involve chlorine and/or incineration.

dispersal   Migration of organisms from a concentrated population into areas with lower population densities.

domestic water   Water used for domestic activities, such as drinking, air conditioning, bathing, washing clothes, washing dishes, flushing toilets, and watering lawns and gardens.

dose equivalent   The absorbed dose times a quality factor.

E

ecocentrism   A theory in ethics that considers the value of ecosystems and larger wholes to be primary.

ecology   A branch of science that deals with the interrelationship between organisms and their environment.

ecofeminism   The view that there are important theoretical, historical, and empirical connections between how society treats women and how it treats the environment.

ecological or environmental economics   An approach to economic accounting that incorporates environmental goods and harms as part of the cost of economic activity.

ecological footprint   The area of the Earth's productive land and water required to supply the resources that an individual demands as well as to absorb the wastes that the individual produces.

economic growth   The perceived increase in monetary growth within a society.

economics   The study of how people choose to use resources to produce goods and services and how these goods and services are distributed to the public.

ecosystem   A group of interacting species along with their physical environment.

ecosystem diversity   A measure of the number of kinds of ecosystems present in an area.

ectoparasite   A parasite that is adapted to live on the outside of its host.

electron   The lightweight, negatively charged particle that moves around at some distance from the nucleus of an atom.

element   A form of matter consisting of a specific kind of atom.

emergent plants   Aquatic vegetation that is rooted on the bottom but has leaves that float on the surface or protrude above the water.

emigration   Movement out of an area that was once one's place of residence.

endangered species   Those species that are present in such small numbers that they are in immediate jeopardy of becoming extinct.

endoparasite   A parasite that is adapted to live within a host.

endothermic reaction   Chemical reaction in which the newly formed chemical bonds contain more energy than was present in the compounds from which they were formed.

energy   The ability to do work.

entropy   The degree of disorder in a system. All systems tend toward a high degree of disorder or entropy.

environment   Everything that affects an organism during its lifetime.

environmental aesthetics   The study of how to appreciate beauty in the natural world.

environmental cost   Damage done to the environment as a resource is exploited.

environmental justice   Fair application of laws designed to protect the health of human beings and ecosystems; that no groups suffer unequal environmental harm.

environmental management systems   Voluntary management processes and procedures designed to aid private and public agencies in establishing policies and practices for identifying, analyzing, and solving environmental problems.

environmental pragmatism   An approach to environmental ethics that maintains that a human-centered ethic with a long-range perspective will come to many of the same conclusions in environmental policy as an ecocentric ethic.

Environmental Protection Agency (EPA)   U.S. government organization responsible for the establishment and enforcement of regulations concerning the environment.

environmental resistance   The combination of all environmental influences that tend to keep populations stable.

environmental science   An interdisciplinary area of study that includes both applied and theoretical aspects of human impact on the world.

environmental site assessment   The standard process of evaluating and assessing potential present, historical, and/or future environmental threats to commercial real estate.

environmental terrorism   The unlawful use of force against environmental resources so as to deprive populations of their benefits or destroy other property.

enzymes   Protein molecules that speed up the rate of specific chemical reactions.

erosion   The processes that loosen and move particles from one place to another.

estuaries   Marine ecosystems that consist of shallow, partially enclosed areas where freshwater enters the ocean.

ethics   A discipline that seeks to define what is fundamentally right and wrong.

euphotic zone   The upper layer in the ocean where the sun's rays penetrate.

eutrophication   The enrichment of water (either natural or cultural) with nutrients.

eutrophic lake A usually shallow, warm-water lake that is nutrient rich.

evapotranspiration The process of plants transporting water from the roots to the leaves where it evaporates.

evolution A change in the structure, behavior, or physiology of a population of organisms as a result of some organisms with favorable characteristics having greater reproductive success than those organisms with less favorable characteristics.

executive branch The office of the president of the United States.

exothermic reaction Chemical reaction in which the newly formed compounds have less chemical energy than the compounds from which they were formed.

experiment An artificial situation designed to test the validity of a hypothesis.

exponential growth phase The period during population growth when the population increases at an ever-increasing rate.

extended product responsibility The concept that the producer of a product is responsible for all the negative effects involved in its production, including the ultimate disposal of the product when its useful life is over.

external costs Expenses, monetary or otherwise, borne by someone other than the individuals or groups who use a resource.

extinction The death of a species; the elimination of all the individuals of a particular kind.

extrinsic limiting factors Factors that limit population size and that come from outside the population.

F

fecal coliform bacteria Bacteria found in the intestines of humans and other animals, often used as an indicator of water pollution.

first law of thermodynamics A statement about energy that says that under normal physical conditions, energy is neither created nor destroyed.

fissionable The property of the nucleus of some atoms that allows them to split into smaller particles.

fixation A form of waste immobilization in which materials, such as fly ash or cement, are mixed with hazardous waste to prevent the waste from dispersing.

floodplain Lowland area on either side of a river that is periodically covered by water.

floodplain zoning ordinances Municipal laws that restrict future building in floodplains.

food chain The series of organisms involved in the passage of energy from one trophic level to the next.

food web Intersecting and overlapping food chains.

fossil fuels The organic remains of plants, animals, and microorganisms that lived millions of years ago that are preserved as natural gas, oil, and coal.

free-living nitrogen-fixing bacteria Bacteria that live in the soil and can convert nitrogen gas (N_2) in the atmosphere into forms that plants can use.

freshwater ecosystem Aquatic ecosystems that have low amounts of dissolved salts.

friable A soil characteristic that describes how well a soil crumbles.

fungicide A pesticide designed to kill or control fungi.

G

gamma radiation A type of electromagnetic radiation that comes from disintegrating atomic nuclei.

gas-cooled reactor (GCR) A type of nuclear reactor that uses graphite as a moderator and carbon dioxide or helium as a coolant.

gene A unit of heredity; a segment of DNA that contains information for the synthesis of a specific protein, such as an enzyme.

genetic diversity A term used to describe the number of different kinds of genes present in a population or a species.

genetically modified organisms Organisms that have had their genetic makeup modified by biotechnology.

genetic engineering Inserting specific pieces of DNA into the genetic makeup of organisms.

geothermal energy The heat energy from the Earth's molten core.

Global Reporting Initiative Guidelines for reporting on the economic, environmental, and social performance of corporations.

greenhouse effect The property of carbon dioxide (CO_2) that allows light energy to pass through the atmosphere but prevents heat from leaving; similar to the action of glass in a greenhouse.

greenhouse gas Gas in the atmosphere that allows sunlight to enter but retards the outward flow of heat from the Earth.

Green Revolution The introduction of new plant varieties and farming practices that increased agricultural production worldwide during the 1950s, 1960s, and 1970s.

gross national income (GNI) An index that measures the total goods and services generated within a country as well as income earned by citizens of the country who are living in other countries.

gross national product (GNP) An index that measures the total goods and services generated annually within a country.

groundwater Water that infiltrates the soil and is stored in the spaces between particles in the earth.

groundwater mining Removal of water from an aquifer faster than it is replaced.

H

habitat The specific kind of place where a particular kind of organism lives.

habitat management The process of changing the natural community to encourage the increase in populations of certain desirable species.

hard pesticide A pesticide that persists for long periods of time; a persistent pesticide.

hazardous All dangerous materials, including toxic ones, that present an immediate or long-term human health risk or environmental risk.

hazardous air pollutants Certain airborne compounds with high toxicity.

hazardous materials or substances Substances that can cause harm to humans or the environment.

hazardous-waste dump A site in which hazardous waste is disposed of in a dump, landfill, or surface impoundment without any concern for potential environmental or health risks.

hazardous wastes Substances that could endanger life if released into the environment.

heavy-water reactor (HWR) A type of nuclear reactor that uses the hydrogen isotope deuterium in the molecular structure of the coolant water.

herbicide A pesticide designed to kill or control plants.

herbivores Primary consumers; animals that eat plants.

horizon A horizontal layer in the soil. The top layer (A horizon) has organic matter. The lower layer (B horizon) receives nutrients by leaching. The C horizon is partially weathered parent material.

host The organism a parasite uses for its source of food.

humus Partially decomposed organic matter typically found in the top layer of the soil.

hydrocarbons (HC) Group of organic compounds consisting of carbon and hydrogen atoms that are evaporated from fuel supplies or are remnants of the fuel that did not burn completely and that act as a primary air pollutant.

hydrologic cycle Constant movement of water from surface water to air and back to surface water as a result of evaporation and condensation.

hydroxide ion A negatively charged particle consisting of a hydrogen and an oxygen atom, commonly released from materials that are bases.

hypothesis A logical statement that explains an event or answers a question that can be tested.

I

ignitability Characteristic of materials that results in their ability to combust.

immigration Movement into an area where one has not previously resided.

incineration Method of disposing of solid waste by burning.

industrial ecology A concept that stresses cycling resources rather than extracting and eventually discarding them.

Industrial Revolution A period of history during which machinery replaced human labor.

industrial solid waste A wide variety of materials such as demolition waste, foundry sand, scraps from manufacturing processes, sludge, ash from combustion, and other similar materials produced by industry.

industrial water uses Uses of water for cooling and for dissipating and transporting waste materials.

infrastructure Permanent structural foundations of a society such as highways and bridges.

insecticide A pesticide designed to kill or control insects.

in-stream water uses Use of a stream's water flow for such purposes as hydroelectric power, recreation, and navigation.

integrated pest management A method of pest management in which many aspects of the pest's biology are exploited to control its numbers.

interspecific competition Competition between members of different species for a limited resource.

intraspecific competition Competition among members of the same species for a limited resource.

intrinsic limiting factors Factors that limit population size that come from within the population.

ion An atom or group of atoms that has an electric charge because it has either gained or lost electrons.

ionizing radiation Radiation that can dislodge electrons from atoms to form ions.

irrigation Adding water to an agricultural field to allow certain crops to grow where the lack of water would normally prevent their cultivation.

ISO 14000 Commonly used name for a group of voluntary environmental processes and practices developed by the International Organization for Standardization, headquartered in Geneva, Switzerland.

isotope Atoms of the same element that have different numbers of neutrons.

J

judicial branch That portion of the U.S. government that includes the court system.

K

keystone species One that has a critical role to play in the maintenance of specific ecosystems.

kinetic energy Energy of moving objects.

kinetic molecular theory The widely accepted theory that all matter is made of small particles that are in constant movement.

K-strategists Large organisms that have relatively long lives, produce few offspring, provide care for their offspring, and typically have populations that stabilize at the carrying capacity.

L

lag phase The initial stage of population growth during which growth occurs very slowly.

land The surface of the Earth not covered by water.

land disposal The placement of unwanted materials on the surface in landfills or impoundments or by injecting them below the surface of the land.

landfill A method of disposing of solid wastes that involves burying the wastes in specially constructed sites.

land-use planning The process of evaluating the needs and wants of the population, the characteristics and values of the land, and various alternative solutions before changes in land use are made.

latent heat Heat transfer that occurs when a substance is changed from one state to another—solid to liquid, gas to liquid—in which heat is transferred but the temperature does not change.

laws Legislative or judicial frameworks determining how members of a particular society should behave.

law of conservation of mass States that matter is not gained or lost during a chemical reaction.

LD$_{50}$ A measure of toxicity; the dosage of a substance that will kill (lethal dose) 50 percent of a test population.

leachate Contaminant-laden water that flows from landfills or other contaminated sites.

leaching The movement of minerals from the top layers of the soil to the B horizon by the downward movement of soil water.

legislative branch That portion of the U.S. government that is responsible for developing laws.

less-developed countries Countries of the world that typically have a per capita income of less than US $5000.

life cycle analysis The process of assessing the environmental effects associated with the production, use, reuse, and disposal of a product over its entire useful life.

light-water reactor A nuclear reactor that uses ordinary water as a coolant.

limiting factor The primary condition of the environment that determines the population size for an organism.

limnetic zone Region that does not have rooted vegetation in a freshwater ecosystem.

liquefied natural gas Natural gas that has been converted to a liquid by cooling to −162°C (−260°F).

liquid metal fast-breeder reactor (LMFBR) Nuclear fission reactor using liquid sodium as the moderator and heat transfer medium; produces radioactive plutonium-235, which can be used as a nuclear fuel.

lithosphere A combination of the crust and outer layer of the mantle that forms the plates that move over the Earth's surface.

litter A layer of undecomposed or partially decomposed organic matter on the soil surface.

littoral zone Region with rooted vegetation in a freshwater ecosystem.

loam A soil type with good drainage and good texture that is ideal for growing crops.

M

macronutrient A nutrient, such as nitrogen, phosphorus, or potassium, that is required in relatively large amounts by plants.

mangrove swamp ecosystems Marine shoreline ecosystems dominated by trees that can tolerate high salt concentrations.

mantle The layer of the Earth between the crust and the core.

marine ecosystems Aquatic ecosystems that have high salt content.

marsh Area of grasses and reeds that is flooded either permanently or for a major part of the year.

mass burn A method of incineration of solid waste in which material is fed into a furnace on movable metal grates.

matter Substance with measurable mass and volume.

mechanical weathering Physical forces that reduce the size of rock particles without changing the chemical nature of the rock.

Mediterranean shrublands Coastal ecosystems characterized by winter rains and summer droughts that are dominated by low, woody vegetation with small leaves.

megalopolis A large, regional urban center.

methane (CH₄) An organic compound produced by living organisms that is a greenhouse gas.

micronutrient A nutrient needed in extremely small amounts for proper plant growth; examples are boron, zinc, and magnesium.

migratory birds Birds that fly considerable distances between their summer breeding areas and their wintering areas.

mining waste Waste from the processing of rock from mining operations. It includes solid materials that are typically dumped on the land near the milling site and liquid wastes typically stored in ponds.

mixture A kind of matter consisting of two or more kinds of matter intermingled with no specific ratio of the kinds of matter.

moderator Material that absorbs the energy from neutrons released by fission.

molecule Two or more atoms chemically bonded to form a stable unit.

monoculture A system of agriculture in which large tracts of land are planted with the same crop.

more-developed countries Countries of the world that typically have a per capita income that exceeds US $10,000; Europe, Canada, United States, Australia, New Zealand, and Japan.

mortality The number of deaths per year.

mulch An organic material that is used to cover the soil.

multiple land use Land uses that do not have to be exclusionary, so that two or more uses of land may occur at the same time.

municipal solid waste landfill A waste storage site constructed above an impermeable clay layer that is lined with an impermeable membrane and includes mechanisms for dealing with liquid and gas materials generated by the contents of the landfill.

municipal solid waste (MSW) All the waste produced by the residents of a community.

mutations Changes in the genetic information of an organism.

mutualism The association between organisms in which both benefit.

mycorrhizae Symbiotic soil fungi, present in most soils, that attach themselves directly onto the roots of most plants. They help the host plants to absorb more water and nutrients while the host plants provide food for the fungi.

N

natality The number of individuals added to the population through reproduction.

National Priorities List A listing of hazardous-waste dump sites requiring urgent attention as identified by Superfund legislation.

natural capitalism An economic approach that pursues market-based profitability while at the same time preserving (or increasing) environmental quality.

natural resources Those structures and processes that can be used by humans for their own purposes but cannot be created by them.

natural selection A process that determines which individuals within a species will reproduce more effectively and therefore results in changes in the characteristics within a species.

nature centers Teaching institutions that provide a variety of methods for people to learn about and appreciate the natural world.

negligible risk A point at which there is no significant health or environmental risk.

neutralization Reacting acids with bases to produce relatively safe end products.

neutron Neutrally charged particle located in the nucleus of an atom.

niche The total role an organism plays in its ecosystem.

nitrifying bacteria Bacteria that are able to convert ammonia to nitrite, which can be converted to nitrate.

nitrogen cycle The series of stages in the flow of nitrogen in ecosystems.

nitrogen dioxide A compound composed of one atom of nitrogen and two atoms of oxygen; a secondary air pollutant.

nitrogen-fixing bacteria Bacteria that are able to convert the nitrogen gas (N_2) in the atmosphere into forms that plants can use.

nitrogen monoxide A compound composed of one atom of nitrogen and one atom of oxygen; a primary air pollutant.

nitrous oxide N_2O, one of the oxides of nitrogen.

nonpersistent pesticide A pesticide that degrades in a short period of time.

nonpersistent pollutants Those pollutants that do not remain in the environment for long periods.

nonpoint source Diffuse pollutants, such as agricultural runoff, road salt, and acid rain, that are not from a single, confined source.

nonrenewable energy sources Those energy sources that are not replaced by natural processes within a reasonable length of time.

nonrenewable resources Those resources that are not replaced by natural processes, or those whose rate of replacement is so slow as to be noneffective.

nontarget organism An organism whose elimination is not the purpose of pesticide application.

northern coniferous forest See **boreal forest**.

nuclear breeder reactor Nuclear fission reactor designed to produce radioactive fuel from nonradioactive uranium and at the same time release energy to use in the generation of electricity.

nuclear chain reaction A continuous process in which a splitting nucleus releases neutrons that strike and split the nuclei of other atoms, releasing nuclear energy.

nuclear fission The decomposition of an atom's nucleus with the release of particles and energy.

nuclear fusion The union of smaller nuclei to form a heavier nucleus accompanied by the release of energy.

nuclear reactor A device that permits a controlled nuclear fission chain reaction.

nucleus The central region of an atom that contains protons and neutrons.

O

observation Ability to detect events by the senses or machines that extend the senses.

oligotrophic lakes Deep, cold, nutrient-poor lakes that are low in productivity.

omnivores Animals that eat both plants and other animals.

organic agriculture Agricultural practices that avoid the use of chemical fertilizers and pesticides in the production of food, thus preventing damage to related ecosystems and consumers.

organophosphate A class of soft pesticides that work by interfering with normal nerve impulses.

outdoor recreation Leisure activities carried out in the natural out-of-doors.

overburden The layer of soil and rock that covers deposits of desirable minerals.

oxides of nitrogen (NO, N_2O, and NO_2) Primary air pollutants consisting of a variety of different compounds containing nitrogen and oxygen.

ozone (O_3) A molecule consisting of three atoms of oxygen that absorbs much of the sun's ultraviolet energy before it reaches the Earth's surface.

P

parasite An organism adapted to survival by using another living organism (host) for nourishment.

parasitism A relationship between organisms in which one, known as the parasite, lives in or on the host and derives benefit from the relationship while the host is harmed.

parent material Material that is weathered to become the mineral part of the soil.

particulate matter Minute solid particles and liquid droplets dispersed into the atmosphere.

particulates Small pieces of solid materials, such as smoke particles from fires, bits of asbestos from brake linings and insulation, dust particles, or ash from industrial plants, that are dispersed into the atmosphere.

passive solar system A design that allows for the entrapment and transfer of heat from the sun to a building without the use of moving parts or machinery.

patchwork clear-cutting A forest harvest method in which patches of trees are clear-cut among patches of timber that are left untouched.

peat The first stage in the conversion of organic material into coal.

pelagic Those organisms that swim in open water.

pelagic ecosystem A portion of a marine or freshwater ecosystem that occurs in open water away from the shore.

periphyton Attached organisms in freshwater streams and rivers, including algae, animals, and fungi.

permafrost Permanently frozen ground.

permissible exposure limit (PEL) Standard acceptable level or limit of chemicals thought to be "safe" for human exposure utilized by the U.S. Occupational Safety and Health Administration (OSHA) as enforceable criterion for the protection of human health.

persistent pesticide A pesticide that remains unchanged for a long period of time; a hard pesticide.

persistent pollutant A pollutant that remains in the environment for many years in an unchanged condition.

personal ethical commitment A determination of ethical right and wrong made by an individual.

pest An unwanted plant or animal that interferes with domesticated plants and animals or human activity.

pesticide A chemical used to eliminate pests; a general term used to describe a variety of different kinds of pest killers, such as insecticides, fungicides, rodenticides, and herbicides.

pH The negative logarithm of the hydrogen ion concentration; a measure of the number of hydrogen ions present.

pheromone A chemical produced by one animal that changes the behavior of another.

photochemical smog A yellowish-brown haze that is the result of the interaction of hydrocarbons, oxides of nitrogen, and sunlight.

photosynthesis The process by which plants manufacture food. Light energy is used to convert carbon dioxide and water to sugar and oxygen.

phytoplankton Free-floating, microscopic, chlorophyll-containing organisms.

pioneer community The early stages of succession that begin the soil-building process.

plankton Tiny aquatic organisms that are moved by tides and currents.

plate tectonics The concept that the outer surface of the Earth consists of large plates that are slowly moving over the surface of a plastic layer.

plutonium-239 (Pu-239) A radioactive isotope produced in a breeder reactor and used as a nuclear fuel.

PM$_{10}$ Particulate matter that is 10 microns or less in diameter.

PM$_{2.5}$ Particulate matter that is 2.5 microns or less in diameter.

point source Pollution that can be traced to a single source.

policy Planned course of action on a question or a topic.

pollution Any addition of matter or energy that degrades the environment for humans and other organisms.

pollution costs The private or public expenditures undertaken to avoid pollution damage once pollution has occurred and the increased health costs and loss of the use of public resources because of pollution.

pollution prevention Action to prevent either entirely or partially the pollution that would otherwise result from some production or consumption activity.

pollution-prevention costs Costs incurred to prevent pollution that would otherwise result from some production or consumption activity.

pollution-prevention hierarchy Regulatory controls that emphasize reducing the amount of hazardous waste produced.

polyculture A system of agriculture that mixes different plant species in the same plots of land.

polyploidy A condition in which the number of sets of chromosomes increases.

population A group of individuals of the same species occupying a given area.

population density A measure of how close organisms are to one another, generally expressed as the number of organisms per unit area.

population growth rate The rate at which additional individuals are added to the population; the birthrate minus the death rate.

porosity A measure of the size and number of spaces in an aquifer.

potable waters Unpolluted freshwater supplies suitable for drinking.

potential energy The energy of position.

prairies Temperate grasslands.

precipitation Removal of materials by mixing with chemicals that cause the materials to settle out of the mixture.

precision agriculture The use of computer technology and geographic information systems to automatically vary the chemicals applied to a crop at different places within a field.

predation The act of killing and feeding by a predator.

predator An animal that kills and eats another organism.

preservation Action to keep from harm or damage; to maintain in its original condition.

preservation approach An approach that seeks to ensure that large areas of nature together with their ecological processes remain intact.

pressurized-water reactor (PWR) A type of light-water reactor in which the water in the reactor is kept at high pressure and steam is formed in a secondary loop.

prey An organism that is killed and eaten by a predator.

price The monetary value of a good or service.

primary air pollutants Types of unmodified materials that, when released into the environment in sufficient quantities, are considered hazardous.

primary consumer An animal that eats plants (producers) directly.

primary sewage treatment Process that removes larger particles by settling or filtering raw sewage through large screens.

primary succession Succession that begins with bare mineral surfaces or water.

probability A mathematical statement about how likely it is that something will happen.

producer An organism that can manufacture food from inorganic compounds and light energy.

profitability The extent to which economic benefits exceed the economic costs of doing business.

proton The positively charged particle located in the nucleus of an atom.

pseudoscience A deceptive practice that uses the appearance or language of science to convince, confuse, or mislead people into thinking something has scientific validity, when it does not.

public resources Those parts of the environment that are owned by everyone.

R

radiation Energy that travels through space in the form of waves or particles.

radioactive Describes unstable nuclei that release particles and energy as they disintegrate.

radioactive half-life The time it takes for half of the radioactive material to spontaneously decompose.

radon Radioactive gas emitted from certain kinds of rock; can accumulate in very tightly sealed buildings.

range of tolerance The ability organisms have to succeed under a variety of environmental conditions. The breadth of this tolerance is an important ecological characteristic of a species.

reactivity The property of materials that indicates the degree to which a material is likely to react vigorously to water or air, or to become unstable or explode.

recycling The process of reclaiming a resource and reusing it for another or the same structure or purpose.

reduced tillage A tillage method that generally leaves 15 to 30 percent of the soil surface covered with crop residue following planting.

reforestation The process of replanting areas after the original trees are removed.

rem A measure of the biological damage to tissue caused by certain amounts of radiation.

remediation A method of providing a remedy or correction for an environmental contamination issue, as in "cleanup" operations.

renewable energy sources Those energy sources that can be regenerated by natural processes.

renewable resources Those resources that can be formed or regenerated by natural processes.

replacement fertility The number of children per woman needed just to replace the parents.

reproducibility A characteristic of the scientific method in which independent investigators must be able to reproduce the experiment to see if they get the same results.

reserves The known deposits from which materials can be extracted profitably with existing technology under present economic conditions.

Resource Conservation and Recovery Act (RCRA) The 1976 U.S. law that specifically addressed the issue of hazardous waste.

resource exploitation The use of natural resources by society.

resources Naturally occurring substances that can be utilized by people but may not be economic.

respiration The process that organisms use to release chemical bond energy from food.

ribbon sprawl Development along transportation routes that usually consists of commercial and industrial building.

risk The probability that a condition or action will lead to an injury, damage, or loss.

risk assessment The use of facts and assumptions to estimate the probability of harm to human health or the environment that may result from exposures to specific pollutants, toxic agents, or management decisions.

risk-based corrective action (RBCA) Scientific process for "cleaning up" environmental contamination sites through health and safety risk assessments.

risk management Decision-making process that uses input such as risk assessment, technological feasibility, economic impacts, public concerns, and legal requirements.

risk tolerance Ability to adapt to potential loss, damage, or injury.

rodenticide A pesticide designed to kill rodents

r-strategist Typically, a small organism that has a short life span, produces a large number of offspring, and does not reach a carrying capacity.

runoff The water that moves across the surface of the land and enters a river system.

S

salinization An increase in the amount of salt in soil due to the evaporation of irrigation water.

saltwater intrusion The movement of saltwater into aquifers near oceans when too much water is pumped from aquifers.

savanna Tropical biome having seasonal rainfall of 50 to 150 centimeters (20–60 inches) per year. The dominant plants are grasses, with some scattered fire- and drought-resistant trees.

science A method for gathering and organizing information that involves observation, asking questions about observations hypothesis formation, testing hypotheses, critically evaluating the results, and publishing information so that others can evaluate the process and the conclusions.

scientific law A uniform or constant fact of nature that describes *what* happens in nature.

scientific method A way of gathering and evaluating information. It involves observation, hypothesis formation, hypothesis testing, critical evaluation of results, and the publishing of findings.

secondary air pollutants Pollutants produced by the interaction of primary air pollutants in the presence of an appropriate energy source.

secondary consumers Animals that eat animals that have eaten plants.

secondary recovery Techniques used to obtain the maximum amount of oil or natural gas from a well.

secondary sewage treatment Process that involves holding the wastewater until the organic material has been degraded by bacteria and other microorganisms.

secondary succession Succession that begins with the destruction or disturbance of an existing ecosystem.

second law of thermodynamics A statement about energy conversion that says that whenever energy is converted from one form to another, some of the useful energy is lost.

selective harvesting A forest harvesting method in which individual high-value trees are removed from the forest, leaving the majority of the forest undisturbed.

sensible heat The heat energy stored in a substance as a result of an increase in its temperature.

septic tank Underground holding tank into which sewage is pumped and where biological degradation of organic material takes place; used in places where sewers are not available.

seral stage A stage in the successional process.

sere A stage in succession.

sewage sludge A mixture of organic material, organisms, and water in which the organisms consume the organic matter.

sex ratio Comparison between the number of males and females in a population.

smart growth Land development that emphasizes the concept of livable cities and towns.

Small Business Liability Relief and Brownfield Revitalization Act ("Brownfields Law") U.S. law (January

2002) designed to limit Superfund liability to allow for the environmental remediation and reuse of contaminated sites that were previously industrial or commercial areas.

small-scale chemistry "Green" approach to that chemistry that minimizes the amounts of chemicals used in the laboratory in order to limit wastes.

social ecology The view that social hierarchies between groups of people are directly connected to patterns of behavior that cause environmental destruction.

soft pesticide A nonpersistent pesticide that breaks down into harmless products in a few hours or days.

soil A mixture of mineral material, organic matter, air, water, and living organisms; capable of supporting plant growth.

soil profile The series of layers (horizons) seen as one digs down into the soil.

soil structure Refers to the way that soil particles clump together. Sand has little structure because the particles do not stick to one another.

soil texture Refers to the size of the particles that make up the soil. Sandy soil has large particles, and clay soil has small particles.

solidification The conversion of liquid wastes to a solid form to allow for more safe storage or transport.

solid waste Unwanted objects or particles that accumulate on the site where they are produced.

source reduction Reducing the amount of solid waste generated by using less, or converting from heavy packaging materials to lightweight ones.

speciation The process of developing a new species.

species A group of organisms that can interbreed and produce offspring capable of reproduction.

species diversity A measure of the number of different species present in an area.

stable equilibrium phase The phase in a population growth curve in which the death rate and birthrate become equal.

standard of living The necessities and luxuries essential to a level of existence that is customary within a society.

steam stripping The use of heated air to drive volatile compounds from liquids.

steppe A grassland.

stormwater runoff Stormwater that runs off of streets and buildings and is often added directly to the sewer system and sent to the municipal wastewater treatment facility.

strict joint and several liability Legal phrase used to describe environmental and other potential liabilities where *strict* liability means liability without fault; *joint and several* liability means that any one of the liable parties involved in a contaminated site may be held liable for the entire cost of cleanup.

strip farming The planting of crops in strips that alternate with other crops. The primary purpose is to reduce erosion.

submerged plants Aquatic vegetation that is rooted on the bottom and has leaves that stay submerged below the surface of the water.

subsidy A gift given to private enterprise by government when the enterprise is in temporary economic difficulty and is viewed as being important to the public.

succession Regular and predictable changes in the structure of a community, ultimately leading to a climax community.

successional stage A stage in succession.

sulfur dioxide (SO$_2$) A compound containing sulfur and oxygen produced when sulfur-containing fossil fuels are burned. When released into the atmosphere, it is a primary air pollutant.

Superfund The common name given to the U.S. 1980 Comprehensive Environmental Response, Compensation, and Liability Act, which was designed to address hazardous-waste sites.

supply Amount of a good or service available to be purchased.

supply/demand curve The relationship between the available supply of a commodity or service and its demand. The supply and demand change as the price changes.

surface impoundment Pond created to hold liquid materials. Some may hold only water, while others may be used to contain polluted water or liquid contaminants.

surface mining (strip mining) A type of mining in which the overburden is removed to procure the underlying deposit.

survivorship curve A graph that shows the proportion of individuals likely to survive to each age.

sustainable agriculture Agricultural methods used to produce adequate, safe food in an economically viable manner while enhancing the health of agricultural land and related ecosystems.

sustainable development Using renewable resources in harmony with ecological systems to produce a rise in real income per person and an improved standard of living for everyone.

swamp Area of trees that is flooded either permanently or for a major part of the year.

symbiosis A close, long-lasting physical relationship between members of two different species.

symbiotic nitrogen-fixing bacteria Bacteria that grow within a plant's root system and that can convert nitrogen gas (N$_2$) from the atmosphere to nitrogen compounds that the plant can use.

synergism The interaction of materials or energy that increases the potential for harm.

T

taiga Biome having short, cool summers and long winters with abundant snowfall. The trees are adapted to winter conditions.

"take back" concept Process that can require the return of goods to the original manufacturer for recycling/reclaiming of usable materials.

target organism The organism a pesticide is designed to eliminate.

technological advances Increasing use of machines to replace human labor.

temperate deciduous forest Biome that has a winter-summer change of seasons and that typically receives 75 to 150 centimeters (30–60 inches) or more of relatively evenly distributed precipitation throughout the year.

temperate grasslands Areas receiving between 25 and 75 centimeters (10–30 inches) of precipitation per year. Grasses are the dominant vegetation, and trees are rare.

temperate rainforest Areas where the prevailing winds bring moisture-laden air to the coast. Abundant rain, fertile soil, and mild temperatures result in a lush growth of plants.

terrace A level area constructed on steep slopes to allow agriculture without extensive erosion.

tertiary sewage treatment Process that involves a variety of different techniques designed to remove dissolved pollutants left after primary and secondary treatments.

theory A unifying principle that binds together large areas of scientific knowledge.

thermal inversion The condition in which warm air in a valley is sandwiched between two layers of cold air and acts like a lid on the valley.

thermal pollution Waste heat that industries release into the environment.

thermal treatment A form of hazardous-waste destruction involving heating waste.

threatened species Those species that could become extinct if a critical factor in their environment were changed.

threshold level The minimum amount of something required to cause measurable effects.

total fertility rate The number of children born per woman per lifetime.

toxic   A narrow group of substances that are poisonous and cause death or serious injury to humans and other organisms by interfering with normal body physiology.

toxicity   A measure of how toxic a material is.

toxic waste   Waste substances that are poisonous and cause death or serious injury to humans and animals when released into the environment.

tract development   The construction of similar residential units over large areas.

transuranic waste   Nuclear wastes of weapons programs that consist primarily of isotopes of plutonium.

trickling filter system   A secondary sewage treatment technique that allows polluted water to flow over surfaces that harbor microorganisms.

trophic level   A stage in the energy flow through ecosystems.

tropical dry forest   Regions that receive low rainfall amounts, as little as 50 centimeters (20 inches) per year, and are characterized by species well adapted to drought. Trees of dry tropical forests are usually smaller than those in rainforests, and many lose their leaves during the dry season.

tropical rainforest   A biome with warm, relatively constant temperatures where there is no frost. These areas receive more than 200 centimeters (80 inches) of rain per year in rains that fall nearly every day.

tundra   A biome that lacks trees and has permanently frozen soil.

U

unconfined aquifer   An aquifer that usually occurs near the land's surface, receives water by percolation from above, and may be called a water table aquifer.

underground mining   A type of mining in which the deposited material is removed without disturbing the overburden.

underground storage tank   Tank located below ground level for the storage of materials, such as oil, gasoline, or other chemicals.

uranium-235 (U-235)   A naturally occurring radioactive isotope of uranium used as fuel in nuclear reactors.

urban growth limit   A boundary established by municipal government that encourages development within the boundary and prohibits it outside the boundary.

urban sprawl   A pattern of unplanned, low-density housing and commercial development outside of cities that usually takes place on previously undeveloped land.

V

vadose zone   A zone above the water table and below the land surface that is not saturated with water.

variable   Things that change from time to time.

vector   An organism that carries a disease from one host to another.

volatile organic compounds (VOC)   Airborne organic compounds; primary air pollutants.

W

waste destruction   Destruction of a portion of hazardous waste with harmful residues still left behind.

waste immobilization   Putting hazardous wastes into a solid form that is easier to handle and less likely to enter the surrounding environment.

waste minimization   A process that involves changes that industries could make in the way they manufacture products that would reduce the waste produced.

waste separation   Separating one hazardous waste from another or from nonhazardous material that it has contaminated.

water diversion   The physical process of transferring water from one area to another.

water table   The top of the layer of water in an aquifer.

waterways   Low areas that water normally flows through.

weathering   The physical and chemical breakdown of materials; involved in the breakdown of parent material in soil formation.

weed   An unwanted plant.

wetlands   Areas that include swamps, tidal marshes, coastal wetlands, and estuaries.

wilderness   Designation of land use for the exclusive protection of the area's natural wildlife; thus, no human development is allowed.

windbreak   The planting of trees or strips of grasses at right angles to the prevailing wind to reduce erosion of soil by wind.

Z

zero population growth   The stabilized growth stage of human population during which births equal deaths and equilibrium is reached.

zoning   Type of land-use regulation in which land is designated for specific potential uses, such as agricultural, commercial, residential, recreational, and industrial.

zooplankton   Weakly swimming microscopic animals.

CREDITS

Design Elements

Standing on a Rocky Peak: © Brand X/Jupiter RF; Students Socializing: © BananaStock/Jupiter RF; Water Ripples: © Comstock/Picture Quest RF; Trail through Shady Forest: © Corbis RF.

Chapter 1

Opener: © Copyright 1997 IMS Communications Ltd./Capstone Design. All Rights Reserved.; 1.2(wolves, elk, dam): © Getty RF; 1.2(willows): © Judy Enger; p. 4: © Getty RF; p. 5: © Stockbyte/PunchStock RF; 1.3: © Getty RF; 1.5(left): © Steve McCutcheon/Visuals Unlimited; 1.5(middle): © Getty RF; 1.5(right): © Corbis RF; 1.6(dust): © Getty RF; 1.6(erosion, pesticides): USDA; 1.6(corn): © Corbis RF; 1.7(grazing): © Digital Vision/PunchStock RF; 1.7(urban): © Getty RF; 1.7(irrigation): USDA; 1.7(grazing): © Brand X/PunchStock RF; 1.8(right): © Corbis RF; 1.8(left): © PunchStock RF; 1.8(middle): © Comstock/PunchStock RF; 1.9(left): © Corbis RF; 1.9(middle,right): © Getty RF; 1.10(left): © Corbis RF; 1.10(middle): © Getty RF; 1.10(right): © Corbis RF; p. 12(top): © Jupiter RF; p. 12(bottom) USFWS (U.S. Fish and Wildlife Service).

Chapter 2

Opener: © PunchStock RF; 2.1: © Getty RF; 2.3(left): Photo by Lynn Betts, USDA Natural Resources Conservation Service; 2.3(middle,right): © Getty RF; p. 20(Emerson, Thoreau): Library of Congress; p. 20(Muir): © Bettmann/Corbis; p. 20(Leopold,Carson): © AP Wide World Photos; 2.5a,b: Courtesy of the North Carolina State Archives, reprinted by permission of Raleigh News & Observer, photographer Harry Lynch; p. 24(left):NASA/Jeff Schmaltz, MODIS Land Rapid Response Team; p. 24(right): Courtesy of U.S. Army/U.S. Coast Guard/photo by Petty Officer 2nd Class Kyle Niemi; 2.7: © Corbis RF; 2.8a: © Getty RF; 2.8b: © Natalie Fobes/Getty Images; 2.9(top left, top right, bottom right): © Corbis RF; 2.9(bottom left): © Getty RF.

Chapter 3

Opener: © Getty RF; p. 39, 3.1: © Getty RF; 3.4: USDA Soil Conservation Service; 3.5: © Dr. Byron Augustin; 3.6(danger): © Jeff Greenberg/PhotoEdit; 3.6(smog): © Vol. 25/PhotoDisc/Getty Images; 3.6(smoking): © PunchStock RF; 3.6(paint,cattle): © Corbis RF; 3.6(towers): © Getty RF; 3.6(visual pollution): © Michael Newman/PhotoEdit; 3.6(junk): © Getty RF; 3.7(buggy): © Corbis RF; 3.7 (boats): © Vol.10/PhotoDisc/Getty Images; 3.7(signs,hiker): © The McGraw-Hill Companies/Roger Loewenberg, photographer; 3.7(canoe): © Vol.10/PhotoDisc/Getty Images; 3.7(street): © PunchStock RF; 3.8: © Vol.31/PhotoDisc/Getty Images; 3.10b: © AP Wide World Photos.

Chapter 4

Opener: © Getty RF; p. 62: © Stockbyte/PunchStock RF; 4.2: © Getty RF; 4.3: © The McGraw-Hill Companies, Inc./Jim Thoeming, photographer; p. 70: © Jupiter RF; 4.10: © Getty RF; 4.11: © Digital Vision/PunchStock RF; p. 76: © Corbis RF.

Chapter 5

Opener: © Corbis RF; 5.2(both): © Creatas/PunchStock RF; 5.3b: © Getty RF; 5.3c: © PunchStock RF; 5.4: © Corbis RF; 5.5(beaver): © Creatas/PunchStock RF; 5.5(dam): © Getty RF; 5.5(grebe): © Masterfile RF; 5.5(gnawing,pond): © Brand X/PunchStock RF; 5.6(left): © Getty RF; 5.6(right): © It Stock/age fotostock RF; 5.6(middle): © Getty RF; 5.7: © Creatas/PunchStock RF; 5.9: National Library of Medicine; 5.10: © Corbis RF; 5.11: © Pixtal/age fotostock RF; 5.13: © National History Museum, London; 5.14, 5.15: © Corbis RF; 5.16(top): © It Stock/PunchStock RF; 5.16(bottom): © PhotoAlto/PunchStock RF; 5.17: © Creatas/PunchStock RF; 5.18: © PunchStock RF; 5.20(left): U.S. Department of Agriculture; 5.20(middle): © Manfred Kage/Peter Arnold, Inc.; 5.20(right): © Malcom Kitto, Papillo/Corbis; 5.21: © Brand X/PunchStock RF; 5.22a: Eldon Enger; 5.22b: © Getty RF; 5.23: © K. Malowski/Visuals Unlimited; 5.24: © Corbis RF; 5.27(father and son, fish, frog, grasshopper, cattails): ©: Getty RF; 5.27(spider): © Digital Vision RF; p.100(top): NOAA Great Lakes Environmental Research Laboratory; p. 100(bottom): © Eldon Enger; p. 105: © The McGraw-Hill Companies, Inc./Pat Watson, photographer.

Chapter 6

Opener: © Alamy RF; 6.2: Eldon Enger; 6.5: © Steven P. Lynch; 6.10b,c: © Getty RF; 6.10d: © Corbis RF; p. 117: Photo by Tim McCabe/USDA; 6.11b: © Creatas/PunchStock RF; 6.11c: © Getty RF; 6.11d: © It Stock/age fotostock RF; 6.11e: © U.S. Fish and Wildlife Service/John and Karen Hollingsworth photographers; 6.12b: © Corbis RF; 6.12c: © Brand X/PunchStock RF; 6.12d,e: © PunchStock RF; 6.13b: © Steven P. Lynch; 6.13c: © U.S. Fish and Wildlife Service/ Lee Karney photographer; 6.13d: © Judy Enger; 6.14b: © Doug Wechsler/Animals Animals/Earth Scenes; 6.14c: © Superstock RF; 6.14d: © David Zurick; p. 123(left): © The McGraw-Hill Companies/Roger Loewenberg, photographer; p. 123(middle): © Harold Hungerford/University of Southern Illinois-Carbondale; p. 123(right): © Carl Bollwinkel/University of Northern Iowa; 6.15b: © Stephen P. Lynch; 6.15c: © Getty RF; 6.15d: © PunchStock RF; 6.16b-d: © Getty RF; 6.17b: U.S. Fish and Wildlife Service; 6.17c: © Corbis RF; 6.17d: © Getty RF; 6.18b: © Stephen P. Lynch; 6.18c: © Creatas/PunchStock RF; 6.18d, 6.19b: © Corbis RF; 6.19c-e: © Getty RF; p.134(all): © Judy Enger; 6.21(left): © Getty RF; 6.21(right): © Corbis RF; 6.22: © Eldon Enger; p. 137: © Digital Vision/PunchStock RF.

Chapter 7

Opener: PunchStock RF; 7.2a: © Getty RF; 7.2b: © John W. Bova/Photo Researchers, Inc.; 7.2c: Courtesy of Monika Al-Mufti; 7.4a,b: © Getty RF; 7.4c: © Corbis RF; p. 147(right): © Julia Sims/Peter Arnold, Inc.; p. 147(right): Photo by Jeff Vanuga, courtesy of USDA Natural Resources Conservation Service; p. 152: © Bettmann/Corbis; p. 159: © Corbis RF; 7.13a: © Peter Ginter/ScienceFaction; 7.13b: © Peter Menzel; p. 163: © PunchStock RF; p. 167: The Hudson Bay Project. Photo provided by Robert Jefferies.

Chapter 8

Opener: © Corbis RF; 8.1: © Corbis RF; 8.2: © Getty RF; 8.3: Library of Congress; 8.5a: © Getty RF; 8.5b: © The McGraw-Hill Companies, Inc./Andrew Resek, photographer; p. 174: © Getty RF; p. 175: © Corbis RF; p. 176: © Getty RF; 8.7a : © Steve McCurry/Magnum Photos; 8.7b: © Galen Rowell/Corbis; 8.8, 8.9: © Getty RF.

Chapter 9

Opener: © Getty RF; 9.8: © William Campbell/Peter Arnold, Inc.; 9.9: © Getty RF; 9.10a: © Matt Meadows/Peter Arnold, Inc.;

INDEX

A

A Horizon, 294–97
abiotic environment, 80–81, 98, 100–101, 109
absorbed dose, radiation, 216
abyssal ecosystems, 131
acceptable risks, 40–41
acid mine drainage, 191–92
acid rain
    cause-and-effect relationships, 70, 375–78
    coal burning, 191
    environmental costs, 44
    politics of, 30
acids, chemical nature of, 67–69
activated-sludge, 357–58
activation energy, chemical reaction, 70–71
active solar power systems, 202–03
acute toxicity, 414
administrative law, 436
aesthetic pollution, 274
Afghanistan, 182
Africa. See also individual country
    agricultural production, 301–02, 312
    climate change, 381–82
    hydropower development, 200
    soil management, 309
    wildlife recovery, 442
African American minorities, 21–23
age distribution, population, 142–43, 154–55
Agency for Toxic Substances and Disease Registry (ATSDR), 415–16
Agenda 21 (Rio Declaration on Environment and Development, 1992), 5, 21, 444
Agent Orange, 442
agricultural chemicals, 315, 318. See also fungicides and rodenticides; herbicides; insecticides; pesticides
agricultural methods
    conventional agriculture, 326
    genetically modified crops, 330–31
    history and development, 312–14
    integrated pest management, 327–30

organic farming, 310, 325–27
    "slash and burn," 312
    sustainable agriculture, 325–26
agricultural waste, 393
agriculture. See also fertilizer/fertility; food supply/production; soil
    biodiversity and, 241
    biomass production, 175, 196–97
    climate change and, 29–30, 381–83
    conversion of land to, 242, 266
    dam construction and, 201
    destruction of ecosystems, 113
    environmental issues, 7–8
    ethanol fuel and, 210
    government subsidies/assistance, 50–51, 300
    groundwater mining, 358–60
    impact of urban growth on, 272
    impact on nutrient cycles, 104–05
    land-use strategies, 276
    as a natural resource, 41–42
    supply/demand economics of, 43
    Third World development of, 30–32
    use of fossil fuels, 103–04, 314
    water pollution, 350
    water use, 29, 342–43
agro-ecology, 325–26
air conditioning, 75
air pollution
    agriculture and, 7
    biomass conversion and, 199
    case study, 369
    catalytic converters, 70
    categories/types of, 367–73
    climate change and, 381–82
    control measures, 373–75
    diesel vs. gasoline engines, 76
    environmental issues of, 4
    indoor sources, 385–87
    making personal choices, 389
    mining and, 190–91
    true vs. perceived risks, 40–41
    urban growth and, 271
    urbanization and, 265
air quality
    atmospheric pollution and, 366–73
    automobiles and, 209

biomass fuels and, 175
    cost-benefit analysis, 47
    ethics and, 15–16
    pollution control measures, 373–75
    U.S. environmental legislation, 435
air stripping, hazardous waste, 422
air toxics, 367, 373
air travel, 177. See also transportation
Akosombo Dam (Ghana), 200
Alaska, 5–6, 126, 195, 251–52, 273, 351
Alaskan National Interests Lands Conservation Act of 1980, 195, 435
alcohol, true vs. perceived risks, 40–41
Aleutian Islands, 290
algae
    agricultural runoff, 7–8, 104–05, 315, 319
    in freshwater ecosystems, 134–35, 191
    in marine ecosystems, 130–31, 133, 260, 351
    in primary succession, 110–11
    role in ecosystems, 95–96
    water pollution and, 100
Algeria, 180–81
alkaline solutions, chemical nature of, 67–69
alpha radiation, 215–16
alpine tundra, 128
alternative agriculture, 325
Amazon River, 134
American College and University Presidents, 446
American Conference of Governmental Industrial Hygienists (ACGIH), 413
American Society for Testing and Materials (ASTM), 37, 412
American Solar Energy Society, 50
American Wind Energy Association, 205
Amherst (Lord), 442
amino acids, 102
ammonia, 37, 102–03, 196, 294, 367
Amsterdam Treaty of 1997, 445
anaerobic digestion, 197–98
Anderson, Ray, 25
Animal Liberation Front (ALF), 33
animal rights/welfare, 18, 38–39
animals
    dung as fuel source, 171, 175

animals—*Cont.*
  invasive/exotic species, 49
  overexploitation of resources, 249–50
  selective breeding issues, 329
  species extinction, 87–88, 235–36
  sustainability management of
    populations, 258–60
  wastes as biomass, 197
Annan, Kofi, 436
Antarctica, 253, 378, 423
Antarctica Treaty of 1961, 443
Antarctic Conservation Act of
  1978, 435
anthracite (hard) coal, 189
anthrax (*Bacillus anthracis*), 442
anthropocentrism, 17
aquaculture (fish farming), 131, 133,
  248–49
aquatic ecosystems. *See also* fisheries
  management
  acid rain effects, 377–78
  freshwater ecosystems, 133–36
  habitat loss, 247–48
  human impact on, 131, 133, 136
  impact of pollution on, 347
  introduced species and, 250–51
  marine ecosystems, 129–33
  marine oil pollution, 351–52
  preservation efforts, 360–61
  thermal pollution and, 227, 344
  tidal energy issues, 207
aquatic succession, 110–11
aquiclude/aquitard, 338
aquifers, 338. *See also* groundwater
Aral Sea, 244
archeological sites
  conservation easements, 117
  loss of, 201
  today's landfills will become, 395
Arctic National Wildlife Refuge
  (ANWR), 195
Arctic Sea, 386
area conversion, metric, 453
Argentina, 157–58, 160, 302, 331
argon gas, 366
Arizona, 75, 261, 274, 285, 358, 405
arsenic, 44, 394, 400, 415
artesian water, 338, 340
asbestos
  carcinogenic effects, 369
  indoor air pollution, 385
  true vs. perceived risks, 40–41
asexual reproduction. *See* reproduction
ash, incinerator, 394, 399, 400, 411, 422
Asia. *See also* individual country
  agricultural production, 301–02, 312

diet and food production, 322
Association for the Advancement of
  Sustainability in Higher
  Education (AASHE), 12, 183
asthenosphere, 289
Aswan Dam (Egypt), 200
AT&T Corporation, 53
atmosphere/atmospheric pressure, 366
atmospheric pollution
  categories, 366–73
  as cause of global warming, 378–83
  control measures, 373–75
  damage from acid deposition, 375–78
  dealing with climate change, 383–85
  ozone depletion, 378
atomic structure, 66–67, 214–15
attitudes, environmental, 18–21
Auburn University, 161
Australia, 30, 106, 119, 121, 151–53, 162,
  181, 189–90, 199, 221, 239, 246,
  327, 336, 384, 406
automobiles
  air pollution and, 49, 367, 389
  biofuels, 197
  carbon monoxide and, 367–68
  catalytic converters, 70
  diesel vs. gasoline, 76
  fuel cells, 209–10
  fuel efficiency, 179
  gasoline taxes, 448
  hybrid electric vehicles, 209
  oil production and, 172–73
  photochemical smog, 371–73
  reducing use in cities, 174
  true vs. perceived risks, 40–41
  urban sprawl and, 268–70
  urban transportation planning, 279–80
  U.S. gasoline prices and, 184
auxins, plant, 317–18

B

baby boom, 162–63
Bacillus anthracis (anthrax), 442
Bacillus thuringiensis (Bt), 329
bacteria
  cyanobacteria, 105
  in nitrogen cycle, 101–02
  nitrogen-fixing, 93
Balsas River (Mexico), 337
Bangladesh, 56, 159, 178, 406
Bank of America, 50
Barbados, 406
Basel Action Network, 419
Basel Convention (1992), 423, 445
bases, chemical nature of, 67–69
batteries

fuel cell technology, 28
  heavy metals contamination, 58, 370,
    400, 415, 424
  hybrid electric vehicles, 209, 434
  nickel-cadmium, 53
  photovoltaic cells, 203
Bay of Fundy (Canada), 360
Bechtel Corporation, 230
Belgium, 414
benthic marine ecosystems, 131
benzene, 368, 373, 417
beryllium, 415
beta radiation, 215–16
beverage container waste, 52, 403
B Horizon, 294–97
Big Rock Point nuclear power plant, 229
bioaccumulation
  DDT case study, 316
  of persistent organic pollutants, 32
  pesticide use and, 320–22
  of waste products, 145
  water pollution and, 10
biocentrism, 17
biochemical oxygen demand (BOD),
  135, 348
biocides, 315
biodegradable materials
  composting, 295
  plastics regulations, 406
  product claims of, 68
  waste disposal, 45
biodiesel, 197
biodiversity. *See also* genetic diversity
  addressing the urgency of, 262
  biomass conversion and loss of, 199
  contributions and value of, 239–42
  defined, 236–39
  economics vs. ecology, 47–49
  ethics and, 15–16
  "Going Green" initiatives, 254
  human impact on, 242–53
  legal protections for, 254–58
  North American losses, 137
  preservation efforts, 253–62
  soil conservation and, 303
  species extinction from loss of, 87–88,
    235–36
  sustainable agriculture and, 325–26
  tropical rainforest biomes, 123–24
biofuels, 160, 197
biogas, 198
biogeochemical cycle. *See* nutrient cycles
biological pest control, 327–29
biological sewage treatment, 358
biological weapons, 442
Biological Weapons Convention (BWC), 442

biomagnification, 320
biomass
    case study on use of, 175
    CO_2 sequestration, 384
    conversion to energy, 196–99, 210
    as energy source, 28–29, 97, 186
    use of wood for fuel, 171, 173, 176
biomes. *See* aquatic ecosystems; terrestrial
        biomes
biotechnology issues, 329–31
biotic environment, 80–81, 98,
        100–101, 109
biotic potential, 143–44
birth control, 152, 155, 157, 161
birthrate, population, 140, 153–54
bismuth, 214, 387
bisphenols, 424
bituminous (soft) coal, 189
black lung disease, 190
blue-green bacteria (cyanobacteria), 105
boiling-water reactors, 219, 229
Bolivia, 57
boreal forest biomes, 126–27
Borlaug, Norman, 28
boron, 218, 314
borosilicate glass, 226
Botswana, 57
bottled water, 340, 403
Boundary Waters Treaty of 1909
        (U.S./Canada), 5
Brazil, 21, 32, 49, 175, 177, 181, 197
breeder reactors, 220
Britain. *See* United Kingdom
Brown, Noel, 30
brownfield development, 51, 276
Brundtland Commission, 21, 54
Bullard, Robert, 24
Bush, George W., 18, 51, 195, 226

C

cadmium, 53, 218, 400, 411, 415, 417, 419
calcium, 292, 377, 383, 389
calcium carbonate, 375–76
calibration of consumer knowledge, 54
California, 9, 20, 30, 55, 121, 126, 167,
        203–04, 270–74, 283, 327, 343,
        345, 359–61, 373–74, 386, 404–05
California Condor (*Gymnogyps
        californianus*), 261
California Urban Conservation
        Council, 341
Cambodia, 423
Canada. *See also* North America
    ecolabeling programs, 446–47
    habitat preservation, 360
    hazardous waste laws, 423

hydroelectric power, 200
loss of biodiversity, 137
Northern Yukon National Park, 195
pattern of land use, 265
population case study, 165
population policies, 157
recycling programs, 53
tidal power use in, 207
use of biomass, 196
waste production, 393
wildlife management, 252
cancer
    particulate matter and, 369
    pesticide residues and, 323–24
    radon gas and, 387–88
    risk probabilities, 39
    secondhand smoke and, 385–86
    true vs. perceived risks, 40–41
Cape Cod National Seashore (U.S.), 360
capitalist work ethic, 18–19
carbamate pesticides, 317
carbon absorption, 422
carbon cycle, 98–101
carbon dioxide (CO_2)
    addressing the effects of, 383–84
    biomass conversion and, 199
    burning fossil fuels, 76, 103–04, 191
    carbon cycle, 100–101
    greenhouse effect, 380
    Kyoto Treaty standards, 179
carbon footprint. *See* ecological footprint
carbonic acid, 375. *See also* acid rain
carbon monoxide (CO), 367–68, 385
carbon sequestration, 104, 384
carcinogens, 369, 414
carnivores, 95–96
carrying capacity, environmental,
        146–47, 167
Carson, Rachel, 2, 20, 316, 434
Case Studies
    air pollution, 369
    Arctic National Wildlife Refuge
        (ANWR), 195
    beverage container waste, 403
    biomass as energy source, 175
    California condor, 261
    determining toxicity, 414
    diet and food production, 322
    Everglades National Park, 354–55
    Grameen Bank and microcredit, 159
    grasslands succession, 124
    Hurricane Katrina, 439
    land capability classes, 308
    North American population, 165
    nuclear waste cleanup, 231
    pesticides, 316

philosophers of nature, 20
pollution prevention, 53
population growth predictions, 152
poverty and natural disasters, 24
risk probabilities, 39
thermoplastics recycling, 400
water scarcity, 345
catalysts, chemical reaction, 70–71
catalytic converters, 70, 368, 371
cause-and-effect relationships, 62, 70
ceiling limit (CL), 413
cell phone recycling, 419
cellular respiration, 75
certified organic foods/labeling,
        326, 332
chaparral (Mediterranean shrubland)
        biomes, 120–21
chemical reactions/bonds, 69–72
chemical sewage treatment, 358
chemical wastes. *See* hazardous waste
chemical weapons, 442
chemical weathering, 290
Chernobyl disaster (1986), 222–23
Chevron Corporation, 55
Chile, 120–21, 126, 248
China
    agricultural production, 301–02, 312
    air pollution issues, 390
    biogas production, 198
    coal production, 51
    diet and food production, 322
    ecolabeling programs, 446–47
    energy use in, 174, 181, 438, 440
    history of fossil fuels in, 172
    plastics regulations, 406
    pollution control programs, 52
    population control, 157
    use of hydropower, 200–201
Chittagong University, 159
chlorinated hydrocarbon pesticides,
        316–17
chlorofluorocarbons (CFCs), 378, 380–81,
        384–85, 443
C Horizon, 294–97
chronic toxicity, 414
cigarette smoking. *See* smoking
Civil Rights Act of 1964, 23
Clark, William, 15
Clausen, A. W., 56
Clean Air Act of 1970, 23, 373–75, 412
Cleantech Network, 50
Clean Water Act of 1972, 23, 344, 351, 412
clear-cutting, forest, 244–45
climate
    biodiversity and, 240
    creates biomes, 114–15

climate—*Cont.*
  relationship of geography to, 372–73
  soil formation and, 292, 297–98
  species range of tolerance, 81–82
climate change
  biomass conversion and, 199
  burning fossil fuels and, 191, 196
  carbon sequestration and, 104
  causes and effects, 378–83
  dealing with the issues of, 383–85
  diesel vs. gasoline engines, 76
  economics vs. ecology, 47–49
  energy consumption and, 28–29
  environmental ethics and, 29–32
  food production and, 27–28
  "Going Green" initiatives, 446
  green economics and, 49
  Kyoto Protocol, 5
  nuclear energy and, 229
  species survival and, 253
  U.S. policies toward, 205
Climate Change 2007, 379
climax communities, 109–13
Clinton, William Jefferson, 23
Clinton Administration, 18
cloud forests, 254
The Club of Rome, 27
Coalition for Environmentally Responsible
    Economics (CERES), 26
coal production. *See also* fossil fuels;
    mining
  electrical energy from, 178
  as energy source, 186–87
  extraction and use, 189–92
  formation process, 186–87
  international geopolitics, 438
  mercury contamination, 58
  removing sulfur from, 375
  as replacement for wood, 171, 181
coevolution, 88–89
co-generation power plants, 376
Colorado River (U.S.), 30, 244, 301, 337,
    345, 358, 431
Columbia River (U.S.), 231, 357
Columbia University, 404
combustion, energy conversion by, 75
commensal relationships, 92–93
commercial energy use, 174–75. *See also*
    energy consumption; industrial
    development
common property resources, 48–49
communication, scientific method, 65
community. *See* ecosystems
community sustainability
  land-use planning and, 268–70
  smart growth for, 281–83, 285

strategies for, 276
  urban growth and, 271–72
competition, species, 90–91
competitive exclusion principle, 91
composting and mulching, 295, 304, 399–401
compounds, relationship of matter in, 67
Comprehensive Environmental Response
    Compensation and Liability Act.
    See Superfund Act of 1980
computer systems, 175, 419
Computer TakeBack Campaign, 419
Conferences, international. *See* United
    Nations
confined aquifer, 338
coniferous forest biomes, 126–27
consequences of risk, 37
conservation. *See also* overexploitation of
    resources
  agricultural water use, 342–43
  ANWR case study, 195
  biodiversity preservation and, 253–54
  debt-for-nature exchanges, 57
  domestic water use, 340–41
  environmental ethics of, 19–21
  "Going Green" initiatives, 361
  groundwater mining and, 358–60
  habitat preservation, 147
  industrial water use, 343–44
  saving energy by, 207–09
  soil management practices, 301–05
  soil tillage practices, 305–07
  U.S. environmental legislation, 435
conservation easements, 117
Conservation Fund, 147
Conservation International, 57
consumerism, 28
consumers
  choices affecting biodiversity, 257
  ecolabeling programs, 446–47
  ecosystem organisms as, 95–96, 101–02
  green marketing principles, 54
  information programs, 51–52
consumption ethics, 27–29, 56
container deposit-refund programs, 52, 403
container laws, 402–04
contingent valuation method
    (CVM), 43–44
contour farming, 303–04
contraception (birth control), 152, 155,
    157, 161
controlled experiment, 64–65
conventional agriculture, 326
conventional tillage, 305–07
Conventions, international. *See* United
    Nations
conversion tables, metric, 452–53

coral reef ecosystems, 131, 133,
    235, 253, 382
corporations, environmental ethics, 23–27
Costa Rica, 52, 446
cost-benefit analysis, 45–47
Council on Environmental Quality, 436
credibility of product claims, 54
criteria air pollutants, 367–68
crop resistance, 329
crop rotation, 327
crust, Earth, 289
Cryptosporidium sp., 349–50
cultural relativism, 16
culture and traditions, population, 155–56
curbside recycling programs, 404
Cuyamaca College, 361
cyanobacteria, 105
Czech Republic, 417

D

DaimlerChrysler AG, 209
Dalupiri Ocean Power Plant
    (Philippines), 207
dam construction, 344–46
dam-removal projects, 43
Darwin, Charles, 2, 85
DDT (dichloro-diphenyl-trichloroethane),
    32, 236, 316, 320–22, 412, 415
death phase, population, 145
deaths
  from accidents, 40
  from air pollution, 367
  cancer, 39–41, 387
  from environmental risks, 38
  infant mortality rates, 158
  from ionizing radiation, 216–17
  population mortality rates, 140–41, 153–54
debt-for-nature exchanges, 57
deceleration phase, population, 144
decibels/decibel scale, 387–89
decision-making. *See also* environmental
    policy
  choices affecting biodiversity, 257
  determining risks in, 37–40
  land use planning, 274–79
  true vs. perceived risks, 40–41
decomposers, ecosystem, 95, 101, 240
deep ecology, 18
DEET (insect repellant), 350
deferred costs, 44
deforestation, 30, 56, 104, 152, 175–76,
    256, 299. *See also* forestry
    management
Delaware River (U.S.), 355
Delta College, 183
demographics, population, 153–55

denitrifying bacteria, 102
Denmark, 178, 204, 229, 406, 423
density, population, 143
density-dependent limiting factors, 144–45
density-independent limiting factors, 145
deposit-refund programs, 52, 403
desalination, 29, 337, 339
desert biomes, 116–18
desertification, 28, 151, 159, 161, 199, 247–48, 256, 299
desert soils, 294–98
Design for Environment program, 53–54
detergents, 348, 350
detritus, 98
developing countries. *See* less-developed countries; Third World development
development. *See* sustainability/ sustainable development
diesel engines, 76
dioxins, 415
dirty bombs, 223–24
disease. *See also* health/health issues
    climate change and, 253, 381–82
    control of pests and, 251
    drinking water and, 164, 335–37
    indoor air pollution and, 282, 385
    insects role in, 316
    introduced species and, 250–51
    population density and, 143
    as population limiting factor, 144–45
    urbanization and, 265
    water quality and, 347–50
Disease Registry (ATSDR), 415–16
dispersal, population, 143
distilled water, 340
DNA (deoxyribonucleic acid), 84, 98, 102, 216–17, 236, 329
domestic terrorism, 33
domestic water use, 339–41
Dominican Republic, 57
dose equivalent, radiation, 216
Dow Chemical (corporation), 50, 324
Drake, Edwin L., 172
drinking water
    agriculture and, 8
    bottled water, 340
    domestic water use and, 339–41
    future world needs for, 29, 164
    groundwater mining, 358–60
    land-use planning and, 274
    lead contamination, 46, 362
    solar purification, 204
drought, water ethics and, 30
Duke University, 18, 361
DuPont Corporation, 50, 378

E

Earth (the planet)
    consumption ethics, 27–29
    future predictions, 32
    geologic processes of, 289–91
    global warming and climate change, 378–83
    relationship of climate and geography, 372–73
    satellite observations, 386
    as seen from space, 15
    water ethics, 30
Earth Day (1970), 2, 430, 434
Earth First!, 33
The Earth Island Institute, 257
Earth Liberation Front (ELF), 33
earthquakes/seismic activity, 265, 273–74, 289–90
Earth Summit. *See* UN Conference on Environment and Development (UNCED)
ecocentrism, 17
ecofeminism, 18
ecojobs, 50
ecolabels, 446–47
Ecoli (*Escherichia coli*), 349–50
ecological footprint, 29, 31, 153
ecology, science of
    concepts of, 80–84
    interactions of community and ecosystems, 94–105
    interactions of individuals, 89–94
    natural selection and evolution, 84–89
economics/economic growth
    addressing climate change with, 384–85
    biodiversity and, 241
    coal environmental issues, 190–92
    consumption ethics and, 27–29
    corporate ethics and, 25–26
    dealing with environmental issues, 49–54
    economics vs. ecology, 47–49
    environmental costs, 44–47
    environmental ethics and, 23
    environmental science and, 4–5
    ethanol vs. gasoline, 210
    green business, 26–27
    human population and, 151–53
    of natural disasters, 439
    natural gas environmental issues, 194–96
    nuclear energy and, 229
    oil environmental issues, 193–94
    politics of energy use, 179–80
    politics of hunger and food, 160–61
    population growth and, 155–58
    recycling programs, 402–06

    risk management, 37–40
    of sustainable development, 54–56
    in Third World development, 57
    urban sprawl and, 269–70
    of U.S. gasoline prices, 184
    value of resources in, 41–44
ecosystems
    benefits of biodiversity in, 239–42
    climate change and, 381–83
    defined, 3–4, 94
    economics vs. ecology, 47–49
    energy flow in, 96–97
    energy sources in, 171–72
    environmental ethics and, 27–29
    food chains/food webs, 97–98
    human impact on, 236
    interactions of individuals, 89–94
    loss of biodiversity, 137
    natural disaster impact on, 439
    nutrient cycles, 98–104
    political boundaries and, 13
    role of organisms in, 95–96
    role of predators in, 3
    succession and climax communities, 109–13
ecoterrorism, 33, 441–42
ecotourism, 47–48
ectoparasites, 92
Ecuador, 180–81, 300, 353
education
    consumer programs, 51–52
    green marketing, 54
    socially responsible investments, 18
    sustainability issues in, 12
    of women and children, 155, 157–58
Egypt, 30, 154, 156, 175, 200, 302, 337, 353
E Horizon, 294–97
Ehrlich, Paul, 27
electricity
    co-generation power plants, 376
    as energy source, 177–79
    from hydroelectric power, 344
    from landfill gas, 395
    nuclear power generation, 219, 230
    power plant emissions, 375
    pricing and consumption, 179–80
    solar energy and, 203–04
    toxic chemical releases and, 418–20
electronics (e-waste), 419
electrons, 66–67, 214–15
elements
    atomic structure, 66–67
    periodic table, 451
    radioactive half-life, 214–16, 387
Elwha Restoration Project (U.S.), 43
Emerson, Ralph Waldo, 19–20

emigration, population, 143
emissions fees and taxes, 51–52
employment, green-collar, 50
endangered species. *See also* U.S.
    Endangered Species Act
  biodiversity preservation and, 253–54
  climate change and, 253
  conservation easements, 117
  forestry management and, 9
  Indian protection of, 4
  international environmental treaties, 32
  legal protections for, 254–58
  from loss of ecosystems, 137
  overexploitation of resources, 248–50
  societal perceptions of, 262
endemism, 244
endoparasites, 92
endothermic reactions, 69–71
energy
  in chemical reactions, 69–70
  ecological stability and, 96–97
  environmental implications of,
    74–75, 80–81
  fossil fuels vs. human labor, 314
  photosynthesis, 70–71
  as population limiting factor, 164
  principles/kinds of, 72–74
  renewable vs. non-renewable, 186–87
  as requirement for life, 145, 160, 171
  resources vs. reserves, 186–87
  U.S. environmental legislation, 435
energy conservation, 207–09
energy consumption
  electricity, 177–79
  environmental ethics and, 28–29
  government policy and, 182
  green economics and, 49
  for heating water, 176
  history of, 171–74
  industrial use, 175–77
  international geopolitics, 438, 440
  politics and economics of, 179–80
  residential and commercial use, 174–75
  taxation and, 448
  transportation use, 177
  trends in, 180–82, 196
energy efficiency, 383
Energy Information Administration
    (EIA), 195
energy production. *See also* biomass;
    fossil fuels
  government subsidies, 50–51
  renewable resources for, 196–207
  resources and reserves for, 186–87
  technology advancements, 49
entropy, 73–74

environment. *See* environmental science
environmental aesthetics and
    attitudes, 18–21
environmental costs/economics, 44–47,
    419–22. *See also* economics/
    economic growth
environmental debt, 431
environmental ethics
  approaches to and definitions, 15–18
  attitude toward the environment, 18–21
  biodiversity and, 241–42
  consumption, 27–29, 56
  corporations, 23–27
  ecosystem restoration, 3
  environmental justice and, 21–23
  globalization, 29–32
  individual, 27
  individual responsibility and, 447
  society, 23
environmental governance, 429–30
environmental impact analysis, 43
environmental impact statements, 46
environmental investment policy, 18
environmental justice, 21–23
environmental law. *See* laws and treaties;
    U.S. environmental legislation
environmental policy. *See also*
    government policies/regulation
  administrative law, 435–36
  challenges for the future, 428–32
  development of U.S., 432–35
  ecoterrorism and, 441–42
  EPA role in, 436–37
  international geopolitics, 438–41
  international organizations, 442–47
environmental racism, 22
environmental regulation. *See* government
    regulation; U.S. Environmental
    Protection Agency (EPA)
environmental resistance, 144–46
environmental risk
  assessment and management, 37–40
  determining tolerances, 40–41
  economics, 41–49
  ethics, 15–16
  hazardous and toxic materials, 410–11
  of mercury contamination, 58–59
  sustainable development and, 54–56
  Third World development and, 57
  tools for addressing issues of, 49–54
environmental science. *See also* science
  concept of ecology, 80–84
  concept of interrelatedness in, 2–3
  economic and political issues, 4–5
  economic tools, 49–54
  ecosystems approach, 3–4

  global and regional nature of, 5–11
  measuring population impact, 151–53
environmental tax systems, 51–52
environmental terrorism. *See* ecoterrorism
environmental treaties, 32
enzymes, 71–72
erosion. *See* soil erosion
estuaries, 131
ethanol, 197, 210
ethics. *See* environmental ethics; personal
    ethics/choices
Ethiopia, 56
euphotic zone, 130
Euphrates River (Middle East), 337
Europe
  agricultural potential, 301–02
  green initiatives, 384
  hazardous waste dumps, 416–17
  noise pollution abatement, 389
European Union, 445–47
eutrophication, 135, 348
evapotranspiration, 337
Everglades National Park, 354–55
evolution and natural selection, 84–89
exothermic reactions, 69–71
exotic species
  fisheries management and, 49
  fish farming, 248–49
  green landscaping and, 304
  impact on biodiversity, 244, 250–51
  introduction of, 100
experiment, hypothesis testing, 64–65
exploring, scientific method, 64
exponential growth phase, population, 144
exposure limits, 413–15
extended product responsibility, 52–54, 419
external costs, 44, 50–51
extinction. *See* species extinction
extrinsic limiting factors, 144
Exxon Valdez oil spill, 25–26, 351, 439, 441

F

family planning programs, 156–57
farming. *See* agriculture
"farming with nature," 325–26
fecal coliform bacteria, 349–50
federal environmental legislation, U.S. *See*
    U.S. environmental legislation
feedback loops, 29–30
female literacy, 155–56
fertility rate, population, 154–58
fertilizer/fertility
  activated-sludge, 357–58
  agricultural waste as, 393
  agriculture and, 7–8, 104–05, 314–15
  Everglades case study, 354–55

government regulation, 333
increased food production from, 27–28, 309
natural gas use, 196
nutrient cycles and, 102
nutrient pollution, 319
soil conservation and, 303
symbiotic relationships, 93
Finland, 178, 180, 197, 226
first law of thermodynamics, 73–75
fisheries management
biodiversity and, 241
dam construction and, 201
dam-removal projects, 43
environmental ethics and, 16
environmental justice and, 22
government subsidies, 50–51
human impact on, 131, 133
introduced species and, 250–51
mercury contamination, 59
nutrient pollution and, 319
overexploitation of resources, 248–49
overharvesting, 48–49
Pacific Northwest salmon, 4–5
preservation efforts, 360–61
sustainability of populations, 260
fish farming (aquaculture), 131, 133, 248–49
fission. See nuclear fission
fissionable material, 218
flood irrigation, 342–43
floodplains, 3, 272–73, 277, 298
floods/flooding
dams and flood control, 43, 201, 244, 274, 344, 346
deforestation and, 56
development in low-lying areas, 11, 24, 30, 275
soil erosion, 301
successional changes, 111–13, 136, 146, 239–40
urban growth and, 272, 339
water use planning, 353
wetlands role in, 360–61
wildlife role in, 83–84, 134
Florida, 11, 30, 131, 269, 277–78, 354–55
fluorescent lighting, 77, 208
fly ash, 394, 399, 400, 411, 422
food chain/food web
in ecosystems, 97–99
impact of bioaccumulation on, 10, 316–17
impact of exotic species on, 100
mercury contamination, 58–59
persistent organic pollutants in, 32
persistent vs. nonpersistent pollutants, 415–16

food supply/production. See also agriculture
biodiversity and, 240
biomass conversion and, 199
climate change and, 253
competition for resources, 90–91
economics case study, 322
environmental ethics and, 27–29
environmental impact of, 158–60
genetic engineering issues, 329–31
human energy pyramid, 160
organic foods/labeling, 332
pesticide residues, 323–24
politics of, 160–61
soil fertility and, 309
wildlife pressures on, 167
forestry management. See also deforestation; wilderness
biodiversity and, 241
biomass conversion and, 196–97
environmental ethics and, 16
environmental issues of, 9, 15
government subsidies, 50–51
habitat loss from, 242–46
international environmental treaties, 32
introduced species and, 250–51
market-based programs for, 52
secondary succession and, 113
sustainability in, 54–55
UNCED efforts, 5
urban forestry, 276
wood as energy source, 171
forest soils, 294–98
The Forest Stewardship Council, 257
formaldehyde, 385
fossil fuels
air pollution and, 366–73
carbon cycle and, 101
consumption ethics and, 28–29
electrical energy production, 177–78
environmental impact of, 103–04
environmental issues, 188–96
fertilizer production from, 314–15
formation process, 187–88
history of development and use, 172–74, 196
landfill gas recovery, 395
particulate matter pollution, 76, 368–69, 385
politics and economics of, 179–80
pollution control measures, 373–75
substitution for human labor, 314
Framework Convention on Climate Change. See Kyoto Conference on Climate Change (Kyoto Protocol, 1997)

France, 174–75, 178, 207, 218, 220, 222, 225, 330, 341, 375, 382, 400
"frankenfoods," 330
Freedom to Farm Act of 1996, 51
free-living nitrogen-fixing bacteria, 102
freshwater ecosystems. See aquatic ecosystems
fuel cell technology, 28–29, 209–10
fuelwood, 196
fungicides and rodenticides, 318. See also pesticides
fusion. See nuclear fusion

G
Galbraith, John Kenneth, 49
gamma radiation, 215–16
Gandhi, Mahatma, 23
Ganges River (India), 4, 56, 336, 353
garbage. See solid waste management; waste disposal
Garrison Diversion (U.S.), 355–56
gas-cooled reactors, 220
gasification, 199
gasoline engines, 76
genes, 84–86
genetic diversity, 236–37, 240, 313
genetic engineering
food production, 27–28
genetically modified crops, 330–31
selective breeding and, 329
true vs. perceived risks, 40–41
geography
climate change and, 381–83
relationship of climate to, 372–73
geologic processes, Earth, 289–91
geopolitics, 438–41
Georgia, 147
geothermal energy, 178, 205–06
Germany
energy policy, 384
energy use in, 174–75, 178–80
green economics, 49
nuclear energy, 220
recycling programs, 52–53
use of biomass, 197, 199
use of solar/wind energy, 204
glaciers/glacial activity, 30, 88, 109, 238, 290–92, 335, 339–40, 381–82, 386
Glen Canyon Dam (U.S.), 200
globalization
consumption ethics and, 27–29
corporate ethics and, 25–26
energy consumption and, 49
environmental ethics and, 29–32
environmental issues and, 13
environmental justice and, 22

globalization—*Cont.*
  environmental policy and, 428
    Principles for Responsible Investment, 436
    World Trade Organization and, 25
Global Reporting Initiative (GRI), 26
global warming. *See* climate change
Gobi Desert (Asia), 116
"God's First Temples" How Shall We
    Preserve Our Forests: (Muir), 20
"Going Green" initiatives
  air conditioning, 75
  biodiversity preservation, 254
  co-generation facilities, 376
  college and university, 7, 12
  composting, 295
  conservation reserves, 119
  consumption ethics, 28
  exchanging surplus chemicals, 421
  global warming commitments, 446
  green design and building, 278
  greenhouse gases, 55
  green landscaping, 304, 320, 361
  integrated pest management, 320
  nuclear energy, 230
  reclamation projects, 81
  recycling, 404
  renovations for energy efficiency, 183
  socially responsible investments, 18
  war on hunger, 161
  water conservation, 361
  what individuals can do, 13
  wind energy, 206
Gonzales, Alberto, 33
Gore, Al, 379
government and governance, 429–30
government policies/regulation
  addressing climate change with, 384–85
  administrative law, 436
  agricultural subsidies, 50–51, 300, 314
  biodiversity protection, 254–58
  bottled water, 340
  certified organic foods/labeling, 332
  cost-benefit analysis and, 45–47
  economic programs in lieu of, 49–54
  electronics (e-waste), 419
  energy consumption, 179–80, 182
  energy deregulation, 205
  environmental ethics and, 16
  EPA role in, 436–37
  fertilizer/pesticide, 323–24, 333
  green initiatives, 384
  hazardous waste, 411–16, 423–24
  household hazardous waste, 424
  industrial water use, 344
  international waste trade, 422–23
  land-use issues, 12, 283–85

managing common resources, 48–49
nuclear energy, 229–30, 233
population control, 156–57
recycling programs, 402–06
risk assessment and, 37–40
state/local agencies, 276–77
urban sprawl and, 269–70
voluntary standards, 425
zoning restrictions, 278–79
government subsidies, 50–51
Grameen Bank (Bangladesh), 159
grants to small business, 51
grasslands management, 246–47, 283–84
grassland soils, 294–98
grasslands succession, 124
Great Lakes region
  environmental issues of, 9–10
  introduced species, 250–51
  PCB contamination, 321
  water use politics, 5, 13, 431
green business
  concepts of, 26–27
  consumerism and, 28
  individual decisions matter, 7
  marketing and product claims,
    54, 68, 384
green-collar jobs, 50
green design and building, 278, 281–83
green development, 21
green economics, 49
greenhouse gases. *See also* Kyoto
    Conference on Climate Change
    (Kyoto Protocol, 1997)
  causes and effects, 379–81
  "Going Green" initiatives, 55
  industrial development and, 49
green initiatives. *See* "Going Green"
    initiatives
green landscaping, 81, 295, 304, 320, 361
"green" politics, 438–41
green power, 206
Green Revolution, 313–14
greenways, 276
gross national income (GNI), 157–58
ground cover, 303
ground-level ozone, 366, 389
groundwater
  agriculture and, 8
  domestic water use and, 340
  hydrologic cycle and, 338
  landfill leachate and, 396–98
  population growth and, 11
  radiation contamination, 225
  sources of pollution, 352
groundwater mining, 358–60
guest-worker program, 163, 169

Gulf of Alaska, 351
Gulf of Mexico, 319, 431

H

habitat management, 258–60
habitat/refuges
  biomass conversion, 199
  ecological concept of, 83–84
  endangered species and, 12, 147
  legal protections for, 257
  loss of biodiversity, 240, 242–48
  population pressures on, 167
  preservation efforts, 360–61
The Hague, 444
Haiti, 438
half-life, pesticide, 316
half-life, radiation, 214–16, 387
Hanford Site (U.S.), 231
Hardin, Garrett, 48
Harvard University, 18
Hawaii land use regulation, 277
hazardous air pollutants (HAP), 367, 373
hazardous materials, defined, 410–11
Hazardous Materials Transportation Act
    of 1975, 412
hazardous waste. *See also* U.S.
    environmental legislation
  environmental issues of, 416
  environmental justice and, 21–23
  handling and disposal, 15
  health risks, 416
  household wastes, 424
  international trade, 422–23
  negligent disposal, 416–19
  nuclear waste cleanup, 231
  regulation of, 411–16, 423–24
  toxicity case study, 414
hazardous waste management
  choices, 419–22
  international waste trade, 422–23
  natural disaster impact on, 439
  regulatory program evolution, 423–24
hazardous waste sites, 416–18
health/health issues. *See also* disease
  air quality and, 366–73
  climate change and, 381–83
  coal mining, 190
  drinking water and, 164, 204, 362
  environmental costs, 44–45
  food production and, 158–60
  hazardous wastes, 416
  indoor air pollution and, 385–87
  mercury contamination and, 58–59
  noise pollution and, 387–89
  personal ethics/choices, 29
  pesticide use and, 323–24

radiation exposure and, 216–17, 224
risk assessment and, 37–40
traditional medicine, 250
true vs. perceived risks, 40–41
water quality and, 346–47, 349–50
heat energy, 72–75
heat pump systems, 205–06
heavy metals
    in electronics (e-waste), 419
    environmental persistence, 415–16
    as hazardous waste, 424
    waste incineration and, 398–99
heavy-water reactors, 219–20
herbicides, 317–18. *See also* pesticides
herbivores, 95–96
high-density polyethylene (HDPE), 400, 405
Hill, Julia Butterfly, 33
Hispanic minorities, 21–23
historic preservation/restoration, 81
Hoechst AG (corporation), 324
Homer (Greek poet), 315
Hong Kong, 406, 423
host, parasitic, 92
household hazardous waste, 419, 424
housing, environmental justice and, 22
Huang He River (China), 337
Hudson River (U.S.), 355
human energy pyramid, 160
human impacts on environment
    biodiversity and species extinction,
        235–36
    climate change, 253
    control of predators and pests, 251–52
    habitat loss, 242–48
    introduction of exotic species, 250–51
    overexploitation, 248–50
human population. *See* populations,
        human
human rights
    cultural relativism and, 16
    Darfur violations, 18
    environmental justice and, 21–23
humus, 292
Hungary, 157, 440
hunger
    consumption ethics and, 27–28
    finding long-term solutions for, 314
    food production and, 158–60, 309
    "Going Green" initiatives, 161
    politics of, 160–61
Hurricane Katrina, 24, 286, 439
Hussein, Saddam, 441
hybrid electric vehicles, 209
hydrocarbons. *See* volatile organic
        compounds (VOCs)
hydroelectric power

as energy source, 178, 186
in-stream water use for, 344–46
technology and potential, 199–201
trends in energy consumption and, 180–81
hydrogen fuel/technology, 28–29
hydrologic cycle, 337–39, 381–83
hydro turbine, 207
hypotheses, scientific method, 64–65
hypoxia, 319

I

Iceland, 178, 205
Idaho National Engineering Laboratory, 230
illegal immigration, 162–63, 165
immigration
    geopolitics and, 438
    government policies, 156–57, 169
    North American case study, 165
    population growth and, 142–43
    U.S. population and, 162–63
incandescent lighting, 77, 208
incineration, waste, 40–41, 398–99, 422
India
    agricultural production, 312
    Bhopal incident, 414
    biogas production, 198
    energy use in, 174–75, 181
    Keoladeo National Park, 4
    plastics regulations, 406
    population control, 157
    resource exploitation, 56
    water pollution, 415
indigenous peoples, 6, 21–23
individual responsibility. *See* personal
        ethics/choices
Indonesia, 52, 150, 154, 156, 180–81, 205,
        322, 423
indoor air pollution, 385–87
Indus River (Pakistan), 30, 337
industrial development
    atmospheric pollution and, 366–67
    automobiles and, 390
    brownfields, 51, 276
    capitalist work ethic, 18–19
    corporate ethics, 23, 25–27
    economics vs. ecology, 47–49
    energy use and, 174–76
    environmental ethics and, 16, 29–32
    environmental impact of, 9–11
    environmental justice and, 21–23
    green-collar jobs, 50
    groundwater mining, 358–60
    habitat loss from, 248
    hazardous waste trade, 422–23
    human population and, 151–53
    problems from lack of planning, 265–66

role of water transportation in, 267
socially responsible investments, 18
toxic chemical releases and, 418–20
water pollution, 350–51
water use, 343–44
industrial ecology, 27
Industrial Revolution, 172, 265, 312–13
industrial solid waste, 393
infant mortality rates, 158
information programs, market-based, 51–52
inner-city redevelopment, 281
innocent landowner defense, 418
inorganic molecules, 69
insecticides, 315–17. *See also* pesticides
in-stream water use, 344–46
integrated pest management, 320, 327–30
Intergovernmental Panel on Climate
        Change (IPCC), 49, 379
International Conference on Population and
        Development (Cairo, 1994), 27
international conventions and conferences.
    See United Nations
international environmental treaties, 32
International Fund for Agricultural
        Development (IFAD), 300
International Joint Commission
        (IJC), 5, 49, 335
International Organization for
        Standardization (ISO), 25, 37, 412
International Paper Corporation, 147
international trade, 22
International Tropical Timber Agreement
        (1983), 32
International Union for Conservation of
        Nature and Natural Resources
        (IUCN), 242, 253, 256
international water disputes, 353, 355–56
International Year of Microcredit (UN,
        2005), 159
Internet, 54, 175
interspecific competition, 90–91
intraspecific competition, 90–91
intrinsic limiting factors, 144
invasive species, 49
involuntary risks, 40–41
ionizing radiation, 216–17
ions, 67
Iran, 154, 156, 180–81, 233
Iraq, 180–82, 441–42
Ireland, 406
irrigation, agricultural, 8–9, 342–43,
        355–56, 358–60
isotopes, 66–67, 214–15
Italy, 142, 142–43, 155, 179, 205, 229
Ithaca College (NY), 295
Ives, J. C., 345

J

Japan
  birth/death rate, 153–54, 157
  demographic transition model, 162
  energy use in, 174–75, 178–81, 194
  geothermal energy in, 205
  hydroelectric power in, 200
  nuclear energy, 220, 222, 225, 230, 232
  plastics regulations, 406
  recycling programs, 52
  solar energy in, 204
  use of biomass, 197
  waste disposal, 398
Jeffords, Jim, 362
jobs in environmental science, 50
Jordan River (Israel), 30
Journals (Emerson), 20

K

Kariba Dam (East Africa), 200
Kazakhstan, 181, 221
Kenya, 120, 142, 157–58, 302, 309
Keoladeo National Park (India), 4
keystone species, 95–96
kinetic energy, 72–74
kinetic molecular theory, 65–67
Kodak Corporation, 53
K-strategists, 148–49
Kuwait, 441–42
Kyoto Conference on Climate Change
        (Kyoto Protocol, 1997), 5, 32,
        179, 384–85, 438, 443, 445

L

labor-intensive agriculture, 312
lag phase, population, 144–45, 148–49
lakes and rivers. *See* freshwater ecosystems;
        water; water management
Lake Victoria (Africa), 244
land capability classes, 308
landfills
  anaerobic digestion, 198
  biomass conversion and, 197
  gas production and recovery, 395
  green building recycling and, 282
  groundwater pollution from, 352
  hazardous waste disposal, 416–18, 422
  municipal solid waste and, 393–98
  regulation, 22
land reclamation. *See* reclamation projects
landscaping, lawn and garden
  environmental impact and cost, 107
  "Going Green" initiatives, 81, 295, 304,
        320, 361

habitat/open space loss, 242, 272
  water use and, 345, 349, 358
"The Land Ethic," 17
land use
  biodiversity protection and, 257
  conflicts among users, 283–84
  economics vs. ecology, 47–49
  environmental costs, 44–45
  environmental issues of, 6–11
  government regulation of, 12, 308
  habitat preservation, 147
  indigenous peoples and, 6
  state/local regulation of, 277–78
  U.S. environmental legislation, 435
  waste disposal, 21–23
  wilderness conversion, 29
land use planning. *See also* urban growth
  environmental policy and, 428
  ethics and, 15–16
  government policies, 283–84
  mechanism for implementing, 276–79
  principles of, 274–76
  protecting nonfarm lands, 307–08
  special issues, 279–83
  strategies for, 276
  urban growth and the need for, 265
La Rance River (France), 207
latent heat, 73
laws and theories, scientific, 65, 73–75
laws and treaties. *See also* United Nations
  biodiversity protection, 254–58
  electronics (e-waste), 419
  environmental assessment case study, 256
  environmental ethics and, 16
  environmental policy, 435–38, 442–44
  fertilizer/pesticide regulation, 333
  recycling programs, 402–06
  shared water resources, 431
LD_{50} (lethal dose 50 percent), 414
leaching/leachate, 296, 396–98
lead (Pb), 370, 385
lead-206/lead-214, 387
lead contamination/poisoning, 414, 421
lead contamination/poisoning, 22,
        46, 362
Leadership in Energy and Environmental
        Design (LEED), 278, 281
Lead-Free Drinking Water Act, 362
Le Grande River (Quebec), 201
Lekkerkerk hazardous site (Netherlands), 410
length conversion, metric, 452
Leopold, Aldo, 17, 20
less-developed countries. *See also* Third
        World development
  economic development in, 151–53

energy consumption, 174–75, 178
  use of biomass, 196
lethal dose, 414
Levitt, William, 267
liability protection, 51
life cycle analysis
  extended product responsibility, 52–54
  solid waste, 401–02
  toxic substances, 410
life expectancy, 158
light emitting diodes (LED), 77
light output per watt of electricity, 77
lightwater reactors, 220
lignite (brown) coal, 189
limiting factors
  competition for resources, 90–91
  ecological concept of, 81–82
  human population, 153–57, 164–66
  plant nutrition, 93, 102–03, 348
  population growth, 144–46
  species reproduction, 148–49
limnetic zone, 135
liquid metal fast-breeder reactors, 220
liquified natural gas, 194
literacy rates, 155–56
lithosphere, 289
littoral zone, 134–35
livestock production
  animal rights/welfare, 18
  certified organic labeling, 332
  economics vs. ecology, 47–49
  environmental issues of, 8–9
  grazing practices, 246–47
  predator control, 251
logging. *See* forestry management
Love Canal hazardous site (U.S.), 410
low-density polyethylene (LDPE), 400
lung cancer, 387

M

Maastricht Treaty of 1992, 445
Maathai, Wangari, 32
macronutrients, fertilizer, 314–15
Madagascar, 57
Malawi, 309
Malaysia, 52, 423
Maldives, 438
malnutrition, 30, 160
Malta, 406
Malthus, Thomas Robert, 27, 152
mangrove swamp ecosystems, 131, 133, 235
mantle, Earth, 289
The Marine Aquarium Council, 257
marine ecosystems. *See* aquatic
        ecosystems

Marine Protection, Research and
    Sanctuaries Act, 23
market-based pricing of resources, 51–52
marriage, 155
marshes (wetlands), 136
Maryland, 277–78, 362
Massachusetts, 20, 269, 277, 410, 421
mass transit, 279–80. *See also*
    transportation
matter
    as abiotic factor, 80–81
    energy and the state of, 72–74
    nature/structure of, 66–68
    organic vs. inorganic, 69
Mauritius, 406
Mead, Margaret, 29
mechanical weathering, 289–90
mechanized agriculture, 312–14
Mediterranean shrubland (chaparral)
    biomes, 120–21
megacities, environmental impact of, 30
megalopolis, 269
Mekong River (Vietnam), 353
mercury contamination, 58–59, 66, 201
mesosphere, 366
methane gas (CH_4)
    from anaerobic digestion, 197–98
    as atmospheric component, 366
    global warming and, 29, 380–81
    as natural gas component, 69–70, 188, 196
    from pyrolysis, 198–99
metric system, 452–53
Meuse River (Europe), 353
Mexico. *See also* North America
    air pollution, 4, 369
    biodiversity, 137, 239, 254
    energy use in, 175, 177
    geothermal energy, 205
    illegal immigration, 169
    loss of biodiversity, 137
    nuclear energy, 218
    oil production, 181, 194
    population case study, 165
    population growth, 153–56, 165
    wildlife management, 260–61
Michigan, 33, 183, 258, 285, 403–04,
    410, 415
microcredit, 159
micronutrients, fertilizer, 314–15
Microsoft Corporation, 254
migration. *See also* immigration
    from city to suburbs, 265–69
    wildlife routes/patterns, 143, 195, 207,
        236–38, 253
migratory birds/waterfowl, 4, 167, 259–60

Millennium Declaration, 256
Minamata Bay hazardous site (Japan), 410
mineral water, 340
mining
    coal as energy source, 172–73
    coal extraction, 189–92
    environmental costs, 44–45
    groundwater, 358–60
    industrial water pollution, 350–51
    land reclamation, 52, 106
    land use issues, 283
    toxic chemical releases, 418–20
    uranium ore, 221, 225
    waste management, 393
Minnesota, 105, 258, 278, 285, 404
Mississippi River (U.S.), 273, 301, 319
Missouri River (U.S.), 273, 355–56
mixtures, relationship of matter
    in, 67–68
Mojave Desert (U.S.), 203, 345
mold, 385
molecules, nature of, 67, 98
monoculture, 313
Montreal Protocol (1987), 32, 378, 443, 445
moral responsibility to environment,
    16–18
Morbidity and Mortality Weekly Report
    (CDC), 251
more-developed countries, 151–53
mortality (death rate), 140–41
Mozambique, 309
Muir, John, 2–3, 15, 19–20, 32
mulching and composting, 399–401
municipal environmental issues
    solid waste, 393–95
    water pollution, 349–50
    water-use planning, 353–54
Murray River (Australia), 30
mutations, genetic, 216, 236
mutualistic relationships, 93–94, 110
mycorrhizae, 93

N

Narmada River (India), 337
natality (birth rate), 140
National Environmental Justice Advisory
    Council, 23
National Environmental Policy Act of
    1969, 23, 46, 436
National Institute of Occupational Safety
    and Health (NIOSH), 413
National Law Journal, 22
National Organic Program, USDA, 332
National Priorities List, 418
national security issues, 440

Native Americans
    cultural impacts on, 43, 242
    historic preservation projects, 81
    impact of European diseases on, 143, 442
    tribal rights, 48–49
natural capitalism, 26–27
natural disasters
    climate change and, 29–30
    environmental effects of, 439
    environmental justice and, 22, 24
natural gas. *See also* fossil fuels;
        methane gas
    as energy source, 181, 186
    extraction and use, 194–96
    formation process, 187
    growth in use of, 173–74
natural ingredients, 68
natural/native landscaping. *See* green
        landscaping
natural pesticides, 315–16, 327–29
natural resources
    assigning value to, 43–47
    debt-for-nature exchanges, 57
    defined, 41–42
    sustainability and, 54–56
natural selection and evolution, 84–89
"Nature?" (Emerson), 20
nature and interpretative centers, 280.
        See also recreation
Nature Conservancy, 117, 119,
        137, 147
Nauru phosphate mining, 106
navigation, waterway, 344–46
Nebraska, 343
negligible risk, 39
Nelson, Gaylord, 55
Nepal, 300, 353
Netherlands, 52, 414, 417, 444
neutralization, hazardous waste, 422
neutrons, 66–67, 214–15
Nevada, 203, 227, 232, 265, 345
New Essentials of Biology (American
        Book Company, 1911), 54
New Hampshire, 375–77
New Mexico, 225, 274, 359
New York (state), 33, 250, 267, 279, 295,
        362, 396
New York City, 223, 240, 270, 339, 353,
        355, 396–97, 406
New Yorker (magazine), 20
New York University, 376
New Zealand, 126, 151, 153, 160, 181,
        205, 229, 246
niche, ecological concept of, 83–84, 91
nickel-cadmium batteries, 53

Nigeria, 154, 156, 180–81, 194, 353
Nile River (Egypt), 30, 200, 301, 337
NIMBY (not-in-my-backyard) syndrome, 397–98, 412
9/11 terrorist attacks, 223
nitrogen cycle, 101–02
nitrogen dioxide (NO_2), 370, 385
nitrogen-fixing bacteria, 93, 102, 294
nitrous oxides (NO_x), 76, 371–73, 380–81
Nixon, Richard, 436
Nobel Peace Prize, 32, 159, 379
Noise Control Act of 1972, 389
noise pollution
    environmental justice and, 22
    health issues related to, 387–89
    land use planning and, 274
    U.S. environmental legislation, 435
nongovernmental organizations (NGOs), 430, 434
nonpersistent pesticides, 315
nonpersistent pollutants, 415–16
nonpoint source, pollution, 7–8, 348–49
nonrenewable resources, 41–42. *See also* fossil fuels; sustainable development
nontarget organisms, 315, 323
nontoxic ingredients, 68
North America. *See also* Canada; Mexico; United States
    conservationism in, 19–21
    consumption ethics, 27
    domestic water use, 339–41
    energy consumption, 174–75, 178
    industrial water use, 343–44
    loss of biodiversity, 137
    regional environmental issues, 6–11
    shared water resources, 431
    use of nuclear energy, 218
North American Free Trade Agreement (NAFTA), 442
North Atlantic Treaty Organization (NATO), 440
North Carolina, 21–22, 361, 420
North Dakota, 206
Northern Arizona University, 404
northern coniferous forest biomes, 126–27
Northern Yukon National Park (Canada), 195
North Korea, 233, 442
Norway, 178, 180, 196, 200, 248
"not in my backyard" syndrome (NIMBY), 23, 397–98, 412
nuclear breeder reactors, 220
nuclear chain reaction, 217, 219
nuclear energy
    characteristics and nature of, 214–17
    electrical energy production, 178

as energy source, 173, 181, 186, 196
    environmental issues, 222–29
    fission reactors, 218–20
    future potential of, 229
    growing energy demands and, 28–29
    history of development, 217–18
    nuclear fuel cycle, 221–22
    nuclear reactor alternatives, 220
    technical trends in, 230
    true vs. perceived risks, 40–41
nuclear fission, 217–20
nuclear fusion, 220
nuclear medicine, 214
nuclear waste
    disposal and storage, 224–27
    Hanford Site cleanup, 231
    plant decommissioning and cleanup, 229
    Yucca Mountain storage, 226–27, 232
nuclear weapons
    history of development and use, 217–18
    terrorism and, 223–24
    true vs. perceived risks, 40–41
    waste disposal, 224–27
    WWII legacy of, 214
nucleus, atomic, 66–67
nutrient cycles
    biodiversity and, 239
    carbon cycle, 98–101
    human impact on, 103–05
    nitrogen cycle, 101–02
    phosphorus cycle, 102–04
nutrient pollution, 319, 348

O

obesity, 160
observation, scientific method, 63
Occupational Safety and Health Act of 1970, 412–13
odors, 25, 45–47, 194–95, 274, 347–48, 366, 397–98
O Horizon, 294–97
Oil Pollution Act of 1990, 25
oil production. *See also* fossil fuels
    ANWR case study, 195
    ecoterrorism and, 441–42
    as energy source, 186
    environmental ethics and, 28–29
    extraction and use, 192–94
    formation process, 187
    gasoline taxes and, 448
    marine oil pollution, 351–52
    OPEC and, 180
oil spills, 25–26, 351, 439, 441
Oklahoma City bombing (1995), 33
oligotrophic lakes, 135
Olympic National Park (U.S.), 43

omnivores, 95–96
open space, 276. *See also* recreation
Oregon, 9, 26, 33, 277, 283, 353, 403
Oregon State University, 230
organic agriculture, 310, 325–27
organic foods/labeling, 326, 332
Organic Foods Production Act of 1990, 326
organic matter
    accumulation in ecosystems, 109–12, 131, 135–36, 260
    as energy source, 80, 95–104
    fossil fuels formation, 172, 187
    mulching and composting, 399–401
    soil fertility and, 313–15, 326
    in soils, 199, 239–40, 291–98, 302–03
    structure and formation, 69, 71
    water quality and, 347–50, 356–57
organic molecules, 69, 98, 367
Organization for Economic Cooperation and Development (OECD), 181
Organization of Petroleum Exporting Countries (OPEC), 180–82
organophosphates, 316–17, 416
Our Common Future (Bruntland report, 1987), 21, 54
overburden, surface mining, 189–90
overexploitation of resources
    fisheries, 248–49
    forest, 242–46
    grasslands, 246–47
    legal protections against, 258
    wildlife and plants, 249–50
overgrazing, grasslands, 246–47, 284
ozone (O_3)
    atmospheric depletion of, 32
    formation of, 367
    as ground-level pollution, 371–73, 389
    importance of, 378
    international agreements, 384–85
    Montreal Protocol (1987), 32, 378, 443

P

P2 costs (pollution-prevention costs), 45, 420–22
Pacific Gaze & Electric Company (PG&E), 55, 205
Pakistan, 30, 154, 156, 164, 300, 337
parasite-host relationships, 92, 94–95
parent material, soil, 292, 294–98
parks and open space. *See* recreation
particulate matter pollution, 76, 368–69, 374–75, 385
passive solar systems, 201–02
patchwork clear-cutting, forest, 245
PCBs (polychlorinated biphenyls), 21–22, 321

pelagic marine ecosystems, 130–31
Pennsylvania, 172, 187, 222, 362, 395, 410
performance bonds, 52
periodic table of elements, 451
periphyton, 136
permissible exposure limits (PEL), 413
perpetual energy, 186
Persian Gulf, 29, 182, 439, 441
persistent organic pollutants (POPS), 32,
        236, 316, 415–16
persistent pesticides, 315–16, 320–22
personal ethical commitment, 16
personal ethics/choices
    air pollution as, 389
    consumption of resources, 27
    ecological footprint, 29
    environmental protection, 447
personal property, environmental
        regulation and, 12
pest control, 315
pesticides
    agriculture and, 7–8
    DDT case study, 316
    genetic mutation from, 236
    as hazardous waste, 424
    impact on natural selection, 86–87
    increased food production from, 27–28
    information programs, 51
    as organic pollutants, 32
    problems with use of, 320–24
    projected changes in use, 324–25
    soil conservation and, 303
    U.S. environmental legislation, 435
petroleum industry. See oil as energy
pheromones, 327
Philippines, 154, 156, 205, 207, 322
philosophy, environmental, 16–18
phosphates, 93, 103–06, 348, 350, 358
phosphorus cycle, 102–04
phosphorus lawn fertilizer, 105
photochemical smog, 371–73
photosynthesis
    carbon cycle and, 100–101
    energy conversion, 96–97, 186
    process of, 71–72
    in soil formation, 294
photovoltaics, 75, 203–04, 384. See also
        solar energy/technology
pH scale
    acid rain effects, 376–77
    measurement, 67–69
    soil formation and, 292
phthalates, 424
physical sewage treatment, 358
phytoplankton, 95–96, 130–31, 133,
        135–36, 250

Pinchot, Gifford, 19
pioneer community, 109–10
Pizarro, Francisco, 442
planning agencies. See government
        policies/regulation
Plan of Implementation, Agenda 21
        (UN, 2002), 21
plantation forestry, 245–46
plant-growth regulators, 317–18
plastax (tax on plastic bags), 406
plastics, recycling, 400, 405–06
plate tectonics, 289–90
plutonium-218 (P-218), 387
plutonium-239 (P-239), 220
point source, pollution, 348, 350
political boundaries, 4–5, 13
politics
    corporate ethics and, 25–26
    of energy use, 179–80
    environmental science and, 3
    food production, 160–61
    immigration policy, 162–63
    international geopolitics, 438–41
    land use planning, 276–77
    nuclear energy future, 229–30
    nuclear weapons disposal, 224
    resource exploitation, 195
    risk management, 37–40
    of world oil price, 181–82
pollination
    biodiversity and, 240
    coevolution relationships, 88
    symbiotic relationships, 94
pollution. See also atmospheric pollution;
        noise pollution; thermal
        pollution; water pollution
    corporate ethics and, 25–27
    economics of, 41
    energy consumption and, 28–29
    environmental costs, 44–45
    fossil fuels and, 190–91, 193–94, 196
    industrial development and, 30–31
    international environmental treaties, 32
    market-based efforts to reduce, 51–52
    personal perceptions about, 27
    risk assessment, 37–40
    thermodynamics and, 75
    urbanization and, 265
pollution prevention. See also recycling
    P2 costs (pollution-prevention), 45,
        420–22
    techniques, 53
polonium-214, 387
polybrominated dephenyletrhers
        (PBDEs), 424
polychlorinated biphenyls (PCBs), 415, 417

polyculture agriculture, 312
polyethylene terephthalate (PET), 400, 405
polyploidy, 87
polystyrene (PS), 400, 405
polyvinyl chloride (PVC), 400
The Population Bomb (Ehrlich), 27
population cycles, 149
population demographics, 153–55
population growth/density. See land use
        planning; urban growth
populations
    characteristics, 140–43
    defined, 84–85
    environmental issues, 8–13
    ethics and, 15–16
    genetic diversity in, 236–37
    global warming impact on, 381–83
    growth curve, 143–44
    limiting factors, 144–47
    pest control impact on, 323
    reproductive strategies, 148–49
    species extinction, 87–88, 235–36
    sustainability of fisheries, 260
    sustainability of wildlife, 258–60
populations, human. See also human
        impacts on environment
    age distribution, 142, 154–55
    biological characteristics, 153–55
    consumption ethics, 27–29
    demographic transition model, 161–62
    economic development and, 151–53
    energy consumption and, 186
    environmental degradation, 158–61
    environmental ethics, 23
    environmental policy and, 428–32
    growth curve, 149–51
    limiting factors, 164–66
    North American case study, 165
    pesticide residues and, 323–24
    political factors influencing, 156–57
    social factors influencing, 155–56, 166
    standard of living, 157–58
    water scarcity and, 335–37
porosity, soil, 338
postwar baby boom, 162–63
potable water, 335
potassium, 104, 292, 309, 314, 401
potential energy, 72–74
poverty
    credit programs to combat, 159
    environmental justice and, 22, 24
    impact on environment, 11
    sustainable development and, 21, 56
    urbanization and, 265
precipitation. See climate
precipitation, hazardous waste, 422

precision agriculture, 326–27
predators
  beneficial insects as, 327–29
  impact of bioaccumulation on, 10
  introduced species as, 250–51
  pest control, 323
  as population limiting factor, 144–45
  role in ecosystems, 3, 89–90, 95, 251–52
  symbiotic relationships, 94
preservation, 19, 21
pressurized-water reactors, 219
primary air pollutants, 367
primary sewage treatment, 356–58
primary succession, 109–11
Princeton University, 18, 361
Principles for Responsible Investment, 436
probability vs. possibility, 37
producers, ecosystem, 95–96, 100–102
product claims/endorsements, 54
property rights, 48–49
proton exchange membrane fuel cells,
    209–10
protons, 66–67, 214–15
pseudoscience, 66
public transportation. *See* transportation
Puget Sound Energy (PSE), 206
purified water, 340
pyrolysis, 198–99

Q

questioning, scientific method, 64

R

racism, environmental justice and, 21–23
radioactive waste disposal, 224–27
radioactivity. *See also* nuclear energy
  biological effects of, 216–17
  half-life, 214–16, 387
  handling and waste disposal, 15
  measuring, 214–16
Radiological Dispersal Devices (RDDs,
    dirty bombs), 223–24
radon gas, 225, 385, 387–88
railroad travel, 177. *See also*
    transportation
The Rainforest Alliance, 257
rainforest biomes, 29, 56, 126–27
Ramscar Convention on Wetlands, 256
rangeland management, 246–47, 283–84
range of tolerance, species, 81–82
raw materials as limiting factor, 145
Razali, Tan Sri, 56
reclamation projects
  coal mining, 190–91
  covering the cost of, 52
  Everglades National Park, 354–55

"Going Green" initiatives, 81
  grasslands succession, 124
  mining, 351
  Nauru phosphate mining, 106
recreation
  biodiversity and, 240
  dam construction and, 344–46
  environmental ethics and, 16
  environmental justice and, 22
  land use issues, 283–84
  managing common resources, 48–49
  urban planning and, 276, 279–80
  water management for, 10
RecycleBank, 404
recycling. *See also* pollution prevention
  agricultural water use, 342–43
  approaches to, 402–06
  biomass conversion and, 196–99
  deposit-refund programs, 52, 403
  economics and technology, 405–06
  ecosystem decomposers, 95
  electronics (e-waste), 419
  extended product responsibility, 52–54
  green design and building, 278, 281–82
  hazardous waste, 422
  hydrologic cycle, 337–39
  marketing and product claims, 68
  in soil formation, 294
  supply/demand economics of, 42–43
  thermoplastic resins, 400
  wastewater, 337, 349
Red River of the North (U.S.), 355–56
Red Sea, 182, 289
reduced tillage, 305–07
reforestation, 52, 245
regional planning agencies. *See*
    government policies/regulation
religion
  birth control and, 155–56
  environmental ethics and, 16
  environmental justice and, 22
  humanitarian aid, 160
renewable energy. *See also* biomass;
    geothermal energy; hydroelectric
    power; solar energy/technology;
    tidal power
  electrical energy production, 178
  green-collar jobs, 50
  green economics and, 49
renewable resources
  defined, 41–42
  as energy source, 186
  sustainability of, 54–56
reproducibility, scientific method, 65
reproduction
  biodiversity and, 235–37

birth control, 152, 155–57
endangered species, 252–53
integrated pest management and, 327
natural selection and, 84–85, 91
population characteristics, 140–43
population size limitations, 166–67
species strategies, 148–49
reserves, fossil fuel. *See* fossil fuels
residential energy use, 174–75
resistance, pesticide, 87, 316, 322–23, 329
resistance, selective breeding, 329–31
Resource Conservation and Recovery Act
    of 1976, 23, 412, 416, 424
resource exploitation
  consumption ethics, 27–29
  cost assessment, 44–45
  cost-benefit analysis, 45–47
  debt-for-nature exchanges, 57
  economics vs. ecology, 47–49
  economic tools to address, 49–54
  ecoterrorism and, 33
  environmental ethics and, 23
  resources vs. reserves, 186–87
  sustainability and, 54–56
  value of resources in, 41–44
  wildlife protection and, 195
resources, energy. *See* fossil fuels;
    renewable resources
reverse migration, 269
R Horizon, 294–97
ribbon sprawl, 269
Rio Declaration on Environment and
    Development. *See* Agenda 21
    (Rio Declaration on Environment
    and Development, 1992)
Rio Grande River (U.S.), 30, 244, 337, 431
risk. *See* environmental risk
risk values/probabilities, 39
rivers. *See* specific river by name; water;
    water management
road salt pollution, 278
Romania, 440
Roosevelt, Theodore, 432
root-fungi mutualism, 93
r-strategists, 148–49
rural-to-urban population shift, 265
Russia/Russian Federation, 30, 126, 154, 156,
    181, 188, 201, 207, 220, 222, 230,
    232, 443. *See also* Soviet Union

S

Safe Drinking Water Act of 1974, 23,
    362, 412
salination, water, 358–59
saltwater intrusion, 358–60
A Sand County Almanac (Leopold), 17, 20

Sahara Desert (Africa), 116
savanna biomes, 119–20
scavengers, 95
science
    basic elements of, 62–65
    of chemical reactions, 69–72
    energy principles, 72–74
    environmental implications of
        energy, 74–76
    limitations of, 66
    of matter, 66–69
scientific method, 62–65
Scotch Brand Magic Tape, 53
Scotland, 246, 248
Sea Grant program, 49
sea level, rising, 39, 109, 379, 381–82
Seattle University, 320
secondary air pollutants, 367
secondary sewage treatment, 356–58
secondary succession, 109, 111–13
secondhand smoke, 385–86
second law of thermodynamics, 73–75
Seine River (France), 337
selective breeding, 237, 329–31
selective harvesting, forestry, 245
self-sufficiency, 160
sensible heat, 73
septic systems, 352
September 11 terrorist attacks, 223
Serageldin, Ismail, 29
seral stage, 110
Serbia, 417
sewage treatment, 356–58
sex ratio, population, 141–42
sexual reproduction. See reproduction
short term exposure limits (STELs), 413
Sierra Club, 20
Silent Spring (Carson), 20, 316, 434
Simon, Julian, 27
Singapore, 123, 174, 446
"slash and burn" agriculture, 312
Slovakia, 417, 440
Small Business Liability Relief
        and Brownfield Revitalization
        Act of 2002, 51, 412
smallpox, 442
smart growth, urban, 281–83, 285
smog, photochemical, 371–73
smoking
    carbon monoxide and, 368
    indoor air pollution and, 385–86
    synergism from, 414
    true vs. perceived risks, 40–41
social ecology, 18
socially responsible investment, 18
social networking, 54

society
    cultural relativism, 16
    culture and traditions, 155–56
    economic development, 151–53
    endangered species awareness, 262
    environmental ethics in, 23
    environmental policy and, 428, 432
    land use decisions, 274–76
    lifestyle and waste production, 394–95
soil
    acid rain effects, 376–77
    components, 291–92
    conservation practices, 301–05
    formation process, 110, 240, 292
    land capability classes, 308
    properties, 292–94
    protecting fragile lands, 307
    tillage practices, 305–07
soil erosion
    agriculture and, 7, 326
    biodiversity and, 240
    deforestation and, 175
    geologic weathering, 290
    mechanized agriculture and, 313
    organic farming and, 310
    types and process of, 298–301
soil profile, 294–98
soil texture, 292–94
Solar Energy Generating System (U.S.), 203
solar energy/technology
    Chinese investment in, 49
    climate change and, 29
    electrical energy production, 178
    energy conversion, 75
    German policies, 384
    "Going Green" initiatives, 55
    growing energy demands and, 28–29
    land use strategies, 276
    technology and potential, 201–04
Solar Tres Power Tower (Spain), 203
solid waste management. See also waste
        disposal
    amounts and kinds produced, 393
    methods of disposal, 395–406
    municipal wastes, 394–95
    pay-by-the-bag systems, 52, 402
    U.S. environmental legislation, 435
source reduction, 401–02
South Africa, 406
South America, 301–02, 312
South Korea
    biogas production, 198
    demilitarized zone, 442
    drinking water, 164
    energy use in, 179–80
    nuclear energy, 218, 220, 230

Southlands Experimental Forest (U.S.), 147
Soviet Union, 214, 217–18, 224–25,
        378, 417. See also Russia/
        Russian Federation
Spain, 155, 178–80, 203–04, 218
spatial distribution, population, 143
speciation, 87–88
species
    biotic potential, 143–44
    defined, 84–85
    interactions of communities, 94–104
    interactions of individuals, 89–94
    population characteristics, 140–43
    range of tolerance, 82
species diversity, 237–39
species extinction
    biodiversity loss, 235–36, 244
    biodiversity preservation and, 253–54
    California condor case study, 261
    evolutionary patterns, 87–88
    societal awareness of, 262
spray irrigation, agricultural, 342–43
stable equilibrium phase, population, 144
standard of living, 56, 157–58, 174
Stanford University, 18
starvation. See hunger; poverty
Statement of Principles for the Sustainable
        Management of Forests, 5
state planning agencies. See government
        policies/regulation
sterilization, 157
St. Lawrence Seaway, 10
stormwater runoff, 278, 353
stratosphere, 366
streams and rivers. See freshwater
        ecosystems; water; water
        management
Streptococcus faecalis, 349–50
strip farming, 304–05
strip-mining, 44–45, 52, 189–91
styrofoam (polystyrene), 400
subatomic particles, 66–69
subbituminous coal, 189
subirrigation, agricultural, 342–43
subsidy programs, 50–51, 210, 314
subsoil, 294–98
succession, 109–13, 124
sulfur dioxide (SO$_2$), 51, 58, 191, 201,
        370. See also acid rain
Superfund Act of 1980, 412
Superfund Amendments and
        Reauthorization Act of 1986, 412
Superfund program, 418
supply/demand economics, 42–43
surface mining, 189–91
survivorship curve, 141

Sustainability, Tracking, Assessment, and
    Rating System (STARS), 12, 183
Sustainability Reporting Guidelines (GRI), 26
sustainability/sustainable development. *See
    also* community sustainability;
    "Going Green" initiatives
    Agenda 21, 5, 21
    Brundtland Report, 21
    carrying capacity, 146–47
    defined, 54–56
    groundwater mining and, 358–60
    Maastricht Treaty of 1992, 445
    Principles for Responsible Investment, 436
    urbanization and, 268–70
sustainable agriculture, 324–25
swamps (wetlands), 136
Sweden, 31, 157, 174, 196
Switzerland, 398
symbiotic nitrogen-fixing bacteria, 102
symbiotic relationships, 92–94
synergism, 414

T

T8 fluorescent lighting, 77
taiga biomes, 126–27
Tanzania, 309
target organisms, 315
taxonomic richness, 237–39
technology advancements
    addressing climate change with, 383–84
    environmental policy and, 428, 432
    green economics and, 49
    precision agriculture, 326–27
    recycling, 405–06
    sustainable development, 55–56
temperate deciduous forest biomes, 124–26
temperate grassland biomes, 118–19, 246–47
temperate rainforest biomes, 126–27
temperature. *See* climate; climate change
temperature conversion, metric, 453
Tennessee, 278, 420
Tennessee River (U.S.), 281
terracing, soil conservation, 304–05
terrestrial biomes
    acid rain effects, 376–77
    climate creates, 114–15
    desert, 116–18
    Mediterranean shrubland
        (chaparral), 120–21
    savanna, 119–20
    taiga, coniferous forest, boreal
        forest, 126–27
    temperate deciduous forest, 124–26
    temperate grasslands, 118–19, 246–47
    temperate rainforest, 126–27
    tropical deforestation, 245

tropical dry forest, 121–22
tropical rainforest, 122–24
tundra, 127–29
terrestrial succession, 109–10
terrorism, 33, 223–24. *See also*
    ecoterrorism
tertiary sewage treatment, 358
Texas, 4, 10, 265, 359, 404
Thailand, 154, 156, 322, 353, 423
theories and laws, scientific, 65
thermal inversion, 372–73
thermal pollution, 227–28, 344, 351
thermodynamics, laws of, 73–75
thermoplastics recycling, 400
thermosphere, 366
Third World development. *See also* less-
    developed countries
    agriculture and, 30–32
    debt-for-nature exchanges, 57
    economics and sustainability, 54–56
    geopolitics and, 438, 440–41
    hazardous waste trade, 422–23
    human population and, 151–53
    urbanization in, 265–66
Thomas, Lewis, 15, 32
Thoreau, Henry David, 2, 19–20
threatened species, 254–58
3M Corporation, 53
Three Gorges Dam (China), 200–201
Three Mile Island (1979), 222–23
threshold level, 414
tidal energy, 178, 206–07
Tide (detergent) "Coldwater Challenge," 54
Tigris River (Middle East), 337
tillage, agricultural, 305–07
timber harvesting. *See* forestry
    management
Times Beach hazardous site (U.S.), 410
Time Weighted Average (TWA), 413
Tittabawassee River (U.S.), 415
Tolba, Mustafa, 442
topsoil, 294–98
total energy, 73–74
tourism
    ecotourism, 47–48
    environmental ethics and, 16
    pollution and, 10
toxic chemical releases, 418–19
Toxic Release Inventory, 51
toxic substances, defined, 410–11. *See
    also* hazardous waste
Toxic Substances Control Act of 1976, 412
toxic-waste incineration
    heavy metals and, 398–99
    true vs. perceived risks, 40–41
    waste disposal by, 422

tract development, 269
tradable emissions permits, 51
traditional medicines, 250
traffic congestion, 174
"The Tragedy of the Commons" (Hardin), 48
transportation
    automobiles, reducing use of,
        174, 369, 389
    energy use in, 177
    environmental justice and, 22
    fossil fuels, 190, 193–94
    fuel cell technology, 209–10
    gasoline taxes and, 448
    government subsidies, 51
    hybrid electric vehicles, 209
    industrial development and, 267
    nuclear fuel, 222
    urban growth and, 271
    urban planning and, 279–80
    water management for, 10
transuranic waste disposal, 224–25
trash. *See* solid waste management; waste
    disposal
tree-spiking, 33
trickle irrigation, agricultural, 342–43
trickling filter system, wastewater, 357
tropical deforestation, 245
tropical dry forest biomes, 121–22
tropical rainforest biomes, 122–24, 254, 298
tropic levels, 96–97
troposphere, 366
Turkey, 154, 156, 353
2,4,5-T (2,4,5-trichlorophenoxyacetic
    acid), 318, 324
2,4-D (2,4-dichlorophenoxyacetic
    acid), 318

U

ultraviolet light, 378
unconfined aquifer, 338
underground mining, 190
underground storage tanks, 352, 411
United Church of Christ, 22
United Kingdom, 179–81, 218, 220, 222,
    225, 377–78, 448
United Nations
    addressing environmental policy, 442–44
    Commission on Sustainable
        Development, 5, 444
    Conference on Environment and
        Development (UNCED, 1992), 5,
        21, 444
    Conference on Human Environment
        (Sweden, 1972), 31
    Convention on Biological Diversity,
        254, 445

Convention on International Trade in Endangered Species of Wild Fauna and Flora (CITES, 1973), 32, 258, 443, 445

Convention on Long-Range Transboundary Air Pollution (Geneva, 1979), 443, 445

Convention on Migratory Species (Bonn, 1983), 256, 445

Convention on Persistent Organic Pollutants (Stockholm, 2001), 32, 445

Convention on the Law of the Sea (1994), 443, 445

Convention on the Prior Informed Consent Procedure (Rotterdam, 1998), 445

Convention to Combat Desertification, 256, 445

Educational, Scientific, and Cultural Organization (UNESCO), 4

Environment Programme, 31–32, 49, 316, 335, 379, 423, 443

Food and Agriculture Organization (FAO), 240–41, 442–43

Framework Convention on Climate Change. See Kyoto Conference on Climate Change (Kyoto Protocol, 1997)

Microcredit Summit (1997), 159

Millennium Ecosystem Assessment Synthesis Report (2005), 5–6, 256

Principles for Responsible Investment, 436

Universal Declaration on Human Rights, 16

World Food Program, 161

World Heritage Sites, 4

World Summit, 21

United States. See also individual states; North America

agricultural potential, 301–02

biodiversity protection, 254–58

ecolabeling programs, 446–47

endangered species, 262

energy conservation efforts, 207–09

environmental policy, 432–38

hazardous waste laws, 423

immigration policy, 169

international environmental treaties, 445

nuclear energy policies, 233

pattern of land use, 265

pesticide regulation, 316–17

plastics regulations, 406

population case study, 165

population characteristics, 162–63

population growth, 432

radioactive waste sites, 225, 228

recycling programs, 52–54

use of biomass, 196

use of geothermal energy, 205

use of hydropower, 200

use of nuclear energy, 218

use of solar/wind energy, 204–06

waste production, 393

University of Arizona, 75

University of Arkansas, 81

University of British Columbia, 421

University of California-Berkeley, 119

University of Kansas, 254

University of North Carolina, 361

University of Tennessee, 278

University of Vermont, 278

University of Western Ontario, 295

uranium-235 (U-235), 218–19, 221

uranium-238 (U-238), 215–16, 220–21, 387. See also nuclear energy

uranium enrichment, 221

uranium mining and milling, 225

urban growth. See also land use planning

migration to the suburbs, 265–69

problems from lack of planning, 270–74

water management and, 8–9

urban growth limit, 275

urban sprawl, 268–70, 281

U.S. Army Corps of Engineers, 354–55

U.S. Bureau of Land Management (BLM), 284

U.S. Census Bureau, 163

U.S. Centers for Disease Control (CDC), 251, 362

U.S. Commission on Civil Rights, 23

U.S. Congress

development of policy, 432–34

government subsidies, 50–51

immigration policy, 162–63

U.S. Corps of Engineers, 273

U.S. Council on Environmental Quality, 436

USDA National Organic Program, 332

U.S. Department of Agriculture, 332, 447

U.S. Department of Defense, 440

U.S. Department of Energy, 205, 224, 231–32, 383

U.S. Department of the Interior, 254

U.S. Department of Transportation, 411

useful energy, 73–74

U.S. environmental legislation

addressing major issues, 435–37

Clean Air Act of 1970, 23, 373–75, 412

Clean Water Act of 1972, 23, 344, 351, 412

Endangered Species Act of 1973, 12, 34, 254, 257

Federal Flood Control Act of 1944, 355

Federal Insecticide, Fungicide, Rodenticide Act of 1947 (FIFRA), 412

Freedom to Farm Act of 1996, 51

Hazardous Materials Transportation Act of 1975, 412

Lead-Free Drinking Water Act, 362

Marine Protection, Research and Sanctuaries Act, 23

National Environmental Policy Act of 1969, 23, 46, 436

Noise Control Act of 1972, 389

Occupational Safety and Health Act of 1970, 412

Oil Pollution Act of 1990, 25

Organic Foods Production Act of 1990, 326

Resource Conservation and Recovery Act of 1976, 23, 412, 416, 424

Safe Drinking Water Act, 23, 362

Safe Drinking Water Act of 1974, 412

Small Business Liability Relief and Brownfield Revitalization Act of 2002, 51, 412

Superfund Act of 1980, 412

Superfund Amendments and Reauthorization Act of 1986, 412

Toxic Substances Control Act of 1976, 412

Water Pollution Control Act. See Clean Water Act of 1972

Water Quality Act of 1987, 353

Wild and Scenic Rivers Act of 1968, 360

U.S. Environmental Protection Agency (EPA)

creation of, 430

Design for Environment program, 53–54

enforcement options, 436–37

environmental justice and, 21–23

green power programs, 206

hazardous substances, 410–11, 413

indoor air pollution research, 385

mercury emissions program, 58–59

noise pollution regulation, 389

nuclear waste cleanup, 231

pesticide regulation, 324

quality standards, 367–73

radon gas recommendation, 387

solid waste reduction, 399

water-efficiency standards, 341

U.S. Federal Bureau of Investigation (FBI), 33

U.S. Fish and Wildlife Service, 3, 137, 147, 262, 284

U.S. Food and Drug Administration (FDA), 413
U.S. Geological Survey (USGS), 350
U.S. Government Accounting Office (GAO), 22
U.S. Green Building Council, 281–82
U.S.-Mexico Water Treaty of 1944, 431
U.S. National Marine Fisheries Service, 137
U.S. National Oceanic and Atmospheric Administration (NOAA), 351
U.S. National Park Service, 284
U.S. Nuclear Regulatory Commission, 226, 229
U.S. Occupational Safety and Health Administration, 411
U.S. Office of Environmental Justice, 22
U.S. Public Health Service, 413
U.S. Soil Conservation Service, 300, 308
Utah, 277, 345

V

Vác hazardous site (Hungary), 410
vadose zone, 338
variables, experimental, 64–65
Varzea Forests, 134
vector, parasitic, 92
Velsicol Chemical Corporation, 324
Vietnam War, 442
Virginia, 285, 395
volatile organic compounds (VOCs), 367–68, 370–73, 385
volcanoes/volcanic activity, 109, 239, 265, 273, 289–90, 366, 375
Volta River (Ghana), 200, 337
volume conversion, metric, 453
voluntary environmental standards/goals, 37, 53, 166, 277, 412, 420, 423, 444

W

Walden; or, Life in the Woods (Thoreau), 20
warfare and the environment, 441–42
war on hunger, 161
Washington (state), 9, 33, 43, 126, 206, 227, 231, 357, 362, 403, 419
waste disposal. *See also* solid waste management
  biodiversity and, 240
  biomass conversion and, 196–99
  composting, 295
  economics of, 41
  environmental costs, 44–45
  environmental ethics of, 15, 25
  environmental justice and, 21–23
  industrial ecology and, 27
  nuclear contamination, 224, 231
  nuclear fuel, 221–22

nuclear weapons, 224–27
population density and, 9–11
as population limiting factor, 165
public littering, 49
true vs. perceived risks, 40–41
waste heat (thermal pollution), 75, 227, 376
waste minimization, 420–22
waste products, accumulation of, 145
wastewater
  geothermal energy, 206
  groundwater pollution from, 352
  pollution control programs, 52
  recycling/reclamation, 337, 349–50
  sewage treatment, 356–58
  water treatment, 340–41
water. *See also* aquatic ecosystems; hydroelectric power
  ecosystem interrelationships, 4
  energy and the state of, 72–74
  energy costs for hot water, 176
  environmental issues, 335–37
  geologic weathering, 289–90
  hydrologic cycle, 337–39
  international conflicts over, 362
  land-use planning and, 272–74
  molecular characteristics, 67, 335
  nuclear power and, 222
  in soil formation, 292, 296–98
  thermal pollution and, 227
  transportation and commerce, 267
  world freshwater supply, 335–37
water disputes, international, 353
water diversion, 355–56
water erosion, 298–301
water management
  agricultural water use, 342–43
  biodiversity and, 239–40, 244
  case study, 345
  dam-removal projects, 43
  dealing with exotic species, 100
  domestic water use, 339–41
  environmental issues of, 7–11
  ethics of allocation and use, 30
  future world needs for, 29, 164
  industrial water use, 343–44
  politics of, 5, 13
  power generation and navigation, 344–46
  reclamation projects, 81
  shared resources, 431
  urban growth and, 272–73, 345
  use planning issues, 353–61
water pollution
  acid rain, 30, 44, 70
  agricultural runoff, 104–05, 315, 333
  bioaccumulation, 10
  Gulf of Mexico "dead zone", 319

impact of shortages on, 335–37
kinds and sources, 346–52
land use planning and, 274
mercury contamination, 59
Water Pollution Control Act. *See* Clean Water Act of 1972
water quality
  environmental justice and, 22
  industrial water use and, 344
  shared water resources, 431
  U.S. environmental legislation, 435
  water-use planning issues, 353–61
Water Quality Act of 1987, 353
water table, 338
"water war," 362
waterways, soil conservation, 304–05
wave energy/technology, 28–29
weathering, geologic, 289–90
weeds, 315, 317–18
weight conversion, metric, 453
Western Washington University, 206
wetlands
  benefits of, 240, 360–61
  Everglades case study, 354
  freshwater ecosystems, 136–37, 249
  losses, 11, 138, 167, 270, 273
  preservation efforts, 4, 260, 276–77, 285, 304, 360–61
  wastewater treatment, 358
Whitman, Walt, 19–20
wilderness
  Arctic National Wildlife Refuge (ANWR), 195
  consumption ethics issues, 29
  environmental issues, 6–7, 9
  land use issues, 283–84
Wildlife Conservation Society, 249
wildlife management
  carrying capacity, 146–47
  for ecotourism, 47–48
  India Wildlife Protection Act, 4
  loss of ecosystems, 137
  overexploitation of resources, 249–50
  population control in, 167
  predator control, 251–52
  resource exploitation and, 195
  sustainability of populations, 258–60
  U.S. environmental legislation, 435
  water-use planning, 360–61
wind, 289–90, 298–301
windbreaks, soil conservation, 305
wind energy/technology
  addressing climate change with, 384
  Chinese investment in, 49
  electrical energy production, 178
  energy conversion, 75

green power initiatives, 206
growing energy demands and, 28–29
technology and potential, 204–05
wolves
  population control, 252
  as predators, 89–90
  reintroduction to Yellowstone Park, 3
women
  education and literacy rate, 155–56
  family planning, 157
  Grameen Bank and microcredit for, 159
wood as energy source. *See* biomass
World Bank, 46, 57, 256, 443
World Commission on Environment and
    Development. *See* Brundtland
    Commission

World Health Organization (WHO), 316, 367
World Heritage Sites, 4
World Trade Organization (WTO), 25,
    442–43
World War I, 442
Worldwatch Institute, 439
World Water Council, 341
World Wildlife Fund, 57

X

Xerox Corporation, 53

Y

Yale University, 18
Yangtze River (China), 200–201
yard waste. *See* composting and mulching

yellow-cake uranium, 221
Yellow River (China), 30, 244, 322
Yellowstone National Park (U.S.), 3, 34
Yucca Mountain nuclear waste storage,
    226–27, 229, 232
Yunus, Huhammad, 159

Z

Zambezi River (East Africa), 200
Zambia, 309
zero pollution, 28–29, 49
zero population growth, 154, 163
zero risk, 39
Zimbabwe, 309
zoning restrictions, land use, 278–79, 308
zooplankton, 130–31, 133, 135–36

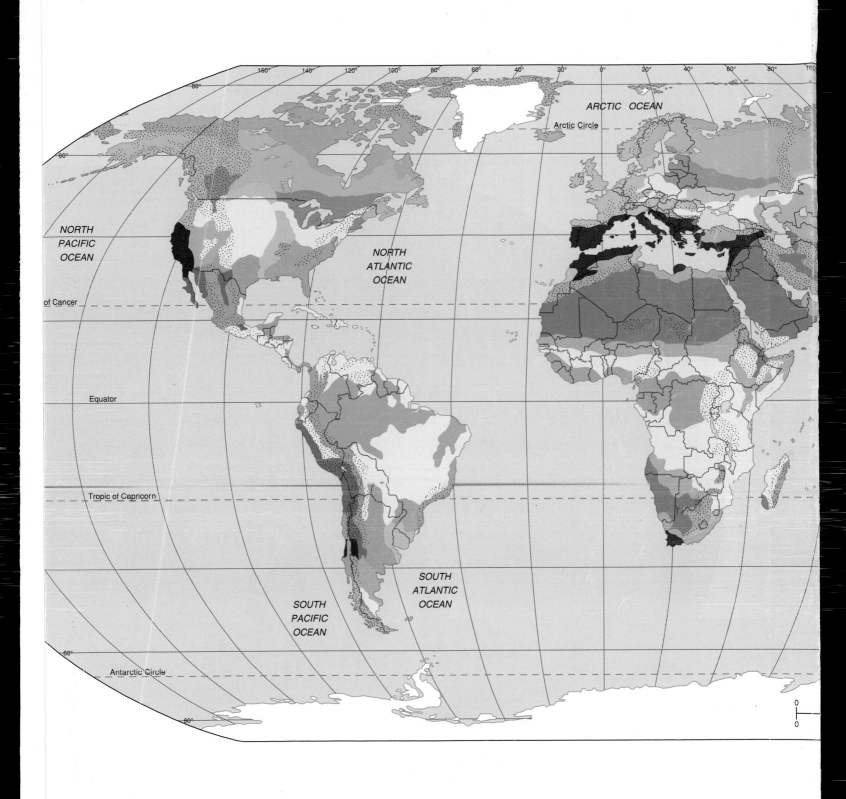

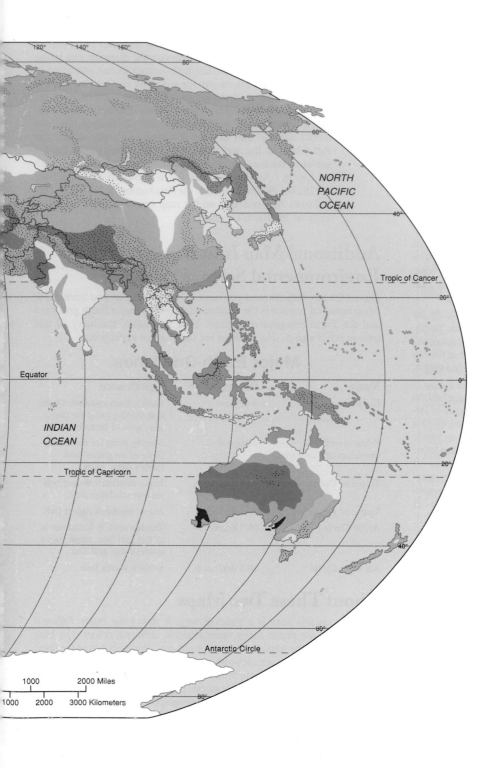

NORTH
PACIFIC
OCEAN

Tropic of Cancer

Equator

INDIAN
OCEAN

Tropic of Capricorn

Antarctic Circle

120° 140° 160° 80°

60°

40°

20°

0°

20°

40°

60°

80°

1000   2000 Miles

1000   2000   3000 Kilometers